Handbook of Practical Astronomy

Günter D. Roth (Editor)

Handbook of
Practical Astronomy

**Extras on
www.springer.com/978-3-540-76377-2**

Springer

Günter D. Roth
Ulrichstraße 43a
Irschenhausen
82057 Icking/Isartal
Germany

This book is a completely revised and updated edition of the *Compendium of Practical Astronomy*, published in 1994 at Springer-Verlag.

Cover photo: Copyright by Franz Xaver Kohlhauf.

ISBN 978-3-540-76377-2   e-ISBN 978-3-540-76379-6
DOI 10.1007/978-3-540-76379-6
Springer Dordrecht Heidelberg London New York

Library of Congress Control Number: 2008944087

© Springer-Verlag Berlin Heidelberg 2009

Typesetting and Production: le-tex publishing services GmbH, Leipzig, Germany
Cover Design: WMXDesign, Heidelberg

Printed on acid-free paper

Springer is part of Springer Science+Business Media (www.springer.com)

Preface

With amateurs, students and teachers of astronomy in high schools and colleges particularly in mind, the *Handbook of Practical Astronomy* comprises an essential source of current trends in astronomy and examines the broad variety of astronomical observations.

Methods used by amateur astronomers have changed significantly in recent years. Classical photography has been superseded by video astronomy and CCD-systems. Telescopes are now computer-controlled. The Internet has become the basis for exchanging knowledge and experience, even providing amateur observers with access to scientific data files. Personal contact can now be made worldwide without delay.

Astronomy online for users of this handbook: To access astronomy data, physical constants, catalogues, maps, and object search visit www.springer.com/978-3-540-76377-2.

The first edition of this book was published in 1960 in Germany with the title *Handbuch für Sternfreunde*, followed by further editions in 1967, 1981 and 1989. The current edition has two English forerunners, *Astronomy: a Handbook*, published in 1975, and *Compendium of Practical Astronomy*, published in 1994.

As the Editor, I wish to thank sincerely all authors for their understanding and friendly cooperation during the planning stages and execution of this work. I would like to add a word of grateful remembrance of Professor Dr. Felix Schmeidler of the University Observatory at Munich, he had been an author in all editions since 1960 and had from the very beginning warmly supported this project, but whose death in the middle of the current edition ended his collaboration prematurely.

I gratefully acknowledge the helpfulness of Springer-Verlag, Heidelberg, where Dr. T. Schneider gave every possible consideration to the authors' and my suggestions.

Irschenhausen, March 2009 Günter D. Roth

Table of Contents

H. Boehnhardt

H. Boehnhardt

Extras on
www.springer.com/978-3-540-76377-2

List of Authors

Dr. Hermann Böhnhardt
Max-Planck-Institut für Sonnensystemforschung,
Max-Planck-Straße 2, 37191 Katlenburg-Lindau, Germany
boehnhardt@mps.mpg.de

Michael Delfs
Waldsassener Straße 23, 12279 Berlin, Germany
m.delfs@arcor.de

Prof. Dr. Hilmar W. Duerbeck
Postfach 1268, 54543 Daun, Germany
hilmar@uni-muenster.de

Prof. Dr. Johannes Viktor Feitzinger
Tewaagstraße 13, 44803 Bochum, Germany
www.johannes-feitzinger.de

Michael Gottwald
Lichsstraße 37, 73529 Schwäbisch-Gmünd/Bargau, Germany
michael.gottwald.bargau@web.de

Dr. Reinhold Häfner
Universitäts-Sternwarte München, Scheinerstraße 1, 81679 München, Germany
haefner@usm.uni-muenchen.de

Prof. Dr. Elmar Junker
FH Rosenheim, Fakultät für Angewandte Natur- und Geisteswissenschaften,
Hochschulstraße 1, 83024 Rosenheim, Germany
junker@fh-rosenheim.de

Prof. Dr. Christoph Leinert
Max-Planck-Institut für Astronomie, Königstuhl 17, 69117 Heidelberg, Germany
leinert@mpia-hd.mpg.de

Dr. Hartwig Lüthen
Behnstraße 13, 22767 Hamburg, Germany
h.luethen@alice-dsl.net

Dr. Martin J. Neumann
Redaktion Sterne und Weltraum, Max-Planck-Institut für Astronomie,
Königstuhl 17, 69117 Heidelberg, Germany
neumann@mpia.de

Dr. Klaus Reinsch
Institut für Astrophysik, Friedrich-Hund-Platz 1, 37077 Göttingen, Germany
reinsch@astro.physik.uni-goettingen.de

Dipl.-Kfm. Günter D. Roth
Ulrichstraße 43a, Irschenhausen, 82057 Icking, Germany

Dr. Gero Rupprecht
European Southern Observatory ESO,
Karl-Schwarzschild-Straße 2, 85748 Garching, Germany
grupprec@eso.org

Prof. Dr. Felix Schmeidler[†]
Mauerkircherstraße 17, 82057 München, Germany

Prof. Dr. Klaus-Peter Schröder
Departamento de Astronomía de la Universidad de Guanajuato,
Apartado Postal 144, Guanajuato, GTO, México, C.P. 36000
kps@astro.ugto.mx

Christine Treichel
Hauptstraße 81, 16727 Oberkrämer/Schwante, Germany

Peter Völker
Weskammstraße 13, 12279 Berlin, Germany

Dipl.-Ing. Bernd Weisheit
Bürgermeister-Langer-Straße 10, 75181 Pforzheim, Germany
weisheit@pro-bw.de

Dr. Volker Witt
Ganghoferstraße 5, 82178 Puchheim, Germany
volkerwitt@t-online.de

Peter Wright
European Radio Astronomy Club, Ziethenstraße 97, 68259 Mannheim, Germany
erachq@aol.com

1. Why Astronomy?

G.D. Roth

1.1 Introduction

Sooner or later each person has a personal experience with the starry sky. When there are no street lights or city dust veiling the stars, and there are still the black nights in the countryside, it is these moments that we find ourselves in a thoughtful mood. By gazing at the stars our thoughts turn to questions about the creation of the universe and the conditions of the Sun, the Moon, the planets, and ultimately our place in the cosmos.

In these moment of thoughtfulness our eyes are open to observe celestial phenomena. It is our chance to become an amateur astronomer. Thousands of people worldwide enjoy astronomy as their hobby. Telescopes have become a good value for the money, and many cities have public observatories and active astronomy clubs. Many activities are presented within current astronomy outreach programs, targeting schools, children and adults, by means of observatories, robotic telescopes, podcasts and more. Global telescopes create a network for astronomical education without worrying about the weather and without even leaving the classroom (Figs. 1–7 show various objects of interest related to the study of astronomy).

Fig. 1. Comet Hale–Bopp

See the chapter "The Social Astronomer", p. 281.

1.2 Astronomy and the Observer

To wonder about the world of the stars is one of the characteristic human endeavors. The observation of the stars transcends all boundaries. The entire universe is the subject of astronomy:

> Astronomy involves the exploration of all phenomena external to and including the Earth. Its realm is the whole of space, out to the remotest distances, and also all of time back to the origin of the Universe as deduced from stellar dating and cosmological theory. Apart from meteorites and the very nearest celestial bodies which can be reached directly by spacecraft, astronomical researchers are spatially separated from the objects of their study and therefore cannot experiment with them at will. They must make observations at the times and under the conditions that nature prescribes. Furthermore, the apparent diameters, separations, and motions which must be measured in astrometric work as well as the amount of light available for astrophysical analysis are usually so small that an appreciable degree of natural uncertainty is inherent in most observational measurements. The analysis and removal of this uncertainty has itself become a primary focus of astronomical research. (W.D. Heintz).

Fig. 2. The Great Red Spot on Jupiter

Fig. 3. Maksutov Telescope

See the chapters "Optics and Telescopes", p. 41, and "Telescope Mountings, Drives and Electrical Equipment", p. 95.

See the chapters "Fundamentals of Spherical Astronomy", p. 5, and "Applied Mathematics and the Computer", p. 23.

See chapter "The Nature of Light and Matter: Fundamentals of Spectral Analysis", p. 175, and chapter "Principles of Photometry", p. 205.

See chapter "Signals from Space: Radio Astronomy for Beginners", p. 239.

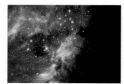

Fig. 4. The Omega Nebula M 17
See the chapter "Astrophotography", p. 133.

Even in the days when professional astronomers were no better suited than the amateur with regards to equipment, the number of amateur observers who managed to make startling scientific advances was comparatively small. Furthermore, it is probably true to say that the lucky few usually had some professional connection with a branch of science, or they had some general scientific grounding, or failing either of these, had a particular aptitude for systematic work. Often they also had two other useful attributes, namely time and money. No doubt there are exceptions to this generalization, and I mention it merely to emphasize the fact that by far the greater part was the personal satisfaction. Nowadays group work can, to some extent, compensate for lack of time and money, although the personal attributes mentioned above are still essential. The pooling of resources has the advantage that a given group can purchase equipment which individually they could not begin to afford. However, that still leaves the question of where and how the instruments are to be housed. Details regarding instruments and the location of observatories will be examined in the course of this book.

A scientific attitude demands methodical thinking and methodical working, in that order. A good observer ought to have more than just a facility with a telescope in his chosen field; he should also be widely read. Only then can he hope to get full value out of his observations and make the best use of the limited means at his disposal. There is really no reason why, even in this day and age, the amateur observer should not render a useful service to science.

Astronomy is closely related to mathematics and other "exact" sciences in that its methods and results are usually quantitative. The reduction of the observations, that is, the processing of raw data to obtain a result, often requires lengthy computations and the use of extensive tables; astronomical theories that depend on observations frequently use the most complex mathematical methods and high-speed computers.

1.3 Specialized Instruments, Methods and Objects

Classic reflecting and refracting telescopes gather all the information about the universe based on ordinary visible light. It concerns the radiation reaching the observer between $\lambda = 300$ nm to around $\lambda = 1000$ nm. With the aid of suitable instruments, this radiation can be decomposed into its spectral constituents, and the spectrum analyzed. Besides the observation of spectra the amateur astronomer can study the radiation of stars and other distant celestial objects by photometry involving the measurements of radiation summed over a more or less extended wavelength region.

With the discovery of non-visible electromagnetic radiation amateur astronomers started observing with radio telescopes, concave disks reflecting radio waves to a focus. Objects in the radio window are the Sun, the planets, especially Jupiter, but also galaxies.

Compared with visual observations astrophotography has the advantage to record fainter structures of celestial bodies and to provide permanent documents of these objects. Moreover these documents are the base for photometrical and astrometric studies. The availability of techniques to produce digital image information has developed the field of astrophotography in a dramatic manner for the amateur astronomer. The application of CCD cameras and webcams is as common today as the use of computers. Digital astrophotography without the assistance of a computer is impossible.

1.4 Major Objects for the Amateur

This book presents the substance of celestial bodies for the amateur observer in groups: the planets, stars and galaxies. The Sun is the local link to the other stars, the nexus of cosmic evolution. The Solar System is comprised of the Sun and all the celestial bodies orbiting it. This system is of special interest for the observing amateur.

The value of the work undertaken obviously depends on how systematic one's approach is and also how much time one is able or willing to devote to it. There is no reason why such a self-imposed task should ever become a drudgery; if this happens, then either the observer is temperamentally unsuited for this kind of work, or his method is wrong. Reading should always complement observations made at the telescope, not substitute these. The right sort of book should make the reader want to go out and see for himself. He can accomplish this with a comparatively modest optical apparatus: a good 3-inch. refractor will show a fair amount of detail on the Moon and the planets.

It is often said that the real value of education lies in its ability to stimulate people to practical activity on their own accounts. The educational value of telescopic observation is as great as the acquisition of sufficient optical and technical know-how to build one's own telescope. Club and school observatories should not confine their activities to a few set observing evenings alone. For example, it is very satisfying to have followed the planet Mars from the first weeks of visibility right through to opposition and beyond. It can clarify all sorts of points that one may have heard or read about the planet. Lunar and planetary observation is a most rewarding study topic for either adult or juvenile education.

Nevertheless within the Solar System are smaller celestial bodies and phenomena worthy of the attention of the observer. Besides the spectacle of lunar and solar eclipses, or the unexpected event of a comet or a nova, keen observers direct their attention towards meteors or occultations of stars. Modern techniques are a help as well, e.g., when registering meteors a video supported system is an advantage in comparison with visual observations.

The space programs of the US and Russia have energized mankind with amazing results about our place in the universe, but they also conduct research into the Earth. Artificial Earth satellites become bright enough to be observed by amateur astronomers, and an object of special interest is the International Space Station (ISS), a multidisciplinary research station for various studies in weightlessness. Besides exploring the formation of the Solar System and the objects of outer space, the missions of these satellites also focus on the Earth and the space near the interplanetary space. But for the amateur, observing satellites both natural and man-made depends on the weather and climate on Earth, where good and bad seeing are the basis of all observations.

By observing stars, nebulae and galaxies the amateur learns about the construction, the creation and the fate of the universe. Star clusters and clouds of glowing gas can tell the observer about the lives of stars. In the huge clouds of interstellar gas and dust, stars are born. The largest objects in the universe are the galaxies, groups of billions of stars, e.g., our galactic system, the Milky Way. Modern techniques for visual and photographical observations put the amateur in a comfortable financial position to obtain powerful equipment.

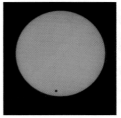

Fig. 5. Transit of Venus 2004

See the sections "The Sun", p. 309, "Modern Sundials", 255, "The Moon", p. 358, "Observation of the Planets", p. 427, and "Asteroids and Kuiper Belt Objects", p. 483 in Chap. 13.

See the sections "Eclipses and Transits", p. 386, "Comets", p. 498, "Meteors and Bolides", p. 526, and "Occultations of Stars", p. 406 in Chap. 13.

See the sections "Artificial Earth Satellites", p. 416, "Noctilucent Clouds, Aurorae, Zodiacal Light", p. 535, and "The Terrestrial Atmosphere and Its Effects on Astronomical Observations", p. 552 in Chap. 13.
See the sections "The Stars", p. 573, "Variable Stars", p. 584, "Binary Stars", p. 625, "The Milky Way and its Objects", p. 645, and "Extragalactic Objects", p. 667 in Chap. 14.

Fig. 6. The Milky Way Galaxy

Fig. 7. Photographic Refractor of Paris Observatory 1885

See the chapter "An Historical Exploration of Modern Astronomy", p. 267.

1.5 "The Universe, Yours to Discover"

This is the motto of the International Year of Astronomy (IYA) declared for 2009 by the International Astronomical Union (IAU) in collaboration with UNESCO and the United Nations. This is also the central idea of this book. Astronomy is the oldest fundamental discipline with a connection to many spheres of human existence. Advances in astronomy during the past 100 years have been sensational. In the near future we can look forward to enjoying the benefits of space technology, where astronomy is essential to developing a perspective on human existence and its relation to the cosmos.

Not every amateur observer will, or should, try to fulfill a specific scientific purpose with his observations. Much real pleasure can be had simply from astronomy as a hobby, though, as with most things, trying to do the job as well as possible adds to the pleasure. The contribution which amateurs make to the advance of science is decreasing year by year, but the role which amateur astronomy plays in satisfying man's scientific curiosity remains as great as ever. It is the opportunity to see and work things out for oneself which is the bug that bites the amateur and makes him spend his leisure time at some self-appointed scientific task.

2. Fundamentals of Spherical Astronomy

M. Gottwald

2.1 Introduction

As its name suggests, the subject of spherical astronomy purports to describe, using the language of mathematics, the positions and motions of phenomena which occur on the celestial sphere. This is a field that today is often neglected, and yet it is still the foundation of many branches of astronomy for observations and research, even if this is not always clearly evident. Every observer is confronted with problems involving spherical astronomy in practical work; for example, to observe a certain astronomical object, it is necessary to find its location, or coordinates, on the celestial sphere. The reduction or processing of observed data also often involves methods which are based on spherical astronomy.

The aim of this chapter is to provide the reader with a brief survey of the fundamentals of spherical astronomy. Because of restricted space in this book, several engaging subjects have been excluded.

In the last edition the reader was assumed to have some knowledge of numerical calculations, but today it is not necessary to compute transformations of coordinate systems and others manually. For example it takes only a click to change between several coordinate systems. Every astronomical software supports the important coordinate systems and several time systems as well.

2.2 The Coordinates

A point on a sphere is defined by two coordinates, assuming that the coordinate system itself is uniquely defined. There are several kinds of coordinate systems in spherical astronomy, and the most important ones will be dealt with in this chapter.

2.2.1 Geographic Coordinates

With the exception of space-based observations, observers work from a point on the Earth's surface defined by the three coordinates:

Geographic latitude, φ

Geographic longitude, λ

Altitude above sea level, H

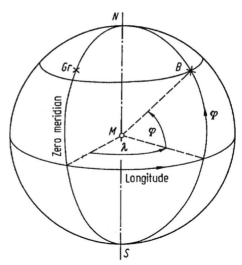

Fig. 1. Spherical coordinates

The geographic latitude is measured from 0° at the Earth's equator up to 90° at its poles, positive to the north, negative to the south. All circles parallel to the Earth's equator are called *parallels of latitude*, and all are smaller than the equator itself.

Any plane which cuts through the imaginary polar axis of the Earth is perpendicular to the equator and intersects the Earth along a *meridian*, which is used to measure geographic longitude (see Fig. 1). By international agreement, the meridian through the Royal Observatory, Greenwich, England, has been taken as the *zero meridian*.

Longitude is measured from 0 to 180°, negative to the west of Greenwich and positive to the east.

The determination of the geographic coordinates and the shape and size of the Earth is one of the tasks of astronomy and geodesy. In reality the Earth is not precisely spherical but rather is shaped somewhat like a flattened, or oblate, ellipsoid. For practical reasons, the so-called *reference ellipsoid* was introduced as a standard surface. This is chosen so that the directions of its normals agree as closely as possible with the directions of a plumb line at points on the Earth's surface. Also by international agreement, the following dimensions have been accepted for this reference ellipsoid:

$$a = \text{semimajor axis} = \text{equatorial radius} = 6378.140 \, \text{km} \, ,$$

$$f = \text{flattening} = \frac{1}{298.257} \, .$$

From the relation

$$f = \frac{a - b}{a} \, ,$$

the value b is obtained:

$$b = \text{semiminor axis} = \text{polar radius} = 6356.755 \, \text{km} \, .$$

The polar radius is therefore 21.4 km smaller than the equatorial radius. The physical surface of the Earth as a rule deviates somewhat from the reference ellipsoid. On the other hand, the term *geoid* refers to an equipotential surface defined by the average sea level, and whose position over the continents can be imagined as being represented by the water level in channels connected with the oceans.

As a consequence of the ellipsoidal shape of the Earth, there is a difference between the *geographic latitude* φ, defined by the angle between the equatorial plane and a plumb line at that point, and the *geocentric latitude* φ' given by the angle between the equatorial plane and the direction to the center of the Earth.

2.2.2 Horizontal Coordinates

If one imagines a plane passing through the place of observation at right angles to the vertical, it will then intersect the apparent sphere in a circle called the *horizon*, or, more precisely, the *apparent horizon*. Vertically above the observer is the *zenith*, and vertically downward is the *nadir* (see Fig. 2).

At any given point on the horizon, one can imagine a circle drawn perpendicular to it. Such a circle is called an *altitude circle*, and all such circles intersect at the zenith and below the horizon at the nadir. The altitude of a star above the horizon is measured along the altitude circle from 0 to 90° at the zenith. All altitudes measured above the apparent horizon are called *apparent altitudes a*; owing to atmospheric refraction, they are slightly greater than the true altitudes.

The *zenith distance* $z = 90° - a$ is the complement of the altitude and is frequently given instead of a.

A plane through the center of the Earth, running parallel to the apparent horizon, intersects the celestial sphere at the *true horizon*. Altitudes referred to it are called *true altitudes*. Since any celestial body is located at a finite distance from Earth, its apparent

The *refraction* of light on its way through the Earth's atmosphere will be considered in many astronomical software packages. More and more freeware include this feature. The parameters to compute it are the altitude of your observation point, the air temperature and air pressure.

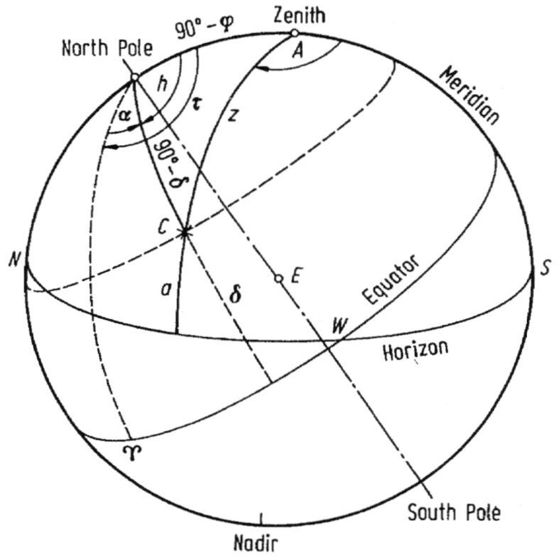

Fig. 2. The horizon coordinate system

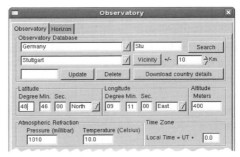

Fig. 3. Consideration of refraction (Cartes du Ciel)

altitude will be smaller, sometimes substantially so, than the true one; in the case of the Moon, the difference can amount to about 1°. On the other hand, astronomical refraction caused by the deflection of the rays from a celestial object in the Earth's atmosphere elevates its apparent position with respect to the horizon; therefore the apparent altitude is increased with respect to the true one.

All celestial observations must be corrected for the effects of *refraction* (see Fig. 3).

The second coordinate in the horizon system is defined by the direction of the point where the altitude circle through the star cuts the horizon and is called the *azimuth* A. In astronomy, it is measured south to west, north, and east, from 0 to 360°. In geodesy, on the other hand, the measurement starts from the north and proceeds in the same sense. The four points on the horizon having astronomical azimuths 0°, 90°, 180°, and 270° are called the *south point, west point, north point,* and *east point,* respectively. The altitude circle passing through the south point, the zenith, and the north point is called the *local meridian*; the altitude circle perpendicular to it through the west point, the zenith, and the east point is called the *prime vertical.* Because of the rotation of the Earth, each celestial body traverses the meridian twice every day, once in its highest position, or *upper transit,* which for observers in the northern hemisphere is usually in the south, and the second time in its lowest position, or *lower transit* in the north. For some stars both transits take place above the horizon; such stars are called *circumpolar.* The condition for a star with declination δ to be circumpolar is given by the relation

$$|\delta + \varphi| \geq 90° ,$$
(1)

Fig. 4. Several coordinates of Deneb (KStars)

where φ is the geographic latitude of the observer. In the northern hemisphere, $\varphi \geq 0$ and the circumpolar condition becomes simply that $\delta \geq 90° - \varphi$.

All stars that do *not fulfill* the circumpolar condition of Eq. (1) must rise and set, provided that they can be seen at all. The positions on the horizon where the star rises and sets and the star's motion above the horizon all depend on the geographic latitude of the observer as well as on the declination of the star.

As viewed from a point on the Earth's equator, all stars rise perpendicularly to the horizon. The further the place of observation is from the equator the smaller the angle between the horizon and the direction of the star's rising or setting. When viewed from the Earth's North or South Poles, all stars move in circles parallel to the horizon and therefore never rise or set; the Sun, Moon, planets, and comets, however, because of their own intrinsic motions in declination, can in fact rise and set as seen from these two specific locations.

2.2.3 The Equatorial System, Vernal Equinox, and Sidereal Time

The major disadvantage of using the horizon system is that the coordinates of each star change continuously during the course of the day, and also that they differ from place to place. This disadvantage is removed in the *equatorial system*. If one imagines an extension of the plane of the Earth's equator, it intersects the celestial sphere at the *celestial equator*. The imaginary rotational axis of the Earth is perpendicular to this plane and

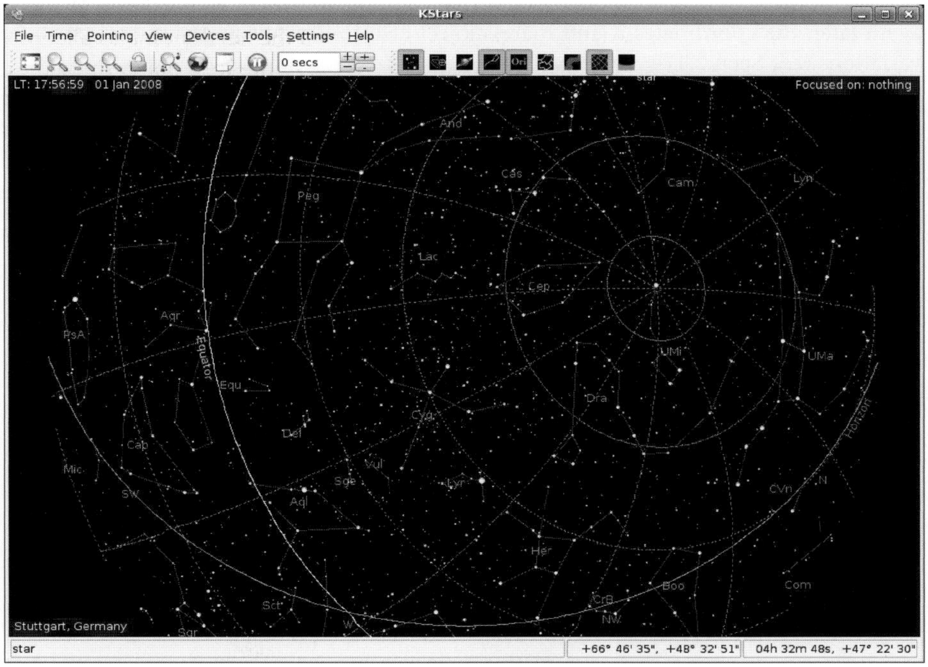

KStars is a part of KDE (*K Desktop Environment* http://edu.kde.org/kstars), a graphical user interface (GUI) for Linux. Meanwhile it is a powerful (and free) desktop planetarium (see Figs. 4–8 for examples of astronomical software).

Users of GNOME (the other widespread GUI for Linux) can also use KStars, by installing the KDE-Education-RPM.

Fig. 5. Horizon, equatorial grid and celestial equator (KStars)

intersects the celestial sphere at the north and south celestial poles. In a system thus defined, it is possible to define more permanent coordinates of the star, analogous with the geographical coordinates of a point on the Earth's surface.

The first equatorial coordinate is obtained by measuring the angular distance of a star from the celestial equator toward the north or south from 0° at the celestial equator to 90° at the celestial pole; this quantity is called the *declination*, δ. It is counted positive toward the north and negative toward the south. Any great circle through the star and perpendicular to the equator is called a *circle of declination*.

The other coordinate is counted along the equator in analogy to the geographic longitude. Here the zero point is the *vernal equinox*, also called the *first point of Aries* (Υ), which is one of the two points of intersection between the celestial equator and the plane of the annual solar motion (the *ecliptic*). When the Sun arrives at this point, spring begins. The diametrically opposite point is called the *autumnal equinox*. The second coordinate of the star corresponds to the geographic longitude and is measured by the angular distance from the vernal equinox to the point of intersection of the circle of declination with the equator; the possible range is from 0 to 360°, and the sense of the angle is from west to east, that is, in the sense of the annual motion of the Sun. This coordinate is called the *right ascension*, α, or simply RA. The position of a star on the celestial sphere is uniquely determined by the two coordinates of right ascension and declination.

The vernal equinox itself therefore has the coordinates

$$\alpha = 0^\mathrm{h}0^\mathrm{m}0^\mathrm{s} \quad \text{and} \quad \delta = 0°0' \,.$$

Because of the slow displacement of the vernal equinox, to which reference will be made again later, the coordinates α and δ are subject also to slow, progressive, and periodic variations (see Sect. 2.4).

It is not difficult to find a relation between the coordinate systems referred to the horizon and to the equator if the concept of the *hour angle h* is introduced. This quantity is the angular distance between the circle of declination through the star and the upper meridian. It is measured from south through west, north, and east from 0 to 360°, or from 0 to 24$^\mathrm{h}$. Since a whole revolution of 360° takes place in 24$^\mathrm{h}$, the angles can always be expressed either in time *or* in degrees by means of the following conversions:

$$1^\mathrm{h} = 15°$$
$$1^\mathrm{m} = 15'$$
$$1^\mathrm{s} = 15''$$

Time	
Julian:	2454467.03950
UTC Date:	1/01/2008
UTC Time:	12:56:53
Sidereal:	19:39:08
TZ Name:	XXX
TZ Offset:	0:00:00
Local Date:	1/01/2008
Local Time:	12:56:53
Delta T:	(Auto) 73.80

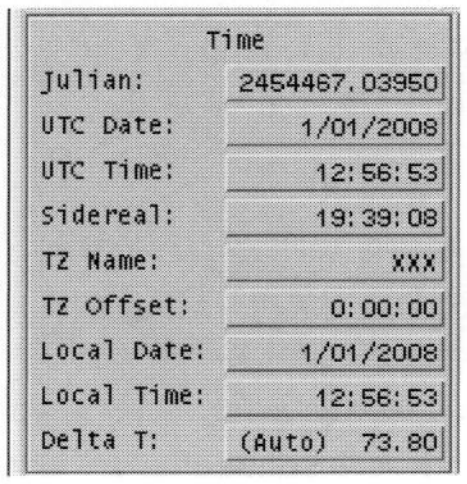

Fig. 6. The local sidereal time and more information for Stuttgart on 01.01.08 (XEphem)

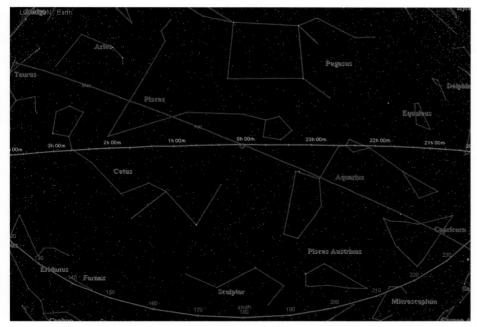

Fig. 7. Upper transit of vernal equinox = $0^h 0^m$ sidereal time (Distant Suns)

At upper transit, the star has the hour angle $0°$ (0^h), and at the lower transit $180°$ (12^h). Of course the vernal equinox, just like a star, also has a certain hour angle. This special quantity is known as the *sidereal time τ*. When the vernal equinox is in upper transit the local sidereal time is $τ = 0^h 0^m 0^s$ (see Fig. 7). There exists a very simple relationship between sidereal time, right ascension, and hour angle (see also Fig. 2):

$$τ = h + α \quad \text{or} \quad h = τ - α . \tag{2}$$

This relation is of greatest practical importance, one wishes to locate an object in the sky: if $τ > α$, the hour angle is west; if $τ < α$ the hour angle is east.

XEphem
http://www.clearskyinstitute.com/
xephem/index.html (Freeware)

Distant Suns
http://www.distantsuns.com
(Commercial)

2.2.4 Transformation of Horizontal Coordinates into Equatorial Coordinates and Vice Versa

The transformation from horizontal coordinates into equatorial ones and vice versa is achieved by applying the formulae of spherical trigonometry. To convert horizontal into equatorial coordinates you need only to change between the coordinate systems by clicking on the buttons of a planetarium software. Or you can use a coordinate converter/calculator, provided by some astronomical programs (Fig. 8).

2.2.5 Other Coordinate Systems

Apart from the systems of coordinates referred to the horizon and to the equator, there are two other coordinate systems which are used in astronomy: *ecliptic coordinates* and *galactic coordinates*.

Fig. 8. The coordinate converter of XEphem

2.2.5.1 The System of the Ecliptic In the ecliptic coordinate system the fundamental plane is the ecliptic itself. The *ecliptic latitude β* is measured at right angles from the ecliptic to the north and the south, running from 0 to ±90°, and the *ecliptic longitude λ* starts at the vernal equinox and runs from 0 to 360° in the same eastward sense as the right ascension. Since ε, the *obliquity*, or inclination of the ecliptic plane, is known, the equatorial coordinates can easily be converted into ecliptic coordinates, and vice versa. The obliquity is slowly variable and given by the formula

$$\varepsilon = 23°26'21''.448 - 46''.82\,T - 0''.0006\,T^2 + 0''.0018\,T^3 \,, \tag{3}$$

with T in Julian centuries from the year AD 2000.

2.2.5.2 The Galactic Coordinate System For investigations of stars and galactic structure it is convenient to use the galactic plane, or Milky Way, as the fundamental plane. The coordinates in this scheme are called *galactic coordinates* and are defined as follows (see, e.g., Mihalas and Binney [1]):

> *Galactic latitude b* runs from 0 to 90°, and is positive to the north and negative to the south of the galactic plane. *Galactic longitude l* is measured from 0° in the direction of the galactic center and runs eastward along the galactic plane to 360°.

The north galactic pole has equatorial coordinates

$$\alpha = 12^h51^m.0 \,, \quad \delta = 27°08' \quad (\text{equator } 2000.0) \,,$$

while those of the galactic center are

$$\alpha = 17^h45^m.7 \,, \quad \delta = -29°00' \quad (\text{equator } 2000.0) \,.$$

The inclination i of the galactic plane against the 2000.0 equator is 62°52'.

Prior to the 1959 definition, a different system was used and can be found in earlier literature. It counted galactic longitude from the ascending node of the ecliptic plane on the equator. In the new system this node has the coordinates

$$\alpha = 18^h50^m.40^s \quad \delta = -0°22'10''$$

and

$$l = 32°31' \,, \quad b = 0° \,,$$

so that apart from a slight change in the inclination, $l(\text{new}) = l(\text{old}) + 33°$. During the transition period, the coordinates l, b in the old and new systems were distinguished by superscripts I = old and II = new.

2.3 Time and the Phenomena of Daily Motion

The simplest and most obvious periodic change that can be used to measure time is the continual passage from day to night and night to day; for longer intervals, the change of seasons is convenient to use. Thus the rotation of the Earth and its revolution about the Sun are the basis for nearly all astronomical time calculations. Since most of the important astronomical quantities are time dependent, each measurement should include the specific date and time when it was made.

Decades ago, it was recognized that the period of rotation of the Earth about its axis, that is, the length of the day, is not strictly constant. Therefore, the day consisting of 24 hours can no longer be considered an accurate measure of time when high precision is needed. Far more accurate units of time defined in terms of atomic oscillations now exists and must be employed instead. For the purpose of most astronomical observations, however, such extreme precision is not required, and the older methods of time determination, most of which were in use before the variability of the day was recognized, are still of value.

2.3.1 True and Mean Solar Time

Originally the position of the Sun provided the interval of the day. However, when observations were made from a fixed location of the intervals between two successive upper transits of the Sun, it was noticed that these intervals are subject to conspicuous variations. Thus the hour angle of the "true" Sun, which gives the *true solar time* (TST), is non-uniform and therefore unsuitable for measuring time with precision. There are two major reasons for this non-uniformity:

1. The Earth's orbit around the Sun is not a circle, but an ellipse. Because of Kepler's second law it is therefore impossible for the Sun to traverse equal arcs in the sky in equal times.
2. Even if the Sun's motion along the ecliptic swept out equal arcs in equal intervals of time, the projection of these arcs onto the celestial equator, along which time is measured, would lead to unequal time segments owing to the mutual inclination between the equator and the ecliptic.

It is for these reasons that the concept of a *fictitious mean Sun*, whose position coincides with that of the true Sun at the vernal equinox and which moves with uniform speed along the celestial equator, is introduced. This defines *mean solar time* (MST) as a uniform measure of time.

The hour angle of this mean Sun is called *mean local time* (MLT) and differs from place to place. Mean solar time and also the true solar time are therefore both *local* times.

The difference between the true and mean solar times, TST − MST, is called the *equation of time* (Figs. 9 and 10). The equation of time becomes zero on four dates each year, namely on April 16, June 14, September 2, and December 25. The extreme values, on the other hand, are $-14^m.3$ on February 11, $+3^m.7$ on May 14, $-6^m.4$ on July 26, and $+16^m.4$ on November 3. The behavior of the equation of time over the course of a year is shown in Fig. 9.

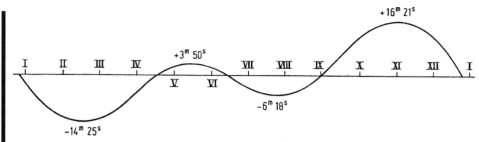

Fig. 9. The equation of time

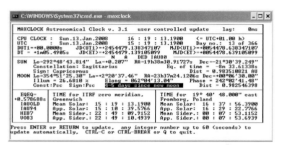

Fig. 10. Several time and date informations (MAXCLOCK)

Before 1930, the definition of the equation of time used to be opposite in sign to that given above. In current nomenclature, the term is mostly avoided and is replaced with the term *ephemeris transit of the true Sun*.

2.3.2 The Relation Between Sidereal Time and Mean Time

Sidereal time has already been defined as the hour angle of the vernal equinox (Sect. 2.2.3). Strictly speaking, it is not a uniform measure of time, since nutation (see Sect. 2.4.2) causes the vernal equinox to perform a small periodic variation, although in practice this effect can be neglected. A sidereal day, that is, 24 hours of sidereal time, is the interval between two upper transits of the vernal equinox. Sidereal time is also a local time, since, from Eq. (2),

$$\tau = h + \alpha .$$

On the meridian, $h = 0^h$, and therefore $\tau = \alpha$. This means that the right ascension of a star at its upper transit is equal to the sidereal time. This can also be put another way: sidereal time is the right ascension of all those stars which simultaneously traverse the meridian at their upper transit. Of course, the fictitious mean Sun also has a certain right ascension at its upper transit, and this is called *sidereal time at mean noon*.

Astronomical almanacs generally give the sidereal time 0^h universal time from day to day. This is valid only for the meridian of Greenwich, and since sidereal time is a local time, it is necessary to add a correction $\Delta\tau$ to obtain the sidereal time at mean midnight at a point on the Earth whose longitude is $\Delta\lambda$ (in hours) measured from Greenwich:

$$\Delta\tau = \pm 9^s.8565\,\Delta\lambda \,, \tag{4}$$

where the + and − signs are for longitudes west and east, respectively, of Greenwich. A sidereal day is somewhat shorter than a mean day since the Sun moves daily some amount toward the east and thus arrives at upper transit later than a fixed star. Dividing a sidereal day into 24 hours of 60 minutes each, with each minute of 60 seconds, it is found that every unit of sidereal time must be somewhat shorter than the corresponding unit in mean time. During a *tropical year* the Sun moves from vernal equinox to vernal equinox. This requires 365.2422 mean solar days. During this time the Sun has completed one revolution; a star, therefore, has one more upper transit per year than the Sun. The relation between the two days, is thus

MAXCLOCK
http://www.ginko.de/user/udomark/
maxclock_main_e.html (Freeware)

$$\text{One mean solar day} = \frac{366.2422}{365.2422}\ \text{sidereal days}\,,$$

$$\text{One sidereal day} = \frac{365.2422}{366.2422}\ \text{mean solar days}\,.$$

From this it follows that

$$24^h0^m0^s\ \text{mean solar time} = (24^h + 3^m56^s.555)\ \text{sidereal time}\,,$$
$$24^h0^m0^s\ \text{sidereal time} = (24^h - 3^m55^s.909)\ \text{mean solar time}\,.$$

Or, per hour,

$$1^h\ \text{mean solar time} = (1^h + 9^s.856)\ \text{sidereal time}\,,$$
$$1^h\ \text{sidereal time} = (1^h - 9^s.829)\ \text{mean solar time}\,.$$

For most purposes the conversion can be achieved with sufficient accuracy using a simple approximate relation given in 1902 by Börgen:

> For every hour of mean solar time, the sidereal time *gains* $(10 - 1/7)^s$.
> For every hour of sidereal time the mean solar time *loses* $(10 - 1/6)^s$.

The error due to this approximation is only $0^s.00067$ per hour in the first case and $0^s.00379$ in the second.

2.3.3 Other Phenomena of Diurnal Motion

During the 24 hours in a day, the apparent motion of celestial bodies causes some particular phenomena which are to be commented on. Upper transit has already been mentioned; there is also *rising* and *setting, passage through the prime vertical*, and **largest digression**. Here you see an example of several diurnal information (Fig. 10), such as setting or rising time of a star (Fig. 11).

Largest digression: The two positions of an object where it moves exactly vertical up or down. These points are the largest eastern and western azimuth of this object.

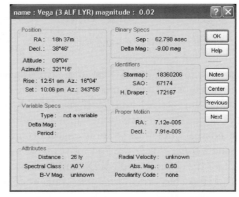

Fig. 11. Diurnal motion and further information of Vega (Distant Suns)

2.4 Changes in the Coordinates of a Star

While the coordinates of a star measured in the horizon system usually undergo very rapid changes with time, the coordinates in other systems vary only slowly. The causes of these variations are discussed in the sections below.

2.4.1 Proper Motion

The Tycho-2 catalog contains 2,539,913 proper motions of the brightest stars in the sky with a magnitude down to $12^m.0$.

Edmund Halley recognized in 1718 that the stars change their positions on the celestial sphere with relation to one another. These continuous displacements are due to the fact that all stars, including the Sun, are actually traveling through space, and the component of the yearly displacement which is perpendicular to the line of sight (i.e., the projection of the annual displacement onto the celestial sphere) is called the *proper motion* μ of the star. Because the stars are very distant this proper motion is very tiny and therefore escaped detection until the eighteenth century. Halley was able to demonstrate the proper motions of only a very few stars; today over 2,500,000 proper motions are known.

The largest annual proper motion is $10''.34$, exhibited by a faint star of magnitude $9^m.6$, named Barnard's Star in the constellation Ophiuchus. Its the fourth star (after the three components of the α-Centauri system), a red dwarf, with a distance of 5.979 light years.

Only about 500 stars are known to have proper motions which exceed $1''.0$. The annual proper motion is resolved into two components, one in the direction of right ascension (Fig. 11) the other in declination; these are denoted by μ_α and μ_δ. The observer will, apart from a few exceptions, have no need to make allowance for the annual proper motion.

2.4.2 Precession and Nutation

In Fig. 12 you see the proper motions of a period of about 50,000 years ($\pm 25,000$ years). The middle of the gray line is the actual position, the red dot marks the final position.

Precession is of the utmost importance since it causes relatively rapid changes in the coordinates α and δ. The gravitational pull of the Sun and the Moon on the equatorial bulge of the rotating Earth causes the polar axis to slowly move, eventually describing a complete revolution about the pole of the ecliptic in about 26,000 years. This period, which is more precisely stated as 25,784 years, is also called the *Platonic year*, and the movement of the polar axis about the pole of the ecliptic, in analogy to the behavior of a spinning top, is called *precession*. This phenomenon was recognized by Hipparchus as

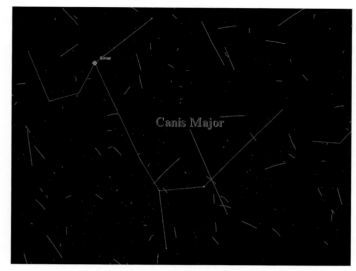

Fig. 12. Proper motions of stars in Canis Major (Distant Suns)

early as the second century BC and appears as a secular retrograde movement of the vernal equinox. The largest part of the precession, so-called *lunisolar precession*, is caused by the inequality of the moments of inertia of the Earth and can be determined only by empirical methods. The annual lunisolar precession, according to international convention, is found to be

$$p_0 = 50''.37 + 0''.000049\, t \,, \tag{5}$$

where t is measured in tropical years from the year 2000. To be added to this is a corresponding effect due to the planets in the solar system, the annual *planetary precession*:

$$p_1 = -0''.12 + 0''.000189\, t \,. \tag{6}$$

The total effect is the so-called *general precession*:

$$p = p_0 + p_1 \cos \varepsilon = 50''.27 + 0''.000222\, t \,. \tag{7}$$

In consequence of this precession the ecliptic longitude of the star increases continuously.

Currently, the north celestial pole is in close proximity to the star α Ursae Minoris; it will come nearest to that star in the year AD 2115, when its angular distance will be only 28′. Around AD 14000, the bright star Vega (α Lyrae) will be the new "pole star" (Fig. 15). The revolution of the Earth's axis about the pole of the ecliptic is not entirely uniform. The term "precession" means the progressive part of this motion; the small fluctuations superimposed on this motion are known as *nutation*. In 1747, the British astronomer J. Bradley discovered an oscillation of the polar axis with a period of 18.6 years. This short period movement is the major contribution of nutation and is caused by the fact that the lunar orbit does not lie in the plane of the ecliptic, and by the retrograde motion of the nodes of the intersection of the lunar orbit and the ecliptic. It therefore has the period of

A good view of precession is seen in Figs. 13 and 14. The software Distant Suns can calculate the precession of stars and constellation outlines separately. Thereby the difference of star positions in a period of time is very good to see.

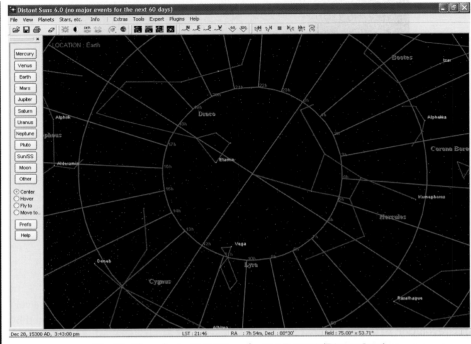

Fig. 13. The pole star 58 Iot Hercules in AD 15300 (Distant Suns)

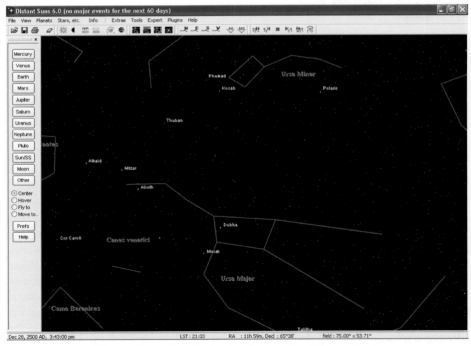

Fig. 14. Precession calculated without outlines (Distant Suns)

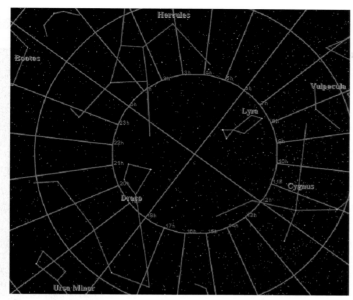

Fig. 15. The famous star Vega as a pole star in A.D. 13500

one revolution of the line of nodes, namely, 18.60 tropical years. Owing to nutation alone, the celestial pole describes an ellipse on the celestial sphere. The semimajor axis of this ellipse (i.e., the amplitude of the nutation oscillation) is called the *constant of nutation* and is equal to $9''.202$.

2.4.3 Aberration

The phenomenon of *aberration of starlight* originates from the fact that the orbital velocity of the Earth is a finite fraction of the speed of light. Because of this, the stars appear displaced a little in the direction of motion of the Earth. This phenomenon was discovered accidentally by Bradley in 1728. The *annual aberration*, caused by the motion of the Earth in its orbit about the Sun, must be distinguished from the much smaller *diurnal aberration* resulting from the rotation of the Earth around its polar axis. Because of the annual aberration, a star changes its position periodically throughout the year, tracing out a small aberration ellipse on the celestial sphere.

> To imagine the aberration you can think of a ride in your car on a winter's night. The snowflakes appear to fall at an oblique angle to the windshield of the car, but as you drive faster, they fly more horizontally in front of you.

The maximum value of the displacement, called the *aberration constant*, is

$$k = 20''.496 .$$

At the pole of the ecliptic ($\beta = 90°$), the aberration ellipse becomes a circle, and on the ecliptic ($\beta = 0°$) it degenerates to a linear segment.

The diurnal aberration originates from the rotation of the Earth. The speed of displacement of a point at the Earth's equator gives the constant of daily aberration in the amount of $0''.31$.

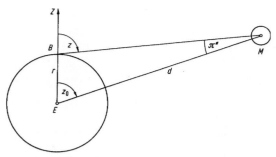

Fig. 16. Parallax

2.4.4 Parallax

Diurnal parallax is due to the fact that angles to a celestial object vary somewhat depending on whether they are measured from a point on the surface of the Earth or from the center of the Earth. In Fig. 16 the angle z_0 indicates the zenith distance of the center of the Moon as it would appear from the center of the Earth; however, the zenith distance observed at the same time from a point B on the surface of the Earth would be z. Thus there exists a small angle π'' subtended at the center of the Moon which is the parallax. The larger the distance between the centers of the celestial bodies, the smaller the value of π''. The equatorial-horizontal parallaxes of the Moon and Sun are

$$\pi''_{\mathbb{C}} = 3422''.44 = 57'2''.44 \,,$$
$$\pi''_{\odot} = 8''.794 \,.$$

In exact calculations the fact that the Earth is flattened at the poles must also be taken into account. The difference between the geographic and the geocentric latitude, $\varphi - \varphi'$, and the definition of the corresponding radius r' have been given in Sect. 2.2.1. The *annual parallax* is defined analogously, except that the baseline is now the semi-major axis of the Earth's *orbit*. The annual parallax of even the nearest fixed star is $0''.772$, but most observers will seldom, if ever, need to take such minute angles into account.

2.4.5 Reduction of Mean Position to Apparent Position

The *mean position* α_0, δ_0 of a star is its position at some chosen instant of time referred to the mean coordinate system at that time from a star catalog and with precession taken into account according to the formulae of Sect. 2.4.2; it also includes the constant term of aberration. On the other hand, the *apparent position* α_{app}, δ_{app} is the position of the star at perhaps a different instant, referred to the true coordinate system at that instant. Thus the apparent position differs from the mean position in that it is affected by nutation, the variable part of aberration, and the displacement due to precession and proper motion between the two instants of time.

References

1. Mihalas, D., Binney, J.: Galactic Astronomy 2nd edn, p. 39. Freeman, San Francisco (1981) (actual edition: Binney, J., Merrifield, M.: Galactic Astronomy. Princeton University Press, Princeton 1998)

The URLs of the websites of discussed software can be found in the marginal column on the corresponding page.

Acknowledgement. I want to thank the following people, organizations and distributors: The other authors of this book for the good cooperation, especially Dr. rer. nat. Oliver Montenbruck and Prof. Dr. Klaus-Peter Schroeder for their helpful hints. Mike Smithwick (Distant Suns), Patrick Chevalley (Cartes du Ciel/Sky Charts), the KStars team and Clear Sky Institute (XEphem) for authorizations of software screenshots.

Based on *Fundamentals of Spherical Astronomy* by Prof. F. Schmeidler, Compendium of Practical Astronomy, Vol. 1, Chap. 2, Springer, Berlin, Heidelberg, New York (1994)

3. Applied Mathematics and the Computer

M. Gottwald

3.1 Introduction

Practical astronomical work requires, to a large extent, the evaluation of mathematical formulae by numerical calculation. For this reason, astronomy was ranked as a subfield of mathematics until well into the nineteenth century. Today, more than ever, the treatment of many astronomical problems is impossible without the use of mathematical tools. But now no observer must master these mathematical techniques, since many astronomical software packages already include the important algorithms, such as ephemeris calculations, coordinate transformation or interpolation. So the numerical performance of calculations is no longer in the domain of logarithm tables or pocket calculators. Furthermore most amateur astronomers now have access to powerful electronic computers. Customary in the trade, the computer now provides enormous possibilities for astronomical software.

These days, you do not have to pay for astronomical software. On the Internet one can find freeware or shareware (the author expects a small amount). The quality now is close to the level of commercial software.

This chapter of the last edition was completely overworked, except for the theory of errors (this is the original text by Prof. Felix Schmeidler).

The interested reader can find such details in standard textbooks on mathematics, and further details in special books [1, 2], or in the previous edition of this book.

3.2 The Theory of Errors

Every measurement is subject to *errors* (more properly called *uncertainties*) because man-made instruments and human senses are incapable of infinite precision. One can distinguish between two different types of errors: *systematic* and *random*. *Systematic errors* depend in a known manner on some external circumstances and tend to shift the observed values consistently toward either higher or lower values than the "true" value. Such systematic shifts can be determined, although usually not without tedious analysis. *Random errors* (also called *accidental errors*), on the other hand, tend to "scatter" the observed data so that some of the values fall above the true value and some fall below; by their very nature they cannot be predicted. It is only the latter with which the general theory of errors is concerned. Its task is to establish the law of frequency of errors and to judge the accuracy of a measurement or of a calculation.

The fundamental principle of the theory of errors was established by Gauss, and may be stated as follows: Suppose a series of measurements have been made to obtain a particular required quantity. Then the "best" estimate of the sought quantity is that value for which the sum of the squares of the individual errors is a minimum. This theory expresses not a fundamental law of nature but rather a very plausible definition. A good introduction to error theory can be found in Taylor [3]; another, older reference is the book by Mendenthall [4].

There is no software known by the author especially for this case. But you can use every statistical software. The usual formulae like *standard deviation* or the *principle of least squares* is mastered by this software.

3.2.1 Direct Observations

Consider the simplest case of a single quantity l that can be measured directly. Let there be n measured values $l_1, l_2, \ldots, l_n$ which would be identical if there were no errors; obviously this is never true in practice. Here, in accordance with the *principle of least squares*, the theory of errors asserts that the "most probable value" of the required quantity is given by the *arithmetic mean*

$$\bar{l} = \frac{1}{n}(l_1 + l_2 + \ldots + l_n) . \tag{1}$$

The *standard deviation*, or *mean error*, is expressible in terms of the differences, also called *residuals*, $v_i = l_i - \bar{l}$ between the individual measurements and the mean value $\bar{l}$. Denoting the sum of the squares of the differences, $\sum_{i=1}^{n} v_i^2$, by $[vv]$, as is customary in the theory of errors, then the *standard deviation of a single measurement*[1] is

$$\sigma_l = \sqrt{\frac{[vv]}{n-1}} \tag{2}$$

and the *standard deviation of the mean* is

$$\sigma_l = \frac{\sigma_l}{\sqrt{n}} = \sqrt{\frac{[vv]}{n(n-1)}} . \tag{3}$$

Thus, by forming the mean of n individual measurements, the accuracy of the results can be improved by a factor of $\sqrt{n}$. These formulae assume that all measurements are of equal reliability, but this is often not the case. For example, an observer with a small telescope of low magnification will be able to measure the distance between the components of a double star less accurately than someone using a large instrument. Such differences in the accuracy can be allowed for, however, by assigning to each measurement a *weight*. The larger the degree of reliability, the larger the weight. The determination of these weight factors p_i depends in each case on an assessment, frequently only a very approximate one, of the quality of the observation. Once the weight factors have been assigned, the mean is no longer equal to the arithmetic mean of the single measurements but instead is given by the *weighted mean*

$$\bar{l} = \frac{l_1 p_1 + l_2 p_2 + \ldots + l_n p_n}{p_1 + p_2 + \ldots + p_n} = \frac{[lp]}{[p]} . \tag{4}$$

---

[1] This definition is more commonly referred to as the *standard deviation of the population*.

In this case, the standard deviation of one measure of unit weight is

$$\sigma_l = \sqrt{\frac{[vvp]}{n-1}} \tag{5}$$

where the term $[vvp]$ denotes the sum $\sum_{i=1}^{n} v_i^2 p_i$. The standard deviation of the mean is

$$\sigma_{\bar{l}} = \sqrt{\frac{[vvp]}{[p](n-1)}} \; . \tag{6}$$

Very often the required quantity cannot be measured directly but is a known function of other quantities that can be measured. For instance, it is impossible to measure the absolute magnitude of a star directly, but it can be calculated if the apparent magnitude and the distance of the star are known. Since these two quantities can be measured only with a certain mean uncertainty (standard deviation), it is desirable to know the expected uncertainty of the computed absolute magnitude. The answer to this question is provided by the *law of the propagation of errors*.

Let the required quantity x be a known mathematical function of n other quantities $x_1, x_2, \ldots, x_n$, i.e.,

$$x = \varphi(x_1, x_2, \ldots, x_n) \; . \tag{7}$$

If each of these quantities x_i is affected by a mean uncertainty σ_i, then the standard deviation of x is given by the formula

$$\sigma_x^2 = \left(\frac{\partial \varphi}{\partial x_1}\right)^2 \sigma_1^2 + \left(\frac{\partial \varphi}{\partial x_2}\right)^2 \sigma_2^2 + \ldots + \left(\frac{\partial \varphi}{\partial x_n}\right)^2 \sigma_n^2 \; . \tag{8}$$

The standard deviation, or mean error, is a measure of the accuracy of the observation. However, one must be aware of the fact that the actual uncertainty exceeds the mean error. The following rule of thumb can be formulated: The discrepancy between a single measurement and the mean could, in unfavorable circumstances, be as much as 2.5 to 3 times the mean error. Therefore, if a previously known quantity is to be re-examined by new measurements and the resulting value turns out to be different, then this difference is meaningful (i.e., real) only if it exceeds the mean error of the measurement by at least a factor of 2.5; otherwise the result of the new measurement should be interpreted as a confirmation of the old value.

3.2.2 Indirect Observations

If one aims at the determination of several unknown quantities whose mathematical connection with the measured values is known, then a least-squares solution of indirect observations is made. In most cases the relation between the measured and the unknown quantities is linear, and if this is not the case the computation can be linearized by using approximations. It is assumed for the sake of simplicity that there are only three unknowns; the same principles are used for any other number of unknowns. Let the measured quantity l be a function $\varphi(x, y, z)$ of three unknowns x, y, z. If, on the basis of some plausible hypothesis, the approximate values x_0, y_0, and z_0 are introduced for the unknowns and their true values denoted by $(x_0 + \xi)$, $(y_0 + \eta)$, $(z_0 + \zeta)$, we are left with the determination of the *corrections* ξ, η, and ζ. According to *Taylor's theorem* a power series can be set up as follows:

$$f = \varphi(x_0 + \xi, y_0 + \eta, z_0 + \zeta) = \varphi(x_0, y_0, z_0) + \xi \frac{\partial \varphi}{\partial x} + \eta \frac{\partial \varphi}{\partial y} + \zeta \frac{\partial \varphi}{\partial z} + \ldots \qquad (9)$$

The numerical values of the derivatives are to be calculated at the point (x_0, y_0, z_0).

Using the notations

$$\frac{\partial \varphi}{\partial x} = a , \quad \frac{\partial \varphi}{\partial y} = b , \quad \frac{\partial \varphi}{\partial z} = c , \quad f - \varphi(x_0, y_0, z_0) = l , \qquad (10)$$

a linear relation between the measured quantity l and the unknowns is obtained:

$$l = a\xi + b\eta + c\zeta . \qquad (11)$$

Of course, at least three such equations connecting the numbers l_1, l_2, l_3 with the three unknowns ξ, η and ζ are required. Usually, however, there are more equations than unknowns, and our task is to determine the most probable values of the unknowns ξ, η and ζ from the measured quantities l_i, each of which carries a random error of measurement.

The various measurements must be performed in such a way that the values of the derivatives of the various functions φ_i differ from each other as much as possible (in order to separate the unknowns). If there are n different measured quantities, then n *equations of condition* are to be satisfied:

$$\begin{aligned}
a_1 \xi \; + \; b_1 \eta \; + \; c_1 \zeta \; &= \; l_1 , \\
a_2 \xi \; + \; b_2 \eta \; + \; c_2 \zeta \; &= \; l_2 , \\
\cdots \quad\;\; \cdots \quad\;\; \cdots & \\
\cdots \quad\;\; \cdots \quad\;\; \cdots & \\
\cdots \quad\;\; \cdots \quad\;\; \cdots & \\
a_n \xi \; + \; b_n \eta \; + \; c_n \zeta \; &= \; l_n .
\end{aligned} \qquad (12)$$

From these are formed the *normal equations*:

$$\begin{aligned}
[aa] \, \xi + [ab] \, \eta + [ac] \, \zeta &= [al] , \\
[ba] \, \xi + [bb] \, \eta + [bc] \, \zeta &= [bl] , \\
[ca] \, \xi + [cb] \, \eta + [cc] \, \zeta &= [cl] .
\end{aligned} \qquad (13)$$

The algebraic solution of Eq. (13) yields those values of the unknowns that are most probable according to the theory of the errors.

The coefficients of the normal equations are symmetrical with respect to the main diagonal of their determinant, e.g., $[ab] = [ba]$, $[bc] = [cb]$, etc. This property facilitates the algebraic process of the solution. First of all the first normal equation is multiplied by $-[ab]/[aa]$ and the result is added to the second equation; next the first equation is multiplied by $-[ac]/[aa]$, and the result is added to the third equation. Each of these two operations leads to an equation which does not contain the first unknown ξ. The same process is applied to the two resulting equations for η and ζ and this leads ultimately to one equation in ζ only. If ζ is known, η and ξ can then be determined from the previous equations.

In order to find the mean errors of the three unknowns, the *mean error of unit weight* and the *weight coefficients* must be known. The mean error m of unit weight is given by

$$m^2 = \frac{[vv]}{n - \mu} \, , \tag{14}$$

where μ is the number of the unknowns (in this case $\mu = 3$) and $[vv]$ the sum of the squares of the residual errors. This sum can be found either by calculating the right-hand sides of the equations of condition and then forming the differences with the observed values l_i, or from the easily verifiable equation

$$[vv] = [ll] - [al]\xi - [bl]\eta - [cl]\zeta \, . \tag{15}$$

The weight coefficients Q_{1i} follow from the normal equations by replacing the right-hand sides by $1, 0, 0$:

$$\begin{aligned}
[aa]\,Q_{11} + [ab]\,Q_{12} + [ac]\,Q_{13} &= 1 \, , \\
[ba]\,Q_{11} + [bb]\,Q_{12} + [bc]\,Q_{13} &= 0 \, , \\
[ca]\,Q_{11} + [cb]\,Q_{12} + [cc]\,Q_{13} &= 0 \, .
\end{aligned} \tag{16}$$

In an analogous manner the coefficients Q_{2i} are obtained by solving the normal equations with the right-hand sides set at $0, 1, 0$, and finally the Q_{3i} by setting the right-hand sides equal to $0, 0, 1$. Usually the greatest labor required for a complete least-squares solutions is that of solving the system of normal equations three times (or μ times in the case of μ unknowns). The standard deviations σ_x, σ_y, and σ_z of the three unknowns can be found from the equations

$$\sigma_x^2 = m^2 Q_{11} \, , \qquad \sigma_y^2 = m^2 Q_{22} \, , \qquad \sigma_z^2 = m^2 Q_{33} \, . \tag{17}$$

Thus the task of the determination of the most probable values of the unknown and of their standard deviations is completed.[2]

In all of these formulae, it has tacitly been assumed that all of the equations have the same weight. If the available measurements are of different precision then it can no longer be assumed that this assumption is satisfied, in which case the individual equations of condition have unequal weights. The coefficients of the normal equations are then no longer $[aa]$, $[ab]$, etc., but must be replaced by the weighted sums $[aap]$, $[abp]$, etc. In practice the weight factors can be very simply allowed for by multiplying each equation of condition by the factor $\sqrt{p}$, that is, by the square root of the assigned weight. The resulting set of equations can be treated as equations of equal weight by application of the previously presented procedures.

The subject of uncertainties described above is one in which familiarity with the concepts involved comes only after considerable practice with numerical cases. An example of the least-squares solution for the case of two unknowns is given below.

---

[2] The normal equations are more commonly solved with high-speed computers using the method of determinants in which case the weight factors are implicitly calculated in that process, since they are a ratio of cofactor over determinant of the diagonal terms in the system of normal equations.

Example: On March 8, 1987 the zenith distances of six stars were measured in Munich, and were subsequently compared with the values, which for these stars, were theoretically derived using the catalogued coordinates of the star. Small differences resulted which may be attributed to three causes:

1. Each of these six measurements, needless to say, possessed random errors.
2. It was to be expected that the measuring instrument itself had a systematic deviation of constant amount.
3. It was known that the telescope had a flexure which resulted in measuring errors that, to good approximation, were proportional to the sine of the zenith distance.

Because of error sources 2 and 3, the theoretical equation of approach for the differences $z - z_0$ between the measured zenith distances z and the theoretical values z_0, the equation

$$z - z_0 = x + y \sin z \tag{18}$$

was assumed; the remaining errors of the individual measurements as mentioned under 1 are then the remaining differences v.

The results are presented in the following list, giving in the first column the name of the star observed, in the second column the approximate zenith distance z, and in the third column the differences $z - z_0$:

35 G. Columbae	$z = +75°$	$z - z_0 = -1''.78$
16 Orionis	$+38°$	-0.48
51 Geminorum	$+32°$	-0.40
κ Aurigae	$+19°$	-0.85
ι Aurigae	$+15°$	-0.12
o Draconis	$-72°$	$+1.19$

Introducing the respective values of $\sin z$, then, according to Eq. (18), the following equations of condition for the two unknowns x and y are obtained:

$$x + 0.97y = -1.78$$
$$x + 0.62y = -0.48$$
$$x + 0.53y = -0.40$$
$$x + 0.32y = -0.85$$
$$x + 0.26y = -0.12$$
$$x - 0.95y = +1.19$$

In this example, all of the coefficients a_i of the unknown x have the value 1, thereby simplifying the numerical evaluation. The sums of the products are now formed; as an example, the sum $[bl]$ is here given in detail:

$$[bl] = -0.97 \times 1.78 - 0.62 \times 0.48 - 0.53 \times 0.40$$
$$-0.32 \times 0.85 - 0.26 \times 0.12 - 0.95 \times 1.19 = -3.6699 .$$

The other sums of the products are calculated analogously, but extreme care must be taken with the algebraic signs! As a safeguard against errors, it is recommended that the sums be calculated twice: once from beginning to end, and the second time in reverse.

The next step in the present example is the formation of the two normal equations

$$6.0000x + 1.7500y = -2.4400 , \quad \| \quad -0.291667$$
$$+ 2.6787y = -3.6699 ,$$

where, by convention, the first coefficient of the second equation is not shown since by definition it is identical to the second coefficient of the first equation (+1.7500). The first of the two normal equations is multiplied by the factor given on the right (−0.297667), which is calculated as the quotient −1.7500/6.0000. Adding the resulting expression to the second normal equation yields an expression without the unknown x:

$$+2.1683y = -2.9582 .$$

Then, with the help of the first equation, x can also be evaluated. The result is

$$y = -1''.3643 ,$$
$$x = -0''.0088 .$$

For the standard deviation of unit weight, and for the weight coefficient, it is found that

$$m^2 = 0.1710 ,$$
$$Q_{11} = +0.2060 ,$$
$$Q_{22} = +0.4612 .$$

Therefore, the final result of the least-squares solution, rounding the result to two decimals, is:

$$x = -0''.01 \pm 0''.19 ,$$
$$y = -1''.36 \pm 0''.28 .$$

The unknown x is practically zero, whereas y is almost five times as large as its standard deviation and is, for this reason, certainly real. Of course both unknowns do not equal their "real" values; what have been found here are simply those numbers which best represent the observations of one night. A longer series of observations would have yielded more reliable values.

3.3 Photographic Astrometry

In photographic astrometry an observer uses photographic films or *CCD* images of a certain region of the sky to determine the coordinates of the stars. Photographic films are rarely ever used, therefore, we only discuss CCD images. It is necessary to base this operation on a sufficient number of stars, at least three, whose coordinates on the CCD image are known and which can serve to define the orientation of the coordinate system. H.H. Turner has given a complete derivation of this method which can be found in the well-known book by W.M. Smart, *Textbook on Spherical Astronomy* [5].

For examples with modern astrometry I will show this with Astrometrica (see Figs. 1–6), a powerful and spread software:

CCD = charge-coupled device. More about CCDs in the chapter "Astrophotography."

Astrometrica
http://www.astrometrica.at (Shareware). Since 1993 many amateur astronomers have used Astrometrica for data reduction.

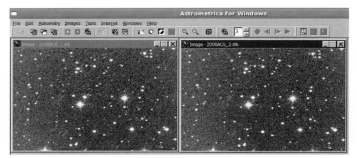

Fig. 1. Two photos loaded into Astrometrica

Linux users can also use the popular astrometry software Astrometrica. Install it with WINE or other Windows emulators.

The *Flexible Image Transport System* (FITS) is a flexible file format for photos and spectra and was developed in 1982. It is one of the approved file formats by the International Astronomical Union. It also contains additional information from the image data, such as the telescope, date and time, equinox, information on the coordinate system and much more.

ST-6 is the file format of "ST-6" CCD cameras from SBIG. Similar to FITS it also contains additional (but less) information.

The *USNO-B* catalog is the successor of the USNO-A2.0 and presents positions, proper motions and magnitudes for more than a billion objects compiled by the US Naval Observatory.

Example: You made two photos of the same region in the sky between two minutes to search the Near-Earth Asteroid 2000 AC6. First you can blink these images to see if there is a moving object. Modern software like Astrometrica provides tools to do this. Now you see a moving object near the bright star in the center (Fig. 2).

The next step is to find reference stars near the moving object to locate the coordinates of the objects on the images. This is called *astrometric data reduction*.

To do this, any software needs the coordinates of the center of the image. Alternatively you can type the designation of the asteroid (or comet) and let the software compute the coordinates.

After you enter "2000AC6" the software will load a part of the USNO-B1.0 star catalog from the Internet. Astrometrica usually finds the reference stars automatically, otherwise you can match them manually.

The result of astrometric data reduction is a lot of colored circles around detected objects (Fig. 3). The green circles are reference stars for astrometric data reduction, blue circles are other stars and yellow circles are "bad" reference stars (too large residuals). Moving objects are circled gray, as well as other detections (very faint objects or image

Fig. 2. The moving object in front of the star field

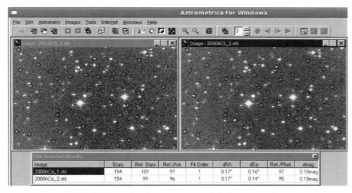

Fig. 3. Found reference stars with Astrometrica

noise). Now you can click on any object (circle) to measure and name it. A dialog window will be opened with a magnification of the area around the clicked object and two graphs: the *point spread function* (PSF) of this pixel data and the residuals of that fit. Other measured values are usually the *signal-to-noise ratio* (SNR), *flux*, *full width half maximum* (FWHM) and the *root mean square* (RMS).

After typing the *packed designation* of the object (for more information see http://cfa-www.harvard.edu/iau/info/packeddes.html) the asteroid is now stored and marked in the image (Fig. 5).

If this is an unknown object, you can send a MPC Report to the Minor Planet Center (http://cfa-www.harvard.edu/iau/mpc.html).

Many astrometry software packages provide a function to do this for you. For example the MPC Report for 2000 AC6:

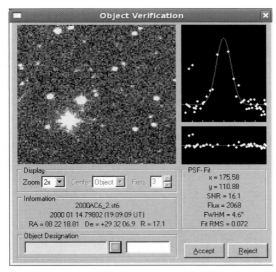

Fig. 4. The graphical analysis of the object

Point spread function: A function that describes the distribution of light from a point source in the focal plane of a telescope, after it has passed through the atmosphere and the optics of the telescope. A Gaussian (bell-shaped) curve is generally a good model for the PSF.

Signal-to-noise ratio: The ratio of the signal to the random noise in a set of data. In general, a higher SNR means that more accurate and more reliable information can be extracted from the data.

Flux: The stream of photons coming from a light source. The number of photons collected by a pixel, and therefore the pixel value, is a measure of the flux, and thus of the brightness of the light source.

Full width half maximum: The width of a curve at half of its peak value. In astronomy, the "diameter" of a stellar image as it appears on a CCD frame is often described by the FWHM of the PSF of the source.

Root mean square: The square root of the average value of the square of residuals. This value generally describes how good a model fits to a set of data.

Fig. 5. The stored and marked asteroid

```
COD 540

ACK MPCReport file updated 2007.10.28 19:51:53

NET USNO-B1.0

K00A06C C2000 01 14.79802 08 22 18.98 +29 32 17.2 17.1 R 540

----- end -----
```

If you have at least three images, the software can detect a moving object automatically.

A further application of astrometry is for stacking images, either to adding the signal of a faint object or to shifting the images to compensate for the motion of an object. The result of stacking the last examples with Astrometrica is shown in Fig. 6. The images are shifted and you see the point near the center is the asteroid. All other trails and strings are stars (blue) and reference stars (green).

The next steps are the same as before.

Fig. 6. Stacked images (Astrometrica)

But astrometry also works with a single image to measure a fixed object, for example a star, nova or supernova. The procedure is the same like for moving objects.

When you want to compare it to your favorite astronomy software, make sure that the orientation of the image equals the sky chart (north–south, east–west). Otherwise you must flip the image horizontally or vertically.

One tip is to save your measured image as a FITS file, so the software includes the WCS coordinates in the file header (if configured).

World Coordinate System (WCS): The keywords in the FITS header define the relationship between pixel coordinates in the image (x/y) and sky coordinates (RA and DEC). Software that can read WCS keywords can display sky coordinates for any pixel in the image without finding the complete astrometric solution for the image.

3.4 Determination of the Position and Brightness of Planets and of the Planetographic Coordinates

Although the ephemerides of the planets are published in various astronomical almanacs, it may sometimes be desirable to calculate them directly, if, for instance, greater accuracy is sought, or if the calculation concerns a minor planet not yet contained in the almanacs. For this purpose six orbital elements are required:

A small selection of websites of widespread software:

Distant Suns
http://www.distantsuns.com
(Commercial)

GUIDE
http://www.projectpluto.com
(Commercial)

Redshift http://www.redshift.com
(Commercial)

Cartes du Ciel/Sky Charts
http://www.ap-i.net/skychart
(Freeware)

XEphem
http://www.clearskyinstitute.com
(Freeware)

T = Time of the passage through the perihelion of the orbit **or**

M = Mean anomaly

μ = Mean daily motion

e = Eccentricity

Ω = Ecliptic longitude of the ascending node
    (distance of the node from the vernal equinox/point of Aries)

ω = Angular distance of the perihelion from the node in the orbit

i = Inclination of the orbit to the ecliptic plane

Today it is not required to calculate the ephemerides manually, some astronomical software can do it for you (Fig. 7). For the planets, Moon and Sun you do not need the orbital

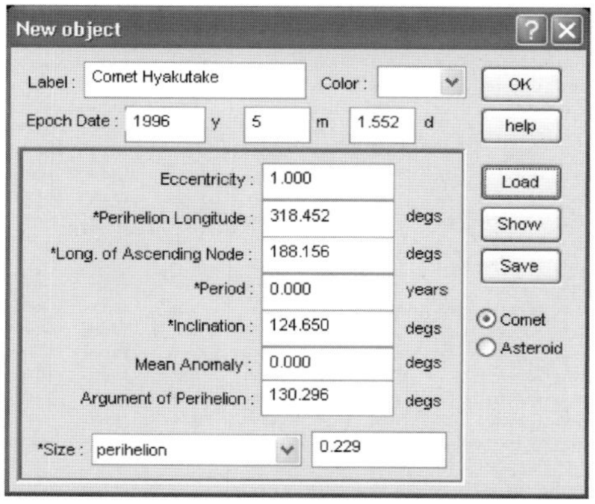

Fig. 7. Window to edit a new orbit (Distant Suns)

elements; they are included. But for asteroids, dwarf planets and comets it is very necessary to have actual data. Therefore these programs provide functions to load actual data files from the Minor Planet Center or Lowell Observatory.

Additionally it is possible to edit the orbital elements manually, e.g., if you want to simulate an orbit by a modified asteroid or comet.

The calculation of ephemerides with modern software is very simple: You only choose the object, time to start the calculation, stepping rate and the period.

The result is an extensive chart with positions, magnitudes, disc diameters, phases (when an inner planet or the Moon) and rise/set times, etc. (Fig. 8). Normally you can also print the chart for later observations.

Frequently the observer is faced with the task of calculating the coordinates of a point on the observed disk of a planet relative to the equatorial planetary plane. Although in most cases this problem can be solved with sufficient accuracy by graphical methods. But any manual computation, formerly also needed, is not necessary today. Software which can display any planet in realistic view (or as a unicolored globe only) provides mostly a coordinate grid. Normally you can also choose the distance between the grid lines. The green line you see on the example screenshot is the central meridian. Some programs also show landscape features and landing sites (Moon, Mars and Venus).

If you need a point between the grid lines, you must interpolate manually; some programs show the coordinates by clicking on the planet disc or selecting a planet feature (Figs. 9 and 10).

Calendar Stuttgart Time zone—00:00:00

Date from 2007.10.01 to 2007.10.27   Refresh   Help   Close

at 21:19:54   by 1 days   Save to File   Print   Reset Chart

Twilight | Planets | Comets | Asteroids | Solar Eclipses | Lunar Eclipses | Artificial Satellites

Sun | Mercury | Venus | Moon | Mars | Jupiter | Saturn | Uranus | Neptune | Pluto

Mars 21h20m UT	Date Coord. RA	DE	Magn.	Diam.	Illum.	Rise	Culmination	Set	Az	Alt
2007-10-01	6h05m34.9s	+23°23'13	-0.1	9.8	0.87	20h45m	4h49m	12h51m	+58°33'	+04°18'
2007-10-02	6h07m29.3s	+23°24'34	-0.1	9.8	0.87	20h43m	4h47m	12h49m	+58°55'	+04°35'
2007-10-03	6h09m22.0s	+23°25'52	-0.1	9.9	0.87	20h41m	4h45m	12h47m	+59°16'	+04°53'
2007-10-04	6h11m13.2s	+23°27'07	-0.1	9.9	0.87	20h39m	4h43m	12h46m	+59°38'	+05°12'
2007-10-05	6h13m02.7s	+23°28'18	-0.2	10.0	0.87	20h37m	4h41m	12h44m	+60°00'	+05°30'
2007-10-06	6h14m50.4s	+23°29'27	-0.2	10.1	0.87	20h34m	4h39m	12h42m	+60°23'	+05°49'
2007-10-07	6h16m36.4s	+23°30'34	-0.2	10.2	0.87	20h32m	4h37m	12h40m	+60°46'	+06°09'
2007-10-08	6h18m20.6s	+23°31'38	-0.2	10.2	0.87	20h30m	4h35m	12h37m	+61°09'	+06°28'
2007-10-09	6h20m02.9s	+23°32'40	-0.2	10.3	0.87	20h27m	4h33m	12h35m	+61°32'	+06°48'
2007-10-10	6h21m43.4s	+23°33'40	-0.2	10.4	0.88	20h25m	4h30m	12h33m	+61°56'	+07°08'
2007-10-11	6h23m21.9s	+23°34'40	-0.2	10.4	0.88	20h23m	4h28m	12h31m	+62°20'	+07°29'
2007-10-12	6h24m58.5s	+23°35'38	-0.3	10.5	0.88	20h20m	4h26m	12h29m	+62°44'	+07°50'
2007-10-13	6h26m33.0s	+23°36'35	-0.3	10.6	0.88	20h18m	4h23m	12h27m	+63°09'	+08°11'
2007-10-14	6h28m05.5s	+23°37'31	-0.3	10.7	0.88	20h15m	4h21m	12h24m	+63°34'	+08°33'
2007-10-15	6h29m36.0s	+23°38'28	-0.3	10.8	0.88	20h13m	4h19m	12h22m	+63°59'	+08°55'
2007-10-16	6h31m04.2s	+23°39'24	-0.3	10.8	0.88	20h10m	4h16m	12h20m	+64°24'	+09°17'
2007-10-17	6h32m30.4s	+23°40'21	-0.4	10.9	0.88	20h08m	4h14m	12h17m	+64°50'	+09°40'
2007-10-18	6h33m54.3s	+23°41'18	-0.4	11.0	0.89	20h05m	4h11m	12h15m	+65°17'	+10°04'
2007-10-19	6h35m15.9s	+23°42'16	-0.4	11.1	0.89	20h02m	4h09m	12h12m	+65°43'	+10°27'
2007-10-20	6h36m35.3s	+23°43'16	-0.4	11.2	0.89	20h00m	4h06m	12h10m	+66°10'	+10°52'
2007-10-21	6h37m52.3s	+23°44'16	-0.4	11.2	0.89	19h57m	4h03m	12h07m	+66°37'	+11°16'
2007-10-22	6h39m06.9s	+23°45'18	-0.4	11.3	0.89	19h54m	4h01m	12h05m	+67°05'	+11°41'
2007-10-23	6h40m19.1s	+23°46'22	-0.5	11.4	0.89	19h51m	3h58m	12h02m	+67°33'	+12°07'
2007-10-24	6h41m28.8s	+23°47'29	-0.5	11.5	0.89	19h48m	3h55m	11h59m	+68°01'	+12°33'
2007-10-25	6h42m36.0s	+23°48'37	-0.5	11.6	0.89	19h45m	3h52m	11h57m	+68°30'	+13°00'
2007-10-26	6h43m40.7s	+23°49'49	-0.5	11.7	0.90	19h42m	3h50m	11h54m	+68°59'	+13°27'
2007-10-27	6h44m42.7s	+23°51'03	-0.5	11.8	0.90	19h39m	3h47m	11h51m	+69°29'	+13°55'

Fig. 8. The Ephemeris Calendar by Cartes du Ciel

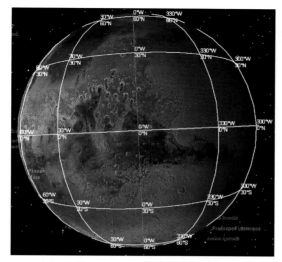

Fig. 9. Coordinate grid of Mars (Redshift 6)

Fig. 10. Coordinates of a planet feature (GUIDE 8)

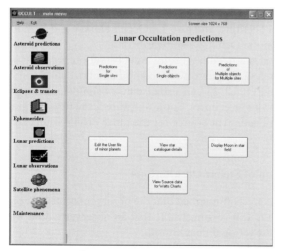

Fig. 11. OCCULT, a powerful and widespread software for occultations and more

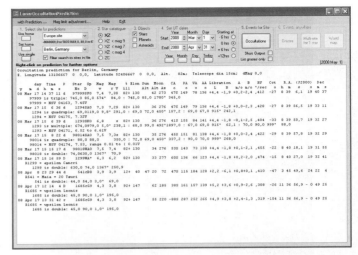

Fig. 12. Predictions of star occultations for Berlin with OCCULT 4

3.5 Star Occultations and Grazes with the Computer

The Moon occults many stars as it passes through the celestial sky (for more information see Sect. 13.4). This part will show you only examples with known software.

3.5.1 Predictions of Star Occultations and Grazes

To make predictions of star occultations by the Moon, planets or asteroids, you will find various software packages, including OCCULT, software (Windows) for predictions, reduction and more (found on the website for the International Occultation Timing Association/European Section), and also LinOccult (Linux and Windows). (Figures 11–17 show examples from the OCCULT software.)

The predictions of star occultations for an observation site is very simple: You choose a city and a date, thereupon the software computes all occultations for this site.

Some software like OCCULT can also display a visibility path on a world map (like the path from a solar or lunar eclipse). Some software also supports graze occultation predictions and Baily's Beads. The source for the lunar profile usually is the chart prepared by C.B. Watts (1963).

3.5.2 The Reduction of Stellar Occultations

Although as a rule it is recommended that the observers of stellar occultations send their results directly to *IOTA*, it is conceivable that an observer would want to carry out the reduction of the observational data himself.

In the last edition this was described with formulae but today's computer software will do it for you.

When the software computes predictions, it reduces automatically. If you have your own observations, type it into a text file and let the software reduce it for you.

OCCULT (Version 4)
http://www.lunar-occultations.com/ iota/occult4.htm
(Shareware)

LinOccult
http://andyplekhanov.narod.ru/ occult/occult.htm (Freeware)

International Occultation Timing Association (IOTA)/*European Section* http://www.iota-es.org

European Asteroidal Occultation Network (EAON) http://www.astrosurf.com/eaon

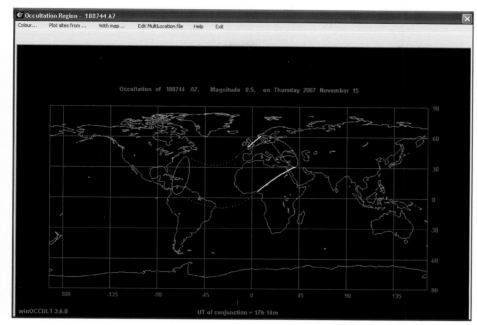

Fig. 13. The visibility path of a star (OCCULT 3.6)

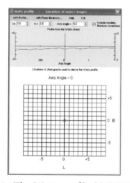

Fig. 14. The Watts profile (OCCULT 4)

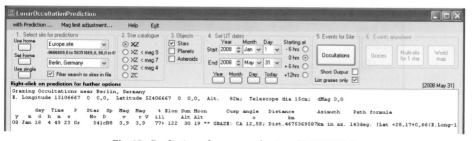

Fig. 15. Prediction of graze occultations (OCCULT 4)

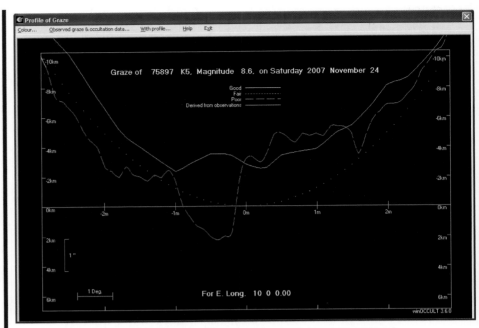

Fig. 16. Plot of lunar profile and a graze occultation (OCCULT 3.6)

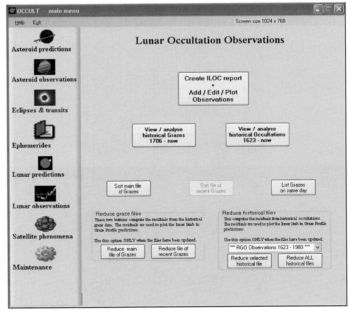

Fig. 17. *Reduce* and more features of OCCULT 4

References

1. Montenbruck, O.: Practical Ephemeris Calculations. Springer, Berlin Heidelberg New York (1989)

2. Montenbruck, O., Pfleger, T.: Astronomie mit dem Personal Computer. Springer, Berlin Heidelberg New York (2004)

3. Taylor, J.R.: An Introduction to Error Analysis. University Science Books, Mill Valley, CA (1982)

4. Mendenthall, W.: Introduction to Probability and Statistics, 2nd edn. Wadsworth, Belmont, CA (1967)

5. Smart, W.M.: Textbook on Spherical Astronomy. Cambridge University Press, Cambridge (1977)

Acknowledgement. I want to thank the following people, organisations and distributors: Herbert Raab (Astrometrica) for his support. The other authors of this book for the good cooperation, especially Dr. rer. nat. Oliver Montenbruck and Prof. Dr. Klaus-Peter Schroeder for their helpful hints. Mike Smithwick (Distant Suns), United Soft Media (Redshift), Bill J. Gray (Project Pluto/GUIDE), Patrick Chevalley (Cartes du Ciel/Sky Charts) and Dave Herald (OCCULT) for authorizations of software screenshots.

Based on *Applied Mathematics and Error Theory* by Prof. F. Schmeidler, Compendium of Practical Astronomy, Vol. 1, Chap. 3, Springer, Berlin, Heidelberg, New York (1994)

4. Optics and Telescopes

B. Weisheit

4.1 Introduction

Most instruments employed for astronomical observations are designed primarily to increase the intensity of illumination on the surface of an image-recording device, be it a conventional photographic plate, a sophisticated electronic detector, or merely the human eye. This may be achieved by enhancing the light-collecting area or by sharper imaging, with the aims of improved angular resolution in order to separate closely adjacent objects and high light-gathering power in order to clearly pick out faint objects. There is no all-round, "universal" instrument available that satisfies all of these requirements simultaneously. Rather, an observing instrument is optimized for a specific purpose and is often named after a particular type of construction or observational mode (e.g., zenith telescope, transit circle, binoculars, rich-field telescope, coronograph, astrograph or Schmidt camera).

Another kind of classification refers to the method by which the optical image is formed:

1. *Dioptric systems* employ refraction of light via lens.
2. *Catoptric systems* employ reflection of light via mirror.
3. *Catadioptric systems* employ both refraction *and* reflection to image objects optically.

This chapter surveys the variety of observational instruments and telescope systems in optical astronomy which are available to the observer. Besides a description of optical systems, a discussion of different imaging errors will be relevant as well. First, the basics of optical imaging are treated, in order to judge the imaging quality. This is of fundamental interest at this juncture insofar as it permits the determination of the value of individual imaging errors. The explanation of the various kinds of errors and their origins is followed by a description of the most widely used telescope systems and their imaging properties. There is no attempt at completeness since the possible variations of telescope types are extraordinarily diverse. Additional literature will be referred to for special optical systems. The description of telescope accessories, by means of which the telescope system supplies "results," is treated at considerable length.

4.2 Pupils and Stops

The components of an optical system, which include lenses, mirrors, diaphragms (stops), prisms, etc., have finite dimensions and therefore the cross-sectional size of a beam will be limited. This is of great significance in matters of image brightness and field size, and

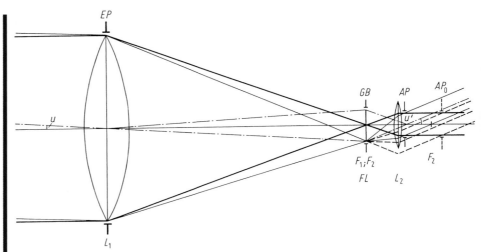

Fig. 1. Image beam (*solid line*) and pupil beam (*dot-dashed line*) in a telescopic system: objective L_1, field lens FL, magnifying lens L_2, field diaphragm GB, entrance pupil EP, exit pupil AP; also pupil beam and vignetting (*dashed line*) with exit pupil AP_0 in the absence of the field lens FL (see text for details)

Fig. 2. This view of the eyepiece clearly shows the diameter of the exit pupil beam that leaves the eyepiece

can contribute to the correction of some imaging errors. Berek has discussed the importance of the proper positioning of a diaphragm or stop. Any consideration of imaging properties of the system as a whole must include not only the imaging rays, which are represented by the cardinal points, but also the illuminating rays, which are limited by the various stops.

The beam size is always limited by two separate stops: the *aperture stop* and the *field stop*. Aperture stops, such as the support frame of a lens or the edge of a mirror, are those stops which appear under the smallest angle when viewed from the intersection of the optical axis with the object or image plane. To determine the real aperture stop, each stop and lens frame of a system is imaged by the preceding or following element into the object or image space. Some of these images can have smaller diameters than the aperture stop, but lying closer to the object or image plane they appear under larger viewing angles. The image of an aperture stop in the object space is called the *entrance pupil*, while in the image space it is the *exit pupil*. Since both pupils are images of the same diaphragm, they are also simultaneously images of each other. In telescopic systems, the aperture of the instrument generally acts as the aperture stop, i.e., the entrance pupil in refracting telescopes is the objective or the corrector plate in catadioptric systems; in reflectors, it is the primary mirror. Each optical element, such as the secondary mirror, field lenses, eyepiece, etc., images this stop until it forms the exit pupil in image space, the eye position in visual observation. This location should be at a suitable distance behind the eyepiece so that the exit pupil and eye pupil can properly coincide. If it is too close behind the lens, then visual viewing will be inconvenient and the field of view will be correspondingly narrowed, giving rise to what is termed "keyhole viewing."

To recognize possible vignetting (i.e., missing of rays by one optical element) in the optical system, both the imaging beam, whose conic points are at the axis points of the windows, as well as the pupil beam, whose principal rays intersect the axis at the pupils, must be graphed (Fig. 1). All frames, stops, plate formats, etc., are thereby imaged with the preceding or following elements, respectively into the object or image space, and the corresponding pupils and windows are graphed to scale in both position and size. The potential vignetting will then be seen. Figure 1 shows the beam in an astronomical Keplerian telescope with three convex lenses: the objective lens L_1, the field lens FL, and the magnifying or eyepiece lens L_2. The entrance pupil EP coincides with the aperture stop (the objective frame), whereas the exit pupil AP_0 gives the pupil position *without* the field lens. The vignetting at the eyepiece lens is immediately seen and leads to light loss for off-axis objects with large inclinations.

4.3 Imaging Errors

4.3.1 Seidel Theory

The *Seidel theory* provides the connection between the ideal paraxial image and the imaging errors of real, wide-aperture systems. Seidel (1821–1896) showed that the coefficients required for the determination of the errors can be found by calculating just *one* paraxial ray.

This section will not deal with the theory of imaging errors of third and higher order but will provide the formalism to calculate the partial coefficients and the individual imaging errors. These are much less complicated than they look and can easily be determined with the aid of any of the now-affordable personal computers. Theory dictates that each of the five primary imaging errors is composed of the sum of the individual errors of each surface. Thus, the five coefficients for each optical surface are calculated and, at the completion of the calculation, finally added. From these sums, the contributions of the individual imaging errors can be found. The formulae are described by Köhler, whereas Berek deals with the conditions that one or more coefficients vanish. The latter author appreciates Seidel's theory less for determining aberrations and more for systematically studying the effect of the single surfaces, and in this way to influence these effects.

4.3.2 The Primary Aberrations

Based on the Seidel coefficients and the determination of their values, the significance of the various types of aberrations will now be mentioned. The *primary aberrations* as obtained from the Seidel theory, including third-order polynomials, are discussed. Worthy of further study are the theoretical roots of aberrations, including diffraction analysis, for which the standard book by Born and Wolf [1] can be strongly recommended. Cagnet, Francon, and Thrierr [2] give a readable and illustrative presentation of the most important optical phenomena, such as images of the primary aberrations, diffraction images, interferograms, etc.

Spherical aberration, also called *aperture error*, occurs, as the name implies, when lenses or mirrors with spherical surfaces are used in optical systems. These spherical elements hold great importance in view of the ease with which they can be manufactured

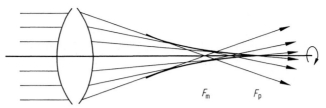

Fig. 3. Spherical aberration and caustic: F_p paraxial focus, F_m marginal focus (i.e., focus for light coming from the edge of the objective); the objective in this example is spherically undercorrected since $f_m < f_p$

Fig. 4. Diffraction images of **a** an ideal objective and **b** one which is affected by spherical aberration. See text for explanation. Adapted from Cagnet, Francon, and Thrierr [2]

commercially. The intrinsic error of the spherical surface is that near-axial and near-limb rays do not fit into one point. Each annular *zone* of the spherical surface possesses its own distinct focal length. This difference in the focal lengths of the individual zones is called *spherical aberration* and leads to the caustic, which, like zones, is rotationally symmetric about the optical axis. The dependence of this aberration on the incidence height and the focal length, and therefore on the focal ratio $1/N = D/f$, is evident from Fig. 3 (see also Fig. 4). Spherical aberration thus can be reduced in two ways: by reducing the aperture, which is obviously not very desirable for astronomical telescopes, or by choosing sufficiently long focal lengths. It can be shown that spherical aberration of an optical system is negligibly small (in the sense of the Rayleigh criterion) if the F-number N exceeds a particular value, namely

$$N \geq 3.4 \sqrt[3]{D} \,, \tag{1}$$

where D is the aperture size in centimeters. Spherical aberration can be partially removed if two lenses of a suitably chosen refractive power are combined. It is removed entirely only by a departure from the spherical shape, that is, by *aspherical* surfaces. For a single mirror this is the condition of a *rotation paraboloid*. Use is made of the defining property of the parabola, namely, that all points on that conic curve have the same distance (or optical path length) to one point and to a straight line. The straight line represents the plane wavefront of an infinitely distant object which is reflected over the entire aperture into one point, the focus. If the objective is free of spherical aberration, then the radius of the diffraction or Airy disk increases symmetrically on both sides of the paraxial focus; otherwise the increase is asymmetrical. This asymmetry hints at the presence of spherical aberration.

Coma is purely an asymmetry error occurring for beams incident at an angle, i.e., in the present case for objects within the field but off the optical axis. Through the inclination of the beam, the rotational symmetry with respect to the optical axis is lost and limited to a plane of symmetry, the *meridional plane*, which contains the optical axis and the inclined principal ray. Since the aperture stop (or entrance pupil) is in this case no longer spherically symmetric, none of the rays of the meridional beam keeps a special position on the basis of symmetry properties. In general, this beam contains not only the spherical aberration which is already present in the axial rays, but also an asymmetry of this aberration. This emerges in a light pattern with some resemblance to a comet whose tail is directed radially toward or away from the optical axis.

Coma increases in proportion to the inclination angle u, and therefore becomes more pronounced with increasing distance from the optical axis. Figure 5 shows a highly mag-

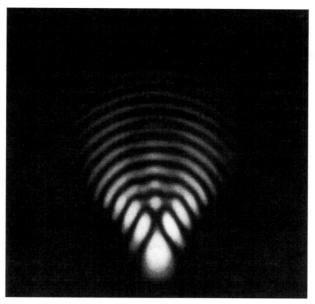

Fig. 5. Diffraction image of an optic with a round entrance pupil affected by coma. Adapted from *Atlas of Optical Phenomena* by Cagnet, Francon, and Thrierr [2]

nified diffraction image of a single point source in the case of *pure* coma, and can be removed by simply restoring the symmetry for beams incident at an angle. Optical systems corrected for spherical aberration and coma are called *aplanatic* (non-deviating) by Abbe (1840–1905). Aplanatic systems enjoy a prominent role in astronomical optics owing to their relatively good correction state.

Astigmatism and *field curvature* are interrelated and should therefore be discussed together. For skew beams, the stigmatic focus is lost and split into the two foci F_{mer} and F_{sag}, as is shown in Fig. 6. This effect (called *astigmatism* for *no* point imaging) occurs, in contrast to coma, even for beams of small inclination. Through the inclination of the beam, the refractive conditions in the two planes, the *meridional* and the *sagittal*, differ. Considering simply the refraction angles for the edge-rays of the meridional and sagittal section, it is seen that these rays, because of different incidence angles, lead to different foci of incident skew beams. Since the sagittal rays are not yet converged in the meridional focus, and the meridional rays have already diverged upon reaching the sagittal focus, the two foci F_{mer} and F_{sag} degenerate into focal *lines*. In the center between the two lines (on the surface of mean field curvature) the distribution of light is circular, but elsewhere, except at the three points named, it is more or less elliptical, and the ratio of the ellipse axis depends on its distance from the focal lines.

Distortion results from the fifth coefficient of Seidel theory, and affects not the sharpness but rather the shape of the image. When present, the optics lack a constant imaging scale over the field of view. This simply means that the effective focal length of the system (responsible for the image scale or magnification) changes with the field angle. Distortion

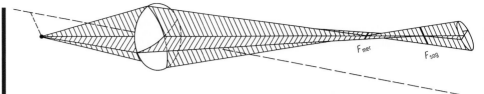

Fig. 6. Astigmatism of skew beams. The graph shows the two main sections, the meridional (*vertical*) and sagittal (*horizontal*) planes with the corresponding focal lines F_{mer} and F_{sag}

is revealed when straight lines not crossing the optical axis are imaged into "lines" which are curved outward or inward. The former case is termed "barrel" or negative distortion, while the latter is called "pincushion" or, as it represents an increase of scale, positive distortion. This can be readily demonstrated by observing a square mesh or lattice with the optics.

In summary, three of the five primary image errors in Seidel theory (spherical aberration, coma, and astigmatism) affect the sharpness of the image, the other two (field curvature and distortion) the position of the image.

4.3.3 Chromatic Aberration

Previous considerations were based upon strictly monochromatic light. The refractive index n, which occurs in numerous formulae, changes with wavelength λ. Thus, in addition to the primary aberrations, lens systems are also affected by color, or *chromatic*, aberrations. In order to determine the refractive index $n(\lambda)$ and the resulting chromatic aberrations for separate wavelengths, monochromatic light of different wavelengths (which is available from the emission lines of various kinds of atoms), is required. Hence it is useful to refer all color-dependent quantities to the light of certain spectral lines as supplied.

The refractive power, $F = 1/f$, of a lens depends on the wavelength, i.e., a lens has a different focal length f for each color. The focus is thereby split into two foci for two different wavelengths; the focus of one color is surrounded by the "defocused" diffraction circle of the other color (and vice versa), the diameter of which depends upon the F-number N of the system and the difference in focal length. The difference in focal length between the red and blue color extremes of the brightest portion of the visual spectral range is of order

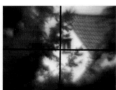

Fig. 7. The direct comparison by a viewfinder telescope with an achromatic optics and a viewfinder telescope with a single objective lens demonstrates the enormous loss of contrast that the chromatic aberration can create

$$f_{\mathrm{C}} - f_{\mathrm{F}} = \frac{f_{\mathrm{e}}}{\nu_{\mathrm{e}}} \ . \tag{2}$$

The indices "C," "F," and "e" refer to the respective wavelengths 656.3, 486.1, and 546.1 nm, and ν_{e} is an important quantity called *Abbe's number*,

$$\nu_{\mathrm{e}} = \frac{n_{\mathrm{e}} - 1}{n_{\mathrm{F}} - n_{\mathrm{C}}} \ , \tag{3}$$

that gives the *reciprocal dispersive power* of the glass used. The subscript "e" in Abbe's number refers to the green e-line ($\lambda = 546.1\,\mathrm{nm}$) of mercury, which is nowadays preferred over the yellow sodium D-line as the "mean" wavelength of the visual spectral range.

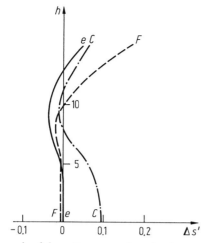

Fig. 8. An example of aberration curves for a simple cemented achromat

As the Abbe numbers are always positive, a two-lens objective can merge two colors into one focus only when composed of a convergent and a divergent lens. Correcting the on-axis displacement in the paraxial focus for two wavelengths does not guarantee that the edge rays of wide-open beams also pass this focus. Even if spherical aberration is corrected for one wavelength, this need not be so for other wavelengths; this leads to the chromatic error of the spherical aberration (aberration in an achromat is shown in Fig. 8). Its correction is difficult, if not impossible, and often the color correction on-axis is abandoned in favor of a chromatic correction of the focal length of an annular zone.

4.4 Methods of Optical Testing

4.4.1 Determination of Focal Length

The focus of a convex lens is found simply by picking up the real image of a distant (by definition, at infinity) object on a screen. The image of a faint object can be inspected using a magnifying lens with reticle; if both the image and the reticle simultaneously appear sharp, then the focus of the objective is at the position of the reticle. The focal length can also be measured in the laboratory by a procedure introduced by Bessel (see Fig. 9). If the distance between object and focusing screen is a constant l, then there are two intermediate positions at which the lens forms a sharp image of the object on the screen. If e is the distance between these two positions of the lens, then the focal length f is given by

$$f = \frac{l^2 - e^2}{4l} \, .$$

(4)

To obtain two focal positions, the distance between object and screen must exceed four times the focal length f of the lens. If it *equals* $4f$, then the image scale is 1 : 1.

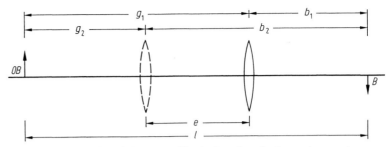

Fig. 9. Focal length determined by the Bessel method: $g_1 = b_2$; $g_2 = b_1$

Table 1. Material constants and derived quality factors of potential mirror materials

Material	Density	Thermal expansion coefficient	Thermal conductivity	Specific heat
	ρ	α	λ	c
Flint glass	4.4	8.5	0.0015	0.13
Crown glass	2.5	7.5	0.0018	0.17
Duran 50 (Pyrex)	2.23	3.2	0.0028	0.18
Fused silica (quartz)	2.2	0.4	0.0033	0.18
Zerodur (Cer-vit)	2.53	< 0.1	0.0039	0.20

The dimensions for the specific quantities are as follows: $\rho \, (\mathrm{g \cdot cm^{-3}})$, $\alpha \, (10^{-6} \cdot \mathrm{grad^{-1}})$, $\lambda \, (\mathrm{cal \cdot cm^{-1} \cdot s^{-1} \cdot grad^{-1}})$, $c \, (\mathrm{cal \cdot g^{-1} \cdot grad^{-1}})$

The focal length of a divergent or negative lens is determined using the scattered circle of a very distant point source. If D is the diameter of the lens, b the diameter of the scattered circle, and l the distance between lens and screen, then the focal length is

$$f = \frac{lD}{D - b} \, . \tag{5}$$

It should be noted that mirrors change shape and focal length with temperature. Therefore, an ideal mirror material should have a very high thermal conductivity in order to reach thermal equilibrium immediately after a temperature change (Table 1).

Now available are materials like *Zerodur* from Schott glassworks and *Cer-Vit* from Owens-Illinois, which have practically no thermal expansion and thus are currently the preferred materials for mirror blanks. These are not glasses but rather *glass ceramics*, a mixture of ceramics and glass with respectively negative and positive expansion coefficients. The production and properties of the outstanding material Zerodur are detailed in an article by Petzoldt [3]. Glass ceramics have now superseded borosilicate glasses such as *Duran* (Schott glassworks) or *Pyrex* (Corning glassworks) of which, for instance, the famous Palomar 5-m telescope was made.

4.4.2 The Hartmann Test

To test astronomical optics by the *Hartmann method*, the Hartmann screen, a mask consisting of holes that are radially symmetric is placed in front of the objective or mirror.

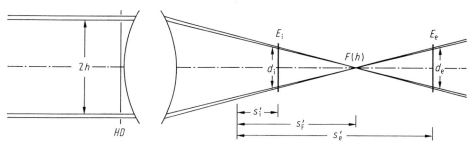

Fig. 10. Objective testing after Hartmann [4]. Only one pair of holes with height h is graphed. The pairs of holes are generally arranged on several diameters of the Hartmann screen HD; intrafocal and extrafocal widths s'_i and s'_e are measured from an arbitrary reference point

This method approximates the rays of geometrical optics by small beams generated by the holes, each representing a certain *height of incidence h* (see Fig. 10).

Illuminated by a plane wavefront (e.g., from a distant artificial point source or a real star), two out-of-focus photographs are taken, one inside and one outside the telescope focus. Measuring the intrafocal and extrafocal distances s'_i, s'_e and the corresponding ray separations d_i, d_e from the geometry of Fig. 10, the focal distance follows:

$$s'_F(h) = s'_i + \frac{d_i}{(d_i + d_e)}(s'_e - s'_i) . \tag{6}$$

The dependence of s'_F on the incidence height h permits conclusions on imaging errors such as spherical aberration or zonal errors. A change of focal distances for holes circular to the optical axis on the Hartmann screen (i.e., the absence of rotational symmetry) reveals astigmatism. The Hartmann test supplies quantitative results, i.e., numerical values of the aberrations, but it is limited by its nature to a pointwise checking of the optics under test.

4.4.3 Foucault's Knife-Edge Test

The knife-edge test encompasses the entire optical surface at once and is also quite sensitive to small-scale deviations which are not recorded by, for instance, the Hartmann test. Its effect is also explained by geometrical optics. All rays emanating from a point source at the center of curvature C of a spherical mirror are reflected back to that center. In the actual laboratory test (Fig. 11) the center of curvature replaces the normal focal point. Thus the terms *intrafocal* and *extrafocal* for knife-edge positions near the center of curvature will be used. If a sharp-edged screen (such as a razor blade) is inserted from the side into the beam, the shadow moves in the same direction as the screen in an intrafocal position but opposite to the screen in an extrafocal position (i.e., behind the center of curvature). Geometrical optics explains this by the obscuring of individual rays. If the edge is inserted exactly at the position of the center of curvature C (the focus), the obscuration of the brightly lit mirror surface occurs suddenly (or at least evenly) over the entire surface. Here is thus a *highly* sensitive method for determining the center of curvature of a mirror or the precise focal position of a real star.

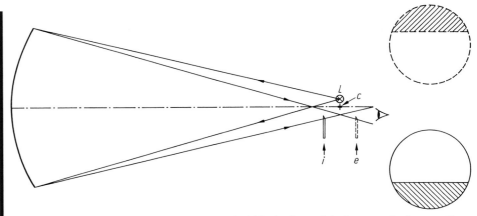

Fig. 11. Knife-edge test after Foucault. Intrafocal (*solid line*): edge and shadow move in the same direction; extrafocal (*dashed line*): edge and shadow move in opposite directions

If the mirror deviates from its ideal form by zones or by small-scale irregularities, these deviations will become visible as bright and dark regions on the shadow and the illuminated parts respectively, as the rays bypass the center of curvature because of their different inclinations. The mirror surface then appears as a "relief" or embossed surface illuminated from the side (opposite to the edge). Foucault's test therefore is primarily a qualitative test to find minute surface errors. The method is sensitive enough to show

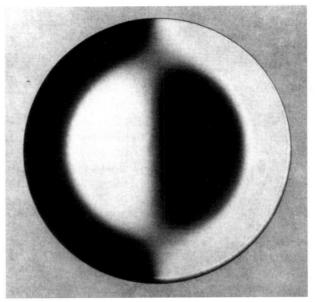

Fig. 12. Foucault graph of spherical aberration. The edge is inserted between the paraxial and the marginal focus into the beam. Adapted from *An Atlas of Optical Phenomena* [2]

deviations down to $\lambda/100$. In order to separate object and image, the light source L is placed slightly off the center of curvature C in the actual laboratory test (Fig. 11). The image point is then axisymmetric to the light source and therefore accessible. Foucault's test also shows imaging errors. When an ideal or error-free spherical mirror shows no deviations when examined in the center of curvature (and thus is brightly illuminated) it is evident that a parabolic mirror in the same setup will show shadows since the spherical wave does not converge at the center of curvature. However, if both mirrors are tested with a plane wavefront (as from a distant star), the parabolic mirror will show no focal deviations, whereas the spherical mirror distinctly displays spherical aberration (Fig. 12). The same holds true for zonal errors. Needless to say, this method also applies to the testing of lens objectives in *transmitted* light.

The test as described includes a point source (spherical wave) which can be replaced by a slit. This results in a strong increase in the image brightness which is very valuable in visual testing. The illuminating slit must be strictly parallel to the edge. The image brightness can be further increased by observing past the edge an extended light source which is partially covered by the edge itself. This ensures that both edges, knife and slit, are parallel. The construction of such a test device is described by Malacara [5], who also gives the diagnostic images of the individual imaging errors as they occur within the Foucault test.

4.4.4 Interferometric Tests

Interferometric tests are based on wavefronts orthogonal to the rays considered thus far. In this sense, the ideal wave surface has a spherical shape after passing through the optical system with its center of curvature at the image point, the focus. All departures from this wave surface are called *wave aberrations*. In interferometric methods, these errors are made visible by the interference of two wavefronts, the wavefront being tested and a reference wavefront. Their advantage consists of permitting the qualitative *and* quantitative determination of imaging errors simultaneously.

From the wide variety of interferometric tests, many of which would be beyond the scope of this book, only a few types will be mentioned here briefly as examples. References such as Malacara [5] and DeVany [6] can be consulted for detailed instructions and interpretations of the test results obtained. The best known test device is probably the one by Twyman–Green, originating from the two-armed Michelson interferometer, but replacing one plane mirror by the optical system to be tested. Superposition of both reflected wavefronts by the beam splitter yields characteristic fringe patterns. Malacara [5] gives a list of the various patterns for perfect and defocused optics and such affected by spherical aberration, coma, or astigmatism. Counting the fringes gives the amount of deviation in units of half of the wavelength ($\lambda/2$). Interferometric methods are quite demanding in setup and are quite sensitive to vibrations, air turbulence in the optical path, adjustment, and so on. To overcome such difficulties, a type of *shear interferometer* has been developed. This device does not require two separate interferometer beams, since the optical system which is under test produces *both* the test and the reference wavefronts and superposes them after a displacement (i.e., shear). The possible variations are so numerous (lateral, radial, rotated, and inverted shear) that Malacara [5] and Bryngdahl [7]

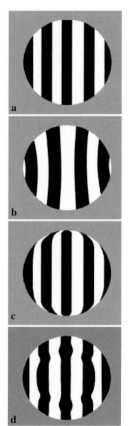

Fig. 13. Optical errors in the Ronchi test: **a** perfect optics, **b** over- or under-correction, **c** edge error, **d** zone error

must be referred to. A relatively simple setup for the amateur is given by Bath [8] with instructions and with interferograms of optical errors.

4.4.5 The Ronchi Test

Finally, there is the *Ronchi test*, which is similar to the Foucault test except that the knife-edge is replaced by a fine grating. In its original form, a point source in (or near to) the center of curvature was used to illuminate the mirror which was then viewed through a grating. For the purpose of better illumination, the point source has been replaced by a slit. For a further increase of the image brightness, the slit can be replaced by an extended light source which illuminates the mirror through the same grating which will be used to observe it. In this way the parallelism of "multiple slit" and grating are guaranteed, which is important for this method too. Malacara [5] shows the Ronchi-grams for all imaging errors and for aspherical surfaces and compares them with the interferograms of a Twyman–Green interferometer. It is then seen that the Ronchi-gram for spherical aberration agrees with the Twyman–Green coma interferogram, and the Ronchi-gram for coma with the astigmatism interferogram. Image generation in the Ronchi test can be explained by geometrical optics, but the interpretation is usually carried out interferometrically. A comparison between Ronchi-grams and Foucault-grams is given by DeVany [6], who also shows an illuminating device for both tests. In order to simplify the interpretation by standardizing the Ronchi test patterns, Schultz [9] proposes a grating whose line number depends on the aperture ratio $1/N = D/f$. These figures show for gratings with $12N$ lines/in. (e.g., $f/8 \rightarrow 100$ lines/in.) the same Ronchi-grams for mirrors of quite diverse apertures and focal lengths. The fine gratings with equidistant bright and dark lines of equal widths can also be produced photographically by amateurs. Thus, for amateurs, the Ronchi test together with the Foucault test will undoubtedly provide the most attractive method for testing astronomical optics.

4.4.6 Star Testing a Telescope

Analyzing the diffraction pattern generated by a telescope is one of the most sensitive methods of assessing the quality of the optics. Only in rare cases will the outside-focus and inside-focus diffraction patterns be absolutely identical, with the focused star being a flawless point. On the other hand, it takes a trained eye to analyze the most subtle details. This is not the aim here. Instead, the goal here is to recognize fundamental optical aberrations based on diffraction patterns, as illustrated in Fig. 14. This is easy to do after some patient waiting for the right night with a steady atmosphere and a telescope that has been optimally collimated beforehand. The following is an effective procedure:

After setting up the telescope, you should wait at least 2 to 4 h for the tube and optics to cool down completely. Otherwise, the star test will always be impaired by residual movements of the air within the telescope itself. Before setting up the telescope, you should first make sure the stars are not twinkling, but are shining steadily and uniformly. Excellent nights for this purpose are those with stable layers of air that are marked by a bit of fog formation or by the hazy sky of a high-pressure region that is starting to lose strength. The optimum site is away from developed areas and surrounded only by meadows or bushes since this is where air turbulence from ground heating is at its lowest.

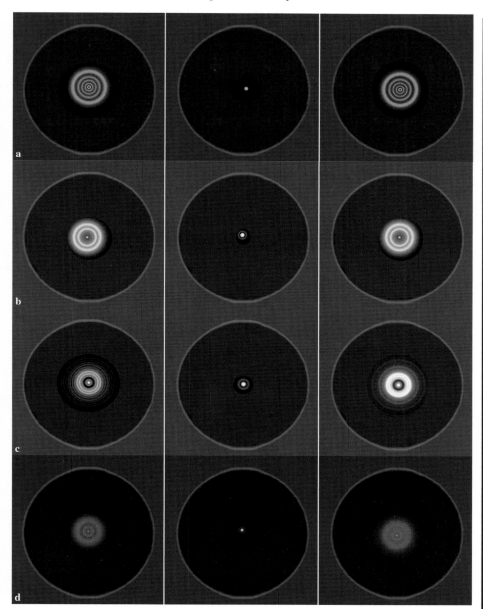

Fig. 14. Diffraction patterns caused by various optical defects. **a** Ideal diffraction pattern. **b** Diffraction pattern with 25% obstruction without secondary-mirror spider vanes. **c** Turned-down edge, classic fabrication defect for reflecting telescopes. **d** Chromatic aberration for a simple optical lens. **e** Spherical aberration, unsharp extrafocal view with undercorrection, and vice versa. **f** Astigmatism, 90° jump when passing through focus. **g** Coma at the center of the field with optics out of collimation, inherent in the system for short-focal-length telescopes. **h** Seeing, characteristic light pattern in focus instead of the Airy disk. **i** Tube currents, warm air rises up and out of the aperture. **j** Optics pinched by the cell

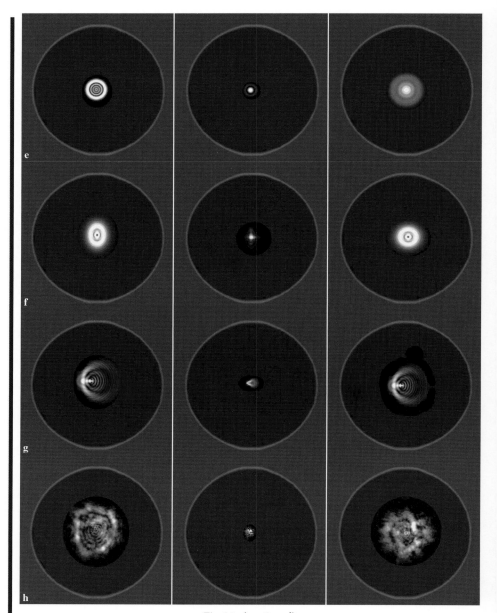

Fig. 14. (continued)

After the telescope has cooled down, aim it at a bright star located as close as possible to the zenith. Magnification should be roughly 2 times the diameter of the objective in millimeters. The brighter the overall diffraction pattern appears, the easier it is to clearly recognize the individual structures. Even after the telescope has cooled down completely and the stars in the sky seem to be steady to the naked eye, a first glance through the

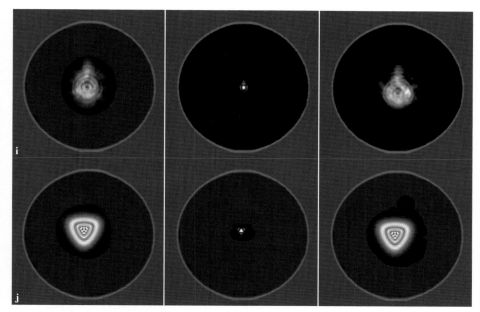

Fig. 14. (continued)

telescope at high power will reveal if the seeing is sufficiently steady. A writhing, dancing blob of light will cast doubt on any subsequent assessments. In this case, it is better to stop the star test and try another day so as not to make false interpretations due to the poor seeing.

Once the seeing is adequate, the first step is a critical assessment of the image at its best possible focus. Ideally, the star appears as a steady, tiny round spot. On closer examination, you discover that the Airy disk may be surrounded by one or two very weak diffraction rings. A slightly out-of-focus image pattern reveals more about the image-formation effects produced by the optical component. Here the focuser is turned slightly outside the focus point. As a result, the stellar point grows into a small disk that ideally consists of a whole series of light-dark-light rings. This is the extrafocal diffraction pattern. Turning the focuser tube back into focus and then inward beyond this point by a comparable distance reveals the intrafocal diffraction pattern visible. The star test consists first of all of analyzing the diffraction patterns, and, critically, in making a direct comparison between the extrafocal and intrafocal image. Figure 14 illustrates the fundamental optical defects and their effects on the diffraction pattern.

4.5 Telescope Systems

4.5.1 Refractors

Perhaps the most distinctive feature of refracting optics is their chromatic error due to the wavelength dependence of the refractive index n (Sect. 4.3.3). An objective corrected for two wavelengths was first computed and ground by Joseph Fraunhofer in the early nineteenth century. This *achromatic*-type of objective, also known as *Fraunhofer-type*, consists of a convex front lens of crown glass and a concave rear lens of flint glass, with a small air space in between. The procedure for calculating and correcting a simple achromat is shown in detail, for instance, in Klein [10], who also addresses the fine correction of the objective. The Fraunhofer achromat corresponds to Zeiss-type E (Fig. 16). The secondary spectrum (see Sect. 4.3.3) of achromats can be reduced by a proper choice of glasses. These somewhat more expensive special glasses lead to the two-lens *half-apochromat* of Zeiss-type AS (Fig. 16).

The addition of a third lens of proper glass type can achieve chromatic correction for three wavelengths. This classical apochromat, corresponding to Zeiss-type B, not only possesses nearly ideal chromatic correction but is also free of coma and spherical aberration. Manufacturing data and aberration curves for various wavelengths of this and a variety of other objectives are given by König and Köhler [11]. Cemented three-lens apochromats of $f/8'$ and faster (e.g., Starfire Triplet, APQ Fluorite of Zeiss) that can be used for astrophotography are also now available. In order to reduce the tube length of the refractor, amateur observers have sometimes used plane mirrors to "fold" the beam. The telescope length is correspondingly reduced to $1/2$ to $1/3$ of the focal length. Needless to say, the quality of the mirrors enters into the imaging performance of the telescope. If they are not quite plane, the result will be increased astigmatism. Also very critical are the stability of the tube and of the support frame of the mirrors, conditions that can nevertheless be fulfilled by amateur telescope makers.

Fig. 15. A 5-inch refractor telescope for amateurs

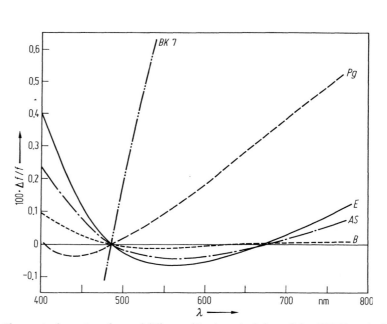

Fig. 16. Chromatic aberration of several different objectives: single lens of glass BK7, Fraunhofer achromat *E*, half-apochromat *AS*, apochromat *B*, example of an objective corrected for photography *Pg*. Adapted from Bahner

Modern lens-type optics with almost complete correction of chromatic aberration have become possible through the use of ED glass. These types of glass are produced by adding fluoride salts to the molten glass. This results in types of relatively low-refractive glass that on the other hand in part exhibit highly varying "anomalous" dispersion.

Normal crown and flint glasses have a higher index of refraction in the blue than in the red. Compared to crown glass, a flint glass nevertheless has somewhat greater dispersion than would be indicated by its index of refraction. This fact enables dispersion of the crown element to be compensated, without completely eliminating the positive refractive effect. The varying but always parabolic dispersion curves of the two glasses intersect at two points. This is how the color correction of an achromat is achieved for two colors. If ED glass is used in place of crown glass and combined with a normally dispersive glass, the various dispersion curves can have three or more intersection points. As a result, at least three colors at the focus are superimposed exactly, while the intermediate values are situated significantly closer to the focal point. In this case, the two-element optical system is called an (ED)-apochromat. An even more effective approach is to use from the start a crystal composed of fluorite (usually CaF_2) as the lens element since this naturally has the strongest anomalous dispersion. For example, the inside lens of a Zeiss APQ triplet is a fluorite element.

Figure 16 indicates that the objectives of telescopes used primarily for photographic studies are in general corrected for different wavelengths than objectives used for visual observations. The vertex of the color curve lies at shorter wavelengths, and this is conditioned by the higher blue-sensitivity of earlier photographic emulsions. Lens objectives are nowadays still in use as astrographs. The demands on these objectives are, of course, much higher than on those used for visual observations, at least regarding field size and flatness of the image. To be mentioned first is the venerable *Petzval objective* which, at $f/3.4$, was a fast system for its day. While its 7° field of view was quite large, it suffered from field curvature. Later, the *Taylor triplet* (up to $f/3$) followed from which the well-known *Ross* and *Sonnefeld four-lens objectives* were developed. These have useful fields of up to 10° × 10°. In contrast to the Petzval objective, which is composed of two two-lens achromats with a large air space in between, these objectives have a negative lens as the central component.

4.5.2 The Newtonian Reflector

All previously mentioned problems with chromatic aberrations of refraction optics do not, of course, occur in purely reflecting systems, as these are, owing to the law of reflection, wavelength independent and are hence *absolutely* achromatic.

A simple arrangement of reflecting optics is the combination of a concave primary and a small secondary mirror to deflect the beam outward (Fig. 17) and making it accessible to the eye. The simplest form of the primary is a concave spherical surface, which is easily manufactured but which contains a severe imaging error: spherical aberration. Despite this serious disadvantage, the spherical mirror can be used for astronomical observations if its f-ratio is chosen sufficiently small (Sect. 4.3.2). If the disadvantages thus caused—the low light power and the large tube length—are not deemed acceptable, a two-lens corrector can remove spherical aberration and coma almost entirely to an

Hint:
Some "ED-APO" labeled telescopes should be called "semi"- oder "half-apochromats". Their color correction is much better compared to achromats, but it is not perfectly colorless.

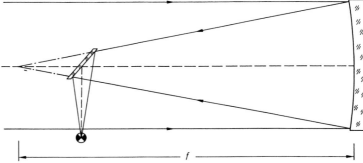

Fig. 17. An $f/3$ Newtonian telescope. The primary mirror is a rotation paraboloid, the secondary is a plane mirror of elliptical shape

Hint:
for the do-it-yourselfer, the free software "Newt for Windows" by Dale Keller is recommended: http://www.dalekeller.net.

f-ratio of $f/2$. More on the corrector-type and the aberration curve are given by Wiedemann [12], who provides the construction data.

Another way of correcting the aperture error or spherical aberration is by using a mirror of *parabolic* shape. This uses the property of the parabola (the conic curve of equal distance from a point and a straight line) to concentrate light parallel to the optical axis at one focus, irrespective of aperture. A plane secondary mirror inclined by 45° against the optical axis moves the focus outside the telescope tube. This construction was conceived back in the seventeenth century by Sir Isaac Newton and is named in his honor (Fig. 17). The parabolization of a spherical mirror can be executed by anyone with the necessary instructions (e.g., Rohr [13]) and with a good deal of patience. After removing the spherical aberration, the chief remaining imaging error of the Newtonian telescope is the *comatic error*, which gives rise to a comet-shaped distortion of stellar images that increases toward the edge of the field (Fig. 5). The coma thus determines the useful field size of a paraboloid mirror.

Two ways of defining the useful field are possible. A maximal comatic lateral aberration of $1''$, for example, may be permitted during visual observations. For telescope apertures larger than 10 cm, this corresponds approximately to the resolution limit set by the Earth's atmosphere. In this case the useful angular field diameter is determined to be

$$\phi \simeq 0°003\,N^2 \ . \tag{7}$$

The field size increases with the square of the f-number N. Still, in order not to be too limited in light-gathering power, a three-lens corrector of the Wynne-type can be used, as it almost entirely removes the coma of a parabolic mirror to an f-ratio of $f/2.5$. Wiedemann [12] gives details and elsewhere [14] specifies a modification of the Baker corrector, which, largely complying with the needs of an amateur, achieves a quite satisfactory correction up to a ratio $f/5$.

The position and size of the secondary mirror depends on the focal length and f-number and can be found graphically. The plane mirror is usually of elliptical shape in order to reduce obscuration, and its two axes are chosen somewhat larger in order to deflect, free of vignetting, not only the axial rays but also the desired image field out of the tube. The path and intersection of these marginal rays with the secondary can be seen by drawing the desired field size in scale at the primary focal plane. To hold the secondary mirror in place, a four-arm support frame (preferred over a three-arm frame) is used as a spider. Since the diffraction of light is always perpendicular to the disturbing edge, the diffraction images of two arms coincide (i.e., four spikes in the case of orthogonal arms), whereas the three-arm spider leads to six radial rays in the diffraction image. The tube should not be sealed at the primary mirror end so that warm air in the tube can mix with the ambient air and flow out of the tube; warm air convecting within the tube would heavily degrade the image enlarged by the eyepiece.

Collimating a Newtonian will be briefly described here. For a comprehensive treatment of the procedure, the reader is referred to Valleli [15]. When starting the aligning, one should be sure that eyepiece tube and optical axis (i.e., the tube axis) are orthogonal. The plane secondary mirror is inserted and moved along the optical axis until its circular shape is centered in the eyepiece tube. Next, the three adjusting screws of the secondary are tuned until the image of the primary appears concentric with the circular rim of the secondary. Finally, moving and tilting the primary mirror can be stopped when the image of the secondary with its spider appears concentric to the primary's rim. This rough collimating, where all rims of the mirrors and their images must appear concentric to each other is carried out in daylight following the test on a star. As in the procedure on the refractor, one observes the defocused image of a bright star in an eyepiece of high magnification. The screws of the primary are fine-tuned until the oval diffraction rings of the star image appear concentric in the eyepiece. In the case of a manufacturing error of the mirror, this will not be successful. To test this one moves the eyepiece to the other side of the telescope focus. If the defect in the concentric diffraction rings turns to the opposite side of the defocused image, the mirror possesses a defect or is stressed in its cell.

Fig. 18. A compact Newton telescope for beginners

4.5.3 The Cassegrain Telescope

The telescope length can be substantially reduced, yet retaining the equivalent focal length, by replacing the plane secondary mirror with a convex one also resulting in a conveniently accessible focus location (Fig. 20). The curved secondary S_2 images the primary focus F_1 into the telescope focus F. The focal length of the telescope is then no longer the primary focal length f_1 but rather the distance between the system's focus F and the intersection of the narrow beam with the incident marginal rays. If $m = f/f_1$ is the magnifying factor of the primary focal length, the tube length of the Cassegrain telescope is less than $1/m$ of the telescopic focal length. The following formulae are useful in relating the Cassegrain system parameters:

$$f = mf_1 = \frac{f_1 f_2}{f_1 + f_2 - e},$$
$$a = mc = \frac{(f_1 - e)f_2}{f_1 + f_2 - e}.$$

(8)

Fig. 19. Classic Cassegrain systems are rarely found in amateur astronomy

For instance, to modify a Newtonian telescope into a Cassegrain system, and with given f_1, m, and g (see Fig. 20), the following equations apply:

$$e = \frac{mf_1 - g}{m + 1} \qquad f_2 = -\frac{m}{m^2 - 1}(f_1 + g).$$

(9)

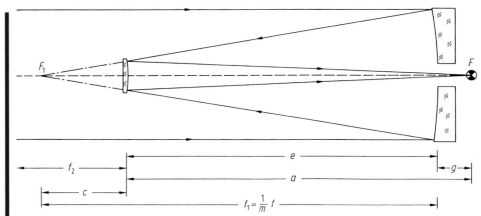

Fig. 20. A Cassegrain telescope ($f/12$). The primary is a rotation paraboloid, the secondary a rotation hyperboloid. The focal length is increased by the factor $m = 4$

The minimum diameter of the secondary follows from the *law of rays*,

$$D_2 = \frac{a}{f} D_1 \, , \tag{10}$$

where an additional amount ($2e \tan u$) must be taken into account for the angle u of the image field radius. The original Cassegrain system consists of a parabolic primary (deformation constant $k_{1,\text{Cass}} = -1$) and a convex secondary. In order to transform the spherical-aberration-free prime focus into the Cassegrain focus, the secondary mirror must have a deformation which ensures the constancy of the light path for both foci F_1 and F. This constraint leads to a hyperbolic meridional cross-section. A test setup to examine the shape of the Cassegrain secondary has been described by Richter [16].

Also noteworthy is the *Gregory telescope*, which accomplishes the magnification of focal length by inserting a *concave* secondary mirror behind the prime focus F_1. This position lengthens the telescope tube by the amount $2c$ (Fig. 20) compared with the Cassegrain system. This may be one of the reasons why the Gregory system has not successfully competed with the Cassegrain system. In this system the spherical-aberration-free imaging of the parabolic primary is maintained by deforming the concave secondary to an elliptical curve in the meridional cross-section. As in the Cassegrain system, coma and astigmatism are retained.

The Cassegrain focus is modified to the *Nasmyth focus* by inserting a tertiary, plane mirror in front of the primary mirror. This third mirror deflects the beam after reflection from the secondary into the elevation (altitude) axis out of the tube. The positioning of the focus to the side of the tube, however, is preferable only in an alt-azimuth mount over the Cassegrain position, because in this case, it is fixed with respect to Earth's gravity and also does not burden the tube with the instrumental weight.

Another distinctive focus is the *Coudé focus*, whose position is absolutely fixed in the hour axis of the equatorial mount or in the azimuth axis of the alt-azimuth mount. After reflection at the Cassegrain secondary, the beam passes several plane mirrors until arriving in the rotation axis of the telescope. Normally the heaviest spectrographs are placed here; the image rotation (which for the Coudé focus also occurs in an equatorial mount) is without significance in the spectroscopy of point sources. If necessary, however, this can be compensated. The Coudé focus has gained significance nowadays for its use when combining several telescopes to form an optical interferometer. This new observing method still requires numerous technological developments, for example, for coatings of highest reflectivity.

4.5.4 The Ritchey–Chrétien System

The mirror systems considered thus far remove spherical aberration by parabolizing the primary mirror. To maintain this state of correction in the system focus (the transformed primary focus), the secondary must be correspondingly deformed into an aspherical hyperbolic mirror, i.e., the focal properties of conic sections are employed. Correction of spherical aberration can also be accomplished by deforming both mirrors in relation to each other instead of correcting each mirror individually. Thus, by insertion of a secondary mirror into the beam, this second surface adds another degree of freedom which can be used for further correction of imaging errors. This gives exactly *one* combination of aspherical mirrors which removes spherical aberration and coma simultaneously. In this system, named after Ritchey and Chrétien, who first elaborated its design, the two mirrors of the Cassegrain are even more strongly deformed according to the two deformation constants

$$k_{1,RC} = k_{1,Cass} - \frac{2a}{em^3} = -1 - \frac{2(f - em)}{em^3} ,$$

$$k_{2,RC} = k_{2,Cass} - \frac{2f}{e(m-1)^3} = -\left(\frac{m+1}{m-1}\right)^2 - \frac{2f}{e(m-1)^3} , \tag{11}$$

that is, both mirrors are rotation hyperboloids ($k < -1$). Of course, the primary itself loses its correction for spherical aberration so that a special corrector is needed when observing at the primary focus. The system focus, however, is free of spherical aberration and coma because of the combination of the two aspherics. The remnant astigmatism and field curvature are somewhat stronger than in a comparable Cassegrain system.

These errors limit the size of the useful field of view. A field diameter of $1/2$ degree can be considered typical, where the astigmatism of the RC system leads to a scattered circle of some $1''$ diameter at the edge of the field. Since the RC system is a further development of the Cassegrain, it has also, by virtue of the similar mirror arrangement (Fig. 20), a tube length of less than $1/m$ of the telescopic focal length. Because of its good optical correction state (absence of chromatic, spherical, and comatic aberrations) while having only two optical surfaces and a short telescope length, many of the existing large telescopes have been designed as RC systems.

4.5.5 The Schiefspiegler

The extreme difficulty in the accurate manufacturing of two strongly deformed aspherical surfaces is likely to keep the RC system just mentioned out of the reach of the amateur telescope maker. The situation is totally different for the *oblique mirror telescope*, or

Table 2. Construction data for the Kutter Tri-Schiefspiegler. All measurements are given in inches and in centimeters (the latter in parentheses)

Primary mirror			Distance e_1	21.77	(55.30)
Free aperture	4.30	(10.92)	Distance e_2	13.19	(33.50)
Radius of curvature r_1	+87.40	(+221.99)	Distance M_3-focus	18.90	(48.01)
Secondary mirror			Distance c_1	3.90	(9.91)
Diameter	2.40	(6.096)	Distance c_2	11.02	(27.99)
Radius of curvature r_2	−127.56	(−324.00)	Equivalent focal length	63.00	(151.19)
Tertiary mirror			Total length	25.00	(56.35)
Diameter	2.40	(6.096)	Aperture ratio	$f/14.7$	
Radius of curvature r_3	+704.00	(+1788.15)			

Schiefspiegler, which is used almost exclusively by amateurs; it can be regarded as a development of the Newtonian design. The simplest variation of the Schiefspiegler consists of a concave primary and the eyepiece at the periphery of the front end of the tube ("Herschelian Telescope"). Even apart from the disadvantage of the long, cumbersome tube, the primary imaging errors now have an enhanced effect since this is a pure "off-axis" observation with inclination angle u without any correction. If a plane secondary is placed adjacent to the tube so that it reflects the light from the primary again, the result is a shortened tube length along with a larger focal length, which, via the increased f-number N, leads to a decrease in imaging errors. To further reduce the errors, the secondary mirror can be deformed or a suitable correction lens inserted (Fig. 21).

The theory and practice of the Schiefspiegler have been comprehensively presented by Kutter [17]. The manufacturing of the plano-convex correction lens may be difficult; therefore Kutter [18] replaced it with a third mirror, creating the Tri-Schiefspiegler (Fig. 22). The specialty of this instrument consists of all three mirrors having spherical surfaces; they are thus of particular interest to amateur telescope makers. Table 2 collects the data of a Tri-Schiefspiegler according to Fig. 22. The constructional data can also be used for larger apertures. The values in Table 2 must then be multiplied by the desired scale factor.

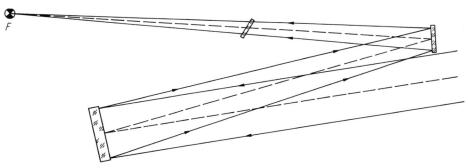

F

Fig. 21. The Kutter Schiefspiegler ($f/20$)

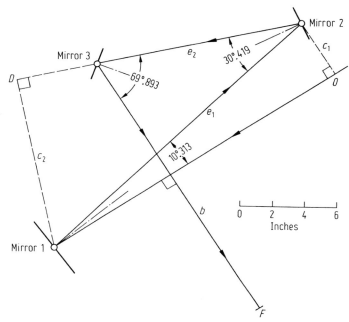

Fig. 22. The Kutter Tri-Schiefspiegler ($f/14.7$), constructed exclusively from spherical mirrors. See Table 2 for the construction data

4.5.6 The Schmidt Camera

The telescopes discussed above possess a rather small field of view of less than 1°. They are usually employed to observe individual objects. The type of telescope to be described in this section is used for the photographic recording of large celestial areas with diameters of several degrees, but it can be modified to another focal position and then be used for other purposes. The ingenious idea of the optician Bernhard Schmidt consisted in removing the asymmetry of the known optical systems for inclined beams (and the resulting imaging errors) and removing the remaining spherical aberration by a correction lens. To eliminate the asymmetry, he used a spherical mirror and moved the entrance pupil of the system from the mirror vertex into the plane through the center of curvature of the mirror (Fig. 22). This eliminates the central role of the optical axis; a beam incident parallel to the optical axis is no longer distinguished from a beam incident at an angle which remains symmetric to its chief ray. Thus the asymmetry errors (Sect. 4.3.2) coma and astigmatism cannot occur.

Schmidt eliminated the spherical aberration of the mirror by inserting a correction plate at the entrance pupil. It still has the imaging errors of a lens (e.g., chromatic aberration) but these remain insignificantly small since the plate is nearly without any refractive power. This so-called Schmidt plate possesses one plane and one aspherical surface. Schmidt also showed how to make the complex aspherical surface; with zonal differences of several tens of µm, it is strongly deformed. Schmidt's procedure (grinding a glass plate that is bent over a cylinder by evacuating it) and other methods are detailed by Ingall [19]. Instructions for the do-it-yourself production of Schmidt plates using a vacuum machine are given by Cox [20]. The adjustment of a Schmidt camera is described, for instance, by Weigel [21]. The field curvature remains as the only optical error. Its radius of curvature R_F corresponds to one-half of the radius of curvature of the spherical mirror, which is equal to the focal length of the telescope, and is given by

$$\text{Field curvature} \quad R_F = \frac{r_1}{2} = f . \tag{12}$$

This is a rather strong curvature, but the thin photoplates used with larger instruments can be bent by a suitable plateholder and adapted to this curvature. Smaller Schmidt cameras must use emulsions on film. The problem of the strong curving of the focal plane can be avoided with the aid of a flattening lens; the image is flattened by a single plano-convex lens with radius of curvature r_P adapted to the telescope focal length f. The radius of the curved front surface of this plano-convex lens is given by

$$\text{Flattening lens curvature} \quad r_P = \frac{n-1}{n} f . \tag{13}$$

The quantity n refers to the refractive index of the lens which is placed immediately in front of the image plane. Shifting the entrance pupil from the vertex to the center of curvature of the mirror requires that the mirror's diameter be larger than the correcting Schmidt plate; otherwise vignetting for inclined incident beams would occur. The amount depends on the size of the field, i.e., on the diameter D_{PH} of the plateholder used. Figure 23 readily shows that the mirror diameter D_1 must be

$$D_1 = D_0 + 2D_{PH} , \tag{14}$$

where D_0 is the size of the correction plate. The angular diameter of the imaged field ($2u$) can thus be found from

$$\tan u = \frac{D_{PH}}{2f} . \tag{15}$$

By virtue of its unrivaled imaging quality, Schmidt systems can be constructed as fast telescopic systems ($f/3$ and less); for example, the Schmidt telescope in Tautenburg, Germany, has $D_0/D_1/f = 134/200/400$ cm with a $3°\!.4 \times 3°\!.4$ field, and the Mount Palomar Schmidt has $122/183/307$ cm with a $6°\!.5 \times 6°\!.5$ field. The latter produced the photographs for the famous and widely used *Palomar Observatory Sky Survey* (POSS).

Drawbacks of the Schmidt system are the large tube length of twice the focal length, the field curvature (which can be compensated for), and the relatively inaccessible focus. Numerous modifications of the Schmidt camera were constructed to eliminate these problems, but only at the expense of image quality. Various forms and further developments of the Schmidt system can be found in Köhler and Slevogt. One such development is the *Wright–Väisälä System*, with the correction plate and entrance pupil in the plane through the focus, which is free of the asymmetry aberration coma through additional deformation of the mirror but readmits astigmatism. Despite this astigmatic aberration, the mean field curvature is plane but the useful field remains less than that of the original Schmidt camera. The manufacture and testing of such a system are described by Waineo [22]. Owing to high production costs, the use of Schmidt systems without correction plates is being contemplated. Schmadel [23] has listed the maximum focal length and f-ratio for which such a system may still produce useful images.

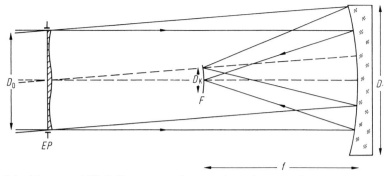

Fig. 23. Schmidt camera ($f/1.5$): Free aperture (corrector) D_0, diameter of spherical mirror D_1, diameter of plateholder D_{PH}, field curvature $= r/2 = f$. The deformation of the Schmidt corrector is much exaggerated in the graph for illustration

4.5.7 Schmidt–Cassegrain Systems

In order to remove the drawbacks of the original Schmidt system, namely the ponderous tube length and the inconvenient focus location, it was soon suggested that the Schmidt system be modified into a *Schmidt–Cassegrain* design by insertion of a secondary mirror. One such Schmidt–Cassegrain system which eliminates all four significant errors (spherical aberration, coma, astigmatism, and image field curvature) comes from Baker. He obtained this excellent state of correction by deforming one of the two mirrors while retaining the spherical shape of the other. Apart from removing the field curvature, the lengths could be reduced to 1.4 to 1.7 times the telescopic focal length and the focal plane placed near the primary's vertex. One disadvantage is that the increase of focal length by the secondary mirror was limited to values less than 2, which resulted in a rather strong central obscuration (15%).

Extremely short Schmidt–Cassegrain systems (tube length less then 1/4 of the system focal length) have been developed for amateur use (Fig. 24). These systems, usually with f-ratios of $f/10$ to $f/12$, have recently become quite popular. They are a compromise between image quality, very short tube length, and easily accessible focal plane outside

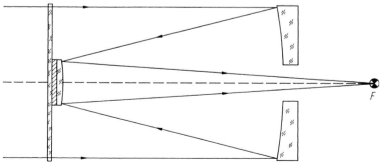

Fig. 24. Schmidt–Cassegrain system

Fig. 25. Schmidt–Cassegrain telescopes enjoy great popularity, due to their compact size

the optical system. Data of one such Schmidt–Cassegrain system have been determined from optical calculations by Rutten and Venrooij [24], who also discuss imaging quality, sensitivity to adjustment, and the different focusing mechanisms of the system. It is assumed that a spherical primary mirror and a correction plate deformed on one side, lying within the primary focal length and defining the entrance pupil of the system, form the optical system. To further reduce the optical errors, the secondary mirror, which increases the focal length, is deformed into a rotation ellipsoid. By far the largest amount of error is contributed by the very strong field curvature. This can be compensated for by curving the receptor surface, so that the remaining imaging errors are largely within the size of the diffraction disk and are thus insignificant.

A Schmidt–Cassegrain system with a plane field has been described by DeVany [25], who also provides hints on manufacturing and testing the optical surfaces. It is particularly suited for amateur telescope makers, since it is based upon two spherical mirrors of identical curvature. The plane image results from the Petzval condition that the radii of curvature of the primary and secondary mirrors must, apart from their opposite signs, be identical. Both mirrors thus result from grinding a single optical surface. The small secondary mirror is cut out of the convex grinding shell and glued underneath the correction plate. Thus, the diffraction spikes of the secondary's spider are removed, a feature shared by all commercially available Schmidt–Cassegrain systems. By abandoning the condition of a plane image field, for instance, by using the instrument visually with a field of 1–2°, a variety of possible Schmidt–Cassegrain systems with aplanatic or anastigmatic imaging can be created. Formulae for calculating the system parameters and the primary imaging errors of such systems are given with related diagrams by Sigler [26]. A similarly detailed treatment of Cassegrain and Gregory systems with one- or two-lens correctors (composed of purely *spherical* surfaces) in the telescope aperture, as well as of Maksutov–Cassegrain systems, is again given by Sigler [27]. He also gives the aberration formulae in closed form, including the imaging errors to be expected for various configurations; they are of great interest indeed to amateur astronomers.

Collimating the Schmidt–Cassegrain is done directly with a star in an eyepiece of low magnification since these commercially available telescopes do not provide any means to adjust the primary mirror itself. For rough alignment one strongly defocuses the star so that its disk fills up nearly one-third the field of view. A dark spot appears within the disk, the silhouette of the secondary mirror. The three aligning screws on the periphery of the secondary are tuned so that the secondary's shadow lies concentric within the defocused stellar disk. One has to repoint the telescope several times since by this procedure the star moves outside the center of the field of view. The fine tuning is done with the diffraction rings of a small defocused star image as described in earlier sections of this chapter. The collimating procedure (also for Maksutov systems) is treated in great detail by Valleli [15].

4.5.8 Maksutov Systems

Since the production of the aspherical correction plate for the Schmidt camera is quite difficult, attempts were soon made to replace it by dioptric elements with spherical surfaces. The optical principle involved was discovered independently and almost simultaneously by K. Penning, A. Bouwers, D. Gabor, and D.D. Maksutov. It is based upon the use of a simple *meniscus lens*, which compensates the spherical undercorrection of the mirror by an overcorrection of the lens. The meniscus is placed concentric with the center of curvature of the spherical primary, and is located (depending on thickness and curvature) closer to the mirror so that the tube length is significantly shortened. A diaphragm (entrance pupil) in the plane through the common center of curvature results in a strictly concentric system without any distinguished axis and with a symmetrically constructed image field, which lies on a spherical surface inside the system. Despite its great similarity to the Schmidt camera, the image quality of the Maksutov camera is slightly inferior, and subsequently the former has prevailed in the design of large professional telescopes. The latter will certainly suffice, however, for high demands by amateurs, and also is simpler to make and less expensive in commercial production owing to the use of only three spherical surfaces. Many modifications of this system are known, some of which have been described by Wiedemann [28] as being especially attractive to amateurs.

As with the Schmidt camera, the Maksutov camera can be modified to a Cassegrain system with an improved state of corrections, resulting in the known advantages of a shorter tube length and a focal plane lying outside the system. The simplest way to introduce the convex secondary mirror into the system is by coating the rear surface of the meniscus with a reflecting layer of corresponding diameter. System data and construction hints for Maksutov–Cassegrain telescopes ($f/15$; Fig. 26) from three spherical surfaces are given by Gregory [29]. The advantages include partial avoidance of light diffraction at the spider (which improves image contrast), and also relative stability against temperature fluctuations (since the tube is closed on both ends). The aberration curves by Wiedemann [12] show that the correction state, even at $f/15$, is not exceptionally good, so a separation of the secondary mirror from the meniscus back is advisable. The freedom thus obtained in the choice of curvature and of positioning of the secondary mirror can again be used for image correction, and Maksutov–Cassegrain systems with very good imaging properties are possible for f-ratios up to $f/7.5$. Comparisons of very different Cassegrain telescope types including spot diagrams have been given by Willey [30].

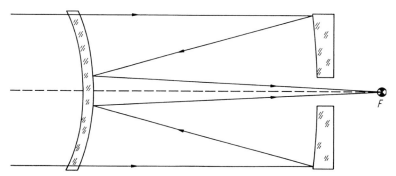

Fig. 26. Maksutov–Cassegrain system ($f/15$). Silver or aluminum vapored onto the back of the meniscus lens serves as the secondary mirror

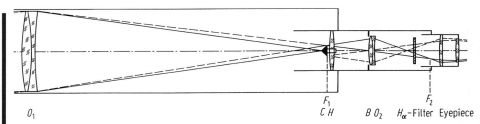

Fig. 27. Schematics of a prominence telescope: objective O_1, conic diaphragm C, auxiliary lens H, scattered light stop B, objective of prominence attachment O_2

Hint:
the color artifact of cheap refractors can be easily removed by the modern Minus Violet filters. The resulting green-yellow tint does not interfere, as experience has demonstrated during solar observations

4.5.9 Instruments for Solar Observations

When observing the Sun, the problem is not the low light intensity which characterizes stellar astronomy. On the contrary, the radiation collected by the instrument is *too* intense for the receiver (such as the eye or a photographic plate), and causes undesirable heating of instrument parts. Notwithstanding the higher price and perhaps remnant chromatic errors, an observer specializing in the Sun would be well-advised to purchase a refractor.

If the atmospheric limited resolving power, which is at best $1''0$, is to be utilized, the telescope aperture cannot be less than 10 cm in diameter according to the diffraction criterion. Smaller instruments with 6- to 8-cm apertures are quite adequate for viewing sunspots and other large-scale phenomena, but for observing finer details such as granulation (see Sect. 13.1), which is of the order of $1''0$, objective diameters of up to 15 or 20 cm are advantageous.

The most preferable (and costly!) light reduction method is by means of *objective filters*, which are plane-parallel glass plates of high surface quality (in order not to spoil the image quality) coated with a partial transparent reflecting layer. Special reflecting foils of the optical quality of glass filters. The transmissivity of the available objective filters ranges from 10^{-3} to 10^{-5}, which corresponds to a reduction by 7.5 to 12.5 magnitudes. The reduction should not be chosen too strong, as there is the possibility of secondary filtering. Objective filters block excess radiation that otherwise could heat the instrument.

Another, undesirable technique is to use an *eyepiece filter*, which diminishes the sunlight *inside* the instrument after it has been collected by the objective. Apart from detrimental effects to the image quality due to the dissipation of heat, there is the ever-present danger that an eyepiece filter, because it is subjected to intense heating near the focal point, will suddenly crack and allow the concentrated solar radiation to reach the observer's retina unattenuated. Such exposure even for an instant can result in permanent eye damage and even blindness. A particular caution should be stated here against the use of soot-blackened glasses, overexposed films, or similar items, since these scarcely diminish the invisible heat radiation which can cause eye damage. More suitable than eyepiece filters are solar eyepieces, or *helioscopes*, whose light-reducing effect is based on refraction, reflection, polarization, or a combination of these properties. The *Handbuch für Sonnenbeobachter* [31] compiles the most important variance of this type of eyepiece, and also treats the entire problem of light reduction in more detail than can be achieved here.

An instrument specially designed for observation of the faint solar atmosphere is the *prominence telescope* , or *coronograph* (Fig. 27). Amateurs can use the coronograph to observe the solar limb and the so-called prominences. Since the hydrogen within the prominences radiates primarily within the Balmer spectral lines, an Hα filter (656.3 nm) with a low spectral passband is used to increase the contrast. The eccentricity of the Earth's orbit causes the apparent solar diameter to increase from 31.'51 in July (at aphelion) to 32.'58 in January (at perihelion), and consequently the conic diaphragm must be changed at various times during the year in order to be adapted to the diameter of the solar disk in the focal plane of the objective O_1 (corresponding, respectively, to 0.00917 or 0.00948 times the focal length O_1 in millimeters). Three diaphragms of small (summer), medium (spring and fall), and large (winter) diameter will probably suffice.

It is possible to equip an existing refractor for observations of the solar limb using a *prominence attachment*. It has the same optical components as the prominence telescope mentioned above and the advantages of relatively short length (25–30 cm), low weight (about 500 g), and improved accessibility for changing the diaphragms; moreover, it can also be added at any time. Construction modes of the prominence attachment are described in the *Handbuch für Sonnenbeobachter* [31] and by Hanisch [32]. Additional shortening and weight reduction are obtained with a so-called *prominence eyepiece* , which is merely a Kellner-type eyepiece with the conic diaphragm on the field lens.

Aside from observing the Sun in the visible spectrum and observing protuberances, interest is growing in observing the entire surface of the Sun in H-alpha light since here the Sun reveals a quite-varied and ever-changing face. Special systems are available for this purpose that are based on the etalon interference filter. Based on the functional principle of a Fabry–Pérot interferometer, these filters allow sunlight to pass only within a very narrow band less than 0.1 nm of the wavelength of ionized hydrogen at 656.28 nm. Functionally, two plane-parallel mirrors of high reflectivity together form an optical resonator. This type of Fabry–Pérot interferometer with a fixed mirror separation is called an etalon. The transmission spectrum of this system reveals multiple narrow transmission maxima for wavelengths that meet certain resonance conditions, while other spectral regions are almost completely extinguished by the interference of partial beams. The maxima of the undesired wavelengths are masked by a block filter. Such combinations of etalon filter and block filter are offered in numerous variants: from the specialized solar telescope to a modification kit for a commercially available refractor. Filter systems using the same technology can also be implemented that allow you to observe the Sun's surface in the short-focal-length light of calcium II at 393 nm and 396 nm. This makes hot-flare regions easily visible.

Fig. 28. The Coronado PST is a compact solar telescope with an Etalon base

4.5.10 Binoculars

Binoculars are certainly the most widespread instruments for visual observations. Depending on their entrance- and exit-pupil size, they may be better or less suited for astronomical observations. Binoculars are of the refracting telescope type, which, since they are frequently used for terrestrial observations (birdwatching, sports, opera, etc.), are designed to provide an *erect* image. Erecting the image with the correct left-to-right orientation can be achieved with various kinds of prisms, among which the com-

Fig. 29. Binoculars mounted on a photo-tripod, attached at the central hinge

bination of two rectangular prisms—the *prism-inverter* according to Porro—is most widespread and achieves a substantial reduction of the tube length by "folding" the beam.

Binoculars are always labeled with designations such as 8×21, 7×50, or 10×50. The first number gives the magnifying factor γ, and the second the free objective diameter D in millimeters; together these provide all the characteristic data for the binoculars. The size of the exit pupil, which is critically relevant for the brightness of extended objects, is found by division as $A_P = D/\gamma$. For observations of extended objects (e.g., comets, gaseous nebulae, galaxies) binoculars with minimum magnification should be chosen, i.e., the diameter of the exit pupil is some 8 mm. Except for observing suddenly occurring phenomena such as meteors and satellites, the binoculars should be as firmly mounted as possible for astronomical observations. Homemade mountings may suffice, but it is preferable to use the readily available camera tripod on which the binoculars can be mounted with a special support cradle (Fig. 29) that can be purchased from most camera shops. With this device, the binoculars can be easily and quickly pointed to any region of the sky, and the vibration-minimized image will show a variety of celestial objects in all their splendor. A bonus is the unique visual impression which can be experienced only by viewing with both eyes. A book full of suggestions for observing with binoculars has been written by R. Brandt [33].

4.6 Telescope Performance

4.6.1 Resolving Power

The resolving power of a good optical system is ultimately limited by the light diffraction at the entrance aperture of the telescope. Figure 4 shows the diffraction image of a circular aperture. When λ is the observed wavelength, and D is the *free aperture* of the telescope, the radius of the first minimum, known as the *Rayleigh criterion* (Fig. 30), can

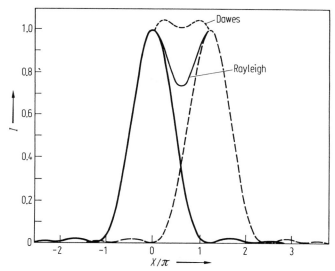

Fig. 30. Intensity distribution in the diffraction disk of a star from a circular aperture. Shown superposed is a second star of the same brightness and the separation between the components (Rayleigh criterion); *dashed line* = intensity sum for the Dawes criterion

be calculated with

$$\rho_0 = 1.22 \frac{\lambda}{D} \quad \text{(radians)}$$

$$= 2.52 \times 10^5 \frac{\lambda}{D} \quad \text{(arcsec)} . \tag{16}$$

Twice this angle, $2\rho_0$, is the diameter of the *Airy disk*. Two point sources are thus barely distinguished when the central maximum of the diffraction curve for one of the sources coincides with the first minimum of the diffraction curve for the second source. Inserting a representative wavelength of 500 nm for visual observations, the Rayleigh criterion mandates a minimum separation of

$$\rho_0 = \frac{12.5}{D(\text{cm})} \quad \text{(arcsec)} , \tag{17}$$

for a telescope of aperture D. Less stringent criteria such as the *Dawes criterion* , which requires an intensity depression between the two maxima of only 3%, may be applied; by this standard, the two images can overlap a bit more before they become indistinguishable, and the constant in the above equation relating to the minimum separation is reduced from 12″.5 to 10″.5. The resolution limit is reached when the overlapping of diffraction patterns no longer gives a dip and the two diffraction maxima merge into one. In this case, 12″.5 is replaced by the limiting value of 9″.77 in Eq. (17).

The resolving power thus obtained holds only for ideal conditions, that is to say, for two point sources of equal brightness, no atmospheric disturbance, and no imaging errors of the optical system.

An almost inevitable limitation of the resolving power arises in the disturbance of the light rays from celestial objects as they pass through the Earth's atmosphere. The

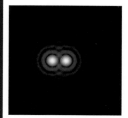

Fig. 31. The Rayleigh criterion specifies the lower limit for an exact separation between two neutral points

Table 3. The astronomical seeing scale as suggested by Tombaugh and Smith [34]

Image quality	Image diameter
−4	50″
−3	32
−2	20
−1	12.6
0	7.9
+1	5.0
+2	3.2
+3	2″0
+4	1.3
+5	0.79
+6	0.50
+7	0.32
+8	0.20
+9	0.13

chaotic motion of turbulent elements there causes a ray passing through them to continually change its direction (directional scintillation), resulting in the ray meandering toward the observer. The amount of broadening depends on the state of the atmosphere and is called *seeing*. At exceptionally good sites, values somewhat below 0″.5 have been observed, while in more ordinary locations, the seeing disk may shrink to about 1″ on excellent nights. Taking 1″ as the order of magnitude resolution permitted by the Earth's atmosphere, then for telescopes with apertures larger than 10 cm the resolution remains about constant while for telescopes under 10 cm the resolution improves with increasing aperture according to Eq. (17). Nevertheless, larger telescopes are still desirable since in addition to their higher light-gathering power, their theoretical resolving power may be realized by the application of special techniques in which the atmospherical image motions are "frozen" in a rapid sequence of very short exposures.

In order to determine the resolving limit set by the atmosphere (seeing) Table 2 Sect. 14.3, p. 627 lists 30 visual double stars with apparent magnitudes m_1, m_2 in order of decreasing separation. As circumpolar stars, they are always above the horizon for most northern hemisphere observers. (A map is given in Sect. 14.3 Fig. 21, p. 626 [35].)

The diffraction disk diameters of stars obtained from double star observations may be represented on a logarithmic *seeing scale* (Table 3) as suggested by Tombaugh and Smith [34]. Apart from the broadening of the stellar images by directional scintillation, there is also *intensity* scintillation, which leads to brightness fluctuations of the star images, depending on the size of the instrument used. It is usually expressed in a five-step scale (1 = best, ..., 5 = worst),[1] but it cannot be measured by eye so objectively as the separation of double stars and is thus strongly governed by subjective impressions.

---

[1] This numbering scheme is reversed in the US.

4.6.2 Magnification and Field of View

During visual observations with a telescopic system, an infinitely distant object is imaged by the objective to the focal plane of the telescope. This real image in the focal plane is inspected through an eyepiece which acts as a magnifying glass (Fig. 1). The magnification γ is then given by the simple relation

$$\gamma = \frac{f_{\text{tel}}}{f_{\text{eyp}}} \,, \tag{18a}$$

or,

$$\gamma = \frac{D}{A_P} \,, \tag{18b}$$

where D is the diameter of the entrance pupil (free aperture) and A_P is the diameter of the exit pupil. Equation (18b) shows that the diameter of the parallel beam emerging from the eyepiece is determined by the choice of the magnification. The telescopic system thus transforms the large entrance beam down to a much smaller "pencil" with the diameter of the eye pupil, and the conditions for the highest and the lowest useful magnification are thereby fixed.

Since all light collected by the telescope objective should reach the eye, the diameter of the exit beam or the exit pupil A_P should not exceed the diameter of the eye pupil. The limiting width results from the maximum pupil size of the eye, about 8 mm (see Sect. 4.8.1). The lowest feasible magnification, or *normal magnification* , and, correspondingly, the longest feasible eyepiece focal length then follow from Eqs. (18) as

$$\gamma_{\text{min}} = \frac{D\,(\text{mm})}{8\,\text{mm}} \,,$$
$$f_{\text{eyp}}^{\text{max}} = N\,(\text{mm}) \,, \tag{19}$$

where $N = f_{\text{tel}}/D$ is called the *aperture number* or *focal ratio*. On the other hand, the resolving power of the eye (about $2'$) determines the highest feasible magnification. Owing to adaptation of the receiving elements on the retina to the diffraction disk, this corresponds, according to Eq. (17), to a pupil diameter of 1 mm. The highest feasible, or *useful*, magnification, and, correspondingly, the shortest useful eyepiece focal length, thus results from

$$\gamma_{\text{max}} = D\,(\text{mm}) \,,$$
$$f_{\text{eyp}}^{\text{min}} = N\,(\text{mm}) \,. \tag{20}$$

Pushing the magnification still higher by shorter-focus eyepieces leads to what is termed *overmagnification* (also called *empty magnification*). More specifically, the optimum seeing disk of only $1''$ is then spread over an apparent angular size and thus becomes discernible to the eye.

The *field of view* is limited by a field stop in the focal plane which is common to the telescope and the eyepiece or by the format of the photographic plate. The image field is

$$2u = 2\arctan(b/f) \,, \tag{21}$$

with b as either the radius of the stop or as one-half the plate diagonal, depending on the particular setup. When observing with an eyepiece, the quantity $2u$ is also called the *true diameter of the field of view*.

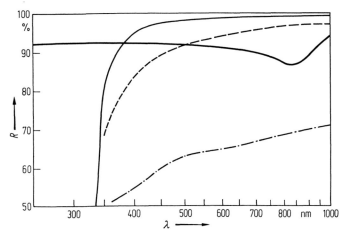

Fig. 32. Reflectivity of the most common mirror coatings. The *heavy line* shows aluminum, freshly vapored-on; the *thin line* shows silver, freshly vapored-on; the *dashed line* shows silver, chemically deposited; the *dot-dashed line* shows polished mirror bronze. Adapted from Bahner

Table 5 Reachable stellar magnitudes

Limit in magnitude	Opening
6	Naked eye
8	Small binoculars
10	2-in. telescope
12	4-in. telescope
13	5-in. telescope
14	10-in. telescope
15	16-in. telescope
16	24-in. telescope

4.6.3 Image Brightness and Limiting Magnitude

The term *brightness* depends on a variety of factors in imaging systems: the apparent brightness (magnitude) and the angular extent of the object in the sky, telescope aperture D, focal length f, f-number $N = f/D$, light losses owing to absorption and reflection in all optical components, imaging quality, and scattering of light from the sky background and the instrument itself. The interested reader should refer to Chap. 8 for a more detailed treatment of some of these items.

Light losses incurred within the optical system of the telescope have several causes. First of all, some light is lost upon transmission through a glass lens, the exact amount depending on the composition, quality, and thickness of the lens. Light losses by reflection at the glass/air boundaries are at least comparable, and occur for each dioptric element twice, once entering and once leaving the lens. As a result, considerable losses can occur for multilens systems. Therefore, the losses by reflection amount (assuming a 4.2% loss per boundary) to as much as 30% for a four-lens objective. They can be reduced only by coating the surfaces with an antireflection layer to below 1% per surface, which, in the preceding example corresponds to a reduction of the total loss from 30% to less than 8%. Another possibility is to cement the lenses together without air spaces; this, however, results in the loss of freedom in the choice of the radii of curvature and the distances between lenses in order to correct imaging errors. An expensive coating is therefore virtually mandatory for high-performance lens optics. Mirror surfaces can also cause losses by absorption owing only to imperfect reflectivity. Coating materials other than silver and aluminum are available but have not gained great acceptance for astronomical mirrors (Fig. 32). A surface upon which silver has been freshly deposited has the highest reflectivity of the common materials, but also two glaring drawbacks. First of all, when exposed to air silver slowly tarnishes and thus loses an appreciable degree of reflectivity. In addition, there is a strong decline in reflectivity for wavelengths shorter than 400 nm.

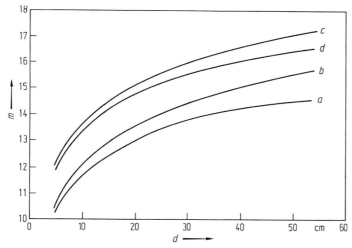

Fig. 33. Limits of visual observation. **a** In dark field (H. Siedentopf). **b** With eye pupil of 7.6 mm (J.B. Sidgwick). **c** After W.H. Steevenson. **d** Including light losses in the optics (after J.B. Sidgwick)

The *limiting magnitude* of stars just barely perceived by the eye depends on the telescope aperture D and on the brightness of the sky background at night. In the following discussion, the light losses in the optical system will be neglected. The following formula is mostly used to estimate the limiting magnitude $m_{\lim}$ for a telescope with aperture D:

$$m_{\lim} = m_0 + 2.5 \log D^2 = m_0 + 5 \log D , \qquad (22)$$

where D is in centimeters, and the value for m_0 is usually given in the range 6.5–6.8 mag corresponding to limiting magnitudes of 6.0–6.3 mag, respectively, for the naked eye depending on which maximum pupil diameter is assumed. The visibility of a point source, however, depends critically upon the background brightness, so the optimum power equals the highest applicable magnification $\gamma_{\max}$ (Eq. 20), and that overmagnification does not improve the limiting magnitude (Fig. 33).

The situation is rather different for extended objects such as nebulae or comets. The light collected by the instrument is spread over a larger area by the magnification of the eyepiece. It is easily shown that the light gain in the case of extended objects is proportional to the ratios of the areas of exit pupil and eye pupil. If a high magnification is chosen, so that the exit pupil is smaller than the eye pupil, then the object will appear less bright in the eyepiece than with the naked eye. If both pupils are of equal size (minimal magnification), then the image will appear as bright through the instrument as with the naked eye (if instrumental losses via absorption and reflection are neglected).

4.7 Collimating a Telescope

"Bad collimation is the number one killer of telescopes worldwide…" is something the well-known American DeepSky observer Walter Scott Houston emphasized at every opportunity. And in fact, criticisms about bad optical performance and fuzzy images are largely due to this fact. Yet collimation of a commercially available refractor or Newtonian reflector is not that difficult as long as you utilize the inexpensive, authoritative aids available. One helpful rule of thumb is that you should always start collimation at the eyepiece and end at the objective optics.

If you want to get the optimum use out of your scope, whether refractor or Newtonian reflector, there is no getting around the need for occasional collimation.

4.7.1 Collimation Tools

Collimating a telescope involves a sequence of steps in which the approach is to assess and then improve the geometric position or configuration. In the past, especially when aligning Newtonian reflectors, all that was needed was a small sight tube that allowed you to precisely center your eye along the optical axis of the telescope. The Cheshire "eyepiece" is a development of these sight tubes. Here a mirror tilted 45° with a central hole is located directly beneath the peephole, while the mirror has a larger opening on the side. As a result, what you see is not the dark alignment barrel but a bright shiny ring with a dark center, the peephole of the observer. The "light ring" is a help, but today an indispensable aid in collimating telescopes is a laser collimator. Nevertheless, lasers are not able to detect flat parallel-displaced secondary mirrors. As long as the deflection angle is right, the laser collimator will give the impression of a perfectly collimated system where even a brief glance is sufficient for you to recognize that the position of the secondary mirror is not correct.

4.7.2 Refractor

A commercially available lens-based telescope, especially the inexpensive starter scope, is in no way as insensitive to misalignment as people often claim. The claim of the supposed "robustness" of long focal lengths and small apertures is also not quite accurate. For example, a focuser tube that is tilted only a few degrees points farther and farther away from the center line of the objective as the distance between focuser and objective increases. Even with a newly delivered telescope, the user cannot be absolutely sure that it is perfectly collimated. With this in mind, it must be emphasized that alignment errors in a refractor do visibly degrade the imaging performance and significantly magnify optical aberrations such as chromatic aberration.

There are two components to align in a refractor: the focuser must aim precisely through the tube at the center of the objective, while the optical axis of the lens must follow the same path back through the exact center of the focuser. Since both elements may be tilted or offset, both must be checked and aligned as necessary. In principle, collimation of a refractor would be simpler than for a reflector if it were not for the fact that often the adjustment screws are missing. The first step is to check for the correct alignment of the focuser and correct this as necessary. It is, however, critical here that the focuser function without any play and cannot be moved back and forth or up and down. When viewing the objective at a slight angle, the reflection of the laser beam coming from the focuser and passing through the objective is visible. Your head should absolutely never be held exactly in line with the beam since the laser beam could enter your eye and cause damage to your retina. Now the focuser is aligned such that the laser beam

Fig. 34. With a small directional eyepiece (*middle*), a Cheshire eyepiece (*left*), and a laser collimator (*right*), a stargazer has a powerful equipment to adjust and control nearly any amateur telescope

passes precisely through the center of the objective. To perform subsequent collimation, you need a Cheshire eyepiece in the focuser. The telescope objective must be covered completely here to prevent any light from entering the scope from the front. A light is now directed into the telescope through the side opening of the Cheshire eyepiece. When looking through the rear alignment hole of the Cheshire in a misaligned refractor objective, two or more light reflections from the scope's front and rear lens surfaces are seen that must be made to coincide.

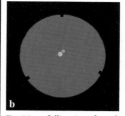

4.7.3 Newtonian Telescope

Since a reflector with its secondary mirror has a second principal optical component, in contrast to a refractor, three alignment operations are required per se: for the focuser, secondary mirror, and primary mirror. Upon initial examination of a newly purchased or just-built telescope, all of the components should, of course, first be checked promptly. A protractor is used to check whether the focuser is really situated perpendicular to the tube. When performing subsequent adjustments, the focuser can usually be ignored. Then the secondary and primary mirrors must be collimated.

Fig. 35. a Collimation of a refractor objective requires a Cheshire eyepiece. **b** Via the view through the rear directional opening in the Cheshire, two light reflections can be seen on the front and the back of the refractor objective, which now need to be aligned

The first step is centering all optical elements that are visible when looking through the focuser. The human eye is surprisingly sensitive to the superimposition of circular surfaces resulting from slight changes in position. All that is required here is a sight tube that assists in keeping the eye exactly at the center of the focuser. If available, of course, a Cheshire eyepiece can also be used since it has a central "peephole." It is also helpful if the primary mirror has a center mark stuck to it in the form of a common "paper hole reinforcement" found in ordinary office supplies. This marks the center of the primary mirror more visibly for your eye and also allows the subsequently required laser collimator through its center.

Now all components are collimated in the proven order "from eyepiece to objective." The first step is done by looking through the focuser at the secondary mirror and adjusting the view of the primary mirror in it such that the primary appears centered in the secondary. After this, the primary mirror is positioned. To accomplish this, the view of the secondary cell and of the cell's struts seen in the primary mirror must move so as to be centered in the image of the primary mirror. This produces a concentric series of images that from the inside out show: (1) image of the secondary mirror image, (2) image of the primary mirror, (3) secondary mirror. All together must also end up at the center of the focuser tube. A laser is very useful in making the fine adjustments to the relative tilt angles of both mirror surfaces, the laser also being used in two steps from the focuser to the primary mirror. When looking into the tube from the front at the primary mirror, the reflection of the laser beam is visible as a small red spot. The goal is to move this into the center of the primary mirror by making fine adjustments. In a second pass, the primary mirror is readjusted so that the laser beam is again reflected back on itself. This is true if all of the optical elements have not moved into improper tilt angles relative to each other along the way; the optics are therefore correctly aligned. Laser collimators that use a small opaque screen to show the position of the returning beam are quite helpful here and facilitate the last alignment steps until the beam disappears within itself. The Newtonian telescope is now collimated.

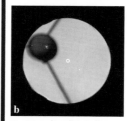

Fig. 36. a In a Newtonian telescope, all components are adjusted in sequence "from the eyepiece to the objective" with a laser. **b** The secondary and main mirrors are adjusted so that the laser beam is reflected back at itself

4.7.4 Other Optical Systems

Good prealignment of a Schmidt–Cassegrain or Maksutov telescope can also be achieved using a laser. However, with almost all commercially available systems using these designs, the primary mirror is not adjustable; as a result, the only adjustment possible is for the secondary mirror. However, the secondary mirror in these systems is not a flat but a highly curved convex mirror. This expands the laser beam noticeably so that the reflected laser point becomes a large diffused disk. Now the secondary mirror can be tilted to make the diffused disk centered with the exit aperture of the laser. This is only a rough collimation due to the imprecision involved. Fine alignment of an SC system must be performed on a star. When collimating more complex systems, the advisable approach is to consult the manufacturer.

4.8 Accessories

4.8.1 Eyepieces

In order to be able to assess the quality and suitability of an eyepiece, the user must consider a number of criteria.

One fundamental usage factor is, of course, a function of the focal length of the eyepiece. It determines the magnification attainable with the telescope. Magnification can be determined simply by dividing the focal length of the scope by the focal length of the eyepiece.

A 60/700-mm refractor using a 10-mm eyepiece thus provides a magnification of 700/10 = 70 power.

The second fundamental parameter is the apparent field of view indicated as an angular field in degrees. Simple eyepieces provide an apparent field of view of around 40° of the viewed sky, while modern wide-angle eyepieces go up to 100°, thereby offering a wider segment of the sky (field of view) at the same magnification (Figs. 37–40). Aside from specialized planetary eyepieces in which the unnecessary field of view is of secondary importance as compared with image quality, today almost all conventional eyepieces have an apparent image field of at least 50°. A field of view measuring 30° to 40° results in an unpleasant "tunnel-vision" effect: the small visible segment of sky appears to be at the bottom of a dark tube. Conversely, in modern eyepieces with a minimum 60 to 70° image field, a much larger segment of the sky floats directly in front of the eye, an impression that is enhanced by the fact that the limiting tube or field stop of the eyepiece around the rim is barely visible.

However, a generous specification for the image field does not always guarantee a correspondingly large field of view. Wide-angle eyepieces of the same focal length can thus have completely different specs for the field of view while having the same real image field size, that is, they show a segment of sky of the same size. This optical effect is called distortion. In an apparently wide-angle eyepiece, it distorts the actual image and stretches it out to the apparently larger viewing angle. While this is almost unnoticeable for small groups of stars, since at most their relative spacing changes slightly, extended celestial objects are impacted significantly. The Moon appears to be stretched out in egg-shaped

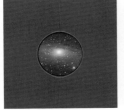

Fig. 37. Kellner eyepiece with a 40° field of view

Fig. 38. Plössl eyepiece with a 52° field of view

Fig. 39. Erfle eyepiece with a 65° field of view

Fig. 40. Ultrawide eyepiece with an 80° field of view

fashion toward the rim in such an eyepiece. Since the distortion varies for each version of an eyepiece, it is all the more important to be able to determine the actual size of the image field yourself. A relevant measurement scale is incorporated as the site of the field stop. If you can read the diameter of the image field in millimeters, you can easily use the geometric relationship

field of view angle = 2 × arctan (measurement value/(2 × focal length of the eyepiece))

to compute the exact value for this specific eyepiece.

Often neglected but just as important is eye relief, which relates to the ability to look into the eyepiece. Normally, the condition for the eye relief needed to view through the eyepiece is an analogous relationship to the eyepiece's focal length. The shorter it is, the smaller the eye relief and the more difficult it becomes especially to see the entire image view. Special eyepieces exist that are appropriate for people wearing eyeglasses since these have an eye relief from about 15 to 20 mm.

It is also interesting to look at the number of lens elements in the eyepiece's construction. Increasing the number of lens elements boosts the eyepiece's performance. This may be an enlarged image field, or greater eye relief, or even better edge correction. Generally, a combination of these optimizations is involved. However, increasing the number of lens elements also has its disadvantages. These turn up first of all in the price since the more elements the system contains, the more expensive the eyepiece becomes. One thing that has to be considered is that despite all of the antireflection coatings and sophisticated types of glass, every lens surface and every lens group consumes some light. As a result, it is in fact the simple eyepieces with few elements that often yield the brightest, most brilliant images.

Antireflection coating of the lens surfaces is critical for contrast and light transmission, as these prevent reflections at the lens surface and ensure that more light can pass through the lens. Simple light-blue coatings do not deserve that high a recommendation. More effective are antireflection coatings that appear deep magenta or green. An important factor to check is whether the lens has not just a single coating but instead is multicoated. Furthermore, all of the lens elements of the eyepiece, or of an objective lens, should be coated ("fully coated"). It is often found that with very inexpensive lens elements in particular only the two visible outer surfaces have been coated.

Scattered light caused by reflections within the eyepiece between lens elements or by the inside of the barrel's not having been sufficiently blackened can be quite objectionable or irritating when observing an object. A good test to check an eyepiece for interior reflections is to place the nighttime Moon or a distant streetlight with the border between light and dark (the terminator) exactly in the center of the eyepiece so that half the image field is brightly illuminated while the other half is black. If the bright moonlight generates reflections within the eyepiece, these usually become visible in the dark half as yellow-white striations, often circular, around the center of the image field.

The makeup and type of an eyepiece is defined primarily in terms of the number of lens elements and their arrangement (Fig. 41). Eyepieces with the same number of lens elements often have very similar properties relating to the correction of aberrations or the size of the image field. Nevertheless, even with this in mind there is a wide variety of subtypes and variations in the order of the elements whereby their radii of curvature or configuration relative to each other is modified slightly. Of the approximately

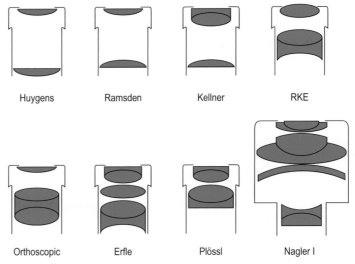

Fig. 41. Different eyepiece designs

100 commonly used eyepiece design types, only around a dozen are relevant to amateur astronomy and deserve further examination.

(a) Two-Element Eyepieces

This simple classic design is usually found in cheap telescopes. Equipped with two simple plano-convex lenses, these eyepieces, named after their inventor Huygens, are only occasionally supplied with very cheap scopes. One modification here using different radii of curvature provides a better view in eyepieces of short focal length and is called a Ramsden eyepiece. Today Huygens eyepieces of long and mid-focal length are relevant only for solar projection since they do not contain cemented achromatic lens elements which can come apart when hot unfiltered sunlight passes through the eyepiece. They are less suited for astronomical observing: both designs have only a small field of view measuring about 30 to 40° and form a sharp image only in a small portion of the field. In addition, this image is visibly degraded by chromatic aberration in all lens elements.

(b) Three-Element Eyepieces

Providing a Ramsden eyepiece with a cemented eyelens produces a better-corrected Kellner eyepiece. The three lens elements also yield a somewhat greater 40 to 45° apparent field of view of very good quality. One appealing aspect of these systems that have fewer elements is their often high contrast and transmission. Modern Kellner eyepieces have an even greater 50° field of view and represent a good solution for beginners who do not want to spend too much money on their first purchase.

(c) Four-Element Eyepieces

Eyepieces having two two-element groups, or a dominant three-element group plus an eyelens, are the ones most commonly found among amateurs. These eyepieces have names such as "Plössl eyepiece" or "Orthoscopic eyepiece," Ortho for short. The design of these eyepieces combines very good performance in terms of field correction and size with a still manageable number of lens elements. As a result, they are very well-suited for many standard applications and have today become inexpensive due to their being produced in large numbers. The result is also a good price-to-performance ratio. The Plössl eyepiece provides a field of view between 50 and 55° that has improved edge sharpness over the Kellner design. They are thus distinguished by a sufficiently large image field, good sharpness, and color correction, as well as an affordable price—clearly the principal reason these eyepieces now represent the most popular design. Sometimes these eyepieces are also sold as "Superplössls." This, however, is simply a marketing name that is not based on any technical or optical modification. Compared with the Plössl, Orthoscopic eyepieces provide especially high sharpness at the center but with a smaller image field. As a result, these eyepieces are popularly used for observing planets, but they represent a niche product and are not as widespread as the Plössl types.

(d) Special Multielement Eyepieces

Improving or enhancing image properties always entails a greater number of lens elements, and usually a generally significantly higher price. For special purposes, multiple designs exist that can be appealing to the amateur astronomer. The relevant performance enhancements relate to increasing the field of view, improving the view near the edge, and increasing eye relief. Often several improvements go hand in hand. For example, it makes sense to optimize edge performance when increasing the field of view since otherwise the added field of view would be useless.

Popular designs for wide-angle eyepieces include the König eyepiece with a field of view ranging between 55 and 65°, and the popular Erfle at 60 to 70°. Both designs tend to have unsatisfactory edge definition in fast telescopes, but here again modifications are available that provide high image quality in almost all telescopes. Extreme-wide-field eyepieces have fields of view of up to 100° and offer the user the most pleasurable observing experience, although at a high purchase cost.

4.8.2 The Barlow Lens

The Barlow lens is a negative (divergent) lens system inserted for the purpose of increasing the telescopic focal length. It consists of an achromatic doublet placed in front of the focus (Fig. 42). The off-axis chromatic errors can be kept to a bare minimum if the Barlow lens is adapted to the objective. This is more easily achieved the smaller the focal ratio D/f and the smaller the lengthening factor B. The latter corresponds to the ratio $f'/f_{BL} = s_2/s_1$, so that the new effective length f' of the telescope follows from

$$f' = \frac{f_1 f_{BL}}{f_{BL} - s_1} .$$
(23)

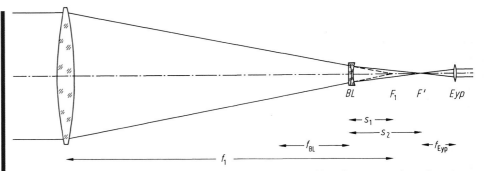

Fig. 42. Position and function of a Barlow lens. The telescopic focal length is increased, resulting in an increased image scale for visual and photographic observation

The distances s_1, s_2 are obtained from the following relations:

$$s_1 = \frac{f_{BL}\,(B-1)}{B}\,, \quad s_2 = f_{BL}\,(B-1)\,. \tag{24}$$

Thus, if the focal length is doubled, the system focus lies at half the Barlow's focal length f_{BL} behind the objective focus. The increase in telescopic focal length is therefore achieved without a significant increase in the tube length. The Barlow lens should *not* be used as an element to shorten the tube length (with a fixed telescopic focal length) as the image quality is inferior to that of a telescope with equal focal length but without a Barlow lens.

4.8.3 Tube Construction

The tube serves as a structure which fixes the optical parts and also protects against rapid temperature fluctuations, humidity, and scattered light. The tube dimensions are given by the focal length and the free aperture of the objective and are found by simply constructing a true-to-scale graph of the entire optical system.

A general statement regarding preference for a closed or an open tube cannot be made. For telescopes that are closed at *both* ends (e.g., refractors, Schmidt or Maksutov systems, astrocameras) the closed tube might be preferred, in contrast to open-ended systems (reflectors), for which the heat inside has direct exchange with ambient air. Either the tube diameter is chosen a bit larger than the mirror so that surrounding air can pass through the tube outside the beam, or else the tube is constructed entirely as an open framework, in which case the heat exchange with the surrounding air proceeds much more rapidly. Upwards of a certain telescope size, an open tube is advisable anyway, as the large cross-section increases the force of the wind on the telescope mount and drive, and the prevention of vibrations caused by wind gusts requires even more attention.

Irrespective of which construction is preferred, the tube must guarantee stability for the adjusted state of the optical system. For small- and middle-sized telescopes this rigidity against flexure can be achieved. The compensation for flexure of large telescopes by *Serrurier struts* is beyond the scope of this chapter. Suitable materials range from metals to wood and carbonfibre, for which Schiffhauer [36] gives useful hints for construction.

All optical components should be attached to the telescope tube in such a way that they may be adjusted. No stress or shift of the components should occur when either the pointing of the telescope is changed or the temperature varies (this is particularly important for primary mirrors). Lens objectives are framed at the edge while small mirrors can be supported on their back by an elastic base (e.g., a felt plate). Adjustment of the mirror is achieved by supporting its back on three points normally located on a circle of radius $r = D/\sqrt{2}$. For large mirrors, the three-point support is insufficient since the mirrors deform under their own weight. In such cases, an *astatic* support, which adjusts itself according to the tilt of the mirror during telescope tracking, must supply an equal counter weight to the mirror at numerous points. Such "flotation cells" are described, for example, by Maksutov [37].

Baffles may be added to the optical components as stops in order to prevent scattered light from reaching the detector (eye, photographic plate, etc.), and should be inserted into a closed tube which has been blackened on the inside. Where and how large these diaphragms should be can also be found from a true-to-scale graphing of the optical path. These *baffles* are especially important for telescopes of the Cassegrain-type, where light can pass the secondary mirror and fall directly onto the receiver. Owing to the conical shape of the baffle, the central obscuration here is somewhat stronger than that produced solely by the secondary mirror.

The *dewcap* can also help to suppress scattered light. It is chosen to be a rather long cylinder but should not vignette the field of view. This cylinder also prevents the formation of dew on the objective. This function can be enhanced by employing a tiny electric resistance heater. If dew forms on the surface of the unheated objective, it should never be wiped dry; it should instead be evaporated by gentle fanning, or dried the next day by exposing it to natural air motion in the open dome. Moisturized optics should not be closed but rather covered by a cloth which is porous to air. In spite of even meticulous care, some dirt will inevitably deposit on the objective so that regular cleaning will be needed from time to time. Specific instructions regarding these procedures have been layed out by, for example, MacRobert [38].

4.8.4 Finding and Guiding

The *finder scope* is a simple refractor with a relatively small (a few centimeters) aperture. It enables the observer to locate faint objects that are invisible to the naked eye and to point the telescope quickly, especially if it is one which lacks accurate setting circles or the amenity of computer-driven coordinate settings. The true field of view of the finder should not be less than $3°$ in order to facilitate finding the object after a rough pointing. In order to avoid unnecessary gymnastics at the telescope, it is useful to have the finder located close to the eyepiece. For exact positioning, the finder is usually equipped with a crosswire eyepiece. The finder must be pointed parallel to the optical axis of the telescope, a task best accomplished with twin holder rings with adjustable collimating screws, which ensures parallel adjustment also after transportation. Unlike classical finder scopes, zero-magnification finders do not have light-gathering lenses but instead are intended to assist the eye of the observer in determining in which direction it is looking. Since the zero-magnification finder is permanently attached to the telescope, the line of sight then also coincides with the direction in which the telescope is aiming.

Fig. 43. While zero-magnification finders excel with their large fields of view, the viewfinder telescopes with their greater screen brightness make it easier to find small and faint objects

This assistance is implemented first of all by projecting a target or an illuminated dot. Another possible approach uses a strong laser beam to show the direction in which the telescope is pointing. The advantage of all zero-magnification finders is that you have the entire sky in view for reference. One disadvantage, on the other hand, is that all of these devices need batteries to power the LED or minilaser. This undoubtedly is not the most economical and ecological solution. Furthermore, it is very possible that the desperately needed batteries will be dead precisely when, after weeks or months of waiting, you have a promising opportunity to observe. In addition, the glass plates with the target markings in general noticeably reduce the detectable limiting stellar magnitude.

One of the simplest arrangements is a pair of sighting rings as used with some rifles. In this scheme, the observer's line of sight is framed by a small ring.

Guide telescopes serve for guiding *during* photographic exposures. There are three requirements: (1) a moderately large aperture in order to see even faint guide stars, (2) a long focal length, and (3) high mechanical stiffness. If the aperture is too small, it may be difficult to find a suitable guide star in sparsely populated sky regions at high galactic latitudes. The focal length of the guide scope must be larger than that of the photographic telescope, but this depends in detail on the "pixel size" of the detector and on the effective focal length of the system. It is essential that the drift which can just be noticed in the crosswire eyepiece or from a guiding camera remains below the angular resolution of the detector. This can be accomplished by means of a Barlow lens and/or the subsequent magnification by an eyepiece in the case of visual guiding.

4.8.5 Eyepiece Micrometers / CCD Astrometry

The *eyepiece micrometer* measures the position angle and relative separation of two objects (e.g., double-star components in the apparent field of view). It uses a crosshair eyepiece with an additional third wire, perpendicular to one of the crosshairs and movable across it by means of a micrometer screw. The entire crosshair unit is rotatable, and its position can be read at a *degree circle* outside the eyepiece. Centering one star on the fixed crosshair and the other on the movable crosshair, their position angle can be read directly at the circle, while the micrometer position—each unit corresponds to a certain displacement in degrees on the celestial sphere—after conversion gives the separation between the components.

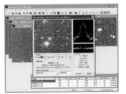

Fig. 44. Screenshot of astrometry software

Hint:
Add-on programs for observers working with astrometry are available for many current astronomy programs. Standalone astrometry programs that can import current star catalogs can also be found in various forms on the Internet.

Another method of position measurement has been made possible by the widespread availability of digital cameras. Whereas an eyepiece micrometer is appropriate primarily for measuring the distance between very close bright objects like double stars or planetary moons, opportunities that were previously the purview only of professional astronomers can now be handled even by relatively small telescopes using CCD cameras aimed at amateur astronomers and powerful personal computers (Fig. 44). The positions of extremely faint individual objects can be measured to fractions of an arc second.

Measurement is divided into two steps: measuring the stellar positions on the CCD and determining the plate coordinates. In the simplest approach, the positions of a star can be determined on the CCD chip by selecting the pixel with the greatest intensity and assuming the associated values for x and y to be the position of the star. The image of the star usually covers several pixels, so determining the position only by the brightest pixel is too imprecise. The analytical programs therefore use other methods to determine the star's position more accurately (i.e., to fractions of a pixel). The center-of-gravity method sums the intensity values by vertical and horizontal lines, then precisely calculates the centers of gravity for the stellar positions to between 0.1 and 0.01 pixel. Comparing the star's image with a mathematical Gaussian distribution also yields precise values for the x and y coordinates of the star's image. The point spread function determines the brightness distribution based on a few bright but not overexposed stars, then applies this to the celestial object to be measured. It is important here that the stellar coordinates be converted from the catalog epoch (e.g., 2000) to the recording date, after which proper motion must then also be taken into account. Effective computer programs are available for this purpose in which only the generated image has to be inputted along with the celestial object to be measured. This approach can be utilized to measure positions for planets, planetary moons, minor planets, comets, fast-moving stars, or binary stars.

The error in estimating the separation increases rapidly for position angles near 0 and 180°. A precision of $1/10$ of a second of time (≈ 1–$2''$) can be obtained in right ascension.

4.8.6 The Photometer

Brightness estimates of stars made with the eye are naturally somewhat subjective, but attempts have in fact been made to increase the precision by certain techniques (e.g., the "step-method" of Argelander (see Sect. 8.4.2) or the "fraction method" of Pickering). To measure stellar magnitudes objectively, a photometer or CCD image are utilized. The collected photons are converted into a measurable electrical current which supplies a quantitative linear measure for the illuminance at the telescope focus.

The most commonly used color system is the *UBV* system of Johnson. It is defined by the combination of a reflecting telescope with a photomultiplier tube of type RCA 1P21 and employing the following filters:

U: 2 mm Schott/UG2; maximum transmission at 360 nm; bandwidth about 55 nm

B: 1 mm Schott/BG12 + 2 mm GG13; maximum transmission at 440 nm; bandwidth about 100 nm

V: 2 mm Schott/GG11; maximum transmission 550 nm; bandwidth about 80 nm

The use of a photometer, however, also entails working with high voltage and complicated filter systems. For the amateur, it is therefore interesting to know that CCD cameras can also be employed for photometric purposes. Those looking to choose a suitable camera for this observing goal should, however, consider a few points that can affect the measuring accuracy and the amount of work involved in the subsequent analysis in their selection of the camera. It is important that the CCD chip be cooled. In addition, dark current and readout noise must be low in order to obtain the greatest possible dynamic range in the images. The images must be read at a data resolution of at least 12 bits, although 14 or 16 bits are the norm today in good CCD cameras. Cameras that contain a chip without antiblooming and without interline structures have a surface with greater light sensitivity per pixel. Additionally, it is advisable to select a chip with good blue sensitivity. Whereas visual estimates achieve an accuracy of around 0.1 magnitude, use of a CCD camera can reduce the mean error for an individual measurement down to mag 0.05. The simplest approach for measuring brightness on a CCD image uses the method of aperture photometry. Newer methods utilize the point spread function (PSF) of stars for the measurement and are similar to the procedures used in professional photometry programs.

4.8.7 The Spectrograph and the Spectroscope

The spectrograph allows a *qualitative* analysis of starlight by investigating characteristic spectral lines. It has three basic components: the collimator, the dispersing element, and the camera. The collimator transforms the convergent rays from the telescope into a parallel beam which falls at a certain angle onto the dispersing element. Having passed this element, the light beam exits at some other wavelength-dependent angle, and is imaged by the spectrograph camera onto the photographic plate or CCD. The collimator ensures that all rays fall at the same angle onto the dispersing element and can be thought of as an "inverted" telescope. The two dispersing media used are prisms and diffraction gratings. The prism spectrograph uses the dispersing effect of glass while in the diffraction grating interference of the individual rays leads to spectral dispersion. The grating produces what is known as a *normal spectrum* since the diffraction angle depends linearly on the wavelength, while the prism response is distinctly non-linear and disperses the blue spectral region more strongly than the long-wavelength red region. Spectroscopy requires high light-gathering power since the starlight, owing to dispersion and the widening of the spectrum, is distributed over a rather large receiving area and normally needs integration on the detector. One exception is the Sun.

To view the solar spectrum and its numerous absorption lines direct with the eye, a *spectroscope* is employed. A *direct-vision prism* (Fig. 45) introduced by Amici (1786–1864) usually serves as the dispersing medium. It consists of up to five individual prisms of two different glasses cemented into one prism. Instructions for building such a spectroscope with a direct-vision prism are given by Gebhardt and Helms [39]. When viewing a point source, such as a distant star, on the other hand, the widening is provided by a cylindrical lens, while photographic widening is best achieved by physically moving the telescope in right ascension or declination depending on which motion is parallel to the slit.

Amateur astronomy also increasingly involves the use of spectrometer equipment that employs diffraction gratings to break up light into its constituents. In addition to the unchanging component (0 order) of light, these provide a spectrum of multiple orders that reveal almost completely uniform color separation and higher intensity in the

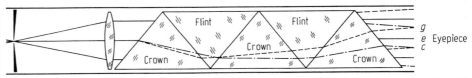

Fig. 45. Spectroscope with direct vision after Amici; the slit is in the focus of the telescope objective and of the collimator lens

blue region. By appropriately designing the form of the grooves, what is known as a blaze grating is able to significantly increase the intensity of a diffraction order in a way that is equivalent to using a much larger telescope with a normal grating. Spectrometers using a blaze grating allow for both visual and photographic observation of spectra. Use of a cylindrical lens enables the viewer to expand the otherwise threadlike spectrum and thus more easily observe the spectral colors as well as the stronger absorption and emission lines of the stars. In addition to spectroscopic analysis of normal stars, spectroscopy of special objects such as Wolf–Rayet stars or planetary nebulae is also possible.

4.8.8 Sun Projection Screens

Instruments for direct solar observation were described in Sect. 4.5.9. Lacking those elements (such as solar filters, absorbing glasses, helioscope, etc.) to reduce the intense solar radiation, the Sun can still be observed by magnified projection onto a white screen. The screen should lie in the shadow of a larger stop surrounding the telescope tube. Owing to their proximity to the telescope's focus and the heat concentration there, cemented eyepieces must *not* be used for solar projection, but only those of the Huygens, Mittenzwey, or Ramsden-type. The diameter of the projected solar disk $P_\odot$ depends on the focal length f_{eyp} of the eyepiece and the distance d to the projection screen. An approximate formula is

$$P_\odot = D_\odot d \left(f_{obj}/f_{eyp} \right) = D_\odot d \gamma ,$$

where $D_\odot$ is is the angular diameter of the solar disk. Owing to the ellipticity of the Earth's orbit, $P_\odot$ varies between 0.00917 and 0.00948 radians at aphelion and perihelion, respectively. Thus, for graphs of high precision, the distance of the screen should be adjustable. A projected diameter of 15 cm is adequate to sketch the solar disk; it is also a recognized "norm" for which graph paper or stencils with division in degrees are available.

4.8.9 Time Signals

The time of an observation, whether visual or photographic, should always be recorded. This applies to predictable phenomena like eclipses or lunar occultations as well as unforeseen events such as fireballs or novae. It should be emphasized that every photographic picture is a document, so recording the exact observation time—beginning, or perhaps end, or total exposure time—should be routine. For most purposes, a quartz watch will suffice; it can be compared before (and perhaps after) an observation with a time signal which permits a much higher degree of accuracy than the times available from radio and telephone service. A number of time signal broadcasts are obtained on the shortwave band (Table 5); they can be readily obtained using simple radios which have the designated band. Thus, the instant of observation and the time signal can be recorded simultaneously on a tape.

In the US, station WWV provides shortwave broadcasts at 5, 10, 15, and 20 MHz and time signals over the telephone (1-900-410-TIME, and, for the Washington DC area, 1-202-653-1800).

Table 5. Some radio stations which broadcast time signals on the shortwave band

Location	Station	Frequency	Power	Location	Station	Frequency	Power
Great Britain	MSF	2.5 MHz	50 kW	Czech Rep.	OMA	2.5 MHz	1.0 kW
		5.0	50	USA	WWV	2.5	2.5
		10.0	50			5.0	10.0
Hawaii	WWVH	2.5	2.0			10.0	10.0
		5.0	2.0			15.0	10.0
		10.0	2.0			20.0	10.0
		15.0	2.0			25.0	10.0

Phenomena observed in the sky are almost always recorded in *Universal Time* (UT) as obtained from the time signals, while for the telescope drive, however, *sidereal time* (ST) is needed (see Chap. 2).

4.9 Visual Observations

4.9.1 The Eye

The properties of the eye, especially where they relate to and limit the performance of telescopes, have been repeatedly addressed. The diameter of the eye pupil is of unquestionable importance in determining resolving power and image brightness. The maximum attainable pupil width is often given in the range 6–8 mm for a person of age 20 down to approximately 4 mm at the age of 80. Thus the minimum or normal magnification should be arranged so that the beam from the exit pupil is not vignetted by a too small an eye pupil.

Another important quantity is the resolving power of the eye, given as about 120″ (= 2′) for high image contrast such as that for double stars. This value agrees surprisingly well with the minimum pupil diameter of 1–2 mm. This means that magnifications with exit pupils under 1 mm do not gain any more angular resolution, since then the smallest diffraction element is spread over several receiving elements on the eye's retina. Likewise, the use of magnifications much in excess of 100× is not practical since then the maximum resolution of about 1″ permitted by the Earth's atmosphere reaches the threshold of perception.

The sensitivity of the eye is wavelength dependent (Fig. 46). The retina of the eye consists of a network of nerve cells that send impulses along the optic nerve to the brain. There are two principal kinds of retinal cells, namely *rods* and *cones*. The rods provide vision in dim light and reach their maximum sensitivity at 507 nm, while the cones respond best to bright light and reach their maximum sensitivity at 555 nm. Color vision is provided by the cones. The rods, on the other hand, are responsible for so-called night vision, and are stimulated at a threshold value which, under ideal conditions, corresponds to a star of 8th magnitude. In reality, however, the background brightness of the night sky diminishes the limiting magnitude for the naked eye to about $6.^{m}0 \pm 0.^{m}5$. In this connection, the adaptation of the eye to the dark is of some interest and proceeds roughly as follows: the retinal sensitivity increases very little during the first ten minutes, then rises

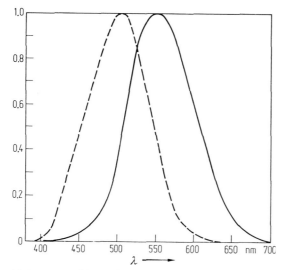

Fig. 46. Spectral sensitivity curves of the eye (normalized so that maximum = 1): *dashed line* represents nighttime, and *solid line* represents daytime vision

steeply from 25% to around 80% of the maximum sensitivity during minutes 15 to 25, and ultimately approaches the maximum asymptotically, so that the maximum sensitivity of the retina is essentially reached after about 1 hour.

The distribution of rods and cones in the retina is of particular interest for visual observations. The color-sensitive cones are concentrated in the center, or *fovea* of the retina. The more sensitive but color-neutral rods, on the other hand, are more numerous toward the outer periphery of the retina. This explains why some very faint light sources (e.g., nebulae) are "invisible" by direct vision, but may be glimpsed by looking slightly to one side of the sought object; this is termed *averted vision.* A comprehensive description of the human eye can be found in Fry [40].

Optimizing your Eyes for Astronomy see Special Report Sky and Telescope, September 2005, pp. 30–42

4.9.2 Sketching What You See

Unlike photography where the celestial object is imaged as faithfully as possible matching the original, drawing celestial objects by hand has a somewhat different emphasis. Here the goal is to reproduce the visual impression at the telescope, not the object itself, as authentically as possible. You do not need to have special artistic drawing talent to sketch celestial objects, although it does help to have a systematic procedure so as to enable you later to compare drawings. This is of special interest when you want to reproduce the impression made by similar objects, such as a series of galaxies, in a telescope.

In terms of tools, the observer should have a stable but easily manageable drawing board illuminated by red light on which to create his or her rough sketches at the telescope. Prepared drawing sheets with a field of view and places to record observing conditions are also very useful to have. The tool set is rounded out by a series of pencils of varying hardness along with a rubber eraser. The sensible approach is to make only rough sketches when observing at the telescope itself. Here the observer should concentrate primarily on capturing the broad outline and essential structural elements of the celestial object, along with field stars located in the image field. Actual observing, however, begins by gradually drawing in individual details. The use of different magnifications also helps in teasing out as many visually detectable details as possible. When observing at the limit of perception, there is always a danger as to whether this or that detail was really recognized or whether one wished to see it, and memories of previously viewed photographs of the object begin to merge. The drawer must always be self-critical and question whether the impression sketched really matches what was observed. In terms of certain special features, a descriptive word can be useful, for example, a note on color or the filter used (Fig. 47).

The finished drawing is then completed at home at your desk. Here you begin by transferring the stars, from brighter to fainter. One possible perfectly legitimate approach is also to use a planetarium program to print out the stars, obviously without the celestial object to be drawn, and then draw in the object. It is advantageous here to begin with the boundaries of the object, then the essential gross features, before finally drawing all of the details recognized. Drawing with a black pencil on white paper is the simplest approach.

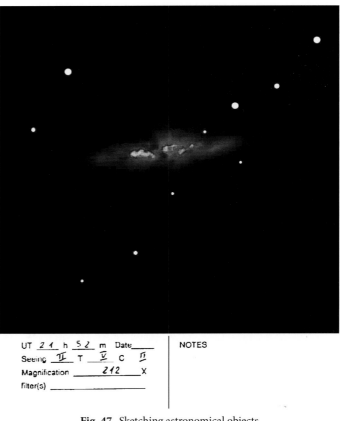

UT _2 1_ h _5 2_ m Date____   NOTES
Seeing _II_ T _V_ c _II_
Magnification _____ _212_ ____ X
Filter(s) _____

Fig. 47. Sketching astronomical objects

To obtain a more realistic view, the method used to involve drawing with a white pencil on black india-ink paperboard. Today, drawings can simply be scanned and inverted in a computer.

4.10 Services for Telescopes and Accessories

Most astronomical magazines such as *Sky and Telescope* and *Astronomy* print suppliers' advertisements on state-of-the-art equipment and instrumentation. Extras on www.springer.com/978-3-540-76377-2.

4.11 Further Reading

Gera, H.D.: Welches Nachtglas? Sterne Weltraum **26**, 167 (1987)

Robb, P.N.: Selection of Optical Glasses. Appl. Opt. **24**, 1864 (1985)

Ulrich, M.H., Kjär, K. (eds.): Proceedings of the ESO Conference on Very Large Telescopes and their Instrumentation, ESO, Garching (1988 and 1992)

Nögel, O.: Ein Fernrohr zur Beobachtung der Protuberanzen für den Amateur. Die Sterne **28**, 135 (1952) and **31**, 1 (1955)

Nemec, G.: Das Protuberanzenfernrohr als Hochleistungsinstrument. Sterne Weltraum **10** (1970) and **11** (1971)

Veio, F.N.: An Inexpensive Spectrohelioscope by a Californian Amateur. Sky Telescope **37**, 45 (1969)

McAlister, H.A.: High Angular Resolution Measurements of Stellar Properties. Annu. Rev. Astron. Astrophys. **23**, 59 (1984)

Hartshorn, C.R., Ingalls, A.G. (ed.): Amateur Telescope Making III, p. 277. Scientific American, New York (1953)

Boulet, D.L.: A Simple Photochronograph. Sky Telecope **68**, 76 (1984)

Harris, C.: Silicon Eye: A CCD Imaging System. Sky Telescope **71**, 407 (1986)

References

1. Born, M., Wolf, E.: Principles of Optics, 6th edn. Pergamon, New York (1980)
2. Cagnet, M., Francon, M., Thrierr, J.C.: An Atlas of Optical Phenomena. Springer, Berlin Heidelberg New York (1962)

3. Petzoldt, J.: 'Zerodur' – ein neuer glaskeramischer Werkstoff für die reflektierende Optik. Sterne Weltraum **15**, 156 (1976)

4. Hartmann, J.: Objektivuntersuchungen. Z. Instrumentenkunde **24**, 1, 33, 97 (1904)

5. Malacara, D.: Optical Shop Testing. Wiley, New York (1978)

6. De Vany, A.S.: Master Optical Techniques. Wiley, New York (1981)

7. Bryngdahl, O.: Applications of Shearing Interferometry. In: Wolf, E. (ed.) Progress in Optics IV, 37. North Holland, Amsterdam (1965)

8. Bath, K.L.: Ein einfaches Interferometer zur Prüfung astronomisher Optik. Sterne Weltraum **12**, 177 (1973)

9. Schultz, S.W.: Standardizing the Ronchi Test Pattern. Sky Telescope **67**, 272 (1984)

10. Klein, W.: Bemerkungen zur Korrektion und Durchrechnung von optischen Systemen. Jahrbuch für Optik und Feinmechanik 1983. Fachverlag Schiele & Schön, Berlin (1983)

11. König, A., Köhler, H.: Die Fernrohre und Entfernungsmesser. Springer, Berlin Heidelberg New York (1959)

12. Wiedemann, E.: Verfeinerte Optiken für Astroamateure. Sterne Weltraum **15**, 366 (1976)

13. Rohr, H.: Das Fernrohr für Jedermann, 5. Aufl. Rascher Verlag, Zürich (1972)

14. Wiedemann, E.: Verfeinerte Optiken für Astroamateure II. Sterne Weltraum, **17**, 374 (1978)

15. Valleli, P.: Collimating Your Telescope. Sky Telescope **75**, 259, 363 (1988)

16. Richter, J.L.: A Test for Figuring Cassegrain Secondary Mirrors. Sky Telescope **39**, 49 (1970)

17. Kutter, A.: Der Schiefspiegler. F. Weichardt, Biberach an der Riß (1953)

18. Kutter, A.: A New Three-Mirror Unobstructed Reflector. Sky Telescope **49**, 46, 115 (1975)

19. Ingalls, A.G.: Amateur Telescope Making III. Scientific American, New York (1953)

20. Cox, R.E.: The Vacuum Method of Making Corrector Plates. Sky Telescope **43**, 388 (1972)

21. Weigel, W.: Justieren einer Schmidt-Kamera. Sterne Weltraum **18**, 272 (1979)

22. Waineo, T.: Fabrication of a Wright Telescope. Sky Telescope **38**, 112 (1969)

23. Schmadel, L.D.: Schmidt-Systeme ohne Korrektionsplatte. Sterne Weltraum **16**, 214 (1977)

24. Rutten, H., Van Venrooij, M.: Die optischen Eigenschaften eines 200-mm Schmidt–Cassegrain-Teleskops. Sterne Weltraum **23**, 274 (1984)

25. De Vany, A.S.: A Schmidt–Cassegrain Optical System with a Flat Field. Sky Telescope **29**, 318, 380 (1965)

26. Sigler, R.D.: Compound Schmidt Telescope Designs with Nonzero Petzval Curvatures. Appl. Opt. **14**, 2302 (1975)

27. Sigler, R.D.: Compound Catadioptric Telescopes with all Spherical Surfaces. Appl. Opt. **17**, 1519 (1978)

28. Wiedemann, E.: Optiken für die Amateur-Astronomie. Sterne Weltraum **19**, 411 (1980)

29. Gregory, J.: A Cassegrainian–Maksutov Telescope Design for the Amateur. Sky Telescope **16**, 236 (1957)

30. Willey, R.: Cassegrain-Type Telescope. Sky Telescope **23**, 191, 226 (1962)

31. Völker, P. et al.: Handbuch für Sonnenbeobachter. Veröffentlichung der Vereinigung der Sternfreunde e.V., Berlin (1982)

32. Hanisch, H.D.: Protuberanzenansatz für kleine Refraktoren. Sterne Weltraum **14**, 370 (1975)

33. Brandt, R.: Himmelswunder im Feldstecher, 7. Aufl.. J.A. Barth, Leipzig (1964)

34. Tombaugh, C.W., Smith, B.A.: A Seeing Scale for Observers. Sky Telescope **17**, 449 (1958)

35. Heintz, W.D.: Doppelsterne in der Polumgebung. Sterne Weltraum **4**, 118 (1965)

36. Schiffhauer, H.: Die Verwendung von Kunststoffrohren im Fernrohrbau. Sterne Weltraum **1**, 158 (1962)

37. Maksutov, D.D.: Technologie der astronomischen Optik. Verlag Technik, Berlin (1954)

38. MacRobert, A.: Caring for Optics. Sky Telescope **73**, 380 (1987); **74**, 573 (1987) and **76**, 5 (1988)

39. Gebhardt, W., Helms, B.: Ein Selbstbau-Prismenspektrograph zum Gebrauch am C8. Sterne Weltraum **15**, 58 (1976)

40. Fry, G.A.: The Eye and Vision. In: Kingslake, R. (ed.) Applied Optics and Optical Engineering, vol. II. Academic, New York (1965)

Based on *Optical Telescopes and Instrumentation* by H. Nicklas, Compendium of Practical Astronomy, Vol. 1, Chap. 4, Springer, Berlin, Heidelberg, New York (1994)

5. Telescope Mountings, Drives, and Electrical Equipment

B. Weisheit

5.1 Introduction

The current trend toward larger and more powerful instruments is mostly limited by the size, type, and quality the mounting. The telescope mounting is a complex structure of many parts and is far less easily comprehended than the optical path in the telescope; some parts are even more difficult for the amateur telescope maker to manufacture than a primary mirror.

Hint:
Most informations could be found in the worldwideweb.

The emphasis of this chapter is placed on design principles. Practice-oriented publications—*Amateur Telescope Making*, *Sky and Telescope*, *Telescope Making*—and many other periodicals and books are available. Innumerable mounting types have been presented over the years in these publications, and in fact there is no type, no concept, no detail of manufacture which has not been described. Mostly lacking, however, has been a treatment of the criteria for the design and layout of a telescope mounting. Amateur ideas regarding telescope vibrations and kinematic principles are similarly vague. But technology needs measurable and calculable quantities. The chapter contains a discussion of basic criteria for statics, kinetics (vibrations), and the kinematics of the structure. From these, the technical principles which are relevant for determining design and layout will be derived. In order also to help the practice-oriented telescope maker with some systematic guidance, the primary consequences of the theory have been condensed into the *fundamental principles*. A similar treatment of the basics of electrical equipment is included. A substantial difference between mechanical and electrical telescope problems is that the latter are much more easily dealt with. In an electrical circuit, components may be replaced simply and at low cost, or even the entire circuit changed, whereas a mechanical structure with serious faults and weaknesses can be corrected only with prodigious effort and expense. It is for this reason that careful designing and layout are supremely important for telescope mountings.

5.2 Types of Mountings

The basic parts of a telescope are the optical system and a mechanical support structure with two rotating shafts arranged at right angles to one another. With these shafts, the telescope optical system can be directed to any point on the celestial sphere above the horizon. Two types of axis orientation systems are distinguished:

Fig. 1. Example of a Dobsonian telescope

5.2.1 The Alt-Azimuthal or Dobsonian Mount

This system has one axis oriented in the vertical direction, normal to the plane of the horizon. In the horizontal plane is the *azimuth circle* with the azimuth A. The other axis is horizontal and its rotation generates the *altitude circle*, with altitude or elevation a. This type is found in theodolites, large radio telescopes, and also in the "new generation" large optical telescopes, since it is mechanically advantageous for such instruments. A typical amateur telescope with an alt-azimuth mounting is the well-known *Dobsonian mounting* (see Figs. 1 and 2).

In a Dobsonian mount, a pivotable connection between the ground board or base and the rocker box creates the azimuth bearing. A pivotable connection consisting of vertical disks on the rocker box forms the altitude bearings. The rocker box is thus the central component of a Dobsonian mount.

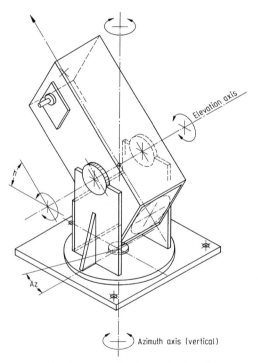

Fig. 2. The Dobsonian mounting is a fork mounting in an alt-azimuth arrangement and is, in static respects, the most favorable orientation of this configuration. J. Dobson gives a fine example for a suitable construction from wood. As such, it can be made with simple tools and at low cost, even for comparatively large Newtonian systems (>300 mm). On the other hand, the use of wood as the construction material and the alt-azimuth positioning limit the application. Its use is predominantly for visual observations. The *altitude* or *elevation axis* is supported in a simple cradle bearing, and the azimuth circle rotates about a pinion. The bearings are coated with teflon, which permits smooth and stick-free motions. The axis system demands no particular orientation. Setting and guiding on astronomical objects is done manually

The bearing between the rocker box and the base enables the telescope to be aimed horizontally. To allow for tracking at high magnifications, this bearing should move easily but also not be susceptible to vibrations. To prevent jerking when tracking, the break-away torque, that is the force needed to start something in motion, must match as closely as possible the force that must act against the frictional resistance (determined from the coefficient of sliding friction and the coefficient of static friction). Having the frictional resistance too low results in allowing the scope to lose its orientation in response to a light breeze or simply to the user touching the scope. The azimuth bearing is usually implemented as a friction bearing, only rarely as a roller bearing. In addition to various other combinations of materials, what is usually used for the friction bearing is an arrangement of three teflon blocks and a hard laminate. To mitigate contamination of the azimuth bearing, the slightly higher teflon blocks are usually attached at the bottom on top of the ground board, while the laminate plate, either in the form of a ring or continuous shape, is glued on the bottom of the rocker box. This allows contamination to fall between the teflon blocks.

Semicircular cutouts are located in the side walls of the rocker box, these cutouts functioning together with the disks attached to the telescope tube as the azimuth bearing. This bearing should also move easily. Making the slip resistance too low here would, for example, cause the scope to move in an uncontrolled fashion with any change in its balance, such as, for example, when changing eyepieces.

One disadvantage of the alt-azimuth mounting is that the tracking of the telescope to compensate for the Earth's rotation requires three complex, non-linear motions: the rotations around the azimuth and the altitude axes, and a rotation of the field of view around the optical axis. For large telescopes, these three motions are computer controlled.

5.2.2 The Parallactic or Equatorial Mount

In this case, one axis, called the *polar axis* (or *hour axis*), is parallel to the rotation axis of the Earth pointing toward the celestial pole and carrying the *hour circle* with the *hour angle h*. At right angles to the polar axis is the *declination axis*, with the *declination circle* showing the angle of *declination δ*. The tracking with this arrangement needs only the polar axis to rotate with the constant angular velocity ω_p of the Earth's rotation.

A movable shaft in two bearings may support the weight either between the two bearings or external to them, as schematically shown in Fig. 2. The first case is known as *center loading* and the second as *cantilever (end)-loading* (see Fig. 4). Mountings with cantilever loading of the declination axis have the system rotatable around the polar axis unbalanced as the center of gravity of the tube is not in the intersection of polar and declination axes. The German and the English mountings thus require a counterweight G_A, which further loads the axes and structure, and also is disadvantageous for vibrational behavior. So the quality of a mounting is ultimately determined upon the carefully considered and solidly founded design.

Fig. 3. Example of a small equatorial mount

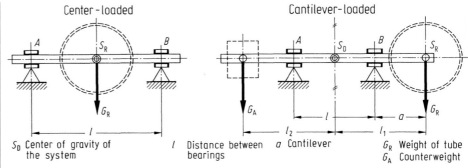

Fig. 4. Arrangement of load on a two-bearing shaft. A load G_R, such as the tube supporting the optical system, may be placed between the bearings A, B (*left*) or outside of them (*right*). These two cases classify the various types of mountings. The former case is *center-loading*; the latter, *cantilever-(end) loading*, requires a counterweight G_A when the system must be balanced with respect to the symmetry axis (*dot-dashed line*). This condition must be met with respect to the declination axis. The balance condition is given by

$$4G_R \cdot l_1 = G_A \cdot l_2 \tag{1}$$

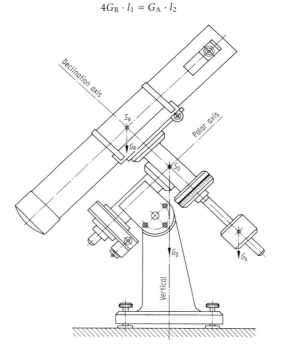

Fig. 5. German mounting. Classical design with a broad range of applications and suitable for long-focus refractors and Newtonian systems, as well as short-tube systems such as Cassegrain reflectors, Maksutov systems, astrographs, etc., covering the entire size range of amateur instruments. The configuration is statically good for medium geographic latitudes, less so for low latitudes. The axis system can be constructed in a compact and rigid way, thus compensating for the disadvantage of the counterweight. The axes and their casings are also simple and in manufacture are unproblematic structures that can be machined to any desired accuracy. The disadvantage is that a long, lower part of the tube is impeded near the zenith by the pier. The telescope must therefore be moved over from the west to the east position

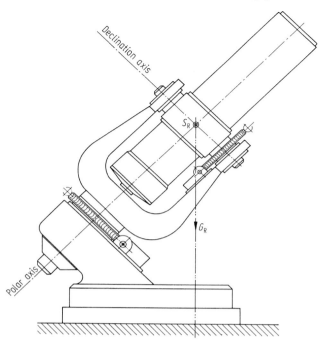

Fig. 6. Fork mounting. Its use on large telescopes has also triggered its increased use by amateurs, despite its drawbacks and problems. It has a favorable static configuration only at high geographic latitudes or as an azimuthal system. Even at middle latitudes, the rather unfavorable load conditions on the polar axis and fork, the low static stability (important for portable instruments), and parts that are problematic in machining are unfavorable forces outweighing the absence of a counterweight. Making the fork mounting with reasonable accuracy and bending-stiffness is exceedingly difficult, particularly in the declination axis bearings. At hour angles $\pm 6^h$, the force component $F_D = G \cos \varphi$ is loaded solely on one arm of the fork. In addition, a truly aligned and stiff design and construction of the two pivots of the axis at the tube is hard to reproduce. Fork mountings may be considered for small, transportable instruments with short tubes and without high demands concerning rigidity and mechanical accuracy. The faultless layout of large fork mountings requires a thorough knowledge of the theory of elasticity and extensive manufacturing facilities

Figures 5–8 graph the basic types; they can be modified in various ways. Pier, fork, and frame mountings can also be modified in many different ways which lead to interesting ideas of layout. Two known variations of the German mounting are the *knee mounting* after Repsold (Fig. 9) and the *Springfield mounting* after W. Porter (Fig. 10).

Especially worthy of mention are the *Folly mounting* and the *horseshoe mounting*, also designed by W. Porter. The latter has become known for its role in the design of the the famous Hale telescope on Mount Palomar in California. It is particularly suited for large telescopes but less so for amateur instruments. Generally speaking, axis arrangements which differ appreciably from the basic types are more prone to problems, and hence are more suitable for experienced designers and advanced telescope makers. The amateur who is not well-versed in matters of mountings may advantageously use the German

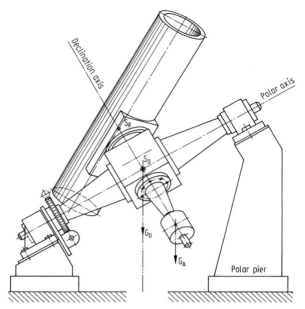

Fig. 7. Cross-axis mounting and Yoke mounting (Fig. 8). Both of these configurations are designed for use at relatively low geographic latitudes and for large reflecting optics of modest focal length. The frame and beam must be assembled from individual parts, which are likely to compromise stiffness at the joints and thus the precision. The beam or frame on two piers requires much more space and thus much more housing than a comparable German mounting. The total effort is considerable, machining not without problems, and applications are limited. Such drawbacks make these mounting types and their variations (horseshoe mountings) less attractive for use with amateur instruments

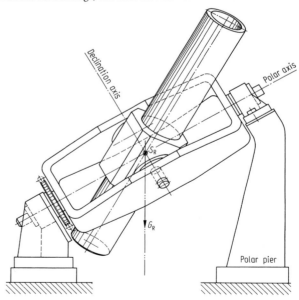

Fig. 8. Yoke mounting (see caption for Fig. 7)

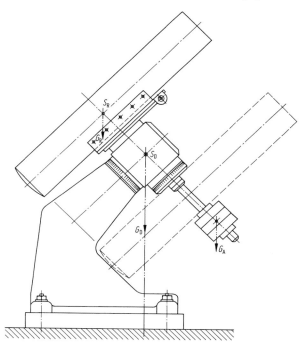

Fig. 9. German knee-mounting. This variant was first conceived by Repsold. Kinematic impediment of the tube's lower end near the zenith is avoided by a bent or "knee-shaped" pier. The arrangement is indicated primarily for photographic work on short-focus Newtonian systems and Maksutov and Schmidt reflectors. It is less useful with optics where the viewing is from the lower end of the tube (refractors and Cassegrains) and for systems with a greatly extended lower tube. The bent pier is more problematic with respect to stiffness, static stability, and construction than the straight pier of classical design. It needs very careful design and manufacture. Besides, anchoring of the pier to the foundation is generally required owing to the position of the center of gravity, so that this type of mounting is less feasible for transportable and unanchored instruments. Its static configuration improves with increasing geographic latitude

mounting. It is well-known, widely used, not too difficult to build in a compact and rigid way, and is suited to optical systems with short and long tubes alike. Less well-known is that the fork mounting, much used in large telescopes, is in some respects very problematic and in fact does not insure a more rigid, vibration-free instrument than do other types.

All types of mounting may confront amateurs with difficult problems. When drawing the layout, one should make certain that these items are concretely fixed and numerically founded. They should also be realistic and realizable within the available means. It is certainly no waste of time to fully explore such limitations before beginning the layout and starting with the construction.

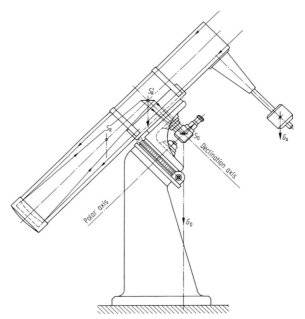

Fig. 10. Springfield mounting. Named after the town of Springfield, Vermont, where it was invented by W. Porter, this is a modified German mounting applicable to long-focus Newtonian systems (over f/10). The optical axis is deflected into the declination axis and is mirrored a second time at the intersection of declination and polar axes (the Coudé principle). The eyepiece thus remains in the same position and makes observing more physically convenient. This axis system is based on a disk arrangement according to Figs. 15e, f or h and can be manufactured quite compactly and rigidly by cast-, weld-, or epoxy-bonding techniques. The peculiarly placed counterweight not only compensates for the weight of the tube but also shifts its center of mass from S_R to S'_R. The double reflection and increased vignetting are optical disadvantages of this interesting mounting for amateur telescopes whose design and construction requires some experience and technical skill

5.2.3 Choosing a Mount

All too often the discussion about which telescope revolves around impressive apertures for the objective or high magnifications. That "thing holding it" under the scope's tube is conveniently ignored in this appraisal of a given telescope. The view that "the only thing the mount has to do is hold the scope" does not do justice here to the real range of functions found in a good mount, with this view sometimes having unpleasant consequences.

The ingenious builder of refractors, Josef von Fraunhofer, is reported to have once said: "A mount must relate to the telescope tube like a clockwork to the hand on the clock." Well, that may be taking the relationship between mount and tube in regard to dimension and weight a bit too far, but there is more than a grain of truth here. This comparison describes in a graphic way the three essential functions of a good mount. It should:

- Hold the telescope in a stable fashion
- Position it accurately
- Enable accurate tracking

In fact, these are the three criteria by which every mount—from a simple one constructed of wood to the computer-controlled GoTo mounts—can be assessed.

The terms "stable" and "accurate" must be taken in their strictest sense here! This is because each little jerking motion and each lack of precision, while hardly noticeable to the eye of the observer, is mercilessly amplified by the magnification used with the telescope. A small jerk measured in millimeters can quickly cause the observed celestial object to disappear from the telescope's tiny field of view. This demand for stability applies not only to the actual mount but equally to the associated tripod or stand which in most cases is the principal weak point of the system.

The requirement of having the greatest functional accuracy possible along with a low-vibration setup of course entails a certain level of mechanical stability. The designer can address a critical aspect such as vibration dampening only by using materials of solid construction, and especially a reasonable weight ratio between mounting and telescope tube.

On the other hand, this requirement conflicts with the desire of the user to have a compact, light, and easily transportable unit. The mount must therefore not be too heavy since otherwise the equipment becomes unwieldy. And this is not the only area where the gap between need and necessity becomes very wide. Since the component "mount" is not very high on the list of performance criteria for the scope, the willingness to invest a significant portion of the purchase price in an effective mount is quite low, especially with first-time telescope buyers. Of course, suppliers then react accordingly by offering telescopes with highly impressive apertures and magnifications, but placed on light and cheaply produced mounts. Although these mounts may no doubt be able to hold your scope, it is often out of the question to speak in terms of stable and accurate handling. What must always be taken into account is that the best optics can never be utilized properly if the mount is not worth anything. For example, the constant backlash motion or lack of tracking can cause every observing session at higher powers to become a war of nerves, or even frustrate it entirely.

The best mount is also not of much use if it does not have the right foundation. This is generally supplied in the form of an extendable tripod made of hollow aluminum. Less commonly used as tripod material is wood, which is more effective in attenuating vibration. The steel post that used to be quite common as the support structure is rarely sold now. In the vast majority of cases, the proud owner of a purchased telescope will thus have to make do with an aluminum or steel-tube tripod. Yet it is precisely these that are frequently the weak point in the stability of the overall telescope design. One reason no doubt for this is in part the excessive trend to minimize weight that is spreading to an ever-increasing degree from the telescope itself to the mount and on to the tripod. In fact, the opposite conclusion would make more sense for reasons of stability, since it is after all only a stable and massive support that can effectively dampen a telescope that has started to vibrate (see Fig. 11). As a result, it is not surprising when a relatively heavy telescope causes the relatively small mount underneath it to vibrate along with it, instead of being attenuated by it. And this whole system is still supposed to be supported by an overloaded tripod made of thin, light hollow sections. Another reason for this no doubt is the potential for the manufacturer to reduce costs, which can best be saved on the tripod after taking into account the expensive optics and mechanically costly mount.

One aspect that is particularly troublesome about this simple aluminum tripod is that suppression of torsional vibration in azimuth is rarely adequate, where the telescope and mount set the top of the tripod into rotational vibration parallel to the ground. Those wanting a stable solution to this problem must often resort to a home-built remedy. Experience shows that using a massive wooden tripod (2 × 4) may significantly reduce the purchased telescope's susceptibility to vibration (Fig. 12).

5.3 Mechanics of Telescope Mounting

In amateur terms, a mounting may be referred to as *stable* or *massive*, but these terms lack quantitative interpretation and measurable values, and thus are not useful as design criteria. Their place is taken by well-defined mechanical quantities from which conclusions on the admissible deviation of images formed in the image plane can be derived.

An image remains fixed when the object does not move and when no forces act on the telescope. The instrument, however, is always subject to its own weight and various other external forces; in addition, the telescope optical system must follow the celestial motion precisely. Apart from kinematic drift effects, this also changes the orientation of components with respect to gravity. All of these factors can in various ways cause images in the focal plane to drift or vibrate.

When designing a telescope for a particular application, the first requirement is to have a firm idea of how much displacement is tolerable. One should not, however, forget that design problems, mechanical demands, and costs increase exponentially with desired precision. Calling the admissible image displacement x_0, the basic criteria for a telescope mounting can be formulated straightforwardly.

The Static Criterion. Static forces shall not cause displacements exceeding a certain value x_0 in the focal plane.

With respect to kinetics, only vibrations are considered since a mounting does not admit to rotary or translatory motions other than the tracking. It is improper to regard a mounting as a *freely movable mass* in order to deduce from Newton's principle of action or the theorem of momentum that it must be "heavy." Even a mounting without anchoring is considered to be a *system at rest*. Its orientation in space must be conserved. A minimum requirement is that after a displacement it returns exactly to its original position.

The Kinetic Criterion. Vibrations resulting from any source shall not excite amplitudes over $x_0/2$ in the focal plane and should be attenuated as quickly as possible.

The Kinematic Criterion. Image drift in the focal plane caused by kinematic inaccuracies shall not exceed the amount x_0 for a given observing time t_b.

5.3.1 Stiffness

Criteria 1 and 2 already imply the quantity *stiffness* (or *rigidity*) c, which is all-important for telescope mountings. A force acting on an elastic body or structure[1] always causes

Fig. 11. Example of a stable German mounting

Fig. 12. Example of a stable wooden tripod system

---

[1] Elasticity, like mass, is an elementary property of matter. All bodies possess some finite degree of elasticity.

a displacement x at the point of contact, or more generally, a *deformation*. Stiffness is the resistance[2] of the body against elastic deformation. The force F, the displacement x, and the stiffness c are connected by the simple relation known as the *force law for springs*, or *Hooke's law*:

$$F = c \cdot x , \qquad (2)$$

where F is the force in Newtons (N), x the displacement in meters (m), and c the stiffness in $N \cdot m^{-1}$.

Stiffness is paramount in the design of mountings for two main reasons:

1. Stiffness determines the layout of every part so as to satisfy criterion 1.
2. It also plays an important role in vibration, and permits the vibrational behavior of the telescope to be controlled in a specified way.

A convenient unit for stiffness is the meganewton per meter (MN/m). Thus, a mounting is said to have a stiffness of 1 MN/m when a force of 10 N causes a displacement of 0.01 mm, as, for instance, for stellar images on a photographic plate.

A telescope mounting is a structure, that is, it consists of a large number of different elements which are connected by discrete joints such as bolted joints, welded joints, slip-joints, bearing, etc. Each element and each joint can be assigned an individual stiffness c_{ki}, where the quantity i in the subscript denotes the ith element. If the elements are connected consecutively in series, the system stiffness c_{ks} can be calculated by the addition theorem of stiffness:[3]

$$\frac{1}{c_{ks}} = \frac{1}{c_{k1}} + \frac{1}{c_{k2}} + \frac{1}{c_{k3}} + \cdots + \frac{1}{c_{kn}} , \qquad (3)$$

where n denotes the total number of elements. Note that the individual stiffnesses c_{ki} are defined in the same state k of load.

The addition theorem of stiffnesses, Eq. (3), predicts two important consequences for the layout of telescope mountings.

Criterion 1. The individual stiffnesses of n series-connected parts and joints must be n times larger than the required system stiffness.

Criterion 2. The stiffnesses of parts should be approximately equal. Equation (3) demonstrates that just one weak element with low stiffness in the structure will suffice to adversely affect the system stiffness, whereas oversized other parts will not improve it appreciably.

The stiffness of an element depends on its geometric form, the material properties (elasticity modulus E, Poisson number v), and the conditions at its fixation point. It does not depend on the acting force. Hence, this stiffness of the parts is primarily significant

---

[2] To be precise, stiffness is a *tensor* represented by a three-dimensional matrix. In the theory of elasticity, stiffness is connected with the stress tensor via the elastic material constants, *elasticity modulus E* and *Poisson number v*. The inverse quantity $1/c$ is called *compliance*.

[3] To be precise, the stiffness *matrices* of the element should be added. Equation (3) neglects the lateral stiffness, whose influence is generally small enough to be ignored. Exceptions are bearings where the lateral effect may be substantial.

Fig. 13. Flexural stiffness in some
important cases of stress for
telescope mountings

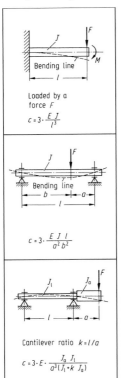

Bending line

l

Loaded by a
force *F*

$$c = 3 \cdot \frac{E\,J}{l^3}$$

Bending line

b ─ *a*

l

$$c = 3 \cdot \frac{E\,J\,l}{a^2\,b^2}$$

Cantilever ratio *k = l/a*

$$c = 3 \cdot E \cdot \frac{J_a\,J_i}{a^3(J_i + k\,J_a)}$$

for the telescope mounting, and should determine their layout. If all elements have sufficient flexural stiffness, then they will not become weak parts under tension, compression, or torsion either. This gives the following important principle:

Criterion 3. The supreme guideline in the design and layout of telescope mountings is high flexural stiffness of the structure.

It remains now to be shown how this may be achieved.

The formulae for flexural stiffness (Fig. 13) include three variables of layout:

1. The flexing length l (a geometrical property)
2. The area moment of inertia J (a geometrical property)
3. The elastic modulus E (a material property)

A structure consisting of elements designed as short as possible will appear in its entirety to be compact and sturdy. The words "compact" and "sturdy" in this context are not meant to be exact physical terms, but they are illustrative words which describe the overall impression of a well-designed telescope mounting. This gives rise to another principle.

Criterion 4. All cradles and cantilever arms in the support structure of a telescope mounting should be made as short as possible.

Criterion 5. Statically speaking, a telescope mounting need not be massive but should be rigid. The amount of weight needed for satisfactory stiffness depends only on the skill of the designer in effecting short cradles and cantilever arms with high moments of inertia.

The *modulus of elasticity*, or *E-modulus*, should not be confused with the *tensile strength*, the *yield point*, or the *hardness* of a material. In numerous technically important metals such as iron and aluminum, alloying constituents increase the strength and hardness, but have practically no influence on E. When it comes to elastic properties, it may be surprising that soft iron and very hard chromium-nickel steel, or soft aluminum and the very hard alloys of light metals, have the same E and therefore identical deflections. It is important to distinguish between, on the one hand, objects for which strength and fracture resistance are the relevant quantities (bridges, cranes, wheel axles, etc.), and on the other hand, where it is the stiffness that is critical (telescope mountings, precision measuring instruments, etc.). As far as stiffness is concerned, materials of high strength and hardness are not needed.

5.3.2 Shafts and Bearings

In every mounting, shafts and their bearings are significant and critical parts. These are consequences of the criterion 3 on kinematics, but they also enter as parameters into the inner stiffness of bearings.

The overall stiffness is composed of, according to the $1/c$ rules, the stiffness of the shaft, the inner stiffness of the bearings and that of the bearing casing. Figure 14 shows in schematic form the individual stiffnesses of the shaft on bearings. Again, it is seen how important short cradles and cantilever arms are since they not only reduce the bending stiffness of the shaft with the third power but also enter into the formulae for the bearing stiffness. The inner bearing stiffness c_L is a rather complex quantity depending on the following factors:

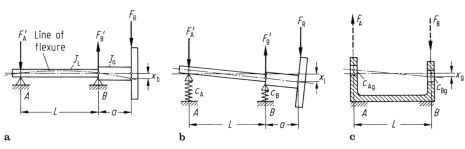

Laverage ratio $\quad k = \dfrac{L}{a}$

Reactive forces
of the bearing $\quad F_A = F_R \dfrac{a}{L} \ , \ F_B = F_R \dfrac{(L+a)}{L}$ $\qquad$ (4)

Inflexibility of shaft c_b $\qquad$ Rigidity of bearing c_i $\qquad$ Rigidity of casing c_g

$$c_b = \frac{3E}{a^3} \frac{J_L J_a}{(J_L + k J_a)}$$

$$\frac{1}{c_i} = \frac{1}{c_B}\left(1+\frac{1}{k}\right)^2 + \frac{1}{c_A}\left(\frac{1}{k}\right)^2$$

$$\frac{1}{c_g} = \frac{1}{c_{Bg}}\left(1+\frac{1}{k}\right)^2 + \frac{1}{c_{Ag}}\left(\frac{1}{k}\right)^2 \qquad (5)$$

Overall rigidity $\quad \dfrac{1}{c_s} = \dfrac{1}{c_b} + \dfrac{1}{c_i} + \dfrac{1}{c_g}$ $\qquad$ (6)

Fig. 14a–c The stiffness of a pivoted shaft. The total stiffness is given by the flexural stiffness c_b, the inner bearing stiffnesses c_A, c_B, and the stiffnesses of the bearing casing frame c_{Ag}, c_{Bg}. In **(b)**, the bearing stiffnesses are symbolically represented as springs. The formulae show the significance of the cantilever arm A not only for the flexure of the shaft but also for the stiffness of bearing and casing. Amateur-made mountings frequently show ample and even much oversized shafts, whereas the stiffnesses of bearings and housings are often ignored

1. The type of bearing (ball bearing, friction bearing, hydrostatic bearing, etc.)

2. The dimensions and geometry of the bearings

3. The internal fit and the amount of bearing clearance

4. The external precision of the shaft fits and the case bores

5. The axial and radial preload (of antifriction bearings)

Figures 15a–h show the various possibilities of the bearing arrangement of the polar axis.

In a balanced system, the weight G of the moving, rotating parts is a spatially fixed, constant vector composed of the force components

$$F_P = G \sin \varphi \quad \text{and} \quad F_D = G \cos \varphi , \qquad (7)$$

where φ = latitude, F_P = component of force in the direction of the polar axis, and F_D = component of force in the equatorial plane. Evidently, the position of the declination axis relative to F_D changes with the hour angle h. When the declination axis is horizontal (at hour angles of 0^h and 12^h), F_D causes radial stresses on the declination bearings, and the axis is stressed in flexure. In the positions $\pm 6^h$, the twisting moment of the component F_D vanishes (but not so for F_P), thus creating an axial load on the bearings. These variable effects cause varying deformations.

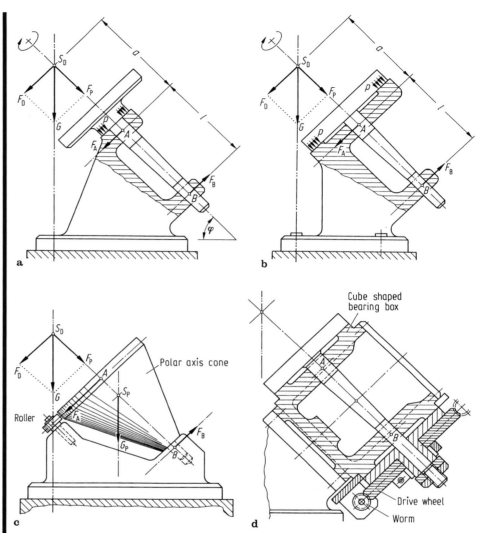

Fig. 15. Various forms of polar axis design. The arrangement of bearings is loaded by the axial force $F_P = G \sin \varphi$ and by the moment $M = aG \cos \varphi$. Primarily M is relevant for the displacement. In (**a**), F_P is transferred through the shaft shoulder onto the support. The bending moment must be absorbed by the shaft. With peripheral support of the flange (**b**), the mounting also absorbs part of the moment. This increases the stiffness. *Porter's Folly mounting* (**c**) possesses very high stiffness of the polar axis cone. Weak parts of this concept are the support rollers. The stiffness of ball bearings in this arrangement is small. Also occurring are *Herz deformations*, bending, and kinematic displacement. The total stiffness is thus not higher than with other optimally designed shaft and bearing arrangements. (**d** and **g**) Cube-shaped casings (the *Badener cube mounting*). The hollow cube has high stiffness in all directions and its favorable shape with respect to machining processes accommodates the rectangular axis system ideally

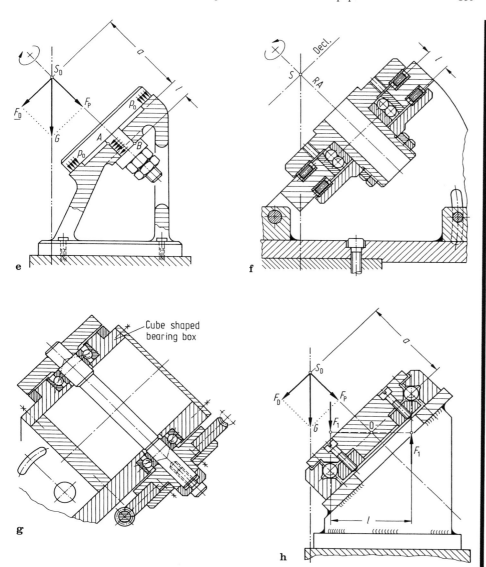

Fig. 15e–h Design arrangements for the polar axis. For reasons of stiffness the distance l between bearings is short. However, it can be reduced further, leading to the disk designs (**e**, **f**, and **h**). These transfer the moment M directly from the disk to the support structure. Since the disks can be made very rigid, such designs can reach very high overall stiffnesses. (**f**) A design with two axial needle thrust bearings and one ball bearing. (**g**) An epoxy-bonded cube construction with preloaded angular-contact ball bearings. (**h**) A do-it-yourself, preloaded four-point ball bearing design which has a highly isotropic stiffness, and can support forces and moments in any direction. Where minimum weight and space are desired, this is the most rigid and vibration-free bearing arrangement which, in addition, can be made to very high precision

Older telescope designs compensated for this effect by the use of elaborate weight/lever "relieving arrangements."[4] The modern design philosophy is to keep deformations as small as possible by a very rigid structure, and to correct the remaining deformation errors automatically with a control computer. Position-dependent deformations are usually quite pronounced in fork-type mountings unless preventive measures are taken. Some hints follow:

At the horizontal position of the declination axis, each fork arm is transversely loaded with $1/2\,F_D$. At $\pm 6^h$, the entire bending load is carried by only one arm. In principle, the axial load component cannot be equally distributed on both fork arms (the statically indeterminate case). Hence, in this position, one arm of the fork can be omitted without harm, thereby resulting in the so-called half-fork mounting. This mounting, which is statically determinate, can, with adequate stiffness, be realized. Also, a fork arm can be designed such that flexure stiffnesses are equal in all orientations of the tube (*isotropy of stiffness*).

There exist design possibilities in which both fork arms are equally loaded or in which a high degree of stiffness is achieved in all orientations of the tube. Briefly, they are:

- A powerful prestressing of the fork arms ($F_{stress} > 5$ to $10 \cdot F_D$), which requires a specially designed bearing arrangement at the axles.
- Forming a rigid frame structure to achieve high integral stiffness. This also requires specially made bearings and a rigid central yoke (crossbar).

The advantage of friction bearings compared with antifriction bearings is their somewhat better ability to damp vibrations. However, it is more difficult to achieve high bearing stiffness and freedom from play, both of which require a very small bearing clearance and a very high accuracy of the sliding surfaces. Steel on bronze slides well, whereas stainless steel, most light metal alloys, and nickel-coated surfaces tend to become pitted.

Antifriction bearings are very attractive for use as elements in amateur telescope mountings. There are various types and versions with different properties. The following criteria may be used when choosing them:

1. Requisite stiffness in radial and axial directions
2. Freedom from play
3. Requisite precision (e.g., the concentricity error)
4. Constraints of price and of design and constructional demands

Telescope shafts can be placed on bearings such that:

- The radial and axial forces are supported by separate bearing units,[5] or
- Roller bearings supporting both radial and axial forces are used. These are normal ball bearings if the axial forces are not too large, but *angular-contact ball bearings* and *tapered roller bearing* otherwise.

---

[4] Such weight/lever relieving systems, such as the refined and elaborate mechanics of F. Meyer of the Zeiss company, were extensively used in older telescope mountings and transit circles. A well-known example was the celebrated *Treptow refractor* on a fork mounting with relieving systems on both axes.

[5] The first variant is at present used only in the very precise, rigid, and low-vibration machine-tool spindles (e.g., lathes, drilling machines, and milling machines) while the angular-contact-bearing arrangement is preferred in high-precision grinding spindles. The latter are at present among the most precise made.

The stiffness of a mechanical structure is, in the *elastic range*, independent of the force, although for the antifriction bearings it is a non-linear quantity depending on the radial bearing load F_r and the initial stressing force. The radial displacement δ of standard ball bearings can be calculated with sufficient accuracy from the *Lundberg–Stribeck formula*:

$$\delta = \frac{1.28 \times 10^{-3}}{\beta} \cdot \sqrt[3]{\frac{F_r^2}{D_0 z^2}}, \tag{8}$$

where δ is the radial displacement (in mm) of the shaft center, z the number of balls, F_r the radial bearing load, D_0 the ball diameter (in mm), 1.28×10^{-3} a dimension factor dependent on bearing shape, and β a dimensionless "spring" coefficient ($\beta = 0.42$ for an axially optimally prestressed ball bearing). The bearing stiffness c_l is then F_r/δ.

The opinion is widespread that in the German mounting (Fig. 5), the bearing distance l should be made very large. Taking the derivative of Eq. (6) the optimum bearing distance to be calculated leads to surprisingly short bearing distances.

5.3.3 Foundation and Stability

The joint between mounting and foundation is the last part in the chain of stiffnesses. This joint should never be the weak one in the system, but in amateur designs it often is. Sometimes the instrument rests on three scantily dimensioned footscrews. In such cases one has only to compare the cross-sections of the shafts with those of the screws in order to see what little sense this scheme makes. Consider also that these joining elements suffer the largest stresses. To achieve a rigid coupling between the mounting and foundation requires amply proportioned footscrews and accurate, even surfaces at the supporting points. The foundation should meet the following requirements:

1. It should go sufficiently deep into the ground, at least below the frost limit.
2. It must have a sufficiently high mass.
3. It should be placed where the ground will be free of vibrations.

Small instruments often cannot be fastened to a foundation if they are to be used on a balcony, a rooftop terrace, or in a portable manner outside on the ground. Such instruments need sufficient static stability, which is achieved when the amount of work W_k needed to topple the system over is large. A structure has a high W_k when its center of mass S is low (h_s is small) and at the same time the distance l_k to the tilting edge is large (see Fig. 16). In principle, a mounting should have its center of mass as low as possible.[6] Therefore, highly elevated parts such as the tube, optics and accessories, and the upper part of the pier should always be designed and constructed to be of low weight. It has been shown that one need not counteract the stiffness criteria in order to satisfy the stability requirements. Larger masses in a mounting are admitted only near the ground, that is, in the lower part of the pier (cf. Sect. 5.3.4 on vibrations).

A typical counterexample is the case of an instrument on a high tripod, since such tripods always have a high center of gravity. Besides, it is nearly impossible to produce

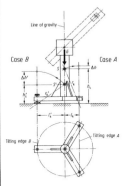

Fig. 16. Stability of a free-standing structure. A measure of the stability of a non-anchored structure is the work W_k required to topple it over. It is given by

$$W_k = \Delta h \cdot G, \tag{9}$$

where

$$\Delta h = r_k - h_s$$
$$= \sqrt{h_s^2 + l_k^2} - h_s. \tag{10}$$

The two cases shown illustrate the situation. Evidently, a "nose-heavy" mounting has a low stability. Large masses are admitted only near the ground (low mass center S, small h_s). It is seen that a three-point base (tripod) has one tilting edge near the line of gravity

---

[6] It makes little sense to place an instrument so high above the ground that steps and ladders are required for observations with it. The lowest eyepiece position should be at eye-height when the observer is in the seated or slightly bent-over position.

a tripod mounting with adequate stiffness; such instruments often display unstable, wiggly images. The unsatisfactory vibrational performance can be improved by attaching braces between the tripod legs and by attaching heavy weights to them. This moves the center of gravity downward while the braces increase the stiffness.

5.3.4 Telescope Vibrations

The image stability of amateur telescope mountings is interfered not so much by static effects but rather by vibrations, and almost always flexural vibrations . Oscillations are rather difficult for the amateur to treat mathematically, as each statement leads to a boundary-value problem of differential equations.

A physical oscillator necessarily contains three elements that determine its behavior. These are:

1. At least two independent energy storage elements
2. A coupling of the storage elements permitting energy to flow in both directions
3. An excitation source which supplies (at least once) energy to the system

Fig. 17. Mounts with long, thin axes are unstable and vulnerable to vibration

In a mounting, the mass m carrying kinetic energy is one of the energy stores, the elasticity of materials represented by the stiffness c, the other. A simple mass-spring oscillator is shown in Fig. 18. Such an oscillator is mathematically characterized by two terms. One term characterizes the vibrational properties of the structure (masses, stiffnesses, and damping properties of the parts), the other the excitation source given by the excitation function $E_F = f(t)$, as shown in the schematic diagram. The structural term supplies an important system characteristic, namely the natural frequency (or frequencies) ω_0:

$$\omega_0 = \sqrt{\frac{c}{m} - D^2} \, , \tag{11}$$

where c is the stiffness of the spring (in N/m), m is the mass (in kg), and D is *Lehr's damping number*. This frequency formula provides two important conclusions, generally valid for structures with elasticity and mass:

1. The mass and the stiffness of a mechanical structure are complementary with respect to the natural frequency.
2. Increasing the stiffness increases the natural frequency of the system, while increasing the mass diminishes it.

A mounting is a complex, three-dimensional, oscillating structure with many degrees of freedom, and for which calculations are difficult at best. For the design, however, relevant results can be also derived from simplified oscillator models in which the mounting is understood as a chain of sequential elements consisting of flexing shafts and masses. Some general statements can be made regarding such oscillator chains, thus allowing us to deduce guidelines for the layout without the necessity to calculate the system numerically.

The normal manipulation forces and pushes incurred during operation and wind-caused oscillations have a frequency spectrum ranging from 0 to 25 Hz or more. In order that these forces, which act primarily on the tube, create only minor oscillatory amplitudes, they must be short-circuited from the structure onto the foundation. The low-pass

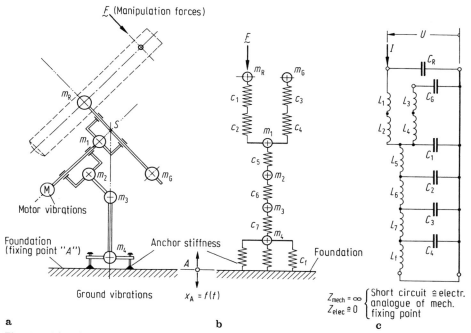

Fig. 18a–c The telescope mounting as a flexural oscillator and oscillating chain, and the electrical analog. (**a**) A telescope mounting can be essentially reduced to flexing shafts with stiffness c and point masses m. (**b**) This flexural oscillator can be interpreted as a branching oscillator chain with an attachment. (**c**) The electrical analog is an LC-circuit with low-pass filter characteristics

filter therefore operates in the transmission range $\omega_a < \omega_s$. For this to occur, the structure must have a high mechanical impedance, which is obtained by designing each part with the highest possible stiffness and smallest mass. This also satisfies the static requirement, and a high-frequency structure has the advantage that any vibrations damp out rapidly. With given damping properties of the material, there is a logarithmic relation between the frequency and the damping time.

So the following important design criteria are required:

Criterion 6. The design of a telescope mounting should aim at high natural frequencies of the system and a high mechanical impedance. High natural frequencies lead also to better damping properties of the structure.

Criterion 7. These properties are obtained by designing every part with stiffness as high and the mass as low as possible and by avoiding all "dead weight."

The subject of kinematics deals with all processes involving motion at telescope mountings. The mere presence of rotatable shafts refers to kinematics. It therefore covers pointing and tracking equipment in declination and right ascension, as well as all adjustment parts. These are for the collimation of the optical components as well as for pointing the axis system toward the celestial pole.

Errors of various drive elements determine how precise the tracking will be. In analogy to the treatment of stiffness, one can speak of an "error chain." In this regard, *system-*

atic errors and *random errors* must be distinguished, as they require different treatment. Systematic errors can be uncovered by a careful investigation and precise measuring of the system in question. They can in principle always be diminished by direct corrections of the system. Not so the random errors, which are treated by the *theory of errors*. This theory makes a statement on the mean error of the chain, and on which error term contributes most to the total error.

The following are examples of random and systematic errors:

1. Parts which are to be combined into a structure are measured with a vernier caliper and a micrometer. Each measurement is affected by random errors (divisions of the scale, reading errors, and parallax errors). This error becomes associated with the length measured. The mean error of the composite structure is calculated by error theory. In this case the total error could be measured directly with a more precise measuring machine. But such high-precision instruments are hardly accessible to amateurs. The theory of errors purports to say something plausible regarding the accuracy of measurements in the absence of direct verification.
2. In the focal plane, a star shows a periodic oscillation with the rotation period of the worm wheel. This indicates a wobble or tumbling error of the worm.

It is caused by a tilt of the axis of the worm which may occur in machining or by concentricity error in the bearings. If the errors in the bearings amount to Δx_1 and Δx_2 and the distance between bearings is l, then they contribute to the tumble, maximally $\tan \psi = (\Delta x_1 + \Delta x_2)/l$.[7] The tumble ψ superposes onto the uniform angular motion of the worm ω_s an oscillation $\psi\omega_s \cos \omega_s t$. The perturbed angular velocity ω_p at the polar axis then is

$$\omega_p = \frac{\omega_s}{u_1} \left(1 + \psi \cos \omega_s t\right) , \tag{12}$$

where u_1 refers to the reducing gear ratio at the worm wheel and ω_s to the angular velocity of the worm.

Although the tumbling error is divided by the reducing gear ratio of the worm wheel it can severely interfere with the tracking if the worms are very inaccurate or have poor bearings. Dividing (indexing) errors of the worm wheel, however, affect the polar axis in undiminished amounts.

Error theory shows that in a structure or chain affected by errors, the part with the greatest error has the most influence on the entire system. There is thus very little purpose in machining *one* part with exceptional accuracy or in measuring only *one* quantity with particular care.

5.4 Drivers for Mountings

Basically, telescope drives can be categorized into three construction parts:

1. Drive mechanics (gears)
2. Drive motor
3. Power source and control unit

---

[7] For this reason, precision worms must be very carefully supported by bearings. The high-precision angular contact ball bearings mentioned earlier are best for this task.

Parts 2 and 3 together comprise the drive system. Drives and tracking train the optical system onto the objects of observation and follow the celestial motion.

From the manually pointed comet finder or Dobsonian telescope to the integrally computer-controlled instrument, the choice of a drive will depend on the intended application, on the amount of technical effort, and on costs. Precision and thus the technical effort for the control electronics should be about the same as that for the mechanical parts.

For the layout of the gearing, the following procedure is advisable:

1. Choose the driving system, by synchronous, stepping, or DC-motor; determine which operations are executed manually.
2. Explore which makes and types of motors are available, and compile the mechanical and electrical data. For instance, the *starting* and *nominal operation torques*, characteristics of the torques, upper and lower limits of the rotational speed in rpm, current, regulating dynamics, requirements of the power source, and control.
3. Explore which gears are available for the chosen type of motor. Note that the gears of such small motors are often only weakly constructed and therefore are not suitable to drive the main worm directly (e.g., for clocks and small instruments).

5.4.1 Drives in Right Ascension and Declination

The drive is usually frictionally connected with the polar or declination axis through a clamp. Instead, it is advisable to insert a finely tuned *friction clutch* between drive and axis. This has two functions: to permit a fast manual motion of the tube, much faster than by any simple motor-drive design, and to protect the sensitive gear parts from damagingly high forces and momenta.

For the mechanical drive itself there are numerous design variants, among them these well-known devices:

(a) *Worm wheel drives.* The most widely occurring in telescopes (see Fig. 20).
(b) *Tangential spindle drives.* A lever or a sector is moved by a tangentially placed worm gear spindle. It permits to realize high gear ratios, but the range of motion is limited. At the end the nut on the spindle is to be returned to the initial position.
(c) *Band drives.* A steel band is wound around a disk, which can also be moved by a threaded spindle. Hydraulic band drives have also been devised. In this type, the friction clutch is already given by the drive principle (i.e., band-disk).
(d) *Friction wheel drives.* A large and precisely machined disk is moved by rollers.

The following general rules apply to designing and building these drive types:

Fig. 19. The worm wheel and snail is a classic drive form for telescope mounts

1. The first drive step connected with the shaft should have a gear reduction ratio as large as possible, $u_1 > 150$. For this to be realized, the worm wheel, the lever (in tangential spindle designs), or the band or friction disk must be made large. Guideline values for the worm wheel diameters are given in Table 1.

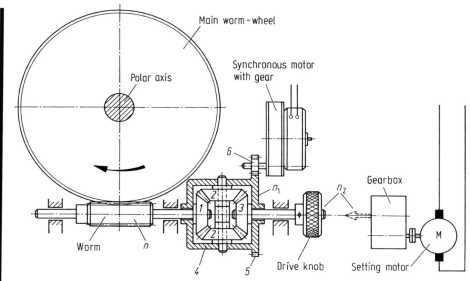

Fig. 20. Drive design with differential gear. Differential and planetary gears are real summing gears in which two input speeds n_1 and n_2 are summed to an output speed n_0. n_1, n_2, and n_0 can be functionally arbitrarily interchanged. For the differential gear, $n_0 = (n_1 \pm n_2)$. On the main worm shaft is the bevel gear (1), which is driven by these two star bevel gears (2). The rotational speeds n_1 and n_2 to be summed are added via the bevel gear (3) and the differential star (4). The drive of this differential star comes from the cogwheel (5) and the pinion (6). In place of the manually operated drive knob, setting motions may also be executed via a servo motor, a DC motor being particularly suitable; see illustration to the *right*

Table 1. Guideline values for the main worm wheel diameters. Wheel pitch circle = Module × Number of Teeth

Refractor D (mm)	Reflector D (mm)	Worm wheel pitch circle (mm)	Module of wheel
60–100	150	130–170	0.70–0.90
100–120	200	180–220	0.80–1.25
120–150	250	220–270	1.00–1.50
150–200	300	270–350	1.50–1.75

2. The first stage largely determines the precision of the drive mechanics. Division errors of the teeth or cogs, rotatory errors, irregularities, and backlash of the subsequent stages are divided by the gear ratio of the preceding drive stages. If high precision is expected from the drive, then attention should be paid to the first stage.
3. When designing the gear mechanics, the stiffness aspect should not be overlooked. No undesired lateral deflections or vibrations should occur in the motion direction of the drive.

In worm drives, the precision is given primarily by the accuracy of indexing (pitch) of the worm wheel, by the concentricity of the worm and worm wheel, and by a precisely adjusted mesh. The gear mesh between worm and wheel should be adjusted carefully. Here, the worm is adjusted centrically relative to the gear teeth, precise in angle so that it is not tilted, and radially so that there is no play. The worm bearings have to provide the adjustment devices needed here. Band and friction wheel drives require precisely worked disks. A recurring problem in friction wheels is dirt grains getting between the driving rollers and the disk. A boxed arrangement with wipers helps alleviate this.

Stiffness in worm drives is conditioned by the *module* of the worm wheel, which should not be too small, and by whether or not the worm is in a rigid bearing housing. For this reason, a worm pressed against the wheel with springs is only a stopgap if the worm sticks and does not revolve smoothly. The stiffness is thereby considerably reduced.

In manually operated drives, it should be noted that the human hand is not capable of arbitrarily small motions. The angle of operation, by which the knob is turned in operation, should not be smaller than 1–2°. When operated over a flexible shaft, this value should be even larger in order to allow for torsion elasticity.

The primary types of motors to be considered for telescope drives are *synchronous*, *stepping*, and small *DC servo motors* (see Figs. 21 and 22). They are available from many manufacturers and have a wide range of characteristics. Many suppliers also have gears fitting these motors, which simplify the design and construction of the drive mechanics considerably.

Synchronous Motors and Stepping Motors. These are motors which employ a rotating magnetic field. The stator contains windings which are fed by AC or by pulses. This induces a magnetic field rotating with the angular velocity ω_m given by

$$\omega_m = \frac{\omega_e}{p} = 2\pi\frac{f}{p} \, , \tag{13}$$

where p is the number of *pole pairs* of the motor. A two-pole motor has $p = 1$, a 16-pole motor has $p = 8$, and so on. Note that in stepping motors a two-pole stator winding is often combined with a multipole rotor. The rotating field "drags" along with it a soft magnet rotor or a permanent magnet rotor. In the first case, this is called a *reluctance motor*, in the second case, a *permanent magnet motor*. In these motors, the rotational speed remains fixed; it is synchronized with the frequency. Thus, the speed can be regulated via the frequency. There is more to the stepping motor: each step corresponds to a precisely defined rotation angle. By simply counting the step pulses, the angular position can be digitally displayed and processed. Thus in a certain sense stepping motors are drive elements and encoders simultaneously.

Stepping motors normally have a two-pole stator winding supplied alternatingly with square-wave pulses of 90° electrical phase shift and of changing polarity (bipolar operation). In addition, there are *half-step* and *microstep* devices.

Modern small DC motors usually have an iron-free drum or disk rotor of small mass moment of inertia and very low inductivity for the purpose of controlling and driving. Their properties differ strongly from those of stepping motors. Their torques are high, are constant over a very large range of speeds, and are free of pulsating torques. For the tracking motion, they can be operated at very low speeds ($n < 20$ rpm). Most of them have speeds ranging from 0 to 5000 rpm, some even as high as 10,000. They can start up and stop very quickly and have no natural frequencies or tendency to-

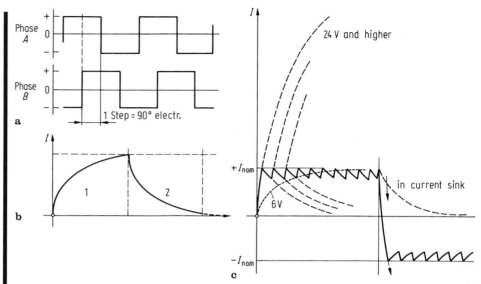

Fig. 21a–c Driver, pulse shapes, and chopper mode. In full-step mode, the driver bridge is controlled by a sequence of pulses (**a**). The inductivity makes the current rise and fall exponentially (**b**). An adequately high voltage at the windings yields a fast increase in the current. When the nominal value is reached, the chopper is triggered via "sense" (**c**). Decline of the field is reached through a current sink switching the windings via "inhibit"

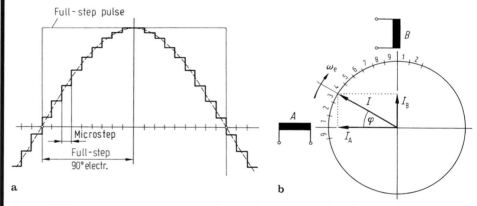

Fig. 22a,b Microstep operation, step curve, and vector diagram. Anomalies of torque of stepping motors in full-step mode can be avoided by microstep operation. The field vector I in the figure (**b**) revolves in microsteps; the graph is drawn for 9 microsteps. The motor phases A and B are excited by the current components I_A and I_B. The current in the windings then follows a step-curve (**a**) approximating the ideal sine curve

ward oscillations. The rotational speed is proportional to the voltage at the armature winding, and is also controlled that way. The disadvantages compared with stepping motors are:

– The commutator (collector) and the brushes suffer wear, especially when continuously operated in the upper range of angular speed and torque.

- The speed depends on the load. If the motor is loaded, then the rpm drops somewhat. Also the temperature has, via the armature resistance, some influence on the speed. If a highly constant speed is required, then the motor should be operated by a feedback loop controlling the rotational speed.
- DC motors are not directly positioning units as stepping motors are. For the positioning task, another control loop with an encoder is needed.

Owing to the large range in angular speed and the excellent controlling dynamics, large telescopes are now often equipped with DC-disc armature and pancake motors. The somewhat larger demand on control technique is insignificant.

Stepping motor gears are designed primarily for instruments which are to be positioned digitally. In such drives, however, considerable costs must be reckoned with. The stepping motor is substantially more expensive than a small synchronous motor. The large span of revolutions ($\omega_p \to \omega_g$) and the characteristic properties of stepping motors require well-harmonized gear and circuit components. To ensure optimal adjustment of stepping motor and control electronics, it is advisable to contact the motor manufacturer, or even to use indexers and gears from the same producer. Many manufacturers of stepping motors supply complete drive systems.

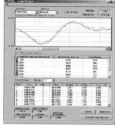

Fig. 23. Due to CCD image technology and computational control, directional orientation errors due to mounts can be digitally captured and analyzed, as shown here with the PEMpro program of CCDWare

The telescope motions can be adequately operated in the full-step or half-step mode. Over the entire range of speeds and for correct positioning, however, it is advisable to operate in microsteps. This is by far the best operating mode for telescope drives. The advantages are:

- Low pulsation of the torque.
- Strongly reduced tendency toward vibrations and eigen-resonances. The motor is practically a system of first order.
- A constant torque over a large range of rotational speeds.
- Increased angular resolution when positioning.
- A better motor efficiency.

However, not all motors are suitable for the microstep mode. The motor torque M, which depends on the turning angle φ, must strictly obey the formula

$$M = k_T I \sin \varphi \,, \tag{14}$$

a condition which is not true for all makes and types. Here I is the current and k_T is a constant.

Many professional telescopes now have all operations (positioning, tracking, guiding, scanning) performed by a single disc armature DC motor. The control is through a control cascade, which is a chain of sequentially connected circuits. The governing functions are performed by a computer or drive controller. Only the tracking speed needs to be run with some precision, and for this a closed circuit with tachometer is needed. All other operations can be performed over uncontrolled circuits.

Note that DC motors have a very low internal resistance R_i. Thus, if the voltage changes suddenly, a very large current will flow through the armature winding (rotor winding) and may damage the commutator and brushes. The control circuit thus needs

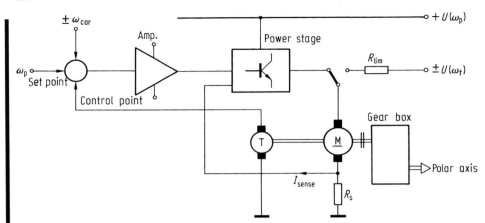

Fig. 24. DC telescope tracking with tachometer. Precise control is needed only for ω_p. The motions ω_f and perhaps also ω_g can be performed uncontrolled through corresponding voltages $U = n \cdot k_e$. The figure shows the n-control circuit with tachometer. A double-throw switch feeds the unregulated voltages to the motor. Current limitation is needed to protect the motor. It is provided, as already noted, by R_s and the "sense"

a current limiter, and the open circuit a current-limiting resistance. These are to be arranged so as not to exceed the nominal current. The schematic of such circuitry is shown in Fig. 24; for this purpose they are commercially available.

Servo motors are used in fast modern drive systems, especially those using GoTo functions. These allow the telescope to move quickly while at the same time providing accurate tracking. The actual motors are DC motors that are driven by an electronic controller within an "automatic control circuit." This means that there is a sensor on the motor that very precisely measures the speed of the motor. The electronics then compares this "actual" speed with the "specified speed," then immediately corrects any deviation by the motor. This drive technique is currently state of the art, but due to the high level of complexity in fabricating the electronic controls—especially combined with the necessary computer interface—this technology can be ignored for home-built systems. Nevertheless, numerous individual solutions have appeared on the market, as is also true for stepper motors, which can be easily connected to home-built mounts.

5.4.2 Connecting Mount and Computer

Interfaces between mount and computer mainly involve two areas. One basic application relates to controlling the telescope with the click of a mouse or using a programmed automatic control to go to a specific site in the sky (GoTo functionality). A second common interface used by astrophotographers involves using the digital signal from a small guiding camera, either installed in the guide scope or attached off-axis on the side of the main camera, to cause the mount to make fine controlling adjustments so as to keep the celestial object exactly in the specified position. This function allows any deficiencies in the mount (such as eccentricities and defects in the worm gear) to be corrected, as well as any atmospheric variations (for example, caused by refraction) that otherwise would have degraded image quality.

Many commercially available GoTo telescopes already include databases containing countless celestial objects in the supplied mount control system that, after an initialization phase, allow the user to move at the press of a button to the desired object. There are several current methods of alignment here. Alt-azimuth-mounted telescopes can be aligned either by using three alignment stars ("sky align") or, on the other hand, by using a "Level North Technology" whereby the telescope must be first aligned so as to be horizontal and aiming true north. The system then moves to an initialization star that must similarly be confirmed by pressing a button as soon as the star has moved to the middle of the image field.

Other types of mounts using stepper motor or servo drives, and even home-built mounts, can readily be made GoTo-capable. For this purpose, you need both the actual control system and power electronics that can activate the motors and handle the electrical current for them. In principle, the control system can generate the requisite signals from software through parallel or serial ports, then send these to the final power stage that runs the motors. However, a completely home-built unit only makes sense for very complex or specialized systems. This is because even with partial solutions, the cost/complexity of development is usually out of all proportion to purchasing a GoTo-ready control system or to adapting a hand-control box with the necessary contacts for connection to guider ports.

The power electronics for the drive motors can be controlled directly by many currently available astronomy programs. A very practical solution for smaller applications and aside from manufacturers' systems is also provided by the ASCOM driver (Fig. 25). The ASCOM initiative has undertaken to standardize interfaces for connecting astronomical equipment. In addition to support for control systems already available, it is now becoming increasingly possible as well to integrate focusers, motorized clutches, or CCD cameras with the computer control system.

Another standard relating to control commands for telescopes and telescope control systems with a serial port is used by the US company Meade. Due to the widespread

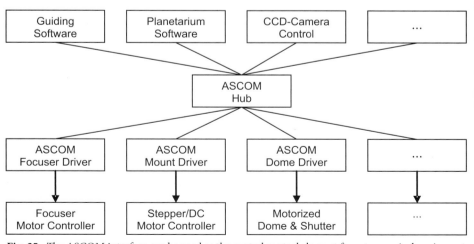

Fig. 25. The ASCOM interface can be used as the central control element for astronomical equipment

popularity of various models of the LX200 series, and support for it by numerous PC applications, the LX200 command set is also employed by other manufacturers for their own telescope controls for reasons of compatibility, at least in reduced scope involving the commands required for the above-mentioned main functions GoTo and autoguiding.

5.4.3 GoTo Mountings

Mounts that are controlled by a computer to find a celestial object are also called GoTo mounts. The main feature of these mounts is that all you have to do is enter the name of the object, or the celestial coordinates for the object in the included computer or computer-aided motor control, and the motors will move the telescope exactly to the desired object. One particular earlier variant involved systems of digital setting circles, which as a kind of forerunner to today's GoTo systems were at least able to detect the axial motions of a mount and display to the observer the positional coordinates of the previously aligned mount. These enabled the observer to move the telescope to a desired object by slewing to a previously calculated celestial position whereby he or she pivoted the scope until the setting circles displayed the searched-for position. Digital setting circles are still being used today by those building their own simple Dobsonian scopes.

Fig. 26. Compact, one-armed fork mounts have experienced a renaissance, with the introduction of GOTO drive technologies

Small Compact GoTo Mounts. Especially in the lower price range, inexpensive mounts are being offered in increasing numbers for the mass market in which the mechanical stability and tracking accuracy are at least sufficient for finding simpler objects. In general, these mounts are not suitable for larger telescopes or adequate to meet the demands of astrophotography. In addition, very small and compact telescope systems are often marketed in which the electronic controls nevertheless right away list more than 10,000 possible observing targets. Since many of the included celestial objects, however, are not even visible with small telescopes, the buyer should carefully consider what type of observing he or she wants to use the scope for. Starting with apertures of around 4 in, a GoTo system makes sense since the number of observable objects reaches a level that makes a positioning system worthwhile. From an aperture of around 8 in and up, the observer will be able to systematically exploit in fuller measure the large range of objects provided by a GoTo system that includes countless small and faint celestial objects.

GoTo for Fork Mounts. The fork mount, which had almost disappeared from the amateur telescope market, is now enjoying a renaissance due to the new opportunities provided by computers. If a fork mount is set up in azimuth mode, the static ratios are very good, while the mechanical stability is sufficiently high for purposes of amateur astronomy. While converting older fork mounts to GoTo operation is rarely encountered, larger fork mounts, especially home-built or made-to-order ones, are frequently equipped with commercially available GoTo systems. In the form of complete systems with permanently installed GoTo controls, fork mounts have recently gained wider popularity once again. Specifically, telescopes of short overall length are often sold on "GoTo forks" and can provide a fairly large telescope aperture in very compact form. There is, however, one limitation with azimuth-based fork mounts: astrophotography is possible only when using relatively short exposure times (up to around 30 seconds, depending on focal length). Exposures any longer than this reveal the effect of image rotation. Such exposures can be achieved only by using a polar wedge that provides an equatorial configuration. But by doing so, you lose the basic advantages of the fork mount in terms of stability and simplicity of operation. A second method of counteracting image rotation uses corotating focusers. However, since the motion sequences are quite complex, such systems have not yet been widely accepted in amateur astronomy.

GoTo for German Equatorial Mounts. German equatorial mounts are increasingly being sold or retrofitted with automatic positioning. This combines the advantages of the German equatorial mount in regard to, for example, flexibility in the type of telescope used, suitability for astrophotography, along with the capability of automatic positioning. There is only one minor caveat. The mount must be precisely polar-aligned; otherwise the system will not be able to move exactly to the object. A polar-alignment scope or permanent mounting is therefore essential for successful GoTo operation.

GoTo: Simple and Inexpensive. Computer-controlled mounts are becoming increasingly popular, shortening the search for objects to observe and significantly lengthening observing sessions. Since complete commercial GoTo systems are sometimes very expensive, individual components for a simple and inexpensive alternative are also marketed. What is needed, of course, is a mount with motors in both axes (right ascension and declination), a telescope controller with an interface for a computer (RS232), a computer (preferably a laptop) with a serial port, and a planetarium program capable of controlling a telescope. Of course the appropriate connector cables are also a necessity.

The control system is connected through an RS232 cable to the serial port of a computer (laptop); operation is also feasible through a parallel port using a control box. Currently available commercial planetarium programs usually support the telescope control system. One very usable freeware program is Cartes du Ciel, installable in numerous language versions, or the demo version of the program SkyMap 9, which also has the capability to control a telescope.

The central linking element between the computer and motor controller of the mount is the ASCOM interface or hub, which does, however, require a motor controller with a serial port to set up and maintain a connection to the PC. On the commercial side, several solutions are available on the market, with mount manufacturers sometimes providing the appropriate controller and requisite interfaces along with the mount. There are also free or inexpensive do-it-yourself products, some commercial.

Examples of programmable controllers for telescope mounts: http://littlefoot.rajiva.de, http://www.astronomy.hu/pulsar.htm, http://www.astro-electronic.de, http://www.boxdoerfer.de, http://homepages.accnorwalk.com/tddi/tech2000/.

Autoguiding functionality is also possible with simple hand controllers where the buttons of the hand controller are actuated electrically and the electrical contact closure is driven by the ASCOM interface through a separately acquired relay box. The control signal in this case must come from an autoguider system; freely selectable positioning (GoTo) cannot be implemented by this route since there is no feedback from the mount motor to the computer.

Very simple GoTo systems are often limited by the positional accuracy attainable, but more significantly by the low positioning speed achievable with them. In such cases, it is useful to use the directional buttons directly and manually to move to a known bright star in the proximity of the searched celestial object, then to place the star exactly at the center of the field of view. Then the connected PC software is told which star has been positioned at the center of the image field. Using the "synchronization" function found in nearly all of the current planetarium programs, the software can now adjust to the position of the mount. From here it is simply a click of the mouse to go to the desired target object. Due to the short slewing path, no fast GoTo function is needed and positioning is of optimal accuracy.

5.4.4 Guiding Systems

Simple astrophotography, say of the Sun, Moon, and planets, using fast exposure times is possible even with undriven telescopes. An equatorially mounted telescope with a tracking motor on the right ascension axis is better, however.

Longer-exposure photographs of astronomical objects through the telescope always required a mount with two motors, and control of tracking accuracy that could be affected by less than optimal alignment with the celestial north pole, or by mechanical or electrical deficiencies in the mount. For this reason, a guide scope is necessary when doing photography using longer exposure times through larger telescopes. The guide telescope is a second scope that is mounted parallel to the imaging telescope. As an alternative to the guide scope, a radial guider can be utilized that is placed in front of the camera and uses a small prism to divert the reflection of a guide star. Also see Chap. 6.

Even when using high-end mounts, very round star images are attainable only with active control of the guiding motion. Either an appropriate standalone guider or a PC with a second camera and a link to the mount controller is used to implement automatic guiding (autoguiding; see Fig. 27). Since the quality of the autoguiding system has a critical effect on the quality achievable in the astrophotographs, careful selection and matching of all components is recommended. This relates to the camera used, the motor controller for the mount, the performance of the computer, and the type of cable connections. The key element here is, of course, the control software. Exposure times of normally 1 to 5 s, in exceptional cases also up to 10 s, are required for the control image of

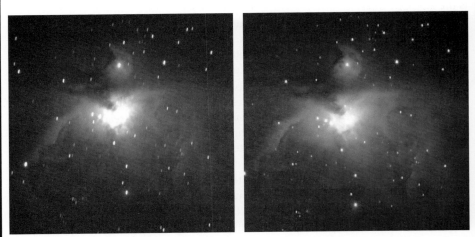

Fig. 27. Even properly aligned mounts are no guarantee for error-free imaging quality (*left*). With autoguiding, any mechanical and electrical deficiencies are effectively compensated (*right*)

the autoguider. However, for tracking a bright star through a guide scope that is not too small, even an unmodified webcam like those used for planetary imaging can be used.

Despite the fact that on first glance this looks quite simple, the control algorithms are quite complex due to the underlying spherical trigonometry and the requisite supplemental functions that, for example, compensate for drive inaccuracies. As a result, commercial control programs often provide better results, yet free software is nevertheless entirely suitable as a means of inexpensive entry into this area and can even yield satisfactory results with a bit of fine-tuning.

The various software programs can control the telescope through different interface systems:

- LX200 (Meade) serial interface. In addition to the ASCOM standard, the LX200 protocol developed by telescope manufacturer Meade for its own mounts has become a widely accepted standard. The telescope mount here is connected to a serial interface on the telescope and activated by easily understandable short commands in ASCII code. The mount control then converts the short commands into the appropriate functions and motion sequences.

- ASCOM (serial interface, http://ascom-standards.org/). The ASCOM initiative is a volunteer group of astronomically interested software developers and manufacturers of accessories that has undertaken to provide an independent, freely available standard for controlling astronomical equipment. ASCOM is a driver system to control telescope mounts, cameras, and even focusers and observatory domes by utilizing astronomy programs such as, for example, Cartes du Ciel (CdC), SkyMap Pro, Starry Night, The Sky, Guide, and others that support this standard. The ASCOM standard is attractive especially for mounts and equipment that do not support the alternative but simpler LX200 command set.

- Parallel port (with control box connected to the computer's parallel port). The parallel autoguider interface with signal cables for each direction can be used for purposes of autoguiding. These autoguiders, usually special types, have a parallel autoguider port. The autoguider and control are connected here to the appropriate autoguider cable or to a multi-interface cable. An adapter is required for autoguiders having an RJ11 connection. The control system then reacts to the control signals coming from the autoguider in the same way it responds to manual confirmation from the directional buttons. The same adjustments and switches should therefore be set to the smallest possible correction speed. Normally, tracking mode is selected for autoguiding. Then you can intervene at any time manually. Older CCD cameras also need a printer port for control purposes. Modern notebook computers unfortunately no longer have this port, which therefore must be retrofitted to them if you still wish to use parallel-port technology.

- GPUSB (Shoestring) USB interface (http://www.shoestringastronomy.com). The GPUSB (Shoestring) USB adapter can be attached directly to a USB port on the computer or connected to a USB hub that is attached to the computer. Using an RJ12 cable, the adapter connects the guider port of the computer to the control input of popular telescope mounts. The interface system is based on optocouplers so as to electrically isolate computers and telescope from each other. The purpose of this is to eliminate any possible sources of electrical interference. This interface adapter does not require any external power supply since it receives its power through the USB bus.

To implement the autoguider function with a webcam or other tracking camera, you need a PC that enables both the actuation capability for the mount control as well as special software to control the camera by evaluating the control images obtained from the camera. Some autoguiders can also communicate directly with the mount control. This is true primarily for systems that come from the same manufacturer where the interfaces can thus be matched to each other.

Once the autoguider is connected, a calibration step is first required. The autoguiding system thereby checks the reaction of the control system to individual control signals, directions, and speeds for the correcting motions on both axes. Autoguiding can then begin. To achieve the best possible results, the correcting speeds for the control system and correction parameters of the autoguider should also be adjusted to account for the imaging focal length. A stepwise test is recommended. Separate power supplies should be used for the control system and autoguider. This avoids a ground loop through the autoguider cable that can corrupt the signals. In addition, this prevents disturbance pulses that could degrade the image from passing from the stepper motors through the power supply to the sensitive camera electronics.

Tip: An ASCOM Telescope Validator is available from amateur astronomer Peter Enzerink at: http://enzerink.net/peter/astronomy/pages/ascomvalidator.htm.

5.5 Basic Adjustments

5.5.1 Direct and Off-Axis Guiding

Long-exposure photographs in cold weather can make guiding at the eyepiece a bit unappealing. Modern electronics now make it possible to carry out this task automatically and with greater precision. The eye is replaced by a photoelectric sensor and corrections are effected by control circuits. Photoelectric tracking has two basic designs:

1. The guide star is sensed with the guide scope which has a *tracking head* supplying the signals for the control circuits.
2. The photographic telescope is used in off-axis guiding (OAG). That is, at the edge of the photographic field, the light of the guide star is deflected by a small prism and fed through a transfer optics into the tracking head.

Advantages and disadvantages of these two modes are summarized in Table 2. It is often a problem to find a sufficiently bright guide star because even for fairly bright stars the signals given by the photoelectric sensors are quite small. For this reason, the aperture of the guide scope should be as large as possible: about 100–120 mm is a lower limit well worth the expenditure of photoelectric tracking.

The deformative displacement between the two optical axes may require attention in long-exposure photographs. The connection and joints between the two systems should be very rigid. Moreover, the optical components in their cells should be free of displacement (or tilt) and the guiding head rigidly fitted to the guide scope.

In OAG mode, the photographic and guide systems use the same optical light path; thus, the hard-to-avoid flexure displacement is not harmful. Of course, flexures in the light ray behind the deflecting prism and in the tracking head should not occur here either. In OAG mode, the mechanics for the plate holder, the prism, and the guiding head are more complex and expensive than with a separate guide scope. On the other hand, only one optical system is needed in OAG mode.

Table 2. Photoelectric guiding systems

Separate guide scope	Off-axis guiding
Advantages	
Guider and photographic (observing) system are independent units and can be designed best for their tasks.	Only one optical system for which the mounting has to be designed.
Useful when off-axis is not feasible: Schmidt, Maksutov cameras, apertures under 300 mm, etc.	Equal aperture and light-gathering, reaches fainter guide stars.
No interference in focal plane of telescope; simple attachment of guiding head at guide scope.	Optical path the same, avoids guiding error caused by axis flexure.
Disadvantages	
Requires two optical systems; guider must not be too small (over 100 mm).	Feasible only for larger telescopes.
More costs for optics and mounting, heavier counterweight.	Not for Schmidt and similar systems.
Deformative displacements between systems may be troublesome; stiff coupling required.	Limitations and complex mechanics near focal plane (prism, special plateholders, etc.).
	Field of telescope not seen while guiding.

Fig. 28. As guide telescopes, refracting telescopes with their low-adjusting vulnerability are particularly popular

5.5.2 Polar Alignment

To align the equatorial axis system with respect to the celestial pole requires adjusting elements in the meridional and horizontal planes of the mounting; in the former, the latitude φ and, in the latter, the azimuth A are adjusted. These important telescope parts are often rather poorly made in amateur telescopes and constitute weak points. Adjusting the telescope precisely with such deficient components can be a tedious task indeed.

The adjusting parts can be placed between foundation and pier (base adjustment), or between pier and bearing case of the polar axis. Instruments on tripods adequately provide an azimuth range over 360°. Here are some basic hints for the design of the adjusting elements:

1. The elements should respond precisely to the operation of adjustment and should not suffer from backlash. The use of the telescope determines the precision $\Delta\sigma_j$ for adjusting it. As mentioned in Sect. 5.4.1 the human hand has a limited accuracy grade of motion σ_m. The adjusting elements should therefore have a gear ratio u of at least $u \geq \sigma_m/\Delta\sigma_j$. This ratio is usually realized by a combination of screw and lever.
2. The elements for adjusting azimuth and latitude must be kinematically decoupled (i.e., independent). Kinematic uncoupling means that when adjusting one coordinate, the other is not also changed. This is a trivial requirement, but one which is not always satisfied in all amateur mountings.

3. The elements should not be in the line of main force, that is, they should not be part of the chain of stiffness. These two elements are thus a coupling link in the chain and can disadvantageously influence the overall stiffness. Adjustment precision and stiffness can make demands on the design which are largely mutually exclusive. A fine adjustment needs fine-pitch screws and long levers; both are elements which are quite unfavorable from the view of stiffness. Therefore, the main flow of force should be in a parallel path past the adjustment parts; otherwise, these should be relieved with respect to forces.

4. The elements should be lockable so that unintentional offsetting is prevented.

Adjustment of the Axis System by the Scheiner Method. Before the adjustment, the placement of the optical axis at right angles to the declination axis should be checked and, if necessary, corrected by aiming the telescope tube alternately at both sides of the pier to a fixed sighting point. The adjustment of the instrument after Scheiner is in four steps, with steps 3 and 4 repeated alternately until the desired precision is reached. A reticle eyepiece with a sufficiently large field of view is needed. The mounting should be adjusted using the primary optics or the guide scope, but *not* the finder.

Step 1. Align the mounting very approximately in the north–south direction. This can be done during the daytime using a compass, or the Sun, or familiar terrestrial marks; at night, the Pole Star can be used. Note that ferrous material in the mounting can deflect the compass needle.

Fig. 29. Many series mountings have pole-finder telescopes inside their polar axes

Step 2. Set the declination wire. This wire in the eyepiece serves as the reference line for the instrumental coordinate system. The declination wire is set as follows: set a star on the wire, clamp the declination axis, move the telescope by its RA drive slightly back and forth and observe the star. The star should run exactly along the wire. If not, rotate the eyepiece until this condition is met. Thereafter keep the eyepiece fixed.

Step 3. Adjust the instrument in azimuth. Point the instrument at a star near the zenith, set the star on the declination wire, clamp the declination axis. With tracking on, the star will gradually move to above or below the declination wire. Apply a correction in azimuth until this deviation is eliminated.

Step 4. Adjust the instrument in latitude. Point the telescope at a star with a declination of about 70° near its eastern elongation. Set the star on the declination wire, then clamp the declination axis. With tracking on, the star will gradually drift off the wire. Correct the latitude until this drift disappears, then repeat Step 3. When the instrument is perfectly adjusted, lock the adjusting elements.

The mounting is considered correctly oriented when the star, after a long tracking period, does not drift off the declination wire. A slow drift along this wire is caused by the tracking speed and not by the adjustment. It should be mentioned that the instrumental axis system is in reality tilted against the true latitude by an angle corresponding to the refraction at the pole, as the instrument has been adjusted via an incident beam refracted in the Earth's atmosphere.

5.5.3 The Setting Circles

Setting or divided circles are a valuable tool for setting the telescope on celestial objects, and should therefore be a component of any large instrument. Since their function on an equatorial mounting is merely to set on a star, and not to ascertain its coordinates (as is the case with a theodolite or transit circle), the divisions need not be extremely fine or accurate. The circles should be readable at very low light levels. The following hints will serve to divide a circle:

1. Recommended angular distances Δ between divisions are, in declination (360° division): $\Delta = 1$ or 2°, labeled: $+90° \ldots 0° \ldots -90°$. In hour angle (24^h division): $\Delta = 2$ or 5 min, labeled: $\ldots 1^h \ldots 12^h \ldots 24^h$. When labeling, be careful of the direction of the polar axis (northern and southern hemisphere), and the sign of the declination, and the arrangement of the circles at the instrument. To use a moving circle with fixed index, simply reverse the labeling compared with a fixed circle with moving index.
2. The minimum required circle diameter D_K is given by the recommendation that the divisions $\Delta = 0.5 D_K \cdot \mathrm{arc}\Delta$ should not be less than 1 mm apart, or better, 2–3 mm.
3. Use thick markings (broader than 0.25 mm). The divisions can also be made as bars 0.5Δ wide. Such a division is readable even in poor illumination and permits an estimate for intermediate values of quarters of Δ.
4. White lines and numbers against a black background correspond better with the accommodation of the eye to the star-studded night sky.

The divisions can be made radially on a flat disk, or on the surface of a cylinder or cone. Ready-made division circles can be obtained from astro-supply centers and in specialty stores. There are also numerous possibilities for making them oneself in an inexpensive and useful way.

Adjustment of the Circles. To set the divided circles to the *instrumental coordinate system* requires that the circles can be finely rotated relative to the pointers, or vice versa. Amateur mountings often have the circles clamped with bolts directly on the shafts. Such an attachment nearly always causes damaged seats on the shafts, and it is hardly conducive to precise adjustments. Tangential and radial clutches (clamps), as mentioned in the discussion of slow motion, are also the best joining elements between circles and shafts.

Adjusting the Circles by the Kolbow Method. In this method, the instrument is first adjusted with respect to the polar axis (Sect. 5.5.2). Next, the instrument is pointed to the zenith and the declination axis made horizontal: the declination circle should then indicate the latitude φ of the geographical position. As the instrument is also in the meridional plane, the hour circle should read 12^h on one side of the pier and 24^h on the other. The adjustment procedure is as follows: Point the instrument in both the east and west positions—or, for fork mountings, in both positions of the fork—to the zenith. Read the circles in both positions and repeat a few times. Find the differences of readings from the nominal φ and 12^h or 24^h for averages, and correct the circles accordingly.

The problem with this method is that the zenith is not a visible reference point, and the zenith position has to be found with mechanical aids of limited precision, such as a bubble level or plumb bob. Also, not every telescope optical system has straight and plane reference surfaces perpendicular and parallel to the optical axis, or is suitable for attaching a level or a plumb bob. Thus this method affords only limited precision.

Setting the Circles from Known Star Positions. This method can, in principle, be made as accurate as desired. Point to several stars near the zenith and find their positions from the circles and the clock. Compare them with the correct positions from a star catalog, allowing for precession if necessary, and determine from the difference the circle corrections. A least-squares calculation (see Sect. 3.2) may be used if it is thought that this will improve the precision.

Digital Position Displays. Professional telescopes are now usually equipped with digital position indicators and computer-operated positioning devices. The signals needed for this are generated by angle encoders coupled with the polar and declination axes. The angle encoders or increment encoders replace the divided circles. Their precision and resolution determine the precision of the positioning and display they control. There are two encoder systems:

1. *Coded-angle encoders (GRAY-code encoders).* Each angle of position is represented by a code number. The numbers are displayed as *bar codes* on a glass circle and are read by reading optics transferred into binary digits or "bits," and the angle can be directly displayed on a simple indicator, or interfaced into a computer. GRAY-code encoders are quite expensive, especially for higher angular resolution.
2. *Incremental encoders.* These devices do not provide absolute angles but rather a defined number of pulses for a change of angle $\Delta\beta$. The original position β_0 of the pulse sequence is therefore entered manually. It is erased when the power is switched off and so always has to be entered anew.

Incremental encoders have a glass circle with ruled grating of division τ which is scanned by photodiodes. Generally, these code tracks have three outputs. One track generates a signal U_1, another an electrically phase-shifted (by 90°) signal U_2, and the third supplies the signal of a reference mark U_0. Signal U_2 is needed for the *direction discriminator* (DRL) which instructs the counter to add or to subtract the angle increments to or from the reference angle β_0. This needs a forward–backward counter VRC. The phase-shifted signal also increases the angular resolution by a factor of 2. Incremental encoders are comparatively inexpensive. Numerous makes and a variety of encoder types are on the market and cover the full range of applications. For reasons of costs, only these should be considered by amateurs.

Figure 30 shows the signals and schematics of digital processing of position. If the circular grating has Z lines, the angle increment is $\Gamma = 360/Z$. The input logic IL can register from the two signals either the front slope, or back slope, or both, and forward them as pulses to the counter. Correspondingly, an angular resolution of $\Gamma/2$ or $\Gamma/4$ is obtained. An encoder with an 1800-line grating therefore has an angular resolution of 3′.

Digital positioning should also include realistic accuracy considerations of the entire instrument. It makes little sense to use precise and expensive encoders of arcsecond resolution if the adjustment and mechanical precision of the telescope are in the range of arc*minutes*. The circuit schematics are simple and need no further explanation. The incremental encoder can be linked in two ways to the astronomical coordinate system:

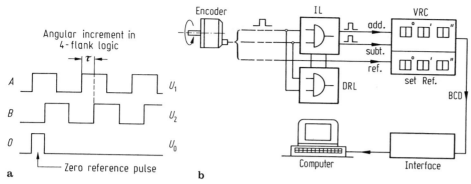

Fig. 30. Digital position indication with incremental encoder. **a** graphs the pulse sequences given by the encoder. The phase-shifted signal B discerns the direction of motion. **b** The pulse logic IL/DRL instructs the counter whether to count forward or backward. If the counter VRC is equipped with a BCD-exit, the position signal can be fed into a computer

1. With the electronics switched on, the telescope is pointed to a star whose coordinates are then input into the counters as reference angles.
2. Another method uses the reference mark U_0. In the first step encoders are mechanically adjusted just as circles are. This assigns certain angle positions to the reference marks. Before use, the telescope is set on the reference position, and the corresponding angular values are then entered into the counters.

5.6 Further Reading

For actual informations on mountings and equipment see astronomical magazines such as Sky and Telescope (USA) and Sterne and Weltraum (Germany). In addition informations can be found in Chap. 12, pp. 283 to have contact with astronomical clubs and societies.

Gera, H.D.: Welches Nachtglas? Sterne Weltraum **26**, 167 (1987)

Based on *Telescope Mountings, Drives, and Electrical Equipment* by H.G. Ziegler, Compendium of Practical Astronomy, Vol. 1, Chap. 5, Springer, Berlin, Heidelberg, New York (1994)

6. Astrophotography

K.-P. Schröder and H. Lüthen

6.1 The Photographic Image: a Brief Introduction

Photographic images are of particular importance to the astronomer. For one, long exposures document galactic and extragalactic deep-sky splendor far beyond the reach of visual observation (see, e.g., Fig. 1). Even more important to science, astrophotography is a means of precise measurements of brightness (photometry) and position (astrometry) of any star, comet, minor planet, supernova, etc.

In the 1990s, cooled CCD cameras were still a specialized tool in deep-sky astrophotography. But after the millennium, affordable digital cameras and new CMOS-DSLRs have been introduced, and digital photography has taken over. While standard consumers cherish the instant availability of digital images and their ease of sharing them, astronomical application has also greatly benefited from their much higher astrometric and photometric accuracy.

Fig. 1. This H_α-filtered photo of IC405, taken by Konstantin von Poschinger under the light-polluted sky of Hamburg, Germany, with a small refractor and a cooled CCD, demonstrates the unique potential of astrophotography

6.1.1 Digital Image Information

Whether digital or conventional, image information simply comprises the illumination intensity at each picture element, registered as a function of two image coordinates. This information is best extracted from film (negatives or slides) and converted into digital format by means of a film scanner. This is preferable to a flatbed scanner used on a print, especially in terms of dynamic range, but the latter approach still gives good results. Either way, conventional photos can thereafter be processed and stored in much the same ways as modern digital photos, on any PC, rather than by wet darkroom chemistry.

CCD and CMOS chips are the electronic sensors which produce the digital image. They consist of arrays of small semiconductor photodiodes or phototransistors, each of which represents one picture element, a *pixel*. During exposure, free photoelectrons are generated, in proportion to the illumination intensity. The readout process first performs, pixel by pixel, an analog voltage measurement. The number of bits used in the subsequent analog-to-digital conversion of each pixel-reading determines how many gray levels there are between 100% white (saturation) and black (no signal). Simple "point-and-shoot" cameras deliver only 8-bit, which is equal to 256 gray, levels in each color channel (JPEG format). DSLRs and other advanced digital cameras offer a RAW image format with uncompressed 12-bit (4096 gray levels) to 14-bit (16,384 gray levels), where some cooled astro-CCD cameras even give 16-bit output (65,536 gray levels). For the astrophotographer, the RAW image format is much preferred to JPEG because it resolves very subtle contrasts and records a larger dynamical range of image intensity. Only after image processing would one want to reduce the relevant intensity range to JPEG format.

6.1.2 Color in Digital Astrophotography

The most common way of producing a color image from a set of, by nature, monochrome (i.e., black and white (B&W)) CCD images taken with different filters is by the RGB-mode. It simply combines three images through red (R), green (G) and blue (B) filters. These three fundamental color channels are required by all monitors and printers to reproduce an image. Likewise, the common single-shot-color CCD is readily equipped with an on-the-chip filter mask (Bayer mask) made of small RGB filter triples. These divide the pixels into red, green or blue pixels in order to simultaneously feed all three channels of the RGB image. As convenient as this is, it costs resolution and sensitivity when compared to a monochrome (B&W) CCD.

But color images can also be represented by a different system, which is more efficient in astrophotography: The Lab system uses three channels, too, but it stores the total B&W image intensity in a luminance channel (L), while color is coded by the readings of the other two channels, a and b. Hence, luminance and color information are separated here. The advantage is a higher-resolution monochrome exposure may replace the original L image. This way, you can use the color information from reduced resolution R, G, and B CCD images, binned to save exposure times, but the final lab image will display the full-resolution and depths of your unfiltered L exposure. This smart method requires R, G, and B exposures (of reduced quality), plus an unfiltered, high-quality B&W image for L, which gave it the name *LRGB imaging* (see Sect. 6.5.3). Its total exposure time can be much less than that of conventional RGB-imaging at the same resolution and S/N (signal-to-noise ratio).

6.1.3 Digital Image Processing

A big advantage of the digital image is, how fast it can be processed on a PC or laptop for optimum display and reproduction. We here briefly introduce software for the general handling of images, like adjusting color balance and contrasts, and which allow basic measurements like intensities and pixel coordinates. Specialized software, e.g., for selecting, registering and stacking images, will be discussed later (see Sects. 6.5.2, 6.7.5 and 6.8).

Photoshop is the flagship of all professional image processing programs. Meanwhile, it handles 16-bit images, RAW images, and there are special plug-ins to open FITS format (as often used in astronomy). Photoshop can do a lot more than basic image adjustment. It can convert standard RGB images into the lab coding, and a very powerful feature is its "layer technique," which allows the superposition of images. Photoshop is available for Windows and Macintosh PCs, with tutorials in almost every language.

GIMP is a freeware Photoshop "clone" with similar functionality. It is an open source project under the GNU public license. Originally designed to run under LINUX, GIMP is now also available for Windows, Mac and some other operation systems, with documentation in many languages. A very basic and fast standard tool under LINUX is *Xview* (xv), which performs only the basic image handling.

IRIS is a powerful freeware package written by Christian Buil, with manuals available in french and english. It offers a large number of high-quality, scientific operations not available elsewhere. It can read and write FITS, RAWs and all relevant image files, does automatic stacking, can employ a large number of digital filters, and it performs photometric image reduction. However, many special functions are not found in its menus but require command-line input. IRIS is a good academic teaching tool, as well as an excellent choice for the advanced astrophotographer.

6.2 Which Type of Camera? An Overview

6.2.1 Conventional Cameras

The old mechanical finder camera of your grandfather's camera can take nice astrophotos (see Sect. 6.3). But the workhorse of the successful astrophotographer was, for decades, the single-lens reflex camera (SLR) for 36 mm × 24 mm format film. A widespread and very useful model was the Olympus OM1 (left side of Fig. 2). It offered exchangeable view-finder screens and allows vibrationless exposures by a manual option to flip up the mirror beforehand. In principle, such conventional SLRs offer the same large potential and versatility as the modern DSLRs (see below), and there is still a rich market for second-hand SLRs, sold at affordable prices. Furthermore, users of conventional cameras do not have to lose out on the benefits of digital image processing: the full image information on film can be digitized with a scanner or by a professional photo service. All this makes the conventional SLR a good choice to embark on astrophotography.

But we cannot ignore a growing disadvantage: Since the market has moved on to digital photography, demand for chemical films has fallen sharply and, consequently, its production has been streamlined. Many emulsions, which once were the astrophotographer's best choice, have already disappeared, and this unfortunate trend will continue!

Photoshop:
http://www.adobe.com/products/photoshop/index.html

IRIS:
http://www.astrosurf.com/ buil/us/iris/iris.htm

Maxim DL:
http:www.cyanogen.com

Fitswork:
http://freenet-homepage.de/JDierks/softw.htm

GIMP: http://www.gimp.org/

Fig. 2. The Olympus OM1 is still a good choice of SLR to embark on astrophotography (*left*). Canon's EOS 300D and further DSLR models are excellent choices for astrophotography for their low-noise CMOS-chips (shown here with focal reducer lens on *right*)

6.2.2 Digital Cameras and DSLRs

Digital point-and-shoot cameras, which revolutionized the camera market around the turn of the millennium, can produce nice lunar photos by afocal photography, even hand-held behind a telescope. But only the more recent development of digital SLRs (DSLRs) has now fully taken the place, held by conventional SLR.

When Canon introduced the EOS 10D camera in 2003, its 18 mm by 24 mm CMOS chip was extremely large compared to any cooled CCDs of that time, and it had an excellent signal-to-noise ratio. The virtually identical, but more economic EOS 300D ("Digital Rebel," Fig. 2 right) soon followed, then the 20D, the 350D (Digital Rebel XT) and the 400D. Of the 20D series, only a limited number of cameras dedicated to astrophotography (20Da) were produced. Although Canon DSLRs still offer the lowest noise in long exposures, there are other useful models offered by Nikon, Pentax and Olympus. In particular, the Pentax Ist DS appeals with its superior view-finder and larger display.

DSLRs make color astrophotography much easier than cooled, monochrome CCD cameras, which require the use of several different exposures through filters, and which either have smaller chips and/or are much more expensive. At the same time, DSLRs can be operated in much the same convenient, laptop-independent way as a conventional SLR. Both camera types offer interchangeable lenses and allow direct attachment of the camera body to the telescope and the use of its prime focus. In addition, like expensive CCD cameras, DSLR photos have a very good astrometric accuracy.

Pixel sizes of DLSRs are around 6 µm or smaller and the digital image background is usually smoother than the one by a film grain. This gives DSLRs a better resolution than achieved with a SLR and film. But for this reason, focusing is much more critical than the apparently sharp image in the DSLR view-finder may suggest. You need to use a magnifier on the view-finder lens, and you should check image quality immediately on the camera display at the highest available magnification. Likewise, any flaws in the optical quality or the precision of guiding is revealed relentlessly!

On the downside, while DSLR cameras are remarkably sensitive, they are not cooled and, hence, cannot fully compete with dedicated, high-end astro-CCD cameras for their low thermal noise in long exposures. They also do not have a strictly linear response, which makes photometry more difficult. A big disadvantage of the off-the-shelf DSLR is its spectral response. In order to limit the chromatic aberration of normal camera lenses, there is an IR block filter directly in front of the CMOS chip. For recording stars and galaxies, this filter is of no concern. But it transmits only a small fraction of the incoming H_α light at 656 nm. This significantly reduces the ability to record emission nebula, where the H_α line contributes greatly to the total luminosity, and a strong color shift is caused! H_α nebulae simply look much too faint and much too pink. If you try to compensate this by giving a larger weight to the R channel, stars and other objects turn out reddish!

The H_α blindness of DSLRs can be overcome by removing the IR filter and, best, by exchanging it for a clear glass or a different filter of the same thickness but of high H_α transmission. Simply removing the filter without any replacement results in a focus difference between the CMOS chip and the matscreen of the view-finder. Unfortunately, the filter exchange requires a partial dismantling of the camera, including unsoldering cables on its electronic circuit boards—such action will invalid your warranty! Courageous astrophotographers have, nevertheless, paved the way and provided us with step-by-step instructions on the Internet. But most readers may simply consider a professional modification, offered by companies like HuTech (California), Baader-Planetarium (Germany), or Canon Germany. After the modification, more light contributes to the R channel of the image, resulting in slightly reddish images. But this effect is easily fixed: You only need to redefine the white balance manually, using a photo of a white paper taken in daylight.

For a detailed and well-illustrated description of how to modify a DSLR for high H_α sensitivity, see http://astro.ai-software.com/articles/mod_350D/mod_350D.html.

Modified DSLRs and modification services are available from, e.g., http://www.hutech.com and http://www.baader-planetarium.de.

6.2.3 Webcams and Video Cameras

Webcams have revolutionized amateur imaging of the planets and lunar detail. They are economic, simple to use, reasonably sensitive, and their lens can be removed quite easily. The apparent downside of a small image size allows, on the other hand, fast streams of short exposures to be recorded. The basic idea is simple: from many hundreds of such images, you select the very best and stack them. Free software, like GIOTTO, does this job for you, and stunning pictures result from further processing of the very low-noise, accumulated image.

Fig. 3. A good choice for planetary imaging (*left*): the Philips SPC900NC webcam, here shown with a flip-mirror assembly for quicker centering and focusing. The Mintron is a low-cost, yet sensitive and integrating B&W CCTV video camera, shown here to serve on an off-axis guider (*right*)

Most planetary imaging today is done with the Philips line of CCD webcams, i.e., the earlier ToUCam Pro models and the SPC900 Pro (see left side of Fig. 3)—and this is for a reason: The cheaper webcams with a CMOS chip (Philips models included) have lower sensitivity and higher noise levels. Other webcam models feature highly compressed or smoothed data streams.

Webcams should be used with an IR block filter to avoid degraded images. Consequently, some astro-dealers offer nice packages which include the webcam, 1.25″ adapter and filter. In addition, there are economic astro-imagers with prolonged exposure times. Their designs started as vastly improved webcams, but the new Meade Deep Sky Imager (DSI-III), e.g., now comes with a very respectable 1.4 megapixel chip.

Digital video cameras for industry application work like a high-quality webcam. They have a larger chip, increased color depth and a much faster frame rate (as required for, e.g., occultation events), but that comes for a much higher price. A remarkable example is the DMK camera, which uses a firewire PC connection. It is available in a very sensitive B&W version and easily handles 30 frames per second; its images are stored in uncompressed AVI format. Exposure times of the DMK astro-version reach up to 60 min.

Some *CCTV analog video cameras* make a very sensitive, reasonably priced choice of astro-imager. Useful models are B&W, have on-board image-integration of up to 2.5 s (Mintron, see right side of Fig. 3), 10 s (Watec) or even 10 min (SBIGs sophisticated STV, which includes autoguider functions). Their output can be conveniently plugged into any TV display (e.g., for public viewing or convenient monitoring of a guide star), and with the help of a USB framegrabber, these electronic eyes can be used in much the same way as their digital counterparts. Besides providing a low-tech approach to deep-sky imaging, these sensitive CCTV cameras are also excellent in recording meteor showers. Analog (as well as digital) camcorders, by comparison, have the disadvantage of a non-removable lens, and their sensitivity cannot compete. But if a camcorder comes with an analog video input plug, it is a very handy item to display the image of a CCTV camera.

6.2.4 Dedicated, Cooled CCD Astro-Cameras

When it comes to lowest noise in long exposures and a linear response for photometric quality images of scientific value, then a cooled, monochrome (non-antiblooming gate) CCD camera with filter-wheel becomes the only option. However, with a large chip size, these dedicated high-end astro-CCD cameras can cost as much as a car! A wide range of quality models is offered by the market-leader, SBIG (California), from small economic CCDs to full 35 mm film format (see Fig. 4). Some models come with a built-in, small off-axis autoguider CCD. But there are also a number of remarkably good designs available from other companies, such as Starlight Xpress (UK), Atik (US), or Apogee (US).

In any case, the superior image quality of a cooled CCD does not come without effort! Focusing, optical and mechanical quality of the telescope, as well as guiding accuracy, all must meet the CCDs relentless resolution. In addition, to obtain a single outstandingly deep astrophoto, you need to add up many long exposures, taking better part of a whole night! Hence, the owner of an expensive CCD will have to be as dedicated as his or her device!

There are many camera choices dedicated to astro-imaging are offered on the Internet, e.g., see http://www.sbig.com (SBIG, California)
http://www.atik-instruments.com (Atik CCDs, US)
http://www.starlight-xpress.co.uk (Starlight Xpress CCDs, UK)
http://www.meade.com/dsi/index.html (Meade DSI, US)
http://www.opticstar.com (Mintron, Watec and more, UK)
http://www.teleskopservice.de (CCDs, Mintron and more, Germany)

Fig. 4. Plenty of choices of cooled CCD astro-cameras, e.g., SBIG's ST10XME (*left*), Starlight Xpress's SXV-H9 (*middle*) or SBIG's ST11000 (*right*)

6.3 Simple Photography of the Stellar Constellations

6.3.1 Camera, Tripod, Cable Release, and Go!

There is quite a number of interesting applications for the simplest photographic exercise: Exposures of up to 30 s of the night sky with a simple camera, at its normal focal length, mounted on a tripod, and with touchless operation by mechanical or electrical cable release (see left side of Fig. 5 for a typical setup). The camera may be of almost any kind: Granddad's mechanical finder-camera, Dad's SRL camera, or a modern digital camera, as long as there is a "B"-setting or any other way of taking longer exposures.

With a nominal (digital or film) speed of 400 ASA and a good lens of f/2.0 to f/2.8, an exposure of several seconds duration already captures the brighter stars and rare planet

Fig. 5. A simple camera on a tripod has many interesting applications (*left*). Crux at dusk, High Andes, 50 mm f/2.8, 12 s on 1000 ASA film (KPS, on *right*)

constellations. About 10 to 15 s are best to show the constellations as they appear to the unaided eye. The optimum exposure depends a lot on the quality of the night sky. A larger physical size of the camera lens records fainter stars. Conventional cameras and DSLRs are better in that respect than the much smaller digital cameras.

With this technique, you can easily do your own "atlas of the stellar constellations." This is a rewarding, yet undemanding first project in astrophotography and very suitable for groups of young students. Very appealing results can be obtained, if the horizon of a remote mountain site and the colors of late dusk or early dawn are included (see Fig. 5 right).

The longest exposure, in which stars would not too obviously become trails, is about 20 to 30 s, depending on the proximity to the celestial pole. Shots with a wide-angle lens may permit 30 to 45 s. In combination with a high-speed (i.e., 1600 ASA), fast lens, and a sufficiently dark location, you can even record the more prominent parts of the Milky Way.

6.3.2 Capture of the Rotating Sky

Longer exposures with a camera on a tripod lead to unpleasant-looking star trails, without reaching any fainter stars: The star-light is not integrated on the same spot of the chip or film. Nevertheless, there is a very attractive exception: Wide-angle exposures of the celestial pole in a dark sky, carried out for several hours, yield a superb "tunnel-effect," which is a dramatic account of the apparent rotation of the celestial sphere, especially if some interesting foreground is included in the photo (see Fig. 6).

This particular technique is difficult to employ with a digital camera, because of its accumulating dark current. The only solution is stacking a rather long series of individual exposures of about a minute, each. Very simple, by contrast, is the use of a conventional

Fig. 6. A 3 h exposure of the celestial pole with 24 mm (f/5.6) wide-angle lens on 400 ASA film, Los Roques, Tenerife (KPS)

camera here: Just choose an ordinary film (i.e., 100 to 200 ASA, not higher) and experiment with lens settings between f/4 and f/8, depending on the residual sky brightness.
The reciprocity failure of conventional film in long exposure times under low light levels
does the rest. It helps to avoid an over-exposure—even 6 h are possible this way, yielding
nice quarter-circle trails around the celestial pole.

6.4 Guided Exposures of Star Fields

6.4.1 Technical Requirements

As pointed out in the previous section, the apparent rotation of the celestial sphere demands that longer exposures of star fields require accurate tracking, in order to reach
fainter stars and high-resolution photos. The mechanical standard solution is a parallactic mount, the right-ascension axis of which must be reasonably well-aligned to the
celestial pole. You could sufficiently improve a very approximate alignment by using the
drift method, but that can take up to an hour. By contrast, a built-in pole-finder makes
this job a matter of just a minute. Other very important accessories are a pulse-driven,
battery-operated motor-drive (in RA) with control box, a camera-bracket for the telescope tube (see Fig. 7 left), and an illuminated reticle-eyepiece (see Fig. 7 right).

Star fields are best captured by SLRs or DSLRs with short (35 mm) to medium
(200 mm) focal length. The respective fields of view range from that of a whole constellation down to detailed views of, e.g., parts of the Milky Way, star clouds or comets.
With a large prism mounted in front of the telephoto lens, you can even perform your
own spectroscopic survey. The exact width W of a field-of-view (in degrees) results from
the length l of the chip or film and the focal length f of the lens (both to be taken
in millimeters). For telelenses, the following simple approximation is sufficiently accurate:

$$W = 57.3° \cdot l/f .\tag{1}$$

The simple camera on a tripod
finds many applications in
astrophotography:
Your personal atlas of the constellations.
Rare planet constellations at dusk
or dawn.
Wide-angle view of the Summer
Milky way.
Dramatic account of the apparent
rotation of the celestial sphere.

Fig. 7. Camera and telelens mounted piggyback on a telescope (*left*). Illuminated reticle-eyepiece as
required for controlled guiding (*right*)

Telephoto lenses for SLRs/DSLRs are usually quite fast. Even when stopped down to improve their image quality, an aperture stop of f/2.8 to f/4.0 is available, which is about 4× faster than the typical focal ratio of a telescope. Hence, even under a dark sky, a sensitivity setting (or choice of film speed) of 400 ASA is sufficient. At 400 ASA you avoid noisier images, which come with higher sensitivity by increased electronic read-out noise or larger film grain, but without getting into uncomfortably long exposure times, either.

Under the above conditions, most DSLR cameras will require only a few minutes to reach the background of even a dark night sky. The optimal exposure time depends, of course, on your local conditions and requires some experimenting. Stacking of a number of exposures (read more in Sect. 6.5.2) then allows you to improve the signal-to-noise ratio (S/N) of your photos and to enhance colors and contrast. As pointed out in Sect. 6.2.2, modified Canon DSLRs presently give the best results in this field of astrophotography.

For the conventional photographer, a very good choice of color slide film used to be the Ektachrome 200 Pro emulsion, push-processed by to 400 ASA. This film used to have a high sensitivity for H_α emission nebulae and a very low reciprocity failure, which keeps exposure times in the range of 10 to 20 min, and maintains a very good color balance, but the astro-relevant performance of an emulsion often changes over the years as its production is modified without notice. You may have to try and compare several promising films.

6.4.2 Follow the Stars

First, mount camera and photolens piggyback on your telescope tube, not too far down from its top end. The use of a second connection, between the telescope tube and the front end of a long telelens, avoids problems with flexure in long exposures. After attachment, the telescope needs rebalancing by sliding it down a bit in its holder.

Short exposures (under a few minutes, as for most DSLR applications) can track the celestial rotation simply by means of an unattended, motor-driven parallactic mount. Long exposures with telelenses, by contrast, require a controlled guiding. For this purpose, aim the telescope at a sufficiently bright star near the center of the intended camera's field-of-view. With the help of an illuminated reticle-eyepiece and the drive-motor control-box, the guide star must be held in the center of the reticle during the entire exposure to assure pin-point star images. Because of residual polar misalignment, you must be prepared to make some small adjustments in declination, too. This requires manual operation of the declination slow motion (unless you also have a motor in declination) by your most sensitive "velvet touch."

The required guiding accuracy can easily be quantified. It results from the focal length of the photolens and the limited resolution of the film (about 20 μm or 0.02 mm on Ektachrome 200) or the chip. Full digital resolution of any image element requires a sampling by, at least, two pixels, according to the Nyquist theorem. Hence, with 6 μ pixels, expect a minimal star size in DSLR images of $d_* = 12$ μm on the chip with the use a moderate speed setting (at 1600 ASA, on-the-chip binning of 2×2 pixels can double the minimal star size). The corresponding angle δ_* in arcseconds ($''$) of the stellar image depends on the focal length of the lens and is given by the simple formula

$$\delta_* = 206'' \cdot d_* [\mu m]/f [mm] .\qquad(2)$$

The standard focal length for a DSLR-lens is $f = 30$ mm. Hence, its stellar images correspond to an angular size of about $80''$ in the sky. The same result is achieved with a fine-grain film like the Ektachrome 200 ($d_* \approx 20\,\mu m$) and the standard SLR focal length of 50 mm. A telelens of 200 mm focal length, by contrast, produces stellar images of only $12''$ (DSLR) or $20''$ (SLR/film), respectively, assuming optimum focus. Many photolenses fail to have a reliable infinity-focus at the end of their scale and you have to apply the focusing techniques described in Sect. 6.5.

The principle idea of guiding is to keep the guide star in the very center of an illuminated reticle, or, at least, not to let it go farther in any direction by more than half of the angular minimum star size discussed above. To know how large that critical deviation appears in your eyepiece, look at Jupiter or Saturn's ring for comparison (40 to $45''$ diameter), or let a star trail (drive switched off) for several seconds. Every second, a star not too far from the celestial equator moves $15''$. At the same time, you want to align the reticle in east–west direction.

After some practice with guiding photolenses of less demanding shorter focal length, you will soon feel quite comfortable with this method, and spectacular results will reward your patience (see, e.g., Fig. 8)! Since only a small, highly portable telescope and mount is required here, you may even travel to the southern sky and return with its highlights, like the Southern Cross with its "Coal Sack" or the Large Magellanic Cloud (left and right side of Fig. 9).

Where to get pole-finders, drive-motors, camera-brackets and illuminated reticle-eyepieces:
http://www.teleskopservice.com (Teleskop-Service, Germany)
http://www.oriontelescopes.com (Orion, US)
http://www.telescopehouse.com (Telescope House, UK)
http://www.celestron.com (Celestron, US)

Fig. 8. Milky Way in Cygnus with the North America Nebula. OM1 with 4.0/100 mm lens, 40 min exposure on Ektachrome 200 film (KPS)

Fig. 9. The southern constellations Centaurus and Crux with the "Coal Sack," OM 1 with 2,0/50 mm lens, 7 min exposure on Fujichrome 400, La Silla, Chile (KPS, *left*). Large Magellanic Cloud, 8 min with 2.0/100 mm lens (KPS, *right*)

6.5 Photography of the Deep Sky

6.5.1 The Classic Approach: Long Focal Exposures

The term "deep-sky objects" refers to all galactic nebulae and distant galaxies. These small and faint objects in the "deep" sky require effectively long exposure times *and* high resolution. Since a guided piggyback photolens does not provide sufficient focal length and resolution, the classic approach is to obtain one (or just a few) long guided exposures with a SLR/DSLR camera body or cooled CCD in the prime focus of the telescope. In order to achieve and maintain precise focus (read more below), your telescope must have a smoothly operating, stable focuser of an ID of 2″. Fast Newtonian telescopes with coma corrector are a very powerful and yet economic alternative to a high-quality, apochromatic refractor. In any case, a rock-solid mount with a fine declination-motor control is more important than a larger telescope (see below)!

As far as the guiding is concerned in principle, we refer to the previous section. But there are several specific points, which need to be discussed in more detail:

1. The focal ratio of a telescope is much less fast than that of a good photolens. It usually lies between f/4 and f/6; for a SCT use a focal reducer to make f/10 become f/6.5. This requires the use of up to 4× higher sensitivity (i.e., 800 to 1600 ASA). Under a dark sky with f/6, optimum exposure times for 1600 ASA film are between 5 and 20 min, while they reach 20 to 90 min with 400 ASA (e.g., push-processed Ektachrome 200). A DSLR set at 1600 ASA requires only about 1 to 4 min for optimum exposure, since it suffers much less from (exposure-time) reciprocity failure than film in low light levels. Cooled CCDs with filters can take between 3 and 20 min. As with DSLRs, the stacking of multiple exposures leads to much improved images. But unless you deliberately use only short, individual exposures (see Sect. 6.5.2), that also means an effort even higher than taking conventional deep-sky photos!

2. Guiding has to be much more precise: If we use Eq. (2), assume a moderate focal length f of 1200 mm and a stellar disk size of the focal stellar image of $d_* = 24\,\mu$, as on fast film or over four $6\,\mu$-pixels (typical with 2×2 binning of most 1600 ASA speed settings), we get an angular size of $\delta_* = 4''$. Hence, to achieve this best-possible resolution with your telescope-camera combination, guiding has to be as precise as $2''$ to either side, declination included! This is hardly more than the image wobbling caused by seeing.

   Obviously, human control of long guided deep-sky exposures is extremely demanding on the skills and patience of the observer! Hence, electronic autoguiders have become very popular in deep-sky photography—either as an integral part of a dedicated, cooled astro-CCD camera (Fig. 10 left) or as a separate accessory (Fig. 10 middle). These are small electronic eyes (small CCD, cooled webcam, or CCTV), which check the position of a suitable guide star on its chip every second or so. To directly engage the motors on the mount whenever the star slips away, the autoguider must be connected with the control box, in parallel with the push-buttons. Modern mounts offer a "ST-4 compatible" connection, a modern autoguider standard introduced by SBIG in the 1990s. Nevertheless, Figs. 11 and 12 show that human guide control can compete to quite some extent!

3. Since the camera now takes the telescope focus, guiding requires an extra telescope rigidly mounted piggyback, or a smart off-axis guider as available from, e.g., Teleskop-Service (see Fig. 10 right). Each of these two alternatives has its downsides: A guide-telescope may bend against the main telescope due to flexure, and it puts extra weight on the mount, making it more sensitive to vibrations. An off-axis guider avoids these problems, but it offers only a very limited, circular field-of-view around the object. You could end up with very faint guide stars when targeting galaxies far off the Milky Way.

4. To achieve the full resolution of your telescope-camera combination, another crucial point is precise focus! The view-finders and mat screens of SLRs and (worse) DSLRs do not provide enough magnification and resolution to achieve that. A lot can be gained with a magnifying angle-viewer added to the camera view-finder. But that may still not prove sufficient, especially with a fast focal ratio. Hence, instant access to the digital image of a DSLR or cooled CCD by a laptop is an important asset. It allows a quick analysis of the width and peak intensity of a stellar image in short

Fig. 10. A built-in (by SBIG, US, *left*) or separate (by Opticstar, UK, *middle*) autoguider helps a lot to achieve long, high-resolution deep-sky exposures. The off-axis guider is a smart way to avoid the need for a second telescope (*right*)

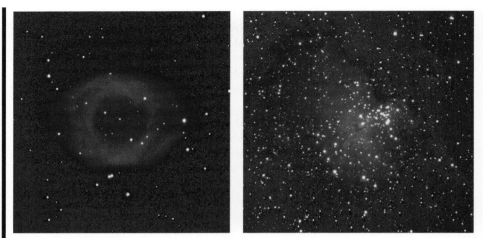

Fig. 11. The Helix Nebula (*left*) and M 16 (*right*), taken with a 10″ Newton f/6, off-axis guider and SLR camera OM1, 45 min and 35 min exposures on push-processed Ektachrome 200 film (KPS)

Fig. 12. North America Nebula, taken with a 10″ Newton f/6, off-axis guider and SLR camera OM1, 65 min exposure on push-processed Ektachrome 200 film (KPS)

test exposures. To accelerate this process, most astro-CCDs have a dedicated focus mode in which a quick, repetitive read-out of just a fraction of the field-of-view is done. If, by contrast, you prefer to use a DSLR stand-alone, careful inspection of a 10 to 13× enlarged image fraction on its display is a sufficient alternative (you may want to use a pocket-lens!). In any case, temperature changes and tube flexure demand refocusing before every new exposure.

A very useful and simple accessory for focusing is an aperture mask with three sub-apertures, which you can easily cut from cardboard. With it, out-of-focus star images become more obvious because of a triangular shape. The knife-edge focuser with a razor blade, as in Foucault testers, is yet another and very precise alternative. However, it must be of exactly the same (within ≈ 10 μ) length as the camera body and its 2″ adapter, and frequent changes between focuser and camera body are not convenient.

6.5.2 The Smart Approach: Stacking Short DSLR Exposures

Summing up a large number of images of short exposure can equal the quality of a long exposure. Still, one single long shot of the same total exposure time yields the better signal-to-noise ratio (S/N), because each individual image of a stack contributes extra readout noise.

However, noise is not the only relevant factor. Long exposures require sophisticated, controlled guiding techniques (see previous section) to yield pinpoint star images, and moving individual clouds may also set limits in some nights. Hence, it may well be simpler to stack 60 (or more) images of 10 s (or less) than to guide a single 10 min exposure. Mintron and Watec cameras are ideally suited for this type of astrophotography. They allow individual integration times of 2.56 or 10 s, much longer even than the DMK firewire camera.

The same smart technique can be used with DSLR cameras, which permit a much larger field-of-view. Most of them offer internal exposure times of up to 30 s, longer exposures are possible with the "B" setting and an electronic cable release. For best results, first find out which is the longest unguided exposure time which still yields pinpoint stars in most pictures. A trained PEC (periodic error control) drive control helps a lot to

Some sources for adapters, coma correctors, focal reducers, autoguiders and off-axis guiders:
http://www.opticstar.co.com (Opticstar, UK)
http://www.teleskopservice.com (Teleskop-Service, Germany)
http://www.lumicon.com (Lumicon, CA)
http://www.celesctron.com (Celesctron, US)

Fig. 13. North America Nebula (NGC 7000), single exposure of 1 min with a 300 mm f/4.5 lens on a modified Canon 350D at 800 ASA, motor-driven without guide-control (*left*). As above, but 35 images stacked with Deep Sky Stacker, final image processed with Fitswork (*right*)

Fig. 14. M 31, technical data as before (*top*). M 8, 14 exposures of 40 s with 10″ f/6 Newtonian (*bottom*), otherwise as before

extend that time from several seconds to about a minute. Less critical telephoto lenses can go without guide-control for even a few minutes. Just delete the few failed exposures before registering (to one or two reference stars) and stacking the rest. The left and right side of Fig. 13 demonstrate the dramatic improvement, while Figs. 14 and 15 show more good results.

Focusing is one of the major problems in DSLR photography, since their view-finders are even less adequate than those of vintage SLR models, and we refer to the previous section for practical advice. In addition, the currently latest DSLR models from Olympus and Canon (EOS 40D) offer a live view mode. Here the user can lock up the DSLRs reflex mirror and monitor the camera view life on its display, or on a laptop—a very convenient feature for focusing. In addition, the 40D software allows for all operation modes

Fig. 15. M 20, as before but 20 × 40 s

required in astrophotography (long exposure times, focusing, programmed images sequences).

Two things are quite important when you set your DSLR for an image sequence: (1) Make it record its 12 or 14-bit RAW images and avoid JPEG modes. (2) If available, choose a mode for "noise suppression on long exposures." In this mode the camera takes a dark frame immediately after each exposure and subtracts it from the image before storing. This gives a much cleaner result than doing your own dark correction afterwards, because it minimizes temperature changes between image and dark. On the downside, you have to stay put until the dark has been done.

Software There is a number of programs which can do registering and stacking of images, among them some excellent freeware. The following are dedicated to DSLR and CCD photography:

Fitswork is a freeware specially designed to stack CCD and DSLR images. One or two reference stars are used to register the images with subpixel accuracy. Results are stored to a 3 × 32-bit FITS format file. Images can be shifted in x and y during the stacking process, enabling stacking on a comet core (the stars will trail, not the comet, see Sect. 6.8.3). Fitswork has a batch mode and can handle hundreds of images in a single run. This is ideal for high-sensitivity CCTV cameras, where you may have to stack a very large number of individual exposures. At the same time, Fitswork can read RAW files of various DSLR camera models. And it provides excellent tools for final image processing, ranging from simple color correction over histogram stretching to advanced sharpening algorithms.

Deep Sky Stacker is freeware designed to automatically stack large numbers of deep-sky DSLR images. It first selects a pattern of stars present in each image, then registers and stacks the images, subtracts darks and/or bias, and does a flat-field correction, all by itself. Recently a comet stacking mode was implemented. From the January 2008 version, Deep Sky Stacker can even operate "on the fly," while images are downloaded from the camera to the laptop. Results can be stored in a 3×32-bit FITS or in various TIF formats. Deep Sky Stacker for stacking, and Fitswork for the final image-processing are a perfect software duo in deep-sky DSLR work. All images shown here, taken by Hartwig Lüthen with Petra and Jörg Strunk, were processed this way.

Other freeware, which does the stacking of DSLR images, is *Giotto* (see Sect. 6.7) and *IRIS* (see Sect. 6.1). Among the commercially available software, we could name *Maxim DL* (including Maxim DSLR), *Images Plus* and *Pixinsight* (free Light Edition available!).

With the color correction and contrast enhancement of the final image, use of a noise reduction routine may also be adequate. Fitswork has a color noise filter and a wavelet noise filter, which could be used in succession. Another software for careful noise suppression (including graininess in scans of conventional film) is Neat Image. But keep in mind that these operations will also reduce image information, so apply them as the very last step of image processing.

6.5.3 The Advanced Approach: Cooled CCD Photography

General. Cooled astro-CCD cameras are capable of longer exposures than DSLRs, as their dark current and thermal noise is even lower than that of the best DSLRs. Exposures of up to 30 or 60 min are possible but risky, in terms of guiding failures and nasty overflows from over-exposed stars. On the other hand, in stacking of very many shorter exposures, as proves to be a good method for DSLRs (see Sect. 6.5.2), an unnecessary amount of read-out noise is accumulated. Hence, the best compromise for cooled CCDs is to add a moderate number (i.e., 5 to 10) of exposures in the range of 3 to 10 min duration.

For optimum results, each exposure series requires a series of *darks* (with the shutter closed, duration and operation-temperature unchanged) and a good *flat-field*. The averaged dark gives the actual dark-current of the CCD, including hot pixels, and it is simply subtracted from each RAW image (the same technique can be applied to DSLR images). Flat-fields are (dark-corrected) exposures against a diffuse, white background, which record uneven illumination over the CCD field and any pixel-to-pixel variation. The dark-corrected image is divided by the flat-field.

Among the cooled astro-CCDs, there is a choice of "single-shot-color" and monochrome (B&W) chips. The latter provide by far the deepest images. Apart from their lower thermal noise and longer possible exposure times, an important advantage of a monochrome CCD is that it uses *all* photons in the visual spectrum. By contrast, single-shot-color CCDs and DSLRs have a mask with filter-triples (RGB) on their chip to separate the color information, and each of their individual pixels receives hardly 1/3 of the photons. The huge sensitivity advantage of monochrome CCDs makes them the best choice for serious astrophotography, even though color-photography is much more effort (see below). Cooled, monochrome CCDs are also the best choice when used in combination with a spectrograph (SBIG offers two smart constructions).

DSLRfocus:
http://www.dslrfocus.com/

Fitswork:
http://freenet-homepage.de/
JDierks/softw.htm

Deep Sky Stacker:
http://deepskystacker.free.fr/
english/index.html

Images Plus:
http://www.mlunsold.com/

Maxim DL and Maxim DSLR:
http://www.cyanogen.com/

IRIS:
http://www.astrosurf.com/buil/us/
iris/iris.htm

NEAT Image:
http://www.neatimage.com/

PixInsight: http://pixinsight.com

Fig. 16. NGC 4038 (Antenna Galaxy), 24″ Hypergraph f/8, ST10 XME, LRGB photo taken from Namibia by Rainer Sparenberg. L: 20 × 5 min (full resolution), RGB: 5 × 5 min (2 × 2 binning), each (From *Sterne und Weltraum* photo archive No. c004461, in landscape format, for 1/2 page)

Monochrome CCDs with a non-antiblooming gate (non-ABG) offer the best linear response to the incoming light. Equipped with Johnson UBVR filters, and by carefully avoiding over-exposure, these CCDs can be used for precise photometric measurements (see Chap. 8 and Sect. 14.1). Antiblooming gates, by contrast, suppress the electron overflow from over-exposed pixels in read-out direction. Nicer-looking pictures are the result, but photometric use is not straight-forward (an accurate calibration of the CCD response-function is required) and less precise. Non-AGB CCDs, on the other hand, can produce really nice pictures, too, provided, you carefully avoid over-exposure and, rather, stack a larger number of shorter shots.

Color photography with a monochrome CCD is done by means of individual exposures through standard R (red), G (green) and B (blue) filters mounted on a filter wheel in front of the CCD. These are specific, defined filters (ordinary color filters would not do) and are different from photometric B,V, and R Johnson filters. To achieve equal signal-to-noise (S/N) levels, B exposures must be longer than R and G by a factor of 1.3 to 1.6×, depending on the individual CCD's blue-sensitivity. In order to avoid overly long exposure times, a 2 × 2 binning may be used.

The best results are achieved, when the image is assembled in the LRGB color system and deep, unfiltered (monochrome) images of the highest resolution (no binning) are taken to provide the intensity information to the extra L channel. To cut off IR-light and for consistent focus, you want to complement your RGB-set with a matching, clear-looking L filter. Stunning, low-noise images result from taking and stacking a whole series of exposures for each LRGB channel (see, e.g., Fig. 16). But be aware that producing a single image of this high quality can easily take more than one clear night!

The LRGB color-system is superior to RGB in S/N because RGB image-intensity is simply added from its three filtered channels, neither of which can reach as deep as the unfiltered L image. In every image processing software (e.g., IRIS, Photoshop, Astroart, Maxim DL) you can switch between these two color systems. Once you have your final L, R, G, and B images ready, (i.e., after dark- and flat-field correction, and stacking), register them to each other and feed them into their respective channels. Finally, tune the color balance of the assembled LRGB image in order to get a neutral to blue-gray background sky, with stars mostly appearing white.

Color photos of faint H_α nebulae present a special case. In order to separate the nebula's line emission from the continuous (in wavelength) background glow, very long (about 30 min) exposures with an narrow-band H_α filter should be taken. Feed these into the R channel (instead of an R-filtered image, but do B and G as normal). The L image is best made from a mixture of the H_α and G images. This method yields astonishing results (see Fig. 1) even under quite light-polluted skies! An R-filtered image may still be taken, to be mixed with the H_α image for a more realistic R channel.

High-quality LRGB and narrow-band filters (Astronomik, Germany) are available from:
http://www.astro-shop.com (Astro-Shop, Germany)
http://www.gerd.neumann.net (Gerd Neumann, Germany)

Other sources of filters and filter-wheels:
http://www.sbig.com (SBIG, CA)
http://www.lumicon.com (Lumicon, CA)
http://www.optecinc.com (OPTEC, US)

Commercial astrophotography image processing software:
http://www.optcorp.com (Maxim DL, CA)
http://www.msb-astroart.com (Astroart, Italy)

6.6 Photography of the Moon and Sun

6.6.1 Focal Images: Capture of Lunar Phases

The focal image of the moon fits best inside the typical DSLR chip-size, when focal lengths of 1.5 m are used (or 2 m for phases before first and after third quarter). For SLRs 24 × 36 mm film format, 2 m (3 m) are best. Near-optimum focal lengths are easily provided by the prime-focus of many telescopes, at least with the additional use of a focal reducer (as with a DSLR on a SC-telescope) or a 2 to 3× magnifying teleconverter lens (as with Newton reflector or short apochromat). In combination with a sensitivity of only 100 ASA for highest resolution, stunning portraits of the entire lunar phase are taken (see Figs. 17 and 18), which cannot be reproduced here in their full detail.

The obvious project, which comes to mind, is to create your personal lunar phase atlas (see also Sect. 13.2). In addition, close conjunctions of the moon with planets and bright stars can be documented. Longer (in the range of 1 to 10 s), motor-driven prime-focus exposures with higher sensitivity (400 to 800 ASA) are capable of recording the faint lunar detail (maria and the brightest craters) in the earth-shine of a thin crescent or the reddish glow of a totally eclipsed moon (see Fig. 19).

The optimum exposure time depends on the lunar phase and the effective f-ratio. It falls in the range of 1/2 to 1/500 s, and a motor drive is not required, if the exposure is shorter than 1/8 s. Integral exposure meters can fail to find the right exposure time, as they get fooled by the dark background around the moon. Hence, you need a good spot-meter. DSLRs allow you to check the image quality immediately after taking. In the "info"-mode you can get an exposure histogram. Its broad peak should be a bit over half-way up the scale.

If you work, e.g., with an entirely mechanic SLR, you might appreciate the following simple formula to calculate the required exposure time t in seconds, for given sensitivity (taken in ASA), f-ratio number $N = f/A$ and a sky of normal transparency:

$$t = \frac{N^2}{\text{ASA} \cdot K} \, .$$

(3)

Fig. 17. Moon 4 days old, 7″ Maksutov–Newton at $f = 2\,\mathrm{m}$, DSLR Pentax Ist, 100 ASA, 1/10 s (KPS)

Here, f is the total focal length of your optical system (i.e., considering any reduction or multiplication factors to the primary focal length), A is the telescope aperture in the same units as f, and K is an object-dependent factor. K ranges from 200 for the full moon down to 2–5 for a thin crescent. For interpolation, match $K = 20$ to the half-moon and 40 to a gibbous phase. The much darker earth-shine and a totally eclipsed moon both require a $K \approx 0.01$, Mars and Jupiter between 40 and 60, Saturn 10, and Venus 200–800 (depending on its phase). Using film, where you cannot check the results until it is

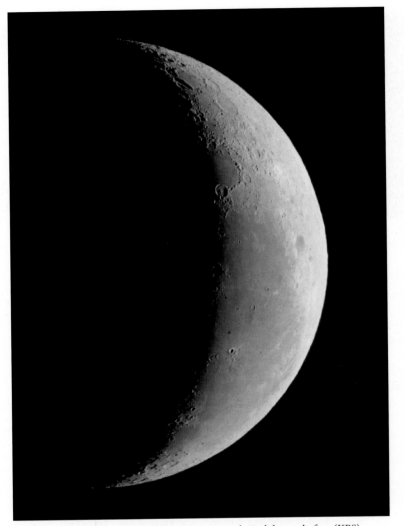

Fig. 18. Moon 4 days before new moon, technical data as before (KPS)

too late, be cautious and do exposures of half and twice the calculated time, since the transparency is difficult to judge.

Focusing is particularly critical for this kind of lunar photography, because of the small image scale. Any slight miss of the exact focus reduces the richness in detail significantly. Hence, extra magnification by a right angle viewer and, if available, a fine matscreen are very helpful. Still, you will find that there is a range within which the image looks fine in the view-finder. Go slowly back and forth with the focuser to either side, where the image starts to blur a little, and then set it right in-between. In most nights, you may also notice the "seeing" from turbulent, inhomogeneous air. Hence, take a series of at least 10 shots, with a new focus every two or three pictures, to be lucky once with optimum focus in a moment of steady air.

Fig. 19. Earth-shine with waning crescent (*left*), 7″ Maksutov–Newton f/6, DSLR, 400 ASA, 2 s, and total lunar eclipse (*right*), 14″ Newton f/6, OM1, Ektachrome 400, 6 s (KPS)

Vibration from the flip-mirror, which normally flips up a split-second before the exposure, can ruin crisp lunar detail despite all precaution taken by using a cable release. But most DSLRs/SLRs have an option to flip up their mirror several seconds prior to the exposure—long enough that the vibration can damp down. It often comes with the time-release function, but the old OM1 has a simple knob to flip the mirror manually. If there is no such option, you can resort to the old "hat-trick": Set the camera to B, hold a dark hat or piece of cardboard in front of the telescope, release the shutter, then expose manually by a fast swing-out of the hat and back, then shut the exposure. In this way, you can do exposure times from about 1/10 to several seconds and suit, at least, a longer focal ratio or a higher magnification (see below).

6.6.2 Detailed Images at Higher Magnification

Obviously, a significant increase of the effective focal length f will lead to a much enlarged image of a only small part of the moon and deliver finest details. You may find, though, that planet photos taken with telescopes under a 5-in. aperture remain a bit disappointing, but the moon is a rewarding target with even the smallest refractor. Hence, photographic studies of individual lunar regions are an obvious application, and time-series of photos taken every hour show nicely, how the shadows gradually change.

Diffraction-limited resolution and the dilution of light both set practical limits to the magnification. According to the Nyquist theorem, there should be at least 2, better 3 pixels (or film grains) across any smallest image detail (the telescope resolution δ, here given in "). This leads to an optimum focal length f_{Nyq} (in meter), which depends on your pixel size d_{px} (in micron):

$$f_{Nyq}[m] \approx 0.6 \cdot d_{px}[\mu]/\delta["] . \tag{4}$$

Hence, for a combination of a 5-in. refractor (1″ resolution) with a DSLR camera (pixel-size 6 μ), you want at least a 3.6-m focal length to get the full resolution. That yields

$N = 28$, since $A = 0.13$ m. About twice that focal length suits, in practice, the inferior resolution of 100 to 200 ASA film.

Precise focusing is rather easier with higher magnification, but exposure times t rise quadratically with f and N (see Sect. 6.6.1)! At the same time, you want to keep t as small as possible, under 1 s, in order to minimize the blurring effects of scintillation ("seeing"). Hence, the best practical choice of f does not exceed f_{Nyq} too much. In any case, the resulting exposure times demand a motor-driven, equatorial mount (with the possible exception of the Sun and Venus). There are two ways to multiply the focal length of the telescope by a desired factor:

Eyepiece-projection. This is the preferred method for any SLR/DSLR, from which you can remove the lens. You need a special projection-adapter between the telescope and your camera body, which in its front part accepts the eyepiece. It offers some reasonable projection distance between the chip or film and the eyepiece (see Fig. 20 left) and is available from most larger astro-dealers. The magnification factor x, by which the telescope focal length is multiplied, is simply given by $x = s/f_{ep}$, where f_{ep} is then focal length stated on your eyepiece, and s is the distance from the outside eyepiece focus (roughly out by half of f_{ep}) to the chip or film.

To determine s, reckon a path-length of 45 mm inside the typical DSLR/SLR camera body and measure the back-side distance of the eyepiece's eye-lens in the projection-adapter (in mm), minus $f_{ep}/2$. In order to get desirable magnification factors of x in the range of 3 to 8, use eyepieces of $f_{ep} \approx 20$ to 8 mm, respectively.

Afocal use of a camera. This is the preferred method for a standard digital camera with integral lens. Use a special clamp (as provided by larger dealers like Teleskop-Service, see Fig. 20 right) to mount it behind your telescope eyepiece. The best focus can be found simply by looking at the camera life-display, but most digital cameras offer an autofocus with on-the-chip verification. Hence, just prefocus the eyepiece visually before mounting the camera and let the autofocus do the rest. In any case, set the camera into the mode

Fig. 20. Camera adapter for eyepiece-projection (*left*). Clamp to mount a digital camera behind the telescope eyepiece (*right*)

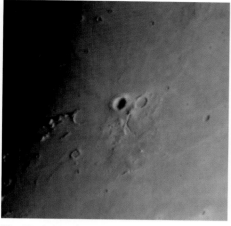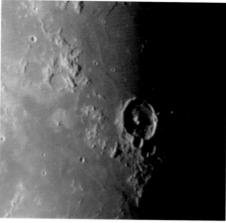

Fig. 21. Aristarch-region on the waning crescent, 7″ Maksutov–Newton at f/25 with DSLR, 200 ASA, 1/4 s (KPS, *left*). Gassendi-region, 11 cm Refractor f/15 with 10 mm Vixen LV 10 mm eyepiece, afocal digital camera w/ f_{cam} = 22 mm to give f = 3.5 m, 200 ASA, 1/6 s (*right*)

in which it automatically chooses only the exposure time, and use manual control to set the aperture wide open.

Here, the magnification factor x is simply given by the ratio f_{cam}/f_{ep}. You will find that ratios of 2 to 4 are already sufficient (as an example, see Fig. 21 right), because the standard digital camera has a much smaller chip and pixel-size than a DSLR. Hence, optimum focal length in terms of resolution and magnification can be shorter, and so are the exposure times.

Digital cameras with a fixed focal length require high-power eyepieces (f_{ep} ≈ 4 to 6 mm) for optimum magnification, but they offer two advantages over cameras with an optical zoom: (1) Their simpler photolens cause fewer reflexes, and (2) their aperture stop is closer to the front lens. Zoom-lenses, by contrast, have the aperture stop deep inside. This causes serious vignetting of the field-of-view, unless you use a special eyepiece with generous eye-relief (20 to 25 mm), as required by observers who need to keep their glasses on.

Needless to say that an absolute vibrationless shutter release (see Sect. 6.6.1) is very important. Because of the seeing, you want to take a series of pictures (a whole film full, or 40 to 60 digital exposures), from which you select the few photos taken in a good moment. Read about more sophisticated techniques to deal with seeing in Sect. 6.7, employing a webcam or video-camera.

6.6.3 Solar Photography

In white light, we observe the solar "photosphere" (see Sect. 13.1). It gives the false impression of a sharp solar limb, because its true extent (< 0.3″) is not resolved. As far as the optical methods are concerned, everything described in Sect. 6.6.2 applies to solar photography, too. A SLR/DSLR camera in the telescope prime focus is best to record the entire Sun, e.g., for positional measurements of sunspots. Such measurements require

Three ways to use a camera with a telescope

1. Focal: DSLR/SLR in prime focus, w/o camera lens; use for your lunar phase atlas. You may use a focal reducer or teleconverter lens to optimize lunar image size.

2. Eyepiece projection: DSLR/SLR and projection adapter, w/o camera lens; use for lunar detail and the planets.

3. Afocal: simple digital camera (w/ camera lens) mounted behind the telescope eyepiece; use for lunar detail.

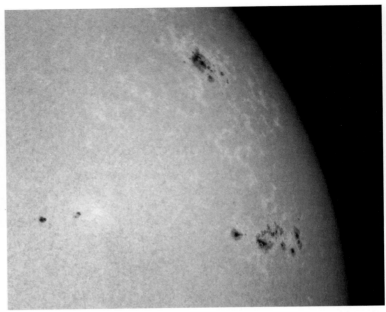

Fig. 22. Sunspots and faculae near the limb (July 10, 1981), 85 mm Refractor at f/50, front-filter 1:1000, 1/250 s, Agfa Ortho 25 (KPS)

precise east-west alignment of the camera, which can be verified by letting the solar image drift without drive. Detailed images of individual sunspot groups (see Fig. 22 and Sect. 13.1), possibly resolving photospheric granulation and filamentary penumbra structure, as well as day-to-day documentation of the evolution of a sunspot group, are done by eyepiece-projection. Any larger magnification is not limited by a lack of light, rather, we need to consider how best to deal with a dangerous over-amount of sunlight!

Fighting the glare. In fact, any glimpse of unfiltered light with the telescope, or the use of an inadequate filter, can permanently damage your eyes! Concentrated sunlight from your telescope can easily set fire to black paper. Hence, simple black filters screwed into the eyepiece barrel are very unsafe because they do burst in the heat! The only safe and sophisticated way to reduce the sunlight is by a reflective front-filter, or by using the partial (4%) reflection from an uncoated prism surface, in combination with a dark filter behind it. Both methods avoid converting solar light into heat. The latter, the "Herschel Prism," may be built by a DIY-enthusiast (mainly, you need to mount a right-angle prism "the wrong way round").

For reflecting front-filters, two types of products are offered on the astro-market: the relatively low-cost Baader filter foil and aluminized glass filters. The latter vary a lot in cost and optical quality but last. The filter foil by Baader Planetarium (see Fig. 23 left) is quite sensitive to rough use, but it is of good optical quality, safe, very economic and easily mounted or replaced. There are two choices of density available: The standard foil reduces the sunlight by a visually convenient factor of 100 000. It is quoted with a logarithmic "density" measure of 5.0. Except for prime-focus photographs, such a high filter factor causes unnecessarily long exposure times. Hence, a photographic version with a factor of 1:6300 (density 3.8) is also available. Used for detailed photographs in eyepiece-projection, exposure times in the desirable range of 1/1000 to 1/250 s result (e.g., for 100 ASA and $N \approx 50$). If your camera exposure meter is not up to the job, you can compute the right exposure time t approximately with the formula given in Sect. 6.6.1, by using $K_{Sun} = 70,000,000$ and multiplying with the filter factor.

Fighting low contrast. The other problem in solar photography is the lack of surface contrast. Fine structure is created by local temperature variations of less than 20% (!). With $I \propto T^4$ that translates into a relatively subtle intensity contrast of under a factor of 2. Hence, ordinary film and DSLR solar images look disappointingly soft and featureless! In the film era, there was a choice of several high-contrast technical emulsions designed to, e.g., record documents, like the Kodak TP 2415 or the Agfa Ortho 25. These superb, extremely fine-grained films have been discontinued since. The best you can do today with a SLR is to use ordinary B&W fine-grain film, such as the Ilford Pan F, and enhance contrast by using a developer for photographic B&W paper. With digital cameras, one soon finds that JPEG images do not take contrast enhancement sufficiently well, either. For improvement, set the camera output to RAW (which requires much more storage space because it provides deeper intensity information). Even better, you stack 10 to 20 selected exposures from larger series, to reduce S/N before enhancing the image contrast.

Special instrumentation for the H_α-chromosphere. The very interesting and dynamic, about 3″ thick solar chromosphere is of insufficient density to be visible in white light. But opacity (absorption efficiency) is sufficiently high, by several orders of magnitude, in the strongest absorption lines of the solar spectrum; best is the red hydrogen H_α line. Chromospheric fine structure is then visible in absorption against its underground (see Fig. 24 left). However, for sufficient contrast, an extremely small bandwidth of under 0.07 nm (0.7 Å, better even 0.5 Å) is required! Such specialized instrumentation borders on that of professional solar observatories and is not cheap. The least expensive options are offered by Coronado, in form of the small PST ("Personal Solar Telescope," see Fig. 23 middle) and Coronado H_α filter systems.

With a much more economic H_α filter of 4 to 10 Å bandwidth, not even a trace of chromospheric fine-structure would be visible. But this bandwidth is still capable of delivering an excellent contrast between solar prominences on the solar limb against the background sky. A sophisticated instrument, with which such a filter is used, is the prominence viewer after the Lyot principle (see Fig. 23 right, and Sect. 13.1). It eliminates a large fraction of the telescope straylight, and a highly reflective metal cone occults the glaring solar disk. The view resembles that of a total solar eclipse (but all in red, and no

Fig. 23. Solar front-filter from Baader foil (*left*). Coronado PST (*middle*). Prominence viewer (*right*)

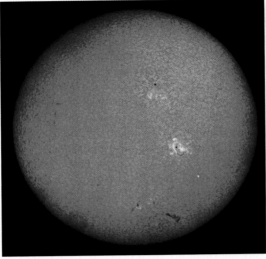
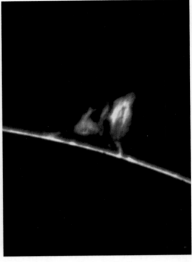

Fig. 24. The solar chromospheric disk in H_α light (*left*). Prominence, taken with 10 cm Refractor, prominence viewer and SLR, TP2415, 1/60 s (KPS, *right*)

coronal detail), higher in resolution and more brilliant than any H_α-disk-view. Unfortunately, the corona is invisible in H_α light and you need the opportunity of a total solar eclipse to see it (see Sect. 6.7).

In order to employ the Lyot construction principles, prominence-viewers must be master-taylored to match a specific telescope, and there is no off-the-shelf solution. But custom-made, high-quality models are available from an experienced German instrument-maker, Wolfgang Lille (http://mitglied.lycos.de/LilleSonne).

Photographic use of both types of H_α instrumentation is straight forward with a SLR camera and color emulsion, also with a *modified* DSLR (see Sect. 6.2.2). Exposure times are in the range of 1/30 to 1/125 s. As in white light, low contrast is a big problem for surface details, but off-limb prominences look great (see Fig. 24 right). Even a webcam can be used, but only with the stacking techniques and series of selected images as described in Sect. 6.7.

Warning 1: Never look at the Sun without a proper filter! That can permanently damage your eyes.

Warning 2: Never use black, absorbing filters in the full prime-focus sunlight. They may burst in the heat while you observe!!
Safe and high-quality Baader-type solar front-filter foil and Herschel prisms are offered by http://www.baader-planetarium.de and http://www.teleskopservice.com.

6.7 High-Resolution Techniques

Low-cost webcams have revolutionized high-resolution imaging of planetary and lunar surfaces; and all this despite low resolution and mediocre S/N of individual webcam images. The trick is this: large frame rate, short exposure time and small data-size allow a stream of some thousand webcam images to be taken within a short time. They are then automatically selected for the best fraction (between about 4 and 30%) and stacked by special software. This method copes well with turbulent air (seeing) and provides, in addition to the maximum resolution of the best moments, a much improved S/N ratio of the final image. The latter point is important for contrast enhancement and sharpening of very subtle planetary detail.

Hence, a well-chosen webcam (see Sect. 6.2.3) can record details close to the resolution limit of the telescope, but, of course, only within a small field-of-view. The webcam technique can also be used well in solar photography, which is plagued with stronger daytime seeing and low contrast.

6.7.1 Best Focal Length

In Sect. 6.6.2, we have given a formula to calculate the minimum focal length f_{Nyq} required to maintain the full resolution of the telescope, according to the Nyquist theorem and a recommended, higher than minimal sampling of 3 pixel per resolution element. Since the theoretical resolution limit of a telescope with aperture diameter A is roughly given by the Dawes limit $\delta[''] \approx A[mm]/120$, we may change this formula into:

$$f_{Nyq} \approx 5.2 \cdot d_{px}[\mu] \cdot A , \qquad (5)$$

where here aperture A is in the same units as f_{Nyq} (e.g., both in meters), and the pixel size d_{px} is, for convenience, to be taken in micron.

The theoretical minimal sampling factor is 2, but oversampling factors of up to 3.5 have been recommended. A longer focal length requires longer exposure times and suffers more of bad seeing, or it needs higher, noisier gain settings. In any case, the larger image cannot reach any better resolution but probably turns out worse.

Hence, for optimum results, a 8″ (0.2 m) telescope and a Philips ToUCam Pro webcam (pixel size 5.6 micron) require around 6 m focal length, or a focal ratio between f/25 and f/35. With an 8″ f/10 SCT, this is achieved by a 3× Barlow lens (or 2× Barlow with additional extension, see Fig. 25). A larger factor and perhaps eyepiece projection is required with a fast telescope (e.g., a Newton).

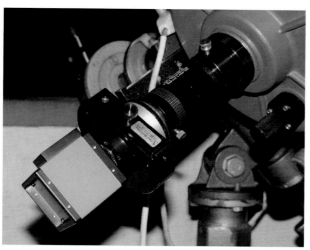

Fig. 25. Setup for high-resolution video-photography: 8″ SCT (f = 2000 mm) with 2× Barlow lens, with additional extension giving an effective focal length of 5.1 m, Astronomik "filter-drawer" and DMK 21AF04 firewire webcam

6.7.2 Filters

IR block filters or UV/IR block filters are routinely used because refractors, as well as any other lens in the optical system (i.e., Barlow or eyepiece) suffer of noticeable chromatic aberration outside the visual spectrum. But webcams and industry cameras like the DMK 21AF04 are quite sensitive to infrared light, which blurs any image in visual focus.

IR pass filters. The impact of seeing has been found to depend inversely on the wavelength of light. This effect can be exploited by using the IR-sensitivity of CCD sensors in combination with an IR-pass filter, which blocks all visual light. Although the theoretical resolving power of a telescope is lower at larger wavelengths, this is overcompensated by the reduced seeing in the near-IR. In addition, some objects display a higher contrast in IR light. This is especially true for Mars, where an IR pass image fed into the R channel or into the L (luminance) channel greatly enhance the image quality. IR pass filters are offered by Baader (650 nm cutoff) and by Astronomik (choice of 742 or 807 nm).

RGB filters. Some industry cameras like the DMK have monochrome sensors. Color imaging requires the combination of three consecutive shots through a red, a green and a blue (RGB) filter. With planets, work as rapidly as possible to avoid shifts from rotation, especially with Jupiter!

6.7.3 Taking Videos

K3CCDops and Giotto are useful webcam image recording programs for any cameras using DirectX drivers, and DMK supplies an excellent recording software of their own with their cameras. After careful focusing, set the exposure time and gain for the camera. As a guideline, ToUCams work well with 1/25 to 1/30 s and a frame-rate of 5 to 10 images per second. A DMK 21AF04 is best used at 1/33 to 1/72 s and a frame-rate of 30 to 60 per second. Be careful when now adjusting the gain: No part of the image should reach saturation, and any underexposure will result in the onion ring phenomenon when sharpened in image processing. For best control, use the live histogram function, as provided by the recording software. The recorded stream of images is then saved as individual BMP files or, more conveniently, as an AVI video file. In the latter case, make sure to avoid any data compression, since that degrades image quality.

Do take a large number of images; a typical series will consist of 2000 or more. But the rotation of the planet and data storage capacity both set upper limits to the duration of the shot. A stream of uncompressed webcam RGB images (about 1 MB each) quickly arrives at several gigabytes. There cannot be enough hard disk space on your laptop! Nevertheless, do not just delete previous videos from your hard disk but regularly burn older raw material on DVD. You may be able extract more from your AVIs in a few years, after further improvement of image processing software.

6.7.4 Image Processing and Analysis

Selection and stacking of many images: The key to top quality! Poor seeing is the most limiting factor in planetary astrophotography. In the days of film astrophotography, a typical Jupiter image was exposed for several seconds. The results were disappointing, because turbulent air blurs the image detail during such a long exposure time. Modern webcams,

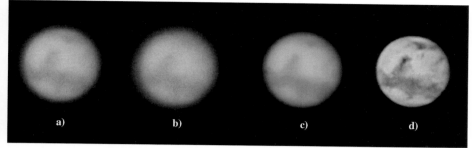

Fig. 26. Processing 2000 video images of Mars, Dec. 15, 2007. (**a**) Best image, evaluated by Registax. (**b**) Worst image (seeing was excellent). (**c**) Stacked image. (**d**) After wavelet sharpening (HL)

however, are sensitive enough to permit short exposure times of 1/25 s, short enough to "freeze" the seeing. Nevertheless, any video will still include a lot of images from bad moments, and the small fraction of virtually unaffected frames (see Fig. 26a), which hold the full resolution of the telescope, has to be identified and separated. In addition, stacking a batch of N images then increases the signal-to-noise ratio by a factor of $\sqrt{N}$. For example, stacking 4, 16 or 100 images reduces the noise by a factor of 2, 4 or 10, respectively.

So, you want to select only those very best images, but at the same time, you want to stack as many good images as possible for reducing the noise. Software like Registax or Giotto does the evaluation and stacking automatically. But, one must try out, how selective the program should be, in order to retrieve the largest possible fraction of fully acceptable images in a particular video.

The resulting stacked image (Fig. 26c) is virtually free of noise. While looking a bit blurred and soft at first, it contains all image information. Sharpening algorithms like unsharp masking, wavelet sharpening (Fig. 26d) and Gauss sharpening are used to extract this information and to yield an enhanced image with a wealth of detail (Fig. 27).

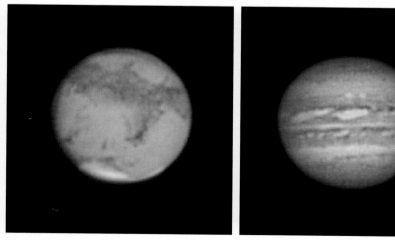

Fig. 27. RGB image of Mars (*left*), from the same sequence as before, mounted in Fitswork, taken with 10″ f/10 SCT at $f = 7.2$ m, DMK 21AF04, Astronomik IR pass (for R), B and G filters (HL). Jupiter, as before but with Philips ToUCam (*right*), processed with Registax (HL)

6.7.5 Dedicated Software

Registax, available in Version 4 at the time of the writing, is a software mainly dedicated to webcam-based lunar and planetary work (Fig. 28). It can load an AVI video file, and one or several reference points for alignment (registering) are selected by the user. The software then registers the images and automatically evaluates the quality of each exposure. In a second step the minimum quality level for stacking has to be set, to ignore all but only the better frames. After stacking, you will be able to sharpen the images with the layer wavelet technique. Wavelet filters are sharpening filters which work on different radii, and you can carefully fine-tune their action on detail of different scale-length. The current version of Registax now supports multipoint alignment, which performs a selective alignment and quality evaluation in individual areas of the image. In this way, information from sharp but deformed (by seeing) images can still be extracted and used.

Giotto was the first software for video-astronomy. It was developed by Georg Dittie, one of the early pioneers who paved the way for webcams and video-astrophotography in the 1990s. This German-language software package addresses all problems of video work, from video recording and stacking to sharpening and image processing.

In contrast to Registax, Giotto can also be used in the recording of videos. In its latest version DirectX is supported, thus virtually any digital video source (webcams, framegrabbers, camcorders and industry cameras) can be handled. This makes Giotto one of the most flexible video recording programs available, and best of all, it is freeware!

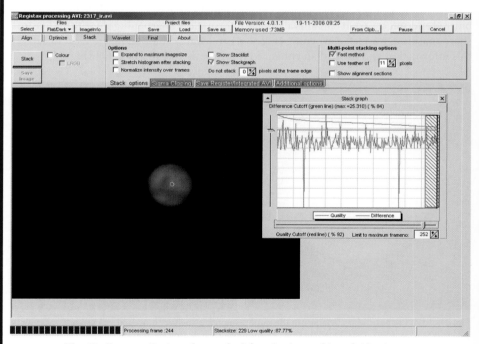

Fig. 28. Freeware Registax (screenshot) for selective stacking of video images

 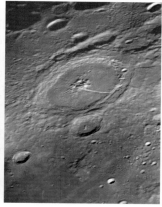

Fig. 29. Mosaic of crater Copernicus from nine individual images, by Autostitch (*left*). 8″ SCT operating at $f = 5200\,mm$, DMK firewire camera (HL). Crater Petavius (*right*), mosaic of three images, otherwise as before (HL)

Stacking is a bit slow in Giotto, but it requires a fewer number of parameters to be set by the user than does Registax. But on the other hand, that gives the user less control over the process. He can specify that Giotto uses only the best, say, 20% of images in a video for stacking, but it is not possible to examine how these look. On the other hand, Giotto is more flexible when it comes to applications off the beaten track; it proves useful in otherwise hopeless cases.

JUPOS and IRIS for final analysis. The final image can be analyzed by software like JUPOS or IRIS, which both can automatically derive Mercator maps from series of high-resolution planetary images, or render a rotating globe animation from the data.

Preparing mosaics with Autostitch. The tiny webcam chips can cover only a small part of the lunar landscape when used at optimum resolution. A mosaic of adjacent images is the way to larger lunar images. Autostitch is a software, which does this job: It analyzes a series of overlapping images, renders precise mosaics (see Fig. 29) and adjusts any different levels of brightness and contrast.

6.8 Documentation of Special Astronomical Events

6.8.1 Solar Eclipse Photography

Many amateur astronomers regard total solar eclipses as major events in their life. They travel across the globe to witness the miracle of those rare moments of daytime-darkness. The astrophotographer's desire is, of course, to document the event. But there are several problems for which a successful eclipse chaser has to be prepared:

1. Careful astrophotography, and focusing in particular, normally requires much more time than the few minutes of totality provided by the swift passing of the lunar shadow. This makes eclipse-photography a real challenge!
2. Air travel restricts the equipment you can bring to the event!

Registax:
http://registax.astronomy.net/
Giotto:
http://www.videoastronomy.org/

K3CCDOps:
http://www.pk3.org/Astro/

Astronomik filters (IR pass and block, LRGB, and more):
http://www.astronomik.com/

Schuler UV pass filters:
http://www.astrodon.com

JUPOS:
http://jupos.privat.t-online.de/index.htm

Autostitch:
http://www.cs.ubc.ca/mbrown/autostitch/autostitch.html

Must-see webpage of high-resolution photographer Damian Peach:
http://www.damianpeach.com/

3. A flimsy tripod is easy to carry but may spoil your images. Also, make sure that your mount or tripod can actually access the position of the sun in the sky at the moment of totality! Consider to have a lightweight, motor-driven mount.

4. Choose an adequate focal length: A simple camera equipped with a wide-angle lens can record the scenery, the movement of the lunar shadow and the sky colors. Corona and prominences at higher resolution require a telephoto lens. But do not use more than 500 mm focal length with a DSLR, or the outer corona may not fit the field-of-view.

5. Focusing is the major hurdle for eclipse photography, especially with DSLRs. Some lenses, like popular Russian Maksutovs, show a thermal focus drift, and a front-filter of glass can cause a general shift of focus. Hence, as temperatures drop rapidly during eclipse, check focus only just before totality, on the remaining thin crescent Sun, with the lens still covered by a Baader filter-foil (see Sect. 6.6.3).

6. Your eclipse equipment and its operation should be thoroughly tested under realistic conditions well ahead of your trip. Try it on the Moon, in a similar position in the sky as the Sun during totality. By this you can identify and eliminate problems with, e.g., balance, vibration, and focusing. Finally, exercise timed rehearsals of your program.

7. When the moment has come, expect the unexpected: Clouds may move through and force you to modify your exposure sequence. Cheering spectators may be so loud that you don't hear the click of your camera. Hence, practice operating your equipment not only in darkness but also without acoustic feedback.

8. Do not forget to remove the solar filter 20 s or so before second contact (one of the most common mistakes in solar eclipse photography)! And put it on again shortly after third contact. Some filters require a stop-down of the aperture for not over-exposing the partial phases. In this case, make sure to also change to full aperture shortly before second contact, to avoid accidental underexposure of the corona!

9. Last, but not least, remember to enjoy the unique event visually! Hence, trim your photographic program to be *shorter* than the totality, and keep it simple. Operating one camera properly is better than failing to use five.

What to Photograph *Partial phases* can be recorded easily. Most important is the use of *safe* solar filters (see Sect. 6.6.3). In this case, the photographic version of Baader sun-filter foil will give images too bright for your purposes. With DSLRs and their higher sensitivity, a sun filter for visual use (factor 1:100 000) is best to cover your telephoto or telescope lens.

Second contact. There is a lot to be photographed in these seconds! The diamond ring and Bailey's beads require short to very short exposure times (w/o filter). At the same time, a quick series of wide-angle shots can record the rapid advance of the lunar shadow across the sky, while a video camera pointed to the ground may detect moving shadow bands ("flying shadows").

Totality. Eclipse photography must cope with an enormous dynamic range, and it is important to take a whole series of different exposures to cover it. Solar prominences and the chromosphere require very short exposure times (e.g., 1/1000 to 1/2000 s at 100 ASA and f/6), but the outermost coronal streamers need several seconds (see Fig. 30)! Still longer exposures will show brighter stars around the Sun or even detail on the lunar surface in earth-shine, if sky transparency is excellent. Hence, a motor-drive is very useful.

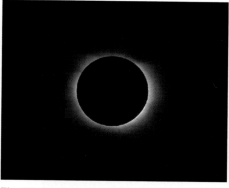

Fig. 30. Inner corona of March 29, 2006, taken with 500 mm f/6.3 lens, Canon EOS 350D, 1/250 s, 200 ASA (*left*). As before, but 1/2 s (*right*) for the outer corona (HL). Inner detail is now heavily overexposed

Plan and practice your sequence with the longest exposures taken around mid-eclipse, shortest exposures at second and/or third contact to cover Bailey's beads and diamond ring phenomena.

Annular eclipses are, by contrast, regarded as boring events. However, nice pictures of Bailey's beads at second an third contact can be taken, through filter foil. When you are placing yourself near one side within an annular eclipse path, these phenomena are less swift. Some daring photographers even took off their filter and recorded solar prominences and the chromosphere near the point of contact, in some cases even the innermost corona. However, be sure you know what you do: Your retina, and even camera shutter and exposure meter may be damaged! And, all such photos have been taken on chemical film; it is not yet known whether CMOS or CCD chips survive this treatment.

Automated Exposure Sequences What a pity, that solar eclipse photographers are so unable to enjoy nature's greatest show while hassling with their equipment. Take pictures *or* watch? Fortunately, modern digital cameras offer remote-control by computer. This provides an attractive option: Take your pictures, ignore the camera *and* observe.

Fred Bruenjes wrote "Eclipse Orchestrator", a nice piece of software for controlling Canon DSLR cameras. First, "Eclipse Orchestrator" determines the precise time and location from a GPS connected to a Windows laptop computer. From this information, precise contact times are computed, even allowing for the lunar limb profile. It then can run scripts for taking a sequence of images relative to the predicted times. The program was tested by various observers during the 2006 and 2008 total solar eclipses and found that it worked very well. There is a similar program for Macintosh computers, Umbraphile by Glenn Schneider.

Coping with the Radial Coronal Intensity Profile No single photograph of the corona could match the visual impression. Unlike human vision, it cannot accommodate for the huge radial intensity gradient (see above). Throughout the 1980s, eclipse photographers tried gradient filters or out-of-focus circular masks in front of the film. But the actual intensity profile of the corona is not predictable and the Sun has to be centered on the filter with high precision.

Fig. 31. As before but stack of different exposures from 1/1000 to 5 s, using Fitswork and its Sekanina Larsen filter. Note faint detail in the lunar disk and stars to 8^m (HL)

The advent of digital image processing made other approaches possible. A sequence of images with different exposure times can be stacked and processed jointly to compress the full range of radial coronal brightness to the small dynamic range of a monitor or paper-print without loosing non-radial detail. Spectacular images have been produced by Wendy Carlos by using Photoshop, and they made it to the front-pages of *Sky & Telescope*. In the 1990s this technique was still based on digital scans of negatives. But since the 2006 eclipse, digital camera RAW images have been used almost exclusively. The use of Photoshop or other packages requires quite a bit of experience in digital darkroom work. There are, however, some very helpful step-by-step instructions on the Web.

Another, probably more straightforward method is the Sekanina–Larsen algorithm (see Fig. 31). Originally, it was designed to reduce the radial density profile in images of comets, to reveal faint jets and streamers in the coma. A free software, which includes this type of processing, is Fitswork. Step-by-step instruction is available on the Web.

6.8.2 Meteors and Meteor Showers

Chemical Photography In deep-sky photography faint objects are given long exposures. But when it comes to meteor photography this approach does not work: The very short time of the meteor visibility and its rapid movement across the sky limit the effective exposure time for a given pixel or film grain. Hence, limiting magnitudes for meteors are much brighter than for stars or nebulae. This problem and the effectiveness of various setups is discussed in the IMO Photographic Handbook. For example, a 28 mm f/2.8 lens is not able to detect meteors fainter than 1.3 mag. For the typical sporadic meteor rate (zenithal hourly rate (ZHR) = 15) this means that this lens (on 24 × 36 mm film) would record a meteor only every 12.5 h! Even in a Perseid maximum night (ZHR = 120) this camera must operate, on average, for 1.6 h before detecting a single event.

Windows software to automate eclipse photography:
http://www.moonglowtechnologies.com/products/EclipseOrchestrator/index.shtml

Umbraphile (Mac-software for controlling cameras during solar eclipses):
http://nicmosis.as.arizona.edu:8000/ECLIPSE_WEB/UMBRAPHILE/UMBRAPHILE.html
http://www.mreclipse.com/ (page for solar eclipse photography)

Technical manual for stacking composite corona images with Fitswork and Sekanina Larsen filters:
http://www.gva-hamburg.de/sofi2006/fitswork/sofi_fitswork_uk.htm

Jerry Lodriguss' technical manual for stacking and filtering composite corona images with Photoshop:
http://www.astropix.com/HTML/J_DIGIT/E_COMP.HTM

NASA Eclipse Bulletins:
http://sunearth.gsfc.nasa.gov/eclipse/SEpubs/bulletin.html

Fig. 32. Leonid meteor storm, Nov. 18, 2001, Corea, ZHR about 2700. 10 min exposure with f/2.8 fisheye lens, Ektachrome 400, push-processed to 700 ASA (HL)

Film meteor photography is still valuable and used today for fireball patrols, since film is able to survey a larger area of the sky for very bright meteors at high resolution. In Germany and in the Czech Republic a camera network of all-sky-cameras is operating. It is able to determine the trajectories of fireballs in the atmosphere and heliocentric meteor orbits. A rotating shutter interrupts the light path every, e.g., 1/50 s, which results in a dash-dotted meteor trail on film and allows the measurement of the meteor speed. Recently, the network successfully determined the impact area of the Neuschwanstein (Bavaria, Germany) meteorite. This effort culminated in the recovery of several large meteorite fragments in 2002.

Despite the limitation of chemical photography to only the very brightest events in a meteor shower, stunning pictures have been taken during meteor storms like the Leonids in 1999, 2001 and 2002, when the ZHR was on the order of several thousands (Fig. 32). With a high fraction of very bright meteors, dramatic images can be taken even with slow wide-angle lenses.

Video Meteor Observation Some video-cameras are sensitive enough to record fainter meteors than could film. A typical setup consists of a Mintron or Watec camera (see Sect. 6.2.3), equipped with a very fast wide-angle lens (e.g., 6–8 mm f/0.8 lens, Fig. 33). Such a system can detect third magnitude shooting-stars, which is 5 magnitudes or so fainter than a still-camera operating on a similar field-of-view with film. Some meteor video-astronomers use image-intensifiers in combination with video-cameras and investigate even fainter meteors. Depending on the population index of the meteors an increase in sensitivity of one magnitude about triples the number of meteors recorded! A disadvantage of image intensifiers is, however, their high price and limited lifetime.

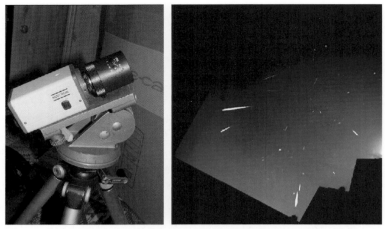

Fig. 33. Mintron with 6 mm f/1.2 lens for meteor patrol (*left*). Perseids and two sporadic meteors, August 8, 2004 (*right*). Same equipment as *left*, images stacked by adding only the brightest pixels of each frame (HL)

And accidentally shining on them with a flashlight may damage or destroy the unit immediately.

Looking at hours of video tape is an annoying procedure, and you will surely overlook many meteors. But there is software which does this work for you. The first one in wide

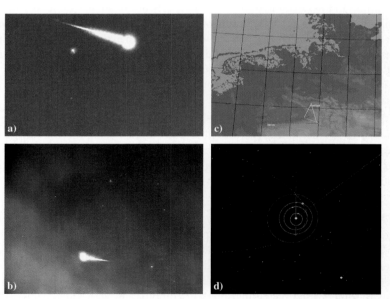

Fig. 34. Bright Orionid fireball, by (**a**) Jörg Strunk (Herford, Germany) and (**b**) Bernd Brinkmann (Herne, Germany), Oct. 22, 2007, using Mintron cameras w/6 mm f/0.8 lens and MetRec detection software. (**c**) Trajectory, computed by UFO orbit, projected on a map of northern Germany, and meteor orbit (**d**)

use is MetRec. It is able to detect meteors in real time, and it even performs astrometry and shower association. There is an international network of MetRec observers. MetRec requires quite old-fashioned hardware (Matrox Meteor frame grabber card) and runs under DOS. However, this hardware is still available second-hand.

The MetRec-output can be fed into and analyzed with a dedicated post-processing tool: Radiant, Rainer Arlt's radiant determination tool, to confirm or investigate new meteor showers. In Japan there is a network of video observers using a similar software called "UFO Capture," which runs under Windows and uses DirectX. UFO Capture comes complete with a post processing tool and an orbit determination for analyzing double station observation.

From the parallax of the meteors, as observed from two stations, the trajectory of the meteor through the atmosphere, and even its heliocentric orbits can be derived (see Fig. 34). However, the accuracy of video double-station work is limited by the low resolution of the footage. Pinpointing a meteorite impact area like in the Neuschwanstein case is difficult to achieve from video double-station data.

International Meteor Organization (IMO):
http://www.imo.net/

Arbeitskreis Meteore (AKM, Germany):
http://www.meteoros.de/meteor/meteore.htm

Fireball network MetRec:
http://www.metrec.org/

Radiant:
http://www.imo.net/software/radiant

UFO Capture:
http://sonotaco.com/e_index.html

6.8.3 Comets

By contrast to deep-sky objects, comets change their appearance from night to night. Surprising leaps in magnitude and rapid changes in structure can occur and make comet photography a thrilling pastime.

Fig. 35. Hyakutake (*left*), April 23, 1996, taken with a Celestron Schmidt-Camera ($f = 225$ mm $f/1.65$) on Kodak TP 2415 film. Hale-Bopp (*right*), April 7, 1997, as before, but on Ektagraphic HC film (HL)

Very bright comets. So-called *great comets* can easily be seen with the naked eye, and their tails stretch over several tens of degrees. However, they are quite rare. The last ones were comet 1975 V1 (West 1976), 1996 B2 (Hyakutake 1996, Fig. 35 left), 1995 O1 (Hale-Bopp 1997, Fig. 35 right), and 2006 E1 (McNaught 2007). Given their angular size and their brightness, a simple fixed camera on tripod with standard or even wide-angle lens (see Sect. 6.3.1) gives good results. Telephoto lenses yield stunning detail of the ion and dust tails, while structures close to the nucleus are best recorded through the telescope.

Reasonably bright comets fail to become naked-eye objects, but can easily be seen in binoculars. Their ion tails may still stretch over $10°$ and more. Relatively short focal lengths can be used; with DSLR cameras fast lenses of 100–300 mm focal length are best. Observing streamers, knots and disconnection events visually is difficult, but they can be photographed quite easily (see, e.g., Fig. 36). By interaction with the solar wind, motion of any feature along the tail is extremely fast and time-series of photographs are very rewarding.

Faint comets are less frequently observed as brighter ones, and their observation is both challenging and of some scientific interest. A 14 mag or even 18 mag comet can still display a tail or show other structure. In 2007 comet 17P/Holmes underwent an outburst of brightness from 17 mag to 2 mag! In such cases, numerous post-outburst shots are available, but the few pre-outburst images are of true scientific value. The example of 17P/Holmes stresses that it can be rewarding for an astrophotographer to document all comets accessible with his/her equipment.

Guiding and stacking comet images. Most comets have a non-negligible angular motion, and their images trail in longer exposures. This is a particular problem when dealing

Fig. 36. Comet 2002 T7 (LINEAR) in May 2004. Mintron with 135 mm f/2.5 lens, 130 images integrated 2.56 s, each, stacked by Fitswork (HL)

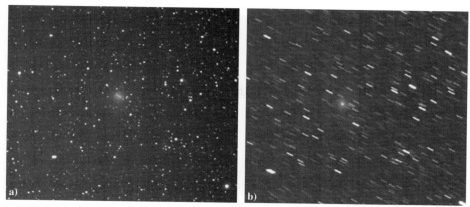

Fig. 37. Comet 8P/Tuttle, Dec. 16, 2007 (1:09–1:33 UT), Canon EOS 350D with 300 mm f/4.5 (HL). Nine images stacked by Deep Sky Stacker, with stars as reference (**a**) and with the comet as reference (**b**)

with faint comets, because these require longer focal length *and* longer integration times. Brighter comets can be guided manually on their false nucleus. Modern autoguiders can track on even modest comets. However, in the last years an alternative technique has emerged: to stack a larger number of comet images of short exposure, using the false nucleus as the reference star for registering provided, the comet appears reasonably bright in the exposures. For fainter comets, a star must be used as reference, but the stellar frame is to be shifted from exposure to exposure according to the comet's proper motion. In either case, stars turn out to be trailed, but the comet is sharp (Fig. 37).

Fitswork and Deep Sky Stacker both can interpret the time stamp in RAW images to precisely derive the comet's apparent proper motion. In the latest Fitswork version, for example, you just click on the comet and on one or two stars, in the first and in the last exposure, and the software computes for each image in-between the comet's displacement in pixels in the image coordinates x and y.

The techniques of median stacking and Kappa-Sigma clipping allow to extract the comet from a maze of star trails. Deep Sky Stacker then allows to refit the comet into the stellar frame, which generates an esthetically appealing image of the comet within pinpoint stars.

International Comet Quarterly: http://cfa-www.harvard.edu/icq/icq.html

VdS Fachgruppe Kometen (in German), lots of images: http://kometen.fg-vds.de/index.htm

Comet Images Yahoo group: http://tech.groups.yahoo.com/group/Comet-Images/

Excellent images by Michael Jäger and Gerald Rhemann: http://www.astrostudio.at

References

1. Covington, M.A.: Astrophotography for the Amateur, 2nd edn. Cambridge University Press, Cambridge (1999)

2. Wodaski, R.: The New CCD Astronomy. Multimedia Madness, Cloudcroft, NM (2000)

3. Rendtel, J.: Handbook for Photographic Meteor Observations. IMO Monograph **3** (1993)

7. The Nature of Light and Matter: Fundamentals of Spectral Analysis

R. Häfner and G. Rupprecht

7.1 Introduction

Our knowledge of celestial bodies results largely from the analysis of the radiation which they emit, absorb, or reflect. Apart from particle emission, there are electromagnetic waves that obey the well-known relation $c = \lambda v$, where $c = 2.99792458 \times 10^8 \, \mathrm{m \, s^{-1}}$ is the speed of light in vacuum, λ the wavelength, and v the frequency. The electromagnetic spectrum spans the range from the ultrashort-wavelength γ rays through ultraviolet (UV), visible, and infrared ranges to the radio waves. The bulk of this spectrum is not observable from the surface of the Earth, because the atmosphere is transparent only within certain wavelength *windows*. This chapter is concerned with the radiation reaching the observer through the *classical* optical window between $\lambda = 300 \, \mathrm{nm}$ to around $\lambda = 1000 \, \mathrm{nm}$. (The much wider *radio window* is discussed in Chap. 9.) With the aid of suitable instruments, this radiation can be decomposed into its spectral constituents, and the spectrum analyzed. The following sections provide a survey of the theory of spectra and the objects, instruments, and several methods of analysis which are accessible to amateur astronomers.

7.2 Theoretical Basics

R. Häfner

7.2.1 Theory of Spectra

The Laws of Radiation The three fundamental laws of spectroscopy attributed to G. Kirchhoff (1859) are:

1. An incandescent solid or liquid body or a gas of sufficient absorption (under high pressure) emits a *continuous spectrum* containing all wavelengths.
2. An incandescent gas under low pressure emits a discrete or *emission-line spectrum*, that is, light at a finite number of specific wavelengths which are characteristic for that gas.
3. A cool gas traversed by white light with a continuous spectrum absorbs those wavelengths from the continuous light which it would itself emit if it were sufficiently hot. These wavelengths are thus missing or weakened, thereby resulting in the appearance of dark *absorption lines* in an otherwise continuous spectrum.

The Sun and most stars display a continuous spectrum. They consist of gases, since their temperatures are much too high for ordinary matter to exist in the liquid or solid state. The spectra also usually show dark absorption lines, indicating that the stars are surrounded by a cooler gaseous *atmosphere*. The latter absorbs from the radiation certain parts characteristic for the composition of the gas. Emission-line spectra are observed in certain objects, and lead to the conclusion that these objects consist of—or are at least surrounded by—rarefied gases.

That stars appear to have distinctly different colors illustrates that the maximum intensity of emission occurs at different wavelengths, a fact which finds its quantitative and qualitative interpretation in *Planck's law of radiation* and in relations derived from it. Strictly speaking, they are valid only for the so-called *black body*, which has certain idealized properties (e.g., thermal equilibrium).

The gases composing a star exhibit a high *opacity*; that is to say, they very effectively absorb the radiation which is generated deep in the interior through thermonuclear reactions and which slowly flows outward. Although the existing temperature gradient within the star and the emission of radiation from its surface, obviously without which the star would not be visible, conflicts with the postulated condition of thermal equilibrium, a star can nevertheless, in first approximation, be considered a black body.

The radiation field of the black body was successfully interpreted theoretically by M. Planck (1900) through the assumption that radiant energy is *quantized*, that is, exists in discrete packages or *photons* of energy $hc/\lambda = h\nu$ (Planck's constant $h = 6.626\,068\,96 \times 10^{-34}$ J·s). The radiated energy emitted by $1\,\text{cm}^2$ of the surface of the black body per second and per unit wavelength at temperature T into unit solid angle is

$$E(\lambda, T) = \frac{2hc^2}{\lambda^5(e^{hc/\lambda kT} - 1)} = \frac{2h\nu^3}{c^2(e^{h\nu/kT} - 1)} \, , \tag{1}$$

where $k = 1.380\,65\,04 \times 10^{-23}$ J·K^{-1} is the Boltzmann constant. Figure 1 shows the energy distribution for several astrophysically relevant temperatures. One can see that the amounts of energy emitted at all wavelengths differ widely depending on temperature. The higher the temperature, the higher the total energy emitted and the greater the shift of the maximum intensity toward shorter wavelengths (hence, hot stars appear blue while cool stars appear red). The latter property is formulated in *Wien's displacement law*, which gives the amount of wavelength shift as a function of temperature:

$$\lambda_{\max} = CT^{-1} \, , \tag{2}$$

where $\lambda_{\max}$ is the wavelength of the maximum intensity. The constant C equals 2897.8 when λ is given in microns and T in Kelvin. Adding the contributions from all spectral regions gives the total energy emitted per second per square centimeter over all wavelengths at a specific temperature T:

$$E(T) = \sigma T^4 \qquad \text{(Stefan–Boltzmann law)} \tag{3}$$

where $\sigma = 5.670\,400 \times 10^{-8}$ J·m^{-2}·s^{-1}·K^{-4} is the Stefan–Boltzmann constant. Both laws can be exactly deduced from Planck's law, although they had been found earlier by empirical studies.

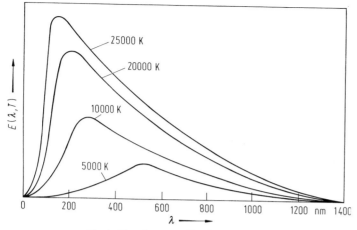

Fig. 1. Planck energy distribution curves

The Line Spectrum J. Fraunhofer (1814) was the first to explore the many dark lines named after him in the spectrum of the Sun and also to investigate them in the spectra of Venus and several bright stars. He determined their positions in the solar spectrum and assigned letters to the strongest lines (e.g., the sodium *D-line*), a notation which is still used today along with more exact spectroscopic codes. The interpretation of the lines was necessarily postponed until the twentieth century when atomic physics supplied the theoretical principles. The atomic model developed by N. Bohr (1913) will here suffice to explain the basic facts. Accordingly, an atom consists of a central *nucleus* surrounded by a system of *electrons*. The entire mass is practically concentrated in the nucleus, which consists of *protons*, each carrying the positive elementary electrical charge, and usually also of uncharged *neutrons*. It is the number of protons which determines the chemical element. The simplest atom, that of hydrogen, for instance, has one proton. The oxygen atom has eight, and that of uranium 92 protons. A non-ionized (see below) atom has zero total charge. The nucleus containing Z protons must therefore have Z electrons, each carrying the one negative elementary charge, somehow distributed around the nucleus. The mass of an electron is about 1/1836 the mass of either a proton or a neutron. The radius of the nucleus is about 10^{-15} m, five orders of magnitude smaller than that of the whole atom.

The features of the simple hydrogen atom (one proton, one electron) will be more closely described here as they may help to understand more complicated atoms. The hydrogen lines appear in certain series whose patterns had already been represented by the *Rydberg formula* (1890):

$$\frac{1}{\lambda} = R \left(\frac{1}{n^2} + \frac{1}{m^2} \right) , \tag{4}$$

where $R = 1.097\ 373\ 156\ 852\ 7 \times 10^{-2}$ nm^{-1} is the Rydberg constant and n and m are integers such that $m > n$. For each fixed n, a series can be generated with m equal to any higher integer. The best-known series, which coincidentally lies in the visible part of the spectrum, derives from $n = 2$. It is named after J.J. Balmer, who, in 1885, first

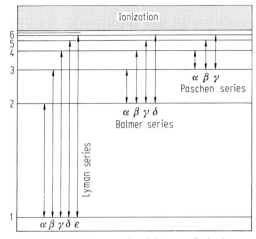

Fig. 2. Schematic energy-level diagram for hydrogen

gave the special formula for it. Following Bohr, the pattern is explained by assuming that an electron can revolve around the nucleus in certain permitted orbits at various distances without emitting energy. Higher energy levels of the electron correspond to orbits farther from the nucleus. Absorption or emission of electromagnetic radiation can initiate the transition of the electron between these permitted orbits or *levels*. As the emitted radiation consists of *photons*, whose energy depends on their wavelength, only those wavelengths whose energy equals the energy difference between permitted levels can be absorbed or emitted. If the permitted energy levels are numbered from 1 to n beginning nearest the nucleus, all transitions beginning or ending at the first or *ground* level ($n = 1$) produce the *Lyman series* in absorption or in emission; transitions starting or ending at the second level ($n = 2$), the *Balmer series*; and so on (see Fig. 2). Even in this simplest of cases, that of atomic hydrogen, quantum mechanics demonstrates that the real features are not quite so simple. In fact, when helium, the next simplest atom with two protons, normally two neutrons, and two valence electrons is considered, Bohr's theory is already inadequate to describe the observations. Nevertheless, the Bohr theory is still an excellent approximation to the full theory so long as only one-electron systems are considered (e.g., alkali metals, or ions with one valence electron, i.e., an electron in the outermost level).

Heavier atoms show more complex features. With increasing nuclear charge number and a correspondingly increasing number of electrons, their various energy levels become gradually filled with electrons. The maximum number z of electrons fitting into a certain level depends on its level number,

$$z = 2n^2 , \tag{5}$$

that is to say, the lowest level ($n = 1$) is occupied by at most 2 electrons, the next ($n = 2$) by 8 electrons, and so on. This so-called *exclusion principle* was empirically derived by W. Pauli in 1925. In heavier atoms, such as iron, the various levels are not filled successively; rather, an outer level may start to build before the lower level is filled. The outer

Table 1. Abbreviations of some chemical elements

H	Hydrogen	Si	Silicon	Zr	Zirconium
He	Helium	Cl	Chlorine	Tc	Technetium
Li	Lithium	Ar	Argon	Ru	Ruthenium
C	Carbon	Ca	Calcium	In	Indium
N	Nitrogen	Ti	Titanium	Xe	Xenon
O	Oxygen	V	Vanadium	Ba	Barium
Ne	Neon	Cr	Chromium	La	Lanthanum
Na	Sodium	Mn	Manganese	Eu	Europium
Mg	Magnesium	Fe	Iron	Gd	Gadolinium
Al	Aluminum	Co	Cobalt	Au	Gold
P	Phosphorus	Ni	Nickel	Hg	Mercury
S	Sulfur	Sr	Strontium	Pb	Lead

electrons may then perform complicated transitions between levels that are not filled or only partially filled, sometimes even within the same level, and thus generate a variety of spectral lines characteristic of the atom considered. Often, special groups of related lines are generated. For instance, sodium always shows two closely neighboring lines (called *doublets*), the best known being the yellow D-lines, while magnesium has singlet and triplet lines.

Many astronomical objects show so-called *forbidden lines* whose origin involves long-lived energy levels. Normally, an electron reaching a higher level will leave it again spontaneously after only $\sim 10^{-8}$ s to return to a lower level. However, there also exist energy states where the electron may dwell for seconds, minutes, or years. Under "normal" conditions, the electron in such a state will, via interactions with other particles or with photons, loose its excitation energy (see below) long before the atom has a chance to radiate the forbidden lines. Only within the extremely rarefied environments of certain gaseous nebulae in space does this collisional interaction become so minimal that the forbidden line radiation can occur. To distinguish forbidden from permitted lines, the elemental codes (see Table 1) are displayed in square brackets (e.g., [OIII], [FeIV]).

The most complicated spectra of all are those emitted by *molecules*, which are aggregates of two or more atoms. Here, in addition to the jumps in the joint electron shell, vibrations between nuclei may occur as well as rotations around certain axes. Quantum theory postulates that, as with the electron transitional energies, the vibrational and rotational energies are quantized. The corresponding energy differences are very low, meaning that there are numerous closely adjacent levels, and transitions between them generate lines in the infrared region. The superposition of electron jumps, rotations, and atomic vibrations generates the observed molecular spectrum in the visible and UV range. Instead of a single line corresponding to the electron jump, there appears a *band* of many closely adjacent lines which are often not entirely resolved. Usually, the lines comprising the band, depending on the type of molecule and the electron transition, crowd together at the short-wave or long-wave end to form the so-called *head* of the band.

Excitation and Ionization At low temperatures, almost all atoms of a gas are in the ground state; their valence electrons reside in the lowest possible energy level. Absorption of suitably energetic photons or collisions with other particles may "kick" the electrons to higher levels. The atoms are then said to be *excited*. To generate, for instance, the Balmer series, some fraction of the existing hydrogen atoms must have their electrons in the $n = 2$ level. How many atoms are in this specific level depends in general on the temperature of the gas.

The formal representation of this relation is based on the results of investigations by L. Boltzmann and is therefore known as the *Boltzmann equation*:

$$N_s \sim N_0 e^{-\chi_s/kT} , \tag{6}$$

where N_0 is the number of atoms in the ground state, N_s that at the excited level, and χ_s the corresponding excitation energy. T is the excitation temperature, which, in thermal equilibrium, coincides with the gas kinetic temperature (proportional to the kinetic energy of the particles) and with the radiative temperature. (In general, however, thermal equilibrium does not hold and these temperatures are not quite equal.) The conclusions are, of course, of a statistical nature. Each electron will, after residing a certain time in the excited state, tend to return to the ground level, although in some other atoms the electrons will occupy these excited levels. Evidently, even at high temperatures, N_s cannot exceed N_0. Only at very high temperatures do the two values become approximately equal. In the solar atmosphere with a temperature of around 6000 K, for instance, only about one hydrogen atom out of 10^8 is excited to the second level and thus able to produce a Balmer line; hence, at any given moment the overwhelming majority of hydrogen atoms does not contribute to the generation of these lines.

As the temperature increases, an atom may gradually reach higher and higher excitation states until finally the electron becomes detached from the parent atom. The atom is then said to be *ionized*. The minimum energy needed to ionize an atom from the ground level is called *ionization energy*. If the energy provided lies above the ionization minimum, the excess energy will, after the ionization process, be converted into kinetic energy of the free electron. The number N_i of ionized atoms (ions) of a kind which exist at a given temperature results from the Boltzmann statistics in connection with quantum theory, and is described by an equation derived in 1920 by M.N. Saha:

$$N_i \sim T^{3/2} (N_0/N_e) e^{-\chi_i/kT} \,(\text{Saha equation}) , \tag{7}$$

where N_0 is the number of neutral atoms, χ_i the ionization energy, N_e the number of free electrons in the gas, and T the ionization temperature, which again coincides with the other temperatures only in the case of thermal equilibrium. The factor $T^{3/2}$ causes N_i to grow without limit as temperature increases, so that at a sufficiently high temperature, most of the atoms will be ionized. Besides the temperature, the electron density is the most important quantity in the ionization process. At a given temperature, the ionization is stronger when the number of existing free electrons in the gas is low. The ionization energies differ for various kinds of atoms; for instance, sodium and potassium are easier to ionize than magnesium or silicon. The expectation is therefore that, at given T and N_e, one element can be multiply ionized (has lost several electrons) while another element remains still essentially neutral. The energy levels of an ion are very different from those of the neutral atom. Specifically, in each ionization state, specific levels exist and generate spectral lines characteristic of that kind of ion when the remaining electrons perform the corresponding transitions.

Unsöld and Baschek [1] provide an excellent introduction to this matter.

7.2.2 The Objects of Spectral Investigation

There are fundamental differences between objects which contain internal energy sources and thereby generate electromagnetic radiation (stars), those which basically reflect the radiation from other sources (planets and moons lacking atmospheres), and those which emit or absorb energy through interaction with radiation (emission nebulae, planets and moons with atmospheres, comets) or with the Earth's atmosphere (meteors). The following section gives a brief survey of basic spectroscopic features of astronomical objects and especially those which may be explored by amateur astronomers.

7.2.2.1 Stars It is found that stellar spectral features, the absorption lines in particular, display substantial differences with respect to number, intensity, and type which are in principle directly correlated with the location of maximum intensity of radiation (and hence with the temperature due to Wien's displacement law). As the shift of maximum intensity is not easy to determine, the classification of stars according to certain features of their spectral lines is a valuable tool.

In view of the concepts mentioned in the previous section, it may be surprising that dark lines appear in stellar spectra at all, as every excited atom reradiates its absorbed energy again after a normally very brief time interval. This is especially true for *resonance lines* (transitions between ground and first excited level), where the electron cannot return to the ground level by emitting the same spectral line previously absorbed. (The atom may, during its brief residence in the excited state, once more be excited by absorption of a photon or by collision with another particle, but this mechanism cannot accout for all absorption features.) The explanation of this paradox is that the spontaneous reemission of the various atoms occurs *isotropically* (i.e., uniformly in all directions), and therefore only a certain small fraction of the energy is directed toward the observer. Except for resonance lines, the electron may also take a path over various intermediate levels and emit different spectral lines. In each case, the net effect is a weakening of the continuum at the position of the absorption wavelength, i.e., a "dark line." Temperature, chemical composition, and, to a much lesser extent, the gas pressure of the stellar atmosphere (extending some 10^{-3} to 10^{-2} of the stellar radius) are the factors determining the features of the stellar spectrum. Assuming that the chemical composition of the atmospheres of all stars is by and large the same, the foremost decisive quantity is the surface temperature.

A basic grouping of stars according to intensities of lines of various elements has long been known under the name *Harvard Classification*. With the aid of Saha's ionization theory, it became possible to interpret intensities of the various lines *quantitatively*. Figure 3, which displays the results of applying Eqs. (6) and (7) for several elements, shows the number of atoms of the element in a certain excited state relative to the total

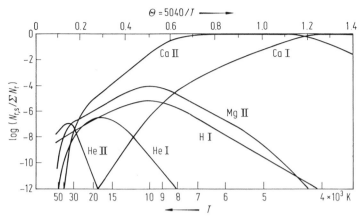

Fig. 3. The dependence of ionization and excitation on temperature (after Unsöld and Baschek [1])

number of those atoms (in all states) as a function of temperature at an average electron density. As is customary in astrophysics, the spectrum of a neutral atom is characterized by a Roman numeral "I" after the elemental code, and higher ionization states by correspondingly higher numbers (e.g., CaI for neutral calcium, CaII for singly ionized calcium, etc.; see Table 2). The occurrence of a higher ionization state is a function, as mentioned, of the ionization energies which differ for various elements (e.g., Ca has low ionization energy, He has high ionization energy). Retaining the traditional letter codes of spectral classes, but re-ordered according to decreasing temperatures (about 40,000 down to about 3000 K), the sequence

$$O - B - A - F - G - K - M$$

was obtained, where each class is further subdivided decimally.

> The still-used terms *early-type* for the hot stars of classes O, B, and A, and *late-type* for the cool spectral types K and M are rooted in astronomy history; they are not related to the true age of the stars.

The basic spectral features and the corresponding surface temperatures are compiled in Table 3.

A graph of the absolute magnitudes of stars versus their spectral types reveals that the fast majority of all stars fall into a narrow band, the so-called *main sequence*, running diagonally from the intrinsically very bright O-stars to the intrinsically faint M-stars. This diagram is named the *Hertzsprung–Russell diagram* (or simply H–R diagram) after the astronomers who first investigated it independently in 1907 and 1913. Figure 4 presents such a graph which is based on data for about 6700 stars. The vertical scatter results from an uncertainty in the luminosity determinations. The *main sequence* and, mostly above it, other groups of data points are present. It is directly apparent that stars of the same spectral class can have different absolute magnitudes and therefore must radiate differ-

Table 2. Wavelengths of selected spectral lines in visible light

Element	λ (nm)	Element	λ (nm)	Element	λ (nm)
Hα	656.282	He I	447.148	Na I	589.594 D$_1$
Hβ	486.133		412.081		588.997 D$_2$
Hγ	434.047		402.619		
Hδ	410.174		396.473	Mg I	457.110
Hϵ	397.007	He II	656.010	Mg II	448.133
H$_8$	388.905		541.152		448.113
H$_9$	383.539		468.568		
H$_{10}$	379.790		544.159	Ca I	657.278
H$_{11}$	377.063		419.983		422.673
H$_{12}$	375.015		402.560	Ca II	396.847 H
H$_{13}$	373.437		392.348		393.366 K
H$_{14}$	372.194				
H$_{15}$	371.197	Hg I	579.006	Cr I	425.435
			576.960		
He I	587.565		546.074	Mn I	403.076
	587.562		435.834		
	501.568		404.656	Ba II	493.409
	471.314		365.014		455.403

ent amounts of energy even at identical surface temperatures. This is possible only if the brighter star has a larger surface area and hence a larger radius. This fact lays the foundation for the following terms used by Morgan and Keenan (MK, [2]) in a finer subdivision of star types into *luminosity classes* characterized by Roman numerals according to the different sizes:

Ia	Brightest supergiants	V	Main sequence stars (dwarfs)
Iab, Ib	Less bright supergiants		
II	Bright giants	VI	Subdwarfs
III	Giants	VII	White dwarfs, also designated
IV	Subgiants		by "D"

Notable examples of this MK classification are: β Ori (Rigel) B8 Ia, the Sun G2 V, α Boo (Arcturus) K2 III, α Ori (Betelgeuse) M2 Iab.

Spectra of main sequence stars, giants, and supergiants of the same spectral class differ from each other in small, but noticeable, details. Luminosity criteria include primarily the width of Balmer lines (narrower in high-luminosity stars) and intensity ratios of lines of certain ions to those of neutral atoms (ionic lines are stronger and neutral lines weaker in high-luminosity stars). As the gas densities in the atmospheres diminish for high-luminosity stars, the *pressure broadening* (Balmer lines) also diminishes, while the degree of ionization increases (metallic lines; compare Saha equation).

Over 99% of the known stars fit into this two-dimensional scheme. The remainder have either anomalous abundances of certain elements in their atmospheres, or different physics of line formation. For example, they may have strong magnetic fields or high rotational velocities associated with mass loss. At the hot end of the spectral sequence appear the *Wolf–Rayet stars* (type W), the *Be* and *shell stars* with anomalous He or N abundances. Among the A-type stars, there are those with strong magnetic fields and

Table 3. Characteristics of the major spectral classes

Spectr. type	Color	Atmospheric temp. (K)	Distinguishing criteria	Typical example	
O	Blue	> 30,000	Lines of highly ionized atoms, primarily HeII, SiIV, NIII; Balmer series very weak	ζ Pup 10 Lac	O5 O9
B	Blue	11,000–30,000	HeII vanishes between B0 and B5; HeI strengthens but disappears by B9; also lines of SiII, SiIII, OIII, MgII; Balmer series intensifies from classes O to A	α Vir (Spica) β Ori (Rigel)	B1 B8
A	Blue	7500–11,000	HeI absent; Balmer series reaches maximum intensity at A2; lines of singly ionized elements such as MgII, SiII, FeII, TiII, CaII, etc., present (maximum at A5); lines of neutral metals very weak	α Lyr (Vega) α CMa (Sirius)	A0 A1
F	Blue-white	6000–7500	Balmer series decreases in intensity; lines of ionized and neutral metals about equally intense; CaII strong	α Car (Canopus) α CMi (Procyon)	F0 F5
G	Yellow-white	5000–6000	Balmer series continues to wane; CaII very strong; lines of neutral metals very strong; first appearance of CH-bands	α Aur (Capella) Sun	G0 G2
K	Orange-red	3500–5000	Lines of neutral metals with low excitation energies very strong; CN- and CH-bands present; TiO-bands appear at K5; Balmer lines virtually absent	α Boo (Arcturus) α Tau (Aldebaran)	K2 K5
M	Red	< 3500	TiO-bands very strong; lines of neutral elements with low excitation energies (e.g., CaI) very strong	α Sco (Antares) α Ori (Betelgeuse)	M1 M2

those with metal anomalies. Among the late classes after type K, the S-types and C-types (carbon stars, formerly designated R- and N-stars) with abundance anomalies branch off as side sequences.

The large group of physically variable stars not fitting the normal sequence will not be discussed here. These objects are often too faint to be investigated spectroscopically with small instruments and with the necessary time resolution. Only the *novae* will be considered here, as interesting spectroscopic observations may on occasion be possible for amateurs. Novae are white dwarfs in close binary systems whose brightness increases by as much as 16 mag in a short time (hours to months) and declines thereafter over years to the original brightness. Though the time scales are different, the spectra of novae in

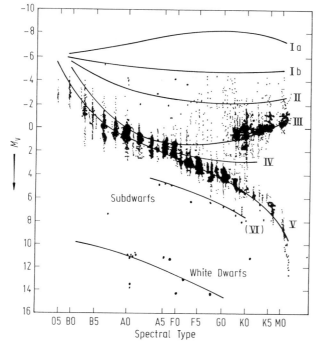

Fig. 4. The Hertzsprung–Russell diagram (after Scheffler and Elsässer [3])

the various stages of eruption are similar enough to be combined in a separate class Q subdivided by the sequence 0 to 9.

The spectrum in the pre-nova state usually resembles that of an early-type star. A maximum halt, about 2 magnitudes below maximum, is followed by a slower ascent showing an A-type spectrum (Q0) with blue-shifted absorption lines (see below) originating in a rapidly expanding shell (up to 1000 km s^{-1}). At the maximum and shortly thereafter, the spectra (Q1, Q2) match F-supergiants. The first decent to about 3.5 magnitudes below the maximum (Q3) shows strong emissions of H and metal ions with absorption edges at the short-wavelength end. Later on, broad, diffuse absorptions of H, OII, NII, and NIII (Q4, Q5) from a second shell ejected at higher speed (sometimes several shells and up to several thousand km s^{-1}) appear. The Q6 spectrum is characterized by emissions of He and [NII]. In the following transition novae behave individually. They may display strong brightness fluctuations or a minimum. All novae return then to a brightness of about 6 magnitudes below maximum. In this transition the *nebular spectrum* Q7, a pure emission spectrum of HeI, FeII, FeIII, CII, OII, etc., develops. The well-known forbidden green oxygen lines [OIII] appear and reach a maximum in the phase of slow descent (Q8, Q9). The post-nova shows a continuous O-type spectrum, superposed by narrow emissions of H, HeI, and [OIII].

7.2.2.2 The Sun The proximity of the Sun permits detailed spectroscopic studies of many individual features, a procedure not possible for other stars. The *Quiet Sun* exhibits a normal G2 V spectrum, very rich in lines. The Rowland Tables [4] list about 24,000 lines over the range $\lambda\lambda$ 293.5–877.0 nm. About 20,000 of them are identified, and their primary data, such as intensity and excitation energy, are compiled. In the red spectral range, absorptions by the Earth's atmosphere interfere increasingly. These *telluric lines* make up about 60% of the lines observed in the solar spectrum between $\lambda\lambda$ 600–900 nm. The tables also list the lines from sunspots, which, as lower-temperature regions (about 1600 K cooler than outside these regions), show a spectrum similar to a K0 star; that is, the lines of neutral elements, particularly Ca, V, Cr, and Ti, are enhanced compared with the normal solar spectrum, and the ionic and Balmer lines weakened. Molecular bands primarily of MgH, TiO, and CaH and lines of Li, Ru, and In also appear. Occasionally, the Balmer lines occur in emission, especially when light bridges cross the spots. Spectra taken of the solar limb also show characteristics typical of lower-temperature stars.

During the course of a total solar eclipse, the *flash spectrum* of the chromosphere is observed, which consists only of emission lines since tangential viewing of the solar limb removes the absorptions and leaves only the re-emissions. About 3000 lines are known, but they are not simply a reversal of the normal solar spectrum. Lines of ions and of neutral atoms with high excitation energy appear enhanced, while neutral lines of low energy are weakened. Also observed are strong HeI lines, in particular the so-called D3-line (587.6 nm). All of this is strongly indicative of a high chromospheric temperature.

Studies show the temperature to rise from the surface discontinuously to about 10^6 K at 10^4 km above the surface, where the transition into the corona begins. The latter appears at total eclipses as an extended, whitish structure several 10^6 km wide and forms the outermost part of the solar atmosphere. In the optical spectrum, the radiation of the so-called K, L, and F coronae are superposed. While the F spectrum is a normal solar spectrum reflected by interplanetary dust some distance away from the Sun (at the transition to zodiacal light), the K spectrum consists of a pure continuum generated by scatter of sunlight at free electrons directly above the chromosphere. The thermal velocities of these electrons are so high, owing to the high temperatures caused (most likely) by shock waves from the chromosphere, that the lines are completely smeared out by the Doppler effect (see Sect. 7.3.2.3). Finally, the spectrum of the L corona represents the "true" light of the corona. It consists of forbidden emissions of highly (9 to 15×) ionized elements such as Fe, Ca, V, Cr, Mn, K, and Co. Over 30 lines, in particular in the short-wavelength range, have been identified. The most notable ons are the forbidden green line at 530.3 nm from [FeXIV], the yellow line at 596.4 nm from [CaXV], and the red line at 637.4 nm of [FeX]. Collisions with fast electrons cause the high ionizations, and the near-vacuum density then allows forbidden lines to appear in the same manner as in emission nebulae.

The classical solar spectrum atlas has been published by Moore et al. [4]. More accessible is the digital High Resolution Solar Spectrum published online by Observatoire Paris-Meudon.

Meudon High Resolution Solar Spectrum: http://bass2000.obspm.fr/solar_spect.php

Other spectral features of solar phenomena will not be dealt with here, as their observation requires large and expensive instrumentation.

7.2.2.3 Planets and Moons   Most planets feature gaseous atmospheres where thick clouds reflect or scatter the sunlight, which subsequently penetrates part of the outer atmosphere. A planetary spectrum consequently consists of a solar spectrum plus additional absorption bands from the gases in the planet's atmosphere. Since these gases may resemble those in the Earth's atmosphere, telluric bands interfere with the planetary spectrum. The observed planetary bands lie mostly in the red and infrared spectral range and can mainly be ascribed to the following molecules (omitting those objects that have no atmospheres in the strict sense, e.g., Mercury, Pluto, Earth's Moon):

Venus and Mars: CO_2, N_2, H_2O ;

Jupiter and Saturn: H_2, CH_4, NH_3 ;

Uranus and Neptune: H_2, CH_4 .

Not all gases can be spectroscopically identified, as many lines of potential constituents do not appear in the accessible range at the temperatures of planetary atmospheres.

Saturn's large satellite Titan is the only moon in the Solar System to possess its own permanent, fully developed atmosphere containing CH_4 and H_2. All other moons (and also Saturn's rings) show, after eliminating telluric features, a purely reflected solar spectrum.

7.2.2.4 Comets   At large distances from the Sun, a comet consists only of a tiny nucleus which shows a spectrum of reflected sunlight. As it approaches the Sun, parts of the nucleus evaporate, are further broken up by photochemical processes, and form a diffuse gaseous shell called the *coma*, which the sunlight excites to fluorescence or resonance light. The spectrum exhibits, apart from sunlight, emission bands of molecules, radicals, and radical ions: OH, NH, NH_2, CH, CN, C_2, C_3, OH^+, CH^+, CO^+, N_2^+, CO_2^+, and, at close approach, atomic lines of Na and perhaps of Ni, Fe, Ca, Cr, and O. For a collection of typical cometary spectra see Swings [5].

In the space roughly inside the orbit of Mars, the comet develops its characteristic *tail*, generated by the radiation pressure of the Sun and by the solar wind (ions and electrons traveling outward from the Sun at speeds of around $400\,km\,s^{-1}$ at the orbit of Earth). Two basic types of tails are distinguished:

1. Type I. A long, narrow tail with emission bands primarily of CO^+, CO_2^+, CH^+, CN^+, OH^+, and N_2^+ generated due to further ionization by short-wavelength solar radiation
2. Type II. A broad, diffuse tail consisting of colloidal particles which merely reflect the sunlight

7.2.2.5 Meteors   Small, solid particles (with masses mostly between 10^{-3} to $10^{-5}\,kg$) penetrating the Earth's atmosphere cause the luminous phenomenon known as *meteors*. Collisions ionize the atmospheric molecules and evaporate the surface material of the intruding body (or even the entire body), with a resulting emission of light. The characteristic spectrum generated depends somewhat on the orbit and the entrance velocity. Basically, it consists of emission lines, sometimes superposed over a weak continuum. Low-excitation lines of FeI, NaI, CaI, MgI, and CaII-H and -K appear, but lines of AlI, SiI, CrI, MnI, NiI, and even H_α and H_β have been observed. Ionic lines have been found in fast meteors with relative velocities of over 30 km s^{-1}, and also in slower meteors toward the end of their path. The N$_2$-bands occasionally seen are of telluric origin. The forbidden green [OI]-line at 557.7 nm sometimes emerging shortly after immersion into the atmosphere for one or two seconds has not yet been explained. Meteor spectra can be divided according to dominant lines and numbers of lines into four types and four classes:

Type X:	Dominated by NaI-D or by MgI (518.4 and 383.3 nm)
Type Y:	Dominated by CaII-H and -K
Type Z:	Dominated by FeI or CrI
Type W:	Fits none of the above cases
Class a:	> 49 Lines
Class b:	20–49 Lines
Class c:	10–19 Lines
Class d:	1–9 Lines

Most spectra obtained are of type Z and class d. For more information see the International Meteor Organization (IMO) and the American Meteor Society (AMS). An in-depth treatise is given by Bronshten [6].

IMO: http://www.imo.net
AMS http://www.amsmeteors.org

7.2.2.6 Luminous Gas Nebulae and Extragalactic Objects

Interstellar gas in the vicinity of O or B stars will be ionized due to the high energetic photons emitted by these stars. Recombination processes of ions and electrons generate then via cascade transitions emission lines of H, HeI, HeII, and of some further ions. Besides this, conditions are optimum to form forbidden lines of several ions (e.g., [OIII], [NeIII], [NII]). One refers to such structures as *diffuse nebulae* or *HII regions*. The so-called *planetary nebulae*, nova and supernova remnants, and shells around *Wolf–Rayet stars* exhibit a nearly spherical morphological structure around a central star. In these cases the ambient gas is due to more or less violent ejections related to instabilities in the outer layers of the stars in question or due to stellar winds. The physics of line formation is the same as for the diffuse nebulae except for the supernova remnants, where the gas is ionized by the high temperatures in shock wave fronts. The wavelengths of the most prominent emission lines of the luminous gas nebula in the optical region are as follows (in nanometers): 658.3 ([NII]), 656.3 (H_α), 654.8 ([NII]), 587.6 (HeI), 500.7 ([OIII]), 495.9 ([OIII]), 486.1 (H_β), 468.6 (HeII), 434.0 (H_γ), 410.2 (H_δ), 396.7 ([NeIII]), 386.9 ([NeIII]), 372.9 ([OII]), 372.6 ([OII]).

Except for special cases where individual stars can be resolved in nearby galaxies, the spectra of galaxies in general are integrated spectra representing the star mixture of the region under investigation. There is a correlation between the integrated spectrum of the central region of a galaxy and its Hubble type, whereas in irregular galaxies the light of O and B stars dominates the spectrum, red dwarfs and giants predominantly influence the spectra of spiral and elliptical galaxies. The spectra of some types of active galaxies (e.g., Seyfert galaxies, quasars) in the optical region resemble those of luminous gas nebulae. The nuclei of these galaxies, called *active galactic nuclei*, presumably contain a supermassive black hole surrounded by an accretion disc where the emission lines originate.

7.3 Practical Applications

G. Rupprecht

7.3.1 Spectroscopic Instruments

This section emphasizes general aspects of spectroscopic equipment, while special instruments will be discussed only when they are accessible to amateurs or professional astronomers at small colleges. We assume the use of electronic detectors (usually CCDs) that are described elsewhere in this volume in more detail.

7.3.1.1 The Methods of Spectral Dispersion The heart of any spectral device is a dispersing element which separates the incident light according to its various wavelengths. Today, various types of gratings are usually employed as dispersers. Prisms are hardly used any more because of their price and the relatively low dispersion they provide. They are however still used as objective prisms by amateur and professional astronomers.

The Prism Light obliquely incident on a glass surface suffers a change of direction owing to the different velocities of propagation in air and in the glass medium; it is thus said to be *refracted*. Refractive behavior is governed by the *law of refraction*, more commonly called *Snell's law*, which states that

$$\frac{\sin i}{\sin i'} = \frac{c_1}{c_2} = \frac{n_2}{n_1} = n \, , \tag{8}$$

where i and i' are the angles of incidence and of refraction with respect to the normal (i.e., perpendicular) to the surface, c_1 and c_2 the propagation speeds in air and in medium, and n_1 and n_2 the corresponding refractive indices. Usually, merely the relative index n of the medium against air is needed. Passage through a prism causes two changes in direction. The amount of the deflection angle φ changes with the angle of incidence i and angle of the refracting edge (apex angle) γ of the prism. The deflection is a minimum when the ray passes the prism symmetrically, parallel to the base B, and thus incident and exiting rays form the same angles with respect to the normals to the prism surfaces (Fig. 5). For small incident angles and small γ, the angle of deflection is in general given by

$$\varphi \approx (n - 1)\gamma \, . \tag{9}$$

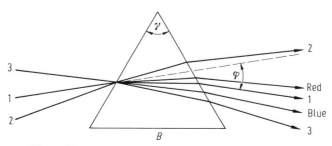

Fig. 5. Transmission and dispersion by a prism (schematic)

Symmetric transmission gives strictly, for any γ,

$$\varphi = 2 \arcsin[n \sin(\gamma/2)] - \gamma . \tag{10}$$

The refractive index depends on wavelength ($n_{red} < n_{blue}$), and blue light is more deflected than red light. The change of φ with wavelength λ is called the *angular dispersion* of the prism. Like the refraction index, it is a material constant which in addition depends on γ. Differentiating Eq. (10) with respect to wavelength yields the angular dispersion for symmetric passage,

$$\frac{d\varphi}{d\lambda} = \frac{2 \sin(\gamma/2)}{[1 - n^2 \sin^2(\gamma/2)]^{1/2}} \frac{dn}{d\lambda} . \tag{11}$$

For a 60° prism of standard glass with $1.4 < n < 1.6$, this relation simplifies to:

$$\frac{d\varphi}{d\lambda} \approx n \frac{dn}{d\lambda} . \tag{12}$$

The change of refractive index with wavelength, $dn/d\lambda$, can be computed with the material data available from the websites of glass manufacturers. Of course, the angular dispersion of a given prism increases for an asymmetric passage, but with the drawback of a decrease in imaging quality as well as in the resolving power A. The latter specifies how much of a wavelength difference $\Delta\lambda$ at a given λ can be separated, and should be as large as possible. In a fully illuminated prism and symmetric transmission, the resolving power is given by

$$A = \frac{\lambda}{\Delta\lambda} = B \frac{dn}{d\lambda} . \tag{13}$$

Note that the resolving power does not depend on the angle γ but only on $dn/d\lambda$ and the base length B of the prism, and is effectively diminished in non-symmetric transmissions. For a standard prism with a base length of 10 cm, A is on the order of 10^4.

Diffraction Gratings The term *diffraction grating* characterizes a large number of fine grooves etched in a parallel and equidistant fashion on a plane or concave surface (e.g. glass). The distance between two grooves or between two "unetched" surface strips is called the *grating constant g*. There are both transmission and reflection gratings, but this distinction is, in theory, immaterial. In practice, reflection gratings are more widely used (see Fig. 6).

When parallel rays of light enter the grating, the unetched surface strips between grooves act as transparent or reflecting diffraction slits. The combined action of many such slits enhances the light of a certain wavelength only in specific directions; this phenomenon is called *constructive interference*. The position of the intensity maximum is given by the condition

$$g(\sin i + \sin \varphi) = m\lambda , \tag{14}$$

where i is the angle of incidence, φ that of diffraction, and λ the wavelength. This formula is known as the *grating equation*. The order m can have the values 0, ±1, ±2, etc.; that is to say, apart from the case $m = 0$ (that is, reflected or transmitted without diffraction for $i = \varphi$), the path difference of neighboring rays is an integer multiple of the wavelength when the light is to propagate into direction φ. A negative value for m occurs when incident and diffracted rays are not on the same side of the normal and $|\varphi| > i$.

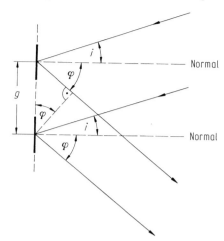

Fig. 6. Geometry of a reflection grating

Since the maximum condition (Eq. 14) holds for any λ, white (i.e., multicolored) light is separated into wavelengths. In contrast to a prism, the deflection is here proportional to wavelength (the so-called normal spectrum), being stronger for red than for blue light. The angular dispersion is nearly independent of λ:

$$\frac{d\varphi}{d\lambda} = \frac{m}{g\cos\varphi} \ . \tag{15}$$

The dispersion is evidently higher the higher the order m used for observation and the smaller the grating constant. In each order, an entire spectrum of practically constant dispersion (except for the change in $\cos\varphi$) is generated, but at higher orders more overlap occurs, with the violet end of the following spectrum falling onto the red end of the preceding one. For instance, 800 nm in the first order coincides with 400 nm in the second order. In order to achieve high resolving power with the grating, the spectral range of interest may be preselected by a prism or a filter. Of course, m cannot be arbitrarily large; the maximum number is given by $m_{max} = 2g/\lambda$. In practice, usually the first or second order is used.

The resolving power of a grating is given by

$$A = \frac{\lambda}{\Delta\lambda} = mZ \ , \tag{16}$$

where Z is the total number of grooves. Commercially available grating dimensions reach resolving powers of about 10^4 to 10^5. The quality of the grating is assumed to be adequate; for instance, grooves which are irregular in shape or spacing do not yield good interference patterns, and periodic division errors produce annoying spurious maxima called "ghosts."

Theoretically, the intensity of spectra should diminish with increasing order. This is not the case for all gratings, but depends on the shape in which the lines are etched. Suitable shapes can even concentrate almost the entire intensity (as in a prism) into one direction, i.e., into *one* spectrum of a particular order, and additionally achieve a high resolving power. Such a device is called a *blaze grating*. There is also the *echelle grating*, which uses the narrow sides of rectangular grooves at a high grating constant and achieves high resolving power (up to 10^6) at the expense of a very narrow, overlap-free spectral range.

7.3.1.2 The Design of a Spectral Instrument This section describes the basic arrangements of spectral devices, while only briefly mentioning the most important special designs.

Slitless Spectral Instruments The simplest spectral instrument, the *objective prism* (Fig. 7) is obtained by mounting a prism of small refracting angle in front of the telescope objective at the end of the tube so as to cover the entire aperture. The great advantage of this design is in obtaining simultaneously the spectra of all objects in the field of view. The angular relation for symmetric transmission gives the angle $\beta = \varphi/2$ between the base of the prism and the optical axis of the telescope. With f as the objective focal length, the wavelength-dependent spectral dispersion d in the focal plane becomes

$$d = f \frac{d\varphi}{d\lambda} \tag{17}$$

$$= f \frac{2\sin(\gamma/2)}{[1 - n^2 \sin^2(\gamma/2)]^{1/2}} \frac{dn}{d\lambda} . \tag{18}$$

It is the reciprocal value of d which is usually given and (not entirely correctly) called "dispersion" (normally in nm/mm or Å/mm). Combining the formulae for resolving power and for angular dispersion gives the resolution Δl of the entire device in the focal plane at wavelength λ:

$$\Delta l = \frac{2f\lambda \sin(\gamma/2)}{B[1 - n^2 \sin^2(\gamma/2)]^{1/2}} . \tag{19}$$

Use of an objective grating is, because of the low intensity of diffraction images, normally restricted to a combination with an objective prism (see below).

Spectral Equipment with a Slit More demanding but advantageous in many ways is the slit design, whose basic composition appears in Fig. 8. At the focus of the objective (aperture D, focal length f) is the slit S onto which the seeing or diffraction disk of the object to be observed is imaged. The *collimator* (D_1, f_1) induces the light to fall parallel onto the dispersing element E, which may be a grating (today the most common solution) or one or several prisms. The slit is at right angles to the direction of dispersion.

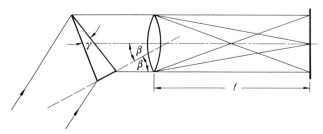

Fig. 7. Principle of the objective prism

The camera lens (D_2, f_2) finally images the spectrally separated light into the focal plane of the camera and onto a detector. The spectral instrument is fitted to the telescope when

$$f/D = f_1/D_1 \,,\tag{20}$$

and hence when the dispersing element is optimally illuminated. Imaging the slit into the focal plane of the camera also makes apparent the imaging properties of the dispersing element in the direction of the dispersion; that is to say, a monochromatic beam changes its diameter Φ_1 after passage through the element into Φ_2 in the direction of dispersion. The ratio of these diameters is the factor V, given by

$$V = \frac{\Phi_1}{\Phi_2} = \frac{\cos i}{\cos \varphi} \,,\tag{21}$$

which is exactly 1 for prisms with symmetric transmission and usually of the order of 1 for gratings. This gives, for the imaging by the slit,

$$b' = bV(f_2/f_1) \,,\tag{22}$$

where b is the slit width and b' the width of the imaged slit in the focal plane of the camera.

The same expressions, Eqs. (18) and (19), are obtained for dispersion d and resolution Δl in the focal plane by a prism as for the objective prism design, but with f replaced by f_2. For a 60° prism and $1.4 < n < 1.6$, the relations are simplified thus:

$$d \approx nf_2\frac{dn}{d\lambda} \,,\tag{23}$$

$$\Delta l \approx nf_2\frac{\lambda}{B} \,.\tag{24}$$

Corresponding formulae for the grating are

$$d = f_2(m/g)\cos\varphi \,,\tag{25}$$

$$\Delta l = f_2(\lambda/g)Z\cos\varphi \,.\tag{26}$$

At perpendicular incidence, the cosine of the deflecting angle is given by the simple relation

$$\cos\varphi = (1 - \sin^2\varphi)^{1/2} = \left[1 - \left(\frac{m\lambda}{g}\right)^2\right]^{1/2} \,.\tag{27}$$

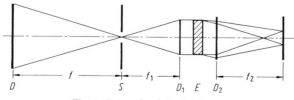

Fig. 8. Principle of the slit design

Special Mountings Various kinds of mountings are feasible for prisms and gratings. The *Littrow mounting* offers a particularly compact design. The *Ebert mounting* is distinguished by low imaging errors, that is, by subdued coma and virtually absent astigmatism. Most mountings with concave gratings use the well-known arrangement of the *Rowland circle* in its many variations. A detailed discussion of special mounting designs can be found, for instance, in Thorne et al. [7].

7.3.1.3 Detectors Since the 1980s electronic detectors, mainly CCDs, have essentially replaced photographic emulsions as radiation detectors in astronomy, professional as well as amateur. They have made the quantitative treatment and analysis of spectroscopic data much easier, and in this chapter we therefore assume the use of CCDs for all applications. Their properties, operation and basic data reduction are described elsewhere in this book.

7.3.1.4 Designs, Hints for Operation, and Accessories

Objective Prisms The use of an objective prism offers the simplest and least expensive way to generate spectra. Reflectors, refractors, or simply telephoto objectives can be used. An upper limit in aperture is reached when the prism, which should completely cover the objective, becomes too large and too heavy for the mounting and generates problems of stability and flexure, or if it becomes too expensive. Equatorial mountings and automatic tracking are advantageous and even indispensable for long exposure times. The seeing disk acts as a quasi-entrance slit, which has the advantage of high light efficiency, but also the drawback that seeing fluctuations combined with tracking errors may substantially diminish the quality of the spectra. The greatest benefit of this method, as mentioned, is in simultaneously obtaining spectra of all objects in the field, although they may be superposed on each other. Since the objects are of different brightness, well-exposed spectra as well as underexposed ones will be obtained. To reach all objects to the limiting magnitude of the instrument may thus require several exposures.

The choice of prism material depends somewhat on the given focal length of the telescope. For short focal lengths, it is desirable (but not always possible) to have an increased refracting angle and also a higher refractive index as well as a higher value of $dn/d\lambda$. With the data of the entire unit, the actual wavelength resolution is

$$\Delta\lambda = d^{-1}\Delta P ,\qquad(28)$$

and the length L of the spectrum is

$$L \approx f\Delta n \frac{2\sin(\gamma/2)}{[1 - \bar{n}^2\sin^2(\gamma/2)]^{1/2}} ,\qquad(29)$$

where d^{-1} is the reciprocal dispersion, ΔP the pixel size of the detector, Δn the difference of refractive indices for the limiting wavelengths and $\bar{n}$ their mean value.

Some experience with amateur observations in stellar spectroscopy using a prism/telephoto objective combination are reported, e.g., by Gainer [8].

Comet spectroscopy requires quite powerful equipment because of the typically low surface brightness of a comet. The dispersion direction of the prism should be oriented approximately perpendicular to the comet's tail to avoid superposition of the coma and tail spectra. Exposure times can be considerable.

Spectra of the solar chromosphere are taken only during total solar eclipses. Shortly before the second and also just after the third contact, the chromosphere flashes on for about one second as a narrow crescent. With the dispersion direction perpendicular to the crescent, the flash spectrum with its characteristically curved emission lines is taken; the crescent acts as an infinitely distant entrance slit of which monochromatic images in the corresponding wavelengths are formed. Because of irregularities of the lunar limb (e.g., lunar valleys), some photospheric light may leak through and manifest itself in the spectrum as continuous bands containing absorption lines. By using a detector mount which is movable perpendicular to the direction of dispersion and a slit which lets only the central part of the crescent-shaped lines pass, one can attempt, by continuous or stepped motion displacement of the detector during exposure, to catch the transition from photospheric to chromospheric spectrum, i.e., a reversal from absorption to emission lines. To obtain the coronal spectrum during totality requires a powerful slit spectrograph.

Slit Spectrographs Slit spectrographs are, in some respects, less economical than objective prisms because, for instance, they produce the spectrum of only one object illuminating the slit and they also require long exposure times to fully utilize the spectral resolution. Since the slit eliminates most of the influence of the sky background, such lengthy exposures become practicable. The slit width b should be fairly small for the purpose of good resolution, but still large enough to cover the seeing disk (diameter approximately $2 \times 10^{-5} f$) in order to avoid unduly long exposures; thus $b \leq 2 \times 10^{-5} f$. Guiding the object during exposure on the slit is facilitated by a slit viewer by which the light reflected by the somewhat slanted slit jaws can be observed. The slit jaws can be composed of polished metal or, even better, glass which has been vapor-coated with aluminum, leaving a correspondingly narrow strip free.

The following considerations assume the availability of a powerful telescope ($D \approx f/2$) and of a suitable dispersing element. In particular, good blaze gratings concentrate most of the dispersed light into a single order. For some work in grating spectroscopy, inexpensive replicas of coarser gratings may suffice (e.g., 100 lines/mm correspond to $g = 10^4$ nm, thus a size of 2.5 cm corresponds to $Z = 2.5 \times 10^3$), but even then some of the theoretical resolving power will be lost.

With a given telescope and dispersing element, the other quantities of the spectrograph may be calculated. The collimator diameter D_1 is

$$D_1 = Zg \cos i = Zg ,\qquad (30)$$

for a grating with perpendicular incidence and

$$D_1 = \frac{B \cos[(\gamma/2) + \beta]}{2 \sin(\gamma/2)} ,\qquad (31)$$

for a minimum of deflection prism.

The condition of optimum illumination gives for the focal length f_1 of the collimator

$$f_1 = \frac{f}{D} D_1 .\qquad (32)$$

Making the slit image b' about equal to the resolution (pixel size) ΔP of the detector fixes the focal length f_2 of the camera objective (normally then $\Delta l < \Delta P$; see Eqs. 19 and 26):

$$f_2 = \Delta P(f_1/f)2 \times 10^{-5} V_{\text{lim}} , \tag{33}$$

where V_{lim} holds for the maximum wavelength (the red limit) of the recorded spectrum. A prism with a symmetric light path has $V_{\text{lim}} = 1$.

The camera lens should be close to the dispersing element as the lens diameter $D_2 \geq D$ increases with the distance. For reflection gratings, it should be outside the incident light path in order to avoid vignetting. The lens diameter can be determined graphically by using the grating equation (Eq. 14) for the limiting wavelength of the range to be recorded. The actual resolution of wavelength again follows Eq. (28). The length of the spectrum produced by a prism is given by Eq. (29) upon replacing f by f_2. For a 60° prism with $1.4 < n < 1.6$, the relation simplifies to become

$$L \approx f_2 \bar{n} \Delta n , \tag{34}$$

again with Δn as the difference of refractive indices for the limiting wavelengths, and $\bar{n}$ their mean value. For gratings, the relation is

$$L \approx m f_2 (\Delta\lambda/g) \cos\varphi_{\bar{\lambda}} , \tag{35}$$

where $\Delta\lambda$ is the difference of limiting wavelengths and $\varphi_{\bar{\lambda}}$ is the deflection angle at the mean wavelength $\bar{\lambda}$.

Provided the seeing disk $f\alpha$ (where α is the angular size of the disk in radians) is contained entirely in the slit, then

$$I = \frac{D^2}{h'd} \tag{36}$$

gives the speed I for this configuration. However, the light losses will be higher owing to the normally larger number of optical parts, and particularly when using a grating. When the slit width is less than $f\alpha$, then the speed is

$$I = \frac{D^2 b}{dh'f\alpha} . \tag{37}$$

Depending on the combination of dispersing element and optical imaging system, the resulting focal surfaces may be tilted (refractor) or curved (Schmidt camera, spherical mirror, concave grating). This has to be taken into account in the optical design of the spectrograph.

There are many descriptions of home-made spectrographs available in the literature and in the World Wide Web. The following appear particularly interesting, but the list naturally cannot be regarded as complete: Tonkin [9] gives a good overview of the matter; C. Buil offers extensive material on design, construction and use of home-built spectrographs; a lot of practical experience is reported by members of the Fachgruppe Spektroskopie der Vereinigung der Sternfreunde e.V. There are also commercial amateur spectrographs on the market.

Because of the extreme intensity of solar radiation, the combination slit-collimator-grating-camera (or eyepiece) will suffice for solar spectra. With the facility to vary the slit width, a spectrograph permits observations of prominences, for example, in the light of the Hα-line and with a wide-open slit.

C. Buil:
http://www.astrosurf.com/buil

Fachgruppe Spektroskopie der Vereinigung der Sternfreunde e.V.:
http://spektroskopie.fg-vds.de
(German; English version available)

Fiber-Coupled Spectrographs The most common problem of a spectrograph directly mounted to the focusing mechanism of a telescope is its weight, which may seriously jeopardize the stability of the setup. Another problem is flexure suffered by the spectrograph itself when it is subjected to a varying gravity vector during a long exposure. An elegant solution to circumvent these problems is a method often applied in professional spectrographs, namely to feed the starlight via an optical fiber of several meters in length to the entrance of the spectrograph that can be conveniently located besides the telescope. In principle this setup would even allow the simultaneous recording of several spectra in a single exposure; however, this requires a mechanism to precisely position several fibers in the focal plane of the telescope. Some disadvantages of fiber spectroscopy should also be noted: a loss in efficiency depending on wavelength due to absorption in the fiber, the loss of light due to the circular fiber entrance as compared to a classical slit, and, most important, the reduction of the focal ratio between the fiber entrance and the exit leading to the so-called focal ratio degradation (FRD).

Two effects are primarily responsible for FRD in step-index fibers: macroscopic bending of the fibers, and stress on the fiber, in particular the microscopic bending caused by the gluing of the fiber ends in the connectors. As an aside: The FRD caused by the refraction index variation along the fiber cross-section does not allow the use of graded index fibers for astronomical applications. For a more extensive discussion of FRD and related effects see Ramsey [10] and Avila [11].

As a general rule FRD requires the injection into the fiber of a telescope beam faster than approximately $f/6$. As a consequence, Newtonian or other "fast" telescopes are especially suited for use with a fiber spectrograph. Alternatively one can introduce a focal reducer optics to obtain a beam, e.g., of $f/4$.

Some fundamental considerations lead to the conclusion that fiber spectrographs are not well-suited for coupling to a small (< 25 cm diameter) telescope, at least not if one wants to perform quantitative spectroscopy at reasonable spectral resolution. Ideally, the fiber diameter should be adapted to the seeing conditions under which the spectrograph is to be used, and to the image scale of the telescope. Assuming $2''$ seeing and an image scale of 6 µm/arcsec (typical for a 25 cm, $f/5$ telescope), then a fiber of $4''$ diameter, corresponding to 24 µm, would pick up essentially all the light of the star. However, the smallest fibers commercially available have a core diameter of 50 µm. This results in a fiber covering approximately $8''$ on the sky and makes a reasonable spectral resolution difficult to reach. Obviously, a larger diameter telescope with the same focal ratio but correspondingly larger image scale is better adapted to the fiber. Fiber-coupled amateur spectrographs will therefore be suitable only for telescopes with diameter above 30 cm.

Fibers can be purchased with both ends already mounted in connectors. As these are usually intended for communications purposes one needs to check, e.g., by means of a laser, that the FRD is acceptable. In practice this means that a parallel beam entering the fiber exits the fiber with a focal ratio slower than $f/8$ and that the fiber ends are perpendicular to the fiber axis within $\approx 2°$. Otherwise the FRD will lead to an unacceptable loss of light. One can also try to polish the fiber ends manually, however this needs some special equipment and extensive experience. The best solution from the optical point of view is the use of dedicated lenses. Their purpose is the proper coupling of the telescope beam into the fiber by imaging the telescope pupil onto the fiber end at an f ratio faster than $f/5$ to minimize FRD. It is advisable to use small lenses to allow high mechanical stability. The image of the star or pupil should not shift by more than a few microns.

Different fiber pick-up systems have been invented and applied that combine the entrance of the fiber with the guiding system. They usually involve an inclined mirror, either with a blank spot in front of the fiber or with a hole behind which the fiber is mounted. Guiding can then be done, visually or with a guide camera, on the reflected actual field.

Finally, for the feeding of the light from the fiber into the spectrograph collimator a trade-off will be necessary: a fast focal ratio helps to reduce the FRD, but a slow beam gives higher resolution.

There are several descriptions by amateur astronomers available in print or on the Internet who have built fiber-coupled spectrographs themselves, like G. Gebhardt or the CAOS group, or who use commercial spectrographs (Dearden [12]).

G. Gebhardt: http://spectros.de

CAOS group: http://www.spectroscopy.wordpress.com

Calibration Following the stage of pure observation and data recording, there are a few crucial steps which should be taken to allow even a comparatively modest processing of the spectra. Such processing will mandate the use of some indispensable accessories, to be described below.

Basically there are three kinds of calibration to be applied to spectroscopic data: removal of the telescope/instrument signature from the data, wavelength calibration, and flux or intensity calibration. The first step is always the removal of the telescope and instrument signature and is described in more detail in the chapter on detectors (see Chap. 6). It is done by applying the bias and flat-field correction.

Although a basic classification of stellar spectral types can be performed by just comparing the spectra to those of standard stars, anything beyond that first step requires a wavelength calibration. In particular the determination of radial velocities (see below) requires that a spectrum of comparison lines of laboratory origin is recorded with exactly the same setup as the object spectrum. For the best results one should take comparison spectra immediately before and after the object spectrum. This allows to correct for possible flexure effects in the spectrograph.

Sources of calibration lines of well-known wavelength are usually spectral lamps emitting a sufficient number of lines (e.g., H, He, Ne, Ar, Th-Ar or Hg). Their light can be guided to the spectrograph slit or the fiber entrance by means of optical fibers. The sharpness of comparison lines can serve as a preliminary indicator of potential errors of the setup.

More advanced analysis, most notably the quantitative study of line changes, requires an intensity calibration of the spectra. This can in principle be done by observing so-called flux standard stars. Their spectra have been measured in absolute units and with high spectral resolution. Lists of flux standards including their measured flux as a function of wavelength are available in the World Wide Web, e.g., in the VizieR "Library of Stellar Spectra" at the Centre de Données astronomiques de Strasbourg (CDS). It allows a search for stars of specific properties (name, spectral type, luminosity class, coordinates, etc.) with flux calibrated spectra.

CDS VizieR:
http://webviz.u-strasbg.fr/viz-bin/VizieR?-source=III/92

7.3.2 The Analysis

This section provides suggestions for an analysis of the recorded spectra. First we describe the basic steps usually undertaken for the reduction of spectral data recorded on a CCD detector.

For the following we assume the availability of all necessary auxiliary data: bias, flat-field and possibly dark frames, comparison spectrum for wavelength calibration. It is crucially important that flat-fields are taken with exactly the same setup of the equipment as the spectral data to be reduced! In case one wants to perform a flux calibration one needs also a spectrum of a flux standard star, its flux distribution in absolute units, all air masses and exposure times of the spectra involved as well as the atmospheric extinction. We note that all the recipes described below are normally performed using one of the dedicated software packages like Visual Spec by V. Desnoux or Prism7 by C. Cavadore and S. Charbonnel for amateurs, or the professional tools ESO-MIDAS or IRAF. They are all (except Prism7) available for free download from the respective websites.

Visual Spec: http://www.astrosurf.com/vdesnoux/
Prism7: http://www.prism7.fr/
ESO-MIDAS: http://www.eso.org/sci/data-processing/software/esomidas/
IRAF: http://iraf.noao.edu/

In a first step one cleans the spectral and calibration data from the (known) bad pixels of the detector and from cosmic ray events.

In the second step one subtracts the bias and (for long exposures) dark frames from the spectra and the bias from the flat-field frames. To improve the statistics of the bias subtraction one should take the average of a large number (usually dozens, depending on the desired signal-to-noise ratio) of bias frames to produce a "master bias" frame.

In the third step one divides all spectral data files by the flat-field. Again, it is advisable to produce a "master flat-field" as the average of many individual flat-field frames.

The next step is the extraction of the object spectrum from the observed data. Apart from the marking of the actual spectrum (usually done with the mouse) one should also mark adjacent strips on the detector uncontaminated by the object spectrum that will subsequently be used to subtract the contribution of the sky brightness.

The wavelength calibration frame is then used to construct the dispersion curve. One first has to identify some of the brighter lines, the programs will then give a first wavelength solution, i.e., λ as a function of position on the detector. This is then refined by incorporating fainter lines and thus reducing the scatter of the wavelength solution to an acceptable value. Once this is done one can apply the wavelength solution to the bias-removed, flat-fielded object and flux standard spectra.

At this stage of the reduction process the spectra still show the influence of the detector's spectral sensitivity curve and the atmospheric extinction. For the determination of spectral types, radial velocities and equivalent widths (see below) it is sufficient to normalize the spectra so that the continuum is at intensity = 1. Again this is a standard procedure in the reduction programs but one can do this also interactively by fitting a spline, Chebyshev or Lagrange polynomial to the observed continuum and then divide the spectrum by the resulting best fitting function.

One is now ready to determine the position, i.e., wavelength of the object's spectral lines, usually by fitting a Gaussian. As an additional result one also gets the value of the equivalent width of the line. In principle this procedure even works for multiple lines (multiple Gaussian fit).

Many quantitative astrophysical problems require a flux calibration of the input data. Apart from the trivial information of exposure time and air mass during the exposure one needs the spectral intensity distribution in the standard star's spectrum in absolute units. These data are available, e.g., from the CDS or even already contained in the reduction program. Note that for a good flux calibration it is necessary to obtain the spectra under photometric conditions and with a slit that is wide enough so that it collects all the flux. Otherwise the flux calibration will suffer from systematic errors.

As was already mentioned, all these reduction steps are usually performed with the help of reduction programs. They also enable the user to do batch-processing, i.e., to reduce many equivalent spectra in an automated way. The details of working with these powerful, partly professional tools should be taken from their respective manuals.

The following sections provide some examples of worthwhile investigations accessible to amateurs with modest equipment. More are given by the Astronomical Ring for Access to Spectroscopy (ARAS), an informal group of spectroscopists promoting cooperation between professional and amateur astronomers in this field.

ARAS: http://astrosurf.com/aras/

7.3.2.1 Spectral Classification The rather low dispersion as usually obtained in objective prism exposures (20–30 nm/mm at Hγ) is sufficient to classify stellar spectra. The first step is to become acquainted with the appearance of stellar spectra, a task most simply done by recording an A-type spectrum with its characteristic Balmer lines (see Table 2) in the blue range. These strong lines are used to determine the dispersion curve, which relates the wavelength of a line to its position in the spectrum, the latter quantity being strongly wavelength-dependent when using a prism. When the dispersion curve (which may depend on the position on the detector) is fixed, fainter lines may be identified with the aid of tables by Adelman et al. [13] or Lide [14]. Comparing the line intensities with those of standard stars from the Digital Spectral Classification Atlas by R.O. Gray or Morgan et al. [15] obtained with the same equipment, the stars under investigation can be classified according to type and perhaps also to luminosity class. For details see Kaler [16]. With sufficient experience, and with the aid of the criteria presented in Sect. 7.2.2.4 and 7.2.2.5, the spectra of comets and meteors can also be classified.

Digital Spectral Classification Atlas: http://nedwww.ipac.caltech.edu/level5/Gray/frames.html

7.3.2.2 Line Changes Drastic changes in spectral lines, such as the transition from absorption to emission lines during a nova eruption, can be easily followed. The observation of periodic line splittings in spectroscopic binary stars usually demands a high dispersion. The quantitative study of these events, however, requires an intensity and a radial velocity calibration.

7.3.2.3 Radial Velocities The radial velocity of a star is the velocity component of its spatial motion along the line of sight from the Earth to the star. It is expressed in km s^{-1} and is counted positive when the star recedes from the observer, negative when approaching. It is found by measuring the displacement, caused by the Doppler effect, of the stellar lines in the spectrum relative to comparison lines. For velocities small compared with the speed of light, the following approximation can be used:

$$\Delta\lambda = v\lambda/c \tag{38}$$

where v is the radial velocity, c the speed of light, λ the laboratory wavelength, and $\Delta\lambda$ the measured shift. Since the motion of the star as well as of the observer enters into $\Delta\lambda$, the measured radial velocities must be corrected for the influence of the Earth's motion.

The corrected velocity is thus referred to the Sun as a point at rest, or, more precisely, to the barycenter of the Solar System. Superposed upon the radial velocity owing to spatial motion may be velocity components caused by relative motions *within* the observed system, such as atmospheric motions in pulsating variables, orbital motions in binary stars, Saturn's rings or even galaxies. Periodic events, repeatedly observed at suitable intervals, provide the necessary data for the construction of the *radial velocity curve*, which shows the relative changes with respect to the systemic velocity. Amateurs (e.g., T. Kaye) have even started to tackle the tiny ($< 0.5\,\mathrm{km\,s^{-1}}$) radial velocity variations caused by extrasolar planets.

T. Kaye:
http://www.spectrashift.com

In the field of stellar astronomy the complex motions which occur in the shells of WR, Be, and P Cygni stars should be mentioned. For stars with high-velocity shell expansions and binaries with sufficiently large radial-velocity amplitudes (see Table 4), already objective-prism spectra permit the determination of *relative* radial velocities, that is, differences of radial velocities of absorption and emission lines, or double lines. *Absolute* radial velocities are best measured with a slit or fiber spectrograph and wavelength calibration spectra. The accuracy of such measurements can be determined by observing radial velocity standard stars of well-known radial velocities (see Table 5). In general, the accuracy improves with higher dispersion and more frequently repeated measurements at the same phase for variable stars. Compromises will often have to be made because of the required exposure times and, in this case, also because of the time variation of the object. For instance, the exposure time should not exceed 1/10 of the period for variable stars. The differences (measured wavelength)–(laboratory wavelength) give from the Doppler formula (Eq. 38) the radial velocities, which are then to be reduced to the Sun. The component of the Earth's orbital velocity in the direction of the star, to be added to the measured radial velocity, is given by

$$v_E = -[(\cos\alpha\cos\delta)\Delta x + (\sin\alpha\cos\delta)\Delta y + (\sin\delta)\Delta z]A \, , \qquad (39)$$

where α and δ are the right ascension and declination of the star, and Δx, Δy, and Δz the daily changes of the rectangular, equatorial coordinates of the Sun tabulated in astronomical almanacs, and which can be considered velocity components of the Earth expressed in astronomical units (AU) per day. The factor $A = 1.731 \times 10^3$ converts this velocity into $\mathrm{km\,s^{-1}}$.

The influence of the Earth's rotation is $0.5\,\mathrm{km\,s^{-1}}$ at most, therefore it only needs to be considered if one wants to investigate smaller effects like in the case of exoplanets. Strictly speaking also the very small correction due to the orbital motion of the Moon should be taken into account, but with an amplitude on the order of meters per second this effect can for the moment be safely neglected here.

An interesting field now becoming accessible to advanced amateurs is spectroscopy of extragalactic objects. Using large telescopes, efficient spectrographs and highly sensitive detectors, it is possible to determine the rotation curves of galaxies seen edge-on. Interestingly, the rotation curves of most galaxies, i.e., the radial velocity as a function of the distance from the center, do not show the decline expected from Newton's law of gravity if one assumes that most of the mass is concentrated in the central region. One rather observes a radial velocity plateau, which is currently ascribed to the existence of dark matter in the galaxies, i.e., matter that only shows gravitational attraction but is otherwise not visible, according to the current standard model of cosmology. More details

Table 4. Spectroscopic binary stars

Name	α(2000)	δ(2000)	V m	Spectral type(s)	Period [d]	V_R [km s^{-1}]	RV ampl.[1] [km s^{-1}]
β Tri	02^{h}09.^{m}5	34°59′	3.0	A5 III	31.39	15.2	67/138
o Per	03 44.3	32 17	3.8	B1 III	4.42	19.8	219/319
λ Tau	04 00.7	12 29	3.4	B3 V + A4 IV	3.42	15.2	111
π Ori	04 52.2	02 27	3.7	B2 III	3.70	24.2	116
η Ori	05 24.5	−02 23	3.4	B1 V	7.98	35.9	290
δ Ori	05 32.0	−00 18	2.2	O9.5 II	5.73	20.1	202
ι Ori	05 35.4	−05 55	2.8	O9 III + O9 III	29.14	26.6	230/392
β Aur	05 59.5	44 57	1.9	A2 IV + A2 IV	3.96	−17.1	215/223
o Leo	09 41.1	09 54	3.5	A2 + F6 III	14.50	27.1	108/126
ζ^2 UMa	13 23.9	54 56	2.3	A2 V	20.54	−5.6	138/135
α Vir	13 25.2	−11 09	1.0	B1 V + B3 V	4.01	0.0/2.0[2]	240/378
α Dra	14 04.4	64 22	3.7	A0 III	51.42	−13.0	94
α CrB	15 34.7	26 43	2.2	A0 V	17.36	1.5	72
π Sco	15.58.9	−26 07	2.9	B1 V + B1 V	1.57	−4.0	262/394
β^1 Sco	16 05.4	−19 48	2.6	B0.5 V	6.83	−1.0/39.2[2]	258/430
ϵ Her	17 00.3	30 55	3.9	A0 V	4.02	−24.2	141/224
θ Aql	20 11.3	−00 49	3.2	B9 III + B9 III	17.12	−27.9	102/127
δ Cap	21 47.0	−16 08	2.8	Am	1.02	−0.2	142
ι Peg	22 07.0	25 20	3.8	F5 V	10.21	−4.6	98

[1] Full amplitude (2K); if known, given for both components.
[2] Radial velocity given for both primary and secondary components instead of system velocity V_R.

are given by U. Geiersbach. Quasars will, due to their apparent faintness, not play a role as targets for amateur spectroscopists in the near future.

7.3.2.4 Equivalent Widths and Line Profiles The term equivalent width is defined as the width of a rectangular absorption feature with central intensity zero and whose area equals that bounded by the spectral line under study. Its formal expression is

$$W = Sd^{-1}I^{-1} , \tag{40}$$

where S is the measured area, I the intensity of the undisturbed continuum at the position of the line, and d^{-1} again the reciprocal dispersion. The equivalent width, usually given in the wavelength scale, characterizes the strength or intensity of the spectral line, which in turn is essentially a function of the number of absorbing atoms and of excitation conditions in the stellar atmosphere. To achieve optimal accuracy, the determination of equivalent widths of spectral lines should be effected with as high a dispersion as possible. In objective prism spectra, only the strongest lines (often only the Balmer lines) can be used.

The amateur can, for instance, monitor changes in the equivalent width, particularly of Balmer and strong metal lines, occurring in many variables such as δ Cephei-type stars. Measuring equivalent widths is simpler and more reliable than determining the true line profiles, since they are independent of the so-called instrumental profile. This term encapsulates all influences which modify the true, original profile of the line. The culprit is primarily the spectrograph slit, whose image superposes on the true profiles. It

Table 5. Radial velocity standard stars. All stars (except δ Lep) are Bright IAU Radial Velocity Standard Stars originally published by Pearce [17]. Measurements are taken from the CORAVEL Radial Velocity Standard Star list by S. Udry/Geneva, http://obswww.unige.ch/~udry/std/std.html

Name	$\alpha(2000)$	$\delta(2000)$	V m	Spectral type(s)	V_R [km s^{-1}]
α Cas	00^{h}40^{m}30^s	+56°32'14''	2.2	K0 II–III	−3.9
β Cet	00 43 35	−17 59 12	2.04	K0III	+13.1
α Ari	02 07 10	+23 27 45	2.01	K2III	−14.6
α Tau	04 35 55	+16 30 33	0.87	K5III	+54.2
β Lep	05 28 15	−20 45 34	2.81	G5II	−14.2
δ Lep	05 51 19	−20 52 45	3.76	G8III–IV	+100.2
β Gem	07 45 19	+28 01 34	1.16	K0III	+3.2
α Hya	09 27 35	−08 39 31	1.99	K3III	−4.7
ϵ Leo	09 45 51	+23 46 27	2.97	G0II	+4.5
β Vir	11 50 42	+01 45 53	3.59	F8V	+4.3
β Crv	12 34 23	−23 23 48	2.65	G5II	−7.6
α Boo	14 15 40	+19 10 57	−0.05	K2III	−5.3
δ Oph	16 14 21	−03 41 40	2.73	M1III	−19.6
α TrA	16 48 40	−69 01 40	1.91	K2II–III	−3.0
β Oph	17 43 28	+04 34 02	2.76	K2III	−12.5
δ Sgr	18 21 00	−29 49 41	2.72	K3III	−20.4
γ Aql	19 46 16	+10 36 48	2.72	K3II	−2.8
ϵ Peg	21 44 11	+09 52 30	2.38	K2Ib	+3.4
ι Psc	23 39 57	+05 37 35	4.13	F7V	+5.6

would be necessary to determine (best empirically) how the profile of an infinitely narrow line finally appears in the recording. In practice, a very faint line of the comparison spectrum is examined, and the measured stellar line can then be mathematically rectified. Such rectified profiles give much more information on the structure of the star's atmosphere than does the equivalent width. But this exceeds the frame of the small college/amateur facilities, and, in particular, requires a high degree of mathematical sophistication.

Several strong effects can be studied qualitatively without rectification procedures. There are, for example, the sometimes very high rotational velocities (up to 500 km s^{-1}) found in particular in many B-type stars, and which lead through the Doppler effect to very broad, dish-shaped line profiles. The width of the lines is a function of the rotational velocity of the star, or rather of the projection of this speed in the direction of the observer, who in general is not in the equatorial plane of the star. With the possibility of working on spectra of high dispersion (better than 3 nm/mm), the analysis can also be made quantitative. From determined equivalent widths of weak lines and of as many elements as possible, the so-called curves of growth can be constructed. These observationally obtained curves are compared with corresponding theoretical curves in order to derive, in a relatively simple way, various stellar quantities such as abundances of the corresponding elements, excitation and ionization temperatures, microturbulence velocities, and damping parameters. This comparison requires extensive information from atomic physics, for instance, excitation energy (Moore [18]) and transition probabilities (Kurucz and Peytremann [19]). Even a summary discussion of this application would exceed the frame of the present introduction. Readers interested in gaining insights into this fascinating subject may wish to consult Voigt [20] or Aller [21], which, among other sources, also survey this subject in an approachable fashion.

References

1. Unsöld, A., Baschek, B.: The New Cosmos: An Introduction to Astronomy and Astrophysics. Springer, Berlin Heidelberg New York (2005)
2. Morgan, W.W., Keenan, P.C.: Spectral Classification. Ann. Rev. Astron. Astrophys. **11**, 29 (1973)
3. Scheffler, H., Elsässer, H.: Physik der Sterne und der Sonne. BI Wissenschaftsverlag, Mannheim (1990)
4. Moore, C.E., Minnaert, M.G.J., Houtgast, J.: *The Solar Spectrum 2935 Å to 8770 Å* (2nd revision of Rowland's preliminary table of solar spectrum wavelengths). National Bureau of Standards Monograph **61**, Washington, DC (1966)
5. Swings, P.: Atlas of Representative Cometary Spectra. University of Liège, Astrophysical Institute (1956)
6. Bronshten, V.A.: Physics of Meteoric Phenomena. D. Reidel, Boston (1983)
7. Thorne, A., Litzén, U., Johansson, S.: Spectrophysics: Principles and Applications. Springer, Berlin Heidelberg New York (1999)
8. Gainer, M.K.: Real Astronomy with Small Telescopes. Springer, Berlin Heidelberg New York (2007)
9. Tonkin S.F.: Practical Amateur Spectroscopy. Springer, Berlin Heidelberg New York (2002)
10. Ramsey, L.W.: Focal Ratio Degradation in Optical Fibers of Astronomical Interest. In: Fiber Optics in Astronomy, ASP Conference Series, vol. 3. Astronomical Society of the Pacific, San Francisco (1988)
11. Avila, G.: Tests of Optical Fibers for Astronomical Instrumentation at ESO. In: Fiber Optics in Astronomy, ASP Conference Series, vol. 3. Astronomical Society of the Pacific, San Francisco (1988)
12. Dearden, S.J.: Spectroscopy with Commercial Spectrographs. In: Tonkin S.F.: Practical Amateur Spectroscopy. Springer, Berlin Heidelberg New York (2002)
13. Adelman, C.J. et al.: Finding List for Multiplet Tables of NSRDS-NBS 3 Astron. Astrophys. Suppl. Ser. **60**, 339 (1985) http://webviz.u-strasbg.fr/viz-bin/VizieR. Accessed 2008
14. Lide, D.R.: CRC Handbook of Chemistry and Physics. CRC, Boca Raton, FL (1998)
15. Morgan, W.W., Keenan, P.C., Kellman, E.: An Atlas of Stellar Spectra, Chicago, Ill., The University of Chicago Press, Chicago (1943) http://nedwww.ipac.caltech.edu/level5/ASS_Atlas/frames.html. Accessed 2008
16. Kaler, J.: Stars and their Spectra. Cambridge University Press, Cambridge (1997)
17. Pearce, J.A.: In: Oosterhoff P. (ed.) Trans. IAU Vol IX B. Cambridge University Press, Cambridge (1955)
18. Moore, C.E.: A Multiplet Table of Astrophysical Interest, NBS Technical Note 36. US Department of Commerce, Washington, DC (1959)
19. Kurucz, R.L., Peytremann, E.: A Table of Semiempirical gf-Values, Special Report 362. Smithsonian Astrophysical Observatory, Cambridge, MA (1976)
20. Voigt, H.H.: Abriss der Astronomie. Spektrum Akademischer Verlag, Heidelberg (2002)
21. Aller, L.H.: Astrophysics, The Atmospheres of the Sun and Stars, 2nd edn. Ronald Press Co., New York (1963)

8. Principles of Photometry

H. W. Duerbeck

8.1 Introduction

8.1.1 General Description and Historical Overview

Radiation from space is practically the only source of information on stars and other distant celestial objects for the observational astronomer. The incident radiation is described completely by its flux, its direction, its color (wavelength), its polarization, and the time of arrival at the location of the receiver. *Photometry* involves the measurement of radiation summed over a more or less extended wavelength region. Depending on the receiver of radiation and the method of observation, one can discriminate between visual, photographic, photoelectric and CCD photometry, point and surface photometry, narrow-band, intermediate-band and broadband photometry, photometry with high time resolution, etc.

The device used to measure the radiation is called a *photometer*. The light-sensitive part of such a device, which transforms the infalling light into another quantity, such as an electric current, is the *detector*. The signal from the detector undergoes a series of transformations before the result is stored permanently. Various kinds of detectors can be used for astronomical observations: the eye, the photographic emulsion, the photomultiplier tube (PMT), the photodiode, and the charge-coupled device (CCD), to name the most important ones (see Sect. 8.1.3).

Photometry has a long tradition in astronomy: in the second century BC, the Greek astronomer Hipparchus introduced the concept of stellar magnitudes and used them for expressing the brightnesses of the stars. This nearly logarithmic scale of stellar magnitudes is still in use today—defined as a purely logarithmic scale, also with fractional magnitudes—in optical as well as in near-infrared astronomy.

An increasing number of professional and amateur astronomers started to engage in visual photometry of variable stars, after an *Appeal to Friends of Astronomy* was made by F.W.A. Argelander in 1844. Amateurs are now organized into many associations all over the world; more than one hundred thousand visual brightness estimates of variable stars are made each year. During the last 25 years, precision photometry has won more and more advocates, first by employing photodiodes and photomultipliers, later by the use of CCDs.

We will attempt to give a general overview of the topic, explaining the possibilities of photometry of different astronomical objects, and will explain the basic facts of photometer types, methods of observation and reduction, and giving hints for useful research programs.

8.1.2 Units of Brightness

This section gives useful relations between astronomical and physical units. Its content is not absolutely necessary for the remainder of the chapter.

A light source (e.g., a star) radiates a certain amount of energy into space; its radiated power L (in astronomy, this is called *luminosity*) indicates the amount of energy per second streaming through an imaginary surface which completely surrounds the star. The radiated power remains the same no matter how large the surrounding surface: an increase of surface (as it surrounds the star at a larger distance) is accompanied by a corresponding decrease in the illumination power, the former quantity increases with the square of the distance between the star and the observer, while the latter decreases by the same amount.

The amount of light per unit time radiated into a unit solid angle (which can be envisaged as an infinitely long cone with an aperture angle of $65.5°$) is called the luminous flux Φ; it is measured in lumens (lm). One lumen is radiated by a light source of radiated power 1 candela (1 cd) into the unit solid angle. One candela is the radiated power which is emitted vertically from an area $1/600\,000\,m^2$ on the surface of a blackbody at the melting point of platinum (2042 K).

Since a light source usually has a certain extent, the brightness or luminance B of the radiation emerging from its surface is the ratio of the luminous flux Φ of the source and its surface area A,

$$B = \frac{\Phi}{A}, \tag{1}$$

and is measured in $cd\,m^{-2}$. The luminance of clear sky is in the range $2000–6000\,cd\,m^{-2}$.

The illuminance E of an object is directly connected with the concept of the stellar magnitude and is defined as the luminous flux Φ which arrives at an area A, i.e.,

$$E = \frac{\Phi}{A}. \tag{2}$$

If 1 lumen arrives vertically at $1\,m^2$, the illuminance is 1 lux (lx). The magnitudes used in astronomy have to be connected with this system of physical units. A luminous flux calibration of the stars α Lyr and 109 Vir by means of a platinum furnace was carried out by Tüg et al. [1]. A star of *apparent* visual magnitude $1^m.0$ has an illuminance of 2.54×10^{-6} lx (if the effect of light absorption of the Earth's atmosphere is neglected). A star of *absolute* visual magnitude $0^m.0$ generates a luminous flux of 2.45×10^{29} cd. The luminance of a "model" sky with one star of 0^m apparent visual magnitude per square degree is $0.84 \times 10^{-10}cd\,m^{-2}$.

Since stars or other objects (e.g., radio or X-ray sources) emit radiation to Earth, which is only vaguely understood as "illuminance," astronomers also use the concept of monochromatic radiation flux (for short, flux), which is the illuminance per unit wavelength (or unit frequency, depending on the subscript). The flux radiated at wavelength 555.6 nm (green light) from α Lyr amounts to

$$\Phi_\lambda = 3.47 \times 10^{-11}\,J\,m^{-2}\,s^{-1}\,nm^{-1}, \tag{3}$$

Wavelength unit: 1 nm (nanometer) $= 10^{-9}$ m; in astronomy the unit $1\,\text{Å} = 0.1\,nm = 10^{-10}$ m is still widely used.

at the Earth. This corresponds to a *monochromatic photon flux* of

$$N_\lambda = 9.7 \times 10^7 \text{ photons m}^{-2}\,\text{s}^{-1}\,\text{nm}^{-1}\,. \tag{4}$$

The above relations are important if physical parameters are derived from astronomical observations; in general, astronomical photometry is a method of comparing "illuminances" E or "radiation fluxes" Φ_λ of different stars.

When two stars observed with the same photometer yield the radiation fluxes Φ_1 and Φ_2, their magnitude difference is

$$\Delta m = m_1 - m_2 = -2.5\log_{10}\left(\frac{\Phi_1}{\Phi_2}\right)\,, \tag{5}$$

where a magnitude difference of 1^{m} corresponds to a ratio of the radiation fluxes of 2.512. The minus sign in the formula takes into account that smaller radiation fluxes correspond to larger stellar magnitudes.

8.1.3 The Receivers

This section will discuss the properties of the human eye, of the photographic emulsion, and of different photoelectric devices as receivers of radiation. While the eye has been used for brightness estimates for millennia, photography has played its dominant role between 1850 and 1950. It was partially replaced by the more sensitive and more accurate photomultiplier tube (PMT), a receiver without spatial resolution, and in the last decades by the charge-coupled device (CCD), which combines the advantage of spatial resolution, sensitivity and photometric accuracy.

8.1.3.1 The Eye In amateur astronomy, a large number of photometric observations, called estimates, have been carried out by eye, and hence we give here a short introduction.

The eye is approximately of spherical shape. An image is formed on the backlying choroid coat, which is composed of two different types of vision cells, about 125 million *rods* and 6 million *cones*. The less light-sensitive cones are employed for color vision during the daytime. These receivers are concentrated in the central region of the choroid coat, the so-called yellow spot or fovea. The angular resolution is about 1/2 arcminutes. The rods are about 10,000 times more sensitive than the cones; they can be employed only for black-and-white vision (indirect vision, extrafoveal vision). Faint astronomical objects are most easily recognized by indirect vision, that is, by looking past the object and thus employing the rods. It is not advisable to examine two stars at the same time to obtain an estimate, because different areas of the choroid coat have different sensitivities.

The power of sensation S for light of different intensity depends, within certain limits, on the logarithm of the strength of stimulus, E (the Weber–Fechner law):

$$\text{constant} \times (S - S_0) = \log_{10}\left(\frac{E}{E_0}\right)\,. \tag{6}$$

The time of adaption from bright to dim is about 15 to 45 min. A dim red light, which just allows the keeping of a logbook or the study of star maps, is recommended when observing in darkness.

More details on the physics of the eye are found at http://webvision.med.utah.edu/.

The reaction of the human eye to red light is different from its response to other colors. When magnitude differences between white and red stars at different intensities (with telescopes of different sizes) are determined, major deviations may occur (the Purkinje phenomenon). Visual magnitude estimates of red stars should be made by choosing other red stars for comparison.

When a 6th magnitude star is observed, 200 photons enter a fully adapted eye every 1/20 of a second. The quantum efficiency of the human eye, which is indeed able to detect single photons, may be as high as 15% and thus has a better efficiency than the photographic emulsion. It does not, however, possess the latter's storage capacity.

The quantum efficiency is defined as the percentage of photons incident on a receiver which produce detectable events.

8.1.3.2 The Photographic Emulsion

The radiation detectors of a photographic emulsion are silver halide crystals with sizes of about 1 μm, embedded in a thin gelatin layer. The emulsion is attached to a base such as a sheet of film or a glass plate. The exposure to light causes the formation of "germs" of metallic silver; during development the exposed crystals are converted completely into black metallic silver (photographic negative). The formation of germs occurs only when the radiation is energetic enough (i.e., of short wavelength or blue light); such emulsions are called orthochromatic. The addition of organic substances increases the sensitivity of the emulsion to light of longer wavelengths (panchromatic emulsions).

For astronomical applications, emulsions for light and infrared radiation up to a wavelength of 1100 nm have been used. At the short wavelength end, the limitation arises because the Earth's atmosphere completely absorbs UV quanta below 320 nm; also the gelatine and the optical elements are not transparent to UV radiation. Generally, the quantum efficiency is below 1%. The photographic plate is less sensitive than the human eye, it can, however, accumulate the light over a long time and thus reach much fainter levels of brightness.

The brightness information of a photographic emulsion is encoded in a scale of density. A density interval corresponds to a small brightness interval when the gradation of the emulsion is high (= high-contrast emulsion). The relation between density and infalling intensity has to be established by a calibration device (a sensitometer where light of different illumination levels is shone on separate sections of an emulsion), or by using, in direct photographs of the sky, fields with stars of known magnitude ("photometric sequences").

8.1.3.3 The Photomultiplier Tube (PMT)

When light hits a cathode (generally composed of alkaline oxides) placed inside a vacuum tube, electrons are released. This is called the external photoelectric effect. The electrons are directed, by a potential difference, onto a dynode, where they are able to release more electrons, which are then directed to a second dynode, and so on. Amplifications by a factor 10^6 are reached with 10 dynodes. The "bursts" of electrons caused by single photons can be counted when fast electronics are used (photon counting). The quantum efficiency can be as much as 20%. PMTs are preferentially blue-sensitive, but special cathodes can extend the sensitivity up to 1.1 μm (S-1 photocathode).

PMTs have been standard equipment in professional astronomy for more than 50 years.

PMTs need a stabilized high-voltage power supply (over 1000 V), and are sensitive to overexposure of light and mechanical stress. The sensitivity in the blue and visual region is very good, the time resolution in photon-counting mode excellent. PMTs can be used either at the ambient temperature or cooled; in the latter case the dark current is reduced by several orders of magnitude. The cooling is achieved with dry ice or by the use of a *Peltier element*.

8.1.3.4 The Photodiode

A semiconductor such as silicon or germanium which has been "doped" with special other sorts of atoms can become conducting because of the released negatively charged electrons (n-type) or the positive "holes" they leave behind (p-type). A *diode* is made up of a combination of a p- and an n-layer, and lets a current pass only in one direction.

Light that is absorbed in the p–n boundary layer releases electrons and changes the resistance. Diodes specially designed for the measurement of light have a layer (I) between the p- and the n-region, where light is absorbed and where the formation of electron-hole pairs by thermal effects is reduced. Such diodes are called *PIN diodes*.

The use of PIN photodiodes is relatively simple. The power is usually supplied by a battery, the costs are low, and the detector is insensitive to overexposure. The sensitivity in the red and infrared regions is good. Disadvantages are that the relatively weak signals must be highly amplified, that the blue sensitivity is low, and that the time resolution is low.

8.1.3.5 Charge-Coupled Devices (CCDs)

CCDs for one- or two-dimensional image recording have been used for photometric purposes in astronomy since the 1980s, and have been used in amateur astronomy since the 1990s. A CCD chip is a wafer of semiconducting silicon divided into little squares ("pixels"). A chip with an area of $100\ mm^2$ carries a grid of about 250,000 light-sensitive elements. Infalling photons produce electron-hole pairs in the silicon layer. Pools of electric charge are formed in each pixel. After exposure, the controlling electronics "reads out" the device by shifting the charge from pixel to pixel to one side of the chip where be measured and the digital output fed into a computer. Like the photodiodes, the CCDs are red-sensitive and can be used for short-wavelength radiation only with special coatings. The CCD has a greater sensitivity than the dark adapted eye, and can accumulate light over a much longer time. The quantum efficiency is at maximum 80%. The time resolution is limited by the read-out electronics; it can be increased by binning (combination of pixels to larger ones), readout of restricted areas on the CCD, or continuous readout of a few CCD columns.

CCDs have revolutionized amateur astronomy. See Howell [2] for more detailed information on CCD characteristics and their application.

8.2 Limits and Accuracies of Photometric Measurements

The radiation of celestial bodies is influenced in many ways while en route to the observer. The radiation of fixed stars is partly scattered and absorbed when it penetrates the interstellar matter. This *interstellar absorption* has different strengths at different wavelengths (i.e., it is selective), which leads to a "reddening" of the light. A similar effect occurs in the Earth's atmosphere, and is there called (selective) *extinction*. Part of the incident radiation is scattered over the whole sky. The scattered light of the Moon increases the sky brightness by several orders of magnitude, and makes the photometry of faint stars less efficient or even impossible.

Turbulent elements in the air randomly deflect incoming radiation from its original path, thereby causing a net widening of the bundle of light rays. The normal bending of light in the atmosphere (the refraction) also tends to widen the incident light bundle. At large zenith distances, each point-like stellar image is widened to a noticeable spectrum owing to the wavelength dependence of refraction.

The light collected by a telescope is also changed by the optical elements. Wavelength-dependent reflection losses are produced at mirror surfaces; radiation is absorbed when passing through glass, and another part is lost by reflection. After an additional weakening and partial blockage by color filters, the remaining light reaches the detector, which registers, depending on its efficiency, only a fraction of the incident radiation. The final result is obtained only after further processing of the signal, where additional information losses occur owing to, for example, amplifier noise. Section 8.5 describes reduction methods where these effects are taken into account. Most effects are internally compensated because the light from the target star is compared with that from one (or more) comparison stars; both rays travel the same path into the measuring device.

8.3 Astronomical Color Systems

Stellar magnitudes can be defined in different wavelength regions. Astronomical photography led to the introduction of *photographic magnitudes*, because the first photographic emulsions were sensitive only to radiation of $\lambda < 490\,$nm. The difference of two magnitudes measured in different wavelength regions is called a color index. The most frequently used color index, $B - V$, is a measure of how much more light a given star radiates in the blue than in the green, compared to a normal star (defined in general as an unreddened star of spectral type A0). Such a color index provides a simple measure of the energy distribution of a star between two wavelength regions. The color index of a star is thus defined quantitatively as

$$m(\lambda_2) - m(\lambda_1) = -2.5 \log\left(\frac{a\Phi_2}{b\Phi_1}\right), \tag{7}$$

where λ_1 and λ_2 are the effective wavelengths at which the radiation fluxes Φ_1 and Φ_2 are measured, and a and b are constants which are characteristic for a given measuring device.

The development of accurate photometers was accompanied by the design of astronomical color systems: Before the light falls onto a photocell, it passes through a color filter, which may be one of a sequence. More precisely, the combination of telescope, filter, and sensitivity function of the receiver defines a bandpass. A set of color filters, in combination with a given telescope type and a characteristic curve of a certain PMT cathode or CCD type, defines a particular color system.

Broadband systems (bandpass widths typically 100 nm), intermediate band systems (30 nm), and narrow-band systems (10 nm) are in use. The most widely known, which will also be discussed here, are the *UBV* system, developed by H.L. Johnson and later expanded to longer wavelengths (R, I, J, K, L, M) (Fig. 1), and the *uvby* system, developed by B. Strömgren.

In the *UBV* system, the brightness in the visual region is roughly equal to the visual magnitude; this magnitude is designated V. The B magnitude is approximately equal to

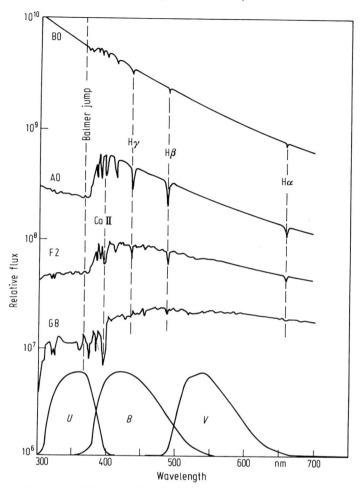

Fig. 1. Transmission curves of the U, B, and V filters of the Johnson UBV system (*bottom*), and radiation fluxes of main sequence stars of different spectral types (*top*). The slope of the stellar continuum, which depends on the surface temperature, is determined by the $(B - V)$ index. The Balmer jump, which depends for a given temperature on the luminosity of the star, is determined by the $(U - B)$ index. From Henden and Kaitchuck [3]

the photographic magnitude, and U is the ultraviolet magnitude. The color indices of the UBV system (see Eq. 7) are $U - B$ and $B - V$.

The original UBV filters were designed for use in a reflecting telescope and a PMT with an S-4 cathode sensitivity; the use of CCDs has made it necessary to change the original colored glass combinations. Suppliers usually offer customized filters for their photometers and cameras.

A few supplementary catalogs are:

- A.U. Landolt (1992): *UBVRI* photometric standard stars in the magnitude range 11.5–16.0 around the celestial equator. A widely used collection of standard stars.

Online version of Landolt:
http://www.ls.eso.org/lasilla/
Telescopes/2p2T/Landolt/

- A.A. Henden, R.H. Kaitchuck (1989): Astronomical Photometry. Table C1/C2, p. 292 contains *UBV* magnitudes of more than 150 stars, some bright, some faint [3].
- B.D. Warner (2006): Lightcurve Photometry and Analysis. Appendix H and I, p. 231, contain finding charts and tables of Landolt *UBVRI* and of Henden *UBVR* photometric standards [4].
- J.-C. Mermilliod, B. Hauck, M. Mermilliod (1997): The General Catalogue of Photometric Data (GCPD). A collection of data of more than 2 million stars.

GCPD on the Web:
http://obswww.unige.ch/gcpd/gcpd.html

In most of the broadband systems in use today, the magnitude of Vega (α Lyr) is set to $0^{\mathrm{m}}.0$ at all wavelengths. The color indices of Vega in such color systems are therefore also equal to $0^{\mathrm{m}}.0$. It should be noted that equal color indices of a star do not mean a constant (flat) energy distribution; it means only that the energy distribution is close to that of an unreddened A0 V star.

From the color indices $(U-B)$ and $(B-V)$ of a normal star, its temperature and spectral type can be determined, and with restrictions, also its luminosity class. The interstellar reddening E_{B-V} changes the color indices of stars. If E_{B-V} is known, the intrinsic or unreddened colors $(U-B)_0$ and $(B-V)_0$ can be calculated:

$$(B-V)_0 = (B-V) - E_{B-V}, \tag{8}$$

$$(U-B)_0 = (U-B) - 0.72E_{B-V}, \tag{9}$$

or vice versa, if the $(B-V)_0$ index is estimated, e.g., from the spectral type, the reddening is calculated. Also the quantity Q (see below) can be used to estimate $E(B-V)$. The interstellar absorption in the V region is

$$A_V = 3.2E_{B-V} . \tag{10}$$

From the measured color indices a reddening-independent quantity Q can be calculated:

$$Q = (U-B) - 0.72(B-V). \tag{11}$$

Possible spectral types and luminosity classes can be derived from the Q-value of a star.

Although not uniquely defined, the following plausibility criteria are of help: bright (and hence nearby) main sequence stars (luminosity class V) are hardly reddened, while faint (and hence distant) supergiants (luminosity class I) are usually heavily reddened.

8.4 The Technique and Planning of Observations

In the planning of a series of observations, one should consider whether the accuracy (spectral resolution, time resolution, photometric accuracy) which can be achieved with a given photometer/telescope combination is sufficient for the tackling of the problem at hand. Second, the use of a special color system should be investigated. One should not forget to observe suitable standard stars or standard fields with known magnitudes and colors together with the program stars. The recording of precise times of measurements is very important when stellar occultations or minima and maxima of variable stars are to be observed. In any case, a time measurement is associated with each photometric measurement, and will later be used to derive corrections for the influence of the Earth's atmosphere.

Additional effects, such as differential extinction or the light-travel time effect (see Sect. 8.6), have to be taken into account as the accuracy of measurements is increased.

The quality of atmospheric conditions plays a role; it is for this reason that the large national and international observatories are located in some of the remotest corners of the Earth. Nevertheless, slight variations of atmospheric transparency are tolerable when spatial receivers (like a CCD or a double-beam PMT photometers) are employed, and program and comparison star(s) are recorded simultaneously with the same receiver.

8.4.1 Point Photometry and Surface Photometry

Two different types of photometric observations can be distinguished, point photometry and surface photometry. When carrying out point photometry (a typical application being photoelectric stellar photometry), a diaphragm is used to isolate a "point" in the sky. The diaphragm can sometimes cover a noticeable sky area without, however, resolving it. When carrying out surface photometry, on the other hand, a region of sky is "scanned" in either one or two dimensions with a certain resolution with a PMT, or recorded with a two-dimensional receiver like a CCD.

8.4.2 Visual Photometry: Differential Observations

The best known method for the visual determination of stellar magnitudes is Argelander's step-estimate method. The magnitude of a star (usually a variable) is determined by comparing it with other stars in the field whose magnitudes are known. In the following, the variable is designated with the letter a, the comparison star with b. The variable whose brightness is to be estimated should appear in the instrument as bright as a star of magnitude $0^m.5$ to $3^m.5$ (the Fechner region) with the naked eye. The required apertures of the instrument are given in Table 1.

The brightness difference of the two stars is given in "steps," which are defined according to Argelander as follows:

0: "If both stars appear to be of equal brightness, or if I estimate sometimes the first, sometimes the second a little bit brighter, I call them of equal brightness and I note *a 0 b*."

1: "If, at first glance, both stars appear of equal brightness, but if careful observing (repeatedly shifting from a to b and back) shows a always, or with rare exceptions, to be just noticeably brighter, I call a one step brighter than b and note *a 1 b*; however, if b is the

Table 1. Necessary apertures for brightness estimates. The limiting magnitude for good estimates is about 1 mag brighter than the (intrinsic) limiting magnitude

Magnitude range of the variable	Aperture of the instrument (in mm)
$0^m.5$ to $4^m.5$	7 mm (Eye)
2.5 to 6.5	20 mm (Opera glasses)
4.5 to 8.5	50 mm (2-in. or binoculars)
6.5 to 10.5	100 mm (4-in.)
8.to 12.5	250 mm (10-in.)

brighter one, I note *b 1 a*, so that always the brighter is in front and the fainter behind the number."

2: "If one star always and without doubt appears brighter than the other, this difference is taken as two steps and designated *a 2 b* if *a* is the brighter one, and *b 2 a*, if *b* is the brighter one."

3–4: "A difference which is noted at first glance is taken as three steps and is noted as *a 3 b* or *b 3 a*. Finally, *a 4 b* denotes an even more pronounced difference."

It is not advisable to estimate in more than 4 or 5 steps. One step corresponds roughly to one-tenth of a magnitude. At least two comparison stars should always be used, one of which, *b*, should be brighter, and the second, *c* fainter than the variable *a*. One then records the date, time, program object and comparison stars, and the relative step-brightnesses as, for example, "*b s_1 a s_2 c*", where s_1 and s_2 are step values. When observing variable stars, special finding charts with the identification of the variable and data on visual magnitudes of comparison stars are quite useful. Such a chart is shown in Fig. 2.

If the variable is "bracketed" between two comparison stars *b, c* of known brightness, and if m_1 is the magnitude of the brighter star *b*, m_2 that of the fainter comparison *c*, then the magnitude of the variable m_V can be calculated according to the following formula:

$$m_V = m_1 + \left(\frac{m_2 - m_1}{s_1 + s_2} \right) s_1 , \tag{12}$$

or, after a simple transformation,

$$m_V = m_2 - \left(\frac{m_2 - m_1}{s_1 + s_2} \right) s_2 . \tag{13}$$

The step value, whose reciprocal value indicates the slope in the equations, is $(s_1 + s_2)/(m_2 - m_1)$.

Important visual photometry on variable stars (pulsating, eruptive, irregular) can be done, especially when results of many observers can be combined. To achieve this task, comparison stars should be chosen carefully. Associations of variable star observers (AAVSO, AFOEV, BAA, BAV, RASNZ/VSS) often supply finding charts (Fig. 2) which are essential for the generation of consistent light curves (Fig. 3).

Collections of charts for the visual observation of variable stars are:

- Charts of the American Association of Variable Star Observers (AAVSO), 49 Bay State Rd. Cambridge, MA 02138 (charts for about 2000 variables with visual magnitude scales for step estimates).
- Charts for Southern Variables, published by The RASNZ Variable Star Section.

  Useful star atlases are:

- R.W. Sinnott, M.A.C. Perryman (2007): Millennium Star Atlas. Sky Publishing Corporation, Cambridge, MA. A collection of 1,058,332 stars down to 11th magnitude on 1548 charts, based on the astrometric and photometric results from the Hipparcos mission.

In order to eliminate later uncertainties, one may note the date in the following way: 1987 June 5/6 Sa/Su.
One records the used time (e.g., Eastern Daylight Saving Time (EDT), or Universal Time (UT)). The use of Universal Time is recommended, because it is not subject to hourly shifts in the course of the year. A count of hours beyond 24 is often useful to remove ambiguities (1 h 00 in the morning of the following day = 25 h 00 of the previous day).

Custom variable star charts can be plotted at:
http://www.aavso.org/observing/charts/vsp/
Magnitudes of suitable comparison stars are also given.

See: http://www.rasnz.org.nz/vss/vss.htm

See also: http://www.rssd.esa.int/index.php?project=HIPPARCOS&page=msa

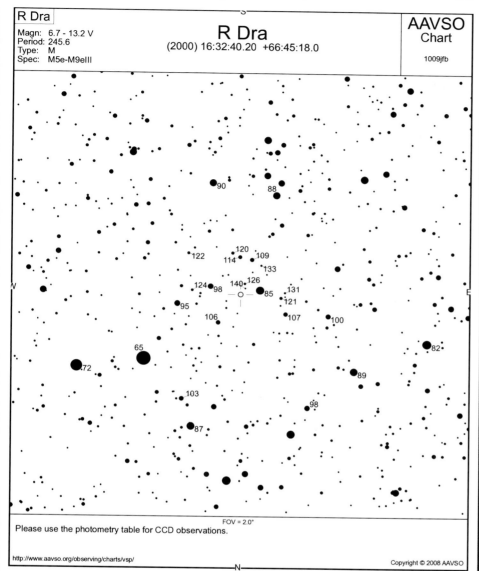

Fig. 2. AAVSO finding chart of the Mira variable R Dra. The field has a size of about 2° × 2°, and south is up. The variable is in the center of the chart. The magnitudes of the comparison stars are given to a tenth of a magnitude; decimal points are omitted to avoid confusion with stars. With kind permission of the AAVSO

- C.E. Scovil (1981): AAVSO Variable Star Atlas. Sky Publishing Corporation, Cambridge, MA. All variable stars brighter than $9^m.5$ (visual) at maximum with amplitudes larger than $0^m.5$ are identified. Magnitude scales are outdated!
- A. Becvar: Atlas Borealis/Eclipticalis/Australis, Sky Publishing Corporation, Cambridge, MA.

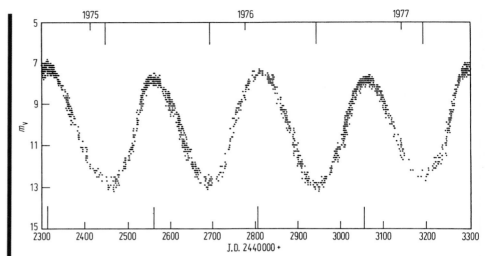

Fig. 3. Light curve of the Mira variable R Dra, derived from visual observations of members of the AAVSO. With kind permission of the AAVSO

The Becvar atlas is useful for selecting comparison stars since it gives the spectra (or at least colors) of all stars. A number of star atlases is also accessible online:

http://archive.eso.org/dss/dss

http://aladin.u-strasbg.fr/

- ESO Online Digitized Sky Survey:
  Digitized scans of Schmidt plates of the northern and southern sky.
- Aladin: An interactive software sky atlas:
  Digitized astronomical images, superimposed entries from astronomical catalogs or databases.

8.4.3 Photoelectric and CCD Photometry: Differential Observations

The method of differential observations is the easiest way to obtain useful observations and results with a CCD or a PMT, especially when observing variable stars of relatively short period and small amplitude. A variable star and two stars in its vicinity are chosen. The two stars should satisfy the following criteria:

1. They should have colors and magnitudes similar to those of the variable to avoid large corrections for second order extinction (see Sect. 8.5.1).
2. They should not be variable. They should not, for instance, be listed in the *Catalogue of Variable Stars*. However, even a star not listed there might be variable, often with quite a small amplitude. The constancy of any of the comparison stars can be checked using two stars whose brightness difference should be monitored. This yields information regarding not only their constancy, but also of the quality of the night, performance of the instrument, etc.
3. Their distance from the variable should be small enough that they can be (a) imaged all three on the CCD simultaneously, or (b) seen together with the variable in the finder of the telescope in the case of single-beam PMT photometers, making it possible to point easily to the various objects.

In the case of single-channel PMT photometry, one proceeds as follows: If the variable is designated with V, the two comparison stars with C_1 and C_2, and the sky alone with (S), the following measuring sequence is carried out (using a UBV filter set for each step, if desired):

$$(S) - C_1 - V - (S) - V - C_1 - C_2 - (S) - C_2 - C_1 - V - (S) - V - C_1 - \ldots$$

The signals are usually recorded as count rates (photon counting or voltage-to-frequency conversion) with respective time marks. Because of the symmetry in the sequence of observations, the following brightness differences can be calculated from the ratios of neighboring measurements:

AAVSO CCD observing manual:
http://www.aavso.org/observing/
programs/ccd/manual/

Sources of image processing
software. For a general overview,
see:
http://www.phy.duke.edu/~kolena/
imagepro.html

MPO Canopus:
http://www.minorplanetobserver.com/
MPOSoftware/MPOSoftware.htm

Astronomical Image Processing for
Windows Software (AIP4WIN):
http://www.willbell.com/aip4win/
AIP.htm

MaxImDL (for Windows):
http://www.cyanogen.com/

Astroart 4.0 (for Windows):
http://www.msb-astroart.com/

Mira AL Software
(for Windows XP):
http://www.mirametrics.com/

Image Reduction and Analysis
Facility (IRAF), software for
the reduction and analysis of
astronomical data (for Linux,
Windows XP):
http://iraf.noao.edu/

ESO-MIDAS: ESO Munich Image
Data Analysis System:
http://www.eso.org/sci/
data-processing/software/
esomidas/ (for Linux)

$$\Delta m(V - C_1) = -2.5 \log_{10}\left(\frac{V + V - 2S}{C_1 + C_1 - 2S}\right) \tag{15}$$

$$\Delta m(C_2 - C_1) = -2.5 \log_{10}\left(\frac{C_2 + C_2 - 2S}{C_1 + C_1 - 2S}\right). \tag{16}$$

The symmetrical arrangement of measurements has the advantage that *slow* atmospheric changes are of little consequence, since they influence the variable and the comparison star by the same amount. Two measurements of two stars yield one magnitude difference. If one calculates, however, in a sequence like $C_1 - V - V - V - \ldots C_1$ from each variable star measurement a magnitude difference (e.g., by linearly interpolating the count rate or deflection of the comparison star), the single magnitude differences would not be independent of each other. It is often better and more convenient to work with fewer measurements of higher precision than to deal with many measurements with larger scatter. There are, however, exceptions, such as the observation of rapidly varying objects (e.g., cataclysmic variables).

In the case of CCD photometry, V, C_1, C_2 and S are recorded simultaneously on a single image frame. Image processing software is used to determine the signals of the stars (and the background) by means of aperture photometry or profile fitting. Before the actual result can be obtained, standard CCD reduction steps have to be carried out (bias and dark current subtraction, flat-fielding). These steps are not described in detail here, the reader is directed to the AAVSO CCD Observing Manual, and to the following sources of image processing software. For each CCD frame, the magnitude difference between the variable and, e.g., the average magnitude of two comparison stars, can be calculated.

The results must be corrected for differential extinction of the first and second order (see Sect. 8.5.1) and for light-travel time (see Sect. 8.6.1).

8.4.4 Absolute Photoelectric Photometry

The field of absolute photoelectric photometry is much less attractive in amateur circles because, in general, no useful results can be obtained. It is important, and difficult, to transform the magnitudes from the local (instrumental) color system to a standard color system. Stringent demands for the stability of sky transparency and the stability of the equipment are required. One possible field of activity is the establishment of magnitude sequences for variable star work.

8.4.5 Photometry of Occultations

When the changes of brightness occur with time scales of seconds to milliseconds and they are not periodic, the observer is forced to omit the comparison star measurements (unless a two-channel photometer is available). Such rapid brightness changes occur when a body of the Solar System occults a star or another object of the Solar System. Instead of a comparison star, one has two reference magnitudes: the combined brightness of occulting and occulted object outside occultation, and the residual brightness of the occulting object during occultation. Possible occulting objects include the Moon, the planets, and planetoids; occulted objects are planets, planetoids, and fixed stars. The reader is also referred to Sect. 13.4 on stellar and planetary occultations by the Moon. The speed of the apparent motion of occulting objects and the diameters of the occulted objects determine the brevity of the observed phenomena. Typical data are given in Table 2.

When a comet or a planet enshrouded by an atmosphere occults a fixed star, the structure of the light curve can be most intriguing, because it shows how different atmospheric layers are transparent to the light of the fixed star. Times of occultation are only roughly defined.

Structured light curves can also be observed when a star disappears or reappears behind an object without atmosphere, such as the Moon or a planetoid. These effects are caused by diffraction effects. For the Moon, the effects begin to occur when the object is 0.01 arcseconds from contact. Twenty milliseconds elapse between the first and the second diffraction minimum if the occultation is a central one. The amplitude of brightness fluctuations is as large as $0^m.5$ (Fig. 4). The smaller apparent motion of a typical planetoid is compensated by its greater distance, so that the time scales of the diffraction effects are of the same order of magnitude.

If a circular planet with an atmosphere centrally occults a star, the so-called central flash of the occulted star can be measured at the instant the star is exactly behind the center of the planet. The flash is caused by light refraction at the planet's limb. An area photometer which can resolve the limb would register a bright ring along the planet's limb during the time of the flash.

In some cases, occultations by planetary objects are accompanied by unforeseen phenomena, such as occultations by unknown planetary satellites or rings. It is therefore advisable to start the observations well ahead of the precalculated "normal" occultation, and to continue them for quite some time after the event.

Some estimates of the obtainable accuracy should be given: Using a 20-cm telescope with a photometer having an integration time of 10 ms, a statistical error of 5% per in-

Table 2. Apparent motion of Solar System objects and angular diameters of occulted objects

Motion of the Moon	$0.''5\,s^{-1}$
Motion of planetoids	$0.01 \times$ Moon motion
Motion of Neptune	$0.001 \times$ Moon motion
Diameter of a planet (typical)	$10''$
Diameter of a bright planetoid	$0.''1$
Diameter of a fixed star	$<0.''05$ (Antares)

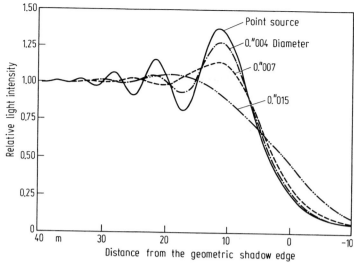

Fig. 4. Theoretical diffraction patterns for lunar occultations of stars of different angular diameters. Higher amplitudes of the oscillations correspond to smaller angular diameters. From Evans [5]

tegration can be expected for stars of $6^m.5$ and occulting planetoids; for objects of $10^m.5$ and an integration time of 1 s, an error of 3% can be anticipated.

Occultations in daylight are easiest to observe visually, since the low contrast makes other observations quite insensitive. If one uses a photometer with a diaphragm diameter of $30''$, and observes with a sky brightness of 2^m per square arcsecond, the observed brightness jump of an occulted star of 0^m is only $0^m.01$, which is at the limit of detectability. The following contain more information on occultations:

- Manual: Richard Nugent, Chasing the Shadow: The IOTA Occultation Observer's Manual
- Occultation prediction software by D. Herald
- Lunar Occultation workbench by E. Limburg
- Occultations by Moon, Planets and Asteroids, with links to predictions: International Occultation Timing Association (IOTA)

Downloadable from:
http://www.poyntsource.com/IOTAmanual/Preview.htm

http://www.lunar-occultations.com/iota/occult4.htm

http://low4.doa-site.nl/index.html

http://www.occultations.org/
http://www.iota-es.de/ (European Section)

http://www.lunar-occultations.com/iota/iotandx.htm
http://www.asteroidoccultation.com/
http://mpocc.astro.cz/

8.5 Reduction Techniques

8.5.1 Reduction of Photometric Measurements

None of the values which are determined by eye, photographic plate, photoelectric and CCD photometry is already a final "stellar magnitude." Each measurement of brightness (and its corresponding time; see Sect. 8.5.3) has to be corrected for falsifying local influences before the true stellar magnitude is obtained.

Starlight is scattered and absorbed in the Earth's atmosphere to a degree which increases when the path in the atmosphere is longer (the astronomer uses the concept of

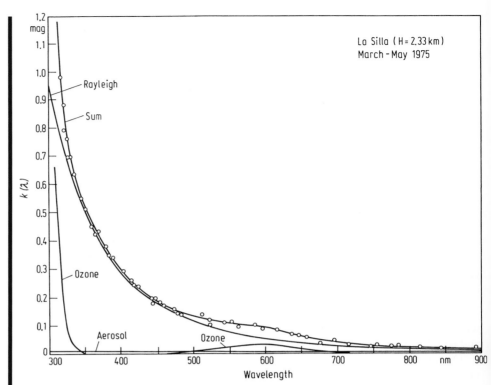

Fig. 5. Coefficient k of the atmospheric extinction as a function of wavelength (in nanometers) for the site La Silla, Chile, 2330 m above sea level. The extinction is composed of different components: small, mutually absorbing dust particles (aerosols), broad molecular absorption bands (predominantly of ozone), and air molecules which cause Rayleigh scatter (in proportion to λ^{-4}). The extinction increases towards small wavelengths, the use of broadband filters yields mean values of k. From Tüg et al. [1]

air mass), and which also increases toward shorter wavelengths. The observer thus measures different magnitudes of the same star, depending on how high it is in the sky. One can determine from a series of observations, taken at different elevations (or zenith distances), how bright the star would appear without the falsifying influence of the Earth's atmosphere. The following formulation is used:

$$m_0 = m - (k' + k''c)X ,\tag{17}$$

where m is the measured magnitude of the star, m_0 the magnitude of the star outside the atmosphere, k' the extinction coefficient of first order (Fig. 5), k'' the extinction coefficient of second order, c the color index of the star, and X the air mass. At the zenith, the air mass is $X = 1.0$, while outside of the Earth's atmosphere it is $X = 0.0$.

A good approximation is:

$$X = \sec z ,\tag{18}$$

where z is the zenith distance of the star and the trigonometric function sec (secant) is the reciprocal of the cosine. From the equation for the zenith distance follows the

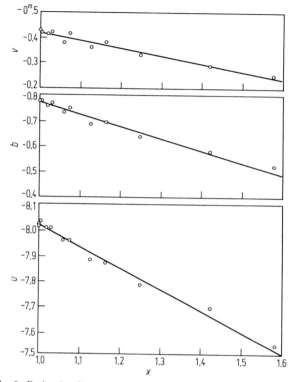

Fig. 6. Extinction (in magnitudes) for the spectral bands U, B, and V as a function of air mass X. From Henden and Kaitchuck [3]

relation

$$X = \frac{1}{\sin \phi \sin \delta + \cos \phi \cos \delta \cos h} \,, \tag{19}$$

where ϕ is the geographic latitude of the observer's station, δ is the declination of the star, and h is the hour angle of the star, expressed in degrees (h is sidereal time minus right ascension of the star). It is apparent that the air mass of the star is smallest when $h = 0$, that is, when the sidereal time is equal to the right ascension. The star is then on the meridian. Observations of "extinction stars" far away from the meridian are also necessary in order to properly determine the extinction coefficients, while program stars (variables) should be observed close to meridian transit, where measurements of highest photometric accuracy are attainable.

To determine true stellar colors and magnitudes, the observer must measure a multitude of extinction stars of various colors under good atmospheric conditions (good, stable transparency of the atmosphere—no haze, no cirrus, no clouds!). One obtains a system of equations of the type Eq. (17) for the single stars, and obtains graphically (Fig. 6) or by the method of least squares the extraterrestrial magnitudes m_0 in the instrumental system. If a series of stars of known color and magnitude is measured in a color system, then transformation equations of the type

$$m_{st} = m_0 + \beta c + \gamma \tag{20}$$

can be established which permit the transformation of the instrumental magnitude m_0 into into a standard magnitude m_{st}. Here, c is the color index of the star in the standard system (since this is not known in the beginning, it has to be assumed and determined by iteration), and β and γ are, respectively, the color coefficient and the zero-point constant of the instrumental system.

The correction for differential extinction between the variable V and comparison star C can be calculated for each hour angle of the variable by determining for both stars the air masses from Eq. (20), the difference ΔX, and, according to Eq. (20), the correction

$$\Delta m = (k' + k''\Delta c)\Delta X . \tag{21}$$

The second term can often be neglected. In some cases, however, this term is not only important, but also a function of time, namely when a variable changes its color noticeably (e.g., during the deep minimum of an Algol system). This correction must be applied to all magnitude differences, which are described in Sect. 8.4.3, if a good precision is attempted.

Transformation programs for *UBV* photometry in BASIC are given in p. 339 in Henden and Kaitchuck [3].
A lot of practical advice on photometric reductions is given in Warner [4].
Photometry software listed in Warner [4] also provides transformation programs.

8.5.2 Reduction of Photographic and CCD Observations

The determination of stellar magnitudes thus requires the observation of several standard stars (photoelectric photometry) or star fields (CCD photometry, photography) according to a particular "schedule." Standard stars can be selected from a list of suitable standard fields, given, e.g., by Landolt (see Sect. 8.3).

Methods of reduction for photographic photometry are sorted below according to increasing accuracy:

1. Visual estimates of the stellar images on the photographic negative, performed with the Argelander method of visual photometry.
2. Determination of diameters of stellar images on a high-contrast, highly magnified copy (or of equidensitometry rings on a copy made with contour film); when preparing such a copy, one must take care to ensure for a homogeneous illumination of the field.
3. Measurement of the plate on an iris diaphragm photometer or on a microdensitometer. Much more effective, although perhaps slightly less accurate, is the digitization by means of a flatbed scanner. It also has the advantage that the digital image is readily available for image processing software.

The reduction of CCD frames is much more straightforward, since digital data are already available. Image processing software usually permits to carry out aperture photometry or profile fitting photometry. A lot of detailed information is available on CCD reductions. The raw image should undergo the following steps: bias subtraction, flat-fielding, possibly normalization due to different exposure times of several frames to be compared (see Sect. 8.4.3). Otherwise, the procedure is similar as described in Sect. 8.5.1.

8.5.3 The Reduction of Time: the Heliocentric Correction

Upon consulting a clock, the time of observation is written down, or, if the data are recorded with the aid of a computer, they are stored concurrently with each datum of measurement. Any deviation of the time of the local clock is determined before and after the observation, and the correction is applied after the observation during the reduction stage. If, for example, the observer intends to determine the time of minimum light of an eclipsing binary, care must be taken to register precisely the times of each brightness measurement. In the final data set, the resulting time of minimum light (or of any other time) is not given in plain universal time: because of the motion of the Earth around the Sun, events in deep space are registered on Earth sometimes earlier, sometimes later. In order to compensate for effects of the finite velocity of light, the time of observation should be transformed to the time of observation of a fictitious observer placed at the center of the Sun. This is called *heliocentric time*.

When observing a variable star for an extended period of time, the time is indicated, instead of by year, month, day and time of the day in the Gregorian calender, by the so-called Julian date (JD). This period started as day number 1 at 12 o'clock universal time on January 1 in the year -4712 (i.e., 4713 BC). The time of the day is given in fractions of a day.

Thus, the reduction of times into heliocentric Julian date (JD hel.) may be performed using one of the following methods:

1. Using tables (the easiest accessible is: A.U. Landolt and K.L. Blondeau: Tables of the Heliocentric Correction. Publications of the Astronomical Society of the Pacific 84, 784 (1972)).
2. Using the scheme published by L.E. Doggett, G.H. Kaplan, P.K. Seidelmann: Almanac for Computers for the Year 1978, Nautical Almanac Office, Washington, D.C. (1978); A.A. Henden and R.H. Kaitchuck [3].
3. Using a script.

The Julian date of the day of observation can also be determined from a script: First calculate the universal time (UT). For observers in Great Britain, through which passes the Prime Meridian, the formula is quite trivially

$$UT = WET \text{ (Western European Time)} ,$$

or, during the summer months when WEST (Western European Summer Time) is in effect,

$$UT = WEST - 1\,\text{h} .$$

In North America, several times zones exist.

http://www.physics.sfasu.edu/astro/javascript/hjd.html

http://www.physics.sfasu.edu/astro/javascript/julianday.html
http://www.go.ednet.ns.ca/~larry/orbits/jsjdetst.html

8.5.4 Determination of Times of Minimum or Maximum and of Periods

The establishment of the times of minimum (or maximum) light of variable stars serves for the exact determination of their periods, and for the determination of possible period changes, which are important for our understanding of the evolution of variables (eclipsing binaries as well as pulsating variables). Organizations of amateur astronomers have been active in this field, although activities have decreased in recent years. The same formalism can be used for the determination of times of minima and maxima.

The visual, photographic, or photoelectric observations yield for the minimum or maximum of the light curve N pairs of values (t_i, m_i), $i = 1, 2, \ldots, N$, representing the heliocentric times and magnitudes (or magnitude differences with respect to a comparison star). All discussed methods are also applicable in the case of step estimates of magnitudes. The interval between two adjacent observations can lie in the range of minutes for short-period eclipsing binaries or pulsating RR Lyrae stars, and is in the range of several days for long-period Mira stars. The series of observations can be analyzed by various methods, which are discussed below and illustrated in Fig. 7.

8.5.4.1 The Curve-Intersecting Line (Pogson's Method) Using graph paper, the magnitudes (Y-axis) are plotted versus time (X-axis). Subsequent points are connected by straight lines. The descending and ascending branches of the light curve are connected by several straight lines (their number should not exceed $N/2$). The central point of each of these lines is determined. If all X-values of the central points scatter around a mean value, this mean value is assumed to be the time of minimum light. Its standard deviation provides a reliable range of the accuracy of the minimum time determination. If the points show a systematic trend to one side, the light curve is asymmetric. Here, the intersecting point of a line, which is drawn "by eye" through the central points or determined by the method of least squares with the light curve, is assumed to be the time of minimum light. A BASIC program of this method is given in p. 64 in Ghedini [6].

8.5.4.2 The Kwee–van Woerden Method This method, which was often employed in earlier times, has somewhat fallen out of favor because it can be applied only to points which are equally spaced in time.

A computer program for the method is presented by A.D. Mallama, available at: http://adsabs.harvard.edu/abs/1987IAPPP..29...33M.

8.5.4.3 The Polynomial Fit A method relatively free of mathematical problems is the "best fit" method, where the measuring points are approximated by a polynomial of nth degree, i.e., a function of the form $y = \sum a_i x^i$, where $i = 0, 1, \ldots, n$. The x-value of the minimum of the approximated curve is assumed to be the minimum time. If the segments of the curve are symmetrical with respect to the minimum time, it is sufficient to approximate it by a parabola ($n = 2$). Asymmetries are taken into account by uneven terms, i.e., the cubes and higher powers.

A BASIC program, which automatically determines the "best" power of the polynomial, and carries out the approximation is given on p. 53 in Ghedini [6].

8.5.4.4 The Polygon Fit This method, which is advantageous if the data points on the two branches of the light curve are unevenly distributed, is discussed on p. 59 in Ghedini [6].

8.5.5 Period Determination

If a minimum time was determined, a comparison with other minima of the same object is warranted. The minima of an eclipsing binary should occur in a very regular way, assuming no physical processes occur which influence the length of a period (mass flow from one component of the system to the other, mass loss from the system, gravitational influence of a third body, etc.).

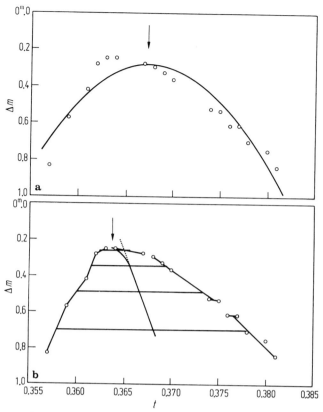

Fig. 7. Determining the time of maximum light by (**a**) approximation by a polynomial or (**b**) curve bisection. The abscissa t is in fractions of a day

Periodically occurring events are also found in other measurements, where, however, they may vanish after some time (from a few minutes to several years). As examples, consider the light changes of asteroids caused by their rotation or the "flickering" phenomenon in cataclysmic variables. If the period is constant and the time of the extreme values of the brightness fluctuation is measured with an error which is small compared with the length of period, the period is given as the greatest common measure of the interval lengths, which results from the combinations of all available minimum (or maximum) times. The period can be determined by trial and error or by the method of least squares. For a single interval, the period is the quotient of the interval length and the number of periods lying in this interval (equal to the difference of epochs). An algorithm, for instance, that of the greatest common measure, is not suitable because of the inherent errors of the minimum times, or because of a possible slight period variation. One should beware of apparent periodicities, especially when the true period and the length of a day, or the true period and the repetition period of the measurement, show a simple ratio. Differences also occur if there are "beat periods," caused by the superposition of several periods (e.g., in pulsating variables). It is sometimes helpful in the period-finding process if one utilizes in the analysis not only the extrema but also some other characteristic parts of the light curve. The accuracy of the period determination is given by the function

$$\text{Total duration} = f \,(\text{number of periods } E) \,, \tag{24}$$

whose slope is equal to the period. It is also dependent on the number of data points (observed minima), on the total length of the interval, and on the accuracy of the individual data points.

If there exist large time intervals during which no observations were made, the connection of different series of observations will pose problems, because the number of periods elapsed in the gap is unknown. Systematic (e.g., computer-aided) tests are suggested. If the deviations, the so-called $(O-C)$ values (see Sect. 8.5.6), of the individual data points for a given period have a periodic structure, the period determination is likely to be erroneous.

On p. 95 of Ghedini [6], two programs for period analysis are given. The first one is based on the Lafler–Kinman method and chooses the best arrangement of observations, i.e., the correct period is the one for which the polygon through the data points, arranged according to phase, is of minimum length. The second one is based on Fourier analysis.

A more modern approach to period finding is provided by software developed at Vienna University Observatory: http://www.univie.ac.at/tops/Period04/.

8.5.6 Period Behavior

When the elements of the light changes of a variable star, the period P and the zero epoch E_0, are known, the times (in JD hel.) of the occurring minima (for eclipsing binaries) or maxima (for pulsating stars) can be calculated for all epochs of the past and the future:

$$E_n = E_0 + nP \,, \tag{25}$$

where E_0 and P can be taken from the *General Catalogue of Variable Stars*. Is it rewarding to observe and determine minimum or maximum times also in such cases? Most certainly, because all "celestial clocks" show irregularities, their minima and maxima can occur earlier or later than anticipated. When planning such observations, one should, however, always consider whether the accuracy obtained is sufficient for a study of possible period changes.

To investigate the accuracy of the "clock rate" of a variable, the $(O-C)$ diagram is used: Compare the observed (O) minimum time with the computed (C) one, according to Eq. (25). If several minimum times have been observed (or collected from the literature), then an $(O-C)$ diagram can be constructed to display the time differences as a function of time (JD should be used) or as a function of epoch E. The course of the differences reveals something of the nature of the elements of the variable:

1. If the sequence of points runs through the origin and shows a rising (or declining) tendency, the zero epoch is correct. The period length is, however, too short (or too long).
2. If the sequence of points does not run through the origin, showing, however, a horizontal course, then the zero epoch must be corrected, the period being correct.
3. If the sequence of points shows a course that cannot be approximated by a straight line, the elements of the light variation of the star are themselves variable, as follows:

- If the points lie approximately on a parabola, the period changes permanently by the same amount, this change can be taken into account by the addition of a quadratic term aP^2 in Eq. (25).
- If the points can be approximated by a polygon, the period shows sudden changes; often, lengthening and shortening of the period follow in an irregular sequence.
- If the sequence of points shows a step-like appearance, the star has a constant period, but exhibits phase jumps.

As an example, Fig. 8 shows an $(O–C)$ diagram of the eclipsing binary RZ Cas. This system shows sudden period changes, as is clearly seen in the $(O–C)$ values derived from photoelectrically determined minimum times. The minimum times which were derived from visual observations do not reveal this behavior because of their low accuracy.

8.6 Photometry of Different Astronomical Objects

8.6.1 Photometry of Solar System Objects

The wide variety of objects in the Solar System demands the use of a correspondingly extensive collection of methods by which to obtain photometric data. Except for the fluorescence of cometary matter and the radiation of meteorites entering the Earth's atmosphere, it is always reflected sunlight which is measured. This radiation carries information regarding the reflection properties of the surfaces.

In this context, the following geometrical quantities are quite important: the orientation of the pole of the object, as given by the coordinates of the planetary north pole, and the planetocentric coordinates of the subsolar and the subterrestrial points. The coordinates which are defined by these three points and the angles between them determine, together with the relevant distances (Sun-object and object-Earth), the geometry of the process of reflection and scattering. The geometry is known for most of the large bodies of the Solar System, and it is described by the position angle of the rotational axis of the planet, the phase of rotation, the phase of illumination, the incident angle of radiation, and the aspect.

For each observation, we can write down an equation which contains known parameters (the geometrical parameters and the measured value) and unknown ones (the physical properties of reflection such as the shape of the object (in the case of a planetoid), the albedo, and the scattering property (the roughness of the surface at all scales)). If, as in the case of planetoids and small, distant planetary satellites, the geometry of reflection is not completely known, additional unknowns, such as the phase of rotation and the position of the polar axis (possibly influenced by precession), enter the equations. These parameters must be determined by a manifold of independent observations. One should try to determine beforehand during the planning stage of observations—and not in the reduction stage—in what sense the planned observations are to cover "new aspects" and which observing times are most informative.

If seasonal or long-term changes of the reflectional properties occur (especially on planets and moons with atmospheres, by sand storms on Mars, or by vulcanism on Io,

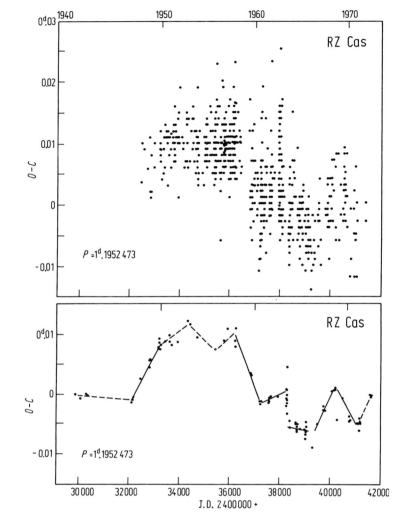

Fig. 8. (O–C) diagram of the Algol system RZ Cas for the years 1940–1973. The *upper* diagram shows the results of visual minimum times, the *lower* one based on photoelectric observations. The period is unstable and shows "jumps," leading to lengthenings and shortenings of the period. From Herczeg and Frieboes-Conde [7]

etc.), a morphological study becomes difficult or even impossible. A series of photometric observations can then be used for meteorological investigations. The most important causes of brightness variations of Solar System objects are:

1. Effects of varying distance.
2. Rotational modulation because of bright-dark surface features and/or irregular shape of the object.
3. Phase effects.
4. "Seasons" and meteorology.

It is important to take into account the influence of the varying distance on the measured brightness. In order to be consistent, the magnitudes are converted into those that would be observed at a *mean heliocentric opposition distance* r_0. This distance is, for Jupiter and its satellites, 5.208 AU, for Saturn 10.529 AU, for Uranus 19.191 AU, and for Neptune 30.071 AU. For the calculation of this correction, the geocentric distance d and the heliocentric distance r for the time of observation are needed. They are tabulated in the annual *Astronomical Almanac*. The value of d is found in Section E under "True Geocentric Distance," $r(t)$ under "Radius Vector." The distance correction of magnitude Δm_d is calculated from

$$\Delta m_d = 5(\log r_0 + \log(r_0 - 1.0) - \log d - \log r) , \qquad (26)$$

where all distances are given in AU.

Phase effects can be noted in the case of the Earth's Moon when comparing the total magnitude at Full and New Moon. However, subtle effects can be detected when comparing the different surface magnitudes of an illuminated region at Full Moon and at first or last quarter.

If the Earth's Moon did not rotate in a bound, synchronous state, we would note a distinct rotational light variation between the "near" side, which is to a large extent constituted by dark mare regions, and the "far" side, which is predominantly bright. Because of similar causes, the amplitudes of rotational light variations of Mercury, Mars, and the large, regularly shaped minor planets amount to several $0^m.01$. The effects are often much more pronounced for Jupiter's moons Europa, Ganymede and Callisto and for Saturn's moon Iapetus because of the large differences of the surface brightness of ice and rock (Fig. 9).

8.6.1.1 Photometry of Minor Planets The largest asteroid, Ceres (1), has a diameter of 940 km and its distance from the Earth at opposition is 1.77 AU. It then appears at an angular diameter of just $0''.7$. In general, planetoids are observed as point-like objects, and thus photometric measurements can be performed in a manner similar to that for fixed stars, aside from the fact that the former change their position in the sky and that they have a different scintillation behavior. The simple empirical fact that "planets do not twinkle" reflects that the statistical fluctuations of scintillation are averaged out across the extended planetary disks. Effects of this sort can be noted even in the case of the major planetoids. The amplitude of scintillation of Ceres amounts to one-third that of a point-like object.

Because of the continuously changing distance of a planetoid from the Earth and Sun, similar corrections, as discussed in the previous section, must be applied to the photometric data. The following parameters are needed, the first of which can be taken from *The Astronomical Almanac*, the others from the Ephemerides of Minor Planets (EMP, Efemeridy Malykh Planyet in Russian):

$$R = \text{distance between Earth and Sun (in AU)}, \qquad (27)$$
$$d = \text{distance between Earth and planetoid (in AU)}, \qquad (28)$$
$$r = \text{distance between Sun and planetoid (in AU)}, \qquad (29)$$
$$r_0 = \text{semimajor axis of the planetoid's orbit (in AU)}. \qquad (30)$$

The online Astronomical Almanac is available at: http://asa.usno.navy.mil/.

Magnitude corrections for the phase angle of the four large moons of Jupiter, of Saturn's moons Titan and Rhea, and of the planets Uranus and Neptune are given by Lockwood [9].

Ceres, as the largest asteroid, has been reclassified in 2006 as a dwarf planet. It is in hydrodynamical equilibrium (round shape), but has not cleared the neighborhood around its orbit.

Some part of the *Tables of Ephemerides of Minor Planets* is available online: http://www.ipa.nw.ru/PAGE/DEPFUND/LSBSS/enguemp.htm.

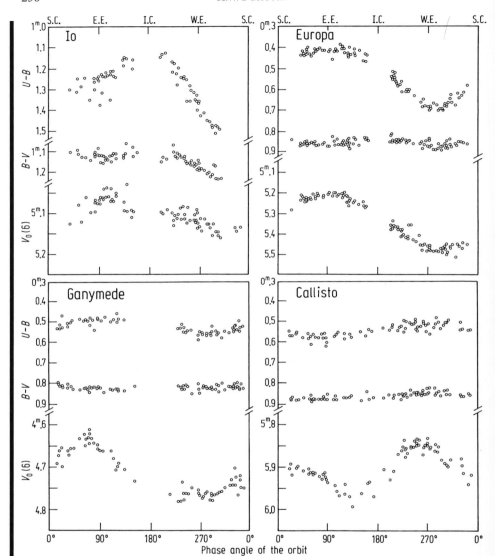

Fig. 9. V magnitudes and $(U - B)$ and $(B - V)$ color indices of Jupiter's four bright satellites, corrected to mean opposition distance and phase angle $\alpha_p = 6°$, given as a function of the phase angle of their orbit. The repeatability of the light curve indicates a bound rotation of the satellites. The measurements were made with different telescopes of the Lowell Observatory (apertures of 0.53–1.07 m). Note the large scatter and the red color of the satellite Io. From Millis and Thompson [8]

The light-travel time must be taken into account. Light travels the Sun–Earth distance of 1 AU in 8.317 min and the planetoid–Earth distance in $8.317 d/R$ min. In order to get the time at which the planetoid had the observed brightness, the value $8.317 d/R$ min must be subtracted from the observed time.

The magnitude must also be "normalized" to the one at unit distance. The observed magnitude of a planetoid is dependent on d and r. In order to obtain comparable magnitudes of a planetoid at different times, or magnitudes among planetoids, the observed magnitude m_{obs} is transformed into the one for which the planetoid has distances $d = 1$ and $r = 1$, both in AU, from the Earth and Sun. This *reduced magnitude* m_r is given by:

$$m_r = m_{\text{obs}} - 5(\log r + \log d) \, . \tag{31}$$

Sometimes the observed magnitudes m_{obs} are reduced to the mean opposition distance $(r_0 - 1)$. The magnitude m_0 is calculated from

$$m_0 = m_{\text{obs}} - 5(\log r + \log d) - \log[r_0(r_0 - 1)] \, . \tag{32}$$

The *phase angle* must also be considered. The phase angle α_p is the angle at the position of the planetoid in the triangle Sun–planetoid–Earth. It is calculated from the relation

$$\cos \alpha_p = \frac{d^2 + r^2 - R^2}{2rd} \, , \tag{33}$$

or, with higher accuracy, from the formula for $\tan(\alpha_p/2)$ (see Chap. 3). If one plots the magnitudes, corrected for light-travel time and transformed to a constant distance, as a function of α_p, a brightness increase is generally noted when α_p becomes smaller. The extrapolation of the brightness curve yields the brightness for the phase angle $\alpha_p = 0°$. The magnitude can be given by linear regression as a function of α_p. The formula is

$$m_0 = 5\log(rd) + H - 2.5\log[(1 - G)\phi_1 + G\phi_2] \, , \tag{34}$$

where

$$\phi_1 = \exp\left\{-3.33\left[\tan\left(\frac{\alpha_p}{2}\right)\right]^{0.63}\right\} \, , \tag{35}$$

and

$$\phi_2 = \exp\left\{-1.87\left[\tan\left(\frac{\alpha_p}{2}\right)\right]^{1.22}\right\} \, . \tag{36}$$

H and G are the absolute magnitude and the slope parameter of a planetoid, respectively. This formula is valid for $0° \le \alpha_p \le 120°$. The well-observed magnitude variation of some planetoids shows that the brightness depends not only upon the phase angle, but also, because of the irregular surface, on rotation. The periods are of the order of several hours, and the amplitudes can be as high as several tenths of a magnitude.

8.6.1.2 Photometry of Major Planets

As a rule of thumb, we give, in Table 3, typical surface brightnesses (visual magnitudes per square arcsecond) and luminances (sb).

The human eye is capable of discriminating between very small differences in the luminances of bright areas and is therefore well-suited for the photometry and structure recognition of small areas on planetary surfaces. The luminances noted by the eye can be magnified or decreased by means of the telescope magnification: the larger the magnification, the lower the luminance which is produced by an illuminated area on the retina. It is therefore advisable to transpose, by a suitable choice of magnification, the luminance into the range of largest contrast sensitivity.

H and *G* are available from the JPL Small-Body Database Browser http://ssd.jpl.nasa.gov/sbdg.cgi

Long-term as well as short-term programs of planetoid photometry are obviously of interest. More extensive information may be found in Binzel [10].

8.6.1.3 Photometry of the Moon Quantitative brightness measurements of the lunar surface by means of photoelectric photometry are still of some interest, despite the fact that very accurate measurements of a few lunar regions had been made during the Apollo landings. Of particular interest is the comparison of intensity and color of known regions with those not yet explored. Lunar photometry entails—if one does not have at one's disposal a two-dimensional detector—the integral measurement of different regions. The diaphragm size should be on the order of $5''$.

Each region has a characteristic brightness function which depends on the phase angle ψ. The radiance of a region (i.e., the radiation of a projected unit surface into unit angle toward the direction to Earth) is thus determined by:

- The albedo, which provides information on the geological and surface characteristics (albedo is equal to total light reflected into all directions over total infalling light).
- The phase angle (the angle between the Sun and Earth as measured on the point observed).
- The longitude (the angle between Earth and the direction normal to the lunar surface, measured at the observed point in the plane determined by the Sun and Earth).

Some of the different projects that can be carried out are:

- General albedo mapping (in different colors).
- Determination of brightness functions of unusual regions.
- Searches for transient lunar phenomena (TLPs).

The brightnesses of the various regions, which are measured with a diaphragm of known diameter (in square arcseconds), can be converted into magnitudes per square arcsecond through observation of a suitable fixed star (of spectral type similar to that of the Sun: G2); thus one obtains absolute lunar photometry. Another possibility is to observe a "standard region" in the vicinity (relative lunar photometry). More details are found in Chap. 13.2.4.

Table 3. Surface brightnesses and luminances of objects in the Solar System

Object	$m_V/\square''$	B (sb)
Sun	$-10^m.5$	
Mercury	1.5 to 4.5	0.6
Venus	0.5 to 1.5	2.0
Moon	3 to 7	0.6 (Full Moon)
		0.1 (First and last quarter)
Earth light	14 (On New Moon)	
Earth light	22 (On Venus at inferior conjunction)	
Mars	4 to 5	0.2
Jupiter	5 to 5.5	0.070
Saturn	6.5 to 7	0.028
Uranus	8	0.0037

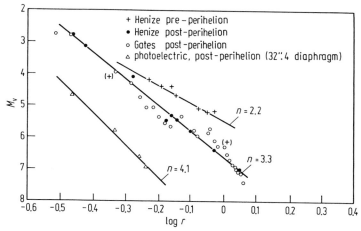

Fig. 10. Absolute magnitude M_v of comet Kohoutek (1973f), reduced to a distance of 1 AU from Earth, as a function of distance r from the Sun. The change of brightness before perihelion (*crosses*) is different from that afterwards (*circles and dots*). *Triangles* indicate photoelectric observations of the cometary nucleus after perihelion. From Angione et al. [11]

8.6.1.4 Photometry of the Sun Solar photometry can be at once both easy and complicated. In general, no comparison stars are available, so that different areas on the Sun can be compared only in a differential way. Since a copious amount of light is available, on the other hand, a small telescope and a relatively insensitive detector (e.g., a PIN photodiode) will suffice for the observations.

The photometry of the intensity variations of sunspots in different color regions can be performed with a telescope with a large image scale and by means of a photometer with a small diaphragm size; the telescope remains stationary while the photometer carries out a scan through the region of interest. An even better way of obtaining data is to employ a two-dimensional receiver like a CCD. The umbra has a surface brightness from 10 to 20% of the undisturbed solar surface. The scatter of light in the solar and the terrestrial atmosphere as well as in the telescope must be taken into account. Similarly, the limb darkening of the Sun can be determined in different colors using a stationary telescope and photometer with a narrow diaphragm.

8.6.1.5 Photometry of Comets and Meteors In contrast to the relatively sharp disks of planets, planetary satellites, and especially planetoids, the photometry of other objects in the Solar System is made difficult because of their extended, often diffuse appearance. This applies to comets, meteors, and the zodiacal light.

Two magnitudes can he defined for comets: the total magnitude and the magnitude of the nucleus. Visual estimates of the total magnitudes are often published in the literature despite their shortcomings. Even when determined by experienced observers, differences of up two magnitudes for the same instant are not uncommon. The estimates of surface brightness depend strongly on the diameters of the entrance and exit pupils of the instrument used, and the tail magnitudes often do not show a regular brightness decline as a function of the distance from the nucleus. Therefore faint parts of the tail are often overlooked, despite the fact that they are non-negligible owing to their vast extent. The total brightness of tailless comets can be estimated quite readily by defocusing a comparison star. When publishing photoelectric magnitudes, the size of the diaphragm used (in arcseconds) must be given, as well as the relevant information on the telescope used to obtain the data. Figure 10 shows visual and photoelectric magnitudes of comet Kohoutek to illustrate the different magnitudes determined.

It is practically unfeasible to determine the magnitude of the nucleus for comets which are near the Earth or the Sun ($r \leq 1$ AU), since the nucleus is always embedded in the bright coma.

The brightness of a meteor is more easily determined, but its interpretation can suffer due to many ambiguities (material, porosity, entering velocity, angle of incidence, fragmentation). Visual estimates yield relatively good results.

Photographs often register only the brighter meteors, a rotating sector allows the determination of its angular velocity. Somewhat fainter objects can be found when the camera is rotated with a typical angular speed comparable to that of the surveyed meteors, away from the radiant, and light from the remaining solid angle is shielded from the camera. More details are found in Sect. 13.9.

> A thorough introduction into the photometry of comets is given by A'Hearn [12], who also gives a list of filters that are especially suited for comet photometry.

8.6.2 Stellar Photometry

As was already mentioned, absolute photometry in well-defined color systems is a field less suited for amateurs. A project to establish (visual) scales for the visual observation of variable stars is mentioned by Henden and Kaitchuck (p. 24 in [3]). The main field of photometric research on stars by amateurs will likely be the investigation of variable stars. Different projects are briefly described in the following sections.

8.6.2.1 Flare Star Patrol The monitoring of flare stars is an activity undertaken by many amateur astronomers. Here, mechanically and electrically stable equipment should be used, in connection with good sky conditions. For most of the time, the brightness of the flare star is monitored with good time resolution by PMT or CCD. The U and B color regions are most suitable. The photoelectric monitoring is interrupted from time to time by short measurements of the sky brightness and the brightness of a comparison star in the same color filter. Flares occur fairly rarely, and one must persist for several nights on a single object to obtain a statistically meaningful result. Unfortunately, most flare stars are quite faint.

8.6.2.2 Short-Period Variables: δ Scuti Stars, Dwarf Cepheids, RR Lyrae Stars, and Cepheids For the first three types of stars listed in this section heading, a few hours of observing time are sufficient to obtain a complete light curve. The objects often show dramatic changes in both the shape and amplitude of the light curve, caused by a beat phenomenon. In such cases, longer observing campaigns are useful.

8.6.2.3 Long-Period Variables: RV Tauri Stars, Mira Stars The main body of data on long-period variable stars is comprised of visual estimates made by a large number of amateurs who, distributed across the world, have been able to obtain fairly complete light curves without the usual gaps caused by poor weather.

Another method of observing such variables is by means of a CCD or IR photometer in the red and infrared regions, where they emit most of their light.

8.6.2.4 Eclipsing Binaries Three groups of eclipsing binaries can be distinguished: the short-period group of W UMa stars, and the medium- to long-period systems of β Lyr and Algol types.

Often it is possible to obtain an "instantaneous" light curve of a W UMa system during a single night. Other systems require weeks, months, or years, before all phases of a light curve are covered by observations. Careful observations show that light curves and periods undergo small but important changes.

Details on these groups are given in Sect. 14.3 on binary stars.

Even if one does not have the time to observe a complete light curve, one can still concentrate on the exciting observation of a complete eclipse: from such data, the time of minimum light can be derived. This is described in detail in Sect. 8.5.5.

8.6.2.5 Eruptive Variables: Novae, Dwarf Novae, and Supernovae These objects are especially interesting, since they show "unique" outbursts: the supernovae, because they undergo only one outburst in their lifetime, the novae (Fig. 11) on historical time scales, and the dwarf novae, because an outburst never completely matches a previous one. Very important contributions to astronomy can be made in all of these cases. Any visual, photographic, or photoelectric observations which can be obtained will often serve as a valuable complement when spectroscopic data or data obtained in other wavelength regions (infrared, ultraviolet, X-ray) are being interpreted by professional astronomers.

8.7 Construction or Purchase of Receivers and Equipment for Reductions

A multitude of observational opportunities are available to the amateur astronomer who is interested in astronomical photometry which can enrich the experience in his or her hobby. He or she can, with the aid of a small or large telescope, either make estimates with the eye or carry out photometric observations. Nowadays, commercially available photometers and CCDs are readily available, as are software packages for image analysis. Furthermore, there are groups of interested amateurs and professionals who publish information on the construction, operation, and astronomical applications of photometers and the associated telescopes.

8.7.1 Advice on the Purchase of Photometers and Cameras

A full range of single-channel photometers for optical and infrared photometry is offered by Optec, Inc., either with photodiodes or photomultipliers.

A choice of CCD imaging cameras is offered by SBIG. Note that for photometry, modest pixel arrays are sufficient, although a larger array is more flexible and can be adopted for readouts of specific regions on the chip to shorten the data acquisition time. The system should also have a five-position filter wheel for, e.g., *UBVRI* filters.

1. Optec Solid-State Stellar Photometer SSP-3 (Fig. 12). The SSP-3 photometer uses a Hamamatsu S1087-01 silicon PN photodiode as receiver. The spectral sensitivity is between 300 and 1100 nm. The blue sensitivity is low; sensitivity maximum is in the

Single-channel photometers are offered by Optec, Inc., 199 Smith Street, Lowell, MI 49331 (http://www.optecinc.com/optec.htm).

CCD imaging cameras are offered by SBIG (Santa Barbara Instrument Group), 147-A Castilian Dr., Santa Barbara, CA 93117 (http:/www.sbig.com).

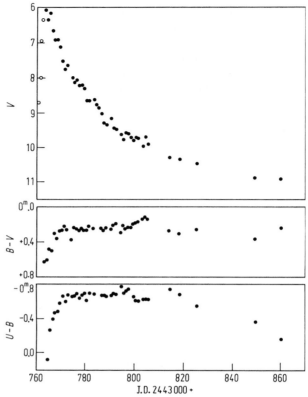

Fig. 11. V light curve and $(B - V)$ and $(U - B)$ color indices of Nova V1668 Cyg (1978), observed with the 0.35-m telescope of Hoher List Observatory and a photoelectric photometer. The brightness maximum coincides with large color indices (i.e., low temperature). From Duerbeck et al. [13]

infrared, at 850 nm. The exit voltage is produced via a shunt resistance of 50 GΩ. It is converted via a voltage-to-frequency converter into a frequency. A counter is built into the photometer housing. By means of a slide, filters can be brought into the beam of light; Johnson *UBVRI* filters are standard. The power is supplied by means of a battery. Stars of blue to infrared magnitude of 10^m (ultraviolet to 8^m) can be observed in combination with a 30-cm telescope. The cost of the photometer is about $1500.

2. Optec Solid-State Infrared Photometer SSP-4. It uses a Hamamatsu G5851 InGaAs PIN photodiode, sensitive from 900 to 2050 nm for photometry in the J (1250 nm) and H (1650 nm) band. The cost of the photometer is about $2500.

3. Optec Photoelectric Photometer SSP-5. The SSP-5 photometer uses a Hamamatsu R6350 miniature photomultiplier with S-5 (i.e., UV-extended S-4) spectral response (or, alternatively, a R6358 extended red response), operating at about 800 V. The PMT signal is fed to a voltage-to-frequency converter and a counter. Manual or motorized 6-position filter sliders can be used. *UBV*-photometry of stars down to 13^m with a 30 cm telescope should be possible. The cost of the standard configuration

Fig. 12. The SSP-3 solid-state photometer of Optec, Inc. From above, the following can be seen: eyepiece and sliding mirror, filter slide, box of the PIN photodiode with switches and digital display

(including Johnson *UBVR* filters and Strömgren uvby filters are available) is about $3000.

Further information on CCD cameras is found in Chap.6.

8.8 Further Reading

The following can be recommended for general reading on astronomical photometry:

Budding, E.: An Introduction to Astronomical Photometry. Cambridge University Press, Cambridge (1993)

Genet, R.M. (ed.): Solar System Photometry Handbook. Willmann-Bell, Richmond, VA (1983)

Sterken, C., Manfroid, J.: Astronomical Photometry: A Guide. Kluwer Academic, Dordrecht (1992)

References

1. Tüg, H., White, N.M., Lockwood, G.W.: Absolute Energy Distributions of α Lyrae and 109 Virginis from 3295 Å to 9040 Å. Astron. Astrophys **61**, 679 (1977)

2. Howell, S.B.: Handbook of CCD Astronomy, 2nd edn. Cambridge University Press, Cambridge (2006)

3. Henden, A.A., Kaitchuck, R.H.: Astronomical Photometry, 2nd edn. Willmann-Bell, Richmond, VA (1989)

4. Warner, B.D.: Lightcurve Photometry and Analysis: A Practical Guide. Springer Science+Business Media, Berlin Heidelberg New York (2006)

5. Evans, D.S.: Photoelectric Observing of Occultations-II. Sky Telescope **54**, 289 (1977)

6. Ghedini, S.: Software for Photometric Astronomy. Willmann-Bell, Richmond, VA (1982)

7. Herczeg, T., Frieboes-Conde, H.: The Period of RZ Cassiopeiae. Astron. Astrophys. **30**, 259 (1974)

8. Millis, R.L., Thompson, D.T.: UBV Photometry of the Galilean Satellites. Icarus **26**, 408 (1975)

9. Lockwood, G.W.: Photoelectric Measurements of Planets and Satellites. In: Genet, R.M. (ed.) Solar System Photometry Handbook. Willmann-Bell, Richmond, VA (1983)

10. Binzel, R.P.: Photometry of Asteroids. In: Genet, R.M. (ed.) Solar System Photometry Handbook, p. 1-1. Willmann-Bell, Richmond, VA (1983)

11. Angione, R.J., Gates, B., Henize, K.G., Roosen, R.G.: The Light Curve of Comet Kohoutek. Icarus **24**, 111 (1975)

12. A'Hearn, M.F.: Photometry of Comets. In: Genet, R.M. (ed.) Solar System Photometry Handbook, p. 3-1. Willmann-Bell, Richmond, VA (1983)

13. Duerbeck, H.W., Rindermann, K., Seitter, W.C.: A UBV Light Curve of Nova Cygni 1978. Astron. Astrophys. **81**, 157 (1980)

Based on *Principles of Photometry* by H.W. Duerbeck and M. Hoffmann, Compendium of Practical Astronomy, Vol. 1, Chap. 8, Springer, Berlin, Heidelberg, New York (1994)

9. Signals from Space: Radio Astronomy for Beginners

M.J. Neumann and P. Wright

9.1 Introduction

The radio-wave region of the electromagnetic spectrum offers amateur astronomers interesting opportunities for observing. Regardless of weather, time of day, or time of year, they can detect astronomical objects like the Sun and Moon, the planets, supernova remnants, and distant galaxies. Radio astronomy is astronomy 24/7.

If our eyes were able to see in the radio-wave region, we would experience the heavens in a completely different way than is revealed to us in the optical region: Radio filaments of the Milky Way, the eggshell-like structures of close-by supernova remnants, and tremendous regions of star formation would dazzle us (Fig. 1). Furthermore, the radio sky would be strewn with countless point sources that appear to us like stars. Yet we would look in vain for the familiar constellations since, unlike the stars that form figures like the great bear of Ursa Major, very few "radio stars" are located within our Milky Way. These "stars" are usually the nuclei of distant radio galaxies and quasars, some being located at the very edge of the observable universe.

After a journey often lasting billions of years through the expanding universe, the radiation from these distant galaxies one day happens to reach the antenna of a radio telescope on Earth, where it generates a tiny electrical current. This must be amplified millions of times by the stages of sensitive electronic equipment in order to then move the indicator of an X–Y plotter or to appear as a curve on a computer screen. It is only due to the unimaginably huge radiative power ranging up to 10^{39} watts that the distant radio galaxies are able to be detected by Earth-based telescopes.

By "stars" of the radio sky we mean the active cores of very distant galaxies.

9.2 What is Radio-Frequency Radiation?

Physically speaking, radio-frequency radiation is similar to optical radiation—in both cases these entail electromagnetic waves. The only difference is that radio waves are of a significantly greater wavelength than visible light. The wavelengths for the spectrum usable for telecommunication extends roughly from one centimeter to ten kilometers. In contrast, optically visible light possesses wavelengths of less than a thousandth of a millimeter. Aside from the optical and radio regions, there are countless other wavelength regions (Fig. 2) that can also be observed.

The radiation from various regions is differentiated either by indicating its wavelength λ, or its frequency v. Often, both pieces of information are used together. The two parameters are interrelated by the speed of light $c = 3 \times 10^8$ m/s: λ [m] $= c/v$ [Hz], or v [Hz] $= c/\lambda$ [m]. Frequencies are usually stated in Hertz, while wavelengths are given

Fig. 1. Seen here is the radio sky over the telescopes of the National Radio Astronomy Observatory in Green Bank, VA. Note the shell-like supernova remnants and irregularly shaped star formation regions. The point-like objects are not stars but mostly distant radio galaxies. Image courtesy of National Radio Astronomy Observatory/AUI/NSF

Radio-frequency radiation is a form of light. The radio region comprises wavelengths from 1 mm to 20 m.

in meters. The wavelength region in which Earth-based radio astronomy can be pursued comprises wavelengths ranging between 15 m and 1 mm. This corresponds to frequencies of between 20 million Hertz (20 MHz) and 300 billion Hertz (300 GHz).

The atmosphere of our planet places natural limits on the radio spectrum of celestial objects detectable from the Earth. The Earth is surrounded by a shell of electrically charged particles. This shell is composed of multiple layers and is called the ionosphere. Radio signals from space at frequencies of less than 15 to 20 MHz normally do not penetrate our atmosphere. They are either absorbed or reflected back into space. As a result, an Earth-based radio telescope must operate at higher frequencies. At very high frequencies, however, the Earth's atmosphere places a limit on radio observations. Radio signals at frequencies above 300 GHz do not reach the ground since they are absorbed by water molecules of the Earth's atmosphere. The wavelength range extending from 20 m to 1 mm is called the "radio window" of the electromagnetic spectrum.

Countless radio sources can be detected within the natural limits of 15 MHz to 300 GHz. The spectra of the radio-frequency radiation shown in Fig. 3 show a curve that depends on the mechanism for generating the radiation. A differentiation is made

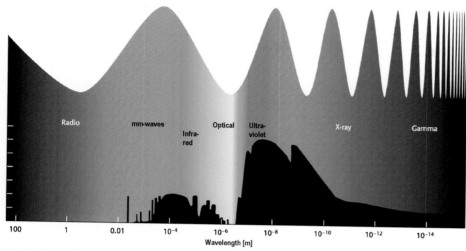

Fig. 2. The radio region observable from Earth occupies a wide range within the electromagnetic spectrum. Radio waves cover an area that is considerably larger than the optical spectrum of light visible to the eye. Image courtesy of *Sterne und Weltraum*

between thermal and non-thermal radiation. Thermal radiation, from the Moon or the Orion Nebula, for example, is produced purely by the temperature of these sources. In other words, these involve heat radiation in the radio region. Here the spectrum rises with increasing frequency or is flat. The situation is completely different for sources involving non-thermal radiation, such as, for example, the quasar 3C 273 and the supernova remnant Cassiopeia A. The radiation from these sources is created by the synchrotron mechanism. Here charged particles moving almost at the speed of light are deflected in magnetic fields and thereby transmit radio waves. A characteristic feature of synchrotron radiation is a decrease in intensity with increasing frequency (Fig. 3).

9.3 The Radio Sun

Anyone wanting to explore the radio universe from his or her own backyard should first try to detect the radiation coming from the Sun. Due to its proximity and high level of activity, the Sun appears to us to be the strongest radio source in the sky. Its radio-frequency radiation originates only to a small extent from the hot gas of the Sun's atmosphere. Much more intensive signals are generated by the interaction of electrically charged particles with magnetic fields, e.g., within the sunspot groups. These are non-thermal radio signals.

Particularly at times of high solar activity, the Sun is an especially interesting radio source for the amateur that he or she can easily observe with easily obtainable equipment. Many amateur radio devotees will have already heard these signals—perhaps without realizing that they are of cosmic origin. The Moon is an entirely different matter. Its temperature-induced radio-frequency radiation is so weak that detecting it is a real challenge for beginners, analogous to observing radio galaxies.

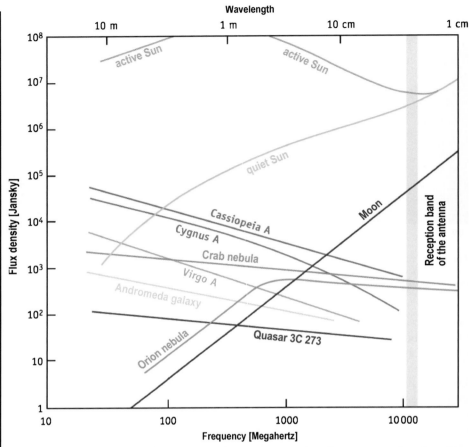

Fig. 3. The diagram illustrates the intensity of the strongest cosmic radio sources as a function of frequency and wavelength. The vertical bar marks the frequency range in which the TV satellite antenna described in this chapter is sensitive. Image courtesy of *Sterne und Weltraum*

The Sun continually emits noise sounding like a hiss from the speaker of a sensitive radio receiver and heard between FM and TV stations. Furthermore, during strong solar activity powerful solar eruptions (flares) can occur that emit intensive pulses of radio noise (radio bursts) that can be received in almost all radio-frequency bands. The radio bursts are in turn grouped together as radio storms that can last minutes or even days. Sometimes they are so strong that they impair radio communication as long as the Sun is located over the horizon for the receiving antenna during the outbreak.

The Sun is the strongest radio source in our sky. Its signals can be detected over the entire radio band.

The Sun emits noise even outside of the active phases—the noise of the "quiet Sun." This can be observed primarily in the centimeter to decimeter region, and therefore in the frequency band of television satellites. The weak radio-frequency radiation from the Moon, which emits thermal radio waves due to its temperature, is located in the same region.

Certain celestial radio sources can also be observed by amateur radio telescopes. Initial forays through the radio universe can be performed in the gigahertz region using a satellite antenna and a satellite finder. In the following section, we will take a closer look at how a simple radio telescope can be constructed using this approach. Then we will tackle a project at the lower end of the radio spectrum: observing the planet Jupiter using a home-made dipole antenna and a sensitive short-wave receiver.

9.4 A Compact Radio Telescope for the Gigahertz Band

Even commercially available parabolic antennas 60 or 90 cm in diameter provide the basis for a sensitive radio telescope that detects radio-frequency radiation from the Sun and the Moon in a frequency range between 10 and 12 GHz. They are inexpensive and also easy to obtain, as is true for the other components of the compact radio telescope described below.

The signal received by the satellite antenna is usually amplified by a so-called low-noise converter (LNC) then converted to a lower frequency, the so-called intermediate frequency, to enable it to then be transmitted through a coaxial cable. The LNC is always supplied along with the parabolic antenna (Fig. 4).

A tiny change in voltage corresponds to the signal leaving the LNC. To detect this, what is needed is a detector that converts it into direct current, as well as an amplifier. Both functions are performed by a "satellite finder" (Fig. 5), which normally functions to align a parabolic antenna with a desired TV satellite. The antenna is moved here until the indicator of the device displays the maximum signal amplitude. The same process is also possible with radio signals from the Sun and from a natural Earth satellite: our Moon.

The Satellite Finder, Model 3735 from Conrad Electronic contains both a broadband detector and an operational amplifier, which boosts the signal voltages from a few millivolts up to 1 V. The gain is manually adjustable by a controller. The satellite finder detects signals in the 950 MHz to 2050 MHz frequency range and operates at normal voltages of 13 V and 18 V.

A small modification lets you easily route the signal voltage from the satellite finder externally to be measured in volts and recorded as a function of time. This is implemented either by an X–Y plotter or a computer connected through the appropriate interface to the voltmeter. Digital multimeters with a computer interface, such as, e.g., the Voltcraft M3610 model, are also available from Conrad Electronic.

The rear panel of the satellite finder is first removed. On the left side of the housing is a connector jack F-Norm (Fig. 6). This is the output for the device. During normal TV operation, this is where the satellite tuner is attached (using an F-connector plug). On the right side is another F jack (F2) that is connected to the LNC of the parabolic antenna to receive a signal. Below this is an operational amplifier chip (OP-Amp) with eight connectors. For our modification, only the connector (pin 1) located on the lower left of the OP-Amp is important: pin 1 supplies the positive potential for the DC voltage sought. A strand of insulated wire is soldered on here.

For little money, a simple radio telescope able to detect radio-frequency radiation from the Sun can be built using a TV dish antenna and a satellite finder.

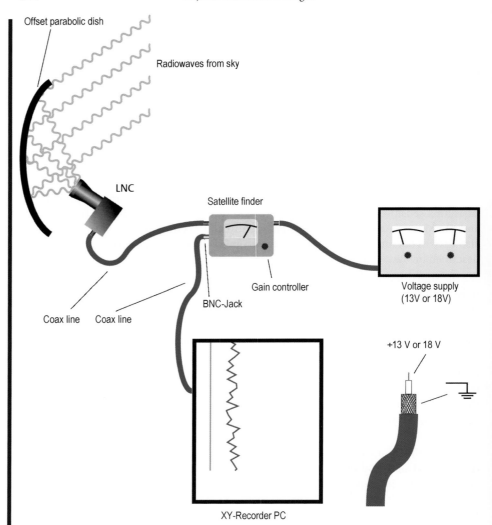

Fig. 4. Diagram illustrating a simple radio telescope. Located at the focus of the offset parabolic antenna is a small horn antenna that sends the radio signal through a short waveguide to the low-noise converter (LNC). From here it goes through a 75-Ω coaxial cable to the terminal F1 of the satellite finder. Terminal F2 is connected to a power supply unit or batteries. Image courtesy of *Sterne und Weltraum*

An additional BNC connector jack must be attached (Fig. 6) to the housing to later enable the signal voltage to be easily measured and recorded. Such a BNC connector jack is available from any electronics mail-order supplier. Its purpose, as is true of the F-jacks, is to connect the coaxial cables; in other words, it has a central wire and an outer conductor that symmetrically surrounds the central wire. The free end of the wire connected to pin 1 is soldered to the central wire of the BNC jack to be installed in the housing panel wall. To enable a voltage to be measured between the central wire and outer conductor of the BNC jack, its outer conductor must be connected to the outer conductor of

A minor modification of the satellite finder enables the observer to record the signals received from the radio telescope.

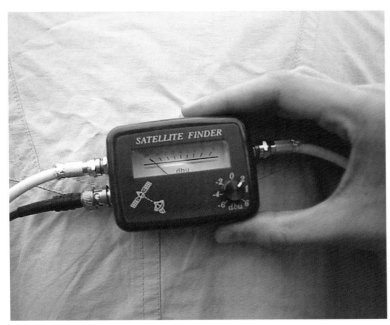

Fig. 5. A satellite finder, such as the Model 3735 from Conrad Electronic shown here, can be easily modified to receive radio-frequency radiation from space. Image courtesy of Peter Wright

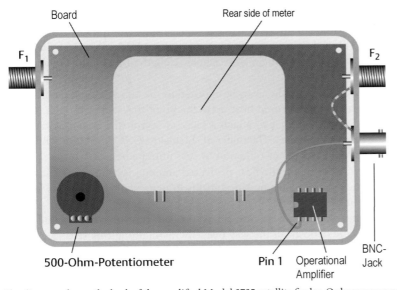

Fig. 6. The diagram shows *the back* of the modified Model 3735 satellite finder. Only components relevant for modification are indicated. At *right* is an operational amplifier (OP) with the connection pin 1, which is connected to the central wire of the BNC connector jack. Image courtesy of *Sterne und Weltraum*

jack F2, since this contains the ground potential. Modification of the satellite finder is then complete.

Now the satellite finder is provided with a power source. For this purpose, terminal F1 is connected to a coaxial cable using an F-plug connector. The cable leads to a stabilized power supply unit that generates voltages of either 14 or 18 V. The outer conductor of the coaxial cable carries the minus while the central wire carries the positive potential of the supply voltage. The power supply unit can be replaced with two 9-V batteries.

9.5 Initial Tests

A good system test consists in connecting the modified satellite finder to a parabolic antenna that has already been aligned with a TV satellite. The satellite receiver is disconnected from the antenna, the incoming antenna line is connected to jack F2, and the power supply is finally connected to F1. Now the needle of the satellite finder should deflect. Adjust the gain control of the satellite finder using the control element "gain control" so as to make the needle deflect all the way over. As soon as the antenna is moved away from the direction of the satellite, the needle should indicate a lower value on the scale. If an X–Y plotter is connected to the BNC jack at the same time, a curve is generated on the plotting paper.

9.6 First Light for Your Radio Telescope

Before you aim your antenna for the first time at the Sun, attach a piece of white paper in front of the waveguide of the LNC. Now slew the antenna until a very bright spot appears on the paper. This is the sunlight reflected toward the LNC by the parabolic antenna. The satellite finder now displays the relative signal strength of the radio-frequency radiation from the Sun.

The LNC should generally never be exposed to the Sun without any protection since otherwise there is a risk it could heat up to the point that its plastic case would quickly melt. The best remedy here is to attach a piece of Styrofoam or, better yet, Styrodur in front of the input of the LNC. The material shields the LNC from heat while at the same time allowing radio-frequency radiation to pass unimpeded.

In our next experiment, the heat-protected antenna is again aimed at the Sun. This should be done around 3:00 p.m. since at this time no artificial satellites are situated in the line of sight whose signals could interfere with reception of the Sun. As soon as the antenna is aimed precisely at the Sun, adjust the gain control of the satellite such that the equipment needle deflects all the way over. By maintaining this setting, along with the alignment of the antenna, you will be able the following day to observe the passage of the Sun through the field of view of the fixed antenna between 2:00 and 4:00 p.m.

If a plotter or a computer, through the interface of a digital multimeter, is also connected, you will obtain a bell-shaped curve of the solar radio-frequency radiation. Observe what happens when the antenna remains aligned in this position over several days: How does the recorded signal change and why?

It is important to protect the LNC of the antenna when aimed at the Sun by a piece of Styrofoam so that it does not overheat!

Are you also able to use the above-described method to aim the antenna at the Full Moon? Can you detect temperature-dependent radio signals? How do they change as the phases of the Moon change?

9.7 Possible Upgrades

Technical enhancements for your radio telescope are possible both in terms of the mechanical system and the electronics. Finding objects is easier if the antenna is mounted equatorially, for example, in parallel with a small refractor (Fig. 7). This allows objects to be observed over many hours. You can, e.g., discover bursts of solar radio-frequency radiation and at the same time observe the Sun optically using the projection method.

You will soon find that the gain control for the satellite finder is too imprecise: It is difficult to reproduce previously set gain factors. It is therefore recommended that you replace the continuous gain control by a rotary switch that selects between multiple trimmer potentiometers with fixed settings, thereby providing reproducible gain factors. The trimmer potentiometers can have values between, for example, 100 and 500 Ω in 100-Ω steps.

We hope your radio telescope will give you much enjoyment and provide you with an interesting introduction to radio astronomy. If you have questions about amateur radio

Mount your antenna in parallel with an optical scope and observe the Sun simultaneously in the optical and radio regions.

Fig. 7. A satellite antenna mounted in parallel with a telescope enables you to easily track the rotation of the sky. This arrangement also allows for simultaneous observation of an object in the optical and the radio region. Image courtesy of Peter Wright

astronomy, the members of the European Radio Astronomy Club are ready and willing to help you (see Sect. 9.14).

9.8 The Giant Planet Jupiter

Even a small telescope shows Jupiter as a flattened sphere of gas with cloud bands running parallel to the equator. And, even binoculars will reveal the four Galilean moons Io, Europa, Ganymede, and Callisto as points of light. Jupiter's moon Io is the most volcanically active body in the Solar System. Its volcanos eject sulfur oxide and dioxide gas as fountains reaching 100 to 300 km in height. The gas leaves the surface of Io and is ionized by colliding with fast charged particles.

The ionized gas, known as the plasma torus, surrounds Jupiter as a thin toroidal cloud at a distance of 5.9 Jupiter radii. Its diameter is about one Jupiter radius. The orbital plane of the plasma torus is not identical with Io's orbital plane but is instead oriented by the effect of Jupiter's magnetic field in the plane of the magnetic equator tilted 10° relative to the equator of Jupiter.

Jupiter's magnetosphere surrounds the sphere of the planet and covers a volume of space hundreds of times greater than the volume of the Earth's magnetosphere. Jupiter's magnetosphere is deformed by the solar wind. The magnetospheric tail generated by Jupiter reaches beyond the orbit of Saturn. The magnetic axis (the connecting line between the magnetic north and south poles) is tilted 10° from Jupiter's rotational axis. In addition, the magnetic field rotates in fixed fashion along with the planet in 9 h 55 m 29.37 s. As a result, during one rotational period we see first the magnetic north pole and then the south pole.

9.9 Jupiter as a Radio Source

The fact that Jupiter is more than a cold gas ball was first revealed in early 1955 when two American radio astronomers came upon its radio-frequency radiation by accident. Franklin and Burke were observing the heavens at 22 MHz using a cross-shaped antenna configuration, the "Mills Cross" near Washington, DC. Mills Cross consists of a total of 64 dipole antennas arranged as a giant cross, each arm of the cross being more than 600 m long.

Jupiter is one of the strongest radio sources in the sky. Radio-interference storms are generated in its magnetosphere and can be detected by a sensitive short-wave receiver.

To test the telescope, the astronomers observed the radio emission of the Crab Nebula. Their measurements were at times disturbed by sporadic signals of unknown origin that appeared each day four minutes earlier. As a result, the signals could not have been terrestrial interference but had to emanate from an astronomical object. It was found that this interference occurred especially often when Jupiter was located within the lobe of the antenna. Subsequent investigation based on older observational data revealed that Jupiter's radio signals had already been recorded in the early 1950s in Australia and were even recorded in the early 1930s by radio astronomy pioneer Karl Jansky.

Figure 8 illustrates Jupiter's radio spectrum. Charged particles move within its magnetic field. They are deflected by the magnetic field lines and in the process emit synchrotron radiation in the decimeter wavelength. This is non-thermal radiation. Thermal

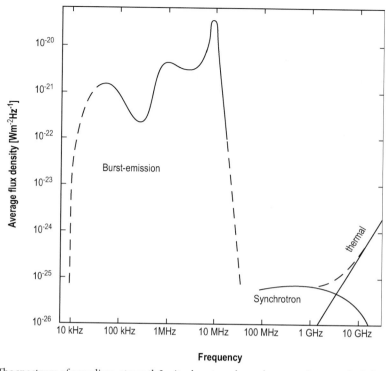

Fig. 8. The spectrum of a medium-strength Jupiter burst produces the curve shown to the left. However, bursts may be ten to hundreds of times stronger. Jupiter bursts can be detected primarily at frequencies around 20 MHz. At higher frequencies, the planet does not emit bursts but synchrotron radiation from its magnetosphere, as well as thermal radiation from its atmosphere. Image based on Carr et al. [1]

radiation can be detected at shorter wavelengths, in the range of a few centimeters, originating directly in the planetary sphere. Observation of this radiation yielded a temperature for the upper cloud boundary measuring 145 K (degrees Kelvin). In addition to the spectral components diagramed in Fig. 8, the Voyager probes detected sporadically occurring radiation at wavelengths of several kilometers.

The strongest component of the radio emission is created by radio-interference storms in the 5 to 39-MHz (decameter) range. They last from a few minutes to several hours and consist of a series of individual noise pulses or bursts. The radio storms occur only sporadically and are of varying intensity. For this reason, Fig. 8 shows only an averaged spectrum.

Earth-based observation of bursts below about 5 MHz is prevented by the ionosphere. Above 10 MHz, their intensity drops sharply. The cutoff frequency of the spectrum at a maximum of 39 MHz is interpreted as being due to the maximum energy of the charged particles being emitted: particles that radiate at this frequency lose their entire kinetic energy through radio emission. Based on this assumption, the site found for the radiation is a magnetic induction of 14 Gauss—a value that is attained in the vicinity of Jupiter's poles.

Two types of bursts can be detected in the short-wave region: L-bursts (long bursts), each of which lasts about 1 to 5 s, and S-bursts (short bursts), which last about 0.1 to 10 ms each. In the speaker of a receiver, L-bursts sound like ocean surf, while S-bursts resemble the crackling noise caused in the radio spectrum by distant thunderstorms. The homepages of the University of Florida Radio Observatory and the Radio Jove project provide sound samples of L-bursts and S-bursts (see Sect. 9.14).

9.10 Cause of Decameter Radiation

For a long period after their discovery, the cause of the radio storms remained mysterious. In 1964, K.S. Bigg noted a connection with the position of Jupiter's moon Io: Whenever the radio storms were most frequent and intensive, Io was located to the right or left of the planet as seen from Earth. These radio storms are called Io-A and Io-B. Furthermore, Bigg found two more types of radio storms independent of Io's position, the so-called A- and B-storms. Io-B- and B-storms consist primarily of S-bursts, while Io-A- and A-storms additionally contain L-bursts.

The radio bursts dependent on Io are created within a magnetic flux tube, a bundle of magnetic field lines that connects Jupiter to Io. Electrically charged particles can move along the flux tube in spiral orbits from Io to Jupiter and back again while emitting radio waves. The flux tube thus acts like a giant transmitting antenna. Since Io orbits Jupiter, the orientation of this "antenna" relative to the Earth is changing constantly. This is why we can receive the radio waves emitted by it only intermittently. In addition, the spiral motion of the emitting particles results in a cone-shaped radiation pattern of the flux tube (Fig. 9).

The engine for Jupiter's radio bursts is the volcanism of the moon Io. Its volcanoes constantly spew out sulfurous gas to altitudes of more than 300 km, where it is ionized by ultraviolet light from the Sun. This is why Io is surrounded by a cloud of electrically conductive gas. In terms of its electric properties, Io thus acts like a conductive sphere.

Jupiter's radio bursts can be observed mainly whenever its moon Io is situated to the left or right of the planet as seen from the Earth.

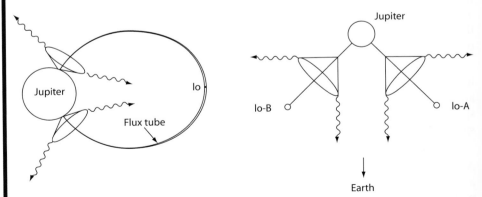

Fig. 9. Jupiter's short-wave radiation is emitted within the surface shell of a cone (*left*). Only when the envelope of the cone passes the line-of-sight to Earth (*right*) can the radio storms be received. When Io is located to the left of the planetary sphere, the right edge of the cone faces Earth and we receive Io-B bursts. When it is situated to the right of the planet, we receive Io-A bursts

Io takes about 42 h to revolve around Jupiter once, whereas Jupiter's magnetic field rotates in about 10 h. As a result, Io is continually overtaken by Jupiter's magnetic field lines. The magnetic field induces a voltage of around 400,000 volts in the electrically conductive moon. This voltage source forces a current of three million amperes along the magnetic flux tube that connects Io to Jupiter. The total power output radiated in the radio-frequency region is about a trillion watts (10^{12} watts).

Not all the radio bursts detected from Jupiter depend on the position of Io. The existence of radio storms not dependent on Io indicates that charged particles are also present away from the flux tube that actively contribute to the radio emission. The active regions are apparently located in the polar regions of Jupiter's magnetic field. Here, where the field lines funnel together, is where Jupiter's magnetic field is the strongest. In fact, the radio storms are observed most frequently when the flux tube has the correct orientation vis-à-vis Earth and at the same time the north or south pole of Jupiters magnetic field is tilted in the direction of the flux tube.

9.11 A Simple Radio Telescope for Jupiter

A receiver system for observing Jupiter consists of a short-wave antenna, a preselector, the short-wave receiver, and a cassette recorder or other tape recorder to record the signals (Fig. 10). In the following segment, we take a closer look at these components.

The antenna. In principle, any antenna that resonates in the 18 to 24-MHz frequency range is suitable for observing Jupiter. Below 18 MHz, the Earth's ionosphere blocks reception of radio waves from space, while above 24 MHz the radio-frequency radiation from Jupiter is too weak. An appropriate unit is, e.g., a $\lambda/2$ dipole for the 15-m amateur radio band (21.0 to 21.45 MHz) suspended at a height of $0.35 \times \lambda$ above the ground and oriented in an east–west direction. Its antenna pattern favors radio-frequency radiation coming from the south and at an altitude of about 40° above the horizon. (This corre-

The key components of a radio telescope for Jupiter are a $\lambda/2$ dipole antenna and a sensitive short-wave receiver operating in the 18 to 24-MHz range.

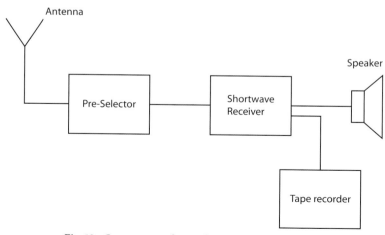

Fig. 10. Components of a simple amateur radio telescope

sponds to the altitude of the celestial equator for a geographic latitude of 50°.) Jupiter remains within field of view of the antenna for about 2.5 h before and after culmination.

What's known as a balun is placed at the center of the antenna. Additional information on constructing dipole antennas can be found in reference books for amateur radio operators and the homepages of NASA's Radio Jove Project, as well as from Radio-Sky Publishing (see Sect. 9.14).

Preselector. Commercially available short-wave receivers are often affected by signals from strong short-wave stations transmitting outside the intended reception band. Short-wave transmitters operating in the lower frequency bands, e.g., in the 49-m band, can cause interference in the upper frequency bands, e.g., in the 15-m band. These undesirable signals are most noticeable at night when the signal level of the short-wave transmitters is increased. The best solution to this problem consists in not allowing the interfering signals to ever reach the receiver. This function is performed by a tunable filter, known as a preselector, that is connected between the antenna and receiver input and allows only those signals that are in the frequency band of interest to pass. Preselectors are commercially available or can be home-built.

Receiver. The receiver should be a sensitive short-wave receiver (manufacturer's specifications: 1 μV for 10 dB (S+N)/N or better) for the 18 to 24-MHz band that has good selectivity and a coaxial antenna jack. This selectivity is needed in order to avoid interfering transmitters. In addition, it must be possible to switch off the automatic gain control, AGC, that functions during the reception of short-wave transmitters to compensate for signal fluctuations (fading), since otherwise Jupiter's radio signal of highly variable intensity could not be detected. Radio signals from space are always amplitude-modulated; therefore, only reception in AM mode is useful.

Suitable receivers do not have to be expensive. They are often sold used at ham-radio flee markets. Radio sets for the short-wave bands used for amateur radio are also suitable for observing Jupiter. NASA offers a complete radio telescope in the form of a kit for observing Jupiter as part of its Radio Jove Project. A free downloadable manual describes all the technical details of the radio telescope in depth.

Recording the signals. The signals received can be recorded on cassette tape, standard audio tape, or a computer. Most cassette recorders are equipped with an automatic volume control (AVC) that causes differences in intensity to become almost inaudible. Therefore it must be possible to switch off the AVC of the cassette recorder. Many recorders allow the recorded tape to be played back at higher speed, thereby enabling suspected signals to be found quickly.

Reception test: When the antenna is connected to the receiver, the noise audible in the speaker must rise significantly.

9.12 Reception in Practice

The following test will show if a receiving system for observing Jupiter is sufficiently sensitive. Set the receiver to a frequency free of interference. When the antenna is connected to the receiver's coaxial jack, the background noise should rise by an audibly significant amount. This additional noise originates mainly in our Milky Way and shows that your receiving system is detecting noise from space. The system's sensitivity is therefore not limited by the internal noise of the receiver but by galactic noise. Jupiter must drown out this natural background noise to be heard from the speaker. In the event the noise does not rise with the antenna connected, the signal must undergo preamplification. The preamplifier must be installed as close to the antenna as possible. To avoid interference from strong short-wave transmitters, you should operate the amplifier together with a preselector.

Reception tests only make sense when Jupiter is in the sky. Information to this effect is provided by various planetarium programs, such as, e.g., Calsky or Easy Sky. The optimum observing period is the second half of the night when the attenuating effect of Earth's ionosphere diminishes. During daytime observations, not only the ionosphere but also the Sun make their presence known: especially during the period of maximum solar activity, the Sun emits intensive radio bursts that resemble the L-bursts from Jupiter.

Before each observing session, the receiver is set to "AM" and the AGC is switched off. Now search for an interference-free frequency and retain this position. Then turn on the receiver; note the starting time or record this verbally on the tape. As soon as you believe you are hearing Jupiter, the receiver is slightly detuned; this is because the presumed L-burst could also be a signal fading in and out from a distant radio station.

To completely exclude the possibility there was interference, you should continually listen to the noise during the observation, then later compare what you've recorded with what Jupiter fans at remote sites have recorded concurrently.

9.13 Future Opportunities

We have attempted in this chapter to describe some initial steps into the invisible universe. Due to their high intensity, radio waves from the Sun and Jupiter are among the strongest signals that an amateur radio astronomer can receive using simple equipment. As we indicated at the beginning, a large number of other projects are also possible beyond these two. An interesting one, for example, is the search for signals from intelligent extraterrestrial civilizations (SETI, Search for Extraterrestrial Intelligence). The US-based SETI League provides guidance for amateur astronomers on how to build your own radio telescope and participate in the search. Additional projects involve observing the Milky Way in light at the 21-cm line of hydrogen and the detection of pulsars. It is our hope that the information of this chapter along with the following references to interest groups, literature, and electronic components for building equipment will assist you in making a successful launch into the realm of radio astronomy.

9.14 Additional Information

9.14.1 Contact Addresses

European Radio Astronomy Club, c/o Peter Wright, Ziethenstr. 97, D-68259 Mannheim, Germany, Tel.: +49-(0)-621-794597, E-Mail: erachq@aol.com: http://www.eracnet.org/

Society of Amateur Radio Astronomers, USA: http://www.radio-astronomy.org/

SETI League, 433 Liberty Street, PO Box 555, Little Ferry, NJ 07643: http://www.setileague.org/

University of Florida Radio Observatory (UFRO), with predictions of Jupiter's radio bursts: http://ufro1.astro.ufl.edu/

9.14.2 Complete Amateur Radio Telescopes and Kit Components

Project "Radio Jove: planetary radio astronomy for schools." Kit for a home-built radio telescope to observe Jupiter bursts: http://radiojove.gsfc.nasa.gov/

Radio Astronomy Supplies, PO Box 450546, Sunrise, FL 33345-0546: http://www.radioastronomysupplies.com

9.14.3 Books on Radio Equipment and Amateur Radio Astronomy

Carr, J.J.: Radio Science Observing, vol. 1 and 2. Prompt, Indianapolis, IN (1999)

Fielding, J.: Amateur Radio Astronomy. Radio Society of Great Britain (RSGB), Cranborne Road, Potters Bar, Hertfordshire, UK (2006)

Flagg, R.S.: Listening to Jupiter. Radio-Sky Publishing, Ocean View, HI. http://www.radiosky.com. Accessed 2008

Heys, J.: Practical Antennas for Novices. Radio Society of Great Britain (RSGB), Cranborne Road, Potters Bar, Hertfordshire, UK (1992)

Lonc, W.: Radio Astronomy Projects, 3rd edn. Radio-Sky Publishing, Ocean View, HI. http://radiosky.com/. Accessed 2008

9.14.4 Textbooks and Scientific Publications

Bagenal, F., Dowling, T.E., McKinnon, W.B.: Jupiter. The Planet, Satellites and Magnetosphere. Cambridge University Press, Cambridge (2004)

Burke, B.F.: An Introduction to Radio Astronomy. Cambridge University Press, Cambridge (2002)

Rohlfs, K., Wilson, T.: Tools of Radio Astronomy. Springer, Berlin Heidelberg New York (2006)

9.14.5 Software

Radio Sky-Pipe: Program to record and analyze received radio signals. Radio-Sky Publishing, Ocean View, HI. http://radiosky.com/. Accessed 2008

Radio-Jupiter Pro: Program to predict radio bursts from Jupiter. Radio-Sky Publishing, Ocean View, HI. http://radiosky.com/. Accessed 2008

References

1. Carr, T.D., Desch, M.D., Alexander, J.K.: Phenomenology of magnetospheric radio emissions. In: Dessler, A.J. (ed.) Physics of the Jovian Magnetosphere, chap. 7, pp. 226–284. Cambridge University Press, New York (1983)

10. Modern Sundials

F. Schmeidler[†]

10.1 Introduction

Modern sundials have become a favorite ornament in houses and gardens. Their design and construction are not difficult with some knowledge of spherical astronomy, and suggestions for improved constructions appear occasionally in the literature (see References).

The basic parts of a sundial are the shadow-casting *style* or *gnomon*, and the dial plate. In principle, the dial can be designed for any surface. Some important and simple cases of plane dials are considered in the following. The gnomon always lies in the plane of the meridian, and is tilted so as to point toward the celestial pole. It is inclined to the horizontal plane by the same angle as the geographic latitude φ.

Any sundial shows directly the *true solar time*, which is obtained simply from the shadow cast by the actual Sun. True solar time is a *local time* and differs from the internationally used system of *standard time* (ST). The latter is the mean local time of the standard meridian, for instance, the 75th parallel west of Greenwich for Eastern Standard Time (EST) in the US. In principle, two different corrections must be applied to observed sundial readings:

1. The equation of time which reduces the true solar time to a mean, uniform time (mean solar time, or MST).

2. The correction for longitude to the standard meridian, which converts the thus obtained mean solar time to standard time. In contrast to the equation of time, the longitude correction at a given location of the dial is a constant to be applied to every reading. This suggests that the longitude correction should already be included in the marking of the dial, see Sect. 10.5 below.

It is not practical, as is sometimes suggested, to find the accurate orientation from an architect's blueprints, since such directions are often not precise enough. Nor should it be obtained from compass readings, which are influenced not only by the magnetic declination but also by unknown local magnetic disturbances which may affect the reading beyond the required precision. An astronomer can and should find the directions only by astronomical observations.

A warmly recommended reference for the actual construction of a sundial is a book by A. Zenkert [2]; it details problems occurring in the construction and contains numerous design graphs and other figures.

International contacts e.g.
British Sundial Society
www.sundialsoc.org.uk

For websites concerning computations, adjustment, equation of time and other information of special interest for the construction, see:
http://cpcug.org/user/jaubert/jsundial.html.

The appearance of different sundials: http://www.sonnenuhr.de.

Links regarding sundials:
http://infraroth.de/slinks.html.

10.2 The Equinoxial Dial

The simplest type of dial is the *equinoxial* or *polar dial*, designed as a kind of armillary sphere. The shadow of the style pointing to the celestial pole is read on a circle in the equatorial plane, which is divided uniformly from 0^h to 24^h, with the 12^h mark at the shadowed position at true noon (i.e., when the true Sun is at the meridian). For the geographic longitude correction of the armillary sphere, the scale is correspondingly shifted. Such equinoxial dials are best placed atop a pillar in the garden.

10.3 Horizontal Dials and Vertical East–West Dials

Also rather simple are the sundials with faces lying in a horizontal plane or on an east–west-oriented wall.

10.3.1 Computations

Figure 1 graphs three planes: the horizontal plane, the east–west plane oriented at right angles to it, and the plane parallel to the celestial equator and passing through the intersection $\overline{AB}$ of the other two. The formulae for dividing the faces in the horizontal plane and in the vertical east–west plane are then readily derived. The style $\overline{NS}$ is perpendicular to the equatorial plane, intersecting it in the point O. At a given hour angle h the shadow on an equinoxial dial meets the east–west line $\overline{AB}$ or its extension at the point P, then $\overline{SP}$ is the shadow on the horizontal plane referring to this hour angle, and $\overline{NP}$ that on the vertical east–west plane. The style is inclined by the geographic latitude, and for an hour angle $h = 0^h$, the shadow makes the lines $\overline{SC}$ and $\overline{NC}$ on the two primary planes. The three triangles $\triangle OCP$, $\triangle SCP$, and $\triangle NCP$ have the line $\overline{CP}$ and the right angle at C in common.

SONNE.EXE is a freeware program for construction different sundials. The program constructs and draws the sundials for different geographical longitudes and latitudes.

The website http://www.shadowspro.com is a proven software for the construction of sundials. Sundials from users and general information of the designer François Blateyron can be found at: http://www.cadrans-solaires.org

The computer program ZW_easy for Windows calculates simple plane sundials for any location on earth. One can view the sundial pattern directly on screen and copy it to the clipboard.

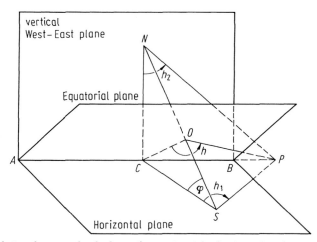

Fig. 1. Relations between the shadows of an equinoxial, a horizontal, and an east–west dial

Therefore, the important time-dependent angles are

$< COP = h =$ hour angle on the equinoxial dial,
$< CSP = h_1 =$ corresponding angle in the horizontal plane,
$< CNP = h_2 =$ corresponding angle in the vertical east–west plane.

We then obtain

in $\triangle OCP$ $\overline{CP} = \overline{OC} \cdot \tan h,$

in $\triangle SCP$ $\overline{CP} = \overline{SC} \cdot \tan h_1,$

in $\triangle NCP$ $\overline{CP} = \overline{NC} \cdot \tan h_2 .$ (1)

And, by division of the equations,

$$\tan h_1 = \frac{\overline{OC}}{\overline{SC}} \cdot \tan h,$$

$$\tan h_2 = \frac{\overline{OC}}{\overline{NC}} \cdot \tan h . \qquad (2)$$

Also, in the right triangles $\triangle COS$ and $\triangle CON$,

$$\frac{\overline{OC}}{\overline{SC}} = \sin \varphi,$$

$$\frac{\overline{OC}}{\overline{NC}} = \cos \varphi , \qquad (3)$$

and hence

$$\tan h_1 = \sin \varphi \tan h,$$
$$\tan h_2 = \cos \varphi \tan h . \qquad (4)$$

These are the two basic formulae which give for any hour angle h the corresponding angles h_1 and h_2, respectively, on horizontal and east–west dials.

10.3.2 Graphical Construction

The graphical solution, if preferred, is also readily obtained (Figs. 2 and 3).

    Draw a circle with radius $\overline{MC} = 1$ and extend the radius to a point O so that $\overline{OC} = \sin \varphi$. Graph a semicircle around O and the joint tangent through C. Mark the angles $15°, 30°, 45°$, etc., on the smaller circle, beginning at $\overline{OC}$, corresponding to whole hours. The desired subdivisions may be added at this point or later. The radii from O are extended to intersect the tangent; the points of intersection connected with M show the dial markings of the horizontal on the larger circle. Label C with 12^h and the other points with increasing hour numbers in the clockwise direction. The style is anchored in M and forms the angle φ with the direction 12^h.

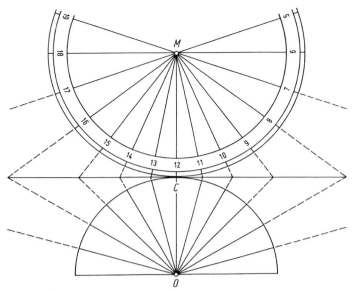

Fig. 2. Construction of a *horizontal* sundial for the geographic latitude $\varphi = 50°$

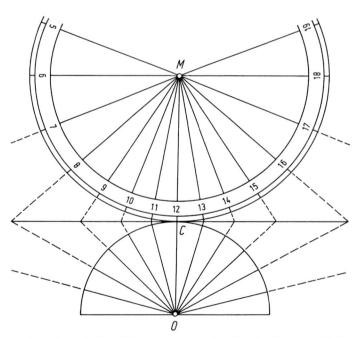

Fig. 3. Construction of a *vertical* sundial in the east–west direction for the geographic latitude $\varphi = 50°$

The dial plate is adjusted horizontally by a level and with 12^h pointing exactly toward north. The easiest way to do this is to calculate the transit of the Sun for any day and rotate the dial accordingly. The longitude correction is not yet allowed for. Figure 2 shows the construction of such a dial for the latitude $\varphi = 50°$.

The vertical east–west dial is constructed in the same way, except that $\overline{OC} = \cos\varphi$ (Fig. 3) and the marking is counterclockwise. The 6^h–18^h line is parallel to the tangent.

The adjustment of the vertical dial in the east–west plane requires an observation. The true solar time corresponding to the position of the Sun at any instant, for example, can be calculated. Adjusting the dial for grazing incidence (at the moment when the Sun crosses the prime vertical) gives the best accuracy, but this can be done only in spring and summer.

10.4 The Vertical Deviating Dial

Usually, the vertical sundial is to be installed at a given wall which in general does not lie exactly in the east–west plane. A vertical dial that is not in this plane is called a *vertical deviating dial*. The first task is to then determine the azimuth of the wall.

10.4.1 Determination of the Azimuth of the Wall

It is probably safe to assume in the construction that the wall is perfectly vertical, although, as was said previously, the azimuthal direction cannot be relied upon. This direction is best found by observing the Sun when its light is just grazing (i.e., parallel with) the wall surface. More specifically, gaze along the plane of the wall toward the east (morning) or west (afternoon). Using a suitable glass filter, observe with one eye the instant when half of the solar disk is covered by the wall. Record this time precisely with a dependable watch which has been synchronized with a time signal. With some experience, the instant mentioned can be found without instruments to about ±10 s; the observation becomes more precise with repeated trials on several days. The recorded times will differ, because the instant of passage through the plane of the wall is affected by the continuous changes of the solar declination and of the equation of time. Each observation gives the azimuth of the wall from the well-known relation,

$$\tan A_\odot = \frac{\sin h}{\sin\varphi \cos h - \cos\varphi \tan\delta} , \tag{5}$$

where

φ = geographical latitude,

h = hour angle of the Sun at the instant of observation,

δ = declination of the Sun at the instant of observation,

$A_\odot$ = azimuth of the Sun and of the wall to be found.

Weather permitting, the readings can be made in the morning or in the afternoon, but, for obvious reasons, only during the spring and summer seasons. With a little practice, the resulting azimuths may differ by about 20′, so that just a few observations will give a sufficiently precise average.

10.4.2 Calculation of the Dial at the Wall

The formulae required for this analysis are more complicated than in the previous cases, but can be deduced from Fig. 4.

Let $\overline{NS}$ again be the style parallel to the Earth's polar axis, meeting the horizontal plane at the point S and the vertical east–west plane, as well as the plane of the wall, at the point N. Besides horizontal and wall planes, the east–west plane, which intersects the wall in a vertical through N, is graphed. The north–south plane through S meets the two other planes at the point M. The direction of the wall is given by the intersection with the horizontal. The angle $< SMT_E = A_\odot$ also gives the azimuth of the wall as defined and determined above.

The points T_E on the morning side (hour angle east) and T_W on the afternoon side (hour angle west) are, for the moment, arbitrary on the line of intersection between wall and horizontal. Then,

$$SMT_E = A_\odot \qquad \text{and} \qquad < SMT_W = 180° - A_\odot .$$

In the horizontal plane, let

$$< MST_E = x_1 \qquad \text{and} \qquad < MST_W = x_2 .$$

The angle $< NSM$ again equals φ, and x_1 and x_2 are the shadow directions of the style in the horizontal plane at hour angles h_1 and h_2, respectively. The angles at N in the plane of the wall,

$$< MNT_E = y_1 \qquad \text{and} \qquad < MNT_W = y_2 ,$$

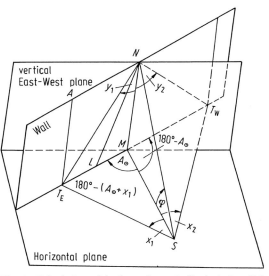

Fig. 4. Calculation of the face of a vertically deviating dial

correspond to x_1 and x_2 at the rotation by the angle $A_\odot$. $\overline{NT}_E$ and $\overline{NT}_W$ are the corresponding shadow directions at the wall for the hour angles h_1 and h_2. To find them, it is necessary only to calculate the segments $\overline{NT}_W$ and $\overline{NT}_E$ or the angles y_1, y_2.

The triangle ΔMST_E gives

$$\overline{MT_E} = \frac{\overline{MS}\sin x_1}{\sin(A_\odot + x_1)} = \frac{\overline{NS}\cos\varphi}{\cos A_\odot + \sin A_\odot \cot x_1} . \tag{6}$$

Because of the relation

$$\cot x_1 = \frac{\cot h_1}{\sin\varphi} , \tag{7}$$

as in the horizontal dial, Eq. (6) becomes

$$\cot y_1 = \frac{\overline{MN}}{\overline{MT_E} = \frac{\overline{NS}\sin\varphi}{\overline{NS}\cos\varphi}} \left(\cos A_\odot + \frac{\sin A_\odot}{\sin\varphi}\cot h_1 \right) . \tag{8}$$

The length $\overline{NS}$ of the style cancels, with the result

$$\cot y_1 = \cos A_\odot \tan\varphi + \frac{\sin A_\odot}{\cos\varphi}\cot h_1 , \tag{9}$$

or, in a more convenient form,

$$\tan y_1 = \frac{\cos\varphi}{\cos A_\odot \sin\varphi + \sin A_\odot \cot h_1} . \tag{10}$$

This equation holds for any true solar time in the morning hours.

For western hour angles, which occur in the afternoon, $A_\odot$ is replaced by $180° - A_\odot$ and x_1 by x_2, which gives

$$\tan y_2 = \frac{\cos\varphi}{- \cos A_\odot \sin\varphi + \sin A_\odot \cot h_2} . \tag{11}$$

Since the position of the points T_E and T_W on the intersection was not specified (except that one is to the left of M and the other to the right), the point at any instant of time can be determined on the line $\overline{T_E T_W}$.

Since the horizontal line $\overline{T_E T_W}$ cannot be arbitrarily long, one turns to the perpendicular to it at a suitable point, for instance, in T_E. Let this be $\overline{AT_E} = \overline{NM}$. The angle $< ANT_E$ equals $90° - y_1$. To find the corresponding points of the time scale on $\overline{AT}_E$, simply change to the co-functions.

10.4.3 Construction of the Calculated Dial Face at the Wall

The next step, which is to transfer the calculated dial markings to the wall, may be executed in the following way.

First mark the foot of the center of the style at point N. Define the vertical with a plumb bob through N, and then graph it. It will generally suffice to make the vertical exactly 1.000 m long to point M (see Fig. 4). At M, the horizontal is determined precisely by a level and graphed toward both sides, also precisely 1.000 m long. At the end points, the vertical upward direction is determined with the plumb bob. Thus, on both sides of the central line $\overline{NM}$, two squares each of 1.0 m side are bisected whose side pairs are oriented exactly horizontal and vertical, respectively.

Graph the points characterized as tan y_1 and tan y_2 from the equations above, horizontally from M to the left and to the right, and do the same with the values of cot y_1 and cot y_2 on the respective verticals (tan y_1 = tan y_2 = 0 lies at the point M, while cot y_1 = 0 and cot y_2 = 0 are diagonally opposite to M).

In this way, all computed points for the dial are at the wall, and subdivisions, perhaps half-hours or even to 10 minutes, can be made according to preference and the size of the dial.

The labeled band can have any arbitrary shape. To obtain the time points on it, merely connect the corresponding points on the rectangular frame of Fig. 4 with the foot of M, where the style will be inserted. Then record the numbers where these lines pass through the labeled band.

Later on, after construction is completed and the wall is painted, the rectangular frame with the original time scale can be removed.

10.4.4 Inserting the Style

The insertion of the style must be performed as precisely as the construction of the dial scale. The accurate fit of the style determines how precise the time readings at the dial are. Two matters in particular should be attended to:

1. When cementing the style, its center should meet the wall as accurately as possible at the point N, since that was the point to which the dial scale referred (cf. again Fig. 4).
2. The style should lie exactly in the plane of the meridian and should form the angle φ with the horizontal.

The second requirement above can best be met as follows: Imagine a plane which passes through the style and can rotate around it as an axis, and consider the angle which the intersection of this plane with the wall forms with the style. This angle will vary with the rotating plane, but reaches a minimum when the rotating plane is vertical to the wall. The intersection of the rotating plane is called a *substyle*, and is the segment $\overline{NL}$ in Fig. 4.

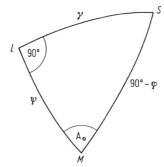

Fig. 5. The determination of the substylar line

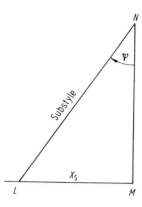

Fig. 6. Determination of the position of substyles

The produced directions of $\overline{NM}$, style $\overline{NS}$, and substyle $\overline{NL}$ point to three different points on the sphere, and form a spherical triangle $\triangle NSL$ with a right angle at L (Fig. 5). The arc $MS = 90° - \varphi$, the angle $< LMS = A_\odot$, and the arc SL marked with the symbol γ is the angle between style and substyle. Formulae for the right spherical triangle MSL give

$$\sin \gamma = \cos \varphi \sin A_\odot , \qquad (12)$$

and

$$\tan \psi = \cot \varphi \cos A_\odot , \qquad (13)$$

where ψ is the angle formed by the substyles with the perpendicular $\overline{MN}$ in the plane of the wall (see Fig. 6).

Call L the point where the substyle meets the horizontal through M, and X_S the distance $\overline{ML}$, and make $\overline{MN}$ again equal to 1.000 m. Then

$$X_S = \tan \psi = \cot \varphi \cos A_\odot . \qquad (14)$$

When $A_\odot < 90°$, X_S is to the left (i.e., west) of M. Otherwise, it is to the right (or east) of M. The positions of the substyles are thus known through X_S.

To insert the style correctly, have a moderate-sized right triangle made of plywood and with one of the acute angles equal to γ. When the style of the desired length is to be inserted, this plywood triangle is placed with one of its sides perpendicular to the substylar line—to support the style—whereby the angle γ comes to lie between the style and substylar line.

If the plywood style has a hypotenuse about as long as the style, the latter can then be placed on the triangle and, if need be, lightly fastened to it. Often, the style is slightly conical, a feature which can be accommodated by a slight correction of the angle γ in the wood triangle, or by a wedged support between triangle and style. The triangle is held in place by the weight of the style or else by a small nail, but the perpendicularity of the wooden triangle with the wall should be exactly maintained. The style can now be cemented, and will be in the proper position when the wood triangle is removed two days later.

10.5 Designs for Higher Accuracy

The methods mentioned so far permit the construction of sundials whose readings may depart from standard time by up to a quarter or half-hour. Higher accuracy may be reached by making additional corrections as described in the sections below.

10.5.1 Correction for Geographic Longitude

This correction, if desired, can be incorporated into the labeling of the scale. For instance, if the sundial is located at a longitude 5° or 20 minutes west of the meridian to which the standard time refers, then the addition of 20 minutes on all marks of the dial will correct for longitude, because standard time is by this average amount ahead of local solar time. The label 12^h, located on the vertical dial directly below the style foot, is now displaced to the left, and below the foot will now be the mark 12^h20^m. A sundial corrected for longitude can be immediately recognized by its having 12^h reading sidewise displaced from directly below the style.

10.5.2 Correction for the Equation of Time

The changes of the equation of time from day to day, and from one year to the next, are below the reading precision of a sundial, which is at best 1 minute. To neglect the leap days creates a maximum error of one day, and the fastest change of the equation of time occurring in late December is only 1/2 minute per day. It is thus adequate to neglect February 29 and to use one permanent graph which reduces apparent to mean solar time for every day (Fig. 7). The date is read on the figure-of-eight, and the corresponding correction of the sundial reading on the top or bottom scale to obtain standard time.

Attempts have recently been made to incorporate this correction in sundials by suitable construction measures. Some successful attempts in this direction have been reported in *Sky and Telescope* [3]. In each case, the principle is to curve the gnomon or the dial or both in such a fashion that the readings include the correction for the equation of time.

Also worth reading is an article in the *Journal of the British Astronomical Association* (JBAA) [4], which discusses under what conditions the dial is equiangular in the sense that equal time intervals are represented by equal labeling intervals. Evidently, a sundial complying with this condition is more precise and convenient to read than would otherwise be the case. On the other hand, the construction of such a dial is necessarily more complex than that of an ordinary sundial. For a simplified theory of such dials, consult H. Lippold [5].

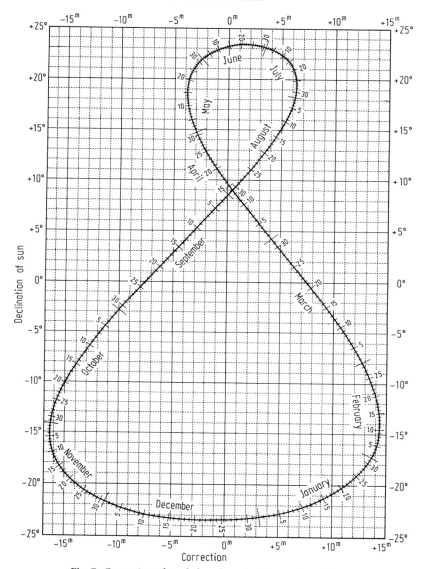

Fig. 7. Correction of sundial readings for the equation of time

10.6 Further Reading

Jenkins, G., Bear, M.: Sun Dials and Time Dials. A collection of Working Models to Cut and Glue Together, Parkwest, 1988

Weis, Ch.: Eine globale Sonnenuhr, *Sterne und Weltraum*, Mai 2005, 87 (www. suw-online.de)

Drinkwater, P.: *The art of Sundial Construction*, Shipston-on-Stour, 1985

References

1. Brunner, W.: Neuartige Sonnenuhr-Konstruktionen. Orion **33**, 44 (1975)
2. Zenkert, A.: Faszination Sonnenuhr. Verlag Harry Deutsch, Thun (1995)
3. Egger, H.: Results of Sundial Competition. Sky Telescope **32**, 256 (1966)
4. Sawyer, F.W.: Bifilar gnomonics. J. British Astron. Assoc. **88**(4):334–351 (1978)
5. Lippold, H.: Zur Theorie der homogenen Sonnenuhr. Die Sterne **61**, 228 (1985)

11. An Historical Exploration of Modern Astronomy

G.D. Roth

11.1 Introduction

There were two principal factors which motivated the people of ancient times to concern themselves with the stars: the link of celestial objects with religious concepts, and the everyday need to determine the time.

First cosmological ideas we see in the geocentric system of Claudius Ptolemy (100–160 BC) based on the physics of Aristoteles (384–322 BC) and the geometric thoughts of Greek thinkers during the fourth century BC. Ptolemy's model was strong enough to survive more than 1400 years.

The instruments and observing methods in astronomy remained basically the same throughout and past the Middle Ages. The study of the motions of celestial objects was paramount to ancient and medieval astronomers. With the invention of the telescope early in the seventeenth century, progress in astronomy underwent a spectacular advance: a detailed physical exploration of celestial bodies could now be pursued.

The evolution of astronomy into a highly specialized science began in the nineteenth century, as the introduction of spectral analysis, photometry, and photography made astrophysics a major part of astronomy. The limitations of optical astronomy were overcome in the twentieth century, when radio telescopes and space-borne instruments made the entire spectrum of electromagnetic waves accessible.

Ancient astronomers believed that the heavens shifted around a stationary Earth.

Ptolemy collected his calculations into 13 volumes called the *Almagest*.

11.2 The Heliocentric System

The era of "modern" astronomy can be said to have come into existence in 1543 with the appearance of *De revolutionibus* by Nicolaus Copernicus (1473–1543, Fig. 1). In this work, the Polish canon introduced a model of the heavens in which the center of the planetary motions is not the Earth but the Sun. This model is known as the Sun-centered, or heliocentric, system. In a nutshell, his thesis was that the observed annual motion of the Sun about the Earth is actually due to the motion of the Earth; the rotation of the Earth about its axis is reflected in the diurnal rotation of the stellar sky, and properties of planetary motion can be understood only by the taking into consideration that observations are made from the moving Earth.

The German astronomer Johannes Kepler (1571–1630, Fig. 2) studied the planetary motions in more detail. In particular, he examined carefully the observations of Mars which had been made by the Danish astronomer Tycho Brahe (1546–1601, Fig. 3) and, as a result, discovered the motion of planets in elliptical orbits about the Sun. (Copernicus had adhered to the venerable tradition of assuming that circular motions are appropriate

Fig. 1. N. Copernicus

N. Copernicus: *De revolutionibus orbium coelestium libri* VI. Nürnberg 1543

Fig. 2. J. Kepler

J. Kepler: *Astronomia nova* 1609

Fig. 3. T. Brahe

Fig. 4. G. Galilei

Visit the following websites for further information: http://www.tychobrahe.com, http://galileo.imss.firenze.it/index.html, http://www.suw-online.de.

Besides his efforts to establish celestial physics on an experimental and mathematical basis Galilei was an excellent artist, who painted the seen lunar objects [1].

Parallax (heliocentric stellar parallax). Apparent shift in the positions of nearby stars due to the changing position of the Earth in its orbit around the Sun.

for celestial bodies.) Kepler's three Laws of Planetary Motion about the Sun are seminal documents of the laws of nature which now pervade all of astronomy:

1. Each planet moves in an ellipse about the Sun, which lies at a focus of this ellipse.
2. The speed of orbital motion is higher in the orbital sections near the Sun than far away from it. The speed changes such that the radius vector from Sun to planet sweeps equal areas in equal times.
3. The ratio between the squares of the periods and the cubes of the mean distances from the Sun is the same for all planets.

Whereas Kepler arrived at his laws without telescopic observations, his contemporary, the Italian scientist Galileo Galilei (1564–1642, Fig. 4), began using the telescope in 1610 for astronomical observations and for providing arguments in favor of the heliocentric system. Galileo discovered the four bright satellites of Jupiter and observed the phases of Venus. His advocation of the Copernican system, however, led to serious conflicts with the Catholic Church. The Roman Inquisition initiated the trial of Galileo in which he was labelled a representative of what seemed to Church authorities to be a preposterous and heretical doctrine.

Early in summer 1609 the Italian artist Adam Elsheimer made first observations of the starry sky with the new telescope. His paintings show craters of the Moon and single stars of the Milky Way.

One objection, in particular, against the heliocentric system remained for nearly three centuries: If the Earth did move about the Sun, then stars should exhibit annual displacements or parallaxes in response to this orbital motion. Critics of the system pointed to the absence of the requisite stellar displacements, while supporters argued that such displacements did exist but were immeasurably minute owing to the presumably enormous distances to the stars. The apparent absence of observable stellar parallaxes became an incentive for observers and telescope builders to design and use better measuring instruments in order to obtain a positive result. The goal was not reached, however, until 1838, when Friedrich Wilhelm Bessel in Königsberg determined the parallax of 61 Cygni, and, almost simultaneously, Wilhelm Struve in Pulkova found the parallax of α Lyrae and Thomas Henderson at the Cape Observatory that of α Centauri.

By that time, the heliocentric system had already gained wide acceptance. But the first stellar parallax measures were not merely the final word in an old dispute; they also marked the beginning of a new advance in the exploration of space with more modern measuring techniques.

11.3 Evolution of the Theory of Motions

The notion that some kind of force originates in the Sun and influences the motions of the planets was already implicit in Kepler's writings. The English mathematician and physicist Isaac Newton (1643–1727, Fig. 5) formulated a mechanical principle which is not restricted to planetary motions only; it is universally valid:

Two bodies attract each other with a force proportional to the product of their masses and inversely proportional to the square of the distance between them.

This law of gravitation permitted an understanding of motions of bodies in the Solar System. Newton could write down the solution of the two-body problem (the motion of one planet about the Sun), but the mathematical treatment of the many-body problem was considerably more difficult. The latter problem occupied many famous mathematicians of the eighteenth century: Leonard Euler (1707–1783) in Switzerland, and Joseph Lagrange (1736–1813) and Pierre Simon Laplace (1749–1827) in France. With the law of gravitation as its base, celestial mechanics developed in the eighteenth and nineteenth centuries into a highly sophisticated science.

The discovery of the planet Uranus by Sir William Herschel (1738–1822) in 1781 and of the first minor planet by the Italian G. Piazzi (1746–1826) in 1801 also stimulated the study of motions of Solar System bodies. Carl Friedrich Gauss (1777–1855) published his *Theory of Motions of Celestial Objects which Revolve about the Sun in Conic Sections* in 1809, thereby furnishing a convenient method for calculating the ephemerides of, for instance, a minor planet from only a few observations. The method developed by the German physician Wilhelm Olbers (1758–1840), *To Compute the Orbit of a Comet from Some Observations*, as a parabolic approximation also found wide application.

The discoveries of asteroids and planetary bodies during the late eighteenth and early nineteenth centuries spawned further searches for planets. The discovery of Neptune in 1846 by Johann Gottfried Galle (1812–1910) is considered the crowning achievement of the celestial mechanics experts Urbain Leverrier (1811–1877) in France and John Couch Adams (1819–1892) in England, who had independently employed their mathematical skills to successfully predict the position of the unknown planet. In 1930, the planet Pluto, now degraded to the status of a dwarf planet, was discovered by Clyde Tombaugh.

11.4 Cataloging the Stellar Sky

Some much-used star catalogs date back to the seventeenth and eighteenth centuries:

1603 Uranometria, by Bayer
1661 Sternverzeichnis, by Hevelius
1679 First southern star catalog, by Halley
1725 List of all stars down to 7th magnitude, visible in northern Europe, by Falmsteed
1762 Star catalog, by Bradley

The list of 103 nebulous "M" objects compiled by Charles Messier (1730–1817) in 1784 and two catalogs by Sir William Herschel in 1786 and 1789, each containing 1000 newly found nebulae and clusters, also date to before 1800. The latter had by 1782 already compiled a catalog of 269 double stars.

Astronomical research relies heavily on the accumulation of a sufficiently large base of observed data, so that the conclusions will be statistically significant. For this reason, sky surveys, data collections, and catalog publications occupied the substantial part of the astronomical working capacity in the nineteenth and twentieth centuries. The prototype of large surveys was the Bonner Durchmusterung, prepared in 1852–1859 and directed by Friedrich Wilhelm Argelander (1799–1875) with co-workers Adalbert Krüger (1832–1896) and Eduard Schönfeld (1828–1891).

Fig. 5. I. Newton

I. Newton: *Philosophiae naturalis principia mathematica.* London 1687.

Gravitation is the tendency of matter to attract matter.

With the decision of the International Astronomical Union (IAU) during the meeting in August 2006 at Prague, Pluto is no longer a major planet but a minor planetary object.

With regard to the discussion about the status of Pluto see: http://kuffner-sternwarte.at/im_brennp/arichv2006/Planeten_Definition.html.

http://www.ipetitions.com/petition/planetprotest

Table 1. Number of known variable stars as a function of year

Year	1786	1844	1890	1896	1912	1970	1983
Number of variables	12	18	175	393	4000	22,650	28,450

Fig. 6. F.W. Bessel

Fig. 7. F.W. Struve

Fig. 8. K. Schwarzschild

For more information see Further Reading p.278.

Improved instruments and observing techniques triggered advances in positional astronomy during the nineteenth century, notably through the efforts of the German astronomer Friedrich Wilhelm Bessel (1784–1846, Fig. 6). The positions of about 32,000 stars which he determined in the years 1821–1835 laid the foundation of many subsequent studies. Better quality and larger telescope apertures also promoted the development of special areas of observation such as double stars. In this regard, an especially assiduous observer was Friedrich Wilhelm Struve (1793–1864, Fig. 7) of Dorpat (Russia), who discovered double stars and also edited catalogs containing their data.

Collection and compilation of data were quickly undertaken in the new fields of photometry and spectroscopy, which were to form the basis for astrophysics. Visual works characterized only the beginning of this "new astronomy," which was to be taken over wholly by photographic methods in the latter half of the nineteenth century. Visual photometric methods, however, were still used to generate some celebrated catalogs:

– Cordoba Durchmusterung of the Southern Sky, begun in 1885, published in 1892
– Harvard Revised Photometry, begun 1879, published in 1907
– Potsdamer-Durchmusterung, begun in 1886, published in 1907

The Göttinger Aktinometrie in the years 1904–1908 by Karl Schwarzschild (1873–1916, Fig. 8) formed the basis for photographic stellar photometry.

There is another special subject which has gained much importance in the wake of these compilations, namely the discovery and description of variable stars. The number of known variables has rapidly increased in the past 200 years, as can be inferred from Table 1.

The advent of spectroscopy led to spectral classifications and catalogs of stars. The first list by the Italian Father Angelo Secchi (1818–1878) appeared in 1867 and contained 316 stars. The monumental spectral survey at the beginning of the twentieth century with 225,300 stars is contained in the Henry Draper Catalogue, which was published under the auspices of Harvard College Observatory in the United States during the period 1918–1924.

11.5 Astrophysics

11.5.1 Stellar Photometry

In the latter half of the nineteenth century, the apparent magnitude system of stars evolved from a foundation of somewhat crude, purely estimated values into a sophisticated, calibrated basis. The transition began in 1861, when Karl Friedrich Zöllner (1834–1882) built the first visual photometer in Berlin. Photographic plates and photoelectric cells dramatically increased the measuring precision, as is revealed in Table 2.

Table 2. Precision of stellar magnitudes by various methods

Year	Method	Precision
Ca. 1850	Visual estimates	$0^m.2$
Ca. 1870	Visual measures	$0^m.1$
Ca. 1912	Photoelectric measures	$0^m.01$
Ca. 1970	Photodetector measures	$0^m.001$

Stellar magnitudes thus came to be very accurately measured in certain standardized wavelength ranges from the ultraviolet into the infrared and have been employed to determine various color indices.

11.5.2 Spectroscopy of the Sun and Stars

When observing sunlight through a slit front of a prism the German optician Joseph von Fraunhofer discovered 1814 hundreds of fine dark lines within the solar spectrum. Today we call them spectral lines (Fraunhofer lines, Table 3).

Fig. 9. J. v. Fraunhofer

The groundwork of modern spectral analysis was laid with the discovery by Gustav Robert Kirchhoff (1824–1887) and Robert Bunsen (1811–1899) of the characteristic spectrum of each chemical element. The amassing of spectral data on stars and other celestial objects characterized the first phase of astronomical spectroscopy. With the invention of the photographic process, permanent records of spectra of the Sun and stars were soon made. The famous principle discovered by Doppler (1803–1853) led to another application in measuring radial velocities of stars. The first photographic measures of this kind were obtained in 1888 by Hermann Carl Vogel (1841–1907).

Among several spectral classification schemes of stars, the Harvard System by Edward Charles Pickering (1846–1919) and Annie Cannon (1863–1941) was the most successful, and was officially adopted by the International Astronomical Union in 1922.

Detailed studies of the solar spectrum continue with unabated intensity to this day. The invention, in 1892, of the spectroheliograph, a device which is used to photograph the solar surface in the light of individual spectral lines, was a major achievement in solar physics. Another milestone was reached when the hybrid technique of spectrophotometry was created by combining the methods of photometry and spectroscopy to measure the intensity distribution in spectra.

Table 3. Early work of exploration of the solar spectrum

1814	Fraunhofer discovered the spectral lines in the solar spectrum
1842	Christian Doppler postulated that wavelength is affected by motion (Doppler effect)
1860	Robert Kirchhoff's laws dealing with the production of absorption lines, emission lines, and continuous spectra
1868	Jules Janssen discovered the solar prominences and helium in the solar chromosphere
1889	George E. Hale constructed the first spectrohelioscope
1897	Henry A. Rowland published his photographical atlas of the solar spectrum

Fig. 10. Total solar eclipse of 1851

See the book by J. Ashbrook, *The Astronomical Scrapbook. Sky watchers, Pioneers, and Seekers in Astronomy*, Cambridge, MA 1984.

Fig. 11. E.E. Barnard

Fig. 12. M. Wolf

11.5.3 Astronomical Photography

The first photograph of the Moon by J.W. Draper on Daguerre plates dates from 1841. The oldest scientifically useful photograph is that of the total solar eclipse on July 18, 1851 (Fig. 10) while the first photographs of star images were obtained by W.C. Bond and J.A. Whipple in 1850. But it was the invention of the dry plate by R.L. Maddox in 1871 that signaled the real breakthrough for astrophotography.

The enormous power of the photographic technique was quickly documented by scores of impressive photographs taken of the Milky Way, star clusters, and nebulae. Much of this pioneering work was performed by the American astronomer Edward Emerson Barnard (1857–1923, Fig. 11) and the German Max Wolf (1863–1932). Photography rapidly established itself in all areas of astronomical research, often in connection with photometric and spectroscopic methods:

1. G.P. Bond took photographs of the double star Mizar at Harvard College Observatory in 1857.
2. Donati's comet of 1858 was the first comet to be photographed.
3. In 1863 Lewis Rutherford constructed a special refractor for astrophotography.
4. In 1880 Draper made the first photograph of the Orion Nebula.
5. In 1887 astronomers of different countries met in Paris to discuss the making of the first star catalog based on photographs, the famous Carte du Ciel.

11.5.4 Large Telescopes

Concurrent with the development of advanced observing techniques was the construction of larger and more powerful telescopes. Progress in astronomical research over the last two centuries is well-correlated with the practicalities and limitations of instrument design. The development of the achromatic objective by Fraunhofer in the nineteenth century initiated the production of refractors with increasingly larger apertures. These range from the moderate-sized Dorpat refractor (objective diameter 24.3 cm, focal length 4.11 m) built by Joseph von Fraunhofer (1787–1826) in 1824, to the immense Pulkowa Observatory refractor (objective diameter 76 cm, focal length 14 m, Fig. 14) constructed by Alvan Clark in 1885.

After 1900, although special photographic refractors were still being constructed, the demand for larger, more powerful telescopes favored the building of reflectors with considerable dimensions. The aperture diameter and year of construction for a few of the most famous instruments are presented in Table 4.

A large field and large aperture ratio are combined in the optical system invented by Bernhard Schmidt (1879–1935) in the years 1930 to 1931 (the Schmidt camera). The largest Schmidt instrument with an aperture of 134 cm went into operation in 1960 in the Karl Schwarzschild Observatory in Tautenburg, Germany.

From the photographic plate to the most recent charge-coupled devices (CCDs), the accessories used to measure and record the radiation emitted by celestial bodies are a vital part of large-telescope research. Detectors of higher sensitivity have improved the information output over shorter recording times. Photoelectric methods in stellar photometry were introduced by Joel Stebbins (1878–1966) and Paul Guthnick (1879–1947).

Table 4. Size and year of construction of some famous telescopes

Observatory	Aperture	Type	Year of construction
W. Herschel (England)	1.2 m	Mirror metal	1787
Dorpat (Tartu, Estonia)	25 cm	Lens	1824
Munich (Bavaria)	28 cm	Lens	1835
Harvard (USA)	38 cm	Lens	1839
Pulkowa (Russia)	38 cm	Lens	1839
Lord Rosse (Ireland)	1.8 m	Mirror metal	1845
Chicago (USA)	47 cm	Lens	1861
L. Foucault (France)	0.6 m	Mirror glass	1862
Washington (USA)	47 cm	Lens	1866
Cambridge (England)	63 cm	Lens	1871
Paris (France)	1.2 m	Mirror glass	1878
Vienna (Austria-Hungary)	68 cm	Lens	1878
Pulkowa (Russia)	76 cm	Lens	1885
Lick (USA)	91 cm	Lens	1888
Lick (USA)	91 cm	Mirror	1895
Yerkes (USA)	102 cm	Lens	1897
Potsdam (Germany)	80 cm	Lens	1899
Mount Wilson (USA)	1.5 m	Mirror	1905
Mount Wilson (USA)	2.5 m	Mirror	1917
Mount Palomar (USA)	5.0 m	Mirror	1948
Selencukskaja (USSR)	6.0 m	Mirror	1976
Hawaii (USA)	10.0 m	Mirror 36 segments	1992
ESO (Chile)	16.4 m	Four mirrors	2000

Fig. 13. Fraunhofer Heliometer 1826

Fig. 14. Pulkowa Refractor 1885

Large telescopes have been designed for specialized purposes (e.g., the 3.8-m Infrared Telescope on Mauna Kea in Hawaii, completed in 1980).

The majority of large telescopes are located on climatically favorable mountaintop sites. Some observatories have been built for the sole purpose of observing the southern skies. One of the first to be established was the Cape Observatory in South Africa (1820). Of more recent construction are the European Southern Observatory (ESO), completed in 1969, and the Cerro Tololo Inter-American Observatory (CTIO), built in 1965, both in Chile.

Fig. 15. Mt. Wilson 2.5 m Mirror

Design of a segmented mirror telescope:
http://www.aura-nio.noao.edu

Large Binocular Telescope with two 8.4-m mirrors
http://www.mpia-hd.mpg.de/LBT

Keck Telescope, Mauna Kea (Hawaii):
http://www.konahi.com-LBT: medusa.as.arizona.edu

ESO Observatory La Silla (Chile):
http://www.eso.org

11.6 Stellar Evolution and Stellar Systems

11.6.1 Stellar Evolution

The matchless discoveries in the areas of stellar structure and evolution are products exclusively of the twentieth century. The recognition of nuclear processes as the source of stellar energy and the identification of the specific nuclear reactions in question paved the way for the present evolutionary models of stars.

In 1904 Johannes Franz Hartmann discovered the stationary K-line of ionized calcium in the spectrum of δ Orions, a binary star. This was the base for the assumption of gas and dust pervading the universe.

The major stepping stones along this path were:

1905 Special Theory of Relativity by A. Einstein Fig. 16
1906 Theory of equilibrium applied to solar atmosphere by K. Schwarzschild; publication of book *Gaskugeln* by R. Emden
1906–1912 Discovery of the relation between surface temperatures and luminosities of stars (the Hertzsprung–Russell diagram)
1915 General Theory of Relativity by A. Einstein
1923 Discovery of mass-luminosity relation by E. Hertzsprung (Fig. 18)
1925 Publication of *Stellar Atmospheres* by C.H. Payne
1926 Publication of *The Internal Constitution of the Stars* by A.S. Eddington
1930 Theory of convective currents in stellar atmospheres first proposed by A. Unsöld
1934 Hypothesis of neutron stars first proposed by W. Baade (Fig. 17) and F. Zwicky
1938 Nuclear reaction as stellar energy sources proposed by H. Bethe and C.F. von Weizsäcker
1951 Formation of carbon in stellar interiors investigated by E.J. Öpik and E.E. Salpeter
1955 Beginning of computer modeling of stellar evolution, allowing for changing interior composition due to nuclear processes, by M. Schwarzschild, F. Hoyle

11.6.2 Stellar Systems

With the opening of observatories with large reflectors during the first half of the twentieth century, the exploration of the Milky Way and other galaxies progressed swiftly. A rich and detailed picture of the Milky Way, further refined with radio astronomical observations since about 1951, began to emerge. The resolution of extragalactic systems (e.g., the Andromeda Nebula) into stars confirmed the fact that most objects commonly found in the Milky Way (e.g., Cepheids, RR Lyrae stars, gaseous nebulae) exist in these distant galaxies as well. Results of investigations into the nature of spiral galaxies and of the Milky Way were merged into a single picture.

Some important stepping stones of exploration of stellar systems are:

1895 Initial studies of the structure of the Milky Way system using the methods of stellar statistics (J. Kapteyn, H. von Seeliger)
1904 Interstellar reddening and absorption found (J. Kapteyn, H. Shapley (Fig. 20))
1918 Determination of photometric distances to systems within the Milky Way using Cepheids (H. Shapley)
1924 Resolution of the outer parts of the Andromeda and other spiral galaxies into stars (E. Hubble)
1926 Theory of galactic kinematics and dynamics developed (B. Lindblad, J. Oort)
1929 Redshift of spectra of extragalactic systems found to be in proportion to their distances (E. Hubble (Fig. 19))
1937 Interstellar CO_2 absorption bands found in spectra
1943 Unusual spectra of certain extragalactic systems found not to be explained solely by thermal emission (C.K. Seyfert)
1951 The 21-cm radio radiation of neutral hydrogen first detected (H. Ewen and E.M. Purcell)

Einstein online:
http://www.einstein-online.info

Concerning the Special Theory of Relativity and the General Theory of Relativity see:
http://www.tempolimit-lichtgeschwindigkeit.de.

Fig. 16. A. Einstein

Fig. 17. W. Baade

Fig. 18. E. Hertzsprung (*left*) H.N. Russell (*right down*)

The Hubble law is the relationship between the redshifts of galaxies and their distances from Earth.

1952 New distance scale of extragalactic systems (W. Baade) removes size discrepancy between Milky Way and other galaxies

1958 Theory of the explosive origin of the Milky Way from a compact protogalaxy proposed (V.A. Ambartsumian)

1963 First quasar discovered and redshift measured (M. Schmidt)

1964 Density-wave theory proposed to explain spiral structure in the Milky Way and other galaxies (C. Lin and F. Shu)

1970 Around the star-like object BL Lacertae diffuse structure was discovered with many spectral lines similar to the spectrum of an elliptical galaxy ("active galaxy")

Exploring the large-scale structure of the universe is so important for a better understanding of the earliest stages of the universe, when matter began to create into structure of stars and galaxies.

Fig. 19. E.P. Hubble

11.7 Observations at Invisible Wavelengths and Space Exploration

Until about 1950, only the visible radiation from celestial bodies reaching the Earth's surface determined the state of astronomical knowledge. This is the so-called optical window in the electromagnetic spectrum, which encompasses ultraviolet, X-rays, and γ-rays in one direction, and infrared, microwave, and radio waves in the other. Most of these ranges cannot be observed from the Earth's surface because of atmospheric absorption. The observations made with the aid of radio telescopes, high-altitude flights, and later space-based instruments have broadened optical astronomy in the past 60 years into a science of all wavelengths. The exploration of the solar system via space missions then led to numerous discoveries regarding the planets and their satellites. The major milestones along this road are:

Fig. 20. H. Shapley

Active galaxies bridge the gaps in our knowledge between galaxies and quasars.

Also see the work by S.L. Jaki [2] and M.A. Longair [3].

1931 Radio radiation from the Milky Way discovered (K.G. Jansky)

1939 Observation of concentrated radio radiation in the galactic plane and toward the galactic center (G. Reber)

1942 Discovery of extragalactic components in radio radiation (J.S. Hey, J. Southworth)

1954 Discovery of radio galaxies

1957 First artificial satellite (Sputnik I, USSR) launched

1957 The 75-m radio telescope Jodrell Bank (England, Fig. 21)

1960 X-ray radiation from the solar corona found by an Aerobee rocket

1961 First unmanned mission to Venus (Venera I, USSR)

1965 Discovery of cosmic X-rays (E.T. Byram, H. Friedman, T.A. Chubb)

1967 Discovery of pulsars (J. Bell and A. Hewish)

1968 The 100-m radio telescope Effelsberg (Germany)

1969 First men land on the Moon (Apollo 11, USA)

1970 160 cosmic X-ray sources discovered (UHURU X-ray satellite, USA)

1976 Unmanned landings on Mars (Viking I and II, USA)

1978 International Ultraviolet Explorer IUE (NASA/ESA)

1979 Flyby of Jupiter and satellites (Voyager 1 and 2, USA)

1979 US satellite Solwind to explore the Sun's outer atmosphere

1980 The 3.8-m infrared telescope Mauna Kea (Hawaii)

Fig. 21. Jodrell Bank Radio Telescope 1957

Fig. 22. Men on the Moon 1969

Fig. 23. Hubble Space Telescope 1990

Selection of information online:
Chandra Orion Ultra deep Project:
http://www.astro.psu.edu/coup/

Huygens Probe. Landing in an Earth-like world, Titan:
http://www.esa.int

SOHO Internet address:
http://soho www.estec.esa.nl

The Orbiting Solar Observatory, a series of satellites OSO 1 to OSO 8 (1962–1975), and the astronauts of Skylab (1973/1974) made spectacular photos of the hot ionized gas on the Sun:
http://www.ksc.nasa.gov/history/skylab.

Hubble Space Telescope (HST) 15 years in space:
http://www.spacetelescope.org, and www.esa.int

Cosmic Background Explorer (COBE):
http://lamda.gsfc.nasa.gov/

1981	Voyager 2 flyby of Saturn
1981	27 parabolic dishes, each 25 m in diameter, linked to the Very Large Array (VLA) radiointerferometer (New Mexico, USA)
1985	Giotto European mission to observe Halley's comet in 1986
1989	Cosmic Background Explorer (COBE, USA)
1990	Hubble Space Telescope (HST, NASA/ESA)
1990	ROSAT European X-rays mission
1995	Soho Solar and Heliospheric Observatory (ESA/NASA)
1995	Infrared Space Observatory (ISO, ESA)
1997	Cassini/Huygens NASA's spacecraft to explore the Saturnian system
1999	X-ray explorer Chandra (NASA)
1999	X-ray observatory XMM (ESA)
2001	Wilkinson Microwave Anisotrophy Probe (WMAP) to explore the dark-matter problem
2005	Huygens visited moon Titan

11.8 Research in Historical Astronomy

11.8.1 Research Problems

The fascination with the history of astronomy can sometimes lead from the mere perusal of historical presentation to the direct pursuit of historical records. The study of the general history of astronomical research and the description of special sections requires some background in both history and science. Biographical presentations often reach beyond purely scientific aspects, and so the historically inclined researcher is encouraged to approach the life work of an astronomer and cast it into a literary form.

There are many books devoted to extensive historical subjects, but magazine articles on historical events and persons often provide the best source of information for a report on the past of astronomy (including amateur astronomy) in both research and education. The report may include material on observational work performed or public activities provided by particular societies, groups, or individuals.

The subject can be approached from various angles. There is, on the one hand, the technical, scientific investigation, and, on the other, one which produces a publication to inform a larger audience on various subjects from astronomy history. The latter may also deal with such matters as community or local schools histories.

11.8.2 Sources

Any historical study depends on the amount and quality of the information available in the accessible sources.

Primary Sources. The focal point of work for the historian is the study of primary sources, which refer to the immediate documentation of the events, and generally date from the time of the events. These include personal documents, foundation documents, diaries and contemporary reports of all sorts (descriptions of institutes, company pamphlets, newspapers, etc.).

Primary sources are often accessible in public and private archives and museums, and also in community public and governmental libraries.

Secondary Sources. Secondary sources include mainly published studies on a particular subject. These include dissertations and theses (on history), book publications and papers, and various manuscripts. These publications generally refer to the primary sources used and include citations from them. In addition to public libraries, university and college libraries may hold important collections of secondary sources.

Topics on astronomy history after 1800 are, apart from books, also documented in serials of technical or general nature (e.g., *Popular Astronomy* since 1893, *Sirius/Die Sterne* since 1868, *Monthly Notices of the Royal Astronomical Society* since 1831, and *Astronomische Nachrichten* since 1821).

Visit for historical reports and helpful tips on books, libraries and archives: the journals *Sky and Telescope* (http://www.SkyTonight.com), *Sterne und Weltraum* (http://suw-online.de), *Astronomy & Geophysics* (http://www.ras.org.uk), *Orion, Schweizerische Astronomische Gesellschaft* (http://www.astroinfo.ch), and *Journal of the History of Astronomy* (http://www.shpltd.co.uk). For associations, observatories and planetaria of Austria visit http://www.austriaca.at/sternwarten.

For news and helpful tips on historical research visit British Astronomical Association online (http://www.britastro.org), also the European Astronomical Society (http://www2.iap.fr/eas/), and the Society for the History of Astronomy (http://www.shastro.org.uk).

The "Arbeitskreis Astronomiegeschichte," an organization of the German Astronomische Gesellschaft (AG) publishes regularly about national and international works dealing with astronomy and geophysics (http://www.astrohist.org).

General Historical Sources. When searching libraries for material, it may be helpful to use the following groups as a guide:

– Bibliographic works (e.g., *Astronomischer Jahresbericht*)
– Introductions (e.g., compendia on history and specifically astronomy history)
– Source studies (e.g., literature on historical sources)
– Document and sources compilations (e.g., printed letters, documents, papers and also collected works)
– Books (e.g., handbooks, textbooks, comprehensive presentations of particular period in time or of an technical subject, such as spectroscopy)
– Genealogies (e.g., of a family)
– Serials (e.g., *Journal of the History of Astronomy*)
– Dictionaries, biographies and reference books (e.g., Allgemeine (ADB) and Neue (NDB) Deutsche Biographie. Encyclopaedia Britannica, http://www.britannica.com)

US mission Solwind: http://lasco-www.nrl.navy.mil/solwind.html

One example of the history of a small telescope maker in Germany is by E. Remmert about Manfred Wachter [4].

For more contact and informations see Chap. 12

The Inter-Union Commission for History of Astronomy is an international body of the International Astronomical Union (IAU) representing the interests of all professional historians of astronomy worldwide (http://www.iau.org).

Searching systems online see, e.g., Google (http://www.google.com) or MSN Search (http://www.search.msn.com/).

11.8.3 Processing

The purpose of an historical investigation, whether for scientific study of for more journalistic general interest information, largely determines the kind of research and studies to be undertaken. All of the general methods of investigation such as interviews, excursions, literature studies, and surveys of local papers may have to be tapped to gain the desired information. Particularly in amateur astronomy, the trails, if any, are difficult to find. Biographical descriptions may be quite general, but cannot be more detailed for lack of primary sources.

For the processing, it is advisable to proceed from the accessible and well-known to the obscure and less-known:

- Preparatory work (inspection of local chronicles, documents, family diaries, etc.)
- Library work (study of technical and documentary literature)
- Archive work (in local or governmental archives)
- Processing (compilation of the previously obtained results in card or computer files)

Any presentation gains by offering the reader well-selected contemporary statements and documents cited, quoted, or reproduced. Footnotes and lists of references are found primarily in scientific studies. A general readership will undoubtedly appreciate some outline of the contemporary historical situation.

Regardless of the particular area in astronomy history, the interrelation of scientific, technological, and societal evolution will ultimately emerge as the work progresses. This interaction of all societal forces has always influenced astronomers, observatories, and research programs, a fact which must be kept in mind in any evaluation or narrative.

Regarding studies of amateur astronomers on the field of history of astronomy, e.g., visit the *Journal für Astronomie*, magazine of the German Association of Amateur Astronomers ("Vereinigung der Sternfreunde," VdS): http://www.vds-astro.de.

11.9 Further Reading

Ashbrook, J.: Astronomical Scrapbook, Sky Publishing Corporation, Cambridge MA (1984)

Brück, H.A., Brück, N.T.: The Peripatetic Astronomer. The Life of Charles Piazzi Smyth. Adam Hilger, Bristol (1988)

Brüggenthies, W., Dick, W.R.: Biographischer Index der Astronomie (Biographical Index of Astronomy). Harri Deutsch, Frankfurt a.M. (2005)

Einstein, A.: „Verehrte An- und Abwesende!" Originaltonaufnahmen 1921–1951, supposé. Köln (2003) http://www.suppose.de

Evans, B., Harland D.M.: Nasa's Voyager Missions. Springer, Berlin Heidelberg New York and Praxis, Chichester (2004)

Evans, D.S.: Under Capricorn. A History of Southern Hemisphere Astronomy. Adam Hilger, Bristol (1988)

Farrell, J.: The Day Without Yesterday. Lemaitre, Einstein and the Birth of Modern Cosmology. Basic Books, New York (2005)

Gruntman, M.: Blazing the Trail. The Early History of Spacecraft and Rocketry. American Institute for Aeronautics and Astronautics, Reston, VA (2004)

Hamel J.: Meilensteine der Astronomie. Von Aristoteles bis Hawking. Franckh-Kosmos, Stuttgart (2006)

Harland, D.M.: The Story of the Space Shuttle. Springer, Berlin Heidelberg New York and Praxis, Chichester (2004)

Harland, D.M.: The First Men on the Moon. The Story of Apollo 11. Springer, Berlin Heidelberg New York and Praxis, Chichester (2007)

Harvey, B.: Soviet and Russian Lunar Exploration. Springer, Berlin Heidelberg New York and Praxis, Chichester (2007)

Hearnshow, J.B.: The Analysis of Starlight. One Hundred and Fifty Years of Astronomical Spectroscopy. Cambridge University Press, Cambridge (1986)

Heck, A.: The Multinational History of Stasbourg Astronomical Observatory. Springer, Berlin Heidelberg New York (2005)

Herrmann, D.B.: Geschichte der modernen Astronomie. VEB Deutscher Verlag der Wissenschaften, Berlin (1984)

King, H.C.: The History of the Telescope. Dover, New York (1979)

Lang, K.R., Gingerich, O.: A Source Book in Astronomy and Astrophysics, 1900–1975. Harvard University Press, Cambridge, MA (1979)

Longair, M.: The Cosmic Century. A History of Astrophysics and Cosmology. Cambridge University Press, Cambridge (2006)

Leverington, D.: Babylon to Voyager and Beyond. A History of Planetary Astronomy. Cambridge University Press, Cambridge (2003)

Leverington, D.: A History of Astronomy from 1890 to the Present. Springer, Berlin Heidelberg New York (1995)

Meÿenn, K. v. (Ed.): Die großen Physiker. 2 Bände, Verlag C. H. Beck, München (1997)

Orloff, R.W., Harland, D.M.: Apoll. The Definitive Sourcebook. Springer, Berlin Heidelberg New York and Praxis, Chichester (2006)

Roth, G.D.: Joseph von Fraunhofer 1787–1826. Wissenschaftliche Verlagsgesellschaft mbH, Stuttgart (1976)

Roth, G.D.: Kosmos Astronomie-Geschichte. Franckh'sche Verlagshandlung, Stuttgart (1987)

Rosenkranz, Z.: Einstein privat und ganz persönlich. Verlag Neue Züricher Zeitung, Zürich (2004)

Van Helden, A.: Telescope Building, 1850–1900. In: General History of Astronomy, vol. 4, part A, p. 40: Astrophysics and Twentieth-Century Astronomy to 1950. Cambridge University Press, Cambridge (1984)

Woltjer, L.: Europe's Quest for the Universe. EDF Sciences, Les Ulis (2006)

Wright, H.: James Lick's Monument. The Saga of Captain Richard Floyd and the Building of the Lick Observatory. Cambridge University Press, Cambridge (1987)

Zinner, E.: Astronomie, Geschichte ihrer Probleme. Verlag Karl Alber, München (1951)

Zirker, J.B.: An Acre of Glass. A History and Forecast of the Telescope. Johns Hopkins University Press, Baltimore (2005)

References

1. de Padova, T., Staude, J.: Galilei, der Künstler. Sterne Weltraum, December (2007) http://www.wissenschaft-online.de/artikel/911841. Accessed 2007

2. Jaki, S.L.: The Milky Way. Isis **66**, 115 (1972)

3. Longair, M.A.: The Cosmic Century. Cambridge University Press, Cambridge (2006)

4. Remmert, E.: Manfred Wachter: Präszisionsmechanik und Optik. Sterne Weltraum **7** (2007) http://www.astronomie-heute.de/artikel/874750. Accessed 2007

12. The Social Astronomer

V. Witt

According to Wikipedia "amateur astronomy is a hobby whose participants enjoy studying celestial objects" [1]. Often devotion to that kind of hobby starts during adolescence with the purchase of a modest telescope through which one merely observes the Moon at first. However, soon the wish for observing more difficult celestial bodies and for sharing one's nocturnal experiences with like-minded people arises. The best way to pursue that wish is to join one of the numerous local astronomy clubs and visit their meetings.

12.1 Astronomy Clubs

Here you can meet people, who are brought together by their joint love of and enthusiasm for astronomy and with whom you exchange experiences. Newcomers benefit from advice offered by the "old stagers" and those with an advanced knowledge of astronomy find a forum where they can discuss their observation results with competent partners. Occasionally organized star parties convey the charm of joint observation.

A worldwide oriented selection of clubs can be found on the "Astronomy Clubs" website or in "The Astronomy White Pages" of the online magazine "AmSky."

A search for the United States leads to "Go Astronomy," where an Astronomy Clubs Directory, arranged according to states, is available, or to the "Astronomy Web Guide," which offers plenty of further links regarding astronomy. Both the astronomy magazines *Sky and Telescope* and *Astronomy* offer a very practical search engine on their homepages, which, after entering the place of residence, generates a selection of nearby clubs or other astronomical facilities.

In order to find addresses of clubs in the United Kingdom, just visit the BBC's website "Science & Nature." Here you receive a selection of clubs in the favored region immediately by clicking on a map. You can find Irish astronomy clubs either on the website of "Astronomy Clubs" or by checking the homepage of the Irish Federation of Astronomical Societies.

Canadian amateurs may find suitable clubs on the homepage of the Canadian astronomy magazine *SkyNews*.

Australian Clubs are listed in "Australian Astronomy", the official website of the Astronomical Society of Australia. Adresses in New Zealand can be obtained on the website of the Royal Astronomical Society of New Zealand (RASNZ).

Astronomy clubs and observatories in Germany, Austria and Switzerland are compiled in the "German Astronomical Directory" (GAD). The directory also contains adresses of planetariums, astronomical societies and research facilities.

Astronomy clubs worldwide
http://www.astronomyclubs.com
http://www.amsky.com/whitepages

United States
http://www.go-astronomy.com/astro-club-search.htm
http://www.astronomywebguide.com/links_clubs.html
http://www.skyandtelescope.com
or http://www.astronomy.com

United Kingdom
http://www.bbc.co.uk/science/space/myspace/localspace/index.shtml

Ireland
http://www.irishastronomy.org/ifas/members.php

Canada
http://www.skynewsmagazine.com

Australia and New Zealand
http://www.astronomy.org.au/ngn/engine.php
http://www.rasnz.org.nz

Germany, Austria and Switzerland
http://www.sternklar.de/gad/index_eng.htm

12.2 Astronomical Societies

Plenty of organizations represent the professional and amateur astronomers' interests on a national scale. Some of the important societies are presented here. A complete compilation is given by AstroWeb (http://cdsweb.u-strasbg.fr/astroweb/society.html).

http://www.aas.org

American Astronomical Society (AAS). Established in 1899 the AAS has approximately 6500 members, the major organization of professional astronomers in North America. The goal of the AAS is to promote the progress in astronomy and related branches of science. Another goal of the organization is to improve both education in astronomy and research on teaching and learning in astronomy.

http://www.astroleague.org

Astronomical League (AL). The Astronomical League is an umbrella organization of over 200 local amateur astronomical societies from all over the US. This organization encourages an interest in astronomy through educational and observational programs for the amateur astronomer.

http://www.rasc.ca

Royal Astronomical Society of Canada (RASC). The society was founded in 1868 and is Canada's leading astronomy organization comprising over 4000 amateurs, educators and professionals. Local astronomy programs and services are offered by 28 RASC centers in every province of Canada. Many of these centers also support programs of public education and outreach.

http://asa.astronomy.org.au

Astronomical Society of Australia (ASA). It was formed in 1966 as the organization of professional astronomers in Australia. Members of the ASA are involved in astronomy education and outreach at all levels of society. The activities range from school visits to university education, media contacts and activities for the general public. Although most members are professionals, the ASA also has established the grade of Associate Member designed for educators or distinguished amateur astronomers.

http://www.rasnz.org.nz

Royal Astronomical Society of New Zealand (RASNZ). The aim of the society, founded in 1920, is the promotion and extension of knowledge of astronomy and related branches of science. Membership is open to all people interested in astronomy. Many observing programs involve collaboration between professional and amateur astronomers in New Zealand and elsewhere. The society contains a number of sections for different interest groups engaged in particular areas of astronomy.

http://assa.saao.ac.za

Astronomical Society of Southern Africa (ASSA). It was a merger in 1922 between the Cape Astronomical Association and the Johannesburg Astronomical Association into one individual organization. It consists of both professional and amateur astronomers. Membership is open to all interested persons, regardless of knowledge or experience.

Astronomical Societies in Europe

http://eas.iap.fr

European Astronomical Society (EAS). The purpose of this society which was founded in 1990 is to promote the advancement of astronomy in Europe by all suitable means. The EAS Newsletter is a biannual publication and is distributed free of charge to the EAS members (http://eas.iap.fr/newsletters.html).

Royal Astronomical Society (RAS). The Royal Astronomical Society (RAS) was founded in 1820 and represents UK astronomy nationally and internationally. It has more than 3000 members (fellows) including scientific researchers in universities and observatories as well as historians of astronomy. The RAS organizes scientific meetings throughout the country, publishes international research and review journals and promotes education through grants and outreach activities.

http://www.ras.org.uk

European Radio Astronomy Club (ERAC). Founded in 1995 the organization has more than 350 members in 19 different countries who share a common interest in radio astronomy. As an umbrella organization ERAC comprises several local radio astronomy groups on a national basis. Every three years there is an international congress for radio astronomy.

http://www.eracnet.org

Vereinigung der Sternfreunde e.V. (VdS). The society is, with more than 4000 members, the major organization of amateur astronomers in Germany and German-speaking countries. The VdS also provides services for public and school observatories as well as for planetariums and local astronomy clubs. The society's mission is to promote amateur astronomy by advancing the amateurs' activities at all levels and by developing public education and outreach. Several subdivisions focus on specific fields of astronomy such as astrophotography, deep sky, planets, comets, solar observation, spectroscopy or history of astronomy. Members of the VdS may use the Volkssternwarte Kirchheim (near the city of Erfurt) as a guest observatory ("VdS-Sternwarte") which is equipped with telescopes up to 20 in. in aperture. They also receive free of charge the periodically published magazine *Journal für Astronomie*. The Bundesdeutsche Arbeitsgemeinschaft für Veränderliche Sterne e.V. (BAV) was founded in 1950 as an independent association of the German variable star observers. As a subdivision of the VdS it supports members in the systematic observation of variable stars.

http://www.vds-astro.de

http://www.bav-astro.de

Astronomische Gesellschaft (AG). The Astronomische Gesellschaft has a long tradition in Germany dating back to the year 1800 when the "Vereinigte Astronomische Gesellschaft" was founded in Lilienthal, a small city near Bremen. The AG itself was established in the year 1863 as an internationally operating society dedicated to the "advancement of science." After World War I most of the global tasks of the AG were transferred to the International Astronomical Union (IAU). In 1947 the society was re-founded and has now more than 800 members, mostly professional astronomers. The AG's activities are devoted amongst others to the organization of scientific meetings, publication of scientific literature, promotion of young astronomers and public outreach. The AG's publication "Mitteilungen der Astronomischen Gesellschaft" include the annual reports of astronomical institutions in Germany, Austria and German speaking Switzerland. The *Reviews in Modern Astronomy* (RMA), published by the AG since 1988, are a compilation of scientific talks given at the society's annual meetings.

http://www.astro.rub.de/ag

Schweizerische Astronomische Gesellschaft (SAG). The society was founded in 1938 and is the umbrella organization of the local astronomy clubs and associations in Switzerland. The mission of the society is to provide a common basis for mutual communication and the exchange of experience for all people who are engaged in the fields of astronomy. The SAG's primary activities focus on the arrangement of astronomical meetings and events, the publication and distribution of astronomical literature and

http://sas.astronomie.ch

the public outreach and education in astronomy. The SAG publishes bimonthly the astronomy magazine *Orion*.

http://www.oegaa.at

Österreichische Gesellschaft für Astronomie und Astrophysik (ÖGA²). The ÖGA² represents in Austria since 2002, as one of the youngest astronomical societies, the interests of astronomers, scientific institutes and persons who are particularly engaged in astronomy.

http://members.eunet.at/astbuero/av.htm

Österreichischer Astronomischer Verein. The society acts today in Austria as the major community of amateur astronomers and people concerned with sky phenomena. It was founded in 1924 by Oswald Thomas. More than 100 years ago Thomas established the Astronomisches Büro which moved to Vienna in 1913 and has been managed by Hermann Mucke since 1963.

http://www.saf-lastronomie.com, http://www.iap.fr/saf

Société Astronomique de France (SAF). Camille Flammarion, the famous popularizer of astronomy, established in 1887 the French society whose bulletin *L'Astronomie* continues until today. The SAF has approximately 2500 members and is open to everybody. Its purpose is to promote the advancement in practical astronomy and to support the amateur–professional partnership. As subdivisions the SAF comprises 13 various commissions which cover different topics of amateur astronomy.

http://www.uai.it

Unione Astrofili Italiani (UAI). The society is the major organization in Italy which offers various services for amateurs and the local astronomy clubs all over the country, since 1969. From the beginning the UAI publishes *Astronomia UAI* as its official bulletin. The activities of the UAI include astronomy courses, scientific workshops and public observing. The members as well as professional scientists participate in an annual congress which usually takes place in September. Various subdivisions of the UAI are devoted to research in special branches of astronomy.

On an international basis several organizations represent specific fields of amateur astronomy:

http://www.iau.org

International Astronomical Union (IAU). Founded in 1919 the IAU promotes astronomy in its scientific aspect through international cooperation. Its members are professional astronomers who are active in research and astronomy education all over the world. The IAU also maintains cooperation with astronomical organizations that comprise amateur astronomers as members.

http://www.aavso.org

American Association of Variable Star Observers (AAVSO). It was founded in 1911 at Harvard College Observatory and is today the world-recognized leader in information and data on variable stars. Membership in the AAVSO is open to anyone, and with members in over 45 countries the AAVSO is the world's largest association of variable star observers. The goal of the AAVSO is to observe and analyze variable stars, to archive observations for worldwide access and to promote the collaboration between amateurs and professionals (see Sect. 12.7).

http://www.amsmeteors.org

American Meteor Society (AMS). The AMS was founded in 1911 by Charles P. Olivier as an individual branch of the American Astronomical Society (AAS). Typical activities are to observe and report on meteors, meteoric fireballs and related meteoric phenomena, to establish a network of radiometeor stations and to publish scientific materials dealing with meteors.

http://alpo-astronomy.org

Association of Lunar and Planetary Observers (ALPO). The association which was founded by Walter H. Haas in 1947 is an international group of observers that study the

Sun, Moon, planets, asteroids, meteors, and comets. Section coordinators correspond with observers, encourage beginners and can supply instructional material to assist in telescopic work. The association provides services for both professional and amateur astronomers, for the beginner and the advanced amateur astronomer. ALPO's work is coordinated by means of a periodical (*The Strolling Astronomer*).

International Dark-Sky Association (IDA). The mission of this association is to preserve and protect the nighttime environment and our heritage of dark skies. Goals of the IDA are to stop the effects of light pollution on dark skies. It helps for example to raise awareness about light pollution, to teach everyone about the values of harmless outdoor lighting and to fight against other threats to our view of the heavens, such as radio frequency interference. IDA was founded in 1988 as a non-profit organization operating exclusively for educational and scientific purposes.
http://www.darksky.org

International Meteor Organization (IMO). IMO is a very young organization and was created in response to an ever-growing need for international cooperation of meteor amateur work. The IMO's main goals are to encourage, support and coordinate meteor observing, to improve the quality of amateur observations and to spread observations and results to other amateurs and professionals.
http://www.imo.net

International Occultation Timing Association (IOTA). The association's mission is to promote the observation of occultations and eclipses. IOTA provides predictions for occultations of stars by the Moon and by asteroids and planets. Timing of asteroid occultations allows for example to study the shape and size of an asteroid. IOTA also provides information on observing equipment and techniques, and reports about successful observations in its *Occultation Newsletter*. Members of IOTA present their observing results during an annual meeting.
http://www.occultations.org

Webb Deep-Sky Society. The society is named in honor of Thomas William Webb (1807–1885), the author of the classic *Celestial Objects for Common Telescopes*, which has been an inspiration to several generations of amateur astronomers. The main goal of the society is to encourage amateurs for observing double stars and deep-sky objects such as star-clusters and nebulae. Observational activities of the society are coordinated in the following sections: double stars, nebulae and clusters, galaxies and southern sky. Results of the society's work are published in the quarterly journal *The Deep-Sky Observer*.
http://www.webbdeepsky.com

Society of Amateur Radio Astronomers (SARA). SARA is an international non-profit society of amateurs, engineers, teachers and non-technical persons who are fascinated in doing their own observations of the radio sky. The membership includes a subscription to *Radio Astronomy*, SARA's official publication. A meeting of SARA members takes place every year.
http://radio-astronomy.org

12.3 Star Parties, Astronomical Events and Astronomy Vacations

A more or less complete compilation is given by a Google search:

http://www.google.com/Top/Science/Astronomy/Amateur/Star_Parties.

A calendar for upcoming astronomy-related events in the US is provided by the magazine *Sky and Telescope* or *Astronomy*, respectively.

Star Parties and Astronomical Events

Astronomy Day and Astronomy Week. Astronomy Day is a worldwide movement "bringing astronomy to the people." Thousands of people will have an opportunity to see first hand why so many amateur and professional astronomers are excited by the joy of astronomy. Astronomy clubs, observatories, universities, planetariums, science museums and other organizations host special events to acquaint people with their activities. Astronomy Day was born in California in 1973. Nowadays events take place at hundreds of sites across the US (see Table 1). Many other countries have hosted Astronomy Day activities such as England, Canada, New Zealand, Finland, Sweden, the Philippines, Argentina, Malaysia and New Guinea. Activities comprise elaborate exhibits at shopping malls, museums, nature centers, libraries, etc. Astronomy Week was founded to give sponsors more time to arrange special events. The Astronomical League maintains the official Astronomy Day webpage, the "Astronomy Day Handbook" by David H. Levy is available for free as a PDF.

http://www.astroleague.org

http://astroday.net

AstroDay in Hilo, Big Island, Hawaii. AstroDay was founded as an offspring from Astronomy Day. Standing date is the first Saturday in May. Exhibitions, demonstrations and activity areas including the observatories on Mauna Kea.

http://www.chicagoastro.org/af/index.html

Astrofest of the Chicago Astronomical Society (CAS), one of the oldest astronomical societies in the Western Hemisphere.

http://www.chiefland.org

Chiefland Star Party at Chiefland Astronomy Village (CAV). Near Chiefland, Florida (Fig. 1).

http://www.socorro-nm.com/starparty

Enchanted Skies Star Party at Socorro, New Mexico. Very popular, one of the leading star parties in the US, dry desert climate (elevation 1400 m).

http://www.tucsonastronomy.org

Grand Canyon Star Party at Grand Canyon State Park, Arizona. Observing facilities on the South Rim (elevation 2135 m). The area offers day trips to Lowell Observatory, Meteor Crater, Painted Desert, Wupatki and Sunset Crater Volcano National Monument.

http://www.nebraskastarparty.org

Nebraska Star Party at Snake Campground, 27 miles south of Valentine, Nebraska (elevation 945 m). High sky transparency, light pollution-free and naked-eye limiting magnitude approaching 7.5.

http://www.okie-tex.com

Okie-Tex Star Party at Black Mesa, Oklahoma (elevation 1515 m). One of the darkest skies anywhere in the Southwest. Service is provided through a whole week including breakfast, lunch and dinner.

Table 1. Dates of Astronomy Day and Astronomy Week

Year	Astronomy Day	Astronomy Week
2008	May 10	May 5–11
2009	May 2	April 29–May 3
2010	April 24	April 19–25
2011	May 7	May 2–8
2012	April 28	April 23–29
2013	April 20	April 15–21
2014	May 10	May 5–11

Fig. 1. Dobson telescope parade at the Chiefland Star Party. Courtesy of Arthur R. Mullis, Brandon, FL, USA

Oregon Star Party (OSP) at Indian Trail Spring on the Ochoco National Forest, Oregon (elevation 1525 m). OSP takes place every August. Speakers program and youth activities.
http://www.oregonstarparty.org

RTMC Astronomy Expo (originally called Riverside Telescope Makers Conference) at YMCA Camp Oakes, five miles southeast of Big Bear City, California (elevation 2210 m). The RTMC comprises all aspects of astronomy such as observing under dark skies, talks on telescope making and using, and a marketplace for astronomical equipment and software.
http://www.rtmcastronomyexpo.org/index.html

Rocky Mountain Star Stare (RMSS) near Colorado Springs, Colorado (elevation 2680 m). Due to recent changes in US Forest Service policies access to suitable observing areas will be more restricted in the future. Therefore the Colorado Springs Astronomical Society plans to purchase its own land for future RMSS.
http://www.rmss.org

Table Mountain Star Party on Table Mountain about 20 miles northwest of Ellensburg, Washington (elevation 1938 m). Each year held in June, July or August.
http://www.tmspa.com

Texas Star Party (TSP) near Fort Davis, Texas (elevation 1537 m). One of the most renowned star parties. Activities comprise meetings with guest speakers, special observing programs, astrophoto contest, astronomy art competition or bus tours to McDonald Observatory. From many locations throughout the area including the observing fields Internet access is available.
http://www.texasstarparty.org

The **Winter Star Party** (WSP) at West Summerland Key, Florida, attracts amateur astronomers from around the world to the warm subtropical climate of Florida. Organized by the Southern Cross Astronomical Society, Miami, Florida.
http://www.scas.org/wsp.htm

Mount Kobau Star Party on Mt. Kobau, located in south central British Columbia, Canada (elevation 1860 m).
http://www.mksp.ca

http://www.nyaa.ca

Starfest is Canada's largest annual star party and attracts over 900 experienced observers and astrophotographers, as well as those people who are new to the hobby. Organized by the North York Astronomical Association, Ontario, Canada.

http://www.vicsouth.com

VicSouth Desert Spring Star Party near the town of Nhill in western Victoria, Australia. The event is hosted by the Astronomical Society of Victoria and the Astronomical Society of South Australia.

http://www.arkaroola.com.au

Star Party DownUnder. Annual star party hosted by the Arkaroola Wilderness Sanctuary and the Astronomical Society of South Australia.

http://www.asnsw.com/spsp/index.asp

South Pacific Star Party at "Wiruna" (Aboriginal for "Sunset") near Ilford, New South Wales. The Star Party, hosted by the Astronomical Society of New South Wales, is Australia's largest international star party, attracting participants within Australia and overseas since 1993.

http://www.qldastrofest.org.au

Queensland Astrofest at Camp Duckadang near Linville, Queensland, Australia. One of the southern hemisphere's most popular star parties, takes place on two consecutive weekends.

http://www.iceinspace.com.au

IceInSpace AstroCamp. Annual star party held in the Hunter Valley of New South Wales, Australia.

Astronomical Events in Europe

http://www.starparty.org.uk

The **Equinox Sky Camp** at Kelling Heath, Norfolk, is the largest star party in the UK, with hundreds of amateur astronomers gathering for a weekend of sky observing. Besides a smaller star party around the spring equinox the main event takes place in autumn.

http://www.richarddarn.demon.co.uk/starcamp

Kielder Forest Star Camp. The camp is organized by the Kielder Observatory Astronomical society and is located in Northumberland which is officially England's darkest county. The Star Camp takes place in autumn, but a smaller scale version is offered during springtime.

http://www.teleskoptreffen.ch/starparty/index.en.php

The **Swiss Star Party** held on the Gurnigel Pass in the Swiss Alps, south of Bern (elevation 1608 m). Accommodation is available at the nearby hotel. This event is rated as one of the world's top ten star parties.

http://www.teleskoptreffen.de/itv

Internationales Teleskoptreffen Vogelsberg (ITV). This meeting is the major star party in Germany which takes place under dark skies in the natural preserve "Vogelsberg." The event is centrally located halfway between the cities of Frankfurt and Fulda.

http://www.embergeralm.info/stella

Internationales Teleskoptreffen (ITT). The popular star party is held annually in September in the Carinthian Alps (Austria). The stargazing takes place at the Emberger Alm, an alpine meadow (elevation 1755 m) with dark skies and a fascinating mountain panorama. Accommodation is available at the hotel "Sattleggers Alpenhof."

Astronomy Fairs in Germany

http://www.astronomie.de/att-essen

– **ATT Astromesse**, one of the major astronomy fairs, is held every year in the city of Essen in the north-western part of Germany.

http://www.astro-messe.de

– **AME Astromesse** was established in the year 2006 and takes place annually in southern Germany in the city of Villingen-Schwenningen.

The **Astronomietag** is, similar to the above mentioned Astronomy Day, a nationwide annual event in Germany which is dedicated to the popularization of astronomy. It is managed by various organizations such as the Vereinigung der Sternfreunde (VdS), local astronomy clubs as well as by public observatories and several planetariums. | http://www.astronomietag.de

Further events are the **Schweizer Astronomietag** taking place every September in various localities of Switzerland as well as the **Österreichischer Astronomietag** which is a presentation of astronomical activities in Austria. | http://www.astronomietag.ch
http://www.astronomietag.at

Important conventions for amateur astronomers in Germany

- The **Astronomische Frühjahrstagung** ("Astronomical Spring Convention") has been held since 1972 in the city of Würzburg. The meeting is a popular forum where amateur astronomers report on their experiences and observational results. | http://sfeu.ebermannstadt.de

- The **Bochumer Herbsttagung** (BoHeTa, "Bochum Autumn Convention") has been a well-established meeting for the amateur community since 1980 with talks on observing, astrophotography, telescope making, astronomical traveling and much more. | http://www.boheta.de

Astronomy Vacations

Astronomy Camp. At the Catalina Observatories atop Mt. Lemmon (elevation 2800 m) near Tucson, Arizona, students of all ages explore the heavens with large telescopes. Astronomy Camp is sponsored by the travel program of the University of Arizona's Alumni Association. It is available internationally to teenagers and adults who experience astronomy by hands-on learning. People become astronomers operating research telescopes, keeping nighttime hours and interacting with leading scientists. A prior background in astronomy is not required. | http://www.astronomycamp.org

Advanced Observing Program (AOP) at Kitt Peak Visitor center (summit elevation 2100 m). It's open to anyone with an interest in observing the night sky using a high quality telescope or using state-of-the-art instruments. No previous experience in astronomy is necessary. The program offers hands-on use of telescopes with an experienced guide and a variety of digital cameras and webcams. Participants in this program are treated as visiting astronomers, dine with other astronomers and have accommodation at the observatory. | http://www.noao.edu/outreach/aop

Astronomical Adventures. The high desert viewing site is located in the middle of a large ranch in Northern Arizona, 1900 m above sea level, and has over 300 clear nights a year. Lowell Observatory, US Naval Observatory and the Navy Prototype Optical Interferometer (NPOI) all selected this area for its dark and transparent sky. Participants of the program may use a 36-in., a 30-in. and various smaller reflecting telescopes, as well as solar and planetary telescopes. | http://www.astronomicaladventures.com

During the day the geologically diverse environment is suited for wanderings across the history of earth.

New Mexico Skies Guest Observatory in Mayhill, New Mexico (elevation 2225 m) is located 32 miles east of Alamogordo. Several well-equipped high-end telescopes are available in six separate permanent observatories. The flagship is a 0.6-m Ritchey–Chretien truss tube telescope mounted on a fork mount. Visitors are welcome to bring their own equipment. | http://www.nmskies.com

Galloway Astronomy Centre in Wigtownshire, Scotland. The center is located in open farmland with a night sky free of light pollution and is open all year. It provides comfortable accommodation and a range of quality telescopes for all levels of experience. In the case of little or no observing experience, personal guidance is offered.

http://www.gallowayastro.com

Astronomy in Chile. QUASARCHILE offers expert guided tours to Northern Chile designed for amateur astronomers. The tours include a visit to the Very Large Telescope (VLT), the Atacama Large Millimeter Array (ALMA) site as well as nights of amateur astronomical viewing in the Atacama desert and a talk by a professional astronomer.

http://www.quasarchile.cl

Astronomy-Travel, LLC. The company offers astronomical trips and specializes in eclipse flights. The flights start from Vienna (Austria) to an international airport located close to the event.

http://www.astronomy-travel.com

TravelQuest International. The company offers astronomy-related tours, mainly to solar eclipses, in partnership with *Sky and Telescope* magazine.

http://www.tq-international.com

Wittmann Travel (Hamburg, Germany) cooperates with the German astronomy magazine *Sterne und Weltraum* and arranges tours to major observatories or to sites of astronomical interest all over the world.

http://www.wittmann-travel.com

Here is a selection of guesthouses which offer bed and breakfast and the use of observing equipment.

Stargazers Inn and Observatory. Big Bear Lake, California. Elevation 2200 m. Guided stargazing tours.

http://www.stargazersinn.com

Jack Newton Observatory B&B. Osoyoos, British Columbia, Canada. From May through early October, use of a 16-in. computer-controlled telescope.

http://www.jacknewton.com

Astronomer's Inn (Vega-Bray Observatory). Benson, Arizona. Reflectors from 6 in. (astrograph) to 20 in. (Maksutov–Cassegrain).

http://www.astronomersinn.com

Fieldview Guest House. North Norfolk, UK. Six telescopes (among them three Dobsonians up to 14.5 in.).

http://www.fieldview.net

Sattleggers Alpenhof. Emberger Alm, Austria. Elevation 1755 m. Roll-off roof observatory equipped with 10-in. Schmidt–Cassegrain reflector and 17.5-in. Newtonian, respectively. Astronomy workshops for beginners.

http://www.alpsat.at

Carlo Magno Zeledria Hotel. Madonna di Campiglio, Italy. 16-in. Ritchey–Chrétien reflector.

http://www.hotelcarlomagno.com/en

SaharaSky Kasbah Hotel. Zagora, South Morocco. Reflectors from 8 to 16 in.

http://www.saharasky.com

For those who want to experience the beauty of the southern sky a wide selection of facilities is available in Namibia. Some guest farms, equipped with astronomical telescopes, are presented here:

- **Hakos** guest farm (http://www.natron.net/tour/hakos)
- **Niedersachsen** guest farm (http://www.natron.net/niedersachsen)
- **Rooisand Desert Ranch** (http://www.rooisand.com/de/ranch.html)
- **Tivoli** guest farm (http://www.tivoli-astrofarm.de/e_tivoli_astrofarm.htm)

New Zealand Astronomy Bed and Breakfast. Richmond Nelson, New Zealand. 150-r refractor in a 4-m dome and 315-mm Newtonian (Dobson).

http://www.astronomybedandbreakfast.com

To plan your own astronomy vacations for the purpose of observing under a dark sky there is a thorough compilation of observing sites in the US and Canada by **ObservingSites.com**. All listed sites are accessible to the public during the night, with few if any restrictions.

http://www.observingsites.com

12.4 Newsgroups, Internet Portals, Mailing Lists and Chat Rooms

The Internet is an inexhaustible source to discuss and exchange experiences, questions, problems or results of observations with other amateur astronomers.

Newsgroups

Newsgroups are in general discussion forums on the Internet in which for a specific topic field so-called postings are exchanged among the participants who have the possibility to post their own comments. Astronomy-related newsgroups are found under the link http://www.cv.nrao.edu/fits/www/yp_newsgroup.html.

Very widely spread are the science groups sci.astro provided by Usenet, among them "sci.astro.amateur" especially enjoys great popularity. But precisely because of this popularity the quantity of daily news has become so overwhelming so that it is difficult to filter out the really valuable and important contributions. Further astronomy-related newsgroups are sci.astro.planetarium, sci.space.news and sci.astro.research. The latter is a moderated newsgroup devoted specifically to research related topics in astronomy and astrophysics.

http://sciastro.astronomy.net

Yahoo Tech Groups

Yahoo! makes a very extensive supply available in the context of its "Tech Groups."

The Yahoo! directory of astronomy groups comprises more than 1100 entries for all imaginable astronomical topics. Among these are groups with several thousand members, which is proof of the popularity of this type of communication. Many of the groups render possible access even if one is not a member of this group. The corresponding archive is labeled "public."

http://tech.dir.groups.yahoo.com/dir/Science/Astronomy

If one opens for example the page "sct-user," a forum for the users of commercial Schmidt–Cassegrain telescopes comprising nearly 4000 members, we first reach the page with the most recent messages. By selecting the menu item "topic list" we find all the topics listed that have been discussed in this forum since 1999.

A very practical search function enables contributions to be selected according to criterions such as date, subject and author. To check it out we enter the search criterion "secondary mirror" and we get in approximately one second more than 170 messages relating to this topic.

"Yahoo! Groups" opens access to more specialized groups as well (http://groups.yahoo.com/). Under the keyword "Astrophotography" one finds for instance 229 groups, which concern themselves exclusively with the various aspects of this special field. In

http://tech.dir.groups.yahoo.com/dir/Science/Astronomy/Astrophotography

a similar manner there are lists of groups which engage in topics such as Auroras, Cosmology, Deep Sky, SETI, Solar System, Telescopes etc. It should be mentioned that Yahoo! Groups are not only limited to the English-speaking community but are practically represented throughout the whole world.

Astronomy Internet Portals

Internet portals offer a large variety of astronomy-related services on the Web. Typical features of such portals are for example astronomical news, discussion forums, image galleries, virtual libraries, observing reports, FAQs, astronomy-related weblinks as well as a market place for astronomical instruments. Target groups are mainly amateur astronomers, but also students, teachers and educators and persons who are generally interested in astronomy. Some popular addresses are listed here:

– Wikipedia Astronomy Portal (http://en.wikipedia.org/wiki/Portal:Astronomy)
– *Sky and Telescope* (http://www.skyandtelescope.com/resources/internet)
– North American Skies (http://home.comcast.net/~sternmann)
– Skylights (http://www.astro.uiuc.edu/~kaler/skylights.html)
– Universe Today (http://www.universetoday.com)
– SPACE.com (http://www.space.com)

Sites in German:

– Astronomie.de (http://www.astronomie.de)
– Astrotreff (http://www.astrotreff.de)
– astronomie.info (Germany: http://www.astronomie.info,
  Switzerland: http://www.astronomie.ch)
– Astro Corner (http://www.astrocorner.de)
– astronews.com (http://www.astronews.com)

Astronomy-Related Mailing Lists

An electronic mailing list (sometimes called e-list) allows for widespread distribution of information to many Internet users by e-mail. A large selection on mailing lists with astronomical topics is provided by the Web server of the National Radio Astronomy Observatory (NRAO). Normally mailing lists allow individuals to subscribe and unsubscribe themselves. College institutes, scientific associations and interest groups, amateur astronomical societies, even the organizers of conferences and conventions forward current information to the subscriber by means of mailing lists.

http://search.nrao.edu

http://listmgr.cv.nrao.edu/mailman/create

Even to create a new mailing list is made easy by entering the relevant information into a given form which is provided by the NRAO Web server. Some mailing lists work as an e-mail forwarding system which allows users to be reached via e-mail anywhere through their username. For example SolarMail gives such a service to all members of the solar physics research community. There is a directory of SolarMail users allowing searches by name, mail ID, or institution.

http://spd.aas.org/SolarMail

A small selection of mailing lists that are of general interest follows here:

- amastro. Amateur astronomy mailing list which is dedicated to the discussion of visual observing, one of the most popular lists. — http://tech.groups.yahoo.com/group/amastro
- The ASTRO and ASTRO-DIGEST is for discussing astronomical topics, exchanging news, results, and experiences related to professional and amateur astronomy. — http://astrored.net/messier/xtra/supp/l-astro.html
- The Amateur Telescope Makers (ATM) e-mail list is a gathering of amateur astronomers who are engaged in building telescopes, grinding their own mirrors, building CCD cameras, and more. — http://www.atmlist.net
- ESO News mailing list serves for the distribution of news about various developments at the European Southern Observatory (ESO). — http://www.eso.org/public/outreach/press-rel/esonews
- HASTRO-L (The History of Astronomy Discussion Group) concerns itself with the history of astronomy in all cultural circles whether Euroamerican, non-Western, or non-literate. The field of interest covers all time periods from prehistoric to contemporary. To subscribe send message to: listserv@listserv.wvu.edu.
- The MeteorObs e-mail list is an open forum for discussions relating to meteor astronomy and observing. — http://www.meteorobs.org
- NetAstroCatalog is a mailing list for deep sky observers of all levels to share their observations of galaxies, star clusters, gaseous nebulae or double and variable stars. — http://www.visualdeepsky.org
- Solar Eclipse Mailing List and Solar Eclipse Newsletter (SENL) provide online information about eclipses. — http://www.williams.edu/Astronomy/IAU_eclipses

Chat Rooms

The usual goal of online chat is to share information via text messages with a group of people in the same (virtual) chat room in real-time. Rather than simply posting and reading messages on a bulletin board chatting means to interact with other users and to enjoy a lively conversation. A very popular species among various systems is Internet Relay Chat (IRC). There are several chat rooms (or groups) devoted to amateur astronomy as for example:

- Astronomy Chat Room (http://www.nightskyinfo.com/chat)
- AstroTips.com (http://astrotips.com/mod-SPChat.phtml)

12.5 Astronomical Education and Outreach

Astronomy education comprises a broad spectrum of learning at various levels of difficulty. The diversity extends from primary school to university, from the kindergarten child to the college graduate and with topics ranging from the planetary system to cosmological problems.

Astronomy Education in the Classroom

http://cdsweb.u-strasbg.fr/
astroweb/education.html

In many schools the handling of astronomical facts takes place within the scope of the physics lessons. For the teachers, who as a rule are not trained astronomers, there is a wide range of help resources available. A large selection can be found in the Yellow Page Service of the AstroWeb, as for example:

http://aer.noao.edu/cgi-bin/new.pl

– The **Astronomy Education Review** is a magazine (or website, respectively) for educators who teach in the field of astronomy and space exploration. Its mission is to encourage teachers to share experiences and innovations, to improve teaching and learning systematically and to take part in workshops and symposia. It provides a forum for publishing the results of research in education and gives advice how to apply these in the classroom.

http://astrosci.scimuze.com

– The webpage "**astrosci**" offers a selection of educational resources in astronomy for teachers and students.

http://nw.pima.edu/dmeeks/
mathsci

– "**Meeksnet**", the homepage of Dennis Meeks from the Pima Community College Science Department in Tucson (Arizona) is a mine of hundreds of astronomy links useful for teachers and students.

http://www.astronomyteacher.com

– The site of "**AstronomyTeacher.com**" is a free resource for astronomy educators at all levels. It contains downloadable papers and essays on astronomy content. It also collects links to interactive projects and for astronomy educators in K-12 schools.

http://btc.montana.edu/ceres

– **CERES** Project (Center for Educational Resources). Through funding from **NASA**, CERES provides a big library of interactive K-12 materials for teaching astronomy including classroom-ready lesson plans and NASA data search engines.

http://www.spacetimetravel.org

– "**SpaceTimeTravel**" offers online papers, images, movies and paper models for the visualization of the theory of relativity.

The Role of Astronomical Societies and Organizations

http://www.astrosociety.org/
education.html

An extensive selection of projects and programs in astronomy education is obtainable through the **Astronomical Society of the Pacific** (ASP) which has developed into the well-known leader in that field of education.

The catalog contains among others a list of workshops and training for teachers of astronomy at K-12 schools and colleges, programs involving amateur astronomers and college students and a variety of educational materials.

ASP offers also an assortment of educational resources such as articles on education and a selected astronomy education bibliography. The "**Universe in the Classroom**" is an electronic educational newsletter for teachers who want to bring the fascination of astronomy to the classroom. In the future the newsletter will be available in many different languages.

http://www.astrosociety.org/
education/astro/
project_astro.html

The **ASP** started in 1994 the project **ASTRO** as a national program in the US that improves the teaching of astronomy by linking professional and amateur astronomers with local educators. The success of this activity is based on a variety of efforts such as training between astronomers and educators at two-day workshops, multiple visits of the volunteer astronomer at his or her "adopted" class and organizing star parties or arranging excursions to local astronomy facilities. The goal of these educator-astronomer partnerships in the classroom is to provide hands-on activities that enable students to act like scientists.

NASA, the **National Aeronautics and Space Administration**, maintains three different education programs for space sciences. The "**Elementary and Secondary Education**" branch applies to high school students. As a consequence NASA supports a network of K-12 teacher resource centers at various locations around the US.

http://education.nasa.gov

The "**Higher Education**" program is designated for students at colleges and universities, whereas the mission of the "**Informal Education**" division is to stimulate public understanding and appreciation of science and technology outside of formal classroom environments.

The **National Optical Astronomy Observatory** (NOAO) which is the head organization for several major astronomical observatories in the US has established an educational outreach program with resources for students and teachers as well as for families.

http://www.noao.edu/education/noaoeo.html

The aim of the **European Association for Astronomy Education** (EAAE) is to develop and to promote the teaching of astronomy at all levels by using the wide network of astronomy educators in Europe. It is managed through the **European Southern Observatory** (ESO) in Garching near Munich, Germany. EAAE has national representatives in 23 European countries.

http://www.eaae-astro.org

Since 1996 the **International Astronomy Olympiad** (IAO) is held every year as an astronomy competition for high-school students classified according to several age groups. The event takes place in one of the astronomical centers of the Olympiad member states.

http://www.eaae-astro.org/eaae/olympiad

As the first all-European network for teachers interested in astronomy the EAAE arranges an annual "Summer School" which is dedicated to the spreading of new educational materials and methods as well as to the exchange of experiences.

In the UK a specific branch of the EAAE is called **Association for Astronomy Education** (AAE). On its homepage one finds information on professional astronomers who will give talks to schools as well as on teacher training events or on activities such as "Ask an Astronomer."

http://www.aae.org.uk

The **European Space Agency** (ESA) and the **European Southern Observatory** (ESO) have produced an "Astronomy Exercise Series" devoted to high school students. The exercises are presented as various small projects that focus for example on measuring the distances in the Universe or on estimating the age of the Universe and its expansion velocity.

http://www.astroex.org

The **International Astronomical Union** (IAU) maintains two commissions that are both engaged in astronomy education and public outreach. Commission 46 occupies itself with "**Astronomy Education and Development**." It seeks to develop and improve astronomical education at all levels throughout the world.

http://iau46.obspm.fr

IAU's Commission 55 is dedicated to the topic "**Communicating Astronomy with the Public**" (CAP). Its mission means to enable and encourage a larger part of the astronomical community to take an active role in the popularization of astronomy. The commission has at the moment six working groups and organizes CAP meetings every second year.

http://www.communicatingastronomy.org

Hands-on Astronomy

Direct access to intensive engagement with astronomical problems is provided through hands-on experiments. For example it motivates the pupils to ask further questions in the classroom when they have immediately observed sunspots or the partially eclipsed Sun through the telescope (Fig. 2). The visual observation of sunspots leads to the study of the solar atmosphere and its structure or the experience with a solar eclipse could be the starting point for the theoretical discussion of celestial mechanics.

http://www.handsonuniverse.org

Hands-On Universe (HOU) is an educational program for students that enables them to solve astronomical problems in the classroom or at home by means of realistic observation data. These are requested via the Internet from an automated telescope or downloaded from a large image archive. The data are then analyzed with the aid of a specific image processing software.

http:// www.schoolsobservatory.org.uk

The **National Schools' Observatory** (NSO) offers schools in the UK the opportunity to observe the sky through the 2-m Liverpool Telescope which is at the moment the world's largest fully robotic telescope. The telescope is sited 2400 m high on the island of La Palma in the Canary Islands and is used worldwide by astronomers for research projects.

http:// www.telescopesineducation.com

The **Telescopes in Education** (TIE) program is conducted by the Mount Wilson Institute and enables students around the world to use a remotely controlled reflecting telescope and charge-coupled device (CCD) camera in real-time. The equipment is located at the Mount Wilson Observatory (see Sect. 12.8) and can be operated remotely by educators and students from their classroom. Images can be downloaded to the user's computer for later image processing and study.

Informal Education Facilities

The learning of astronomy is not limited to the formal classroom education, on the contrary much work of public outreach is done through facilities such as observatories, planetariums and science museums.

Public Observatories offer for many people the first and in many cases the only possibility to view celestial objects through the telescope. Already at a very early age a look at the Moon or the rings of Saturn through a telescope can create excitement which in later years can be the basis for more involvement with astronomy (Fig. 3). The Web provides numerous programs for younger children, for example NASA's StarChild,

Fig. 2. Observing the Sun at the schoolyard. Courtesy of Lutz Clausnitzer, Geschwister-Scholl-Gymnasium Löbau, Obercunnersdorf, Germany

Fig. 3. Even at an early age, a look through the telescope can create enthusiasm for astronomy. Courtesy of Peter Stättmayer, Volkssternwarte München, Germany

"Astro for kids" by the well-known magazine *Astronomy* or KidsAstronomy.com which is operated by the KidsKnowIt Network.

A number of major observatories have established visitor centers at their sites where interested people can get an idea about the astronomer's work through specially trained staff. Some of these observatories are very popular as for example the Kitt Peak National Observatory (see Sects. 12.3 and 12.8) which is the most visited astronomical observatory site in the world.

Planetariums are favorite facilities for the general public to get familiar with the starry sky and its changes during the seasons. For millions of people living in light-polluted big cities this is probably the only chance to experience the beauty of a dark night sky at all (Fig. 4). In the US alone there are approximately 1500 planetariums, half of which do school as well as public shows. The sci.astro.planetarium newsgroup permits the exchange of planetarium-related topics among planetarium professionals. Directories of planetariums are available from the websites of both *Sky and Telescope* and *Astronomy* magazine (see Sect. 12.1).

The major **science museums** have astronomy exhibits where the visitor by means of historical objects such as antique telescopes experiences the evolution of astronomical research in history. Several museums also offer educational resources for the understanding of facts related to astronomy or space exploration. Here follows a small selection of science museums in the US and in Europe:

- Adler Planetarium and Astronomy Museum, 1300 S. Lake Shore Drive, Chicago, IL 60605-2403 — http://www.adlerplanetarium.org
- American Museum of Natural History, 79th Street and Central Park West, New York City — http://www.amnh.org
- National Air and Space Museum, National Mall, Washington, DC 20560 — http://www.nasm.si.edu

Fig. 4. Planetariums are for millions of people the only opportunity to experience the beauty of a dark night sky. Courtesy of Peter Stättmayer, Volkssternwarte München, Germany

http://www.sciencemuseum.org.uk

http://www.deutsches-museum.de

http://www.iayc.org

– Science Museum London, Exhibition Road, South Kensington, London SW7 2DD, UK
– Deutsches Museum München, Museumsinsel 1, D-80538 München, Germany

Every summer the **International Astronomical Youth Camp** (IAYC) takes place in Europe for young people who are between 16 and 24 years old and share a common interest in astronomy. Participants from about 20 different countries live together for three weeks working on astronomical projects in one of several working groups which are led by young scientists.

http://
www.sidewalkastronomers.us

Since 1968 the **Sidewalk Astronomers** bring astronomy to the public. They set up their telescopes in urban areas of California, on busy malls or street corners, to show people passing by the Moon and the planets. The telescopes they are using are of Dobsonian type, named after John Dobson who was the founder of this kind of astronomy outreach (see Sect. 12.6 and [3]). Now the expression "sidewalk astronomy" is used as a general term which means that amateur astronomers take out their telescopes to the streets and show and explain passers-by various sky objects.

12.6 Famous Amateur Astronomers

A lot of well-known names exist in the field of amateur astronomy. The reasons for their widespread reputation may be various. In any case these men have accomplished a great deal for our hobby. Whether it was their substantial contribution to the popularization of astronomy, their development of innovative telescopes for amateurs or their numerous discoveries of comets and supernovae that made them famous.

The following profiles of some prominent amateur astronomers represent merely a minute selection, which is naturally very subjective.

Lunar crater "Porter"

Mars crater "Porter"

Russell Williams Porter (1871–1949). Russell W. Porter has acquired renown among amateurs as founder of the legendary Stellafane Observatory in Springfield, Vermont [2]. Porter majored in architecture at the Massachusetts Institute of Technology (MIT). He embarked on six adventurous expeditions to the Arctic Zone within thirteen years. A lot of paintings and drawings that he depicted during that time, reflected the impressions he gained in this world of ice proving even back then his enormous talent for drawing.

Approximately from 1910 onwards Porter dedicated himself to the drafting and construction of telescopes, in which he was occupied throughout his life. At his friend's James Hartness's request he returned to Springfield where he both designed and constructed the clubhouse and observatory called "Stellafane" for a group of enthusiastic amateur telescopes craftsmen. This group, that later called itself the Springfield Telescope Makers, met for the first time in August 1920. It was Stellafane (originally "Stellar Fane," meaning "shrine to the stars") which triggered an amateur movement, that soon spread throughout the US. Albert G. Ingalls, editor of the magazine *Scientific American*, published in collaboration with Porter the standard work *Amateur Telescope Making*, that became the guiding book for constructors of telescopes. Countless sketches of telescopes and their mountings were—mostly three-dimensionally—drawn by Porter with an inimitable perfection. Every year amateurs from all over the US and Canada go on a "pilgrimage" to the Stellafane Convention bringing their homemade telescopes. Stellafane grew to become a sort of astronomy's mecca.

Hartness–Porter Museum: http://www.hartnesshouse.com/vermont-museum/hartness-porter-museum.shtml http://www.stellafane.com

In 1928 Porter was hired to go to California by George Ellery Hale, in order to participate in the construction of the 200-in. telescope on top of Mount Palomar.

Patrick Moore (b. 1923). Sir Patrick Moore is a living legend in the British amateur scene. Hardly any other amateur astronomer is as equally renowned as Moore in the public at large. The release of more than 60 books and his presence on TV for fifty years explain his fame. In addition he was knighted "for services to the popularization of science and to broadcasting." In April 1957 the BBC series "The Sky At Night" was aired for the first time and has been shown monthly ever since. In its more than 600 episodes it managed to get generations of stargazers interested in astronomy and revealed to them the starry sky's beauty. His continuous appearance on TV for 50 years even bestowed on him an entry in the Guinness Book of Records as the longest serving television presenter. Being an active amateur observer he was concerned with a precise mapping of the Moon's surface, so that in 1959 the Soviet Union used his charts to correlate the Lunik 3 pictures of the far side of the Moon. He was also involved in the lunar mapping preceding the NASA Apollo space missions.

Asteroid (2602) Moore

http://sirpatrickmoore.com

John Dobson (b. 1915). John Dobson (Fig. 5) has acquired renown in the amateur community for developing a telescope design, which was named after him and which has proved very popular with amateur astronomers all over the world in the meantime [3]. The first drafts for the "Dobsonian" telescope were developed even during his stay at the Vedanta Society monastery in San Francisco, where he lived for 23 years. Due to the limited possibilities this place provided for the construction of telescopes, he was forced to keep his drafts as simple as possible—the result of which was the revolutionary design of the Dobson telescope which is basically a Newtonian reflector telescope installed on a very simple alt-azimuth mounting. Dobson's construction method enabled the amateurs to build portable and inexpensive telescopes with large apertures, 20 in. and more. At around 1968 Dobson founded together with two other amateur astronomers the San Francisco Sidewalk Astronomers. Members of this organization simply set up their tele-

Asteroid (18024) Dobson

San Francisco Sidewalk Astronomers: http://www.sfsidewalkastronomers.org

Fig. 5. John Dobson. Courtesy of Elke Schulz, Stuttgart, Germany

scopes on sidewalks and show passers-by craters of the Moon, the moons of Jupiter, the rings of Saturn or the spots of the Sun.

John Dobson is well-known for rejecting the currently popular Big Bang model of the universe. Instead he sticks to his own cosmological theory, a kind of "recycling" model, according to which the universe regenerates itself repeatedly from the border.

David H. Levy (b. 1948). The Canadian amateur astronomer David Levy (Fig. 6) is one of the most successful discoverers of comets. Altogether he has discovered or co-discovered a total of 22 comets. He found them using his own backyard telescopes as well as highly professional equipment. One of his most spectacular discoveries was the comet Shoemaker–Levy 9 which he discovered together with Eugene and Carolyn Shoemaker at the Palomar Observatory. The collision of the disintegrated comet with Jupiter resulted in the most powerful explosions ever witnessed in the planetary system.

Levy has published or written more than 35 books, among them three famous astronomers' biographies. As contributing author he regularly writes the column "Star Trails" for the *Sky and Telescope* magazine. Also he has given more than a thousand lectures and interviews, mostly during 1993 and 1994 when "his" comet was colliding with Jupiter.

Asteroid (3673) Levy

http://www.jarnac.org

Fig. 6. David H. Levy. Courtesy of Wendee Levy, Vail, AZ, USA

Furthermore Levy has cooperated in many television programs and thus made a substantial contribution to the popularization of astronomy.

Levy has launched the "National Sharing the Sky Foundation", an organization that aims at using astronomy to encourage young people to develop an interest in science and engineering. For this purpose he has initiated several school and camp programs.

12.7 Amateur–Professional Partnership

When the exploration of the heavens was strongly stimulated in the second half of the eighteenth century the astronomers involved were often literally "amateurs." An outstanding example would be F. William Herschel (1738–1822) who started out as a celebrated musician and is nowadays reputed to be considered one of the greatest astronomers of his time. Even the foundation of a major observatory was in some cases due to an amateur's enthusiasm. Percival Lowell (1855–1916), a businessman from Boston, had been interested in the idea of searching for intelligent life on Mars and finding a new planet (Planet X). He founded the famous observatory in Flagstaff, Arizona, which is named after him and is one of the most important observatories throughout the North American continent.

The Amateur's Role Today

Today professional astronomy has reached a level of development which is characterized by the employment of elaborate (and therefore expensive) state-of-the-art technology. Earth-orbiting observatories whose observations of the sky range from gamma rays into the far infrared, reflectors with gigantic mirrors (for example the future ESO 100-m mirror OverWhelmingly Large Telescope (OWL)) and high-precision interferometers exceed the amateur astronomer's technological conditions by far. The conventional wisdom was that the amateur could not contribute anything further to professional astronomy. However, numerous recent examples prove that there are still a number of niches left where a collaboration between professional and amateur (pro-am collaboration) can have a considerable impact on science [4].

ASP Conference Proceedings:
http://www.naic.edu/~nolan/amateur.html

The scope of duties that can be carried out by amateurs in the context of research projects is multifaceted. First and foremost a pro-am collaboration would seem the right strategy for the discovery and measurement of variable stars. Observing long-period variable stars or discovering a sudden increase of brightness of a flare star, dwarf nova or supernova typically belongs to the amateur observer's specialities. Such tasks frequently require long-term usage of telescopes or spontaneous reactions to unexpected astronomical events. Neither the former nor the latter can normally be accomplished by professional observatories due to their strict compliance with observing schedules which are predetermined months in advance. Instead thousands of amateurs round the globe position themselves with their comparatively modest telescopes ready to take over such tasks with great enthusiasm.

The American Association of Variable Star Observers

http://www.aavso.org

The American Association of Variable Star Observers (AAVSO), which was founded in 1911, is devoted to the study of variable stars and is one of the largest pro-am astronomy organizations in the world. It has 1200 members (about 20 % professionals) in over 45 countries.

Every year approximately 900,000 observations are archived, globally there is a database of 13 million observations available, dating back partly over 150 years. Hundreds of requests from scientists for variable star data have been filled annually from the AAVSO International Database. The AAVSO focuses its efforts on data archiving, to motivating and mediating pro-am collaborations as well as to education. Activities include public speeches and astronomy classes, star parties and events on Astronomy Day, book releases and articles and the production of radio and TV programs. Furthermore amateurs seek contact with schools and teachers and arrange astronomic workshops. With regard to the dialog mode between members themselves the AAVSO discussions have proved helpful means of communication.

AAVSO Discussion Group:
http://www.aavso.org/mailman/
listinfo/aavso-discussion

The American Astronomical Society (AAS) has established a "Working Group for Professional Amateur Collaboration" (WGPAC), which provides a forum for pro-am collaboration.

Working Group for Professional
Amateur Collaboration:
http://www.aas.org/wgpac

Besides his preoccupation with variable stars an earnest amateur astronomer can make himself useful for professional astronomy in plenty of other fields: comets and minor planets, surfaces of planets and their temporal alteration, active processes on the Sun or the precise timing of occultations of stars by bodies of our Solar System. Among amateur organizations that deal with such topics are the Association of Lunar and Planetary Observers (ALPO) and the International Occultation Timing Association (IOTA).

ALPO:
http://alpo-astronomy.org
IOTA:
http://
www.lunar-occultations.com/
iota/iotandx.htm

While back in the early days amateurs placed their observation data at the organizations' disposal for further elaboration, today amateurs increasingly address themselves to professional astronomers personally with their results. This trend was propelled also thanks to the above-mentioned AAS Working Group for Professional Amateur Collaboration.

Gamma Ray Bursts

"NASA Needs You"

Some time ago the NASA directed an appeal to the amateur community with the headline "Calling All Amateur Stargazers: NASA Needs You".

AAVSO International High Energy
Network:
http://www.aavso.org/observing/
programs/hen

This cry for help was directed at all amateur observers and requested to actively monitor the sky for optical counterparts of gamma-ray bursts. These bursts last only for a short time, typically less than a few hours, and occur randomly from all directions of the sky. To obtain safe optical data of the bursts is difficult and requires instantaneous follow-up observations by optical telescopes. The AAVSO initiated the High Energy Network which is dedicated to the optical monitoring of high energy sky events. The goal of this network is to alert the (amateur) observers in cases of an event such as a gamma ray burst (GRB) afterglow, a blazar or polar (magnetic cataclysmic variable). Meanwhile several amateur observers have been able to monitor GRB afterglows when alerted by the AAVSO Network.

Gamma Ray Burst Afterglows: http://www.aavso.org/observing/ programs/hen/grb.shtml

So Berto Monard from South Africa was the first amateur observer to discover a GRB afterglow before any professional could do it. On July 25th, 2003, he detected the optical signature of GRB 030725 with a 12-in. telescope from his home observatory.

Yet two additional examples of a successful pro-am collaboration shall be mentioned. Figure 7 shows the light curves for an outburst of the cataclysmic variable SS Cygni taken for visible, extreme-ultraviolet and X-ray wavelengths, respectively. In this case the flux measurements by the satellite observatories Extreme Ultraviolet Explorer (EUVE) and Rossi X-ray Timing Explorer (RXTE) have been triggered by amateur astronomers

Wheatley et al. (2003)

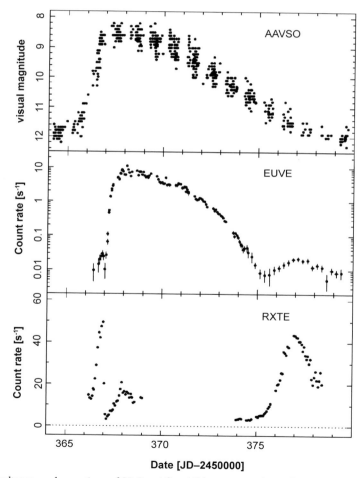

Fig. 7. Simultaneous observations of SS Cygni for visible, extreme-ultraviolet and X-ray wavelengths, respectively. From Wheatley et al. [5]

(mainly AAVSO members) who discovered the dwarf nova outburst visually in its initial phase. It is obvious from Fig. 7 that the EUV outburst as well as the X-ray outburst has been delayed in comparison to the optical outburst by approximately one day [5].

Microlensing

Discovering exoplanets is a new branch in astronomy. The number of exoplanets which have been detected by measuring the light curves when transiting their "parent" stars undergoes an enormous growth. Also in this field of research amateurs can make significant contributions to science, as particularly smaller telescopes with typical apertures between 20 and 40 in. are best qualified to observe exoplanets during their transit across a star. Photometric accuracy requirements are increased approximately by a factor of 20 compared to the usual observation of variable stars. In spite of that even instruments with apertures between 10 and 14 in. allow for successful observations as well, as convincingly demonstrated by Bruce L. Gary in his book *Exoplanet Observing for Amateurs* [6].

In recent time gravitational microlensing is used as a powerful technique for detecting extrasolar planets. The presence of a planet orbiting the lens star modifies its gravitational field in a characteristic manner which influences the light from a more distant star in a predictable way. Two New Zealand amateur astronomers, Grant Christie (see his work [7]) from Auckland and Jennie McCormick from Pakuranga, contributed with their small telescopes essential data to detecting an exoplanet during the microlensing event OGLE-2005-BLG-071. Both amateurs share co-authorship on the announcing paper published in *The Astrophysical Journal Letters* [8].

http://arxiv.org/abs/
astro-ph/0505451

12.8 The Telescope Tourist

In May 2007 the online edition of *The Wall Street Journal* wrote: "Stargazing conjures up staid images of field trips to the local planetarium or puttering with a backyard telescope. But a new breed of obsessed fans is crisscrossing the globe to visit legendary observatories. Destinations include remote deserts, mountaintops—even the South Pole. Many of these travelers are engineering buffs, people who love nothing more than to gawk at the one-of-a-kind quirks of multimillion-dollar machines. Even the smell of old telescopes' lubricants—a bit like crayons—intrigues them. Others are astronomy pilgrims retracing great moments in scientific history."

More and more astronomy enthusiasts are, indeed, being drawn to the big observatories. Whether they travel all the way to the Mauna Kea observatory in Hawaii or to the Cerro Paranal in Chile and are amazed at the modern telescopes which can be referred to as impressive high-tech machines. Or they seek to take in the nostalgic flair emanated from historically notable telescopes on Mount Wilson or at the Yerkes Observatory. Anyway, enthusiasm for telescopes as technical marvels also constitutes an important basis for awakening understanding of astronomy's phenomena.

The observatories' visitor centers attend to curious guests by informing them about accomplished as well as every-day work at the observatory. One should consider that these observatories are oftentimes located on top of high mountains which are difficult to access (warm clothes recommended!) and for the most part do not offer any gastronomic services. Below some selected observatories are listed which are open to the public.

United States

- **Palomar Observatory**, Palomar Mountain, CA. Open to the public from 9 am to 4 pm daily, admission free. The 200-in. telescope was formerly the world's biggest optical telescope, completed in 1949, elevation 1680 m.
  http://www.astro.caltech.edu/observatories/palomar

- **Mount Wilson Observatory**, San Gabriel Mountains, CA. Open daily 10 am to 4 pm from April to November. Public viewing through the 60-in. telescope, advanced booking required. Founded in 1904 by George Ellery Hale, the 100-in. Hooker Telescope was during 40 years the world's biggest telescope, Edwin Hubble worked here, 150-foot solar tower, elevation 1740 m (Fig. 8).
  http://www.mtwilson.edu

- **Mauna Kea Observatories**, Island of Hawaii. Visitor Information Station at 9000 ft, the Keck observatory and the UH 2.2-m telescope have visitors galleries. Keck open 10 am to 4 pm Mon.–Fri.; UH 2.2-m 9:30 am–3:30 pm Mon.–Thu. The world's largest astronomical observatory, 13 working telescopes including those for radio astronomy; the road up the mountain is difficult and dangerous, only 4WD allowed or commercial tour, serious health risks because of the high altitude, elevation 4200 m.
  http://www.ifa.hawaii.edu/mko/maunakea.htm
  http://www.keckobservatory.org
  http://www.ifa.hawaii.edu/88inch

- **Kitt Peak National Observatory**, Tucson, AZ. Open daily from 9 am to 4 pm, Visitor Center (admission free), guided tours (fee), snacks and beverages available at the Visitor Center, nightly observing program available by reservation. Founded in 1958, 4-m Mayall Telescope, 3.5-m WIYN Telescope, McMath–Pierce Solar Telescope (the world's largest solar telescope), elevation 2097 m.
  http://www.noao.edu/kpno

- **Yerkes Observatory**, Williams Bay, WI. Free public tours every Saturday, monthly 24-in. telescope observing program (fee). Founded in 1895 by Hale, five research telescopes including the largest refractor in the world.
  http://astro.uchicago.edu/yerkes

- **Very Large Array**, Socorro, NM. Open daily from 8:30 am to dusk, Visitor Center, gift shop, guided tours twice per year. Radio observatory consisting of 27 radio antennas in a Y- or A-shaped configuration.
  http://www.vla.nrao.edu

- **US Naval Observatory**, Washington, DC. Limited public tours on selected Monday evenings from 8:30 pm until 10:00 pm, admission free, advanced booking mandatory. Founded in 1830, 26-in. Clark refractor. Time Service for the GPS.
  http://www.usno.navy.mil

Fig. 8. Dome of the Hooker telescope at Mount Wilson Observatory. Courtesy of Volker Witt, München, Germany

South America

http://www.eso.org/public/
outreach/visitors/paranal/
tourist.html

– **European Southern Observatory (ESO)**, Cerro Paranal, Chile. Tours on the last two weekends of every month, except December, admission free, booking months in advance required. Very Large Telescope (VLT) consisting of four 8.2-m telescopes, VLT interferometer, elevation 2635 m.

Europe

http://www.rog.nmm.ac.uk

http://www.mpifr-
bonn.mpg.de/english

– **Royal Observatory Greenwich**, London, England. Open daily 10 am to 5 pm, admission free. Founded in 1675, site of the world's Prime Meridian, historic 28-in. refractor (Fig. 9).
– **Radio Telescope Effelsberg**, Bonn, Germany. Public talks at the Visitor Center from April to October (Tuesday to Saturday). With a diameter of 100 m one of the largest fully steerable radio telescopes on earth (Fig. 10).

Africa

http://www.salt.ac.za

– **Southern African Large Telescope (SALT)**, Sutherland, South Africa. Open from Monday to Saturday, day and night tours, booking is required (fee). The 11-m SALT is the largest single optical telescope in the southern hemisphere, elevation 1798 m.

Australia

http://
www.sidingspringexploratory.
com.au

http://www.parkes.atnf.csiro.au

– **Siding Spring Observatory**, Coonabarabran, New South Wales. Visitor Center "Siding Spring Exploratory," 2.3-m Advanced Technology Telescope, 3.9-m Anglo-Australian Telescope, 1.24-m UK Schmidt Telescope.
– **Parkes RadioTelescope**, Parkes, New South Wales. Visitor Center open daily 8:30 am to 4:15 pm, free entry, café and shop, 64-m diameter fully steerable radio dish.

Fig. 9. Flamsteed House at Greenwich Observatory. Courtesy of Volker Witt, München, Germany

Fig. 10. Radio Telescope Effelsberg, Germany. Courtesy of Norbert Junkes, Max Planck Institute for Radioastronomy, Bonn, Germany

12.9 Further Reading

Christensen, L.L.: The Hands-On Guide for Science Communicators. A Step-by-Step Approach to Public Outreach. Springer, Berlin Heidelberg New York (2007)

Heck, A. (ed.): StarGuides Plus: A World-Wide Directory of Organizations in Astronomy and Related Space Sciences. Kluwer Academic, Dordrecht (2004)

Heck, A., Madsen, C. (eds.): Astronomy Communication. Astrophysics and Space Science Library. Kluwer Academic, Dordrecht (2003)

Pasachoff, J., Percy, J. (eds.): Teaching and Learning Astronomy. Effective Strategies for Educators Worldwide. Cambridge University Press, New York (2005)

Pasachoff, J., Ros, R.M., Pasachoff, N.: Innovation in Astronomy Education. Cambridge University Press, New York (2008)

References

1. Wikipedia, The Free Encyclopedia: Amateur Astronomy. http://en.wikipedia.org/wiki/Amateur-astronomy. Accessed 2007

2. Willard, B.C.: Russell W. Porter, arctic explorer, artist, telescope maker. Bond Wheelwright Company, Freeport, ME (1976)

3. Dobson, J.L., Sperling, N. (ed.): How and Why to Make a User-Friendly Sidewalk Telescope, 1st edn. Everything in the Universe, Oakland, CA (1991)

4. Percy, J.R., Wilson, J.B. (eds.): Amateur–Professional Partnerships in Astronomy, ASP Conference Proceedings, vol. 220. Astronomical Society of the Pacific, San Francisco (2000)

5. Wheatley, P.J., Mauche, C.W., Mattei, J.A.: The X-ray and extreme-ultraviolet flux evolution of SS Cygni throughout outburst. Mon. Not. R. Astron. Soc. **345**, 49–61 (2003)

6. Gary, B.L.: Exoplanet Observing for Amateurs. Mira Digital Publishing, St. Louis, MO. http://brucegary.net/book_EOA/x.htm (free download as PDF). Accessed 2007

7. Christie, G.: Detecting Exoplanets by Gravitational Microlensing using a Small Telescope, The Society for Astronomical Sciences 25th Annual Symposium on Telescope Science, 23–25 May 2006, Big Bear, CA. Society for Astronomical Sciences, Rancho Cucamonga, CA, p. 97 (2006)

8. Udalski, A. et al.: A Jovian-Mass Planet in Microlensing Event OGLE-2005-BLG-071. Astrophys. J. **628**, L109–L112 (2005)

13. Earth and Solar System

13.1 The Sun

K. Reinsch, M. Delfs, E. Junker, and P. Völker

13.1.1 Introduction

No observer can avoid being fascinated by the Sun, whose appearance changes from day to day, sometimes from hour to hour. On no day in the future will the Sun look exactly as it does today. A wide variety of phenomena are within reach of small telescopes, including those located in urban sites [1, 2, 3]. Special equipment required for observing the spectacular and dynamic phenomena in the upper solar atmosphere have become available for amateur use at affordable prices. As a welcomed feature, solar observations can be pursued without losing nighttime rest. The enormous surface brightness of the Sun requires special equipment for observation and photography—*never* use a telescope without sufficient filtering!

The vast amount of light visible to the human eye (*white light*) emerges from a thin layer of the solar atmosphere, termed the *photosphere*. It radially expands a few hundred kilometers only and has an effective temperature of 5800 K. Below the photosphere, the Sun becomes opaque due to the rapidly increasing density and temperature of the gas. At the center of the white light image, lower (and hotter) layers of the photosphere are visible while towards the limb we are looking into its upper (and cooler) layers. This can easily be recognized by a *limb darkening* of the solar image. Sunspots (Sect. 13.1.3) and faculae (Sect. 13.1.4) are the most prominent manifestations of *solar activity* in the photosphere (Fig. 1a) and can be viewed with basic equipment (see Chap. 4). Under ideal observing

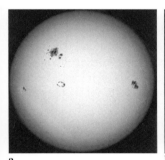  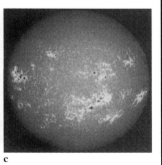

a   b   c

Fig. 1. The Sun in (**a**) white light, photographed October 27, 2003 by R. Buggenthien, and in the monochromatic light of (**b**) the Hα-line, photographed May 4, 1978 by W. Paech, and (**c**) the Ca II K line, photographed September 6, 1981 by G. Appelt

conditions, the photosphere shows a cellular brightness pattern on scales from a fraction of an arcsecond to several arcseconds, the solar *granulation*.

Immediately above the photosphere lies the solar *chromosphere* ("chromos," Greek for "color") in which the density decreases rapidly with height. Due to the highly diluted gas chromospheric emission is produced in the strongest atomic lines of the solar spectrum only, most noticeable the red Hα line and the blue Hβ line of hydrogen and the violet H and K lines of ionized calcium. The chromosphere is outshone by the enormous brightness of the photosphere, and can, unless special filters are used (see Chap. 6), be observed only during the brief moments of solar eclipses when the lunar limb has completely covered the photosphere but not yet the chromosphere. Then it appears as a thin pinkish-red jagged seam of variable thickness between roughly 2000 km and 8000–10,000 km. When viewed through a PST (see Chap. 6) in the brightest hydrogen line, Hα (λ = 656.3 nm), the chromosphere is visible also on the solar disk and gives a visually fascinating picture (Fig. 1b). In the light of the magnetically sensitive Ca II K line (λ = 393.4 nm), chromospheric areas with a moderately strong magnetic field show up bright, but areas with a strong magnetic field (such as in sunspots) remain dark. This line, however, lies so far in the violet spectral range where the sensitivity of the eye is low that only photographic observations are useful (Fig. 1c).

Above the chromosphere, the extremely thin but hot (several million K) solar *corona* ("crown") radially extends over several million kilometers into interplanetary space. The corona itself has a most exciting appearance which, however, becomes visible only during solar eclipses or with specially designed devices ("coronagraph"). Using monochromatic filters, the spectacular prominences and filaments (Sect. 13.1.5)—material with chromospheric-like conditions embedded in the corona—can be observed (Fig. 1b, c) and, occasionally, flares (Sect. 13.1.6) can be witnessed as violent outbursts of radiation.

13.1.2 Observations of the Sun

13.1.2.1 Site Selection The location of the observing site has a large influence on the quality of viewing. If there is a choice of sites at all the observer should consider a few basic selection criteria in order to make optimum use of the investment in instruments and accessories.

The experiences of professional observers have been summarized by Kiepenheuer [4] and interpreted by Müller [5] for amateur astronomers. Basically, the line-of-sight should lead over terrain which warms and cools homogeneously like, e.g., lakes or forest areas. Very unfavorable are built-up areas (houses, streets, etc.) as they heat and cool rapidly and, moreover, the heated air from smokestacks contributes to the notorious "twinkling" phenomenon, particularly in winter. Similarly disadvantageous are farm fields because of their low heat capacity; variations in the solar irradiation causes rapid changes in the ground temperature.

Radiative temperature changes refer particularly to the lowest air layers. On a tower 10 or 20 m above ground, a substantial fraction of the ground turbulence is eliminated and the conditions improve markedly. A balcony on a high-rise can also be used as a tower as long as the objective extends beyond the hot air rising at the wall.

Much turbulence originates in the immediate vicinity of the telescope, and therefore the observing area should be carefully shaped in order to obviate the heating. The vicinty should contain continuous growth; walkways and flat roofs should be graveled, if possible, or even covered with grass or ivy growth.

Most turbulence arises at and in the telescope. The best and most expensive solution is a vacuum telescope; the air is "simply" evacuated from an airtight tube. The experienced telescope maker will take into account the deformation of the tube and optical parts by the pressure gradient. A broad ring placed around the tube in the vicinity of the objective is useful during observations, as it automatically shadows the telescope and also increases contrast on the projection screen. The dewcap should be removed so that the objective may be cooled by air breezes. A complete covering of the telescope during observation every 20 to 30 minutes for about 10 minutes has proven advantageous.

3.1.2.2 Observing Conditions Of the atmospheric effects interfering with telescopic observations, one distinguishes directional scintillation or "image motion" from "blurring," which makes the image unsharp or diffuse. The temperature differences and turbulence eddies in the atmosphere change the optical properties of air as if a lens were changing the focal length several times each second. In obtaining data on atmospheric conditions, which are important to judge the quality of a solar observation, the Kiepenheuer scale [4] (modified by Dreyhsig and Reinsch) has proven useful. The image motion (M) measures the scintillation, the sharpness (S) the amount of "blurring."

Modified Kiepenheuer Scale

Image Motion M

1. No noticeable image motion on the disk nor at the solar limb.
2. Image motion of $\leq 2''$ seen at the limb, not on the disk.
3. Image motion $\leq 4''$ distinctly visible at the limb and on the disk; "boiling" limb.
4. Image motion $\leq 8''$ almost prohibits distinction of umbra and penumbra, with strongly "boiling" limb.
5. Image motion over $> 8''$ reaches spot diameters, with strongly "boiling" limb.

Sharpness S

1. Granulation easily visible; fine structure in penumbra visible.
2. Granulation and penumbra easily seen but with very little fine structure; umbra-penumbra and penumbra-photosphere boundaries are sharp.
3. Granulation visible only in traces, but surface structures are easily found when moving the solar image; umbra and penumbra well-separated without fine structure; boundary to photosphere difficult.
4. Granulation not visible; umbra and penumbra separated only in large spots; boundary to photosphere diffuse.
5. Granulation not visible; umbra and penumbra not distinguished even in large spots.

For some purposes, it is useful also to record the deviation from the average conditions, for example, on long-term sunspot counts (relative numbers), since here the deviations from a series are more important than its absolute quality. Using the following scale, turbulence, sharpness, transparency, and other factors affecting the image quality are considered. "G" (for good) refers to the average conditions at a site; note that "G" for one observer who works at a particular site may correspond to a "P" (for poor) for another who works at a more favorable site.

Scale for Observing Conditions (Seeing)	
Quality Q	
Excellent	Reserved for days when details are observed unusually clearly.
Good	Average visibility of solar surface details corresponding to the conditions for that observer.
Fair	Substandard seeing but not yet considerable interference.
Poor	Considerable interference strongly limiting the value of the observation.
Worthless	Conditions so poor that evaluation of the observation is not useful.

13.1.2.3 Records Proper recording of the solar phenomena and circumstances of their observation (e.g., date, time, instrument, observer, location, conditions) is mandatory for analyzing and evaluating the data later on. For ambitious observers it is not sufficient just to rely on the data stored with digital photographs. Single recording sheets have been proved in solar observations and can be easily archived, evaluated, and copied. An example for a general observing form is shown in Fig. 2. Special observations are recorded using other forms.

Record forms for solar observations: www.springer.com/978-3-540-76377-2 and http://www.vds-sonne.de/index_e.html

13.1.3 Sunspots

13.1.3.1 Morphology and Development of Sunspots Sunspots are extended, strongly magnetic regions on the Sun which, owing to their reduced temperatures (3000–3500 K) radiate less visible light than the undisturbed photosphere (5800 K). The sizes of spots range from 2000 to over 100,000 km. (An angle of $1''$ corresponds to a length of about 700 km on the Sun; $1'$ is about 43,000 km.) Spots with angular diameters under $10''$ are called *pores*. Larger spots usually consist of an *umbra* and *penumbra* (Figs. 3 and 4). A detailed introduction to the various phenomena within sunspots can be found in the textbook by Bray and Loughhead [6] and in the overview article by Wilson [7]. An excellent review on sunspots including state-of-the-art research results has been written by Solanki [8].

The *umbra* is the dark core of a spot. In direct viewing (e.g., using an objective filter) with low turbulence, brightness differences can be seen in the umbra. Larger telescopes show bright spots with a diameter of only about 500 km but almost as bright as the photosphere. Diffraction in the Earth's atmosphere and at the objective makes them appear larger than they really are. Their lifetime is 15 to 30 minutes.

The *penumbra* surrounds the umbra, and consists of bright and dark filaments connected radially to the umbra. The width of filaments is about $0.3''$ or about 200 km, their duration about 2 hours. Changes and divisions can be observed within just a few minutes.

Frequently, umbrae or penumbrae are divided by bright areas resembling photospheric conditions. These so-called *light bridges* are further discussed in Sect. 13.1.3.3. The

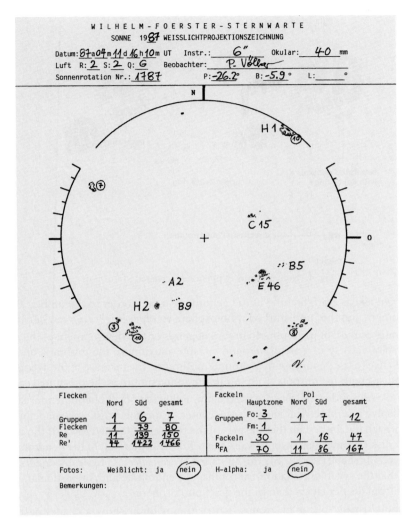

WILHELM-FOERSTER-STERNWARTE
SONNE 19**87** WEISSLICHTPROJEKTIONSZEICHNUNG

Datum: **87** a **04** m **11** d **16** h **10** m UT   Instr.: **6"**   Okular: **40** mm
Luft R: **2** S: **2** Q: **6**   Beobachter: **P. Völler**
Sonnenrotation Nr.: **1787**   P: **-26.2°**   B: **-5.9°**   L: _____ °

Flecken	Nord	Süd	gesamt
Gruppen	1	6	7
Flecken	1	79	80
Re	11	139	150
Re'	11	1422	1466

Fackeln	Hauptzone	Pol Nord	Süd	gesamt
Gruppen Fo: **3** Fm: **1**		1	7	12
Fackeln	30	1	16	47
R_FA	70	11	86	167

Fotos: Weißlicht: ja (nein)   H-alpha: ja   (nein)

Bemerkungen:

Fig. 2. Example record form for observation of the Sun. It contains all relevant data for amateurs and also permits the detection of changes in the graphed sunspot groups over the course of time

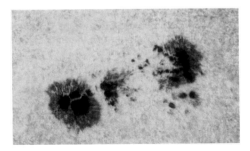

Fig. 3. Sunspot group on April 27, 1986, photographed by C.H. Jahn using a 200/3250 mm refractor with 12.5-mm eyepiece, solar prism, and green filter

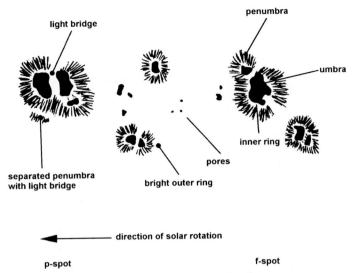

Fig. 4. Nomenclature of phenomena related to sunspots

inner bright ring is a brightening of the penumbra at the border to the umbra. Between the penumbra and the "undisturbed" photosphere an *outer bright ring* can be observed.

The ratio of radii of penumbra to umbra depends on the development of the sunspot and on solar activity in general. Around sunspot maximum, the umbra is on average larger relative to the penumbra than around minimum. The increase of the average magnetic field strength in sunspots from minimum to maximum is the presumed cause.

In 1769, Alexander Wilson discovered the following solar phenomenon: sunspots which display a symmetrical shape at the center of the solar disk appear deformed as they approach the limb, such that the part of the penumbra located towards the center of the solar disk becomes narrower or vanishes, while the other half of the penumbra more or less keeps its shape (see Fig. 5). This so-called *Wilson effect* is physically understood by the fact that the (visible) radiation from sunspot penumbrae and, in particular, umbrae emerges from deeper layers than in the quiet photosphere.

Sunspots usually occur in groups. Large groups often show a clustering of spots around two primary spots, which are termed *bipolar*. The western part of a bipolar group, i.e., the one preceding in the direction of solar rotation, is called *p-spot* (preceding) and the eastern part *f-spot* (following).

The occurrence of a spot group is linked to other phenomena tied to solar activity such as *faculae*, *prominences*, and *flares*, all of which are caused by magnetic fields. The development of these phenomena is of different speeds and also different in each spot region (see, e.g., Bray and Loughhead [6], Wilson [7], and McIntosh [9] for more detailed information). While the sunspot group in the example on www.springer.com/978-3-540-76377-2 has a lifetime of only 60 days, the magnetic field, which is the cause of solar activity, can be traced for about 250 days.

A bundle of magnetic field lines from deeper zones reaches into the photosphere, and there expands (owing to the lesser pressure) into the form of an arch. The two points of penetration in the photosphere mark the two magnetic poles of the spot group. The

Example for the development of an activity region on www.springer.com/978-3-540-76377-2.

Fig. 5. The Wilson effect in a sunspot and photospheric faculae, photographed on June 4, 1985 by C.H. Jahn using a 200/3250 mm refractor with 12.5-mm eyepiece, solar prism, and green filter

details of the developments of spots, such as the formation of sharp boundaries between umbra, penumbra, and photosphere have not yet been clarified. Also, the origin of solar magnetic fields is still much under discussion (Giovanelli [10]).

Sunspots are short-lived phenomena. Ninety percent of all groups disappear after 10 days or less, 50% after only 2 days. The most stable groups, however, have lifetimes of several months. The average lifetime of a group, about 10 days, varies periodically with the *long cycle* (see Sect. 13.1.3.5).

Only in about 10% of all groups does the eastern spot live longer, while in 40% the western spot dominates. Also, the area of the western spot averages 40% larger than that of the eastern spot.

As a bipolar spot group develops, the angle between the line joining the main spots (the axis) and the latitude circles shows a typical change: with increasing spot area, the angle decreases, reaches a minimum at the time of maximal development, and then increases as the number of spots in the group diminishes.

13.1.3.2 Classification of Sunspots The typical evolution of a large group led Waldmeier in 1947 to introduce the following sunspot classes (Fig. 6).

Waldmeier Classification Scheme for Sunspot Groups

A   Single spot or group of spots without penumbrae and without bipolar structure.

B   Spot group without penumbrae in bipolar structure.

C   Bipolar group, one of the two main spots has a penumbra.

D   Bipolar group, both main spots having penumbrae, and at least one spot shows some structure. Maximum extension of the group less than 10° on the Sun (about 1.2×10^5 km).

E   Large bipolar group, the main spots are surrounded by penumbrae and have a complex structure. Numerous smaller spots between the main spots. Extension of the group at least 10°.

F   Very large bipolar or complex spot group, extension at least 15° (or 1.8×10^5 km).

G   Large bipolar group without small spots between main spots. Extension at least 10°.

H   Unipolar spot with penumbra, diameter greater than 2.5° (about 3.0×10^4 km).

J   Unipolar spot with penumbra, diameter under 2.5°.

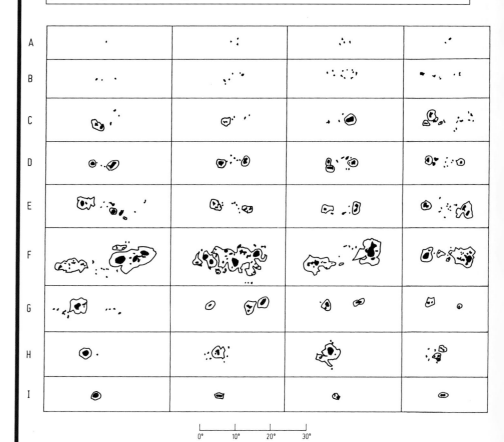

Fig. 6. Waldmeier classification scheme for sunspot groups

Only about 2% of the spot groups pass through all Waldmeier classes. Most groups reach maximum extension in one of the early classes and then regress through classes C, G, or J. The largest number of sunspots is usually reached by the end of the first third of the lifetime of the group. This time asymmetry in the development increases with the maximum class of development, in surprising analogy with one of the Waldmeier rules of sunspot cycles (see Sect. 13.1.3.5).

A useful task for observers is to count the frequencies of spot groups in the separate Waldmeier classes, and to note the change of that distribution during the course of an activity cycle.

For complex structured groups, the Waldmeier scheme does not suffice for classification. The specialists have thus modified and extended it, so that it is now known under the name *McIntosh classification* [9] (see Fig. 7).

McIntosh Classification Scheme for Sunspot Groups

The *first letter* (upper case) corresponds to the Waldmeier class but with classes G and J omitted. The class G, depending on extension, is counted among classes E or F; class J merges with H, which then contains all unipolar groups with penumbrae. If a magnetogram is not used, unipolar groups are characterized by the largest distance between two spots (or between edge of the penumbra of the main spot and any other spot of the group) not exceeding 3°. When new large spots appear in the vicinity of an H-spot, this generally indicates the formation of a new bipolar group which should be treated as a group in its own right. When the diameter exceeds 5°, it can be considered certain that both magnetic polarities exist within the penumbra (bipolar group), and the group is to be classified as Dkc, Ekc, or Fkc.

The *second letter* (lower case) in the McIntosh scheme expresses the appearance of the penumbra of the largest spot in the group:

x   No penumbra (corresponds to Waldmeier class A or B).
r   Rudimentary penumbra partially surrounds the largest spot. This penumbra is incomplete, granular rather than filamentary, brighter than a mature penumbra, and extends as little as 3 arcseconds (2200 km) from the spot umbra. Rudimentary penumbrae may be either in a stage of formation or dissolution.
s   Small, symmetric (corresponds to Waldmeier class J). The largest spot has a mature, dark filamentary penumbra of circular or elliptical shape with little irregularity to the border. There is either a single umbra, or a compact cluster of umbrae, mimicking the symmetry of the penumbra. The north-south diameter across the penumbra is $\leq 2.5°$.
a   Small, asymmetric. Penumbra of the largest spot is irregular in outline and the multiple umbrae within it are separated. The north-south diameter of the penumbra is $\leq 2.5°$.
h   Large symmetric (corresponds to Waldmeier class H). Same structure as type "s," but the north-south diameter of the penumbra is $> 2.5°$. The area, therefore, must be ≥ 250 millionths of the solar hemisphere.

Fig. 7 McIntosh classification scheme for sunspot groups

k   Large, asymmetric. Same structure as type "a," but with north-south diameter >2.5°, and surface area ≧ 250 millionths of the solar hemisphere. This type sometimes contains spots of opposite polarity, and may indicate potential for proton flares.

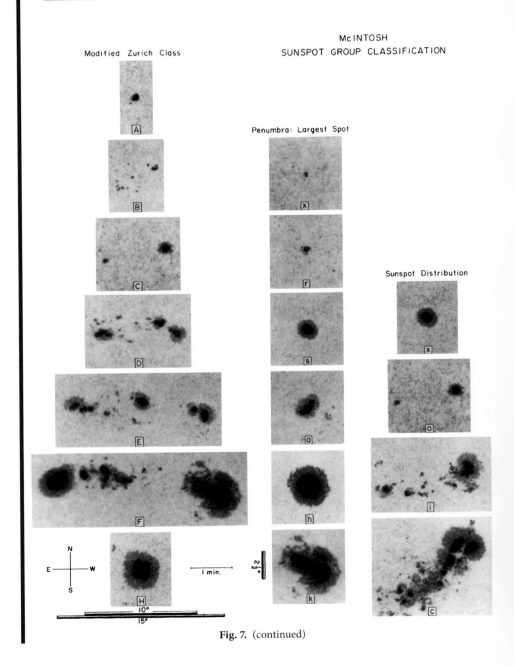

Fig. 7. (continued)

The *third letter* (lower case) distinguishes the distribution of spots within a group:

x   Undefined for unipolar groups (classes A and H).
o   Open. Few, if any, spots between the leader and follower. Interior spots of very small size. Classes E and F of the *open* category are equivalent to class G.
i   Intermediate. Numerous spots lie between the leading and following portions of the group, but none of them possesses a mature penumbra.
c   Compact. The area between leading and following ends of the spot group is populated with many strong spots, with at least one interior spot possessing a mature penumbra. In the extreme case of a compact distribution, the entire spot group is enveloped in one continuous penumbral area.

Fig. 7 (continued)

The second and third letters of the McIntosh code contain information on the magnetic field structure in the group. Statistical data using this classification thus gain in significance. The McIntosh scheme is also recommended for use by amateur observers.

Another relevant quantity which describes the evolutionary state of a sunspot group is the *area* A_i occupied by the spots. It is expressed in 10^{-6} of the visible hemisphere of the Sun (MH), considering the geometrical foreshortening in the vicinity of the limb:

$$A_i = 10^6 \times \frac{A'_i}{2\pi R^2 \cos \rho_i} , \tag{1}$$

where A'_i is the area of the spot group in the solar image, R is the radius of the solar image, and ρ_i the angular distance from the center of the image, given in good approximation by:

$$\rho_i = \arcsin(r_i/R), \tag{2}$$

where r_i is the linear distance of the center of the spot group from the central point of the image. The *area number A* is the sum over the A_i of all groups. On the average, the number A is about 17 times larger than the *Wolf number R* (see Sect. 13.1.3.4) [11].

The area of a spot is intimately related to the magnetic field strength, and thus the development of the spot area reflects the evolution of the field. There is a direct connection between maximum area of the group and its lifetime. The faster the area of the group grows in the early development, the higher the maximum will be; the same also applies for the *number* of spots in the group. After maximum development, the number of spots drops rapidly, but the area is determined by the long-lived western main spot, which decreases rather slowly. Also, the magnetic field diminishes only slowly.

Spot areas can be measured most expediently by graphing the solar image precisely on a firmly mounted and shaded projection screen, and then reading the areas with transparent millimeter paper. A stable mount, precise tracking, and a moderate-size telescope are required for good area measurements. The accuracy can be improved using photography. On digital images sunspot areas can be determined with image processing programs, like, e.g., ImageJ. Solar photographs of adequate quality may be obtained at the primary focus of a 2-in. (5-cm) refractor with a good solar filter (see Chap. 6).

Image processing and analysis in Java: http://rsb.info.nih.gov/ij/index.html

Every feature of a spot group (class, inclination of axis, spot number, area, size and brightness of the associated faculae, number and type of light bridges, frequency of flares, etc.) can be graphed against time as a curve of development characterized by the following parameters:

(a) The time interval (lifetime of the spot group)
(b) The height of maximum
(c) The area under the curve
(d) Mean slopes of ascending and descending branches

Statistical processing should start with these parameters. Long-term series of observations permit the study of the following questions:

(a) Do the parameters of the development curves change within the 11-year activity cycle? If so, which ones?
(b) What connections exist between parameters of the mean development curve of a feature?
(c) What connections exist between the mean development curves of different features?

During the lifetime of a large group of spots, rapid changes in their appearance occur. Spots appear, others vanish, new penumbrae form, light bridges materialize, and spots change relative positions. These changes can be observed within hours or days. Small spots sometimes show evolution within minutes.

As the eye is subject to illusions, photographs of rapid changes are desirable (see Chap. 6). Caution is also advised here because atmospheric turbulence may feign structural changes. The best material from which to study rapid developments are series of photographs taken at intervals of 10 minutes to 1 hour—the ideal observing program for weekend observers!

13.1.3.3 Light Bridges Nearly all sunspots contain bright areas resembling the photosphere. Since they extend usually as narrow tongues into the spots, they are referred to as *light bridges*. Their formation is characterized, first, by the appearance of indentations on opposite sides of the spot. These indentations develop by moving toward each other until connecting to form a bridge. The spot appears divided at this point (Fig. 3). Apart from these "classical" light bridges, bright islands are noted particularly in older, stable sunspots (classes H and J; see Fig. 6), which, with time, may break out into the photosphere.

Light bridges are closely connected with the general development of sunspot groups, which can be only incompletely understood without observing the bridges. Since 1978, the classification by Hilbrecht [1, 2] has been applied, extended in Fig. 8 by the type "n" introduced by Seebörger-Weichselbaum.

When studying the light bridge development with this encoding, it is found that the early and late phases of spots (classes C, D, H, and J) show, per area and number of spots, more light bridges than active groups do (classes E, F, and G). In active groups bridges penetrate from the photosphere far into the spot and evolve rapidly, but otherwise mostly slow changes are observed. In H and J groups, in contrast to C and D groups, "classical" light bridges are less pronounced than islands, as the bridges delineate primarily irregularities in spot contours.

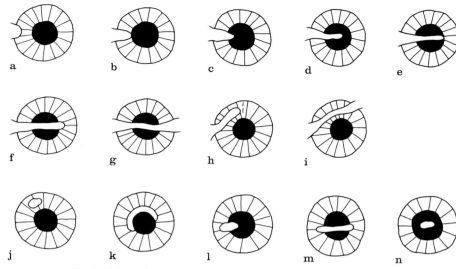

Fig. 8. Scheme for assigning types to light bridges. After Hilbrecht [1, 2]

From long-term series of observations, amateur observers can obtain and process valuable data on light bridges. Since 1977, the VdS Working Group on the Sun has collected amateur observations and introduced observers to the subject. A description (with references) of observing programs with results from professional observatories was published by Hilbrecht [1, 2].

13.1.3.4 Numerical Expression of Spot Activity The degree of sunspot activity is generally described by a single quantity which has been internationally accepted: the *Wolf sunspot number*. It was introduced by R. Wolf in 1848 and is the basis of many investigations of solar-terrestrial relations. Determining Wolf numbers is a favorite amateur observing program as it can easily be performed using small telescopes.

The Wolf sunspot number includes the number of observed single spots s as well as that of entire spot groups g, as the appearance of spots in groups is an important feature of activity. Wolf multiplied the number g of groups with a factor 10 in order to express the fact that the appearance of a new group is weighted 10 times higher than a new spot within an existing group. An isolated spot is regarded as a separate group. The Wolf number R is combined from spot and group numbers according to the relation

$$R = 10g + s .$$
(3)

Example: What is the Wolf sunspot number R from the observing data in Fig. 2? The following groups are observed: an A group with 2 spots, a B group with 5 spots, a B group with 9 spots, a C group with 15 spots, an E group with 46 spots, an H group with 1 spots, and an H group with 2 spots.

Thus, there are 7 groups with a total of 80 spots. The Wolf number is therefore

$$R = 10 \times 7 + 80 = 150 .$$

The ordering of spots into groups may be difficult during times of high activity when several groups are closely adjacent or show complex structures not immediately fitting into Waldmeier's classification scheme (Sect. 13.1.3.2). In this case, the following rules from Künzel [12] may be of help:

1. Spots within an area covering up to $5° \times 5°$ are counted as *one* group if no bipolar structure is noted. Bipolar groups, however, may extend to lengths of $20°$ or more.
2. Single spots separated in heliographic longitude by up to $15°$ are counted as one group when they are the remnants of a larger, formerly contiguous group.
3. A bipolar clustering of spots is considered one group when its western part has the same or a lesser latitude than the eastern part. The average inclination of the axis is $1°$ to $2°$ at latitudes $\pm 10°$, and about $4°$ at latitudes $\pm 30°$.
4. An isolated spot is counted as one group.

There may also be problems in counting the separate spots. Since the development is a smooth transition from granulation-free photosphere to pore to spot, the exact moment of the birth of a spot cannot be given. Under very good atmospheric conditions, the Sun's surface appears bespattered with small A groups, but these are mostly enlargements of the intergranular area with a lifetime of some minutes. These should *not* be counted in the Wolf number. For the statistics of numbers, Waldmeier defines a sunspot as having a diameter of at least $3''$ and a minimum lifetime of 30 minutes.

Very large sunspots often show several umbrae in a joint penumbra. Here, every umbra is counted as a spot when completely separated from other umbrae. Thickening of penumbral filaments are not counted as spots. Numerous circumstances of observation influence whether or not spots are visible and recorded, and thus the Wolf number:

- Atmospheric conditions (air turbulence, sharpness, wind, clouds, haze, etc.)
- Instrument and observing methods (aperture, focal length, optical quality, magnification, filter, mounting and its stability)
- Observer (vision quality, care in observing, experience, physical and psychological conditions)

In order to intercompare the daily Wolf sunspot numbers obtained under various conditions, all values must be reduced to a standard scale. The observations made for over a century at the Zürich Observatory using its Fraunhofer refractor with an aperture 8 cm, focal length 110 cm, and magnification 64× were used as international standards until the end of 1980. In 1981, the collection, evaluation, and publication of international Wolf numbers R_{int} was transferred to the Solar Influences Data Analysis Center (SIDC) in Brussels, Belgium, and since then these values have been used as the standards; they are published, for instance, in *Sky and Telescope* and other magazines.

The reduction factor k must be found for each observer and each instrument from extended simultaneous observations in order to convert the individual Wolf numbers to the international scale. This factor k is found from the ratio of average Wolf numbers over a certain period of time (at least one year) between the standards and the observed series:

$$k = \overline{R_{int}}/\overline{R} \ . \tag{4}$$

Solar Influences Data Analysis Center (SIDC):
http://sidc.oma.be/

Of course, the averages $\overline{R_{\text{int}}}$ and $\overline{R}$ are formed only over days covered by simultaneous observations.

The value of the k factor says *nothing* about the quality of a series, but the scatter of k values from day to day or from year to year indicates the long-term homogeneity of observed series. Every observer should try, after an initial trial period, to maintain the observing method once chosen and to keep the observing conditions as constant as possible. Once the k factor of an observer is determined, his/her observations are reduced to the international scale by the relation:

$$R_{\text{int}} = k(10g + s) = kR .\tag{5}$$

The Wolf number is a simply determined measure of activity, but in practice it has its disadvantages; for instance, the size of individual spots is not considered. A tiny A-spot contributes as much to the number as a giant H-spot of diameter 50,000 km.

Several measures of solar activity deviating from Wolf's method have been suggested to remedy these drawbacks, and are mentioned in the following discussion.

A physically meaningful expression of solar activity as observed in white light should be closely connected with magnetic field strength on the Sun. This could be expressed by the size of a sunspot. As for stable spots, the maximum field strength at the center increases with the area of the spot. Measuring areas, however, is much more demanding than simply recording Wolf numbers.

The new Wolf number (or, more correctly, new area number) R' after Beck [13] purports to combine the advantages of the area number with the simple counting of the Wolf number. It uses the knowledge that for each group type of the Waldmeier classification there is an average ratio of spot area to spot number. When the observer determines the spot number and the Waldmeier class of the ith group, he/she can estimate the area of the group by multiplying the spot number s_i with a weight factor G_i representing the average area/number ratio for the group type. The new area number R' is obtained by adding the area numbers of all groups i:

$$R' = G_1 s_1 + G_2 s_2 + G_3 s_3 + \ldots\tag{6}$$

The following weight factors G_i for the Waldmeier classes are used:

Waldmeier class:	A	B	C	D	E	F	G	H	J
Weight factor G_1:	4	4	8	18	25	36	50	44	37

The highest weights occur in the late classes G, H, and J, with a much-diminished number of spots but still a large area covered by the main spots.

Example: What is the new area number R' for the observation in Fig. 2?

Calculation yields

$$R' = 4 \times 2 + 4 \times 5 + 4 \times 9 + 8 \times 15 + 25 \times 46 + 44 \times 1 + 44 \cdot 2 = 1466 .$$

The new area number R' has been determined by amateur astronomers since early 1977 and has been compared with Wolf sunspot numbers and with area numbers (Fig. 9).

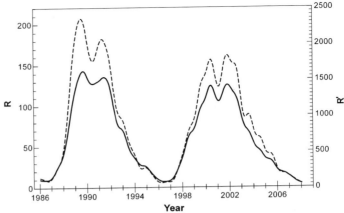

Fig. 9. Comparison of Wolf spot number R (*solid line*) and new area number R' (*dashed line*). After observations by the Working Group *Sonne* made during the period 1986–2008. All data has been smoothed using the P17 method (graphic courtesy of A. Bulling)

Another sunspot number, one which considers the formation of penumbrae as signs of high solar activity, was suggested by Pettis [14]:

$$SN = 10p + s \,, \tag{7}$$

where p is the number of spots with penumbrae and s that of spots without penumbrae. A penumbra with more than one umbra within it is counted as $p = 1$.

Considerations similar to those leading to the definition of the Beck area number have also led the Norwegian Astronomical Society (NAS) to introduce *classification values* (CV) [15]. The CV numbers give larger and more complex groups higher weight than small spots. The system is based on the McIntosh spot classification (see p. 317) which admits 60 possible classes. These are arranged in a sequence giving the highest weight (60) to an F group with large, symmetric penumbrae and with complex spot distribution. A single A-spot is assigned a CV weight of 1. The CV values of all groups are added and thus form the measure of spot activity. Notwithstanding the somewhat arbitrary definition of weights for the spot classes, the CV system has the advantage that spot counts and area measurements may be dispensed with; it needs merely the classification of all groups in the McIntosh scheme.

The most readily determined spot value is the number A of spots visible to the unaided eye [16]. (This observation requires eye protection.) It permits a connection between present observations and records from pre-telescopic times. Keller [17] has performed naked-eye sunspot observations from the Zürich Observatory for over 10 years, and claims good agreement with the Wolf numbers.

All of the aforementioned measures of solar activity have in common that they are subject to strong external influences (e.g., observing conditions and selection effects) and random effects (e.g., the invisibility of the far side of the Sun), and hence their daily values are essentially meaningless. It is only from the *average* of daily numbers over one *apparent* rotation period of 27 days that a fairly dependable description of solar activity can be obtained.

A single observer cannot determine the spot activity without experiencing gaps due to weather and personal commitments. Thus, in order to achieve continuous supervision of solar activity and follow any long-term patterns over years and decades, it is advisable to cooperate with several observers distributed in locations as widely separated as possible. Such observing networks exist in several countries; the oldest amateur group is the solar division within the American Association of Variable Star Observers (AAVSO), which publishes their results monthly in the *Solar Bulletin* and in *Sky and Telescope*. In Germany, a network of observers of Wolf numbers was founded in 1977, and today counts over 130 contributers worldwide and is assisted by the Solar Section of the Vereinigung der Sternfreunde. Results and lists of participants are published online and in the circular *Sonne*. Other observer groups report the new area number R', the Pettis number SN, and the number A of spots observed with the naked eye.

A sample list to record observations of Wolf numbers is available on www.springer. com/978-3-540-76377-2. Every observation should include, in addition to the number of groups G, the number of spots s, the Wolf number R, the new area number R', the time (UT), the image motion M, the sharpness S, and the quality Q of the solar image (Sect. 13.1.2.2). Division of activity into the northern and southern hemispheres of the Sun (g_N, s_N, g_S, s_S) is useful in conjunction with position determinations of spots. The heading includes month and year as well as data on observer and instrument.

Solar Section of the Vereinigung der Sternfreunde (VdS): http://www.vds-sonne.de/index_e.html

13.1.3.5 Sunspot Cycles The face of the Sun changes from day to day, as individual sunspots appear and disappear unpredictably. Long-term studies of sunspots, however, reveal regular patterns, the most obvious being the cyclic variation of solar activity—noted as early as 1843 by the amateur astronomer Heinrich Schwabe—in which the appearance of sunspots varies with a mean period of 11.1 years. This *11-year sunspot cycle* can be traced back, by means of ancient naked-eye observations, to pre-Christian times. Since the beginning of regular telescopic observations in 1749, the sunspot cycles are consecutively numbered. Figure 10 shows the annual averages of Wolf numbers over the cycles 0 to 23. The current cycle is number 24, which began in 2008/09.

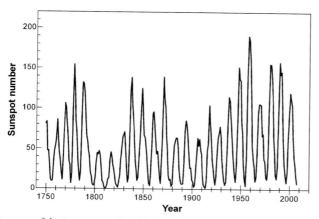

Fig. 10. Annual means of the international Wolf relative numbers from 1749 to 2008 (cycles 0–23) provided by the Solar Influences Data Analysis Center (SIDC) (graphic courtesy of A. Bulling)

Underlying the phenomenon of solar activity is a *magnetic cycle* during which the magnetic polarity of the sunspot groups is reversed; its mean period of 22 years comprises exactly two visible spot cycles.

Individual sunspot cycles vary between 9.0 and 13.6 years in length, and the height of the maxima can also be quite different. The lowest maximum thus far was recorded for the sixth cycle, with a smoothed monthly mean of 48.7, while the nineteenth cycle reached a maximum height of 201.3. Often no spots can be found on the Sun for weeks during minimum, while at maximum often 10 to 20 spot groups may be seen simultaneously. The steeper the rise from minimum to maximum, the higher the maximum becomes. The ascent takes on the average 4.4 years, while the descent to the next minimum averages 6.5 years in duration. Only the very shallow low maxima display any symmetry with respect to time. These rules regarding the pattern of sunspot cycles were formulated by Waldmeier [18] and can be used to predict solar activity. Bendel and Staps [19] revised the coefficients of the Waldmeier rules by including the cycles numbered 19 through 21, and by applying a new smoothing method (*P17 averages*; see below).

In addition to the 11-year cycle, a longer spot cycle of about 80 years, as inferred from the heights of the maxima, is suspected (Gleissberg [20]). However, the limited number of the long cycles observed so far is not sufficient to conclusively prove its existence or to determine its duration. A reconstruction of solar activity from geological records over the past 11 000 years indicates that sunspot numbers have been unusually high during the past 70 years (Solanki et al. 2004 [21]).

The distribution of solar activity between northern and southern hemispheres also shows asymmetries which are potentially connected with the Gleissberg cycle. Observers able to determine the positions of spots should therefore record spots for the two hemispheres separately.

The monthly averages of Wolf numbers scatter too much to permit fixing the exact epochs of sunspot minima and maxima. The curves need to be smoothed to find the instances of time where solar activity was lowest and highest. Smoothing is usually performed by the calculation of 13-month means (*A13 averages*): The means of 13 consecutive months are added, counting the months 2 through 12 twice, and then the sum is divided by 24. This yields the smoothed mean for the seventh month included. The procedure is continued for the next month by shifting the 13 averages included by one.

The calculated epochs and heights of maxima and minima of cycles 0 to 23 are given on www.springer.com/978-3-540-76377-2.

One serious drawback of this method is that the number 13 (months) is not an integer multiple of the solar rotation. It is therefore preferable to average over an interval of either 9 months (10 rotations) or 17 months (19 rotations). The discontinuous weighting function of the 13 months is also a disadvantage. An alternative method which is better suited to obtain a smooth curve and yet includes fairly short-period variations is to use smoothed weighting functions, which assign the highest weight to the mean month of the period to be smoothed and continuously decreasing weight to the adjacent months. The *P17 averaging* proposed by Karkoschka [22], where the monthly averages of Wolf numbers over a 17-month interval are included and are weighted with a polynomial of the form $(1 - x^2)^3$, has proven advantageous. Figure 11 compares the smoothing procedures A13 and P17 for monthly means of cycles 21 to 23.

Example: Calculate smoothed monthly means from the table below according to the "A13" and "P17" prescriptions.

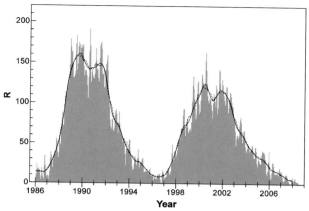

Fig. 11. Comparison of monthly means of the international Wolf numbers during cycles 21–23, smoothed by different methods. *Vertical lines*: unsmoothed monthly means. *Solid curve*: P17 averages. *Dashed curve*: A13 averages. It is seen that the P17 values follow the run of solar activity more accurately than do those of A13 (graphic courtesy of A. Bulling)

Observed SIDC monthly means for 1985–1986

1985		1985		1986		1986	
Jan:	16.5	Jul:	30.8	Jan:	2.3	Jul:	17.8
Feb:	16.1	Aug:	10.4	Feb:	23.6	Aug:	7.4
Mar:	11.9	Sep:	3.9	Mar:	15.7	Sep:	3.9
Apr:	16.1	Oct:	18.5	Apr:	20.4	Oct:	35.7
May:	27.4	Nov:	16.6	May:	13.1	Nov:	14.7
Jun:	24.2	Dec:	17.2	Jun:	0.8	Dec:	6.4

The A13 mean for December 1985 is calculated from the averages from June 1985 to June 1986, counting the 11 inner numbers double:

$$\overline{R_{A13}}(\text{Dec. 1985}) = [24.2 + 2 \times (30.8 + 10.4 + 3.9 + 18.5 + 16.6 + 17.2$$
$$+ 2.3 + 23.6 + 15.7 + 20.4 + 13.1) + 0.8]/24$$
$$= 15.4 \ .$$

The P17 mean for December 1985 is computed from the months April 1985 through August 1986 with the aid of the weighting factors in Table 1.

$$\overline{R_{P17}}(\text{Dec. 1985}) = [0.009 \times 16.1 + 0.062 \times 27.4 + 0.171 \times 24.2$$
$$+ 0.963 \times 16.6 + 1.0 \times 17.2 + 0.963 \times 2.3$$
$$+ 0.009 \times 7.4]/8.226$$
$$= 15.0 \ .$$

The "P17" average calculated for the month $M = 0$ includes the mean for that month and also those of the 8 preceding ($M = -8 \ldots -1$) and the 8 following months ($M = +1 \ldots +8$)

Table 1. Weighting factors for the P17 means, calculated from the formula $[1 - (M/9)^2]^3$

M	Coeff.	M	Coeff.	M	Coeff.
±8	0.009	±5	0.330	±2	0.859
±7	0.062	±4	0.517	±1	0.963
±6	0.171	±3	0.702	±0	1.000

with the tabulated weights. The weighted monthly means added are divided by the sum of the coefficients (= 8.226).

The determination of minimum and maximum epochs will depend on the averaging procedure used. The difference between methods is maximally half a year.

13.1.4 Photospheric Faculae

13.1.4.1 Appearance Faculae are almost always visible on the solar disk. They are distributed over the entire surface, although in white light they are generally visible only near the solar limb due to their low contrast. There they appear brighter than the surrounding upper photospheric layers and occur in various shapes and combinations: as areas, veins, or networks, contiguous, distributed, or point-like (Fig. 12).

Photospheric faculae are activity regions in which sunspots can also appear. Therefore, experienced observers have a close look at faculae areas. Faculae are the first signs of magnetic activity. They appear prior to spots and outlive them by several weeks and months. On average they exist three times longer than the associated sunspot group. The average lifetime of photospheric faculae is 90 days. After a certain age faculae area are frayed out due to convection and lose their original shape. Its magnetic flux bundles are concentrated between the honeycomb-shaped cells of supergranulation making this large area network visible. Subsequently, all fibers are distributed over an increasing area until the faculae region finally becomes invisible.

Using monochromatic filters, such as in the hydrogen Hα line or in the calcium H and K lines, faculae are observable on the entire Sun. Such filters show the continuation of faculae into layers above the photosphere, the *chromospheric faculae* (Figs. 13 and 14),

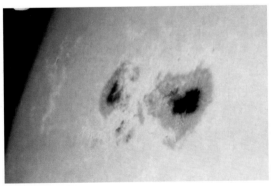

Fig. 12. Photospheric faculae and sunspots near the solar limb photographed by Michael Delfs

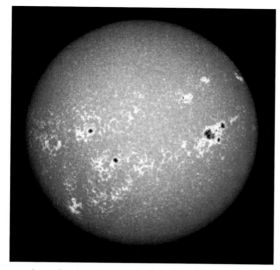

Fig. 13. Chromospheric faculae photographed in the calcium K line by Günther Appelt

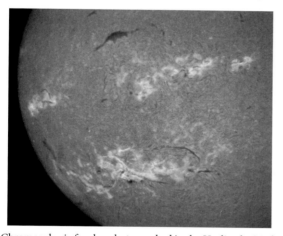

Fig. 14. Chromospheric faculae photographed in the Hα line by Michael Delfs

sometimes also referred to by the French term "plages." Due to the widening of the magnetic field lines they appear larger than in the photosphere. The temperature of the faculae depends on the height above the solar surface. Maximum values (about 1000 K above photospheric temperature) are reached in between the cells of the granulation. In the upper and cooler photospheric layers visible near the solar limb, the temperature in faculae is several hundred Kelvin higher and the brightness about 10% higher than that of the quiet, undisturbed photosphere.

13.1.4.2 Observation of Photospheric Faculae Any telescope or binocular with an aperture of at least 30 mm is suitable for observing photospheric faculae (provided appropriate means for safe solar viewing are followed). Measuring faculae positions and

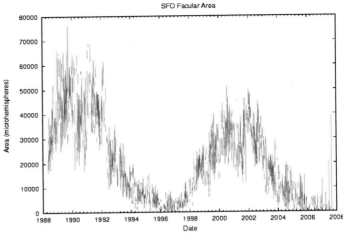

Fig. 15. The total area of all faculae on the Ca II K line image, in millionths of the solar hemisphere, from 1988 until the end of 2007. The solar activity cycle is clearly visible. Courtesy of San Fernando Observatory, California State University, Northridge, CA

areas reliably or drawing faculae using the projection method requires a stable telescope and projection system. Using an automatic tracking system facilitates observing significantly and further enhances the accuracy.

A large variety of observing programs for faculae are within reach of amateur astronomers. Upon planning and performing of observations and their evaluation it should, however, always be taken into account that in most cases each particular faculae area is visible only three to four days while it is not too far away from the solar limb. Only occasionally and under excellent seeing conditions are bright and high contrast faculae of the upper photosphere visible on the entire solar disk.

San Fernando Observatory (SFO): http://www.csun.edu/sfo/index.html

Professional observations of faculae are only performed for area measurements by the San Fernando Observatory in California.

For amateurs, several areas of activities are open: counting area-shaped and point-shaped faculae, measuring faculae areas and positions (Fig. 15).

For each of these and also for other observations it should be kept in mind that seeing variations may be non-negligible. It is important to evaluate seeing conditions critically according to the values of motion (M), sharpness (S), and observing quality (Q) (see Sect. 13.1.2.2).

13.1.4.3 Facula Activity As the visibility of photospheric faculae is restricted to near-limb regions facula activity may, e.g., be presumed to be low if facula areas are only located in the central part of the disk. Therefore, only mean values over sufficiently long time intervals (e.g., one solar rotation, one month, or longer) can provide reliable information on facula activity. Daily numbers are not sufficient for this.

Facula activity can be measured by counting the areas containing faculae. For this, during daily observations the numbers of facula areas with spot(s) Fw and facula areas without spot(s) Fwo are determined. Similar to counting sunspots a single isolated facula is counted as one area. Several faculae within a spacial neighborhood are connected to one area.

The combination of Fw and Fwo leads to the daily sum of facula areas F:

$$F = Fw + Fwo .$$

According to evaluations of observations by Delfs F is clearly related to the sunspot number R. As expected, the main reason for this is the relation of Fw to the occurrence of spots. Fwo, on the other hand, varies less than Fw with the sunspot cycle but dominates during sunspot minima when Fw frequently is zero (see Fig. 16).

A second method to determine facula activity is counting individual area-shaped faculae (ASF) (latitude measurement see 13.1.7).

Usually, ASF appear only in the main zone of solar activity at heliographic latitudes $|b| < 50°$ (for latitude measurements see Sect. 13.1.7). Single point-shaped faculae occur all over the Sun. At heliographic latitudes above $|b| \geq 50°$ faculae are counted separately and are called *polar faculae*. Single point-shaped faculae are primarily formed during the decay of area-shaped faculae but may also occur far away from active regions. In the main zone single point-shaped faculae are added to ASF.

Separated faculae are counted individually as are directly connected faculae.

Counting may, however, be problematic. Even experienced observers with large telescopes may have difficulties with this. Especially with bad seeing the separation of faculae is barely recognizable. With smaller instruments, only larger faculae are observable. It is recommended to note the number of faculae for each area. Count at least twice and take the average value to improve reliability! At the end the sum of all individual faculae is calculated.

By counting facula numbers over the time the coarse of the activity cycle of faculae becomes apparent. The ASF reflect the cycle. During the minimum mainly small or sometimes even no ASF occur. With the onset of increasing solar activity, also the number of ASF increases clearly (see Fig. 17). Reports on faculae counts are collected and analyzed by the facula network SONNE.

Facula network SONNE:
http://www.vds-sonne.de/
index_e.html

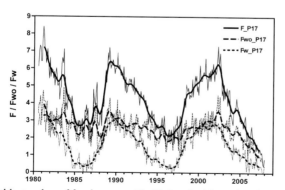

Fig. 16. Mean monthly number of facula areas with (Fw) and without (Fwo) spots and total number of facula areas (F) derived from observations of the facula network of the working group *Sonne* during the years 1980 to 2007. The smoothed course of the indices using the P17 method is indicated by the *continuous line*

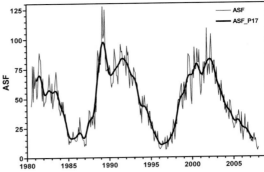

Fig. 17. Mean monthly number of individual area-shaped faculae (ASF) between 1980 and 2007 as observed by the SONNE facula network. The smoothed course of the indices using the P17 method is indicated by the *continuous line*

13.1.4.4 Facula Areas Facula areas are well-correlated with the frequency of sunspots. For precise measuring of facula areas images obtained with good seeing or accurate drawings of the Sun are required. Such drawings can only be obtained using a firmly mounted instrument with a stable projection screen and as possible automatic tracking.

Several methods are available for measuring areas. Image processing tools can offer good support for this. There is, however, a principal difference between area measurements of sunspots and faculae. Spots have much better defined outlines and high contrast against the photosphere. In comparison faculae have a much lower contrast which strongly depends on the distance to the solar limb. The outline of faculae is often diffuse because seeing substantially influences their visibility. Therefore, during image processing the contrast of the faculae should be enhanced. Finally, area measurements have to be corrected for perspective distortion according to their angular distance to the center of the solar image in order to obtain their true area.

13.1.4.5 Latitude Distribution of Faculae The distribution of faculae in heliographic latitude is similar to that of spots. The latitude zones of faculae on both sides of the solar equator are, however, about 10° wider. The migration in latitude of the zones of spots and faculae are closely related. At high latitudes faculae occur prior to spots and outlast them at the end of the solar cycle. In addition, the live span of faculae increases with their size. Investigations in this field are also open for amateurs.

Methods for determining positions are explained in Sect. 13.1.7. These methods are in principle suitable for spots as also for faculae. Aided by position measurements, a facula area may be recognized as the remnant or the precursor of a sunspot activity region. Migration of the facula zones can be found and a butterfly diagram for faculae can be drawn.

13.1.4.6 Polar Faculae While sunspot activity occurs almost exclusively in regions up to about 45 heliographic degrees northwards and southwards of the solar equator the activity of faculae reaches up to the solar poles and therefore extends over the entire Sun. Faculae at heliographic latitudes above $|b| \geq 50°$, the polar caps, are called *polar faculae* (latitude measurement 13.1.7).

Like all other faculae, polar faculae are also observable in the chromosphere, e.g., in the light of the Hα line. They are well-visible in the blue wing of Hα (+0.1 nm). Except for their location, polar faculae differ from other faculae also by their temporal occurrence. Polar faculae are very small and frequently point-shaped. Typical sizes range between 1″ and 5″ corresponding to 2300 km on the Sun.

Determining the lifetime of polar faculae is difficult because unavoidable variations of the seeing can mimic their disappearance or appearance. Professional observers have determined lifetimes of more than 90 minutes.

Polar faculae are most frequent during the minima of sunspot activity and vice versa (see Fig. 18). The polar facula cycle is the second, weaker component of the magnetic activity cycle and comprises 7 to 12 years. The cycle starts with the inversion of the polar magnetic fields around sunspot maximum when the number of polar faculae is lowest. The first new polar faculae appear between heliographic latitudes of 50° and 70° on the northern and southern hemisphere. The zones of activity then start extending towards the poles until the polar caps are covered with numerous polar faculae. The faculae points are magnetically unipolar and have opposite polarity at both polar caps.

Amateur observers should concentrate on registering the activity of polar faculae by daily counting and position measurements. Upon evaluation of the data it has to be taken into account that the orientation of the solar equator varies periodically during the course of the year. For some period the northern solar pole is within view while during other periods the southern pole is visible. Some observers deal with the problem by counting polar faculae on the southern hemisphere between March 15 and April 15 only and on the northern hemisphere between August 15 and September 15. On average over a whole

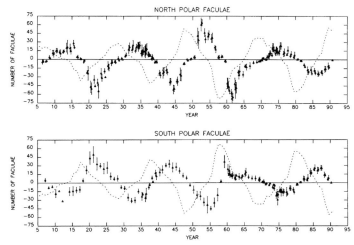

Fig. 18. The numbers of north and south polar faculae observed at the times that Earth had its greatest heliographic latitude in the respective hemisphere during 1906–1990, after Sheeley [23]. The numbers have been assigned the polarities of the associated polar magnetic fields. For comparison, each set of facula measurements has been accompanied by a *dashed line* indicating the full-disk sunspot number with a polarity corresponding to that of the following spots in the respective hemisphere and with an amplitude that is scaled to the maximum number of faculae

year influences caused by variations of the apparent axis orientation are, however, largely smoothed out. Over longer periods it is safe to work with average values.

13.1.5 Prominences and Filaments

13.1.5.1 Introduction Prominences and filaments are clouds of cool and dense material in the solar corona which are trapped and supported against gravity by magnetic loops in active solar regions and resemble physical conditions similar to those of the chromosphere. They are manifestations of the same phenomenon: when viewed at the solar limb in front of the dark background, it appears bright and is called a prominence, but when seen as a dark structure silhouetted against the bright solar disk, it is termed a filament. With a PST (see Chap. 6), the transition from one to the other in the vicinity of the limb can be directly visually observed, but less easily with photography as disk and limb require different exposure times (Fig. 1).

See text on CD-ROM concerning the historical development of instruments and techniques

Prominences are always directly connected with the magnetic field of an activity center, and even provide a visible representation of the magnetic field lines.

Prominences may be broadly subdivided into quiescent and active prominences, which in general is correlated with the age of the feature in the life of an activity center (see below). Prominences are only about 5000 km wide, but differ greatly in height and length, ranging from the size of spicules ($\sim 10^4$ km) up to an extent of several 10^5 km.

13.1.5.2 Classification and Types Scientists have repeatedly presented prominence classification schemes (e.g., Tandberg-Hanssen [24]). Their drawback is that they rely chiefly on physical parameters, which are not easily obtainable, or require extensive series of observations in order to classify the prominence in question. A simple scheme as that of Waldmeier for spots has not been found.

From his own observations, Völker in 1969 derived a classification scheme which is suitable for amateurs [25]. It permits every visible prominence to be immediately (i.e., at the moment of observation) assigned a type according to its appearance. This scheme is shown in Fig. 19. It distinguishes with a letter the bar-shaped, arch-shaped, and area prominences, even those which, due to the effects of foreshortening (which produces an apparent visual "density") or the resolution limit of the instrument, cannot be further resolved. The Völker classification scheme also considers the relative apparent size and the frequent case of a prominence detached from (i.e., "hovering" above) the solar limb.

13.1.5.3 Recording In his standard work "Ergebnisse und Probleme der Sonnenforschung" [26] Max Waldmeier (1912–2000) describes how solar researchers determine profile areas of prominences. The unit is a rectangle of 1° (heliocentric) times 1" (geocentric) (Fig. 20).

Völker found this method of measuring profile areas to be too time consuming for the amateur and, therefore, introduced the prominence number in 1969. It is based on the type categorization of repeatedly observed basic shapes of prominences which he called individual phenomena (Fig. 19).

Usually several individual prominence phenomena are located within a prominence group similar to several individual sunspots forming a sunspot group. The prominence

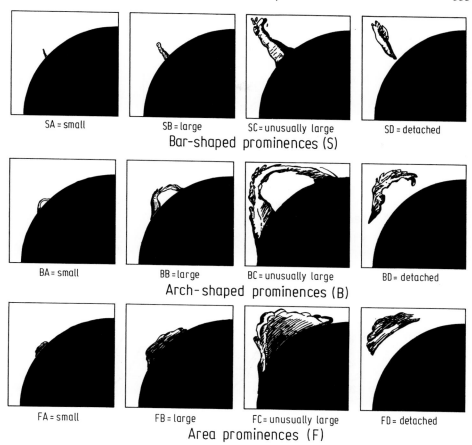

SA = small
SB = large
SC = unusually large
SD = detached
Bar-shaped prominences (S)

BA = small
BB = large
BC = unusually large
BD = detached
Arch-shaped prominences (B)

FA = small
FB = large
FC = unusually large
FD = detached
Area prominences (F)

Fig. 19. Prominence types after Völker

number R_P is determined by counting the number of prominence groups around the entire Sun, multiplying their number with 10, and adding the number of individual prominence phenomena: $R_P = 10h + e$. (For details see [1, 27 (Vol. 2, The Sun)].)

Hugo Stetter demonstrated the suitability of the prominence number by providing a statistic which compares his own observations with the sunspot curve (Fig. 21 and [28]). Observers using a prominence telescope or attachment may continue working with this method.

The PST and Lunt solar instruments provide—in their basic configuration—an overview image of the entire solar disk. Counting individual phenomena is not feasible with this as they can hardly be separated.

From sunspot observations it is known that it is sufficient to count the number of sunspot groups in order to describe the solar cycle. Analogous to this, Völker introduced the Hα number in 2008 [29]. The new number is based on counting the activity centers consisting of spots, faculae, flares, and prominences/filaments (Fig. 22).

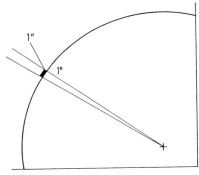

Fig. 20. Unit of prominence area

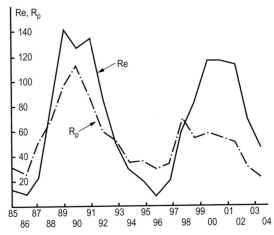

Fig. 21. Comparison of the sunspot number *Re* with the prominence number R_P after observations by H. Stetter, 1985–2004

Not until the middle of the last century did scientists succeed in describing the discrete observations obtained with different special equipments (white light, Hα, calcium, coronagraph, and magnetometers) as related phenomena of a solar activity center [30]. The complete development of an activity center is illustrated in the following example: The birth of the center is indicated by facula points. After two to three days the facula area is larger and brighter and the first spots form. Out of these a fully grown sunspot group may develop within five to ten days. Meanwhile short-lived filaments are observed which are orientated at a sharp angle with respect to the axis of the group. At the solar limb they appear as fast active prominences and are sometimes mistaken for spicules if they remain small. After 10–12 days the sunspot group reaches its maximum state of development. In complex activity centers which are densely populated by large sunspots and in which, therefore, opposite magnetic polarities can approach, energy outbursts may now occur. These are called flares and are in general brighter than faculae and considerably more short-lived, from minutes to hours. After 27 days (one solar rotation period), an average sunspot group has already begun to degenerate. And after two rotations the spots

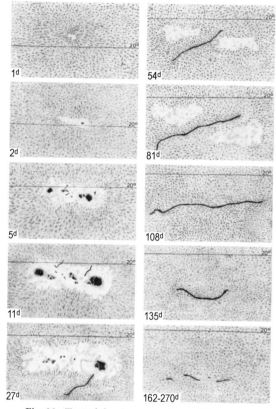

Fig. 22. Typical development of an activity center

have disappeared. The facula herd, however, remains and is decorated by a quiet filament which has turned its orientation now parallel to the equator. Frequently, the filament divides the facula area into two halves. After more than 100 days, i.e., four solar rotations, the faculae have disappeared. The quiet filament, however, will frequently reside four to six further solar rotations while it slowly dissolves. Then the undisturbed Sun is visible at this location again. The activity center has ceased without leaving any trace.

It is beyond the means of an amateur to observe the corona (outside solar eclipses) and to measure magnetic fields. Therefore, they term an activity center as an "activity herd."

For determining the Hα number the activity herds h are counted and multiplied by 10—similar to the well-known sunspot number. Thus, leading to the simple formula for the Hα number: $R_{H\alpha} = 10\,h$.

An example of an observations is shown in Fig. 23. Further evaluation (comparison of observations by several people, k factor, averaging, etc.) is done using the same methods as for sunspots.

13.1.5.4 Long-Term Observing Programs Once statistics have been prepared according to one of the methods described, the prominence activity can be compared with that

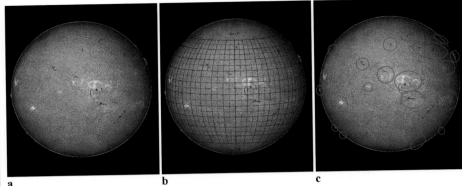

a   b   c

Fig. 23. (**a**) Photograph of the Sun in Hα light taken on August 20, 2004 by M. Weigand, (**b**) with a solar grid superposed, (**c**) with activity herds marked (see Sect. 13.1.7.2-4, Fig. 37)

of sunspots and plages. Attention is to be given to the activity not being synchronous in the main spot zone and in the polar zone. In a statistical comparison with sunspots or plages, it is advisable to use only the main-zone prominences; otherwise the curves may show strong deviations from one another. In addition, a comparison between polar prominence and polar faculae activity may be useful. (For details see [1, 27].)

Long-term observing programs for recording chromospheric activity require, as is the case for spots and plages, long series of observations and, if possible, coordination of observations with other observers in order to obtain meaningful statistics.

13.1.5.5 Short-Term Observing Programs Prominences and filaments also show interesting short-term variations, some of which are well worth following. The often bizarre shapes sometimes perform fantastic movements with ballet-like grace, as has been documented in cinematographic time-lapse films and from scientific satellites like SOHO, TRACE, and STEREO.

Amateur astronomers can also impressively record such events in series of photographs or as a sketched presentation in phases (Figs. 24 and 25).

Besides following, representing, and evaluating events of chromospheric activity, there are additional possibilities which cannot be mentioned here due to space constraints. (Most of these are usually reserved only for very experienced and well-equipped solar observers.) Some of these include:

- Spectroscopy
- Photometry (e.g. light curves of chromospheric details)
- Cinematography (on spectroscopic material or Super-8 film—there are still fans of these techniques!)
- Videography (video monitoring of chromosphere or recording of active features, e.g., for educational use)

13.1.6 Flares

13.1.6.1 Introduction During flares magnetic energy stored in active solar regions is violently released in the solar corona, chromosphere, and photosphere leading to the

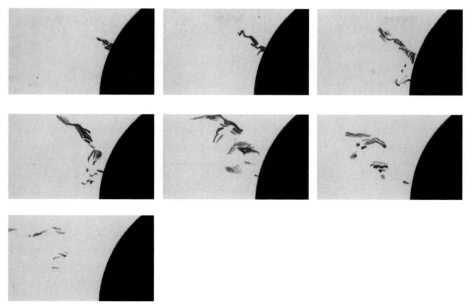

Fig. 24. Ascending and dissipating prominence (spray), observed May 24, 1969 (selected from 14 sketches) by P. Völker. 13:30 UT: beginning of active phase; 14:05 UT: complete dissipation. The phenomenon was 6×10^5 km high at maximum, with no trace left at the solar limb!

ejection of high-energy particles and intense radiation at radio to X-ray energies. In the light of the $H\alpha$ line, flares can be noticed as a sudden local enhancement of radiation but this is only the "tip of the iceberg."

Flares occur in complex sunspot groups and can strongly excite or change the surrounding material (Fig. 26).

Various forms of flares are distinguished: plage flares, two-ribbon flares, homologous flares, etc., whose explanation can be found in [1, 31]. A specialty is the extremely rare white light flare, which is limited to very large sunspot groups and can be observed in integrated light. Here, small points with brightnesses greater than that of the photosphere constitute parts of a very strong flare viewed in monochromatic light.

13.1.6.2 Classification Flares are nowadays usually classified by the order of magnitude of their peak burst intensity measured at the Earth in the 1 to 8 Å X-ray energy band.

Table 2. Importance classes of solar flares

Class	Intensity (in Watt/m^2)
B	$< 10^{-6}$
C	$10^{-6} - 10^{-5}$
M	$10^{-5} - 10^{-4}$
X	$\geq 10^{-4}$

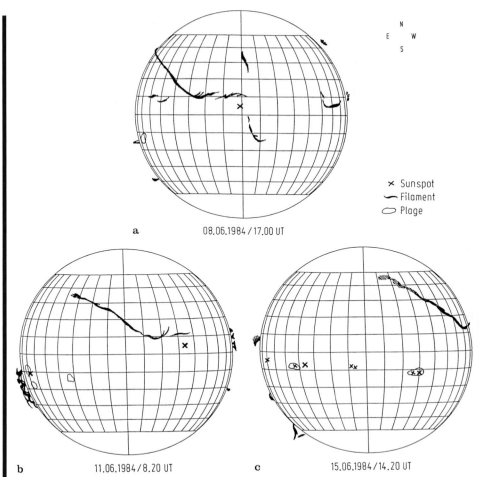

Fig. 25a–c The development of long-lived filaments is well-presented. A giant filament, observed by I. Glitsch and shown in excerpts of a phase sequence covering 21 days (June 2–22, 1984). Owing to the height of the filament, it can be followed as a prominence to the rear side of the Sun

For optical observations, the flare classification approved by Commission 10 of the IAU in 1966 should be used. Flares are divided into classes according to a quantity called their importance. Formerly, the code values were −1, 1, 2, 3, and 3+, but since January 1, 1966, the international code is S (for subflare), 1, 2, 3, and 4. Table 2 gives the pertinent quantities. As Table 2 shows, flares are, compared with other active phenomena, very short-lived.

In addition to the importance class of a flare, the flare intensity is given in the steps f = faint, n = normal, and b = bright or brilliant. There is no quantitative definition. The total description of a flare is thus "Sf" for the smallest features and "4b" for the largest. It may be worth mentioning that the classification should apply to the time of maximum intensity, and not to that of greatest areal extension.

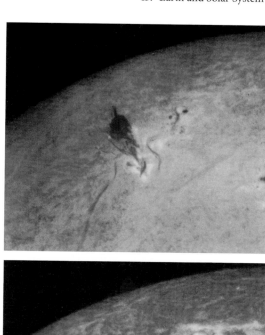

a

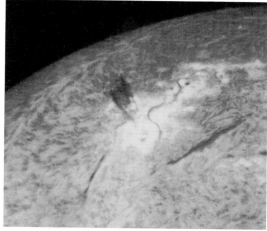

b

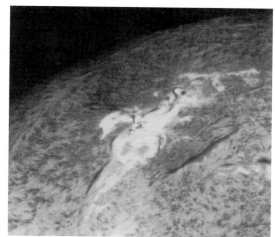

c

Fig. 26a–c Matter is ejected by a flare and returns to the Sun after the activity phase (surge). Observed June 10, 1979 by G. Appelt with a refractor (150/3000 mm) with DayStar filter 0.5 Å: (**a**) 8:12 UT, (**b**) 8:18 UT, (**c**) 8:44 UT

Table 3. Classification and properties of solar flares

Area (degree^2)	Area (10^{-6} solar A)	Class	Typical corresponding SXR Class
≤ 2.0	≤ 200	S	C2
2.1–5.1	200–500	1	M3
5.2–12.4	500–1200	2	X1
12.5–24.7	1200–2400	3	X5
> 24.7	> 2400	4	X9

13.1.6.3 Recording of Data It is virtually impossible to obtain sketches of flare phenomena, as their complex structure changes within a very short time. Therefore, the best approach is to take rapid sequences of photographs.

The following data should be recorded:

Dat: Date

ph: Photographic observations or series

v: Visual observations or series

I: State of the studied heliogram (1 = very poor …5 = excellent)

TB F: Time of beginning of flare (UT)

TB O: Time of beginning of observation (UT)

TE F: Time of end of flare (UT)

TE O: Time of end of observation (UT)

TM: Time of maximum of flare (UT)

LAT: Heliographic latitude of center of phenomenon

L: Heliographic longitude of center of phenomenon

IMP: Importance class according to Table 2

O P: Partial observation

O C: Complete observation

T: UT of measurements

AP: Maximum projected area in millionths of the solar disk

AC: Corrected area in square degrees (if the heliocentric angle ρ is smaller than 65°)

F: Maximum intensity relative to the local undisturbed chromosphere

Two flare phenomena are considered as independent if the distance between the flare knots is over 3° or if their time of appearance differs by more than 5 minutes.

13.1.6.4 Long-Term Observing Programs Two questions regarding the activity cycle are of special interest:

1. Does the total number of flares noted distinctly follow the cycle?
2. Which flare classes occur at the different phases of the cycle and with what frequency?

13.1.6.5 Short-Term Observing Programs One still-important study involves the appearance of the fine structure of the chromosphere and the sunspots before a flare occurred. This so-called pre-flare phase is difficult to catch, since there is no way of knowing when and where a flare will occur. After a flare has occurred, however, the observer can backtrack to locate these structures on a photograph obtained previously during a routine survey.

When a flare occurs, there are numerous ways to describe the event. Figure 27 shows the variety of shapes or forms which can occur.

A surge produced by a flare is shown in Fig. 26. Such effects on solar material have often been observed and are worth studying. In addition to surges, there are also sprays,

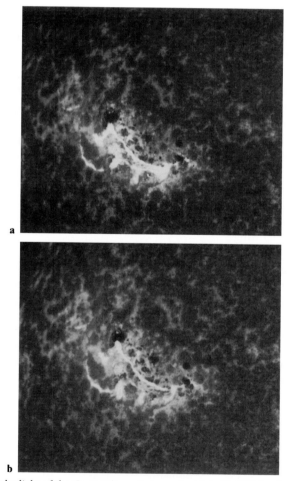

a

b

Fig. 27a,b Flare in the light of the Ca II K line, photographed July 12, 1982 by G. Appelt, using a refractor (125/3000 mm) with DayStar filter 1 Å. It is evident that dramatic changes have occurred in only 25 minutes

the flare loops, and even the Moreton waves and the "disparition brusque" phase of filaments [1, 31]. Changes in photospheric structures are also observed in the wake of flares.

There are many other interesting studies on flares, and a brief listing of some key subjects will be used to complete this discussion:

- The many "knots" which are noticeable in flares can serve to derive data on motions.
- Light curves (see [1]).
- Spectroscopy.
- Cinematography (time-lapse films).
- Videographs for flare monitoring, but again also for sharing with larger groups (instruction).
- Radio observations:
  (a) Monitoring of flares (called bursts with radio telescopes).
  (b) Interference with radio broadcasts (particularly short wave) after large, optically observed flares.

13.1.6.6 Other Chromospheric Phenomena The preceding sections have dealt with the chromospheric features and events that are most evident and most accessible to the observer. These include plages, prominences/filaments, and flares.

The following section will reveal that the structure of the chromosphere is much more detailed, and presents a greater variety of forms than the photosphere. Again, for detailed discussions, reference should be made to the technical literature [31]; a compilation from several sources is found in [1].

Astronomers distinguish between (1) phenomena of the quiet chromosphere and (2) phenomena of the active chromosphere. The best-known features of the quiet chromosphere are the following:

- Spicules, the jagged, flame-shaped structures directly above the solar limb. They reach heights of around 10^4 and diameters of only $1''$, and have a lifetime of 5–10 minutes.
- The chromospheric network (Fig. 1c), a large-scale, chromospheric "superstructure."

The best-known phenomenon of the active chromosphere may be the superpenumbra, where the fibrils continue from the photospheric penumbra into the chromosphere. Its appearance is similar to that of the white light penumbra, but need not correspond with the photospheric fine structure of the penumbra.

13.1.7 Position Measurements of Solar Phenomena

Determining the positions of the previously mentioned solar features on the Sun permits a detailed study of the properties of individual objects as well as those of the entire activity cycle of the Sun. In the following sections techniques available to amateur astronomers are described for measuring and processing solar positions.

13.1.7.1 Targets of Position Measurements Some of the results supplied by position measurements of features on the Sun will be mentioned first. For many observing programs, cooperation with other observers in a network is strongly recommended. The *rotation of the Sun* is noticeable in the motion of active regions across the solar disk, observed on several consecutive days (see also Stix [32], Howard [33], Balthasar et al. [34]). A sunspot appearing at the eastern limb of the Sun moves, as seen from Earth, by about 13.2° westward per day. Measuring this displacement of the spot on the Sun gives the *synodic solar rotation* as: $360°/(13.2°/d) \approx 27.3$ d.

However, the gaseous sphere of the Sun rotates non-rigidly; the polar regions need over one week longer to complete one rotation than those near the equator. This *differential rotation* is portrayed in Fig. 28 using a symbolic trace of spots and faculae located at various latitudes.

It is of interest to follow the motion of sunspots continuously, because in this way the average rotational velocity or even the variation of rotation with latitude can easily be found. Determining the law of differential rotation requires numerous accurate position measurements of stable spots at different latitudes (Zerm [35] and Joppich [36]). The differential solar rotation is expressed by the relation

$$\omega(B) = a - b \cdot \sin^2 B , \qquad (8)$$

where $\omega(B)$ is the angular velocity as depending on latitude B, a is the angular velocity at the equator, and b the coefficient of deceleration with B. The average values for a and b for the time interval 1874 to 1976 (Balthasar et al. [37]) are:

$$a = 14.551(9)° \, \text{day}^{-1} , \quad b = 2.87(6)° \, \text{day}^{-1} . \qquad (9)$$

The values for the two coefficients may vary a lot though, they depend, e.g., on the type of spots used, the time interval within the cycle and the hemisphere observed.

Studying the *distribution* of spots requires a compilation of many observations over months and years. Ideally groups of observers merge their data.

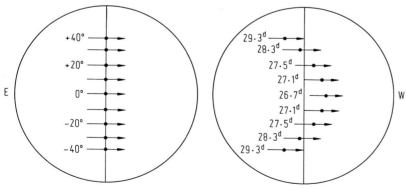

Fig. 28. An illustration of the differential rotation of the Sun using a symbolic trace of sunspots. At the start (*left*), all spots are on the central meridian, a situation which never occurs in practice. One rotation later (*right*), at the "finish line," the latitude dependence of the duration of rotation is seen. The recorded synodic rotation times are derived from over 200 observations of sunspots, calcium clouds, prominences, and metallic lines (from spectroscopy)

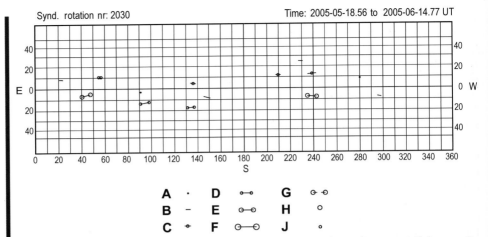

Fig. 29. Synoptic map of the Sun for spots in various Waldmeier classes (Sect. 13.1.3.2) for synodic rotation number 2030 in heliographic coordinates (Möller [38])

A *synoptic map* graphs positions of certain events (e.g., spots, faculae, filaments, etc.) observed during one solar rotation. Figure 29 shows for instance the superposition of all spots observed within 27.3 days (0–360° solar longitude). The actual lifetime of the spots is not considered in the recording.

A synoptic map gives a variety of information: active zones are evident; mean latitudes of spots can be found from a frequency diagram; the north-south asymmetry of a cycle is seen; groups whose p- and f-spots (preceding and following (Sect. 13.1.3.1, p. 312) are on different hemispheres are noted; the beginning of a new cycle is shown by the appearance of spots at high latitudes, and the distribution of the spot group types can be studied; the overall number of groups is a quantitative measure of solar activity. By superimposing several synoptic maps onto an annual chart, the positions of activity centers can be revealed.

The *butterfly diagram* (Fig. 30) illustrates the zonal migration of spots (*Spörer's law*). The mean latitude of spots or groups is graphed versus the time of their observation. The sunspots migrate from higher latitudes during the course of a cycle toward the solar equator. Even before the spots of the old cycle disappear near the equator at the time of minimum, the spots of the new cycle appear at high latitudes. This overlapping of cycles, the occurrence of spots in two zones symmetric with the equator, the interval of maximum activity, the intensity of a cycle, and the differences between hemispheres can all be read from a butterfly diagram (see also Hammerschmidt [40], p. 249 and Yallop et al. [41]).

Properties of Sunspot Groups Revealed by Position Measurements

The *length of a spot group* (i.e., maximum extension) is given by the relation:

$$s = 12148 \cdot \sqrt{(\Delta L \cdot \cos B)^2 + (\Delta B)^2} \qquad (10)$$

(*continued on p. 347*)

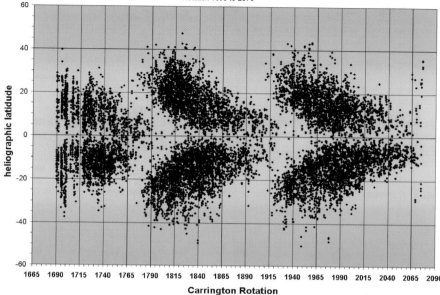

Butterfly Diagram
198.890 Positions
1979-12-27.84 to 2008-11-20.29 UT
Rotation 1690 to 2076

Fig. 30. Butterfly diagram for cycles 21 to 23, composed of 198,890 individual positions the time interval of 1979 to 2008. Data and graphics from Möller [39]

(continued)
where s is the extension of the group in kilometers, B the mean heliographic latitude of the group in degrees, ΔL the east-west extension in degrees, and ΔB the north-south extension in degrees (1° of heliographic coordinates equals 12,148 km at the solar equator).
The *inclination* α *of the group* is given by:

$$\alpha = \arctan\left(\Delta B/(\Delta L \cdot \cos B)\right) \tag{11}$$

where α is positive when the p-spot (preceding in the sense of solar rotation) has a smaller distance from the equator than the following f-spot. In most groups, α is positive, but the numbers depend on the latitude of the group and on the time within the cycle. The literature on this subject is controversial.

For studies of *individual motions* within sunspot groups extreme care has to be taken during the measuring process in order to obtain positions with sufficient accuracy, as the individual spots move usually only a few tenths of a degree per day. Relative measurements, however, are more precise than absolute measurements, and therefore the required precision can be reached (see [1, 2, 42, 43, 44]), e.g., drift motions, formation and dissipation of spots are distinctly visible in the F group of June 1983. Figure 31 illustrates parts of the development of that group. Mehltretter [45] observed substantial motions in spots entering the penumbra of a larger spot. It is neither clear whether

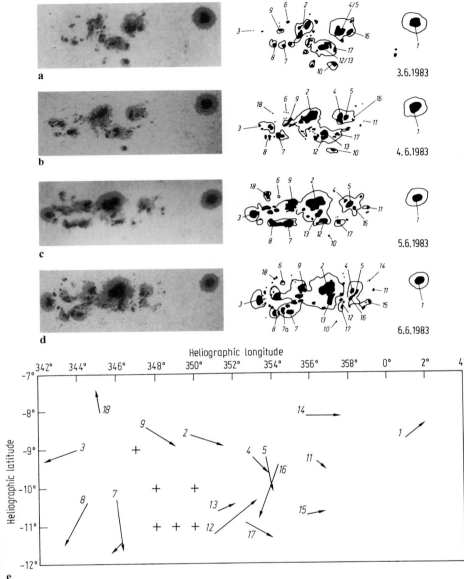

Fig. 31a–e *Left column*: Photographs of the F group in 1983 June. All photos were taken on TP2415 by C.H. Jahn using a 200/3250 mm refractor (with 12.5-mm eyepiece, solar prism, green filter, and exposure time 1/1000 s) of the Astronomical Station of the Universität Hannover. Dates of exposures are (in UT): June 03 $09^h 15^m$, June 04 $10^h 00^m$, June 05 $13^h 34^m$, June 06 $10^h 43^m$. *Right column*: Sketches of the group. The *spots* marked have been used to determine their individual motions. (**e**) Graph of motion of the F group in 1983 June, showing the motion vectors over 4 days between 02. and 11.06.1983. *Crosses* indicate the position of the largest flares

the motion of smaller spots near large penumbrae is accelerated or decelerated, nor which evolutionary phase of a group the largest motions occur. The connection between motions and flare frequency (Sect. 13.1.7.3) has not yet been well-studied. Photography is preferred to direct visual observations for these and similar studies of the Sun (Fig. 10) since interpretive errors which may strongly influence sketches are eliminated.

The methods for measuring solar positions are not limited to sunspots, but are applicable to plages, flares, prominences, filaments, and other phenomena of the solar photosphere and chromosphere. The polar faculae in latitudes over ± 50° should be given more attention during time of sunspot minimum (see Brauckhoff, Delfs, and Stetter [46], Sect. 13.1.4.5–13.1.4.6, Fig. 18, Fig. 23).

13.1.7.2 Heliographic Coordinates The system of the *heliographic reference frame* must first be defined. As in the case of the Earth, the *poles of the Sun* are the points where the rotation axis penetrates the surface. The north pole is that point which, as seen from the Earth, points toward the northern sky. Each plane through the solar poles intersects the surface in a *meridian* of the Sun. All planes at right angles to the solar axis intersect the surface in latitude circles, and the largest of them, which defines the plane through the center of the Sun, is called the *solar equator* (Fig. 32). The *heliographic latitudes B* are counted from the solar equator towards north (+) or south (−). Since there is no permanent fixed point on the Sun which could define a zero meridian—as Greenwich is on the Earth—and the Sun rotates at various speeds in different latitudes (Fig. 28), the definition of *heliographic longitude L* is more difficult:

Depending on the kind of evaluation which is planned, the chosen meridians may be one of the following:

1. The *central meridian* at the moment of observation, from which longitudes are counted west (W) or east (E), e.g., used for studies of individual motions of solar phenomena (Fig. 31).

2. A *zero meridian*, internationally defined as that meridian which passed on January 1, 1854 at 12 h UT through the ascending node of the solar equator on the ecliptic. From this so-called *Carrington zero meridian*, heliographic longitude is counted toward west from 0° to 360°. The instant of transit of Carrington's zero meridian through the central meridian characterizes the beginning of a synodic solar rotation, whose average duration is defined to be 27.2753 d (at $B = \pm 16°$, about the latitude of maximum spot activity). At the beginning of a synodic solar rotation, all active regions lying west of the central meridian are counted to the old Carrington rotation. Rotation 1 began on November 9, 1853 and by December 07, 2009 exactly 2090 rotations (transits of the zero meridian through the central meridian) have elapsed. For example synoptic maps (Fig. 29) refer to the Carrington zero meridian. L_o in Fig. 32 refers to the longitude of the central meridian in the Carrington rotation system. The times of the beginning of the synodic solar rotation are given in the *Astronomical Almanac* and also in some magazines.

Figure 32 illustrates the relative quantities for a position measurement on the Sun and takes into account the actual position of the solar rotation axis relative to the Earth (see

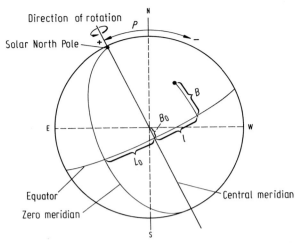

Fig. 32. The heliographic coordinate system (seen in perspective view)

box). P, B_0, and L_0 are tabulated in almanacs and special periodicals for every day of the year, and must be interpolated for the specific time of observation. These seasonal changes of the coordinate system preclude an immediate estimate as to whether a particular spot belongs to the northern or southern solar hemisphere, or whether a facula is or is not in the latitude range of the polar faculae (Sect. 13.1.4.4). In order to make these simple distinctions, a sketch of the Sun including the solar axis and equator regarding B and P_0 is required.

> *Apparent Motion of Sunspots Across the Solar Disk*
> The *apparent motion* of spots depends on the position of the Earth in its orbit. Since both the solar equator and the Earth's equator are inclined (by $i = 7.25°$ and $i = 23.43°$, respectively) against the ecliptic, sunspots as seen from the Earth usually move in ellipse sections and not in straight lines across the solar disk. They perform the latter only when the Earth is located in one of the two points of its orbit which intercept the plane of the solar equator. Figure 34 graphs the annual variations in position of solar axis and equator for the terrestrial observer. The heliographic latitude B_0 of the disk center M varies between $+7.25°$ and $-7.25°$. The *position angle* P (sometimes called P_0) of the solar axis against the north-south direction in the sky varies between $+26.4°$ and $-26.4°$ (+ indicates the tilt of axis is toward east, – toward west).

13.1.7.3 Methods of Position Measurement In the following the two methods of *direct marking* and *trailing* will be presented to measure positions of solar phenomena. In both cases the *orientation of the solar image* must first be clarified (see box).

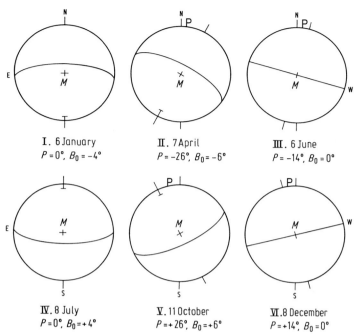

Fig. 33. Position of solar axis and apparent motion of spots in various seasons (schematic). *Graph I:* Around January 6, the position angle of the axis is $P = 0°$, and the rotation axis coincides with the north-south direction in the sky; B_0 is at this time about 4°, the (imaginary) spots on the solar equator thus travel 4° above the center of the disk and across the central meridian. The south pole of the Sun is tilted toward Earth. *Graph II:* Around April 7, the position angle reaches its greatest western elongation, with $P = -26.4°$. The center of the disk has the heliographic latitude $B_0 = -6°$. *Graph III:* Around June 6, P diminishes to around $-14°$. As $B_0 = 0°$, sunspots travel in straight lines across the Sun, and the solar equator symmetrically halves the northern and southern parts of the visible disk. *Graphs IV–VI* show the corresponding situation when P and B_0 are positive

---

Determination of the Orientation of the Solar Image

The orientation of the solar image depends on the lenses and viewing methods used (Fig. 34). With the telescope drive switched off, a spot travels across the sky from east to west by about one solar diameter in 2 minutes. Moving the telescope toward the north celestial pole (i.e., toward higher declinations), the northern hemisphere of the Sun is last to disappear from the field-of-view. It is important to use an equatorially mounted telescope. While an alt-azimuthal mount can provide a survey of the daily spot picture, the lack of tracking and the substantial change in the parallactic angle during the day makes it virtually impossible to obtain reasonably accurate position measurements. (The only exception would be position photographs taken with alt-azimuthally mounted telescopes having focal lengths of less than 1.3 m.) Portable instruments require exact alignment of the equatorial mount with great care. Permanent markings made on the ground will help to get the instrument into the same carefully checked alignment every day. But a control of this adjustment is still needed before each measurement is taken, otherwise a rotation of the solar image might affect the measurements.

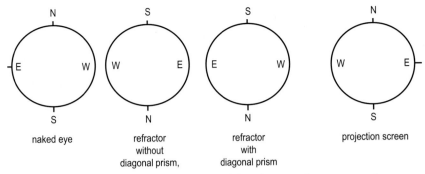

Fig. 34. Position of the cardinal points of the solar image for some examples

The *direct marking technique* is, owing to its simplicity, the most frequently used procedure. It requires a projection screen which is attached to the telescope very rigidly and precisely at right angles to the optical axis. Commercially available screens are usually far too unstable, and should therefore be strengthened. On the screen a prepared sheet of white paper is placed which contains only a circle of diameter 12 to 15 cm and a right-angle cross through its center, the latter serving to mark the north, south, east, and west directions. For the orientation of the cross on the paper see the box. It is important that the diameter of the solar image matches exactly the diameter of the circle on the form. With the telescope drive switched on, the positions of the individual sunspots and faculae are recorded using a light touch with a soft, well-sharpened pencil, without touching the screen with the hand.

A more precise variant of the direct marking technique is the position photography: it requires a double exposure of the focal image of the Sun at the primary focus (Fig. 35), the time difference between the two photographs should be around 100–120 seconds (while the telescope drive is switched off). This can also be achieved by two separate digital photographs superimposed afterwards. One image of the complete disk will suffice. In case of a double exposure the camera must be held rigidly at the eyepiece tube, because the shutter must be cocked between the two exposures.

Like this quite satisfying results can be obtained with very little effort using a camera as all spot positions are recorded simultaneously in a short time. The atmospheric turbulence which has been "frozen" onto the film may, however, distort the recorded position. This method gives the most precise positions, but it must be remembered that it is not necessary to obtain "pretty" pictures in order to determine the positions.

For the *trail method* (see Junker and Reinsch, p. 339 in [1]; Vogt [47]), a grid with the x-axis (east-west direction) and y-axis (north-south) and the diagonals in between, is drawn on a graph sheet on the projection screen (Fig. 36). After carefully adjusting the instrument in the east-west direction, the solar image is allowed to trail across the screen, and the time T_W between the transits of the western limb of the Sun (limb S_1, spot F_1 in Fig. 36a) and of the spot (S_2, F_2 in Fig. 36a) through the y-axis is stopped. The second step is to allow the southern limb of the Sun to trail along the x-axis. Here, the time T_S between the transits of the spots, for instance through the diagonal $y = x$ (S_1, F_1 in Fig. 36b), and through the y-axis is stopped. The spot then has the following position

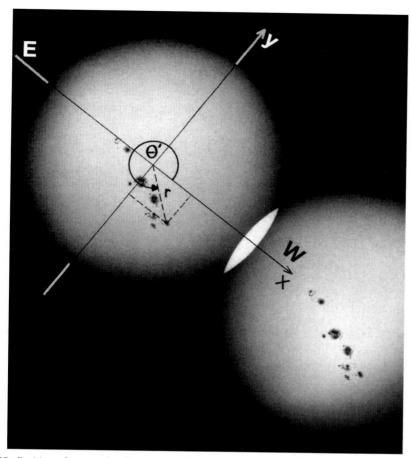

Fig. 35. Position photograph taken on April 29, 1984 by C.H. Jahn on Kodak TP2415 with the 200/3250 mm refractor of the Astronomical Station of the University of Hannover. Graphics by G. Schwaab and E. Junker

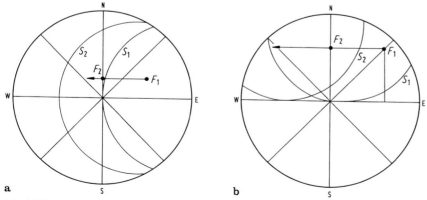

Fig. 36. (**a**) The Sun trails in an east-west direction across the grid. (**b**) The southern edge of the disk moves along the east-west direction. After p. 339 in [1]

on the projection screen (in Cartesian coordinates):

$$x = T_W - R_t \quad \text{and} \quad y = T_S - R_t \tag{12}$$

where $R_t = \cos \delta \cdot (\Gamma'/15)$ is the radius of the Sun in units of time, Γ' its angular radius, and δ its declination (R_t in minutes when Γ' in arcminutes or R_t in seconds when Γ' in arcseconds) (see Sects. 13.1.7.1 and 13.1.7.4).

The method has also been used on fine grids. Also, the field distortion can be compensated for by incorporating it into the grid.

Applying the trail method with a suitable *eyepiece micrometer* provides distortion-free measurements, and thus no accessories, like a screen or camera, whose attachment imposes high demands on the stability of the telescope mount, are needed.

The trail method gives the sunspot position not in a graph but rather in a numerical form which can be directly converted into positions (Sect. 13.1.7.4). One should not expect an accuracy of better than ± 1° using the methods described here. A stable instrument and the exercising of care on the part of the observer may improve the results.

Fritz and Treutner in [1, 48] and Jahn [42] have, with some improvements and corrections, achieved mean deviations of 0.1° and less, which should also be possible using digital position photography, but so far no similar results have been published.

13.1.7.4 Calculation of Heliographic Positions

A special grid or template with coordinates (Fig. 37) will serve to provide an expeditious determination of heliographic positions. The correctly scaled template, when placed in proper orientation (position angle P and B_0) over sketches and photographs, permits the direct reading of heliographic latitude B and the longitude difference l against the central meridian of solar phenomena (Fig. 32; p. 355 in [1]).

Eight transparent grids will suffice for the values $B_0 = 0, 1, 2, \ldots, 7°$. For negative B_0, the grids are rotated 180°.

Construction of Position Grids

Points (x, y) on the grids are calculated according the the formulae:

$$x = R \cdot \cos B \cdot \sin l$$
$$y = R \cdot (\sin B \cdot \cos B_0 - \cos B \cdot \sin B_0 \cdot \cos l) \tag{13}$$

where x is the distance in millimeters of longitude circles from the central meridian at a given latitude B, y the distance in millimeters of latitude circles from the equator ($B = 0$), and R the radius of stencil in millimeters.

For the special case $B_0 = 0$, Eq. (13) simplifies to: $y = R \cdot \sin B$.

Example: For the coordinates $B = 40°$, $l = 20°$ for $R = 50$ mm, x and y are calculated as:

$$x = 50\,\text{mm} \cdot \cos 40° \cdot \sin 20° = 13.10\,\text{mm} \quad \text{and} \quad y = 50\,\text{mm} \cdot \sin 40° = 32.14\,\text{mm} . \tag{14}$$

Because of the higher precision afforded, a numerical technique is preferable to a graphical method (Junker and Reinsch [1, pp. 221–261]). Calculating coordinates from a sketch

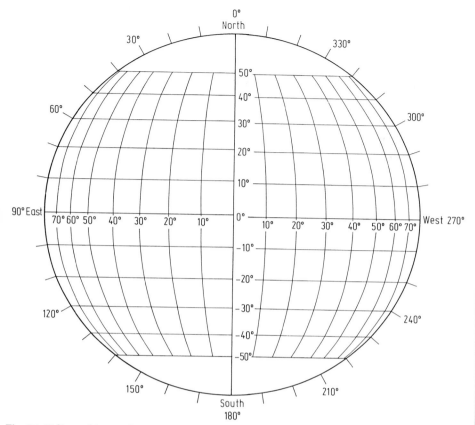

Fig. 37. Heliographic coordinate system with $B_0 = 0°$. Graphed along the limb of the Sun are position angles, which run from north via east from $0°$ to $360°$

or photograph is simplified through the use of a spread sheet program in the reduction which also allows processing and storing larger amounts of data. The position of a spot on the Sun is recorded either in Cartesian coordinates (x, y) or in polar coordinates (r, θ') (Fig. 38).

Polar coordinates can either be directly measured from a sketch or photograph of the Sun (Fig. 35) or are calculated from Cartesian ones via the relations

$$r = \sqrt{x^2 + y^2}$$
$$\theta' = \arctan(y/x) \tag{15}$$

where particular attention must be paid to the quadrant where the spot is located: θ' is

A set of grids for the 8 different B_0 values with diameters 11 cm and 15 cm can be obtained from the VdS Supply Center, Thomas Heising, Clara-Zetkin-Straße 59, D-39387 Oschersleben, Germany (http://www.vds-astro.de/vds-leistungen/06_materialzentrale.htm) or print your own grid with the program from http://www.sonnedeveloper.de/down/gnd_eng.exe.

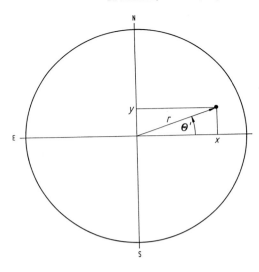

Fig. 38. Plane coordinate system for position measurements (polar and Cartesian coordinates)

counted from west over north; in quadrants 2 and 3, 180° is added, and in quadrant 4, 360°. The angular distance ρ of the spot from the center of the disk is obtained with sufficient precision (approximately 0.1°–0.2°) for most applications via

$$\sin \rho = \frac{r}{R} \qquad (16)$$

where R is the radius of the solar image.

The angular distance of the spot from the solar east-west direction follows as

$$\theta = \theta' - P \qquad (17)$$

as Fig. 39 illustrates.

The heliographic latitude B and the longitude difference l of the spot off the central meridian (Fig. 32) are then found using the relations

$$\sin B = \cos \rho \cdot \sin B_0 + \sin \rho \cdot \cos B_0 \cdot \sin \theta$$
$$\sin l = \cos \theta \cdot \sin \rho \cdot \cos B . \qquad (18)$$

The longitude L of the spot in the Carrington rotation system is found from

$$L = L_0 + l \qquad (19)$$

where P, B_0, and L_0 are obtained from tabulated values by interpolating to the exact moment of observation. L_0 is the heliographic longitude of the central meridian referred to the Carrington rotation (Fig. 32). The pair (B, l) gives the position of the spot relative to the instantaneous central meridian, whereas (B, L) refer to the Carrington system.

Formulae for calculating heliographic positions which differ in detail from those presented here can sometimes be found in the literature, because the trigonometric functions may change with the definitions from which axis and angle is counted. Therefore, formulae from various publications (which are self-consistent) should not be mixed (Junker [49]); the definitions should be studied from illustrations.

An engaged German amateur astronomer provides some software for solar observers analyses sun spot positions and provides graphics and sun spot number statistics: http://www.strickling.net/wstsoft.htm

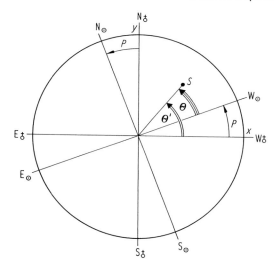

Fig. 39. Explanation of the relation $\vartheta = \vartheta' - P$. N_{δ}, W_{δ}, S_{δ}, E_{δ}: directions on the Earth. $N_{\odot} - S_{\odot}$: projections of the solar rotation axis onto the sky. S: position of a sunspot

A more detailed discussion on position determination has been provided by Junker and Reinsch in [1, pp. 221–261].

13.1.8 Conclusion

The preceding sections have shown that there is a wealth of possibilities for observing the Sun, even though some items are mentioned only briefly. Evaluating one's observations, whether individually or within working groups, cannot be described in general recipes because it would challenge the creativity of the individual. Solar physicists and other active observers are usually glad to help in this regard.

13.1.9 Further Reading

Beck, R., Hilbrecht, H., Reinsch, K., Völker, P. (eds.): Solar Astronomy Handbook. Willmann-Bell, Richmond, VA (1995)

Bruzek, A., Durrant, C.J.: Illustrated Glossary for Solar and Solar-Terrestrial Physics. Reidel, Dordrecht (1977)

Giovanelli, R.G.: Secrets of the Sun. Cambridge University Press, Cambridge (1984)

Solanki, S.: Sunspots: An Overview. Astron. Astrophys. Rev. 11, 153 (2003)

Tandberg-Hanssen, E.: The Nature of Solar Prominences. Kluwer Academic Publishers, Dordrecht (1995)

Taylor, P.O.: Observing the Sun. Cambridge University Press, Cambridge (1991)

Further information can be obtained from one of the following organizations:
(1) The Vereinigung der Sternfreunde Working Group on the Sun. Contact: Steffen Janke Sternfreunde im FEZ e.V., An der Wuhlheide 197, D-12459 Berlin. http://www.vds-sonne.de/index_e.html
(2) The American Association of Variable Star Observers, Solar Division. Contact: Paul Mortfield, c/o AAVSO, 49 Bay State Rd., Cambridge, MA 02138, USA. http://www.aavso.org/observing/programs/solar/
(3) The British Astronomical Association, Solar Section. Contact: Lyn Smith, Dunglass House, Ayr Road, Rigside, Lanark ML11 9TU, Scotland, UK. http://www.baa-solarsection.org.uk/
(4) The Association of Lunar and Planetary Observers, Solar Section. Coordinator: Ms. Kim Hay, 76 Colebrook Rd, RR #1 Yarker, Ontario K0K 3N0, Canada. http://www.alpo-astronomy.org/solar/

13.2 The Moon

G.D. Roth

13.2.1 Problems and Ideas for Lunar Observations

Charles A. Wood is a long-time
Moon explorer and author
of the monthly column "Exploring the Moon" in *Sky and
Telescope* (monthly). His Lunar Photo of the Day website
(http://www.lpod.org) features the very best in lunar
photography.

View the Moon online,
Inconstant Moon:
http://www.inconstantmoon.com
http://www.starcity.freewire.co.uk

13.2.1.1 The Moon as a Test Object for Telescopic Work As the nearest celestial body, the Moon is justifiably a favorite target for amateur observers. Even the naked eye is sufficient to have interesting impressions of the Moon. The so-called earthshine you see when observing the thin crescent Moon, is best seen in the early evening with a waxing crescent and the early morning when you see a thin crescent waning (see Fig. 40). The dimly luminous maria and highlands are objects for binoculars and telescopes with low power.

Binoculars reveal to the observer the richness of observable features which are displayed in larger telescopes in even more detail. These features provide an almost inexhaustible test field for observers and their instruments. Concerning instrumentation, see also Sect. 13.6. And even uninitiated observers using moderate-quality optics can find something to be seen on the surface of the Moon. So, when testing, one should always aim for the small and fine detail. This is most advisable when the observer has a definite program in mind.

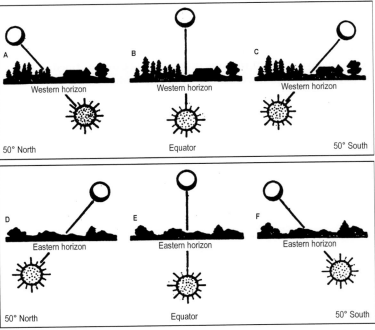

Fig. 40. The crescent of the Moon in different geographical latitudes. *A–C* Crescent of the waxing Moon over the western horizon after sunset. *D–F* Crescent of waning Moon over the eastern horizon before sunrise

13.2.1.2 Previous Studies and Space Missions The Moon is the first heavenly body upon which humans have landed (Neil Armstrong and Edwin "Buzz" Aldrin with Apollo 11 on July 20, 1969). For over 50 years now, the Moon has been explored by space missions, the first close-up study having taken place in 1959 with the USSR's Luna 3 probe. Apart from the direct inspection of the lunar surface by humans, there are also the rock samples, radiation measurements, and photographs at the site itself, which have not only supplied new materials for research, but given the term "lunar exploration" new meaning. Within a few years, the Ranger, Surveyor, Luna, and Apollo missions proceeded to explore the Moon so as to render it now a domain of geoscience (morphography, geology, petrography, mineralogy). Numerous books and magazine articles report on this fact, and the amateur observer interested in the Moon should consult them.

The observer is encouraged to study the lunar surface, not because of the hope of immediate scientific reward, but rather to gain personal experience during and after work on an observational project. Systematic series of observations, such as monitoring possible changes over parts of the lunar surface, can also bring about scientifically useful results. To better one's own understanding, the serious observer should try to know and understand methods of lunar geology. According to Guest and Greeley [50], "To the year 1960, the bulk of work on the Moon executed according to purely geological principles, was not large. But since then, geologists have shown that such principles, established during a century of research on Earth, can readily be applied to the Moon."

The status of scientific knowledge about the Moon as achieved in the frame of the Apollo program is important today for the better understanding of solar-terrestrial relations, and is indispensable for the serious observer. The lunar surface is exposed to a constant bombardment by high-energy ultraviolet and X-rays, and to the solar corpuscular radiation (chiefly electrons and protons). According to Z. Kopal [51, 52], "The Sun represents the hot cathode and the Moon the anode in an ion tube of cosmical dimensions and whose glass walls are formed by Earth's atmosphere." A more detailed knowledge of the structure of the lunar surface was also of value for the early exploration of the Solar System, particularly in the study of crater-shaped surface structures on the planets Mercury, Venus, and Mars. Finally, the origin of the Moon can be understood only in the context of the origin and structure of the entire planetary system.

In spite of this, lunar observations are still attractive to the amateur, even after the "Apollo shock." There are so many objects on its surface that "only amateurs will have time to monitor the Moon accordingly" (H. Hilbrecht). One may even succeed in photographing the short flash when a meteorite strikes the surface. Caution is advised as a blink of sunlight reflected from the surface of an artificial satellite may mimic a flash.

Representation in Maps In topographic studies, the highest perfection is found in the preparation of special maps of certain lunar regions and in overall visual representations of the entire visible surface. The latter has reached, in the work of amateur observers Philipp Fauth and H.P. Wilkins, a height of cartographic artistry which will not easily be surpassed. It will not be necessary to surpass them, however, because excellent-quality photographic maps of the entire lunar surface have meanwhile become available. A particularly important work is the *Photographic Lunar Atlas* by G.P. Kuiper [53]. A huge aggregate of lunar maps based on visual and photographic observations was complied in the US and in Russia in preparation for the space missions to the Moon.

See the definitive sourcebook *Apollo* by Richard W. Orloff and David M. Harland, published by Springer (2006).

The photos of the mission Clementine, started 1994 are the base for a very special map of the Moon (22 CD-ROMs published 1999).

Regarding the future of manned space flight to the Moon and to the planets see weblinks: http://www.suw-online.de/artikel/859577.

The ideal reference guide for beginning Moon-gazers and expert lunar observers alike is the *Atlas of the Moon* by master lunar cartographer Antonin Rückl. *The Modern Moon: A Personal View* by Charles A. Wood is an authoritative guidebook with both drawings on traditional telescope observations and the modern explorations of the Apollo, Clementine, and Lunar Prospector missions. The same author also published *Sky and Telescope's Lunar 100 Card*, a two-sided card which provides information on 100 lunar features, as well as the corresponding chart numbers from A. Rükl's *Atlas of the Moon*. As an exemplary work of an amateur observer with small optics the *Fotografische Mondatlas* by W. Schwinge [54] should be mentioned.

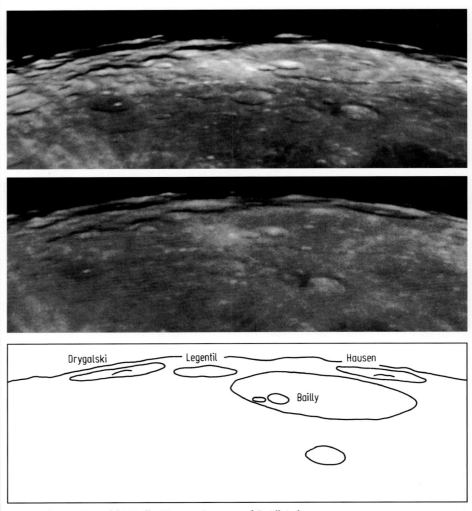

Fig. 41. Craters Drygalski, Bailly, Hausen. Courtesy of C. Albrecht

	71:11:02 – 23.00	73:09:12 – 22.00
Libration in longitude	+2°.2	−4°.8
Libration in latitude	−6°.5	−6°.5

For every charting, good photographs should supplement the visual results. Each chart, even when showing only a very tiny lunar region, should have a scale and should be ordered into a degree grid. Every observer who systematically monitors a certain lunar region should attempt to compile his sketches into a chart (see Fig. 50). This is needed to help understand more deeply the two primary events on the Moon: impacts and volcanism.

Observational tasks whose results suggest the preparation of a special chart are:

– Observing fine structures in the maria
– Observing a ray system
– Observing a region at the lunar limb, and discussing libration
– Observing a lunar rill and its accurate shape
– Observing the distribution of small craters in the vicinity of a walled plane or mountain ring
– Observing yet-unmapped regions (Luna Incognita, Fig. 42)

Such work will be based primarily on visual observations, but these are usefully supplemented by control photographs if suitable equipment is available.

From the Internet one can download the "Virtual Moon Atlas V 2.1" by Christian Legrand and Patrick Chevalley (http://www.astrosurf.com/avl), which is helpful for the observer.

The monthly column by Charles A. Wood in *Sky and Telescope* contains a wealth of information on special observational tasks. See his website http://cwm.lpod.org.

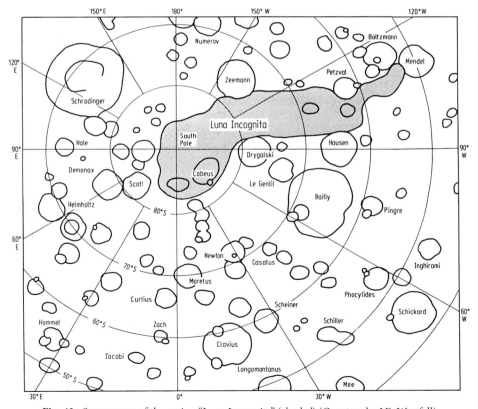

Fig. 42. Survey map of the region "Luna Incognita" (*shaded*) (Courtesy by J.E. Westfall)

Critical Lunar Topography The Moon shows numerous formations whose real shape and extension are inadequately known or which are suspected to show apparent or real changes. It is now known that the Moon's surface is everywhere covered with a fine layer of dust and debris 1 to 20 m thick, and every fresh meteorite impact smashes rock and reshuffles the lunar soil. Thus changes of some formations are conceivable.

It is interesting to observe lunar features produced in a combination of impact cratering and volcanism. There are landforms purely impact or purely volcanic. But also there are lunar features which are combinations of both. Ejecta from impacts are slung across the lunar surface. Volcanism floods and destroys craters blocking its way.

Critical lunar topography therefore includes the comparison of recent with older observations in order to detect possible changes. This may include the use of photographs. In this context the craters Plato, Aristarchus, and Alphonsus are especially interesting. The occurrence of dark spots and luminescence phenomena should also be considered (see bottom of p. 362).

Yet critical lunar topography consists not merely in demonstrating changes, but rather in most cases simply in describing each object more accurately; it requires intense observing and solid knowledge of previously published series of observations. The effects of the light and shadow that are prevalent especially on the Moon may play tricks with the observer's eyes, causing some surface formations to appear, when viewed under certain angles of illumination, different from how they actually are. Of course, the quality of the optics may contribute their own problems when one works near the limit of their resolution. An observer should not be discouraged from working with 2- or 3-in. apertures, although these instruments are in the final analysis quite limited when it comes to perceiving fine and hyperfine detail. In fact, an instrument of 6-in. aperture is the lower limit for critical topographic observations.

New detailed information from such topographic observations is also the foundation by which special maps are made (see p. 359).

Observations Relating to Physical Conditions These include work on the following surface properties:

– Brightness changes
– Color changes
– Polarization of the light scattered by the surface
– Occurrence of luminescent phenomena (transient lunar phenomena, abbreviated TLP or LTP, also called moonblinks)

Luminescence features are caused by gases emanating from the interior and triggered by solar radiation (photoluminescence). The observer will have various perceptions, as J. Classen [56] says: "The emanating gases veil the view of the surface by absorption or scattering, and give the observer the impression of a darkening or a 'grey cloud'. On the other hand, the gases may display luminescence. Colored spots will be seen on the illuminated part of the surface. When observed in the 'ashen' moonlight, whitish patches are perceived as the foveal color vision is then replaced by the quite sensitive, but uncolored, extrafoveal vision."

TLPs have been observed in the following craters: Alphonsus, Aristarchus, Atlas, Censorinus, Copernicus, Eratosthenes, Eudoxus, Gassendi, Grimaldi, Herodotus, Kep-

See the April 2004 issue of *Sky and Telescope*, p. 113, or point your Web browser to SkyTonight.com/lunar100 (information on impact cratering and volcanism of the craters Schickard and Wargentin).

The Aristarchus Plateau is a unique region of the Moon. For more about color and history see the February 2005 issue of *Sky and Telescope*, p. 64.

An example for observing a region at the lunar limb, and discussing libration, also for observing a yet-unmapped region ("Luna Incognita") see [55].

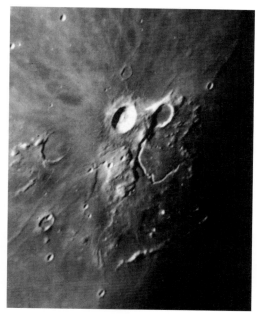

Fig. 43. Aristarchus and Schröter Valley. One of the suspected TLP regions. Photograph taken on 1980 November 19, 20^h 55^m UT, Cotonou (Benin) by J. Dragesco with a Celestron C14, $f/d = 50$, exposed 2 s on TP2415, developed in Ilford Microphen

ler, Lichtenberg, Linné, Menelaus, Peirce, Picard, Pickering–Messier A., Piton, Plato, Posidonius, Procius, Ptolemaeus, Theophilus, Tycho (image of Aristarchus in Fig. 43).

In some craters (e.g., Alphonsus and Atlas), dark spots can be observed which show changes in intensity. Small craters are observed in their centers. Dark spots can be monitored using even a small telescope. Like TLPs, they provide the opportunity to collect data on changes on the lunar surface, which is helpful when interpreting the origins of formations.

Observations of the Dark Side of the Moon Luminescence phenomenon are also observed on the night side of the Moon. A good 5-in. telescope and sufficient magnification (around 150×) will enable the experienced observer to do the job. In this context, the monitoring of the "secondary light" (ashen light) is advised. It can be observed up to 6 days of lunar age (sometimes longer) and again after 22 days. W.M. Tschernow states that brightness fluctuations of an amplitude of about 0.6 mag can be found with maximum brightness in March to May and minimum in June to August. Also, the mean annual brightness varies by about 0.8 mag, which can be ascribed to, among other things, dimming and turbidity in the Earth's atmosphere. Color changes (green tints) have also been noted.

13.2.2 Conditions of Visibility

The orbit of the Moon about the Earth is elliptical and is also subject to large perturbations. The average distance between the centers of Earth and Moon is 384,402 km, but

Much has been published in recent years on TLPs, dark spots, colorations, and similar phenomena. The British Astronomical Association in their *Guide to Observing the Moon* [57] recommends the use of a sequential color glass equipped with red, blue, and neutral filters, which is to be inserted between a Barlow lens and eyepiece. This is known as a crater extinction device (CED). See also *The Moon Observer's Handbook* by F.W. Price [58].

Overall, the objects dealt with here are of interest because of their photometric behavior. In addition to visual observations, photographic studies offer many opportunities, which are increased when color and polarization filters are used to take the photographs.

Moonblinks also include the bright flashes originating at the site of an impact of meteorite on the lunar surface and which are observable from Earth.

Persons interested in an observing project should contact an amateur observatory or working group in their area. See also p. 383.

The ultimate reference book for the serious observer is *The Astronomical Almanac*, compiled and published by the US Naval Observatory and Her Majesty's Nautical Almanac Office in England (annual). With precise ephemerides and phenomena for the Moon and the other objects of the Solar System.

Helpful tips on observing the Moon you will find in regional astronomical calendars and annual observer's handbooks.

the actual distance varies between 356,410 and 406,740 km, which causes the apparent diameter of the Moon to vary between 33′30″ and 29′22″. Thus, whereas the apparent diameter of the Moon is about the same as that of the Sun, the variations in the former are larger. Figure 44 illustrates the difference between the apparent diameters at perigee and apogee (respectively the points nearest to and farthest from the Earth).

The Moon moves quite rapidly against the backdrop of stars, describing an arc of about 13° per day from west to east. Observers should be aware that the apparent motion does not coincide with the geocentric one given in almanacs, because it depends significantly upon the observer's specific location on Earth (i.e., on the position of the Moon relative to the observer). As astronomers who study the occultations of stars well know, an occultation event differs with respect to time, duration, and orientation for every place on Earth. Lunar observations are similarly affected. The instantaneous ap-

Fig. 44. Change in apparent lunar diameter owing to the changing distance from the Earth. Adapted from W. Schwinge [54]

Table 4. The most and least favorable times to observe the Moon at various phases over the year in the Northern Hemisphere

Phase	Three days	First Quarter	Full Moon	Last Quarter	25 Days
Favorable	End of April	Spring Equinox	Winter Solstice	Autumnal Equinox	End of July
Unfavorable	End of October	Autumn Equinox	Summer Solstice	Spring Equinox	End of January

Observers of the Southern Hemisphere may interchange July with January and April with October.

parent semidiameter must be calculated whenever specific observations are made; for example, measuring heights of feature at the lunar limb requires a measure of the lunar semidiameter in order to convert them to meters.

13.2.2.1 The Phases of the Moon

The visibility of lunar formations depends on their illumination by the Sun. The lunar phases are well-known under the names New Moon, First Quarter, Full Moon, and Last Quarter (see Table 4). The phase refers to the illuminated arc of the visible lunar hemisphere. An approximate angle g for the phase is given by

$$g = \lambda - \lambda_\odot ,$$

where the ecliptic longitudes of the Moon and Sun are λ and $\lambda_\odot$, respectively. Thus, for instance, $g = 0°$ corresponds to New Moon, and $g = 180°$ to Full Moon.

More often, the lunar phases are expressed by the time elapsed since the New Moon, called the age of the Moon. For instance, an age of about 7.5 days corresponds to the First Quarter.

In order to determine the lunar phases for given dates in past and future years, one can refer to special tables. The well-known tables by J. Meeus use the fact that 251 synodic months equal approximately 269 anomalistic months: $7412.^d1776$ and $74124.^d1741$, respectively. If a Full Moon takes place at perigee, then it will be in that same position again after 251 synodic months. More precisely, if the difference between the Full Moon (opposition) and perigee is exactly $0°0'$ at the beginning, then it will be $0°.^m76$ at the end of the 251-month cycle. The deviations between actual and mean lunar phases are attributable to the eccentricity of the lunar orbit, and repeat in each cycle at the same positions with only very small shifts.

See LunarPhase Pro Version 1.5 for software on information for any date on the Moon's phase, rise and set times, and libration (Sky Publishing Corp., Cambridge, MA)

13.2.2.2 The Terminator

Of particular interest to the observer is the position of the terminator, which forms the boundary of the illuminated portion of the Moon. Waxing and waning phases taken together, the terminator passes through every object on the surface 25 times per year, corresponding to at least 25 periods of changing illumination angles. Many formations can be observed only when at the terminator. The recording of the position of the terminator is an essential part of any observation of the Moon.

Table 5. Relations between selenographic longitude of the Sun, colongitude, and terminator. The + and − signs distinguish east and west longitude (sunrise and sunset terminator), respectively

Lunar phase	$L_\odot C_\odot = L_\odot + 90°$	Terminator
New Moon to First Quarter	180° to 270°, e.g., 200°	270° to 360° then $C_\odot = 290°$ 360° − 290° = +70°
First Quarter to Full Moon	270° to 360°, e.g., 295°	0° to 90° then $C_\odot = 25°$ 0° − 25° = −25°
Full Moon to Last Quarter	0° to 90°, e.g., 60°	90° to 180° then $C_\odot = 150°$ 180° − 150° = +30°
Last Quarter to New Moon	90° to 180°, e.g., 120°	180° to 270° then $C_\odot = 210°$ 180° − 210° = −30°

It is defined as the selenographic longitude at which, during the instant of observation, the Sun rises (waxing moon) or sets (waning moon), and is tabulated in almanacs and calendars for each day of the year.

The western or eastern longitude L of the terminator can be graphed in the rectangular (ξ, η) coordinates of Table 5. At the western (+) or eastern (−) terminator

$$\xi = \cos b \cos L \, ,$$

where b is the selenographic latitude of the parallel observed.

See also J.E. Westfall's *Atlas of the Lunar Terminator* [59].

The terminator can also be defined by the colongitude of the Sun at the center of the Moon, which is the complement of the selenographic longitude of the terminator at the lunar equator. It is found in various almanacs under the heading "Sun's Selenographic Colongitude." It equals approximately 270° at New Moon, 0° at First Quarter, 90° at Full Moon, and 180° at Last Quarter.

13.2.2.3 Libration Although the Moon always keeps the same face toward the Earth, up to 60% of the entire lunar surface can be observed. The cause is the libration phenomenon, which is composed of three effects: (1) the libration in longitude (maximally ±7°.9), which results from the ellipticity of the lunar orbit; (2) libration in latitude (maximally ±6°.9), caused by the tilt of the Moon's rotation axis with respect to the plane of the orbit (one can sometimes see past the north pole and sometimes past the south pole of the Moon); and (3) the parallactic libration (maximally ±1°), caused by the fact, that an observer on the surface of the rotating Earth views the Moon from changing directions and always from a point which is spatially separated from the geocenter. These three effects may add up to a shift of the lunar limb by as much as 10° and which is particularly pronounced at position angles 45°, 135°, 225°, and 315° of the limb (see Fig. 46). However, the foreshortening of the perspective impedes observations at the limb and in the libration zones.

Astronomical almanacs and some amateur yearbooks list for every day the selenocentric longitude and latitude of the Earth, which are also the longitude and latitude of the center of the lunar disk as seen from the center of the Earth. If the latitude of the

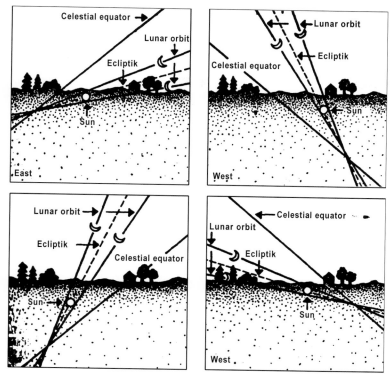

Fig. 45. Extraordinary positions of the Moon on the horizon. (*Top left*) Springtime in the morning. (*Top right*) Springtime in the evening. (*Bottom left*) Autumn in the morning. (*Bottom right*) Autumn in the evening. The figures show the phenomena for observers in the Northern Hemisphere

center point is positive, the observer sees more from the northern limb, and if negative, more from the southern limb; similarly, when its longitude is positive, he sees more of the western limb, and if negative more of the eastern limb. C. Albrecht published photographs of libration regions and showed that 4- to 6-in. refractors are well-suited for such tasks (see Figs. 41, 47 and 48).

13.2.2.4 The Lunar Coordinate Grid An observer's view of the Moon closely matches the orthographic projection, which is a parallel projection centered on the intersection of the planes of lunar equator and zero meridian (Fig. 49). Latitude parallels are projected as line segments parallel to the equator, dividing the circular limb meridian into equal parts; the zero meridian bisects the disk at right angles to the equator, and the meridians of other longitudes become ellipses with the zero meridian as the common major axis. The selenographic latitude b or β is the angular distance of, say, a crater from the lunar equator; north is up for a naked-eye observer in the northern hemisphere, down in an inverting telescope. The selenograph longitude l or λ is the angle between the places of the crater and the zero meridian. As for planetographic coordinates, its direction is now defined (by IAU decision in 1961) in the "astronautical" sense: a person on the Moon facing north has west to the left, for the terrestrial observer west is the side toward which

One region which is much affected by libration is the "Luna Incognita" (Fig. 42), whose cartographic processing began in 1972 by the Association of Lunar and Planetary Observers (ALPO) in the US [55]. This area was not photographed during the space missions to the Moon, and thus it provides amateur observers with the opportunity to contribute to the completion of the lunar map.

Lunar Map Pro 4.0 Deluxe (RITI, PO Box 418, Reading, MA 01867, USA, http://www.riti.com) allows the observer to accurately visualize the Moon's phase with full libration corrections and is a very helpful lunar mapping software.

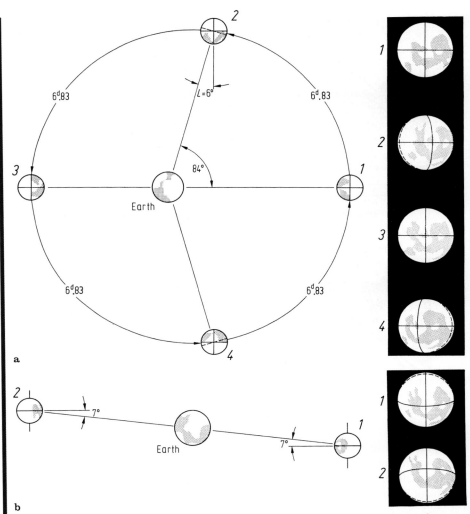

Fig. 46. The rotation of the Moon about its axis is very regular, performing in 27.32/4 = 6.83 days a rotation of 90° (**a**). The motion of the Moon around the Earth is non-uniform, owing to the elliptical nature of the orbit. At the orbital point nearest to the Earth (perigee), the Moon moves fastest, and at the most distant point (apogee) slowest. Position 1 is the apogee. Then, 6.83 days later (position 2), the Moon has rotated 90° but moved 84° with respect to the Earth, thus making visible a small portion of the far side of the Moon. This is called libration in longitude. As the eccentricity of the lunar orbit is subject to perturbations, the maximum longitudinal libration varies. (**b**) The rotation axis of the Moon is not perpendicular to the orbit, which results in a libration in latitude. Sometimes more of the north polar region (position 1), sometimes more of the south polar region (position 2), can be seen. Adapted from B. Koch, *Sternenführer*, Treugesell Publishing, Düsseldorf 1987

the terminator moves but it is seen as east in the sky (Fig. 50). (Older maps are often labeled differently.) When ± signs are used for longitudes, "+" denotes east.

Strict orthographic projection maps the surface details of the Moon (or a planet) as if projected by a parallel beam onto a plane.

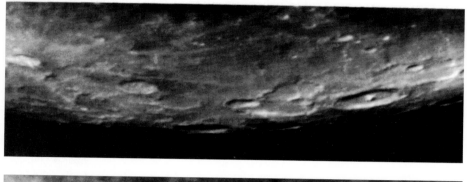

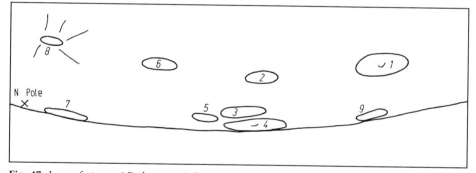

Fig. 47. Lunar features. *1* Pythagoras, *2* Carpenter, *3* Pascal, *4* Brianchon, *5* Mouchez, *6* Anaxagoras, *7* Hermite, *8* Cremona. Photos courtesy of C. Albrecht

	73:02:16 – 22.00	71:05:11
Libration in latitude	+3°.9	6°.3
Libration in longitude	+4°	−3°.7

13.2.3 Lunar Formations

13.2.3.1 Maria A mare (latin for "sea," plural maria) is a large, dark surface region which can be seen even without a telescope. About 40% of the side of the Moon that faces the Earth is occupied by maria. They are the essentially flat areas formed in the belt of maria in one contiguous system: Mare Imbrium, Mare Serenitatis, Mare Crisium.

The greater part of the Moon's big maria are circular because they fill ancient impact basins. The exception proves the rule: Mare Frigoris, an arc of lava, 1500 km long and

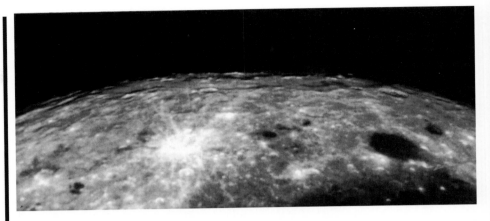

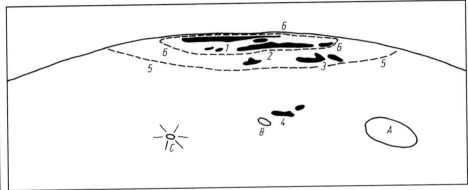

Fig. 48. Mare Orientale. 72:09:22–21.30. Libration in longitude −4°.2; libration in longitude −5°.8.
1 Mare Orientale, *2* Lacus veris, *3* Lacus autumnale, *4* Lacus aestalis, *5* Mts. Cordillera, *6* Mts. Rook,
A Grimalsi, *B* Crüger, *C* LaPaz. Photo courtesy of C. Albrecht

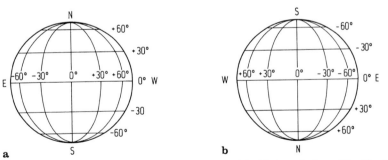

Fig. 49a,b Lunar grid showing the four points of the compass: (**a**) with the naked eye; (**b**) with a telescope

200 km wide, extending from Oceanus Procellarum in the west to Lacus Somniorum in
the east. It is interesting that it is concentric with the impact basin of Mare Imbrium. The
moat of lava that encircles Mare Imbrium is continued on Mare Vaporum, Sinus Medii,
Sinus Aestuum, and Mare Insularum.

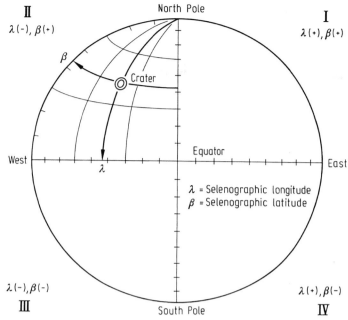

Fig. 50. Selenographic coordinates (east-west in the astronautical sense)

The Moon's best-preserved large impact basin is Mare Orientale (see Fig. 48), surrounded by concentric rings of mountains. Because the situation near the limb of the object is difficult to observe, the rings of mountains are partially seen only, especially the Cordillera Mountains, when the western edge of the Moon is tilted toward the Earth by libration. In general the regions limbwards are very interesting for the observer. The visibility depends on a favorable libration.

The surfaces of maria are not without structure. Three kinds of objects will interest observers: (1) the wrinkled ridges, which are slightly wavy, dam-like mounds or elevated regions on the mare surface; (2) the cone- or dish-shaped craterlets found in maria which are interesting objects with respect to the evolution of the Moon; and (3) the domes, which are observed only at low illumination and which possess holes perhaps similar to volcanic calderas. The rock samples from maria returned during the Apollo space missions have proven to be of the basaltic type; apparently maria consist of large lava flows which have inundated low-lying regions primarily on the Earth-facing side of the Moon. The view of solidified lava streams is particularly evident in the Mare Imbrium (view of Mare Humorum in Fig. 51; also see Fig. 52). The streams show light differences of tint, the older being more reddish, the younger bluish [60].

In a small telescope maria appear to be simple with nothing added. With a larger telescope, especially under low solar light, a lot of structure is visible in the surface, swellings, smooth undulations.

For further reading see the book by G. North *Observing the Moon: The Modern Astronomer's Guide* [61].

13.2.3.2 Terrae Terrae, or highlands, refer to the topography of the brighter areas of the lunar surface which are, as a rule, more elevated than the maria. The terrae occupy more than one-half of the front and almost all of the far side of the Moon's surface. By contrast with the maria, they are mountainous regions which lie over a crust many kilometers thick.

Fig. 51. Mare Humorum and the walled plain Gassendi. Photograph taken on 1981 October 09, 18^h 24^m UT, Cotonou (Benin) by J. Dragesco with a C14, $f/d = 60$, exposure 2 s on XP 400 film

Most craters are found in highlands. Craters are considered the outcome of huge bombardments by meteorites in whose course older craters have been hit and damaged or destroyed by more recent impacts. Thus, recent craters are often found within or near old craters.

Ring Structures

Larger images are wonderfully detailed: *New Atlas of the Moon* by T. Legault and S. Brunier (Firefly, 2006), which has the Moon from day to day, lunar cartography, and lunar movements. Spiral binding allows the book to fold back on itself.

1. Walled plains are craters whose ringed wall enclose a plane of diameter 50–200 km. Examples include Abulfeda, Archimedes, Clavius, Fra Mauro, Grimaldi, Maurolycus, Plato and many others.
2. Ring mountains are similar to the walled plains, but with higher walls and smaller extent of the lower enclosed plain. They are often found to contain central peaks. The best-known examples are Eratosthenes, Copernicus, Petavius, Theophilus, and Tycho.
3. Craters in the narrow sense include all circular forms on the Moon which do not distinctly show any elevated features around them. Their diameters are smaller than those of walled plains and ring mountains. There are also the tiny minicraters and craterlets, which, according to Lunar Orbiter photographs, have diameters of 50 cm and even less (crater holes). Some groups of craters are arranged in string-of-pearls fashion, and in a small telescope can often be mistaken for rills. Terraced inner slopes, the existence of one or several mountains, strata of ejecta, secondary craters, and bright rays in the vicinity make many craters interesting targets of observation. The visibility of the different structures, however, depends strongly on the state of illumination.

Fig. 52. Some typical formations on the lunar surface. *A* Young crater Aristarchus, diameter 40 km, with rays, *B* small, dish-shaped crater, *C* Crater Prinz, partially inundated by mare lava, *D* mare surface, *E* bent rill, *F* straight rill, *G* mare ridge, *H* premare material standing out from mare lavas. Some bent rills follow older ditches and graben structures. Photo courtesy of J.E. Guest, NASA/ULO Planetary Image Center

Mountain Chains These are large mountainous ranges with substantial elevations, the best-known examples of which include the Alps (heights up to 4000 m) and the Apennines (up to 6000 m). Concerning the history of formation, however, they are not to be compared with mountain ranges on Earth. There are convincing reasons to suggest that the lunar structures have been "piled up" as a result of powerful impact events.

Linear Formations

1. Rills are indentations up to 1000 m wide and up to several hundred kilometers long (Fig. 53). Their floors often show many craterlets. Examples are the Ariadeus and Hyginus rills.
2. Crevaces are cracks in the lunar crust which cut into the surface up to several hundred meters deep.

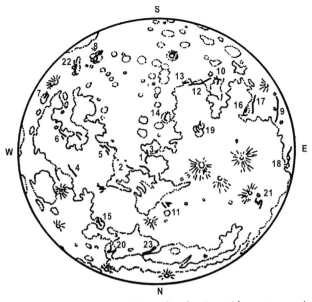

Fig. 53. System of rills (clefts) on the Moon (oriented to the view with an astronomical telescope)

1 Hyginus Rill	9 Sirsalis Rill	17 Mersenius Rills
2 Ariadaeus Rill	10 Hippalus Rills	18 Hevelius Rills
3 Triesnecker Rills	11 Archimedes Rills	19 Parry-, Bonpland and
4 Cauchy Rills	12 Hesiodus Rill	Fra Mauro Rills
5 Sabine and Ritter Rills	13 Pitatus Rills	20 Lacus Mortis Rills
6 Goclenius Rills	14 Alphonsus Rills	21 Schröters Valley
7 Petavius Rill	15 Posidonius Rills	22 Rheita Valley
8 Janssen Rill	16 Gassendi Rills	23 Alpine Valley

3. Valleys are narrow notches in mountains ranges. The most striking example is the Alpine Valley.

4. Bright rays are particularly distinct at Full Moon when no shadows interfere with their visibility (Fig. 54). They are luminous stripes, usually directed outward from a crater. The most conspicuous crater with rays is Tycho; other notable rayed craters include Kepler, Copernicus and Olbers.

Unusual Formations

For the observer, of interest are many objects with incomplete rims, such as impact features with "half craters" only, for example Fracastorius, a crater with a subsided, fractured floor, or Sinus Iridum a big crater without a southern rim. This rim was overflowed by lavas flowing out of Mare Imbrium.

1. Ghost craters appear only shadowy in the dark soil of ma mare. These are craters which have been covered by lava flows. A particularly remarkable object is the half-inundated crater Guericke in the Mare Cognitum.

2. Central peaks are conic mountains or systems there of lying inside walled plains and mountain rings, often appearing similar in structure to shield volcanoes on Earth.

13.2.3.3 On the History of Lunar Nomenclature The names of many lunar regions and formations originate largely from the early days of visual observations of the Moon. A detailed historical survey is given by Z. Kopal and R.W. Carder in [62]. Langrenus in the

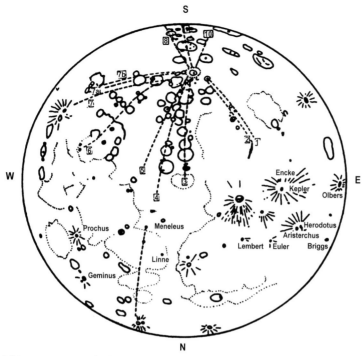

Fig. 54. Bright ray systems on the Moon. Numbers concern rays of the system starting from the crater Tycho (oriented to the view with an astronomical telescope)

seventeenth century was the first to assign names to lunar formations, taking names from the Bible or religious stories about saints. His younger contemporary Johann Hevelius (1611–1687) used names of terrestrial landscapes, of which "Alps" and "Apennines" have remained. The most popular naming scheme in current use was introduced by the Italian astronomer G.B. Riccioli (seventeenth century), who assigned names of known scientists and philosophers to craters and mountain rings. The names given to lunar maria show some astrological influence. The German observers Schröter (1745–1816), Beer (1797–1850), and Mädler (1794–1874) further refined the nomenclature of formations. Beer and Mädler were the first to use letter codes for craters. Under the auspices of the International Astronomical Union (IAU), an internationally accepted nomenclature was prepared in 1935 by Mary A. Black and Karl Müller. The naming after personalities and after geological formations on Earth remains to this day, for instance when naming the recently discovered features of the far side of the Moon (Soviet Mountains, Lomonossow, Joliot-Curie, etc.).

The cartographic recording of the far side and, in recent years, very detailed mappings of the near side of the Moon (scales of 1:250,000 and larger) have created new demands on nomenclature. The large-size maps are divided into 144 regions, each of 16 subregions, each one of these being referred to by the name of a prominent local crater. This organization, however, is not needed for the maps used by amateur observers.

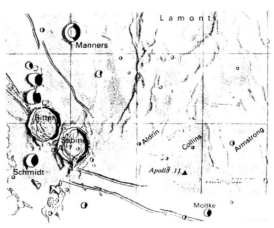

Fig. 55. The soft-landing site of the Apollo 11 mission

The American astronaut Neil A. Armstrong (Apollo 11, 1969) was the first man to set foot on the Moon. To find the place, begin observation at the southern edge of Mare Tranquilitatis, east of the craters Sabine and Ritter. A short distance eastwards the lonely and small crater Moltke is the Apollo 11 landing site's nearest significant lunar object. East of Sabine and north of Moltke you see a chain of three tiny pits named Aldrin, Collins, and Armstrong. The Apollo 11 landing site is between Moltke and Collins, the middle of the three tiny pits.

It is still customary to assign names to lunar formations. In the past decade, a number of small craters, formerly coded by letters, have been given names; for instance, Messier G became "Lindbergh." The old letter codes, however, are still in use.

Since 1976, additional names for structures other than craters have been used. For example, *Mare Insularum*, meaning sea of islands ($b = 7°$ north, $L = 22°$ west) or *Sinus Amoris*, meaning Bay of Love ($b = 7°$ north, $L = 38°$ east). Updates on the nomenclature are published in the triennial transactions of the IAU, which is, by endorsement of the UN and the International Council of Scientific Unions, in sole charge of naming astronomical objects or of changing names. The appearance of many small and very small lunar formations required also new generic names which are taken from the Latin, for instance: *anguis* meaning curved rill, *catena* meaning crater chain, *dorsum* meaning ridge, *fossa* meaning straight rill, *mons* meaning mountain, *montes* meaning mountain range, *promontorium* meaning cape, *rima* meaning rill, *rupes* meaning ditch, and *vallis* meaning valley.

13.2.4 Observational Projects

13.2.4.1 Visual Observations For visual work, long-focus refractors and reflectors (f/10 to f/20) have been used to advantage for optimal image definition. The brightness of the Moon sometimes dazzles the observer. Sensitive eyes will find help in neutral-density (gray) filters or by observing during twilight. Other observational tools include color and polarization filters and also, for instruments with 6-in. or larger apertures and clock drive, filar micrometer.

The situation concerning the proper magnification to use is similar to that in planetary observations. The noted lunar observer Philipp Fauth wrote, on this subject, "I work best on the perception of finest surface features at my 163-mm objective with magnification 160×, at the 176-mm apochromat with 176×, at the Medial at 300-mm aperture with 300× and 350×, and at 385-mm aperture with 350× and 430×. Younger observers with more sensitive eyes may wish to increase magnifications by 50% over these, image steadiness permitting. To go still further has little purpose; although the image grows larger,

Fig. 56. Folded refractor (Schaer refractor by M. Wachter). Aperture 125 mm, f = 2300 mm, length of construction 850 mm

it also becomes paler and more diffuse, and on the Moon clarity is more important than image size."

Clarity is aided by binocular observations, which are particularly well-suited for lunar observation and provide not only a lasting impression but also relaxes viewing. According to the author's experience with Baader binoculars (Munich), details on the Moon are recorded not only faster but also more safely and less ambiguously when using both eyes. G. Miller writes in a report: "It is confirmed that fine details were seen better and apparently larger when using both eyes in binoculars than by using only one of the two eyepieces."

Of interest too are folded refractors because of the long focal length, the good achromatic correction, and a short construction (see Fig. 56).

Drawings at the Telescope According to Philipp Fauth, "A patient acquaintance with a limited region, taking every available opportunity, makes one so engrossed in its structure that ultimately one perceives every detail and records it on paper."

A prerequisite for any topographic work is the knowledge of the basic surface structure. Skeleton maps for orientation and drawings of the outlines (silhouette maps) can serve as a starting pint. Both are obtained by tracing reliable lunar maps or, better still, photographs. Low-contrast copied and enlarged photographs can be used as templates for sketching. Every silhouette map should show exact markings of position so that all

Special instruments for lunar and planetary observers are systems with unobstructed optics, for example an eccentric pupil Newtonian reflector (see *Sky and Telescope* April 2005, p. 88) or an off-axis reflector called a Schiefspiegler, constructed by Anton Kutter of Germany. Even with small apertures of about 4 in., unobstructed scope's are of interest for the Moon watcher. Considering a high-priced 4-in. apochromatic refractor, the unobstructed reflector is an alternative worth looking at.

Fig. 57. Example of a topographic representation of the lunar landscape. The picture is from P. Fauth and represents the walled mountain named after him. Adapted from P. Fauth [63]

details can be entered with their true orientation. The scale to be chosen should not be too small. Fauth drew his map at a scale of 1:1 million, which corresponds to a lunar diameter of 3477 mm. This is an appropriate order of magnitude which should, in the interests of clear presentation, be used as a guide. K.W. Abineri, who observed for many years with a 200-mm reflector, adds in instructions for observers [57] this advice: a relaxed posture (seated, if possible), a stable support for the drawing pad, an adjustable red lamp for illumination, and a soft pencil (see also [60]).

The technique of sketching depends on the ability of the observer. One distinguishes between simple line sketches and the more expressive full-tone drawings, in which one

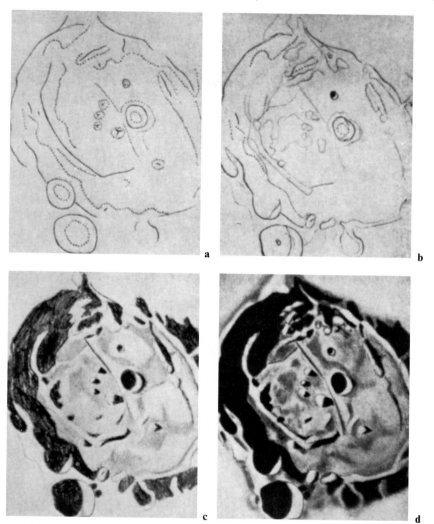

Fig. 58a–d Drawing of the walled plain Posidonius by C.R. Chapman (ALPO). (**a**) The outline taken from a photograph. (**b**) Illumination and libration. (**c**) Addition of shadows. (**d**) Completes the contours and intensities. Courtesy of *The Strolling Astronomer*, Vol. 17, p. 47 (1963)

tries to incorporate a range of light and dark densities into the picture. Fauth drew in contour lines (Fig. 57), and expressive hachure.

Figure 58a–d illustrates the transition of a simple line drawing into a tone drawing which clearly shows shadows and light intensities.

As an example of a plan for systematic observations, several points recommended by F. Billerbeck-Gentz for the observation of lunar ray systems will be mentioned here. The observer should try to answer the following questions: (1) Where does the ray begin? (2) What is its course? (3) Where does it end? (4) How wide is it? (5) How bright is it? (6) Does the brightness of the ray vary along its length and width? If so, at which points

See the book edited by G.D. Roth *Planeten beobachten* [64] for helpful tips on observing and photographing the Moon and planets, and advice on using instruments and accessories.

and when? (7) What is the general surface structure of the ray? Could the structure of the ground perhaps be simulated by a series of bright points, "white" craters, or small surface warps? (8) What is the shape of the ray? Is it straight or does it bend, and if so, where? Is it interrupted, and if so, by how much? (9) Do any parts of the ray show a change of color or tint at times when the ray itself is not visible? (10) At which times do the rays appear and disappear?

Measurements at the Telescope   Direct telescopic measurements make use of the micrometer to determine relative lunar dimensions (crater diameters, mountain heights) and positions. The method of deriving elevations on the Moon was already well-known in the eighteenth century. When the length of the shadow is measured and the altitude of the Sun known, the height of a lunar formation relative to the surrounding elevation are enlarged by a factor of 100. The accuracy reached is only 200 to 300 m. These heights are called "relative" heights because they refer to the terrain onto which the end of the shadow falls. These are to be distinguished from the absolute heights referred to the radius of a chosen reference sphere.

Most altitude data on lunar maps before 1960 are derived from shadow measures, and are relative. Better accuracy of relative heights is possible with photographs taken on space missions and evaluated according to photogrammetric principles. For instance, the Lunar Orbiter maps, which appeared after 1967, have relative heights with precisions of a few meters.

The measurement of the shadow lengths of certain lunar features in order to derive heights is a useful exercise for the observer. Shadows are always measured at right angles to the line joining the horns (or cusps). The fixed horizontal wires are oriented along the shadow, which also defines the mountain peak exactly. The precision obtained will depend on the length of the shadow; the measurement of shadows which are too short leads to unreliable results. The altitude of the Sun during such measurements should be in the range 5° to 15°. If it is less than 5°, there is the danger that the very long shadow will merge with the shadows of other, formations or run into the unilluminated part of the Moon.

For determination of the heights of lunar surface features see *The Moon Observer's Handbook*, p. 276 by F.W. Price [58] and *Astronomisches Praktikum I* (in German), p. 98ff. by O. Zimmermann [66].

See the November 2007 issue of *Sterne und Weltraum*, p. 52, to become acquainted with a very simple method. Helpful information is also provided by the "Virtual Moon Atlas" (http://www.astrosurf.com).

J. Hopmann has determined with a 21-cm refractor (2.92-m focal length) 175 heights of 163 points on the lunar surface [65]. He reported as follows on the measuring technique: "To measure the shadow lengths, the fixed 'horizontal' wire was oriented along the position angle of the terminator from the ephemerides for physical observations of the Moon in the Almanac. Then it can be always clearly identified which part of a crater, ridge, etc., casts the shadow. Because of overexposure, these places appear so bright in photographic atlases (Paris, Kuiper) that identification of the shadow-casting point is not possible. For objects a few arcseconds wide, the eyepiece still shows the shadow peak, whereas on the photographic copy, and probably also on the original negative, the entire region over $2''$ to $3''$ appears white or dark. When observed at the same phase, the peak may be equally well-marked visually and photographically, but then the micrometer measuring procedure follows that of double-star observations: double distances are determined with 3 settings on either side. The time of observation at the midpoint is recorded to the nearest minute."

Visual Photometry   Changes in the relative brightness of the Moon are distinctly seen in the natural course of the phases. However, the reflectivity varies over the different regions of the lunar surface. Therefore the change of brightness of the Moon does not occur quite symmetrically with the phases. Some recommendations for observers concerning the "integrated" photometry of the whole Moon have been published and are referred to in [66]; see also E.A. Whitaker "Lunar Photometric and Colorimetric Properties" in [57]. For advice regarding measurements to be taken during a lunar eclipse, see Sect 13.3.

One task for visual estimates is to find the change in brightness of a certain formation, for instance the surface inside a crater, as a function of the position of the terminator. All estimates are carried out on a relative scale which is in principle similar to that of the step-estimation used for variable stars (see Sect. 8.4.2). Of course, the objects with which the scale is calibrated must be absolutely invariable during the entire course of a lunation.

On a scale recommended by H.J. Klein and H.K. Kaiser, the darkness of the shadow is expressed by $0°$, while $10°$ corresponds to the brightest point on the visible lunar surface. This gives the following other degrees of calibration:

$0°.0$ = Shadow on the Moon
1.0   = Darkest part of the inside of the walled mountains Grimaldi and Riccioli
1.5   = Inner surface of De Billy
2.0   = Inner surface of Endymion and J. Caesar
2.5   = Inner surface of Pitatus and Vetruvius
3.0   = Sinus Iridum
3.5   = Inner surface of Archimedes and Mersinus
4.0   = Inner surface of Ptolemy and Guericke
4.5   = Sinus Medii, surface around Aristyllus
5.0   = Surface around Archimedes, walls of Landsberg and Bullialdus
5.5   = Walls for Timocharus, rays of Copernicus
6.0   = Walls of Macrobius and Kant
6.5   = Walls of Langrenus and Theatetetus
7.0   = Kepler
7.5   = Ukert and Euclid
8.0   = Walls of Copernicus
8.5   = Walls of Proculus
9.0   = Censorinus
9.5   = Interior of Aristarchus
10.0 = Central peak of Aristarchus

The very similar "Elger's Albedo Scale" is also cited in the literature, and was refined by P. Hedervari [67]. Observers may find it worthwhile to verify the scale and amplify it by using other objects and intermediate steps.

The scale for visual estimates can be extended by linking the observed brightnesses to photographically determined photometric values of certain objects. Aside from visual estimates, visual intensities can be measured by photometric means with the aid of neutral wedges and polarization filters which dim an artificial light source to match the brightness of a lunar formation.

Relevant objects here are transient lunar phenomena (TLP). For successful observations it is necessary to work with a telescope of large enough aperture, 8-in. and more. Reflectors should be preferred because the primary image is entirely color-free.

13.2.4.2 Photographic Observations Reflectors as well as refractors are suitable for lunar photography. A telescope of 1 m focal length produces an image of the Moon of about 10 mm in diameter; the image is correspondingly reduced or enlarged for shorter or longer focal lengths.

For information dealing with video cameras, image intensifiers and CCDs see Chap. 6. The Moon is a large and bright object. For imaging any kind of camera can be used (film, digital, video, CCD).

The essentials of techniques and of the selection of imaging is discussed in Chap. 6. As in the procedure for planetary photography, the eyepiece collar carries either a special Moon or planet camera or a single-lens reflex camera (Nikon, Olympus, Pentax, etc.), but minus the optics. For focal photography, the eyepiece of the telescope is not used. If, however, it is desired to obtain an enlarged image of the Moon, then the eyepiece is used to project the primary image of the Moon onto the emulsion. The essential differences between the two methods can be summed up nicely as follows:

– Focal photography gives a bright but relatively small, image requiring a short exposure time and subsequent enlargement.
– Projection photography yields a dimmer, but larger, image (again see Chap. 6), and hence requires longer exposure and less subsequent enlargement.

Excellent photographs have been published of late in various sources, for example by J. Dragesco (see Figs. 43 and 51), B. Flach-Wilken, G. Nemec, G. Thérin, G. Viscardy [68], and W. Schwinge.

The changes in the apparent brightness of the Moon over the phase cycle pose substantial problems to the photographer. The brightness does not increase or decrease proportionally to the extent of the illuminated surface. In comparison with pictures taken of the Full Moon, the exposure time at first and last quarter should be increased by a factor of 4, and for the 3- or 24-day-old crescent by as much as a factor of 12. This is done with the understanding that the "correct" exposure succeeds only near the terminator, while the regions with steeper light incidence are considerably overexposed.

There is no fixed rule regarding the exposure times, as much depends upon the instrument used, the lunar phase, the altitude of the Moon, and the type and speed of emulsion. The following may be used as a crude approximation: a picture of the 50/18° emulsion with 1/50 to 1/10 s, a moderately large (10–15 cm) projected image with 1/2 to 3 s. Use of color filters requires application of the corresponding factors of increase. Any photograph whose exposure exceeds 1 s demands faultless tracking of the telescope.

The successful photographer of the Moon and the planets, G. Nemec, writes, "It is absolutely necessary to work very precisely, and time pressure is always detrimental to the results. In the case of the Moon, this may be the major reason why photographs of the waning Moon are often more successful than those of the waxing Moon, which culminates in the early evening when the astronomer is still occupied with his instrumental preparations and the telescope has not yet cooled down sufficiently. Atmospheric conditions often change irregularly during the night, so that one can never be sure that the second half of the night will guarantee the best seeing. Besides scintillation, it is the altitude of the object on which the definition essentially depends. Photographs taken more than one or two hours before or after transit yield less satisfactory results."

For expert advice on photographing the Moon and planets see Vol. 1, *Compendium of Practical Astronomy*, published by Springer.

For further reading, see *Handbook of CCD Astronomy*, by S.B. Howell, Cambridge University Press, Cambridge, 2000.

A special publication written by Jean Dragesco, *High Resolution Astrophotography* (Cambridge University Press, 1995) reports about specialists in lunar and planetary photography. Jean Dragesco is an internationally well-known pioneer in this matter.

Tasks for Lunar Photography The lunar photographer with a skilled hand and good optics may want to supplement or revise visual observations with photographs. Naturally, he or she will also compare his or her plates with those of other observers. Good negatives can be investigated by photometric methods. The *Fotografische Mondatlas* by W. Schwinge [54] has demonstrated that amateur instruments (apertures 50–200 mm) are capable of high-resolution photographs and that a good-quality atlas can be prepared based on photographs taken over almost an entire lunation. Newly developed objectives such as the high-efficiency optics APQ 100/1000 (Carl Zeiss, Jena) or Starfire EDF, 160 mm, $f = 7.5$ (Astro-Physics, USA, http://www.astro-physics.com) remove the difficulty that visual and photographic foci differ for the observer photographing with a refractor.

Pictures taken of a particular region, say one suspected of luminescence activity, can be taken with color filters, for instance green (545.0 nm) and red (672.5 nm) interference filters. When performed systematically, they supply useful material for comparative processing. This may be achieved with a good 5-in. refractor.

The photographic determination of heights of lunar formations is of special interest, but affected by the drawback that shadow measurements on a photograph are rather inaccurate owing to the influence of bad seeing, diffuseness of the images, and so on. For such research, visual observations are preferable (see p. 380).

13.2.4.3 Photoelectric Observations
Recent technological advances have rapidly made photoelectric observations accessible to the amateur (see Chap. 8 for information on equipment and techniques); such methods can be applied to the Moon. Measurements described in the following sections are also suited for the search for transient lunar phenomena (TLPs) as mentioned on p. 362; see also [57, 67].

Photometric Measurements The scattering of sunlight off the lunar surface does not follow Lambert's law of reflection (Fig. 59). Photometry of comparable surface formations on various parts of the disk and at various solar altitudes permits the determination of the backscattering behavior of the lunar surface as a function of the angles of incidence and reflection (phase curve). The characteristics of scattering are determined primarily by three surface properties:

To undertake an observing project see the website of the Lunar Observing Club, Astronomical League
http://www.astroleague.org/al/obsclubs/lunar/lunar1.html

British Astronomical Association: http://mysite.freeserve.com/lunar/index.html

Association of Lunar and Planetary Observers (ALPO)
http://www.lpl.arizona.edu/alpo/

Vereinigung der Sternfreunde e.V. (VdS), Germany
http://www.vds-astro.de

Schweizerische Astronomische Gesellschaft (SAG),
http://www.astroinfo.ch

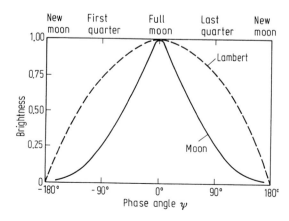

Fig. 59. Photometric phase curve of the Moon. The *dashed curve* is the predicted total brightness from Lambert's law of reflection. The *solid curve* is the actual brightness. Adapted from R.H. Giese, *Erde, Mond und benachbarte Planeten*, Mannheim, 1969

1. Distribution of scattering, and unevenness and cavities on the ground. Sunlight may reach the surface features from very different angles.
2. The albedo of the surface material. Maria have a lower albedo (about 0.05) than the highlands (about 0.10 to 0.15).
3. Scattering functions of various components of this material such as rocks, grains, and dust particles.

Photometry has a long tradition in astronomy. For more about the principles of photometry, see Chap. 8, Vol. 1, *Compendium of Practical Astronomy*, written by H.W. Duerbeck and M. Hoffmann [69].

Observers may wish, for instance, to monitor the phase curves of selected objects such as craters, valleys, rills, bright rays and dark spots. Regions of interest include, for instance, the Long Wall, the Alpine Valley, Vallis Rheita, and the "Cobra Head" of the Schröter Valley.

Color Measurements Differences of color and tint within the Mare Imbrium have already been mentioned (p. 369). The view of the Full Moon makes it difficult to gain an impression other than that of a silvery white disk with gray spots. But spectrophotometric studies indicate that the lunar surface reflects about 8% of incident blue light and 12% of red light. The backscattered light thus has a brownish tint, and various regions of the Moon show small, but distinct changes of tint. By comparison, for example, Mare Tranquilitatis has a more steel-gray coloring, Mare Serenitatis more yellow-brown. Spectrophotometry uncovers the fact that both Maria have about the same blue backscatter (8%) but differences in red: Mare Serenitatis reflects red slightly better (about 12%) than Mare Tranquilitatis (about 11%).

Gray formations in brown surroundings include, for instance, the floor of the crater Boscovich and the floor and central peak of Aristarchus. Examples of brown formations with gray surroundings are the craters Moltke and Plinius, and the floor of Bullialdus. Other good examples can be found in [57].

Polarization Measurements The available measurements indicate that the dark maria polarize light more strongly than the bright terrae. Polarized radiation can be found with the aid of a polarization filter rotatable by 135° and placed directly in the photometer beam. The individual measurements should be accurate to 0.01 mag, which is achievable only with a photoelectric photometer.

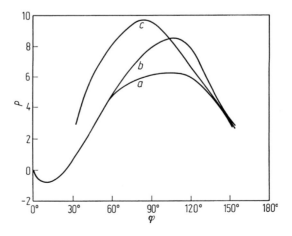

Fig. 60. Change of the degree of polarization depending on the lunar phase (after Lyot): *a* waxing moon, *b* waning moon, *c* ashen light. Adapted from D. Böhme [70]

Various albedo patterns of the waxing and waning moon are manifested in the polarization behavior. Some formations show distinct differences in polarization over a lunation [62] (see also Fig. 60).

A polarimeter connected to a photoelectric indicator permits polarization studies of individual regions on the lunar surface, with the intention of detecting changes of the degree of polarization P, defined as

$$P = \frac{I_1 - I_2}{I_1 + I_2} \times 100\% .$$

Here, I_1 is the light intensity polarized in the plane Sun-Moon-Earth and I_2 the intensity in the plane perpendicular to it.

Observations of Luminescence Certain areas on the Moon fluoresce when subjected to solar irradiation. This luminescence can be identified visually, photoelectrically, and by filter photography. The photoelectric measures are preferably carried out in three colors: green, red, infrared (p. 362 on transient lunar phenomena). For further reading, the section "Luminescence of the Lunar Surface" in [51] is recommended.

Studies of Surface Structures A detailed knowledge of the shape of the lunar surface can be achieved only with a variety of techniques. A combination of the various measurements mentioned in Sect. 13.2.4.2, if carried out under as many different illuminations of the Moon as possible, can lead to the determination of a kind of "microrelief" of the lunar surface. Unfortunately, the requisite accuracy admits only photoelectric measures.

Also suggested are albedo measures of the Moon taken in the infrared, and the search for formations which, when located in the unilluminated part of the Moon, radiate infrared radiation [50]. The peak spectral sensitivity of the measurements should be in the range 700–900 nm. The heat absorbed by the lunar surface in the daytime is reradiated and can be observed in the infrared most distinctly during the lunar night. Lunar eclipses (Sect. 13.3.2.2) are especially useful in this regard. At the onset of the eclipse, the visible surface is more or less evenly heated, but as the eclipse progresses, the incident solar energy is gradually blocked off. The degree of cooling indicates a range of physical properties of the surface material.

When imaging the Moon, the use of selected interference filters when searching for luminescent gas clouds from suspected volcanic eruptions (TLPs see p. 363) has been suggested.

Gray or reddish filters can help to reduce seeing problems because the brightness of the image. Some observers noted a darker impression of lunar basalt when using color filters (orange, red).

13.2.5 Further Reading

Cameron, W.: Report on the ALPO Lunar Transient Phenomena Observing Program. Journal of the Association of Lunar and Planetary Observers **25**, 1 (1974)

Cameron, W.: Lunar Transient Phenomena Catalog. NASA-Goddard Spaceflight Center, Greenbelt, MD (1978)

Cherrington, E.H.: Exploring the Moon through Binoculars and Small Telescopes. Sky, Cambridge, MA (2005)

Cohen, J.: The Face of the Moon: A Descriptive Guide. Melbourne (1998)

Cook, J. (ed.): The Hatfield Photographic Lunar Atlas. Springer, Berlin Heidelberg New York (1999)

Dunlop, S., Gerbaldi, M. (eds.): Stargazers. The Contribution of Amateurs to Astronomy. Proceedings of Colloquium 98 of the IAU, 20–24 June 1987. Springer, Berlin Heidelberg New York (1988)

Fauth, H.: Phillip Fauth and the Moon. Sky Telescope **XIX**, 20 (1959)

Grego, P.: Moon Observer's Guide. Sky, Cambridge, MA (2005)

Hill, H.: A Portfolio of Lunar Drawings. Cambridge University Press, Cambridge (1991)

Janle, P.: Das Bild des Mondes. Vom Altertum bis zum Beginn der Weltraumfahrt. Sterne und Weltraum **38** (1999)

Koch, B.: Handbuch der Astrofotografie. Springer, Berlin Heidelberg New York (1995)

Lunar Flash Revisited. Sky and Telescope **79**, 590 (1990)

Mondkarten von Philipp Fauth. Nachrichten der Olbers-Gesellschaft Bremen **60** (1964) and **63** (1965)

Mysterious Flash Photographed on the Moon. Sky and Telescope **77**, 468 (1989)

Oberst, J., Jaumann, R., Hoffmann, H.: Von den Apollo-Landungen bis heute. Was wir über die Mondoberfläche gelernt haben. Sterne und Weltraum **38** (1999)

O'Meara, S.O.: The Lunar Straight Wall. Sky Telescope **73**, 639 (1987)

Roth, G.D.: Die Mondkarte von H.P. Wilkins. Die Sterne **31**, 180 (1955)

Roth, G.D. (ed.): The Moon, Compendium of Practical Astronomy, vol. 2. Springer, Berlin Heidelberg New York (1994)

Rükl, A.: Atlas of the Moon. Sky, Cambridge, MA (2004)

Taylor, S.R.: Lunar Science: A Post-Apollo View. Pergamon, Oxford (1975)

Wlasuk, P.T.: Observing the Moon. Springer, Berlin Heidelberg New York (2000)

13.3 Eclipses and Transits

C. Treichel

13.3.1 The Nature of Eclipses

Geometry of solar or lunar eclipses.

Solar or lunar eclipses in the Earth–Moon system are due to projection of sunlight on one of the bodies and the resulting shadow on the other of the bodies. This geometrical configuration is only possible when Sun, Earth and Moon are exactly located in one line (Fig. 61). In the case of a Sun-Earth-Moon configuration, as it appears at Full Moon, the result is a lunar eclipse. In the other case of a Sun-Moon-Earth configuration, as it appears at New Moon, we get a solar eclipse. As the Moon's orbit is inclined about 5° with respect to the ecliptic, not every New or Full Moon will cause an eclipse.

Per year, a maximum of five solar or lunar eclipses can occur. The sum per year cannot be higher than seven

The number of eclipses differs from year to year. Eclipses typically occur in pairs, with a solar eclipse immediately preceding and/or following a lunar eclipse by about 15 days.

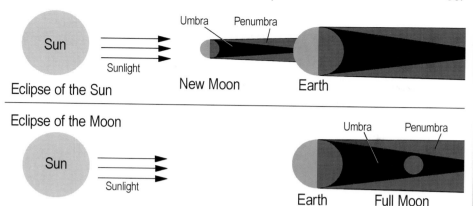

Fig. 61. The geometry of solar and lunar eclipses. Only, when Full or New Moon is near one of its nodes, which are two intersections between the Moon's and the Earth's orbit, can eclipses take place

There is a relationship between the periodicity of the Moon's synodic month and the eclipse year. Two-hundred twenty-three synodic months nearly have the same period of time as 19 eclipse years, which are 6585 years and 0.3 days or 18 years and 11 days and 8 hours. This is called the Saros period. After this period, eclipses repeat in a similar manner and will share many common characteristics, such as duration, type, length, time of the year and so on.

Saros period

The maximum totality for a solar eclipse cannot exceed 7 min 31 s

During a solar eclipse, the Moon's shadow covers a part of the Earth's surface. The Moon's shadow consists of two parts, the dark central umbra and the lighter penumbra. An observer located outside the penumbra will not see any eclipse. An observation site inside the penumbra will show a partial solar eclipse (Fig. 62). Those located within the umbra will see a total solar eclipse (Fig. 63).

Partial and total solar eclipses

On maps which show the narrow path of the Moon's umbral shadow on the Earth's surface during a total eclipse, there is also a center line plotted. At these positions, observers have a centered view to the total eclipse, that means that the Moon's disk crosses the Sun's disk without any deviation to the north or the south. Observations at the center line offer a maximum duration of totality in comparison to positions near the northern or southern limit of the shadow's path.

Center line

Because the Moon surrounds the Earth on an elliptic orbit, the apparent size of the Moon's disk differs from eclipse to eclipse. When the Moon is near the perigee, its disk is larger as it is in the case of apogee. As the Earth's orbit around the Sun is also elliptically, the apparent size of the Sun's disk varies too. This leads to different combinations of the Moon's and Sun's apparent sizes during eclipses.

Moon's apparent size

Sun's apparent size

If the Moon's disk is smaller than the Sun's when it passes centrally, an annular eclipse with a bright ring of sunlight around the Moon appears (Fig. 64). Some annular eclipses are hybrid, which means in the middle of the eclipse path, the distance between Moon and Earth reaches a point where the Moon's apparent size is again big enough to totally eclipse the Sun.

Solar eclipse contacts, see Fig. 65

Annular eclipse

Annular/total eclipse

The fraction of the Sun's diameter obscured by the Moon defines the eclipse magnitude. This is also valid respectively with the Earth's shadow for lunar eclipses. For partial or annular eclipses, the magnitude is always less than 1.0. The magnitude for total

Magnitude of solar eclipse

Fig. 62. Partial solar eclipse near the western horizon on May 10, 1994 in Berlin, Germany. Photo courtesy of Christine Treichel (500 mm, f/8, Gray filter, automatic exposure time, Kodak Chrome Elite 100)

Fig. 63. Total solar eclipse. Photo courtesy of Klaus-Peter Schröder (March 29, 2006, Turkey, Pentax Ist, 700 mm, f/11.6)

Greatest eclipse

eclipses isalways equal to or greater than 1.0. Greatest eclipse is the point on the Earth's surface where the Sun reaches its greatest obscuration from any location on Earth. This is the moment when the axis of the Moon's shadow passes closest to the Earth's center. This location offers the longest duration of totality along the whole path of the Moon's shadow.

Lunar eclipses

Lunar eclipse contacts, see Fig. 66

A lunar eclipse always addresses a larger public because it can be seen from all over the world where the Moon is visible above the horizon. For lunar eclipses also differ-

Fig. 64. An illustration of an annular eclipse near the western horizon

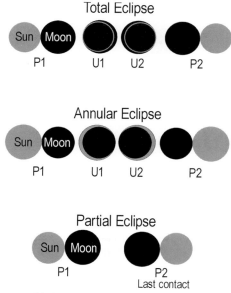

Fig. 65. A solar eclipse consists of different stages, divided by contacts between the solar and lunar limbs. The first contact (P1) is defined as the moment when the eastern lunar limb first touches the western solar limb, the second contact (U1) is the moment when the eastern lunar limb first touches the opposite solar limb. The third (U2) and fourth (P2) contacts are the inverse events at the western lunar limb, respectively. One exception are annular eclipses, where the second contact (U1) is the moment when the western lunar limb first touches the western solar limb, and the third contact (U2) is the moment when the eastern lunar limb first touches the eastern solar limb. For partial solar eclipses, only a first (P1) and last (P2) contact is defined

ent sceneries exist as shown in Fig. 66. The Earth's shadow likewise consists of a light penumbra and a dark umbra. When the Moon crosses the penumbra, a penumbral lunar eclipse occurs which is very dim. But when the Moon transits the umbral shadow,

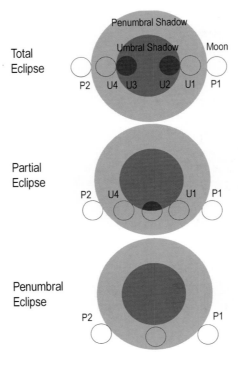

Fig. 66. Also for lunar eclipses, contacts are defined, respectively. A lunar eclipse progresses between stages P1, U1, U2, U3, U4, and P2

Complete reference of all oncoming eclipses and transits, see Fred Espenak's Eclipse Homepage: http://sunearth.gsfc.nasa.gov/eclipse/eclipse.html

Eclipse eyeglasses manufacturer:
http://www.baader-planetarium.de
http://www.rainbowsymphony.com/soleclipse.html

either a partial or a total lunar eclipse occurs. In the case of a partial lunar eclipse, the Moon will not be covered completely by the Earth's shadow as it is in the case of a total lunar eclipse.

13.3.2 Telescopes and Further Equipment

13.3.2.1 Solar Eclipses Watching the Sun without any optics needs special "eclipse eyeglasses" to protect the eyes. The glasses consist of a paper frame clamping astronomical sun blocking filter foil instead of lenses (Fig. 67).

Fig. 67. Eclipse eyeglasses consist of an astronomical foil that protects the eyes against dangerous solar radiation. Image courtesy of Christine Treichel

Fig. 68. The filter must be attached with a frame. Photo courtesy of Christine Treichel

Binoculars offer a huge field-of-view, especially important during totality when the extended corona or the star field are of most interest.

In case of a telescope, a refractor should be preferred over a reflector because focal length is more important than aperture. In addition, a reflector is more sensitive to heat, to which it is subjected during the observation. In order to diminish the impact of atmospheric turbulence, an aperture is useful below 20 cm. It might be necessary to dim out the lens with a whole mask in order to reduce the incoming light and resulting heat inside the tube.

For direct observation with binoculars or telescopes, an astronomical sunblocking filter foil must be attached in front of the optic. These filters have a density of 4 or 5 and reduce the sunlight by about 10 to 12.5 mag. It is very important that the frame of the filter is tightly connected to the optic and cannot fall down during observation (Fig. 68). For the best optical results it is recommended to have the filter stretched in the frame without any wrinkles.

Sun projection with the telescope needs robust oculars like Ramsden, Huygens or Mittenzwey, because they cannot be damaged by the heat. Either a commercial projection screen connected to the ocular rear or just the inner side of a shoe box can be used to show the Sun's image (Fig. 69).

Binoculars

Telescopes

Solar filter manufacturers:
http://www.baader-planetarium.de
http://www.coronadofilters.com

Manufacturers list:
http://www.skyandtelescope.com/equipment/vendors/

Projection

Fig. 69. Telescope projection screen. Photo courtesy of Christine Treichel

Magnification

Hα, IR or polarization filters, objective prism, diffraction grating

Weather stations manufacturer: http://www.lacrossetechnology.fr/en/heavyweather.html

Instrument protection

Plain white sheet

Magnification

Tape recorder

Stopwatch

Time signal

GPS

Photometer

Photographic equipment

Exercising

A magnification factor of 60 is recommended for observation of fine structures in the corona or for prominences. For special programs (see Sect. 13.3.6.2), hydrogenic alpha, infrared and polarization filters are necessary. An objective prism or a diffraction grating is necessary to observe the emission spectrum of the corona.

It makes sense to record meteorological data during the eclipse day and also the preceding and following days in order to estimate weather prospects for future eclipses at similar regions and climates. Small transportable and automatic weather stations can be installed at the observation site. Connected to a notebook, these stations transfer the recorded data to the hard disk. Usually outside temperature, air pressure, humidity, wind direction, wind speed, dew point and rainfall are recorded.

Telescopes and cameras need protection from heat while waiting for the eclipse by covering them with a white or reflective blanket. A plain white sheet or tablecloth laid on the ground can be used to detect "shadow bands."

13.3.2.2 Lunar Eclipses For lunar eclipses, smaller aperture telescopes are also recommended, because they show the penumbral, and respectively, the umbral shadow in higher contrast in relation to the light part of the Moon. Using a magnification of about 150×, an estimation of the penumbral or umbral shadow color is possible. The same magnification is practical for observing the progression of the Earth's shadow over the Moon's surface and for measuring crater contact times. Some observers prefer using a Moon filter, e.g., for better identification of the penumbral shadow on the Moon's surface.

13.3.2.3 Common Equipment A battery-powered tape recorder can be used to record comments or time relevant events during the eclipse without having to write things down on paper. The recorder should be close enough to the observer to pick up his or her voice without having to hold a microphone in hand.

For time measurements a stopwatch and a short-wave or long-wave radio that picks up time signals from a time broadcasting radio station are useful. These signals can be received from nearly any point on the globe. A list of time signal broadcasting stations can be read in Sect. 13.4. Also a GPS receiver with a 1PPS synchronization can be used.

The brightness of the sky throughout the solar eclipse can be measured with a photoelectric cell or exposure meter. Solar and lunar eclipses are also of interest for photography. For more information, see Chap. 6.

13.3.3 Devising the Program

Planning and exercising the observation is essential for having the best results on eclipse day. For solar eclipse photography, focal length and exposure series for totality can be estimated in advance using the Full Moon at night because it has the same apparent diameter as the Sun and the same brightness as the corona at totality. In general, the handling of instruments, filters and other equipment but also the best spatial arrangement of tripods and telescopes should be exercised. For solar eclipses this is practical in twilight, for lunar eclipses in darkness. In the case of a single observer, especially of a solar eclipse, he or she should not try to cover all the opportunities offered by the eclipse. It is more rewarding to have enough time to view the totality in a relaxed manner than to rush from one observation part to another. In the case of several observers it would be helpful to assign different activities to every participant.

13.3.4 Site Selection

In most cases total solar eclipses will only be visible by traveling to a place inside the path of totality, as close as possible to the center line. Such a journey can be planned individually or in cooperation with an astronomical club which offers programs to participate. Preceding to the journey it is important to study the meteorological situation in several target areas. As the weather depends on the shape of the landscape, sites on the leeward side of mountains are preferred as they prevent cloud forming.

In most cases, lunar eclipses can be observed from a place not far away from home. The night-time quality of the observation site should be inspected in advance, e.g., the seeing, the horizon view or sources of light pollution.

In general, the observation site should be even and in order to tower over the dust, at the highest elevation possible. The place should be reachable by a road and also be safe from disturbance from non-observers.

Weather prediction:
http://
www.lunar-occultations.com/
iota/iotandx.htm
http://www.worldweather.org
http://www.intellicast.com
http://www.wunderground.com

13.3.5 Table of Solar and Lunar Eclipses

The following tables (Tables 6 and 7) list solar and lunar eclipses up to the year 2020.

13.3.6 Solar Eclipses

13.3.6.1 Observation Methods A solar eclipse is a wonderful natural spectacle but it is not harmless. Even if the eyes are watching the Sun directly for a few seconds, permanent eye damage or even blindness may result. Besides the visible light, the Sun also radiates intense infrared and ultraviolet light. Those damage the eye's retina faster than it causes sunburn on our skin and it is undetectable because there is no heat-sensitive neurotransmission inside the eyes. **Therefore it is of utmost important to protect the eyes during observation**. The only stage of a solar eclipse where the scenery can be observed without eye protection is totality, as the Moon covers all parts of the Sun where dangerous radiation originates.

Eye Safety

Do not try to form a sunblocking filter on your own, because it is not safe!

For watching the Sun with no optics, the best way to protect the eyes is to wear special eclipse eyeglasses (see Sect. 13.3.2.1). For binoculars or telescopes, an astronomical sunblocking filter foil must be attached in front of the optics.

Filters that screw into eyepieces should never be used! They are fixed near the focus. The heat can break them and incoming sunlight would damage the observer's eye.

Furthermore the handling of a projecting telescope (see Fig. 69) during the partial eclipse must be handled with care because for projection there is no filter foil in front of the optic: Before adjusting the telescope to the Sun, the finder scope has to be removed. In order to center the Sun in the field-of-view the telescope's shadow must be used for adjustment—the smaller the shadow of the tube, the closer the Sun will be visible to the center. If the observer does not use the telescope for a specific time, he or she must always be aware not to allow other people, especially children, to look through the projecting eyepiece.

Be cautious when working with projecting telescopes

Simple Projection Methods A simple way to watch the Sun's image is to construct a pinhole projector from two pieces of white cardboard or strong paper. In one of them a hole with the diameter of about a pencil is cut. Then both cardboards are adjusted in series at right angles to the Sun with the hole in the front (Fig. 70). The sunlight goes through the hole and projects a small dimmed image on the second cardboard. Another pinhole projection can be manufactured from a thin mailing tube. Into one of the tube's end caps

Pinhole projection

Table 6. Solar eclipses. Eclipse table courtesy of Fred Espenak [71], NASA/Goddard Space Flight Center

Date	Type	Saros	Magnitude	Duration	Region
2009 Jul 22	Total	136	1.080	$06^m\,39^s$	India, Nepal, China
2010 Jan 15	Annular	141	0.919	$11^m\,08^s$	Africa, India, Malymar, China
2010 Jul 11	Total	146	1.058	$05^m\,20^s$	Pacific, Easter Is., Chile, Argentina
2011 Jan 04	Partial	151	0.857	–	Europe, Africa, Asia
2011 Jun 01	Partial	118	0.601	–	Asia, America, Iceland
2011 Jul 01	Partial	156	0.097	–	Indian Ocean
2011 Nov 25	Partial	123	0.905	–	Africa, Antarctica, Tasmania
2012 May 20	Annular	128	0.944	$05^m\,46^s$	China, Japan, Pacific, USA
2012 Nov 13	Total	133	1.050	$04^m\,02^s$	Australia, Pacific
2013 May 10	Annular	138	0.954	$06^m\,03^s$	Australia, Solomon Is., Pacific
2013 Nov 03	Hybrid	143	1.016	$01^m\,40^s$	Atlantic, c Africa
2014 Apr 29	Annular	148	0.984	–	Antarctica
2014 Oct 23	Partial	153	0.811	–	Pacific, N. America
2015 Mar 20	Total	120	1.045	$02^m\,47^s$	Atlantic, Faeroe Is., Svalbard
2015 Sep 13	Partial	125	0.787	–	Africa, India, Antarctica
2016 Mar 09	Total	130	1.045	$04^m\,09^s$	Sumatra, Borneo, Sulawesi, Pacific
2016 Sep 01	Annular	135	0.974	$03^m\,06^s$	Atlantic, c Africa, Madagascar, India
2017 Feb 26	Annular	140	0.992	$00^m\,44^s$	Pacific, Chile, Argentina, Africa
2017 Aug 21	Total	145	1.031	$02^m\,40^s$	Pacific, USA, Atlantic
2018 Feb 15	Partial	150	0.599	–	Antarctica, America
2018 Jul 13	Partial	117	0.337	–	Australia
2018 Aug 11	Partial	155	0.736	–	Europe, Asia
2019 Jan 06	Partial	122	0.715	–	Asia, Pacific
2019 Jul 02	Total	127	1.046	$04^m\,33^s$	Pacific, Chile, Argentina
2019 Dec 26	Annular	132	0.970	$03^m\,39^s$	Saudi Arabia, India, Sumatra, Borneo
2020 Jun 21	Annular	137	0.994	$00^m\,38^s$	Africa, Asia, China, Pacific
2020 Dec 14	Total	142	1.025	$02^m\,10^s$	Pacific, Chile, Argentina, Atlantic

Online tables:
http://sunearth.gsfc.nasa.gov/
eclipse/SEcat/SEdecade2001.html
http://sunearth.gsfc.nasa.gov/
eclipse/SEcat/SEdecade2011.html

Manufacturers sell construction
sets for high-quality projection, e.g., the Solarscope from
http://www.science-shop.de

a small hole is punched with a nail. The other cap is replaced by a thin piece of a tissue or nearly transparent sandwich paper. The tube is adjusted with the hole towards the Sun like a telescope and a small image is projected on the diffusing screen at the end of the tube (Fig. 71).

13.3.6.2 Special Astronomical Programs

The Partial Phase Partial solar eclipses are always visible in a larger area on the Earth's surface than total eclipses, and thus an observer will more often be able to view a partial eclipse, especially from home.

Contact times

Measuring of Contact Times With access to a short-wave radio and a stopwatch it is possible to time the moment of first contact [72]: With the time signals of a broadcasting station in the background and the tape recorder switched on, the stopwatch is started at an exact minute mark. The watch is stopped and a voice signal is spoken into the tape recorder when the Moon has covered the first bit of the Sun's limb. The exact time of the first contact results from the addition of the measured time to the hour

Table 7. Lunar eclipses. Eclipse table courtesy of Fred Espenak, NASA/Goddard Space Flight Center

Date	Type	Saros	Umbral Mag.	Duration	Region
2009 Jul 07	Penumb.	110	−0.909	–	Australia, America
2009 Aug 06	Penumb.	148	−0.661	–	America, Europe, Africa, Asia
2009 Dec 31	Partial	115	0.082	$01^h 02^m$	Europe, Africa, Asia, Australia
2010 Jun 26	Partial	120	0.542	$02^h 44^m$	Asia, Australia, America
2010 Dec 21	Total	125	1.262	$01^h 13^m$	Asia, Australia, America, Europe
2011 Jun 15	Total	130	1.705	$01^h 41^m$	America, Europe, Africa, Asia, Austr.
2011 Dec 10	Total	135	1.110	$00^h 52^m$	Europe, Africa, Asia, Austr., America
2012 Jun 04	Partial	140	0.376	$02^h 08^m$	Asia, Australia, America
2012 Nov 28	Penumb.	145	−0.184	–	Europe, Africa, Asia, Austr., America
2013 Apr 25	Partial	112	0.020	$00^h 32^m$	Europe, Africa, Asia, Australia
2013 May 25	Penumb.	150	−0.928	–	America, Africa
2013 Oct 18	Penumb.	117	−0.266	–	America, Europe, Africa, Asia
2014 Apr 15	Total	122	1.296	$01^h 19^m$	Australia, America
2014 Oct 08	Total	127	1.172	$01^h 00^m$	Asia, Australia, America
2015 Apr 04	Total	132	1.006	$00^h 12^m$	Asia, Australia, America
2015 Sep 28	Total	137	1.282	$01^h 13^m$	America, Europe, Africa, Asia
2016 Mar 23	Penumb.	142	−0.307	–	Asia, Australia, America
2016 Aug 18	Penumb.	109	−0.992	–	Australia, America
2016 Sep 16	Penumb.	147	−0.058	–	Europe, Africa, Asia, Australia
2017 Feb 11	Penumb.	114	−0.031	–	America, Europe, Africa, Asia
2017 Aug 07	Partial	119	0.252	$01^h 57^m$	Europe, Africa, Asia, Australia
2018 Jan 31	Total	124	1.321	$01^h 17^m$	Asia, Australia, America
2018 Jul 27	Total	129	1.614	$01^h 44^m$	America, Europe, Africa, Asia, Austr.
2019 Jan 21	Total	134	1.201	$01^h 03^m$	America, Europe, Africa
2019 Jul 16	Partial	139	0.657	$02^h 59^m$	America, Europe, Africa, Asia, Austr.
2020 Jan 10	Penumb.	144	−0.111	–	Europe, Africa, Asia, Australia
2020 Jun 05	Penumb.	111	−0.399	–	Europe, Africa, Asia, Australia
2020 Jul 05	Penumb.	149	−0.639	–	Americas, Europe, Africa
2020 Nov 30	Penumb.	116	−0.258	–	Asia, Australia, America

Online tables:
http://sunearth.gsfc.nasa.gov/eclipse/LEcat/LEdecade2001.html
http://sunearth.gsfc.nasa.gov/eclipse/LEcat/LEdecade2011.html

and minute that the watch has been started. Measurements can be taken for second to fourth contact, respectively. After totality the measurements can be controlled by replaying the tape.

Observations During the Partial Phase While the Moon crosses the Sun's disk (Fig. 62), the silhouette of the black lunar profile can be studied. Using a telescope at magnification of 100× or higher, deep valleys and tall mountains against the Sun's photosphere can be identified. To achieve this, the value of the current lunar libration angle and a high-resolution Moon map are necessary.

Shape of the lunar profile

The eastward moving lunar limb might cover sunspots [72]. In the case of a large sunspot, first the spot's penumbra is covered, then the sunspot's umbra merges into the Moon's profile and disappears. The spot's diameter (if its position is near the center of the Sun's disk) can be estimated by measuring the time it takes for the Moon to cover a sunspot completely. During the eclipse, the Moon passes over about 400 km of the Sun's surface every second. If, for example, the Moon takes 5 s to completely cover a sunspot,

Occultation of sunspots

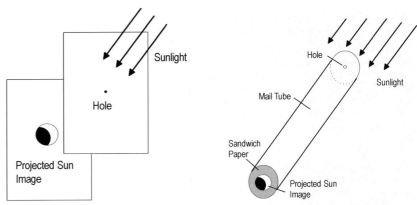

Fig. 70. Pinhole projection paper   **Fig. 71.** Pinhole projection tube

the spot's diameter is about 2000 km. This method does not work properly for sunspots far away from the Sun's center because of the disk's curvature.

Furthermore the limb darkening of the Sun could be examined with long-focus optics. As the limb darkening depends on the wavelengths of light, using different color filters makes it easier to observe it with higher contrast.

Phenomena Near the Second or Third Contact Just before the second contact, any filters attached to the optic should be removed in order to enjoy totality. From this point and until the third contact, the observation is safe for the naked eye. The lunar limb is not perfectly round and consists of several uneven landscapes caused by mountains or craters. Thus, just before the second contact the solar crescent becomes so narrow that parts of it already have contact with mountains at the lunar limb while other parts are still visible as bright spots. The solar crescent looks like a "string of pearls", popularly called "Baily's beads" (Fig. 72).

Seconds later, more and more bright parts of the solar crescent are covered by the Moon until only one spot remains. This is called the "diamond ring" because the already visible inner corona outside the Moon's disk looks like a ring which is jeweled by the last visible bright part of the Sun at a deep blue sky (Fig. 73).

The diamond ring disappears at the second contact. The chromosphere, which is a layer of the Sun's atmosphere outside the photosphere, has a prominent emission spectrum, visible at the same time as Baily's beads or the diamond ring. The chromosphere's emission spectrum, or "flash spectrum," can be observed with an objective prism or a diffraction grating. This is also the moment when prominences at the eclipsed solar limb become visible for the first time. They stay visible during totality and can be seen with the naked eye or with telescopes as deep red, pink or pale violet extensions. The phenomena described appear in reversed order at third contact. At the end of totality, sun-specific observation filters must be reattached to the eyepiece in order to protect the eyes.

Limb darkening of the Sun

Online Moon maps:
http://www.lpi.usra.edu/
resources/mapcatalog
http://www.lunarrepublic.com/
atlas/index.shtml

Baily's beads

Diamond ring

Chromosphere's flash spectrum

Prominences

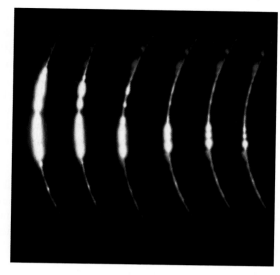

Fig. 72. Baily's beads. Photo courtesy of Rahlf Hansen (11.8.1999, Nikon F3, 500 mm, f/8, series exposure at intervals of some 1/10 s with 1/4000 s exposure time for each single image on an ISO 100 slide film)

Fig. 73. Diamond ring covered by cirrus clouds. Photo courtesy of Rahlf Hansen (11.8.1999, Nikon F3, 500 mm, f/8, automatic exposure time, ISO 100 slide film)

The Corona The shape of the corona depends on the phase within the cycle of solar activity. It varies between an extended corona that stretches along the equatorial plane at minimum (Fig. 63) and a roundish one with pronounced rays extending all around into space at activity maximum. The brightness of the outer corona decreases rapidly with increasing distance from the solar limb. Outside the first solar radius, just one thousandth part has left.

The solar corona consists of three different parts that can be observed by the amateur. The inner and hottest part is called L-corona and exceeds the Sun's surface by about one solar radius. Its spectrum only consists of emitting lines, which can be observed even with a simple spectroscope. The next part extents the Sun's surface by about two solar

The shape of the corona

The corona's spectrum and polarization

Examine the corona in infrared light, because this part of the solar spectrum is still not completely investigated.

radii and is photospheric light scattered by free electrons in the highly ionized gas of the coronal plasma, called K-corona. This light is highly polarized and has a continuous spectrum. The outer part is photospheric light diffracted by the interplanetary dust (the same particles which form the zodiacal light) and shows a "Fraunhofer" spectrum as known from the sun's surface.

Drawing the corona

By drawing, observers are able to detect fine structures of the corona. In advance, the Moon's disk should be positioned in the center of the paper. Outside the Moon, enough space for the full extent of the corona should be left (up to four solar diameters in all directions). After finishing the outline, the corona's fine shadings and fluctuations should be drawn.

Looking for comets, meteors or satellites

The Star Field If the sky is completely clear, the brightest stars and planets will be visible. For identification it would be helpful to have a starmap plotted for the current moment. Some observers look for a comet nearby the sun as comets in perihelion generate tails which might be visible to the naked eye. It also may be possible that a bright artificial Earth satellite (see Sect. 13.5) or meteor appears in the sky.

Baily's beads

Annular Eclipses During an annular eclipse (Fig. 64) Baily's beads are visible at the western limb at second contact and at the eastern at third contact. If the observer's position is at the edge of the Moon's umbral shadow on the Earth's surface, Baily's beads are also visible during the complete annular phase at the northern or southern limb of the Sun.

13.3.6.3 Special Terrestrial Programs

Fading of light

Approaching shadow

Brightness and Color of the Sky With increasing stages of the eclipse, the brightness of the sky reduces significantly. Especially at second contact, the sky darkens rapidly. The Moon's shadow quickly approaches the observation site like a spectacular thunder front from the western horizon. Immediately the sky turns to a deep blue, while the surrounding horizon is still bright due to the light of the partial photospheric light of the Sun. After totality the Moon's shadow escapes to the eastern horizon. Observers may have the impression of daylight already being back. But this is subjective because the eyes have adapted to the darkness during totality and realize the increasing brightness with higher sensitivity.

Measuring the brightness of the sky

The current brightness can be measured with a photoelectric cell or exposure meter attached to a 2-m high stick directed toward the zenith. The sensitivity range of the instrument needs to be adjusted accordingly.

Natural projection of the Sun's crescent

Projection of the Narrow Solar Crescent Approaching totality, the solar crescent becomes very narrow and works like a slit light source. Therefore shadows of any object have a very sharp outline and sunlight falling through leaves or holes is projected as narrow crescents on the bottom or a wall.

Detailed information and observation hints about shadow bands: http://www.strickling.net/shadowbands.htm

Shadow Bands Just before second contact and just after third contact, if the sky is completely clear, another phenomenon called shadow bands occurs [72]. The bands result from the light of the thin solar crescent shining through inhomogeneities in the atmosphere. These faint grayish streaks move across the ground and are best visible on plane surfaces like walls or white sheets laid flat on the ground. They are about 10 to 15 cm wide and move about 10 m/s. During the last seconds before second contact or after third contact, respectively, the shadow bands' contrast and speed increase. Shadow bands can be observed for longer times in the zone of the grazing eclipse at the border of the Moon's shadow path. Changes in the solar crescent's orientation in the sky at this location will have an impact on the shadow band's direction.

Meteorological Observations During totality, a significant drop in temperature can occur. This can cause sudden cloudiness as the temperature reaches the dew point. Another phenomenon is the "eclipse wind," which on the one hand is able to blow away clouds covering the scenery and on the other hand disturbs the observation preparations by blowing away any papers or shaking instruments. With an automatic weather station connected to a notebook, it is possible to record a variety of meteorological data and plot them as diagrams on the screen or later on the printer.

Falling temperatures

Eclipse wind

Recording weather data

13.3.7 Lunar Eclipses

13.3.7.1 Penumbral Lunar Eclipses Although penumbral lunar eclipses are very dim, significant changes in brightness and color are possible to observe with or without optical aid when the penumbral eclipse magnitude is greater than 0.7 or at least half of the Moon's surface is covered by the Earth's penumbral shadow [72].

Penumbral eclipse

Observing the Penumbral Shadow The most difficult part is to exactly measure the contact time, as the penumbral shadow does not have a sharp contour. Some observers try to catch the moment by using a lunar filter attached to the telescope in order to have a dimmed image of the Moon. Another method is to use a medium magnification of the Moon's image (about 150×) and to move the telescope's field-of-view over the Moon's surface.

Contact times of the penumbral shadow

The progression of the penumbral eclipse can be drawn on paper by outlining the penumbral edge together with the exact time on a Moon sketch. The penumbral shadow can change the brilliance of surface details on the Moon, like a crater's rays. Advanced observers are also able to realize a color of the penumbral shadow.

Drawing the penumbral shadow progression

13.3.7.2 Partial Lunar Eclipses

Enlargement of the Earth's Shadow In 1707, measurements of contact times during partial lunar eclipses (Fig. 74) demonstrated that the Earth's umbral shadow in space is enlarged by 2% compared to its theoretically calculated size. Further examinations have shown that the size and shape of the umbral shadow depends on high-altitude clouds, moisture, ozone, volcanic dust and the atmosphere's oblateness.

Contact times help to prove the enlargement of the Earth's shadow

Contact Times As the above theme has not been investigated completely, amateur astronomers can deliver more detailed data by measuring exact contact times of the umbral shadow. Data of scientific value needs measurements within a precision of at least 6 s [72]. This assumes that the borderline between penumbral and umbral shadow is known accurately, which is not easy to determine:

Fig. 74. Partial lunar eclipse on October 16, 1997 in Berlin, Germany. Photo courtesy of Christine Treichel (1000 mm, f/16, 2 s on Fujichrome 100)

Precisely estimate the borderline between penumbral and umbral shadow

Time signal broadcasting

Telescope apertures under 20 cm deliver a better contrast difference between penumbral and umbral shadow. The magnification should be small in order to have the whole disk of the Moon and surrounding sky within the field-of-view. To boost the shadow's contrast, a neutral Moon filter is recommended. As for solar eclipses, the exact contact times should be recorded with a tape recorder using a broadcast time signal in the background.

Crater times

Crater Times In order to be familiar with the Moon's surface at eclipse day, especially for crater timings, the observer should start to get to know it a few Full Moons in advance.

The Association of Lunar and Planetary Observers (ALPO) publishes crater timing maps and report forms: http://www.alpo-astronomy.org

By using a magnification of about 100×, the crossing of the umbral shadow over a crater has to be measured with a precision of at least 6 s. For small craters this is easier than for large craters. The latter need one measurement when the shadow first touches the rim, and another when it touches the opposite site of the crater. The average is equivalent to the point in time when the shadow covered the center of the crater.

Data reports

Observers are encouraged to send their measurements to astronomical institutes or professional astronomical magazines for further calculation of the enlargement of the Earth's shadow.

Color and shape of the umbral shadow

The Color of the Umbral Shadow With a magnification of about 150×, the color of the umbral shadow can be estimated if only the eclipsed portion of the Moon is visible in the field-of-view [72]. The color of the leading and trailing edges of the umbral shadow often show a coppery red, but sometimes a distinct blue or green fringe. Also the shape of the umbral shadow should be noted. Sometimes it is not uniformly curved but rather blunted. This is known as "cusp extensions" [72] (Fig. 75), where the shadow is apparently flattened over the central portion of the Moon's disk and curving off suddenly near the lunar limb. Frequently one cusp extension will appear more prominent than the other.

"Cusp extension"

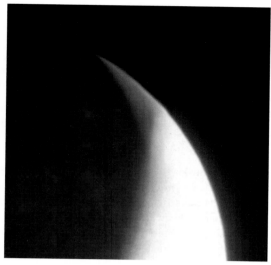

Fig. 75. "Cusp extension"

13.3.7.3 Total Lunar Eclipses

Estimating the Brightness of Totality The brightness and color of totality (Fig. 76) is unique to each eclipse as the Moon's position in relation to the umbral shadow center or the clarity of the Earth's atmosphere varies from eclipse to eclipse. In order to compare different total lunar eclipses, each totality should be evaluated at specific times using the Danjon scale (see Table 8). This is a guideline for estimating the brightness and color of totality with the naked eye. If the occurrence does not fit exactly to one of the scale's levels, a medium value, e.g., 2.6, can be specified.

Fig. 76. Total lunar eclipse on October 16, 1997 in Berlin, Germany. Photo courtesy of Christine Treichel (1000 mm, f/16, 15 s on Fujichrome 100)

Danjon scale

Table 8. Danjon scale [72]

L value	Description
0	Very dark eclipse. The Moon is almost invisible, especially at mid-totality.
1	Dark eclipse, gray or brownish tint. Details are difficult to distinguish.
2	Dark red or rust-colored eclipse, with a very dark spot at shadow center, rather bright outer portions of shadow.
3	Brick-red eclipse. Umbral shadow usually has a bright or yellow rim.
4	Very bright copper-red or orange eclipse. Umbral shadow has a bluish, very bright rim.

Moon's apparent magnitude

Estimating the Moon's Apparent Magnitude Another exercise is to estimate the Moon's apparent magnitude at totality. In order to reduce the 0.5 degrees expanded disk to a point-shaped light source, the most practical way is to watch the Moon through a backwards-oriented binocular. This will shrink the Moon's disk by the inverse of the binocular magnification. Using a 7× binocular will show the Moon's diameter at about 4 arcmin. Its apparent magnitude is estimated as follows: One eye is looking at the Moon through one of the binocular lenses. After that, the other eye is looking for stars at the same altitude as the Moon with similar color and brightness. For the best results, one star should be a little darker, another a little brighter than the Moon. After determining the star's apparent magnitudes, the Moon's apparent magnitude can be interpolated.

Correction factor for binoculars

But as the backwards-oriented binocular diminishes the Moon's brightness, a correction factor has to be considered for each specific kind of binocular (see Table 9).

F is the correction factor,
P is the binocular's magnification, constant 0.3 considers the loss of light in the optical system.

The factor can also be estimated as follows (recommended by ALPO):

$$F = 5 \log P + 0.3 \ .$$

For testing the accuracy of the correction factor, bright stars like Capella or Vega should be observed through the backwards binocular and compared with other stars which appear for the naked eye with the same brightness.

The resulting apparent magnitude of the Moon is calculated as follows:

M is the resulting magnitude,
m is the Moon's observed magnitude,
F is a correction factor.

Totality indicates the current state of the Earth's atmosphere.

$$M = m - F \ .$$

Influences by the Earth's Atmosphere The visible light of totality represents the refraction of sunlight in the Earth's atmosphere at all locations of dusk or dawn. The brightness and color of totality indicates how clear or polluted the Earth's atmosphere is at the moment. When the atmosphere is clear, totality is rather light and of coppery red color with

Table 9. Correction factors for binoculars [72]

Binocular magnification (P)	Correction factor (F)	Binocular magnification (P)	Correction factor (F)
6×	4.2	7×	4.5
8×	4.8	10×	5.3
11×	5.5	12×	5.7
16×	6.3	20×	6.8
25×	7.3	30×	7.7

a Moon's apparent magnitude of about −2. When it is polluted, e.g., by volcanic aerosols, totality is dark, like on December 9–10, 1992 due to the eruption of Mount Pinatubo in the Philippines, when the apparent magnitude was just +3 and the Moon was nearly invisible. But the brightness of totality also depends on the distance between the Moon and the center of the Earth's umbral shadow. The closer the Moon is to the center, the darker the eclipse will be. Another reason for anomalies during totality is the changing position of clouds, haze or dust at the Earth's terminator.

Further Observations The moment of totality can be used for observations that would not have been possible during Full Moon, which are occultation of stars (see Sect. 13.4) or comet searching.

Occultations

Comets

13.3.8 Transits

Transits of planets in front of the solar disk are similar to solar eclipses, but the obscuration by a planetary transit is negligible. During transit, the planet is visible as a small black disk moving westwards for hours across the solar disk (Fig. 77). As the planet disk is smaller than the Sun, contact times are to be understand like that for annular eclipses, but in the reverse direction. The equipment and methods for observing a transit, especially for eye safety, are similar to those for solar eclipse viewing.

Transits are miniature versions of solar eclipses.

Only Mercury and Venus are able to cross the solar disk because they are inner planets and thus can take a position at lower conjunction, where they build a straight line with the Sun and the Earth. Mercury's and Venus's orbits have an inclination in relation to the ecliptic, therefore transits are only possible when Mercury or Venus are crossing their nodes in the moment of lower conjunction (see Tables 10 and 11). While Mercury's transits always happen in May and November in this age, Venus's transits always appear

When does a transit happen?

Fig. 77. Transit of Venus on June 8, 2004 in Schwante, Germany. Photo courtesy of Christine Treichel (1000 mm, f/16, projection screen photography)

Table 10. Mercury Transits Contact Times (UT). Eclipse table courtesy of Fred Espenak, NASA/Goddard Space Flight Center

Date	I h:m	II h:m	Great. h:m	III h:m	IV h:m
2016 May 09	11:12	11:15	14:57	18:39	18:42
2019 Nov 11	12:35	12:37	15:20	18:02	18:04
2032 Nov 13	06:41	06:43	08:54	11:05	11:07
2039 Nov 07	07:17	07:21	08:46	10:12	10:15
2049 May 07	11:03	11:07	14:24	17:41	17:44
2052 Nov 09	23:53	23:55	02:29	05:04	05:06
2062 May 10	18:16	18:20	21:36	00:53	00:57
2065 Nov 11	17:24	17:26	20:06	22:46	22:48

Table 11. Venus Transits Contact Times (UT). Eclipse table courtesy of Fred Espenak, NASA/Goddard Space Flight Center

Date	I h:m	II h:m	Great. h:m	III h:m	IV h:m
2012 Jun 06	22:09	22:27	01:29	04:32	04:49
2117 Dec 11	23:58	00:21	02:48	05:15	05:38

Detailed table of Mercury's transits:
http://sunearth.gsfc.nasa.gov/eclipse/transit/catalog/MercuryCatalog.html

Detailed table of Venus's transits:
http://sunearth.gsfc.nasa.gov/eclipse/transit/catalog/VenusCatalog.html

Venus's transit in 2012:
http://sunearth.gsfc.nasa.gov/eclipse/transit/venus0412.html

For more information on transits, see Fred Espenak's Eclipse Homepage:
http://sunearth.gsfc.nasa.gov/eclipse/eclipse.html.

Simultaneous transit of Mercury and Venus?

Frequency of transits

Influences of Mercury's orbital eccentricity

Frequency of transits

in June and December. As the nodes of both planets are moving slowly along the ecliptic, the occurrence of transits will adjust to later months in future centuries.

In our age, a simultaneous transit of Mercury and Venus is not possible because of their different node's longitudes, and their nodes will not have moved to the right position before the year 69 163. Grazing or partial transits of Mercury and Venus are extremely rare. They will not take place before the year 2391.

13.3.8.1 Mercury Transits

Mercury transits occur about 14 times per century. They appear in intervals of 3, 4, 6, 7, 10 or 13 years. The characteristics of Mercury transits reoccur after 46 years.

As Mercury's orbit has a significant eccentricity, the distance to Earth is longer when it passes the node during a lower conjunction in November instead of May. This implies that transits in November are more often than in May, because there is a larger tolerance for the deviation from the node to make a transit visible. On the other hand, transits in May have a longer duration than in November, because of Mercury's longer distance from the Sun. Mercury's apparent diameter measures about 10 arcsec during transits in November and about 12 arcsec during transits in May.

13.3.8.2 Venus Transits

Venus transits are very rare. They occur in an unusual pattern: At first, two transits in December (eight years apart) occur. At 121 and a half years later, two June transits (also eight years apart) occur. Another 105 and a half years later the pattern repeats from the beginning. In summary, there is a 243-year cycle with 4 transits.

Venus's apparent diameter during a transit measures about 60 arcsec. It may be possible to see it crossing the Sun without optical magnification (but using solar eclipse eyeglasses, see Sect. 13.3.2.1).

Fig. 78. Black drop effect   **Fig. 79.** Lomonossow effect

13.3.8.3 Observations A few seconds after the first contact, assumed the position angle of the planet is exactly known, a small dent at the solar limb is visible. Using Hydrogen-alpha filter, the planet will already be visible against either prominences or the chromosphere outside the solar limb. Measuring of contact times is done accordingly to solar eclipses (see p. 394).

Black Drop Effect Just after the second and just before the third contact, the "black drop" effect is a curious occurrence (Fig. 78). This is an optical phenomenon that can be seen like a black teardrop that connects Mercury's or Venus's disk to the solar limb. This makes an exact measuring of the second and, respectively, third contact difficult. But the effect for Mercury is not so significant as for Venus.

Lomonossow Effect Only for Venus, near to the second and after the third contact, a phenomenon called "Lomonossow effect" occurs (Fig. 79). This is visible as a slim ring of light against the sky at the outer limb of Venus. The cause of this is the diffraction and scattering of sunlight in Venus's atmosphere, visible as a twilight zone. As this effect is even visible with amateur telescopes, observers are able to conclude without any other technical aid, that Venus has an atmosphere.

Determination of the Astronomical Unit The distance to the Sun and planets today can be measured accurately using radar. In the past, Venus transits were used to estimate the solar parallax and the astronomical unit (AU). But because of measurement errors at the second and third contact due to the black drop effect, the results were not quite exact. But to learn about calculating distances in the Solar System it is still interesting to use this method.

13.3.9 Further Reading

Haupt, H.: Lunar Eclipses, Compendium of Practical Astronomy. Springer, Berlin Heidelberg New York (1994)

Petri, W.: Observations of Total Solar Eclipses, Compendium of Practical Astronomy. Springer, Berlin Heidelberg New York (1994)

Hydrogen-alpha filters

First, Venus's atmosphere was made responsible for the "black drop." But also Mercury shows this effect. As it is more significant with small telescopes apertures, the limitation of the telescope's optical resolution could be the reason. Current analyses covered that it is the result of seeing effects due to Earth's turbulent atmosphere.

Lomonossow effect

Calculating the AU:
http://sunearth.gsfc.nasa.gov/sunearthday/2004/2004images/VT_Activity3.pdf

Calculating Venus's distance using the transit duration:
http://eclipse.astronomie.info/transit/venus/theorie/index.html

Observation hints:
See *Sterne und Weltraum*, vol. 8, 2001, p. 656, "Die Messung der Astronomischen Einheit" by Ulrich Uffrecht

Treichel, C.: Kosmische Schattenspiele, Sterne und Weltraum Basics 1. Spektrum der Wissenschaft, Heidelberg (2002)

Treichel, C., Staps, D.: Die totale Sonnenfinsternis am 11. August, Ahnerts Kalender für Sternfreunde (1999)

Eclipses

http://de.wikipedia.org/wiki/Sonnenfinsternis

http://en.wikipedia.org/wiki/Solar_eclipse

http://de.wikipedia.org/wiki/Mondfinsternis

http://en.wikipedia.org/wiki/Lunar_eclipse

Transits

http://de.wikipedia.org/wiki/Durchgang

http://en.wikipedia.org/wiki/Astronomical_transit

http://sunearth.gsfc.nasa.gov/eclipse/transit/transit.html

13.4 Occultations of Stars

C. Treichel

Occultations of stars appear because objects in the Solar System, especially the Moon, but also asteroids, planets or their moons move with a significant motion in front of the stars and therefore change their positions in relation to the stars day by day. When a solar system object crosses exactly the line-of-sight to a star, the star will be covered, and a so-called occultation results.

Occultations by the Moon are visible in principle at every lunar phase, but events at its dark limb are more favored and more promising, because the ingress or egress is observable definitely instead of in the case of an event at the light limb (Fig. 80). For this reason occultations during a total lunar eclipse are also of interest. At the moment of occultation the star disappears suddenly, as if it was switched off. In case of a central occultation of a star by the Moon, the duration takes about 70 min at terrestrial midlatitudes or even more at terrestrial tropical latitudes. An occultation closer to the northern or southern lunar limb, respectively, results in a shorter duration. Occultations of stars by planets or asteroids just take a few minutes or even seconds. Due to the great distance of the star, its light can be assumed to have parallel rays. Thus, the occultation shadow path on the Earth's surface has the same diameter as the occulting body itself.

The Moon's orbit has an inclination of about 5° relative to the ecliptic. Additionally its nodes move backwards along the ecliptic and take 18.6 years for a complete turn. Thus, stars or other objects, e.g., star clusters, within a distance of about 5° north and south

Cause of occultations

Ingress and egress

Duration of occultations

Shadow path

Candidates and frequency of occultations

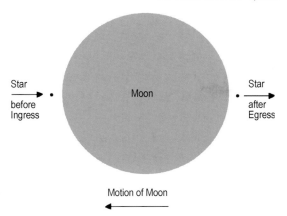

Star
before
Ingress

Moon

Star
after
Egress

Motion of Moon

Fig. 80. When the Moon crosses the line-of-sight to a star, it covers the star always from west. Ingresses take place on the eastern lunar limb, egresses on the western lunar limb

of the ecliptic are candidates for occultations. The number of occultation candidates increases with descending apparent magnitude of the stars [73]. Down to 3.0 mag only 23 stars are available for occultation, down to 6.0 mag there are already 616 stars and about 14,400 stars are candidates with an apparent magnitude down to 9.0 mag. Worldwide there are about 5000 to 7000 visible occultations by the moon. After 18.6 years the characteristics of star occultations by the Moon repeat. Prominent candidates are Aldebaran, Regulus, Spica, Antares and of course the planets. But also faint stars can be observed.

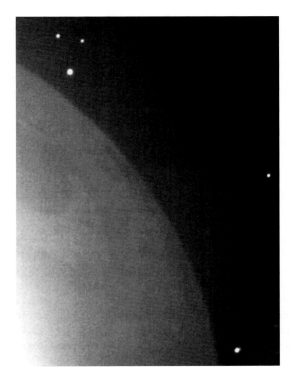

Fig. 81. Stars of M45 after egress at the dark lunar limb on September 12, 2006 in Schwante, Germany. Photo courtesy of Christine Treichel (2000 mm, f/10, 16 s at ISO 100, Canon EOS 20Da)

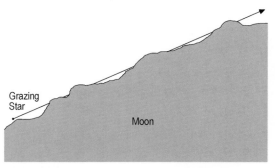

Fig. 82. Schema of a grazing occultation at the lunar limb

A highlight is always the occultation of the Pleiades or Hyades because of the various number of events within one night (Fig. 81). A simultaneous occultation of two planets by the Moon is a very rare event. The last time this happened was on April 23, 1998 when Venus and Jupiter were covered at the same time. The next event of this kind will take place on February 16, 2038 and March 16, 2038 when the Moon will cover Jupiter and Uranus.

A special type is a grazing occultation (Fig. 82). The star streaks the lunar limb profile, which is shaped by high mountains and deep valleys, and alternately disappears and reappears many times. The shadow path of a grazing occultation always relates to a narrow zone on the Earth's surface with just a few kilometers in width.

Also occultations by asteroids, planets or their moons project a narrow shadow path on the Earth.

Occultations are visible already with simple instruments and thus observations are possible for almost every observer. As we will see further, observations obtain information of scientific interest, which may be an important reason for amateurs to take part in observation campaigns.

13.4.1 Implications of Occultation Scenarios

What kind of characteristics or parameters can be retrieved from occultation observations?

13.4.1.1 The Lunar Limb Profile The moment and duration of a star occultation by the Moon enables a determination of the lunar limb profile exactly for these positions where the ingress and egress happened. Grazing occultations especially, but also an occultation of the Pleiades or Hyades, are a basis for measurements of the lunar limb profile. A complete lunar limb profile (Fig. 83) can be registered by participation of several observers at the same event but with different positions of their observation sites between the northern and southern limit of the shadow path.

Having optimal measuring accuracy for occultation times, an average accuracy for the lunar limb profile of about 0.1 arcsec can be reached. Based on the data, the exact

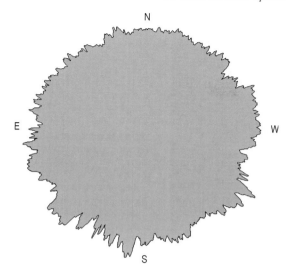

N

E   W

S

Fig. 83. The lunar limb during the total solar eclipse on March 29, 2006, 10:30 UT. The shape is enlarged 40-fold. Eclipse figure courtesy of Fred Espenak, NASA/Goddard Space Flight Center

lunar diameter depending on it's current profile can be retrieved. The shape of the lunar limb profile is not static. Subjected to the libration in longitude and latitude, the lunar limb profile varies respectively.

During a total eclipse of the Sun, the lunar limb profile and also the solar diameter can be determined by use of time measurements during the Baily's beads phenomenon. If observers have their positions at the northern and southern limit of the shadow path, Baily's beads timings deliver the solar diameter with high precision. It is of special interest to understand how the solar diameter changes over the years, and if our Sun is currently expanding or contracting.

13.4.1.2 Shape and Size of Asteroids
In the same way that the lunar limb profile is determined, the profile and size of asteroids can be retrieved when they cause an occultation of a star.

A successful event was the occultation of the star TYC 1259-00984-1=FK6 1115 by (345) Tercidina on September 17, 2002 [74]. Updated calculations of the shadow path on the Earth's surface, and an observer campaign which was in part coordinated via the Internet, made it possible to obtain more than 70 observations of the occultation, which were spread over the whole asteroid profile (Fig. 84). The amount of submitted measurements has made (345) Tercidina one of the best-observed asteroid in the history of occultations.

13.4.1.3 Asteroid Atmospheres
Although asteroids or planetary moons are small objects, it can be relatively easy to determine if they have an atmosphere during an occultation of a star [74]. An atmosphere causes a significant brightness characteristic during the occultation: If the star does not disappear immediately but gradually or the star's brightness increases during the occultation, the asteroid or moon definitely has an atmosphere, but why?

A sunrise or sunset is very similar to this phenomenon. The Earth's atmosphere diffracts the sunlight and therefore it is possible to see the Sun for a moment longer, even

---

For more information on solar and lunar eclipses, see Fred Espenak's eclipse homepage: http://sunearth.gsfc.nasa.gov/eclipse/eclipse.html.

Libration

Solar diameter

Coordination of Tercidina observer campaign and publishing of results: http://mpocc.astro.cz/results

Asteroidal occultation observers in Europe: http://www.euraster.net

Brightness declines delayed

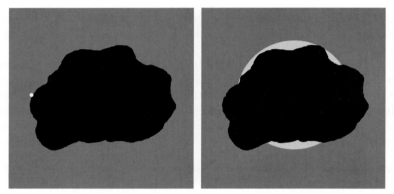

Profile and size of asteroids

Fig. 84. The profile of (345) Tercidina obtained from many observations within the shadow path. Every line corresponds to one observer. The resulting size of the asteroid is about 99 × 93 kilometers. Its irregular shape is clearly seen. Image courtesy of E. Frappa, http://www.euraster.net

The higher the density of the atmosphere, the stronger the diffraction, which means the spreading of starlight around the occulting object, will be. This can even result in brightness increase during a central occultation, because the starlight will shine around the asteroid or moon as if it was spread by a lens ring (Fig. 85, right).

Atmosphere diffracts starlight

Fig. 85. (*Left*) The star exactly before ingress. (*Right*) The starlight is shining around the occulting object due to diffraction of its atmosphere

if it has disappeared behind the horizon. An asteroid or moon with an atmosphere does not block the starlight completely during the occultation. Instead of that the starlight is also diffracted and therefore visible in a larger area of the atmosphere. This is the reason why the brightness decreases gradually during the occultation.

13.4.1.4 Corrections to the Lunar Orbit Occultation times, especially of grazing occultations, are used for exact calculation and correction of the lunar orbit, which means the lunar position in declination or ecliptic latitude. With help from timings for ingresses and egresses, the task is done to empirically connect the astronomically measured universal time (UT), the international atomic time, and the dynamical time—a strictly uniform measure of time on which the calculations of the motions of celestial bodies are based. The definition of time touches upon some very deep yet subtle problems in astronomy and physics, including general relativity theory. Occultations of stars form a direct link

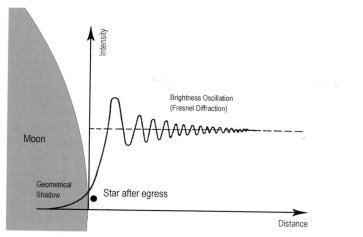

Fig. 86. The lunar limb causes a diffraction of starlight (Fresnel diffraction) [75]. The geometrical lunar shadow is located where the star's brightness has fallen down to 25%

between the Moon as a celestial clock and the system of space coordinates represented by stars.

13.4.1.5 Detection of Binary Stars Occultations by the Moon are useful to discover the binary nature of a star. On April 13, 1819 Professor Johann Tobias Bürg discovered Antares as a binary star during its visual occultation observation.

When the point-like starlight is shining on the lunar limb from behind, its wave character makes the lunar limb shadow appear unsharp because an expanded diffraction pattern with a characteristic sequence of minima and maxima is generated (Fig. 86). This is called Fresnel diffraction. This wave pattern is carried by the Moon like a bow wave over the field-of-view with 0.5 arcsec per second. At the preceding limb the waves are compressed, at the succeeding limb they are extracted. At ingress or egress, the occulted star shows a typical brightness oscillation when it crosses the diffraction pattern. With significant amplitude, it lasts usually less than 0.1 s and this is not visible to the eye, but for photometric devices. Its maximum reaches a value of 40% higher than the unocculted star. The measurement must be carried out with very high time resolution in order to resolve the frequency of the diffraction pattern (30 ms). If the star is a binary, another wave pattern will be generated by the second component independently, see Fig. 87 [75].

13.4.2 Equipment

Occultations can be successfully observed with even small telescopes. A magnification of about 50 to 100× is adequate. For egresses it is useful to have low power so that the whole of the Moon is visible within the field-of-view in order to make it easier to determine the position angle of the star. Also, the scintillation will, at so low a power, interfere with the perception of the contact only in very poor seeing. A larger magnification is necessary in hazy sky with much scattered light, or when a bright star ingresses at the light lunar limb. A track for the telescope is advantageous but not mandatory.

Fresnel diffraction

Very close binary stars and even giant stars can be detected due to their effects on the Fresnel diffraction. Differing from an exact point source, stars of finite diameter blur the wave pattern. This will cause a moderate decrease of measured light intensity at ingress. For binary stars, corresponding to the component's distance, position angle and individual brightness, the measured light intensity will decrease delayed or in two phases at ingress. In this way, binary star component distances below 0.01 arcsec and star diameters down to as small as 0.003 arcsec have been measured. This high resolution is possible because the natural image scanning is done by the lunar limb at a moment before the starlight dives into the Earth's turbulent and scattering atmosphere.

Magnification

For estimation of the star's position angle at egress, a cross-line eyepiece would be helpful.

In order to eliminate a good portion of the bright and annoying moonlight during grazing occultations, a focal diaphragm (i.e., stop) may be used when the lunar phase is bright.

For documentation of observation data it is important to note the exact position and height above sea level of the observation site. Today, on one hand, GPS devices are affordable and a good source for position determination. With a reception of as many GPS satellites as possible, the position accuracy is within a few meters. Another helpful source are satellite images of the Earth's surface found on the Internet. Populated areas can be zoomed in with the PC so that single trees or even cars can be identified from above. The record date of these images may be some years in the past, but as geographic coordinates and elevation are also delivered, it should be possible to exactly determine the position of the observation site on the zoomed satellite image so to say with the computer mouse. For elevation, the height of the telescope above ground level always has to be added by the observer.

For time measurements, a tape recorder with new batteries or a power supply should be used to record contact time and relevant voice signals. The recorder should be close enough to the observer to pick up his or her voice.

One or two stopwatches with at least 0.1 s accuracy and a short-wave or long-wave radio that picks up time signals from a time broadcasting radio station are also necessary. These signals can be received from nearly any point on the globe. Also a GPS receiver with a 1PPS synchronization can be used as a time signal source.

Beside the moment of time of an ingress or egress, the characteristic of measured starlight intensity during occultations by asteroids or planet moons is of interest. For this purpose CCD cameras or video cameras (webcams) are necessary. They record the progression of light intensity and afterwards, a photometric analysis with image processing software is possible. The camera should be able to record several images per second and have a wide grayscale or color scale in order to have a high brightness resolution. The refresh rate should be at 100 Hz or higher. The video image should display the GPS satellite time signal with help of a video time inserter. Those cameras have the advantage to record the progression of an occultation in real-time with a high time resolution.

13.4.3 Occultation Prediction Tables

There are several sources from which to obtain star occultation prediction information. In general date and time, the duration, the apparent magnitude, the coordinates, and the position angle at the Moon's limb, are mentioned for the star. Due to inaccuracies of the lunar or asteroid's orbits, long-term predictions of occulations are always updated closer to the event. Observers who plan to travel to observation sites far away should always study the updates in advance in order to be sure to choose the correct position within the shadow path.

Here is a list of some important prediction information websites:

IOTA: International Occultation Timing Association
       http://www.lunar-occultations.com/iota/iotandx.htm
       http://www.iota-es.de (European Section)

Download of Google Earth PC Client:
http://earth.google.de

Time signal broadcasting stations:
BPM China:
2.5, 5, 10, 15 MHz
CHU Canada:
3330; 7335; 14,670 MHz
DCF77 Germany: 77.5 kHz
JJY Japan: 60 kHz
MSF Great Britain: 60 kHz
RWM Russia:
4996; 9996; 14,996 MHz
WWV USA:
2.5, 5, 10, 15, 20 MHz
WWVH Hawaii:
2.5, 5, 10, 15 MHz
YVTO Venezuela: 5 MHz

Time via the Internet:
http://nist.time.gov/
timezone.cgi?Eastern/d/-5/java

Mintron camera:
http://www.mintron.com

IOTA occultation camera:
http://www.iota-es.de/ioc.html

The IOTA site is a detailed collection of know-how, tables, software downloads, etc., and also provides some guidance for submission of observation data.

ILOC: International Lunar Occultation Center (Tokyo):
> http://www1.kaiho.mlit.go.jp/KOHO/iloc/docs/iloc-index_e.htm

Steve Preston: Updated website
> http://www.asteroidoccultation.com

Of course, upcoming events are also published in well-known almanacs or astronomical magazines.

13.4.4 Occultation Observation

Most observed occultations are ingresses at the dark lunar limb during the first half of night, which observers prefer for observing. Instead of this, lunar orbit data resulting from observations during waning Moon (Fig. 81) are rare, as egresses at the dark limb fall into the second half of night. Egresses are also often missed, if the wrong location at the lunar limb is observed. *Ingresses vs. egresses*

Contacts at the bright lunar limb are more difficult to observe and need more experience. In prediction tables, events are only mentioned for bright stars, because faint stars cannot be observed at the bright lunar limb with much success.

For a visual binary star, the contact times of each component can be recorded. Even if the two are close and not separated by the telescope, their binary nature is often revealed by a distinctly "slow" or two-step contact. The difference of contact times is given by: *Θ, ρ Polar coordinates of binary star*

μ, m Absolute value resp., position angle of apparent motion of moon,

$$\Delta t = \rho \cos(\Theta - p)/\mu \cos(m - p) .$$

p Position angle of contact at lunar limb

Noticeably delayed or stepped contacts which last 1 s thus can occur even for close pairs with $\rho < 0.5$ arcsec. *Occultation of binary stars*

For occultations of binary stars as for occultations by asteroids or small moons, beside the exact contact time, the precise recording of the brightness characteristic and gradient is of interest. For this purpose, a video record of the occultation is separated into single images (frames). Every image is photometrically measured by an image processing software and as a result, a diagram with a brightness curve can be generated. For binary stars, the light curve has a typical two step brightness decrease at ingress and increase at egress (Fig. 87).

In the case of an asteroid with an atmosphere, the light curve shows obvious light boosts during the occultation, e.g., the "central flash" in the middle of the occultation *Occultations by asteroids*

Central flash

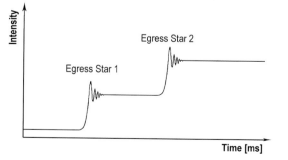

Fig. 87. During an occultation of a binary star, the light curve increases or decreases in two steps. This illustration is for an egress of a binary star

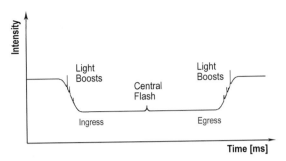

Fig. 88. During a star occultation by an asteroid or moon with an atmosphere, the light curve can show light bursts during ingress and egress and another light burst centered

phase, which are caused by diffraction of the occulted star's light in the asteroid's atmosphere (Fig. 88).

Unevenness at the lunar limb distorts the recorded brightness curve. For this reason, in order to capture more useful data, measurements of binary stars or asteroids should be made several times by either repeating the recording of the same star during the next occultation or by simultaneously recording the same occultation from different positions within the shadow path.

Occultation of planets

Although occultations of planets have less scientific impact, they are still attractive to observe, but occur more seldom. Near-grazing occultations of Venus with durations of ingress of half an hour, and also some partial occultations, have been observed. Predicted contact times are valid for the center of the planetary disc. All measured contact times should be recorded to full seconds, as the uncertainty for extended objects is much more invariable.

Preparations

13.4.4.1 Measuring Contact Times The observation is started long enough in advance to the occultation in order to have enough time to identify the star and to center it in the field-of-view. When the approaching disk of the Moon and the star to be occulted are both in the field-of-view, it can clearly be seen, how the Moon moves across the sky. A track for the telescope is not necessary but an advantage for observing the reappearance of the star: the telescope moves on the star centered when it is behind the Moon and thus it is easier to determine the place and time when the star will be visible again. In addition, a cross-line eyepiece, oriented by a star transit to the N-E-S-W directions, will suffice to estimate the predicted position angle from the center of the Moon. Beside obtaining the occultation duration, time measuring is also necessary for an exact determination of the occultation's beginning and end in UTC.

Time measurement suitable for different equipment

There are several methods for making time measurements at the moment of contact. These methods should anyway be practiced in advance by the observer. If possible, a watch should be synchronized with an accurate time signal:

Stopwatches

1. Start a stopwatch on an exact minute mark and stop it when the star disappears. Add the time on the stopwatch to the time it was started. This will result in the time of the beginning of the occultation.
2. Alternatively start a stopwatch when the star disappears and stop it accordingly when it reappears. This will show the occultation duration.
3. If you have two stopwatches, listen to a time signal broadcast sender and start the first when the star disappears and the second when the star reappears. Stop both

watches on an exact minute mark and note this time. Also note the time recorded on each stopwatch. The occultation beginning and end can be determined by subtracting the appropriate stopwatch time from the UT on which both watches were stopped. Although these first three methods are not quite exact, they can be managed by every amateur or beginner and they even deliver useful data when sent to, e.g., IOTA scientists.

4. Observers who like to record their measurements should use a tape recorder, a stopwatch and a short-wave radio with access to a time signal broadcast sender. With the time signals recorded by the tape recorder, the stopwatch is started on an exact minute mark. The watch is stopped and a voice signal is spoken to the tape recorder when the star disappears. The exact time of occultation beginning results from the addition of the measured time to the hour and minute that the watch was started. The same procedure is used for the star's reappearance. The tape recorder should be kept warm so it does not record at an abnormal speed. — *Tape recording*

5. During a grazing occultation, several measurements have to be recorded. The accuracy should be about 0.1 s. Therefore a tape recorder is switched on definitely. Exactly at disappearance of the star, the observer says, e.g., "in" and at reappearance he or she says "out." It is important that the words sound different in order to distinguish between case 1 and 2 later on replay. Meanwhile the audible time signal in the background has to be loud enough to also be recorded on tape. — *Techniques for grazing occultations*

6. With a CCD camera or webcam connected to a short-focus telescope, the procedure can even be made much simpler and the measuring accuracy can be increased to 0.02 s [73]. This is quite useful especially for grazing occultations. For this purpose, a time signal is transmitted to the camera and displayed in real-time during recording. Alternatively the audible time signal in the background can be recorded on the audio channel of the camera. — *Video recording*

After the occultation, the tape or video-recorded measurements can be verified by replaying the medium. Observers who are interested in sending their measurements for scientific analysis have to use standard forms to fill in their data. Forms are published via the Internet, e.g., by IOTA or ILOC, (see Sect. 13.4.3). — *Verification of measurements / Submission of results*

13.4.4.2 Teamwork

Occultations of stars are predestined for teamwork. When the members of a team are distributed within the shadow path, the diameter and a complete limb profile of the Moon or an asteroid can be obtained by contact time measurements. It is important to have observers positioned perpendicular to the shadow path (in order to have measurements from north to south) as well as along the path (in order to have the same measurement several times). Some observers should also be located just outside the shadow path to definitely have it that no occultation could be observed from there. In this manner a team is sure to have covered the whole profile with measurements. — *Determination of limb profile*

Updates of occultation predictions have to be studied close to the campaign in order to find the correct observation positions. The actual positions of the group members can be determined with GPS devices on the spot. — *Positions within occultation path*

Simultaneous measurements from different positions in the shadow path also deliver significant data to correct calculations of the lunar or asteroid's orbit or for exact determination of the star's coordinates.

13.4.5 Further Reading

Harrington, P.S.: Eclipse!. Wiley, New York (1997)

Heintz, W.D.: Occultations of Stars by the Moon, Compendium of Practical Astronomy. Springer, Berlin Heidelberg New York (1994)

North, G.: Advanced Amateur Astronomy. Cambridge University Press, Cambridge (1997)

Wikipedia: The Free Encyclopedia: http://en.wikipedia.org/wiki/Occultations, http://de.wikipedia.org/wiki/Sternbedeckung.

13.5 Artificial Earth Satellites

C. Treichel

There are several satellites that can become bright enough for observation by amateur astronomers. Although they all are tracked by several ground stations all around the world, they are still rewarding objects for visual or photographic observers (Fig. 89).

13.5.1 Populations of Satellites

On October 4, 1957 the first artificial Earth satellite was launched—the Soviet Sputnik 1. Its height varied between about 230 and 950 kilometers and a complete orbit took about 100 min. After 92 days, Sputnik 1 burned up in the atmosphere on January 4, 1958. Its "beep-beep" sound can be heard at: http://history.nasa.gov/sputnik.

The population of artificial objects is surveyed and cataloged as extensively as possible in order to minimize the risk of collisions, damage and accidental re-entries. In the US, objects are cataloged by the US Space Command (USSPACECOM). Its very sophisticated phased array radars can detect objects as small as 10 cm in low Earth orbit (LEO) and 1 m in geosynchronous orbit.

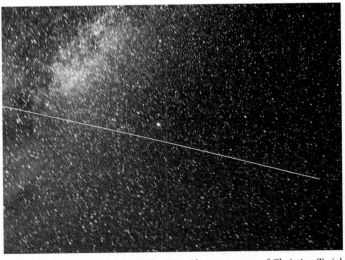

Fig. 89. A polar satellite track crosses the Milky Way. Photo courtesy of Christine Treichel, Schwante, Germany (Canon EOS 20 Da, 6.9.2007, 9:49 p.m., 17 mm, f/4.5, 102 s, ISO 1600)

The first category comprises all satellites that are active or at least under control (operational payloads). The second category contains everything else, which is the vast majority of all artificial objects in Earth orbit (space debris), which means thousands of objects that are big enough to be observed by visual means or radar.

All spacefaring nations are obliged to announce each satellite launch to the International Committee of Space Research (COSPAR). Based on the year of launch, an official designation is assigned to all satellites and fragments.

13.5.2 Classes of Orbits

It is possible to distinguish several classes of orbits which reflect the requirements of some special satellite applications.

Low Earth Orbits A satellite in a low Earth orbit (LEO) surrounds the Earth in an altitude between 300 and 1000 km. A complete orbit takes about 100 min. Therefore observations of these satellites are possible for about 15 min before they disappear again. Members of this class are communication satellites, e.g., the ISS (Fig. 90), Iridium or Globalstar net or some weather satellites.

Medium Earth Orbits The altitude of medium Earth orbits (MEO) is about 10,000 km. A complete orbit can take up to 12 h. Typical members of this class are GPS satellites.

Polar Orbits Satellites with Earth observation or reconnaissance missions require to overfly almost every region of the Earth in a low altitude. To achieve this, they are positioned in a low Earth orbit (LEO) with an inclination of approximately 90° (Fig. 91). If the inclination is exactly 90°, the orbital plane does not precess around the Earth.

In a retrograde orbit with an inclination of approximately 97°, the ascending node moves with a rate of 0.984° eastward. This results in a heliosynchronous orbit, because it compensates for the angular velocity of the Earth's orbital motion around the Sun. Heliosynchronous orbits are essential for satellites that require permanent illumination

Cataloging of satellites

Operational payloads
Scientific/technological
1. Satellites for special programs: telecommunications, Earth observation, navigation, meteorology
2. Satellites for specific military use: photographic reconnaissance, electronic detection, anti-satellite satellites (ASATs)
3. Manned space stations and spacecraft

Space debris
1. Deactivated satellites, spent rocket stages, fragments of satellites and rockets.
2. Objects which are too tiny to be under surveillance from Earth, but can cause a considerable threat to manned and unmanned spacecraft.

LEO

MEO

Heliosynchronous

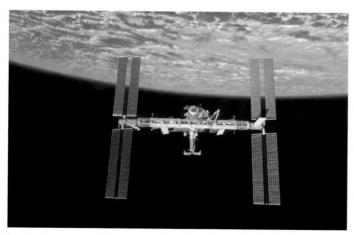

Fig. 90. The ISS photographed from Space Shuttle Atlantis on June 19, 2007 (courtesy of NASA)

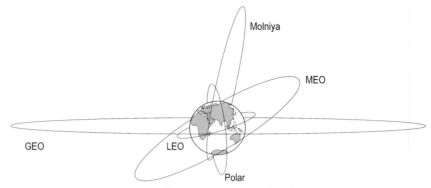

Fig. 91. Illustration of several orbit classes

of their solar arrays or if their mission demands illumination of the overflown areas. The heliosynchronous orbit is used for Earth exploration satellites like the European ERS-1 and ERS-2 or solar satellites like the US satellite ACRIMSat.

Geostationary

Geostationary Orbits Modern telecommunications, such as television broadcasting or the provision of telephone links, demand an uninterrupted connection between the satellites and ground stations involved. Therefore, most communication satellites are positioned in a geostationary orbit (GEO), which is a circular orbit that lies in the Earth's equatorial plane (Fig. 91). Thus, the satellite remains virtually fixed with respect to the Earth. The orbital radius is about 42,164 km and thus the orbital period is $23^h\ 56^m\ 04^s$. The altitude above the Earth's surface is about 35,830 km.

Highly Elliptical Orbit
HEO

Molniya Orbits Ground stations in extreme northern or southern latitudes unfortunately cannot be reached by satellites in geostationary orbits. To solve this problem, satellites are placed in highly elliptical orbits (HEO) (Fig. 91). This orbit allows a satellite to stay for more than 8 h in northern latitudes. The critical inclination of 63.4° is crucial because it avoids a rotation of the line of apsides, which would soon render the orbit useless. Typical members of this orbit class are the Russian Molniya satellites which are also the origin for this orbit name.

13.5.3 Orbit Perturbations

There are certain mechanisms that disturb the ideal two-body motion specified by Kepler's laws and cause the orbital elements of satellites to vary with time.

Perturbations by the Earth's gravitational field

Gravitational Perturbations Due to the deviation of the Earth's gravitational field from that of a homogenous sphere and the Earth's oblateness, the two most important perturbations are the precession of a satellite's orbital plane and the rotation of the line of apsides in the orbital plane. The real mass of the Earth, the geoid, causes perturbations that cannot be predicted exactly by simple formulae, as it is an inhomogeneous and asymmetric body. Therefore, observations of satellite orbit anomalies yield indispensable data for determining the precise shape and mass distribution of the Earth. The Earth's gravitational field is also disturbed by the gravitational attraction of the Sun and the Moon. The resulting perturbations are very difficult to predict, and depend strongly on the relative orientation of the orbit and the attracting bodies. In general, near-equatorial orbits are much less affected than near-polar orbits. If an orbit has a large eccentricity, the gravitational force of the Sun and the Moon can cause substantial changes in the perigee height.

Perturbations by Sun and Moon

Non-Gravitational Perturbations The density of the upper atmosphere is highly variable. It depends on the degree of solar activity, the season of the year, the latitude, and the change from day to night. For LEO satellite orbits, atmospheric drag is the most important perturbation. It is caused by collisions of a spacecraft with atmospheric molecules that are abundant even at these altitudes. This decreases the orbital energy of the satellite and also its semimajor axis. Paradoxically the period shortens and the satellite speeds up. In an orbit with a notable eccentricity, atmospheric friction is a factor mainly near the perigee. Therefore, the drag changes with the orbital position of the spacecraft and tends to decrease the eccentricity of the orbit. As a result of the energy lost, any satellite with a perigee height of less than a few thousand kilometers will sooner or later decay in the atmosphere. Another perturbation arises from the solar radiation pressure. Any satellite that is illuminated by the Sun is disturbed by the same solar force that also forms the well-known comet tails. The solar radiation pressure tends to increase the orbital eccentricity, while leaving the semimajor axis nearly constant. Thus, the perigee height decreases continuously, which can in turn cause a rapid decay due to atmospheric friction.

Atmospheric drag and friction
Solar radiation pressure

13.5.4 Criteria for Visibility

An artificial Earth satellite has to fulfill two criteria to be observable: it has to be above the observer's horizon and it has to exceed a certain minimum brightness to be photographed or recognized. The satellite must reflect sunlight to be visible, unless it has a built-in light source. Its recognition depends on the brightness of the sky background and on the sensitivity of the eye, the photographic emulsion or device. Some of the brightest satellites are the ISS (-2 mag) or Iridium satellites (bright flares up to -8 mag, Fig. 92).

Height above horizon
Minimum brightness
Sensitivity of camera

Fig. 92. Iridium flashes can be observed in twilight due to their extreme brightness. Photo courtesy of Christine Treichel, Schwante, Germany (September 19, 1997, 7:43 p.m., Canon EOS 20Da, 17 mm (detail magnification), f/22, 12 s, ISO 100)

Subsatellite point

Inclination and latitude of launch pad

Influence of the Inclination The inclination i of the satellite's orbital plane limits the area on Earth that can be flown over. The point on Earth exactly under the satellite, the so-called subsatellite point, always moves between the latitudes $+i$ and $-i$. Therefore, the area from which the satellite is visible does not extend much further. To gain the most from the Earth's rotation, most satellites are launched eastward. Thus, the geographic latitude of the launch pad equals the orbital inclination of the satellite. As the latitude of Cape Canaveral is 28.5°, many American satellites are unobservable from mid-northern latitudes. Instead, Russian spacecrafts are launched from different sites with considerably higher latitudes and thus have extended visibility.

13.5.5 Accurate Predictions

Observing a special satellite needs a more or less precise prediction or ephemeris for it. It is most efficient to obtain these predictions with a home computer and a program that is capable of calculating positions from a set of orbital elements. But as these elements do not remain constant for a long time (because of the perturbations mentioned in Sect. 13.5.3), programs in use by amateurs can only predict with sufficient accuracy for a few days or weeks. Thus, current orbital elements are absolutely indispensable. They can be obtained from several Internet sources with the possibility to display resulting satellite trails and time tables for the local observation site.

Internet Links

Satellite trails and time tables for the local observation site

- http://science.nasa.gov/realtime (NASA, Satellite Tracking Homepage)
- http://www.calsky.de: (Visibility, Transits in front of Solar Disk, of ISS and other artificial Earth Satellites)
- http://www.heavens-above.com: (Visibility of ISS, Shuttle Flights, Iridium Flares)
- http://www.satscape.co.uk (Satellite Tracking Software)

Prediction Bulletins per Mail

- NASA Prediction Bulletins, Project Operation Branch (Code 513), NASA-Goddard Space Flight Center, Greenbelt, MD 20771, USA
- British Satellite Prediction Center, Earth Satellite Research Unit, University of Aston, St. Peter's College, Saltley Birmingham, UK
- Satellite Section, The British Astronomical Association, Burlington House, Piccadilly, London WIV ONL, UK

List of geostationary satellites visible from the Eastern and Western hemispheres: http://www.geo-orbit.org http://en.wikipedia.org/wiki/ List_of_broadcast_satellites

Geostationary Satellites Because of their circular orbit with an inclination close to 0° and a period of 23 h 56 min, geostationary satellites remain virtually fixed in the sky (Table 12). But as a consequence of their great distance, they never become brighter than magnitude 10 and are therefore difficult to find. The best way to find such a satellite is to retrieve its equatorial position for a given instance (see satellite tracking links above) and to point the telescope with respect to the stars.

Table 12. Some geostationary satellites that are visible from North America (*left*) and Europe (*right*). The first column gives the nominal longitude (west), the second column the name, and the third column the COSPAR designation of the satellite. The actual and the nominal positions can differ by up to 0.5°. All of these satellites have inclinations of less than 0.1° and should become brighter than magnitude 12

Longitude (West)	Satellite	COSPAR designation	Longitude (West)	Satellite	COSPAR designation
137.0	Satcom C1	90100A	30.0	Hispasat 1A	92060A
135.0	Satcom C4	92057A	19.0	TV-Sat 2	89062A
128.0	ASC-1	85076C	19.2	TDF 1	88098A
125.0	Galaxy 5	92013A	19.2	TDF 2	90063A
125.0	GStar 2	86026A	8.0	Telecom 2A	91084A
123.0	SBS 5	88081B	5.0	Telecom 2B	92021A
111.0	Anik E1	91026A	−3.0	Telecom 1C	88018B
107.2	Anik E2	91026A	−5.0	Tele-X	89027A
105.0	GStar 4	90100B	−13.0	Eutelsat 2 F-1	90079B
103.0	GStar 1	85035A	−13.4	Italsat 1	91003A
101.0	ASC-2	91028A	−16.1	Eutelsat 2 F-3	91083A
99.0	Galaxy 6	90091B	−19.2	Astra 1A	88109B
99.0	SBS 6	90091A	−19.2	Astra 1B	91015A
93.0	GStar 3	88081A	−19.2	Astra 1C	93031A
90.9	Galaxy 7	92072A	−21.6	ECS 5	88063B
87.0	Spacenet 3R	88018A	−23.6	DFS 1	89041B
70.0	Brasilsat 2	82026B	−28.6	DFS 2	90063B
62.0	TDRS 3	88091B	−30.7	DFS 3	92066A
53.0	Intelsat 5a F-1	88040A	−31.2	Arabsat 1C	92010B

13.5.6 Visual Observations

For satellites which are too faint to be visible to the naked eye, a pair of binoculars with an aperture of 50 to 60 mm and a magnification between 7 and 10× is the ideal choice for observation. The telescopic aid also helps to attain an acceptable positional accuracy due to its magnification. Geostationary satellites can be observed with higher magnification and a fixed telescope, as they do not move significantly.

To record the time, an accurate stopwatch is needed. It should have a digital display, readable to 0.01 s, and a "lap time" facility. Also, access to a time signal broadcast radio station is advised.

For more details about stopwatches, time signals and GPS, see Sect. 13.4.

Since the equatorial coordinates of satellites are topocentric, it is important to know the precise geographic position of the observing site.

In advance, the satellites prediction data and its predicted track plotted in a star atlas has to be obtained (see p. 420).

Preparations

13.5.6.1 Determining the Actual Satellite Track
As air drag can cause the orbital period to decrease, it is recommended to start observing the satellite several minutes before its predicted arrival time. If the satellite arrives earlier or later than predicted, it is likely to deviate from its predicted track. This is mainly a consequence of the Earth's rotation. Depending on the satellite's height, the actual apparent track may be several degrees east

Deviation of the actual track

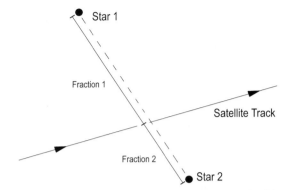

Fig. 93. Fractional distance of the satellite relative to each of the stars

of the predicted one for every minute the satellite is early. If it arrives late, the track will deviate to the west.

Observation program

More information about observing satellites:
http://satobs.org

- Once the satellite has been found, the observer has to wait until it passes between two stars that lie nearly perpendicular to the apparent track and are separated by less than one degree.
- When the satellite crosses the line joining them, the stopwatch must be started. In the same instant, the fractional distance of the satellite relative to each of the stars has to be estimated as shown in Fig. 93.
- Next, identify and mark the two stars in a detailed atlas.
- Finally, the watch is stopped against the signal of a radio or telephone time service. If it has a split-time function, the accuracy of the time recording can be improved by checking the fraction of a second at several preceding time signals.

Reference stars

The precise positions of both reference stars can be taken from a good star catalog. With these data and the observed fractional distance to each of the stars, an accurate position for the satellite is calculated by linear interpolation.

The ESA "Spacecraft Tracker" combines realtime data of satellite positions with Google Maps material and generates a dynamically updated display of the flight track for several ESA and partner missions over the ground:
http://www.esa.int/SPECIALS/
Track_ESA_missions_GE/index.html.

Instead of crossing a line between two reference stars, a satellite can occasionally have a very close approach to only one star. At the moment the satellite appears to coincide with the star, their positions can be assumed to be identical and so a second reference star is not necessary. The time is recored as described above. This technique is easier and more accurate to apply.

13.5.6.2 The Brightness Behavior of Satellites Satellites are, like the Moon or the planets, illuminated by the Sun and thus only visible by reflection of sunlight. Two general kinds of reflections can be distinguished: mirroring and diffuse reflection. (The satellite is assumed here as a simple sphere illuminated by the Sun.)

In the case of mirroring reflection, which can be realized with a polished metal coating, the luminous intensity depends on the illuminance of the incident sunlight, the reflection coefficient and the radius of the sphere. It is independent of direction.

In the case of diffuse reflection, which can be realized with a dull coating, the luminous intensity depends also on the phase angle α, which is the angle subtended at the satellite between the Sun and the observer. For phase angles smaller than approximately 90°, a diffuse reflecting satellite is brighter than a mirroring reflecting one with the same reflection coefficient. The most famous example of a diffuse reflecting satellite is the Moon.

The apparent brightness of a satellite depends strongly on its distance from the observer. If the distance is doubled, the light flux from the satellite is reduced to one-quarter of its initial value. This corresponds to a difference of 1.5 mag.

Influence of the Earth's Shadow When the sky becomes dark enough for satellite observations, the Earth's shadow rises quickly above the observer. The Earth's shadow has the shape of a cone with a diffuse boundary. This is due to the finite angular diameter of the Sun. The Earth's shadow limits the visibility and tracking of satellites considerably, because it covers the satellites with darkness.

When the Sun is 18° below the horizon, the height of the shadow directly over the observer exceeds 300 km. All satellites in lower orbits (i.e., the brightest ones) have already penetrated the shadow cone. But the situation depends strongly on the season. In winter, the useful time for observations in mid-northern latitudes is less than one hour. In summer instead, depending on the observer's latitude, it may be possible to observe satellites throughout the night.

Geostationary satellites are much less affected by the Earth's shadow. For most of the year, the entire geostationary orbit is sunlit, since the orbit is tilted out of the Earth's shadow. Only when the Earth approaches the vernal and the autumnal equinoxes, parts of the orbit move into the shadow and cause satellites to be eclipsed for more than 72 min. This is possible only from February 27 to April 12, and from August 31 to October 16.

13.5.6.3 Light Variations The shape of most artificial satellites deviates considerably from the idealization as a sphere. Cylindrical rocket stages are quite numerous, and many payloads are prismatic or box-shaped. The majority is equipped with solar panels and antenna reflectors. Consequently, their brightness depends not only on the distance and phase angle, but also on their altitude and orientation with respect to the observer. The increase or variability in brightness can be high enough to view the satellite with binoculars or very small telescopes.

The tumbling rotation of many spent upper stages can be observed as a periodic light variation with an amplitude of several magnitudes. The tumbling period of the satellite can be estimated from visual observations.

Spacecrafts which are stabilized with respect to the Earth's limb often show a light variation because they change their orientation to the observer in the course of a passage.

Sometimes satellites exhibit very conspicuous glints of light. These are caused by specular reflections from glazed surfaces on solar arrays or radiators. The increase in brightness can be of more than 10 mag. In the case of the Iridium satellites, the apparent magnitude of a flash can reach up to -8 mag. Soviet Molniya satellites are also candidates to

Mirroring reflection

Diffuse reflection

Phase angle

Height of Earth's shadow

Seasonal influences

Geostationary satellites

Shape of satellites

Rotation period

Orientation change

Flashes

Fig. 94. The ISS dives into the umbral shadow of the Earth (from *right* to *left*) and its light fades away. Photo courtesy of Christine Treichel, Gören-Lebbin (Canon EOS 20Da, 12.08.2007, 11:11 p.m. 17 mm, f/5, 34 s, ISO 800)

Three-axis stabilized geostationary satellites such as Olympus can also cause bright specular reflections. A control mechanism keeps their solar arrays oriented to the point on the celestial equator that has the same right ascension as the Sun. The brightness due to this reflection would reach its maximum when the satellite and the Sun have a difference in right ascension of 12 h (180°), but this always coincides with a penetration of the Earth's shadow. Therefore, the spacecraft is brightest just before and after the shadow passage.

Penumbral phase

Umbral phase

produce flashes, because they remain for hours in high northern declinations and are sunlit almost throughout the night. Originally, several of these events, like the famous Perseus and Aries flashes, were misinterpreted as originating from a then-unknown class of variable stars. In addition, satellite flashes seem to be the explanation for some remarkable phenomena on lunar photographs.

13.5.6.4 Changes in Color When a satellite enters the Earth's shadow, it does not disappear abruptly (Fig. 94). It first passes the penumbral part of the shadow, when, as seen from the satellite, the Sun is partially occulted by the Earth. Progressing further into the shadow, the intensity of direct sunlight on the satellite decreases continuously, but instead, the amount of sunlight that has been refracted and reddened in the Earth's atmosphere is growing. In the umbral part of the Earth's shadow, the direct sunlight is completely obstructed by the Earth and the satellite is invisible. The egress of a satellite is accompanied by a change in color from red to white. Some observers have reported a green tinge due to the ozone layer in the upper stratosphere.

13.5.7 Photographic Observations

Another way to observe satellites is to record them with a photographic camera (SLR or DSLR) or a CCD camera.

Focal length

Aperture

Sensitivity

Exposure time

A 35-mm camera with a manual exposure and an attachable cable release is quite adequate for photographing satellites. The camera lens should have a short focal length and a fast aperture to be able to record a fast-moving object. It also should provide a wide field-of-view in order to record a long satellite trail on the photograph. A standard 50-mm lens fulfills these demands and can be regarded as a good compromise. Only high-sensitivity photographic film, respectively high ISO settings on a DSLR (e.g.,

Fig. 95. Trail of ERS-1 with an unexpected light reflection. The bright star near the *upper right corner* is Polaris. Photo courtesy of Christine Treichel, Schwante, Germany (September 19, 2007, 8:27 p.m., Canon EOS 20Da, 17 mm, f/5.0, 87 s, ISO 400)

800) should be used. The aperture must be fully opened. Exposure times have to be long enough to produce sufficient reference star trails.

To achieve best results, the satellite should be photographed when it is approximately opposite to the Sun. In this position, the sky background is darkest and the satellite is brightest due to its small phase angle. Because of the rapidly rising shadow of Earth, most satellite observations are carried out before the end of the astronomical evening twilight. Thus, the remaining brightness of the sky background has to be taken into account to avoid overexposures. Point in time for photographs

There are two different methods for photography:

1. The camera remains fixed during the exposure, and the satellite and the reference stars produce trails on the film. Only the brightest satellites can be photographed this way (Fig. 89). Fixed camera

2. The camera has an equatorial mount or is fixed on a telescope tube with such a mount, and follows the stars. Again, only the brightest satellites can be recorded, but the reference stars produce sharp points. This makes the evaluation of the photograph easier. Tracking camera

With a CCD camera connected to a short focal length telescope it is possible to record geostationary satellites [76]. Although named like this, their positions on the sky are not static (Fig. 96). Due to gravitational perturbations, their positions vary in an irregular manner. These movements can be recorded with the CCD camera by making a series of images with the same exposure time over a long period, e.g., 12 h during the night. After the images have been added with an image processing software, the deviation from the geostationary position is visible as a curved trail. This method is quite interesting for Exposure series with CCD camera
Satellite Families

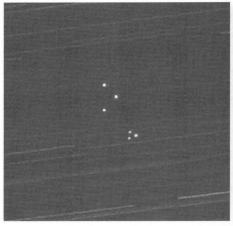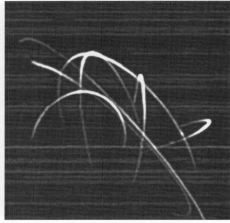

Fig. 96. (*Left*) The ASTRA family, photographed with a fixed camera. The parallel trails belong to stars. Photo from Christine Treichel, Schwante, Germany (September 15, 2007, 9:38 p.m., Canon EOS 20Da, 200 mm (cutout magnified), F/3.2, 199 s, ISO 800). (*Right*) The ASTRA satellite tracks during 12 h and 40 min in the night from January 10th to January 11th 1998. The satellite's brightness varies between 12 and 14 mag. Photo from Stefano Sposetti (C8 fixed, f/6.3 and a Hi-SIS22 CCD camera in 2 × 2 binning mode, simple superposition of 368 CCD images with 120 s exposure time). Courtesy of Stefano Sposetti, http://web.ticino.com/sposetti/index.html

groups of satellites which share nearly the same position on the sky, e.g., the ASTRA family, where all are visible in the recorded field-of-view with different trails [76], (see Fig. 96).

Images of Spacecrafts Many amateurs have recorded amazing photographs or videos of an ISS passage or Space Shuttle Mission with their telescopes. The key is to capture and afterwards continuously follow the satellite with a high magnifying telescope. This work can be done manually with an unlocked telescope or computer based. In either way, in preparation the finder scope must be exactly adjusted in parallel to the main optic and for correct focusing, the video camera or webcam connected to the main optic has to show a well-focused image of a bright star or a planet.

ISS tracking:
http://www.tracking-station.de

Two line elements (TLE):
http://science.nasa.gov/Realtime/
rocket_sci/orbmech/state/2line.html

For manual tracking, the observer looks through the finder scope and keeps the spacecraft in the center, while the telescope must be moved manually to follow the ISS track. As the ISS is very bright, short exposure times can be used to obtain sharp images of it. Some cameras are able to capture 60 images per second, thus with some exercise, it should be possible to have at least 100 to 200 images with many of them sharp. Alternatively, computer-based tracking is done by a video camera on the finder scope, a computer program that receives the recorded images, reads the actual satellite position from a database, controls both axes of the mount in order to point the telescope to the correct position in the sky and then keep the satellite image exactly on the same position in the finder scope. At least, a second video camera on the main telescope records the magnified and highly resolved satellite images. Software used for this case in general requires up-to-date TLE (two line element) formatted satellite position data. Further, the telescope mount or hand controller needs an interface (e.g., RS232) to the computer for

Fig. 97. The ISS photographed with a telescope. Courtesy of Hartwig Lüthen

axis control. With an image processing software the video film has to be evaluated frame by frame, where only the best frames are worthy for the final image. With success, the observer can have a result like Fig. 97. For more information about video capturing, see Chap. 6.

13.5.8 Further Reading

Kresken, R.: Artificial Earth Satellites, Compendium of Practical Astronomy. Springer, Berlin Heidelberg New York (1994)

Wikipedia: The Free Encyclopedia: http://en.wikipedia.org/wiki/Satellite, http://de.wikipedia.org/wiki/Satellit_(Raumfahrt)

13.6 Observation of the Planets

G.D. Roth

13.6.1 The Purpose of and Tasks for Planetary Observations

13.6.1.1 The Amateur Observer and the Planets The nineteenth and twentieth centuries have had a distinguished history of amateur observers who achieved scientific renown as planetary observers, both as visual and high-resolution photographers. The pioneers include C.F. Capen, T.A. Cragg, and W.H. Haas (USA), M.B.B. Heath, B.M. Peek and T.E.R. Phillips (UK), C. Boyer and G. Viscardy (France), S. Cortesi, and M. du Martheray (Switzerland), and W. Löbering, G. Nemec, and W. Sandner (Germany). To name just a few, some of today's experts are: D.C. Parker (USA), T. Platt (UK), J. Dragesco (France), I. Miyazaki (Japan), and B. Flach-Wilken (Germany). Despite the fact that nowadays astronomical research is based primarily on large-scale astronomical equipment and the exploration of the Solar System with space missions, the amateur interest in planetary observations remains. The ambitious observer will attempt to employ physical measuring methods in addition to merely visual observations. Moreover planetary observations are a solid base for astronomy education in schools.

13.6.1.2 Observational Tasks With the dawn of the space age in 1957, new life was breathed into planetary research. In some respects, considering the results of the recent flyby missions to as far out as Neptune, one can even speak of the "rediscovery of the planets." Such findings include the exciting verifications that planetary surfaces show craters like those on the Moon, and that Saturn is not the only planet to possess a ring system. A better knowledge of the solid surfaces of planets and moons in the Solar System supplies new ingredients for interpreting the origin of these bodies and of the entire system, not to mention important deductions regarding planetary geology. The amateur devoted to planetary observations should be acquainted with these new developments. Easy-to-read presentations and recent updates appear regularly in numerous, readily accessible magazines.

The results of the large programs, such as the International Planetary Patrol Program, should be studied. The observer will thereby not only keep informed of the focus of research, but will also receive hints concerning the evaluation of his or her own observations. Even the traditional topographic survey—still the center of most amateur work—gains more weight when processed according to the state-of-the-art methods.

The Voyager flybys of Jupiter and Saturn were accompanied by worldwide organized simultaneous observing programs for assiduous amateur observers, namely the International Jupiter/Saturn Voyager Telescope Observations Program. The national working groups have reported on this and other joint efforts.

There are several observational tasks which, when performed with care and perseverance, may aid the scientific exploration. They include, for instance:

1. Long-term monitoring of planets with respect to the appearance of special phenomena (Martian dust storms, bright or dark markings in the atmospheres of Jupiter and Saturn, a new Red Spot 2006 in Jupiter's South Temperate Belt, and a Great White Spot on Saturn, October 1990).

2. Position measurements of individual objects (e.g., clouds on Mars, the Great Red Spot on Jupiter) on planets in order to derive rotation periods or other motion patterns.

3. Spectrophotometry, visual as well as photographic (observations with filters of known spectral transmission).

4. Total and detailed photometry, visual and photographic (e.g., intensity estimates of bright zones and dark bands on Jupiter, estimates of brightnesses of minor planets).

5. Special observations, e.g., of planetary satellites, occultations, or transits of planets across the solar disk.

The following pages will outline these tasks in connection with the individual planets to which they apply.

When performing a serious observational project, however, it is always of fundamental importance that the observer pay attention not only to the personal reliability and precision of the equipment but also to the relevant literature and care in the processing of

See W. Löbering, *Jupiterbeobachtungen von 1926 bis 1964* [77] for examples of a visual long-term program done by an amateur astronomer.

ESO, the European Organisation for Astronomical Research in the Southern Hemisphere, operates a wide range of education and public outreach (EPO) activities (http://www.eso.org).

Find material for addresses online, periodicals and books in the book edited by G.D. Roth, *Planeten beobachten* [78].

Discover Exoplanet. Amateur astronomers assist professional researchers. See *Sky and Telescope* issue October 2005, p. 96, and September 2006, p. 90. Visit Center for Backyard Astrophysics network (http://cba.phys.columbia.edu).

the results. It is advisable to join a working group or a planetary section of an astronomical association. Knowledge of the literature also permits highly interesting comparisons, for instance, between photographs taken from space and early drawings.

13.6.2 Observing Equipment

13.6.2.1 The Telescope As successful planetary observations depend on very good image definition, even with medium and higher magnifications, an instrument of long focal length is favored (aperture ratio f/10 or larger). In principle, there is no preference between refractors and reflectors (see Chap. 4), although in reflectors, the silhouetting effect of the secondary mirror limits the definition. The disadvantage of silhouetting can be surmounted by choosing a larger aperture.

The high-quality telescope should be accompanied by similarly good eyepieces and by a stable mounting (Chap. 5). For medium and high magnifications, orthoscopic and eyepieces of similar quality are preferred. Eyepieces containing thick meniscus lenses, which diminish the off-axis imaging errors, have proven useful. The Astro-Planokular of Zeiss, for example, and also the wide-angle eyepieces of Baader, Meade, and Tele-Vue deserve attention. Binocular adaptors (supplied by, e.g., Baader, BW Optik, Denkmeier, Lumicon and Tele-Vue) have also been found useful.

The magnifications necessary to see the planetary disk in the telescope at the same apparent angle as the Full Moon viewed with the unaided eye are provided in Table 13. The experienced observer M. du Martheray has compiled experiences concerning the choice of an optimum power for planetary observations into a graph, on which Table 14 is based.

Not to be underrated are personal qualities and the experience of the observer, as well as the quality of the telescope and the atmospheric conditions. Figure 98, which is based on a study by H. Wichmann, considers different atmospheric conditions on the magnification.

The experiences of numerous observers in both Europe and the US emphasize the fact that useful powers can be employed with telescopes of small and moderate aperture (100 to 150 mm) even under fair or poor atmospheric conditions, and thus these apertures can be considered "optimal" for amateur observers over a wide range of atmospheric circumstances.

Table 13. Magnifications for which the planets would appear the same angular size as the Full Moon in the sky

Planet	Apparent diameter	Magnification
Mercury	$4''.8$–$13''.3$	$280\times$ at elongation ($6''.5$)
Venus	$10''$–$64''$	$70\times$ at elongation ($25''$)
Mars	$4''$–$25''$	$70\times$ at opposition
Jupiter	$31''$–$48''$	$40\times$ at opposition
Saturn	$15''$–$21''$	$100\times$ at opposition
Uranus	$3''$–$4''$	$500\times$ at opposition
Neptune	$2''.5$	$750\times$ at opposition

Table 14. Optimal magnifications for planetary observations

Objective Aperture	Magnification for observing detail					Saturn	Uranus	Neptune
	Mercury	Venus	Mars	Jupiter	Jupiter's moons			
75 mm	150	150	175	150	150	175	175	–
135 mm	200	225	275	200	375	250	300	300
250 mm	300	300	325	275	400	350	350	350
300 mm	350	350	350	300	500	375	450	500

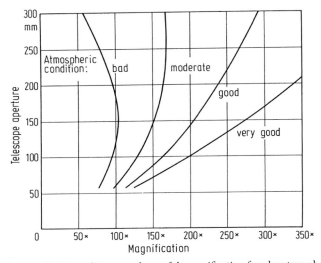

Fig. 98. Effect of atmospheric conditions on the useful magnification for planetary observations as dependent on aperture size

A great optician once wrote: "The worst part of a telescope is the atmosphere…"

In the discussion of reflector versus refractor by W.P. Zmek, the conclusion was that a 100 mm refractor is able to show the same planetary details as a 200-mm reflector [79, 80].

In his book *High Resolution Astrophotography*, J. Dragesco wrote about his experience that his 250-mm Cassegrain is both more convenient to use and has a higher photographic resolution than his 178-mm apochromatic refractor [81].

13.6.2.2 Accessories In order to reach more diversified physical methods of observing and measuring, certain accessories will be indispensable for the amateur observer. Visual as well as photographic observations may be upgraded by the use of additional instruments, the most important of which are:

1. Color filters (Chap. 6)
2. The micrometer (Sect. 4.8.5)
3. The photometer (Chap. 8)
4. The spectroscope (Sect. 4.8.7)

Certainly the color filters are the easiest to work with. They can be advantageously combined with small instruments to make observations of planetary surfaces in various well-defined selected spectral ranges. As a rule, these ranges include the red (e.g., Schott RG2 and RG5), yellow (e.g., Schott GG10) and blue (e.g., Schott BG12 and BG23). The individual filters are framed so as to cap the eyepiece. A micrometer can be used to measure, for instance, the diameter of the crescent of Venus, to determine the time of dichotomy and the cusps (see p. 443), or to measure the positions of striking details on Mars (clouds, polar caps, etc.) and in the atmospheres of Jupiter and Saturn (bright and dark features, widths of zones and bands); see Sect. 13.6.5.4 for further details. With a photometer one can, in particular, perform area photometry of various regions on the planetary surface (Chap. 8) and point photometry to obtain the total integrated brightness of a planet or minor planet (see Sect. 8.6.1.1). The spectroscope enables the observer to find the rotational period of a planet via the Doppler effect and to gain information regarding the planetary atmosphere. The observer may even contemplate the use of a CCD camera, a device particularly suited for photometric tasks (Chap. 6).

13.6.3 Visibility of the Planets

13.6.3.1 Apparent Diameter, Phase, and Oblateness The conditions of visibility planets differ greatly. Information on their visibility during a year can be found in various astronomical calendars. The orbit and size of a planet, as well as the location of the observer, determine its visibility. Specifically, the determining factors are:

1. Rising-culmination-setting of the planet as viewed at the observing site
2. Right ascension and declination of the planet
3. Apparent diameter (see Fig. 99)
4. Apparent brightness
5. Phase angle

The outer planets (Mars, Jupiter, Saturn, Uranus, Neptune) are closest to the Earth at opposition, when they are visible all night long. The inner planets (Mercury and Venus), on the other hand, appear either in the evening or morning sky (at, respectively, the times of greatest eastern or greatest western elongation). The position of the ecliptic relative to the horizon and the seasonally varying duration of twilight, depending on the geographic latitude of the observer, influence the visibility of the planets, in particular the inner ones. But also for the outer planets, the observing conditions are most favorable when the planet's opposition occurs during the winter, when the ecliptic band is high above the horizon. At lower geographic latitudes, the ecliptic is steeper with respect to the horizon, and visibility conditions are better. Even for a planet with a substantial orbital eccentricity, the altitude above the horizon outweigh the perhaps less-favorable orbital conditions (i.e., nearness to the Sun for Mercury, small apparent diameter for Mars).

The phase angle of the planet (which is listed in astronomical almanacs) will have a significant influence on the observations, as it determines the degree of apparent illumination of a planet. At a phase angle of 0°, the disk of the planet is fully illuminated, and corresponds to the Full Moon phase, while at 180° it is invisible (corresponding to New Moon phase). For the inner planets, the phase angle varies from 0° to 180°, while for the outer planets, it plays a significant role only for Mars (where it is 46° when one-eighth

Atmospheric Dispersion Corrector can help coax the best possible performance from the telescope when viewing planets low in the sky. See issue of *Sky and Telescope* June 2005, p. 88.

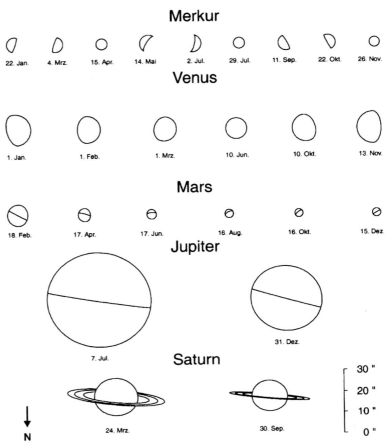

Fig. 99. Appearance of the planets during 2008

of the planetary disk is invisible). Some almanacs give, instead of the phase angle, the degree of illumination of the planetary disk.

In addition to the changes of apparent diameter and illumination, the observer must also be concerned with the position of the rotation axis and the oblateness, information on which is to be drawn from the annual almanacs.

13.6.3.2 Atmospheric and Environmental Influences The effects of Earth's atmosphere on astronomical observations is detailed in Sect. 13.11. For planetary observations, the atmosphere over the observing site should be as free from turbulence as possible. The medium and high magnifications required for planetary observations are corrupted by atmospheric unsteadiness. Haze, on the other hand, is usually less of a problem as it is often associated with very stable conditions.

In order to objectively rate the atmospheric conditions at the observing site, the observer should use an estimated scale and record the result in the observing protocol. Two important data are estimated: the seeing, or image stability, which depends on the magnification employed, and the transparency, which is best estimated without a telescope.

Seeing Scale

Seeing 5   Excellent. Even at high magnification, the image of the planet in the eyepiece is steady and sharp.

Seeing 4   Good. General impression as in 5, but with very occasional turbulence.

Seeing 3   Fair. The turbulence is higher over time, but there are sufficiently long time intervals during which moderate magnification can be used to obtain a useful overall impression of the planetary disk.

Seeing 2   Poor or mediocre. The turbulence interferes perceptibly; only momentarily can details be seen.

Seeing 1   Very poor or useless. It is impossible, even at low power, to see a sharp image of the disk.

Transparency Scale

T5   Very clear sky. 5th mag stars visible.

T4   Clear sky. 4th mag stars visible.

T3   Hazy sky. Visibility reaches to 3rd mag stars.

T2   Thick haze. Only 1st mag stars are seen; planetary observations marginally possible.

T1   Haze, fog, clouds so thick that the planet is intermittently visible with the naked eye. In the telescope the image is too dim to perceive details.

The atmospheric turbulence is best seen on a sharply focused image of a bright star. The image wiggles around its central position. This "turbulence disk" has a radius of $5''$ or more in bad seeing. In quiet air, the disk size diminishes to less than $0''.5$.

Atmospheric turbulence is caused not only by climate and weather at the observing site, but also by other circumstances in the vicinity of the telescope which set into motion the layers of air directly overhead (e.g., heating, smoke updrafts, airplanes). Artificial light sources are not as detrimental to visual planetary observations as they are for other kinds of astronomical work. Therefore, opportunities exist even for planetary observers who live in densely populated areas. However, localized urban heating does favor the generation of small turbulence cells.

13.6.3.3 Personal Qualities   In the final analysis, the perception of planetary details depends on the approach of the individual observer. Here, experience gained from long-term, well-planned observing runs plays an important role. Every observer should be aware of the following rules:

1. Always observe with rested eyes.
2. Avoid light glare (adapt to darkness, dim all light sources needed in the vicinity).
3. Do not overmagnify images.
4. Practice monocular vision, perhaps (at the beginning) by covering the unused eye with the a hand or patch. Note, however, that the experienced observer does not close the unused eye while observing, but merely subconsciously suppresses its signal. Binocular adaptors may be tried.
5. Telescope operation, such as guiding, must be performed effortlessly in the dark.
6. Several breaks should be performed without sketching.
7. No observation should be performed without sketching, as this forces the eye to carefully examine every detail.

The diameter of the shimmering disc can be estimated by means of the distance between the components of a double star pair. C. Tombaugh and B.A. Smith compiled the following scale.

Diameter (arc-second)	Image quality	Diameter (arc-second)	Image quality
50	−7	2	+3
32	−3	1.3	+4
20	−2	0.79	+5
12.6	−1	0.5	+6
7.9	0	0.32	+7
5	+1	0.2	+8
3.2	+2	0.13	+9

"Observing experience is generally acquired and maintained through extensive and continuous activity. Interruptions thereof, even if of short duration, can lead to a loss of practice" (W.W. Spangenberg).

13.6.4 The Representations of Planetary Observations

13.6.4.1 Drawings When making drawings, the directions on a planet must first be defined. Following the IAU decision of 1961, planetary maps have the same "astronautical" orientation used for Earth maps in atlases: north on top, south at the bottom, east to the right (in the direction of planetary rotation), and west to the left. Thus the image is upside-down, or rotated by 180° for observing a planet south of the zenith with an inverting telescope (from most northern locations on Earth), or for a southern observer using a standard, non-inverting pair of binoculars. Surface detail, such as the Great Red Spot on Jupiter, is carried by the planet's rotation to the east (to the left in a northern inverting telescope). Formerly, the terms "preceding end" and "following end" were used, where preceding was in the sense of the diurnal displacement of the planet due to Earth's rotation as well as to the shift due to the planet's rotation. In any event, observers are advised to avoid confusion by proper labeling of graphs, maps, or photographs.

Stencils for graphs can be obtained from various amateur supply companies. There are circular stencils for Mercury, Venus, Mars, while those for Jupiter include the flattening, and for Saturn both flattening and various inclinations of the rings. The two most important things to be kept in mind for Mercury, Venus, and Mars—phases and axial orientation—will be explained in the coming sections.

As a basic technique in drawing, every graph should be developed from coarse contours into fine detail. A medium-soft pencil should be used. For the fast-rotating planets Jupiter and Saturn, the coarse contours should be completed quickly lest substantial longitude errors occur.

Apart from general presentations of the surface, drawings may be used to determine positions. In this case, the emphasis is not on innumerable detail but rather on the careful representation of certain selected features, such as bright and dark spots on Jupiter, and their arrangement into a coordinate system (Fig. 100). With some practice, carefully drawn detail near the central meridian can yield good results.

13.6.4.2 Photographs Techniques and photographic materials are detailed in Chap. 6. Good planetary photographs are also a useful supplement for comparison with the visually obtained sketches. Color filters will show the varying visibility of some surface or atmospheric details on planets as depending on wavelength (Figs. 101–104). Planetary photographs with amateur instruments serve primarily to measure positions of prominent details with the aid of grids. Photographic positions on Jupiter are impeded by the substantial limb darkening across the disk; the true limb of the disk may be difficult to locate. It is therefore advisable to photograph a prominent object near the central meridian in each series, as the position of the central meridian is better determined than that of the limb.

The increased speed of CCD cameras is a valuable asset for planetary photography. CCD cameras are about 75 times faster than the conventional photographic technique. Moreover during the past years immense progress in amateur planetary photography took place with inexpensive webcams. They can produce images with extremely short exposure times, compared with conventional cooled CCD cameras with one-tenth of the duration.

"Under particularly good circumstances, standard errors of ±0.3° to ±0.5°, (jovicentric) were reached, while under unfavorable conditions (turbulence, personal indisposition, etc.) the deviation reached ±0.01°″ (W. Löbering on observations of Jupiter).

Sketching is still the best way to record what you actually see through a telescope eyepiece. See *Astronomical Sketching. A Step-by-Step Introduction*, by R. Handy et al., Springer, 2007.

To process webcam images one of the most powerful programs for sorting, registering, stacking, and sharpening is RegiStax 4: http://registax.astronomy.net (freeware).

Donald C. Parker writes about his "digital darkroom" in *Sky and Telescope* January 2007 issue, p. 129. He uses webcams and a 16-in. Newtonian reflector.

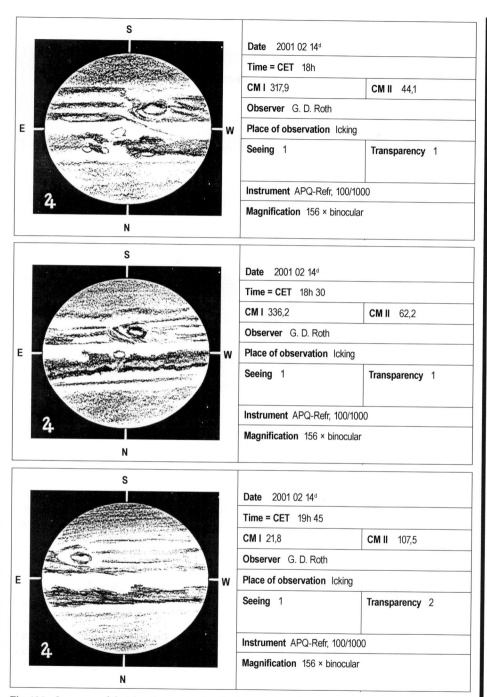

Date 2001 02 14^d	
Time = CET 18h	
CM I 317,9	**CM II** 44,1
Observer G. D. Roth	
Place of observation Icking	
Seeing 1	**Transparency** 1
Instrument APQ-Refr, 100/1000	
Magnification 156 × binocular	

Date 2001 02 14^d	
Time = CET 18h 30	
CM I 336,2	**CM II** 62,2
Observer G. D. Roth	
Place of observation Icking	
Seeing 1	**Transparency** 1
Instrument APQ-Refr, 100/1000	
Magnification 156 × binocular	

Date 2001 02 14^d	
Time = CET 19h 45	
CM I 21,8	**CM II** 107,5
Observer G. D. Roth	
Place of observation Icking	
Seeing 1	**Transparency** 2
Instrument APQ-Refr, 100/1000	
Magnification 156 × binocular	

Fig. 100. Sequence of sketches demonstrating the fast rotating of Jupiter when observing the Great Red Spot within 105 minutes on February 14, 2001 with a 4-in. refractor, and magnification 156× (binocular)

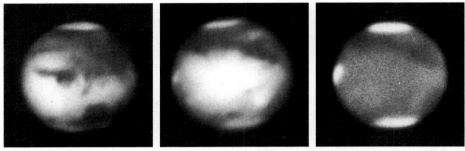

Fig. 101. (*Left*) Mars, photographed on July 20, 1986, 22^{h}18^m UT with a 106-cm reflecting telescope at the Pic-du-Midi Observatory by J. Dragesco. Central meridian 11°.3, apparent diameter 23″, $f/D = 52$, exposure time 1/4 on TP 2415 with W29 filter. (*Center*) Mars, photographed on July 12, 1986 04^{h}52^m UT with a 32-cm Newtonian reflecting telescope by Donald C. Parker, Coral Gables, Florida, US. Central meridian 186°, $f/D = 198$, exposure time 5 s on TP 2415, no filter, developed in Rodinal 1:1000. (*Right*) Same observer and date, but photographed at 04^{h}42^m UT using a violet filter (W47), $f/D = 198$, exposure time 2.5 s on ED 200, developed E-6 process

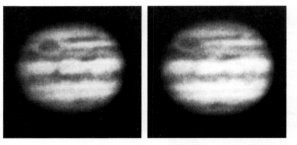

Fig. 102. Jupiter photographed with a refractor of 200/4000 mm with an effective focal length of 36 m, using 14/10 DIN, by G. Nemec, Munich

13.6.4.3 Maps and Planispheres

In principle, planetographic coordinate systems are always the same: the equator is the basic circle, and its position depends on the orientation of the rotational axis. Meridians intersect both poles at right angles to the equator. The position on a planet is expressed by the planetocentric longitude L and the latitude B (e.g., "aerographic" longitudes and latitudes for positions on Mars, or "jovicentric" ones on Jupiter). Longitudes are counted from a zero-meridian at 0° westward to 360°. In the course of time, the meridians of increasing longitude traverse the central meridian. Latitude is counted from the equator to the poles (0° to +90° north, 0° to −90° south).

It is not difficult to construct such a grid for spherical planets. Normally, the so-called orthographic projection is used. For Mars, the tilt of the axis with respect to the Earth is allowed for, which converts the latitude circles into ellipses (as the meridians already are). Orthographic horizontal projections permit the construction of a grid of longitude and latitude at the given tilt. The axis orientation may be found in an astronomical almanac. For the construction by computer see Chap. 3.

In order to compile the readings into a map, cylindrical projection (Fig. 105) may be used, as it is suited to present extended equatorial regions. The projection is true to area

Fig. 103. Mars, October 11, 2005, 22h47m UT, taken with 20 cm SCT. IR-R-G-B. The camera was a Philips ToUCam, Astronomik 742 nm IR-Filter. Photo by H. Lüthen, Hamburg

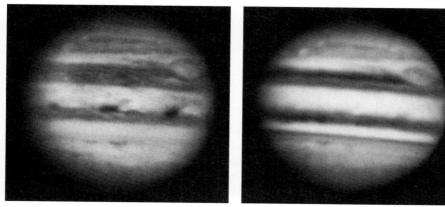

Fig. 104. Examples of photographs taken using color filters. (*Left*) Jupiter, photographed July 20, 1986, $03^h 14^m$ UT with a 106-cm reflecting telescope on the Pic-du-Midi Observatory by J. Dragesco, $f/D = 50$, exposed 2 s con TP 2415, using a Wratten W29 red filter. (*Right*) Same, but taken at $03^h 15^m$ UT at $f/D = 32$, exposed 8 s on TP 2415, using a Wratten 49 blue filter

everywhere, but true to length only at the equator. Parallel circles are stretched to the length of the equator, and graphed at $x = \sin b$, while the vertical meridians are spaced at $y = \text{arc } l$. The area of a zone from the equator to latitude b is then correctly represented as $2\pi \sin b$.

For Jupiter and Saturn, the oblateness, in addition to the substantial axial tilt, complicates the construction. The definition of longitude is not changed by the flattening, but the planetocentric and planetographic latitudes are distinguished.

The preparation of grids helps to determine the positions of surface and atmospheric details, and can in principle be used to process drawings as well as photographs. It also

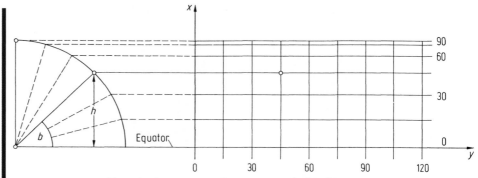

Fig. 105. Construction of a map using cylindrical projection

allows the combination of several observations into a synopsis of the surface. The rapid change of observable detail on some planets, particularly Jupiter, requires a combination of results of several consecutive nights. To represent extended equatorial regions, a mercator projection is advised. A computer is a useful tool for storing and processing planetary observations, and it can also expedite the preparation of maps and planispheres. Examples and programs for PCs and pocket calculators can be found in various astronomical magazines.

13.6.5 The Planets

13.6.5.1 Mercury

Visibility The planet Mercury travels around the Sun in an orbit which lies inside that of the Earth. During its synodic period, it shows phases similar to those of the Moon, as it can move between the Earth and Sun, but never into opposition. Its largest angular distance from the Sun (maximum elongation) is 28°. Beginning with superior conjunction, the pattern of visibility is as follows (with several days' variation owing to the eccentricity of the orbit):

Day 0:   Superior conjunction, "full Mercury"
Day 12:   Mercury appears in the evening sky
Day 36:   Greatest eastern elongation, "last quarter"
Day 47:   Retrogression of Mercury begins
Day 53:   Mercury disappears from the evening sky
Day 58:   Inferior conjunction, "new Mercury"
Day 63:   Mercury appears in the morning sky
Day 69:   Retrogression ends
Day 80:   Greatest western elongation, "first quarter"
Day 104:   Mercury disappears from morning sky
Day 116:   Next superior conjunction, "full Mercury"

Astronomical almanacs supply the following data:

1. Right ascension and declination (needed for finding objects during the daytime)
2. Apparent diameter, which varies between 4″.8 and 13″.3
3. Apparent magnitude, varying between +3.0 and –1.2 during visibility
4. Phase angle (0° = full, 180° = new) or phase (1.00 = fully illuminated, 0.00 = unilluminated)
5. Planetographic coordinates of the center of the disk (sub-Earth point), which may be ±7° in latitude off the equator during periods of visibility

Northern observers at midlatitudes will have their best opportunities to view Mercury in spring in the evening sky and in autumn during the morning. Better observing configuration can be found near the Earth's equator, owing to the steeper angle of the ecliptic to the horizon, or in the southern hemisphere, as Mercury at times of maximum elongation is farther south than the Sun.

Mercury can also be observed in the daytime sky. Its apparent brightness is highest around superior conjunction.

Low light scatter near the Sun is evidently the most essential condition for such observations.

Visual Observations The crater-covered surface of the planet revealed by photographs taken from spacecraft (1974/1975 NASA mission Mariner 10, and the 2004 launched US mission Messenger) cannot be recognized as such by Earth-based observers. Small and middle-sized telescopes show surface detail only in the form of somewhat difficult shadings. The work is also impeded when the planet is observed near the horizon. While it helps to make more frequent observations during the daytime, the best advice is to observe during twilight.

This experience also holds for observations of Venus. The use of filters is advised in order to enhance contrast. The detail on Mercury is, at least in principle, not fuzzy or diffuse, but rather comparable with the district outlines on Mars. Apparent "changes" of dark areas and possible brightening may depend on the phase angle.

The times of greatest elongation and of dichotomy differ by up to ±6 days, owing to the substantial orbital eccentricity of the planet.

Photography When Mercury appears in the evening or morning sky, it can be easily captured on film using a conventional camera with a telephoto lens. Very nice "constellation photographs" may result, showing, for instance, Mercury near the Moon or near Venus. It may also prove worthwhile to try color film or the webcam.

Photographs taken through the telescope require conditions of low light scatter. Photographs of phases or surface details are very difficult to obtain with the telescopes of the size normally used by amateurs. On the other hand, photographs of an occultation of Mercury by the Moon or the rare event of a transit of Mercury across the face of the Sun are within the reach of modest equipment.

Find maps of Mercury in the chapter "Merkur" by Detlev Niechoy in the book *Planeten beobachten* [78], p. 135 and p. 136.

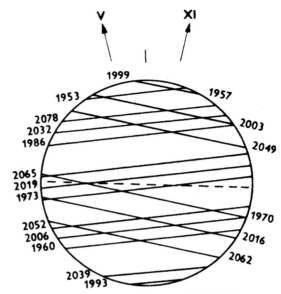

Fig. 106. Transits of Mercury 1920–2080

Transits of Mercury Two transits of Mercury across the face of the Sun will occur in the near future: May 9, 2016 (duration ca. 7.5 h), and November 11, 2019 (duration ca. 5.5 h). Figure 106 shows all the transits occurring between the years 1920 and 2080 (J. Meeus). The first observer to record seeing a transit was the French astronomer Gassendi 1631 in Paris. They can be observed either directly or by projection. The tiny, deep-black planetary disk is a difficult object to observe against the brilliant solar light, as certain optical effects may lead to visual errors. For instance, the occurrence of the "black drop" will interfere with determining the times of second and third contacts.

13.6.5.2 Venus

Visibility As with Mercury, the planet Venus travels in an orbit which lies inside that of the Earth; it is the second "inner" or "inferior" planet. Although still always within the proximity of the Sun as viewed from Earth, Venus is much more conspicuous in the morning or evening sky than Mercury; its maximum angular distance from the Sun reaches 47°. Thus, at the time of optimum visibility Venus can be seen for several hours after sunset as an "evening star" or before sunrise as a "morning star." The planet shows phases just as the Moon and Mercury do. Counting from superior conjunction, the visibility pattern is as follows:

Day 0:   Superior conjunction, "full Venus"
Day 35:   Venus appears in evening sky
Day 221: Greatest eastern elongation, "last quarter"
Day 271: Retrogression of Venus begins
Day 286: Venus disappears from the evening sky
Day 292: Inferior conjunction, "new Venus"

Day 298: Venus appears in the morning sky
Day 313: Retrogression ends
Day 362: Greatest western elongation, "first quarter"
Day 549: Venus disappears from morning sky
Day 584: Next superior conjunction, "full Venus"

The data supplied by almanacs are the same as for Mercury. The apparent diameter of Venus varies from 10″ to 64″, and the apparent magnitude between –3.9 and –4.7. Northern-hemisphere observers have the best evening viewing conditions when eastern elongation falls in the springtime and the planet then remains visible until around midnight.

Venus can be observed over a period of about 7 months at each elongation. The planet reaches greatest brilliancy 35 days after the eastern elongation, and 35 days before western elongation. The high apparent brightness facilitates finding and observing the planet during the daytime. Such observations around maximum elongation are possible with just the naked eye. The observer should stand in the shadow of a house or tree in order to avoid the glare of the Sun. The approximate position of the planet in the daytime sky can be found using the almanac coordinates and a rotatable star map.

An equatorially mounted telescope with circles (see Chap. 5) will facilitate the setting on the planet. It is advisable to have focused the eyepiece on a star beforehand (e.g., on the previous night) in order to immediately secure a sharp image of the planet in the field in the daytime.

Although observation of Venus near conjunction is by no means easy, it should be noted that the planet can, under excellent conditions, be seen with the naked eye about two weeks before or after a superior conjunction. If, moreover, binoculars are used, then Venus may be viewed almost up to or just after the time of conjunction.

During inferior conjunction, the apparent diameter of the narrow crescent of Venus reaches about 60″ from cusp to cusp. Since the planet may be found up to 9° north or south of the Sun, the overlap of the cusps can be seen around inferior conjunction with a 2- or 3-in. telescope under reasonably good viewing conditions. Some observers even report that around the time of inferior conjunction the planet can be seen with the naked eye as an "oblong triangle." When Venus is far north of the Sun at inferior conjunction, it is for a few days both a morning and an evening star for northern hemisphere observers.

Visual Observations The Venera space probes of Russia (former USSR) and Mariner and Pioneer probes of the USA obtained new information on the atmospheres and surfaces of the inner planets. The solid surface of Venus rotates in 243 Earth days (Fig. 107) and thus the rotation is longer than the Venusian year of 225 Earth days. The rotation is also retrograde from east to west, so that on Venus the Sun rises in the west and sets in the east. More than half of the desert-like surface is quite flat. Almost one-fifth lies below zero-level. Elevations reach up to 10,000 m. When the NASA space probe Magellan arrived at Venus on August 15, 1990 the first pictures it took of the surface revealed many more craters and other structures. The atmosphere is composed of several cloud layers with the upper layer rotation distinctly faster than the lower ones. Virtually all of the incoming solar energy is absorbed near the upper edge of the cloud deck. This triggers heating, which in turn drives the circulation patterns in the atmosphere. Ultraviolet photographs taken by the probes (Fig. 108) have confirmed fine structure in the clouds.

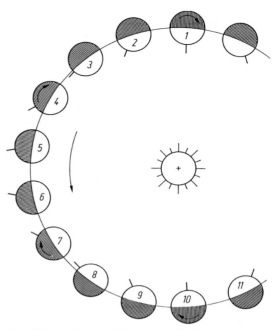

Fig. 107. Part of the orbit of Venus during a full Venusian day (between two consecutive noons). The sense of the rotation is retrograde (*clockwise short arrow*), while the sense of the orbital motion is prograde (*counterclockwise long arrow*). Noon occurs at position 1, sunset at 3, midnight between 5 and 6, sunrise shortly before 8 (in the West!), and 10 marks the next consecutive noon. Adapted from P. Ahnert, *Kalender für Sternfreunde*, Johann Ambrosius Barth, Leipzig 1981, p. 166

In particular eddies in the shape of the letters "Y" or "C" have been repeatedly observed by amateurs.

Since April 2006 the European mission Venus Express orbits the planet with special interest on the atmosphere, the weather and the climate till 2013.

Indeed, the Venusian atmosphere is very dense, nearly structureless in visible light, and conceals the surface from the viewer. Several visual observers working with filters have reported the impression that the structure of the dense atmospheric shell is better recognized. Blue filters apparently offer more detail than yellow and red glasses (see page 430, and Chap. 8).

The visual observer is left with the following tasks:

1. To follow the change of phase and to determine dichotomy
2. To make observations of the shape of the terminator
3. To make observations of the extension of cusps of the narrow crescent and of the "ashen light"

Monitoring the phases and the dichotomy shows the difference between computed and observed times owing to the action of the atmosphere. Many observers have studied the dichotomy and its deviation from the geometric prediction.

The phase angle represents the unilluminated arc of the planetary hemisphere toward Earth. It is by no means a simple task to estimate the phase correctly; the image of Venus

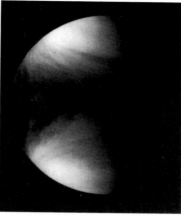

Fig. 108. UV photograph of the planet Venus taken by the Pioneer Venus spacecraft. (*Left*) January 14, 1979 at a distance of 65,000 km. (*Right*) December 30, 1978 at a distance of 43,000 km. Photographs courtesy of NASA

is generally too pale in the daytime, but too bright even during twilight. Moreover, the presence of Venus's atmosphere makes the terminator noticeably less sharp than the limb.

Observations on the shape of the terminator concern its possible deformations. Numerous observers have reported protrusions or indentations at the terminator during the phase cycle. It may here be helpful to use an eyepiece with a cross-wire. This task can also be facilitated by the application of various color filters. It may be assumed that most of the reported anomalies in the terminator are not optical illusions but are indeed real effects in the Venusian atmosphere. This becomes particularly clear in the case of the extension of the cusps of the narrow crescent beyond the poles of illumination. Such observations fall around inferior conjunction. The observer may simply estimate the extension using a cross-wire eyepiece and recording the data using stencils with position marks.

Micrometer measurements will give more precise results. Near-inferior conjunction, the angular cusp extension and its change with the approach of the planet to the Sun are measured according to the scheme of N. Richter: for small angles φ according to Fig. 109a, or for large angles according to Fig. 109b, the lengths $2a$ and b, or $2a$ and $2d$

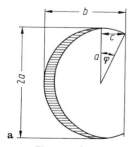
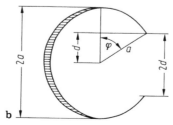

Fig. 109a,b Measurement of the extension of the cusps of Venus

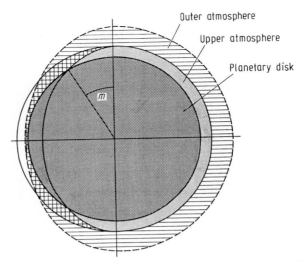

Fig. 110. Refraction by the low and high atmospheric layers of Venus at or near-inferior conjunction

are measured, respectively. Since $c = b - a$, the angular extension is found from

$$\sin \varphi = \frac{b - a}{a} \quad \text{and} \quad \cos \varphi = \frac{d}{a} \, .$$

Once φ is known, the observer can calculate the height of the atmosphere of Venus. E. Schoenberg provided the following considerations and formula.

If the planet Venus is imagined projected onto the center of the solar disk, the dark planetary disk would be surrounded by the bright ring of its atmosphere (although this cannot be seen because of the glare of the Sun). The dark disk appears somewhat reduced in size because solar rays grazing the planet are refracted inward (Fig. 110, right side). The outer ring represents the light-permeated outermost atmospheric layer, the inner ring the still-illuminated part which generates the refraction. The left side of Fig. 110 then represents the situation before inferior conjunction and the phenomenon of extended cusps due to a high atmospheric layer. In the relation

$$\frac{h}{R} = \frac{1}{2} \tan^2 V \sin^2 m$$

where h is the height of the atmosphere, R the radius of the planet, V the angle of elongation, and m the angle of the extension of the cusps.

One noteworthy phenomenon which is seen around inferior conjunction when Venus displays a very narrow crescent is the secondary or ashen light: the unilluminated part of the disk appears in a pale glow resembling the earthshine on the Moon. This phenomenon was first observed by Kircher in 1721 and has since been confirmed by many observers. Unless it is an optical illusion, its cause must again be attributed to the atmosphere of Venus. Observers report different colorings ranging from brownish to gray and violet.

As a summary of visual and filter-visual observations, the persevering observer may learn something of the effects of the dense atmospheric shell which surrounds Venus. As is well-known, and continuously published, recent research results come from the flybys, flights into the atmosphere, and soft landings of numerous space probes.

Photography Apart from radar scans of the surface, the Earth-based exploration of Venus is restricted to atmospheric and cloud observations in the ultraviolet. The overwhelming brightness of the planet makes it suitable for photographic observations by amateurs. Many observers have obtained interesting "constellation photographs" showing Venus near the Moon, other planets, or bright stars. For most photographic work, a single-lens reflex (SLR) camera with exchangeable lenses will suffice. Photographs of, for example, Venus and of the lunar crescent require only very short exposure times (between 1/50 and 1/100 s).

Another task is the photographic recording of the phase change (Fig. 111) and the determination of the dichotomy.

Webcams and other digital cameras capable of recording movies have surpassed in the last years CCD cameras because of low cost and high speed. Up to 60 frames per second with webcam equipment increases the chance to have moments of excellent atmospheric conditions, compared to that with a single-frame camera.

The French amateur Christophe Pellier made photos of Venus with an 8-in. Dall–Kirkham–Cassegrain reflector using a webcam and a near-ultraviolet filter. Elusive details in the cloudtops of the planet were the result. A similar result was recorded by the Italian observer Paolo R. Lazzarotti on September 20, 2004 with a 10-in. Newtonian, a Lumenera LUO75 camera, and a Wratten 47 filter (deep blue).

Cloud motions in the upper layers of Venus's atmosphere were noticed on photographs taken with an 8-in. scope in ultraviolet light. Numerous photographic series secured using large-aperture instruments, primarily at the Pic du Midi Observatory, have subsequently confirmed the clouds and their motions. Attempts with large (10-in. and above aperture) amateur telescopes should prove rewarding. Along with the ultraviolet filter UG 5, blue filters are also suitable for such photographs. If it is desired to extend the focal length of the system with the aid of a Barlow lens, the transmissivity of that lens

Filters, which are optimized for imaging applications, transmitting light mostly outside the visible spectrum, are of special interest for the visual and photographic observations of Venus. Details in the cloudtops of the planet are visible in the near-ultraviolet. A photographical Venus Filter is for example the Baader Planetarium U Filter (Alpine Astronomical, http://www.alpineastro.com).

Photos by Pellier and Lazzarotti and tips by Sean Walker can be found in the *Sky and Telescope* October 2005 issue, p. 115 ("Planetary Masters").

Fig. 111. Photographic example of a sequence of Venusian phases in 2004 by S. Kowollik. Courtesy of *Sterne und Weltraum*, September 2007, pp. 72–75

Fig. 112. The 8-in. Newtonian of S. Kowollik with adapter for projection photography and video module SK 1004-X (Lechner)

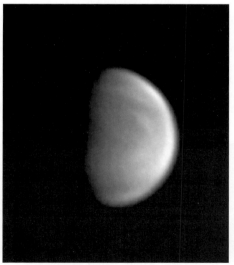

Fig. 113. Venus photo by S. Kowollik of April 5, 2007

in the ultraviolet should be checked. Photographic tests may also be performed using yellow filters, with which clouds in the high atmosphere have been recorded.

A large telescope is needed to secure detailed photographs from which the extension of the cusps near-inferior conjunction can be determined. It may also be worthwhile to use filters such as RG 5 and OG 2. Simultaneous exposures in two colors (e.g., red and

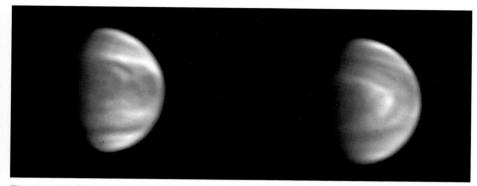

Fig. 114. UV photographs of the planet Venus by B. Gährken. Instruments used were 80-cm reflector Volkssternwarte München, Schüler-UV filter, and webcam

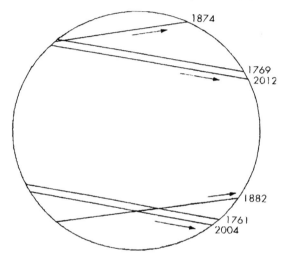

Fig. 115. Transits of Venus 1761–2012

Results of the latest research see Tilmann Althaus "Venus – Die Eigenwillige Schwester der Erde" in *Sterne und Weltraum*, July 2006, p. 32 (http://www.suw-online.de)

orange) of Venus when near-inferior conjunction can be used to determine its angular diameter. G.P. Kuiper suspected that the apparent angular diameter depends on the wavelength of light used in photographing.

Transits of Venus The next transit of Venus will occur on June 6, 2012 (duration ca. 6.5 h, see Fig. 115), followed by transits in 2117 and 2125!

13.6.5.3 Mars

Visibility Mars is the nearest of the "superior" planets which travel around the Sun outside the Earth's orbit. During opposition a superior planet can be observed throughout the night. The visibility after a conjunction with the Sun runs on the average as follows:

Day 0:   Conjunction with the Sun
Day 54:   Mars appears in the morning sky
Day 353: Mars begins retrogression

Day 390: Mars in opposition with the Sun
Day 427: Retrogression ends
Day 726: Mars disappears from evening sky
Day 780: Next conjunction

An almanac will supply the observer with the following data:

1. Right ascension and declination
2. Apparent diameter (varying between 4″ and 25″)
3. Apparent magnitude (between +1.50 and –2.52)
4. Phase angle (also phase = illuminated fraction of disk) and the defect of illumination
5. Central meridian at 0^h UT
6. Position angle of axis of rotation
7. Latitudes of Earth and of the Sun with respect to the Martian equatorial plane
8. Distance from Earth in AU
9. Light-travel time from Mars to Earth
10. Time of transit

Some calendars list the heliocentric, in addition to the geocentric, coordinates of the planet.

The conditions of observation strongly depend on the position of Mars relative to the Earth. Owing to the eccentricities of both orbits, the distance between Earth and Mars at nearest approach varies from under 56 million kilometers, when Mars is near perihelion at opposition, and 101 million kilometers in an opposition near aphelion. Accordingly, the apparent diameter in a perihelion opposition is over 25″, but only 13″.8 in an aphelion opposition (Fig. 116).

Perihelion opposition occur in late summer, when the planet is situated in the southern part of the zodiac and therefore at an unfavorable elevation for northern-hemisphere observers. Aphelion oppositions which take place after the beginning of a year will find the planet in the northern part of the zodiac. The observer should be aware that an apparent diameter of at least 10″ is required to see substantial detail.

Since the atmospheric conditions, which improve with altitude, play a major role in the observation, northern-hemisphere observers will find the most favorable oppositions to be those which occur when Mars lies between its aphelion and perihelion positions.

The equatorial plane of Mars is inclined with respect to its orbital plane by 25°12′. This means that seasons occur on Mars as they do on Earth, and also that the observer sees more of the northern Martian hemisphere at oppositions occurring in springtime, whereas the autumnal oppositions show more of the southern hemisphere (Table 15).

For special information regarding oppositions of Mars see the work by William Sheehan, *The Planet Mars, A History of Observation and Discovery* (http://www.uapress.arizona.edu/ onlinebks/mars/appends.htm).

Table 15. The seasons of Mars

Heliocentric longitude	Northern hemisphere	Southern hemisphere
88°	Spring begins	Autumn begins
178°	Summer begins	Winter begins
268°	Autumn begins	Spring begins
358°	Winter begins	Summer begins

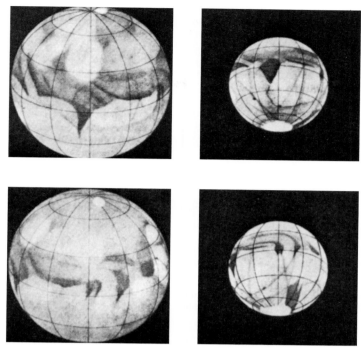

Fig. 116. Comparison of apparent sizes and of axis orientations of Mars at oppositions occurring at perihelion (*left*) and aphelion (*right*)

When the rotation axis of Mars is nearly perpendicular to the line-of-sight (this happens at oppositions between aphelion and perihelion), the rotation of the planet carries surface details across the disk in straight lines. The axial tilt substantially influences the appearance of most details visible on the planetary surface at oppositions near aphelion and perihelion (see Fig. 116).

In addition to the apparent diameter and the tilt of the axis, the observer should also keep in mind the influence of the phase angle. It reaches a maximum of 46° and must be allowed for. Observations made at most one month before and after opposition may neglect the phase angle, and circular stencils may be used for sketches.

The rotation period of Mars is almost the same as that of Earth: one Martian day equals 1.025 Earth days. An observer viewing the planet on consecutive nights at the same time will find very little displacement of surface detail. Only after 38.5 days will the observer see the same central meridian in the eyepiece when observing at the same time. The annual rate of change of the central meridian on Mars is 14°.6. The central meridian must be determined for every observation (on a graph), since the identification of surface detail depends on it.

The rotation of Mars can be measured even with a 2-in. telescope, by way of transit observations of prominent dark markings (e.g., Syrtis Major, Solis Lacus, Titanum Sinus).

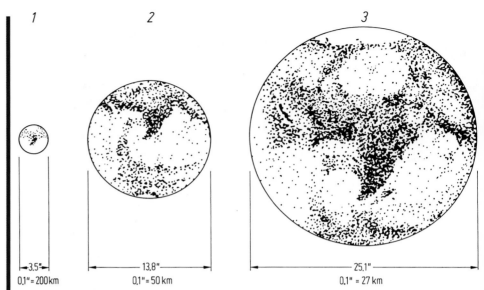

Fig. 117. The apparent diameter of Mars as seen during a conjunction at aphelion (*1*), and at opposition at aphelion (*2*) and perihelion (*3*). From E. Freydank and H. Freydank

Daytime observations of Mars are of little significance although they have occasionally been made and surface detail seen.

Visual Observation Images sent back to Earth from the Mariner 4 spacecraft during its Mars flyby in 1965 gave the surprising result that, instead of "canals" (Fig. 118) the Martian surface is covered with craters and in many ways resembles that of Earth's moon. This similarity was confirmed by all subsequent space missions to Mars, including the two Vikings in 1976. In addition to craters, the spacecraft pictures showed grabens, meandering valleys, and volcanos. The atmosphere was found to be thinner than expected. The polar caps are composed of both water ice and dry ice.

The cameras aboard the US Mars Global Surveyor and the European Mars Express have given scientists and the public an entirely new look at the Red Planet. We have new insights into Mars's recent past: images of the summit calderas, giant shield volcanoes. The High Resolution Stereo Camera (HRSC) of Mars Express also discovered glacial deposits in equatorial and mid-latitudes. This confirms a long-standing suspicion that Mars has been glacially active.

"Mars is not Earth-like. It's difficult to understand the planet because we doesn't know much about its interior. Most of this glacial activity happened over the last few tens of millions of years, which is recent history on Mars" (Gerhard Neukum, team leader and camera inventor of HRSC, Free University, Berlin, *Sky and Telescope* December 2005, p. 30).

Despite the astronautical lead in the exploration of Mars, the amateur observer should continue to observe Mars telescopically under a variety of conditions and acquire experience. Some feasible observing tasks include the following:

1. Finding the period of rotations (see above)
2. Monitoring seasonal changes in the polar caps and the atmospheric cover
3. Observation of bright and dark contours on the surface, and also changes in albedo patterns
4. Observation of atmospheric phenomena; white clouds, ice haze, yellow dust clouds, blue or violet clearing (see below)

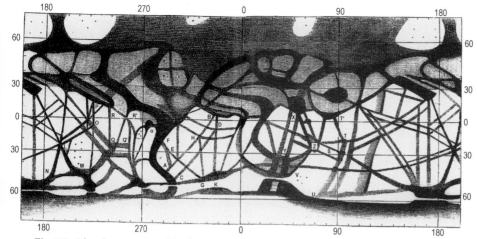

Fig. 118. The phenomenon of the " Martian canals." Observations by G. Schiaparelli in 1886

The appearance and extent of the polar cap follows the seasonal changes on Mars: the cap develops during autumn as layers of ground fog are allowed to form in its surrounding bright zones. During the Martian winter, the cap grows; the observer may also see cloud-like veils in the vicinity of the polar cap. The cloud layer is often especially prominent in spring, when it has the largest extension. With the approach of summer, the cloud veils disappear, and the appearance and increasing prominence of a dark fringe indicates the shrinking of the polar cap (Fig. 120). Owing to the longer winter in the southern hemisphere, more CO_2 becomes frozen near the south pole, and thus a residual southern polar cap remains over the summer.

Observers have reported seeing both caps simultaneously. However, the "cap" of the winter hemisphere actually is an atmospheric cloud layer consisting of CO_2 and H_2O ice crystals and covering the real polar cap. While the caps change in size gradually, the polar cloud veil can change its appearance within a very short time. When observing the polar caps, it is advisable to use a blue filter.

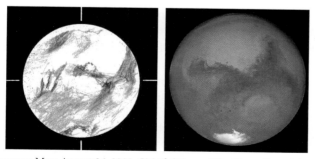

Fig. 119. Dark areas on Mars August 26, 2003, CM 0° (Margaritifer Sinus, Sinus Sabaeus, Hellas). *Left*: Observation with a 6-inch refractor, magnification 274× by G.D. Roth. *Right*: A photo taken with Hubble Space Telescope at the same time

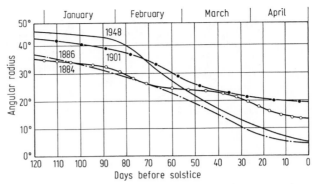

Fig. 120. Sublimation of the northern polar cap of Mars in various oppositions. *Abscissa* represents days before solstice, and the *ordinate* represents angular radius. After G. de Mottoni

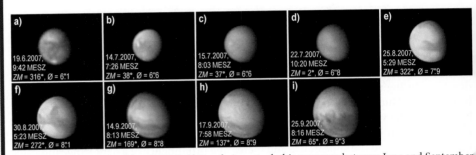

Fig. 121. Dust storm on Mars 2007. Ralf Gersheimer took this sequence between June and September with a 15,5″ Newton 1:36 and a videocamera DMK 21AF04 of The Imaging Source. The peak of the storm see Fig. 121b–d. Further information visit R. Gersheimer http://www.astromanie.de

At the end of the observing season, the observer may wish to measure all of the drawings so that the extent of the polar cap over time can be graphically represented. Observations of the appearance and the evaporation rate of the caps spanning many years may provide useful information regarding possible correlations with solar activity.

The dark areas on Mars are labeled with such terms as sea (mare), bay (sinus), or lake (lacus). But the terms insula (island), fonts (spring), pallus (marsh), ponts (bridge), and regio (land) also occur in the nomenclature, and often relate to legendary (hellenisitc) or biblical names. Noachis, for instance, refers to Noah; Syrtis Major was named after the Syrtis of Libya by the astronomer Schiaparelli, who initiated a major part of Martian nomenclature, including the "canali."

The dark features sometimes show greenish or bluish tints, while the extended bright regions (the "continents") have the typical rust-red or salmon-red coloring to which Mars owes its alternate name, "the Red Planet." The telescope reveals N-shaped connections between the dark areas; it was the often straight-line appearance of thesefeatures that triggered the furor over "Martian canals" (see Fig. 118). However, observations made decades ago using large telescopes resolved these bands into individual spots and pieces. Whether or not Earth-based observers have seen craters on Mars through the telescope is a question of interpretation of earlier graphs and records.

Drawings of the development of the polar caps are published on p. 196 of *Planeten beobachten* [78].

About the great Martian dust storm of 2001 see article by D.C. Parker in *Sky and Telescope*, vol. 102 (2001).

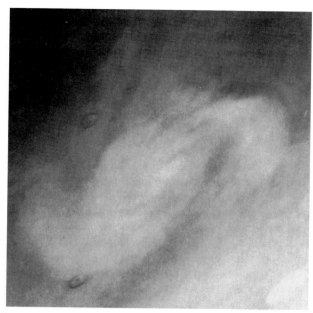

Fig. 122. Cloud formations on Mars in a cyclonic storm. This indicates a mixture of air masses of different temperatures. The cyclone was captured by the Viking Orbiter 1 on August 9, 1978 at latitude 65°N. At the lower right in the picture is the crater Korolev (196° W, 73°N). Photograph courtesy of NASA

A comparison of various maps of Mars shows that there exist several dark regions whose shape changes either very little or not at all (e.g., Syrtis Major, Mare Sirenum, Sinus Sabaeus, Mare Tyrrhenum, Mare Acidalium). Where changes have been repeatedly observed, the cause may be temporary variations due to changes in albedo (resulting from, for example, changes in moisture) or to atmospheric superposition of clouds and veils. This overlying matter can also be dust which has been lifted into the upper atmosphere but which sinks back to the surface after some time. There are always surprises which await the observer from one opposition to the next. In particular, the larger clouds can be seen with only a 4- to 6-in. telescope. There are three basic types of clouds:

1. White clouds, which are commonly visible over the wintery polar regions, but can also be found at mid-latitudes and near the equator. They are cyclonic currents, suggesting the mixing of air masses with different temperatures (see Fig. 122).
2. Blue clouds, which strongly reflect blue and ultraviolet light. The Martian atmosphere contains a "violet layer" which significantly limits the visibility of the surface through blue filters. This layer can split open, for instance, in the vicinity of large dust storms, and the observer can seize the opportunity to make visual trails with filters. The Mariner and Viking missions to Mars have not positively confirmed the existence of the violet layer, and yet the phenomenon impedes Earth-based observations. Temporary clearings have been observed, that is, visibility of dark surface areas, through a blue filter, and this is termed a "blue" or "violet clearing."
3. Yellow clouds, generally caused by dust storms and which obscure major portions of the atmosphere, as, for instance, occurred in 2001.

The International Astronomical Union (IAU) has named an impact crater on Mars after Japanese amateur observer Tsuneo Saheki (1916–1996). Crater Saheki is located at latitude −21.8° and longitude 287° west (north of Hellas) and measures 85 km across.

Any visual observations can be enhanced by the use of color filters. The "classical" Mars filter is the orange filter (e.g., OG 550, 2 mm thick). Red filters (e.g., RG 1 and 2) also may enhance contrasts and are particularly good at showing clouds. Although the transparency in blue light is low, monitoring with a blue filter (e.g., BG 12 and 23) is worthwhile owing to the violet disturbances mentioned. Observers in the US often use the Kodak Wratten filters: W 38 and W 80A are blue filters, W 58 green, W 21, W 23A, and W 25 orange and red.

A micrometer is also helpful for observing Mars (see Chap. 4; Chap. 13.6, Figs. 119 and 123) for the following tasks:

1. Measuring the positions of dark features. Such measured points on the surface can serve as a basis for preparing a map. Changes in position or of size or extension of an object during an observing season may thus also be obtained.
2. Position measurements of cloud-like phenomena and the determination of their sizes. This assumes that the cloud to be measured shows reasonably well-defined edges, which as a rule occurs only in small, compact cloud formations.
3. Measuring the polar caps. The most important point is the position of the center of the cap, which does not always coincide with the rotational pole of the planet. Another task is to measure the diameter of the cap and follow its increase or decrease

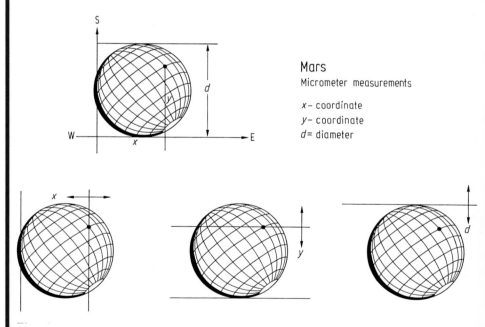

Fig. 123. Illustration of the use of a micrometer to determine the aerographic longitude and latitude of a feature. With the telescope drive switched off, a micrometer wire is oriented by the transit of a star. In the telescope (which is inverting but not mirror-imaging), the x-axis points to the right (east in the sky) and the y-axis upward (south). The axes are tangent to the disk and the coordinate origin lies outside the disk. It is unimportant whether or not Mars has a phase at one of the limbs. The phase, position angle, and inclination of the rotation axis must all be considered when computing the coordinates. The apparent diameter of the disk is measured in the y-direction, again disregarding the phase

in size. Finally, the observer may measure the width of the dark fringe, if present, and its variations.

To determine the position of an object, the observer measures its x and y-coordinates. At least five settings are desired in order to form an average. For each setting, the time is recorded to about ±15 s.

Preparing a Map of Mars The opportunity to observe Mars daily over at least one month supplies the basis for constructing a surface map of the planet. All observations are recorded in drawings. Owing to the changing orientation of the Martian axis, the stencil must be reconstructed as needed. Only detail near the central meridian (to ±10°) is transferred to the map. The old maps by Schiaparelli, Lowell, and Antoniadi have frequently been published. An entirely new generation of maps and globes has originated from the flyby missions of Mariner 6 and 7.

Photography Apart from the wide-field "constellation photographs," which show a planet as merely a bright point of light against the background stars, Mars, when it is close enough to Earth to display a considerable disk, can also be photographed through a small telescope. When the apparent diameter is 10″ or over, a 6- to 8-in. telescope will readily secure useful photographs of the surface and of prominent features such as polar caps, dark areas, and clouds (see Fig. 103, page 437).

To be sure, planetary photographs cannot take the place of direct visual observations, but they can be used to supplement them. In order to make comparisons between the photograph and a map, the central meridian at the time of exposure must be determined.

An advisable task for the photographic observer is to systematically monitor the planet during the time interval when its apparent diameter exceeds 10″. The purpose of the monitoring is primarily to secure a record of unusual events in the atmosphere, such as dust storms and prominent bright clouds. Photographs taken in red and blue light are highly desirable, as are ones in ultraviolet light.

Photometry Careful observations may also detect atmospheric events and cloud formation. This global photometry proceeds according to step estimates (see Chap. 8) and can be made with binoculars. For telescopic observations, the area photometer after Gramatzki may be used, as it permits more accurate brightness measurements. High precision is demanded for photometry of selected bright and dark areas on the surface. It is simplest to make visual estimates by memory scale, as is used for other planets. For global and areal photometries, the photoelectric photometer (see Chap. 8) can be used, and it is also advisable to employ filters. Changes in apparent brightness also depend on the blue and violet transparency of the Martian atmosphere. When no dark surface areas are visible in the blue filter, the global apparent brightness is measured to be 0.2 to 0.3 mag higher.

The Moons of Mars The two satellites Phobos and Deimos were discovered in 1877 with a 65-cm refractor. Spacecraft have now supplied excellent close-up photographs, especially of the satellite Phobos, revealing details of 10 m and smaller.

The apparent magnitude of both moons is around +12. Being so close to the planet, they are very difficult objects to locate. The amateur can at best attempt to find Deimos

For actual maps of Mars visit the Association of Lunar and Planetary Observer (ALPO) website http://www.lpl.arizona.edu/alpo, also the homepage of *Sterne und Weltraum* http://www.suw-online.de

See the historical collection of maps published by G. de Mottoni y Palacios, *The Appearance of Mars from 1907 to 1971, Graphic Synthesis of Photographs from IAU Center at Meudon, Icarus* vol. 25 (296/1975).

Mars photographers have good results with specialized filters improving contrast in the near-infrared, where CCD cameras and webcams are extremely sensitive (e.g., Baader Planetarium IR Pass).

For a detailed report about photographical results with video camera and webcam of the opposition 2005 by an amateur, visit the homepage of the German observer Ralf Gerstheimer: http://www.astromanie.de.

as it is the moon more distant from Mars. In any event, the observer needs a telescope with an aperture of at least 10 to 12 in. It may help to mask out the planet by means of a disk or conic diaphragm placed in the focus of the telescope, a setup similar to that of a prominence telescope.

13.6.5.4 Jupiter

Visibility By contrast with Mars, which becomes a rewarding object to view for only a few brief weeks every 25 to 26 months, the planet Jupiter is, after the Earth's moon, the easiest target for an amateur. Its apparent diameter exceeds 40″ for several months each year. Jupiter, like the other superior planets, experiences opposition at which time it can be observed all night. The pattern of visibility for Jupiter, beginning with the conjunction with the Sun, is as follows:

Day 0:   Conjunction with the Sun
Day 13:   Jupiter appears in the morning sky
Day 140:   Jupiter begins retrogression
Day 200:   Jupiter in opposition with the Sun
Day 260:   Retrogression ends
Day 386:   Jupiter disappears from evening sky
Day 399:   Next conjunction

A good almanac will provide the following data:

1. Right ascension and declination
2. Apparent equatorial diameter (between 31″ and 48″)
3. Apparent polar diameter
4. Apparent magnitude (ranging between −1.3 and −2.7)
5. Position angle of the axis of rotation
6. Jovigraphic latitudes and longitudes of Earth and Sun
7. Distance form Earth in AU
8. Phase angle and the defect of illumination
9. Light-travel time from Jupiter to Earth
10. Jovigraphic longitude of central meridian in rotation systems I, II, and III
11. Times of transit
12. Satellite phenomena

In addition to the geocentric coordinates, some calendars also list the heliocentric coordinates of the planets.

Jupiter's impressive size makes it a favorite object for viewing in even a small telescope. Moreover, the observer always sees essentially the same projection of coordinates since Jupiter's rotation axis is only slightly tilted ($2°.1$) against the ecliptic axis. The maximum of tilt of the planet is $3°.5$. Only then and with high power will any curvature in the parallels be noted. When observing with smaller instruments, the bright zones and dark bands parallel to the latitude circles appear as straight lines.

The oblateness of Jupiter, however, is a respectable 1/16, which can by no means be neglected when making drawings.

The phase angle is at most $12°$, which means that the illumination defect reaches at most 1% of the diameter. The phase angle may thus be neglected. The dark segment will

Fig. 124. Photograph of Jupiter taken by the Voyager 1 spacecraft on February 1, 1979 at a distance of 32.7 million km. Photograph courtesy of NASA

be noted when viewing occultations of the moons of Jupiter, and especially if a large telescope is used.

Atmospheric Phenomena and Events All of the features which the observer sees or photographs on Jupiter (see Fig. 124) are clouds! Measurements of Jupiter's size and density reveal that it consists predominantly of light elements (hydrogen, helium), which are in the gaseous state in the upper atmosphere, but liquid in the deeper-lying layers. Essential information on phenomena and events in the atmosphere were provided by the US space mission Pioneer 10 (flyby date 1973 December 04), Pioneer 11 (1974 December 03), Voyager 1 (1979 March 05), and Voyager 2 (1979 July 09). Further detailed information was provided by the US mission Galileo. The satellite orbited Jupiter from December 1995 to September 2002. Flyby of Saturn mission Cassini in December 2000 resulted in the best photographs of Jupiter. The features created by atmospheric currents were found to be quite diverse: waves, eddies, cyclones, lightning storms, short-lived (a few days) and long-lived cloud formations. The meteorological events on Jupiter are triggered by its rapid rotation and internal heat sources; solar radiation does not play the same role as on Earth. The space missions confirmed a strong magnetic field which had been previously discovered in 1955 by radio astronomical observations. These missions also detected a system of Jovian rings, which can be observed from the Earth only in the infrared and with large telescopes.

The visible details are characterized by bright and dark stripes called zones and bands, respectively; the nomenclature is shown in Fig. 125.

Attention is to be paid to the rapid rotation of Jupiter, which takes place in two visible systems, I and II. The EZ and all of the bright and dark clouds which occur within it rotate as System I, while the NEB and SEB, usually the most prominent dark bands on both sides of the EZ, represent System II. The shortest (fastest) rotational period of $9^h\,50^m\,30^s$ is observed at the equator (I), the slowest in jovigraphic latitudes between $0°$ and $\pm20°$ with a period of $9^h\,55^m\,40^s$ (II). The angular speed increases again toward the poles. The diminution of the rotational speed between latitudes $0°$ and $\pm20°$ is very conspicuous,

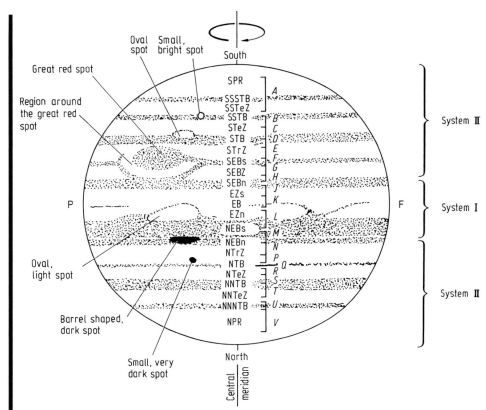

To avoid an ambiguity the word preceding (P) refers to the direction of an object drifting across the planetary disk. Following (F) means the contrary side. These points coincide with east and west on the planet Jupiter, opposite celestial east and west.

Fig. 125. Names of bands and zones on Jupiter. The abbreviations have the following meanings: preceding (P), following side (F) of an object in the Jovian atmosphere, for instance the Great Red Spot (GRS); other designations include north (N), south (S), equatorial (E), tropical (Tr), temperate (T or Te), arctic (A), polar (P), band (B), zone (Z), region (R), BAR = barrel-shaped region, dark cloud, OVAL = plume, bright spot, SPOT = small, bright, round marking

and must be allowed for when observing and charting other details such as bright and dark clouds.

Radio astronomical observations have defined a third rotation system (III) with a rotational period $9^h\ 55^m\ 30^s$, which some researchers expect to be that of the solid planetary body, in contrast to Systems I and II which are linked solely with the atmosphere. Astronomical almanacs publish the central meridians referring to the two systems. They can be interpolated with the angular rotations:

System I   Rotation 877°.9 in 24^h or 36°.6 per hour
System II   Rotation 870°.3 in 24^h or 36°.3 per hour

Visual Observations Visual observations of the planet Jupiter are still superior to those made via photography, particularly concerning fine detail.

For the variety of observed atmospheric features, there is not only the previously mentioned, internationally unified terminology on bands and zones (Fig. 125), but also a computer-adapted code of observable object types to be described as follows (see also Fig. 126).

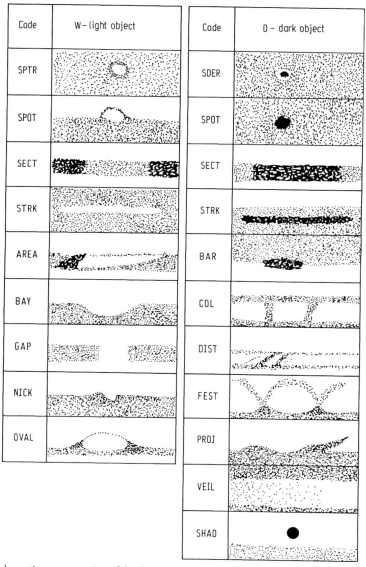

Code	W—light object	Code	D – dark object
SPTR		SDER	
SPOT		SPOT	
SECT		SECT	
STRK		STRK	
AREA		BAR	
BAY		COL	
GAP		DIST	
NICK		FEST	
OVAL		PROJ	
		VEIL	
		SHAD	

Fig. 126. Schematic representation of the designated codes for observed features in the atmospheres of Jupiter and Saturn

The abbreviated name of a feature consists of six or seven letters or numbers:

1. The first letter divides the feature into two categories: "W" for bright or shining features and "D" for dark features.
2. The second letter refers to the center "C," the preceding limb in the sense of rotation, or eastern limb "P," the following or western limb "F," or, for micrometer measurements, the northern "N" or southern "S" limb.
3. The third digit gives a rough designation regarding the visibility of the object: from "1" (prominent) to "3" (difficult to see or very small).
4. The following three or four letters characterize the type of object. Their short code and description is (after Haug and Kowalec [82]):

   (a) Bright (W) objects

Code:	Description
SPOT:	A small, bright round spot.
SPTR:	A small, shiny spot surrounded by a dark ring.
OVAL:	An oval feature of medium or large size, fairly bright, well-bounded, often found in the equatorial zone.
BAY:	A large indentation, usually half oval at the edge of a dark band.
NICK:	A small, semicircular notch at the edge of a band, often somewhat brighter than the adjacent zone.
GAP:	A rather wide, weakened, or missing part in a band.
STRK:	A bright, very oblong spot; when appearing in a dark band, it may look like part of a tear.
AREA:	An extended, bright, and irregularly bounded region.
SECT:	A particularly bright section of a band or a zone.

   (b) Dark (D) Objects

Code:	Designation
SDER:	A small, very dark spot surrounded by a bright ring.
SPOT:	Any single spot not oblong.
SECT:	A noticeably darker section of a band or zone.
BAR:	Dark oblong or barrel-shaped spot.
PROJ:	Something akin to a prominence at the edge of a band, which may also be darker than the main part of the band. Shapes vary, the dents may be rounded or peaked.
VEIL:	An extended, homogeneously smooth, dark region, sometimes occurring in the zones or in polar regions.
FEST:	A dark "fiber" or garland crossing a zone. One end of it or both may originate from a dark condensation in the band.
COL:	A column-shaped dark area in a zone, either perpendicular to it or somewhat tilted. Such columns are occasionally seen in STrZ or SEBZ.
DIST:	Disturbance is a dark, extended area, more or less well-defined containing often very fine detail in irregular distribution and unusual shapes.
STRK:	A very oblong, dark, stripe-shaped object.
SHAD:	The shadow of a moon.

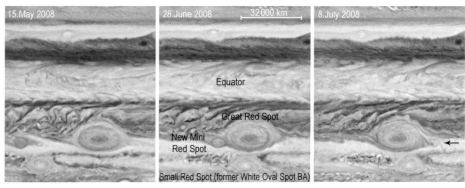

Fig. 127. Jupiter's three Red Spots. Besides the Great Red Spot a second one reddish object was discovered during 2006 by an amateur observer, assigned to the White Oval Spot BA. Called now the Small Red Spot. During May 2008 observers discovered a third reddish spot. This object, also of the "family" of White Oval Spots, was captured weeks later by the Great Red Spot. Hubble Space Telescope documented the event. NASA Photo/Sterne und Weltraum

(c) On Jupiter there are a few other long-lived or frequently appearing features with special designations:

WOS-FA, -BC, -DE: White oval spots FA, BC, and DE.
GRS: Great Red Spot (GRS).
RSH: Red Spot Hollow.
STRD: Disturbance in the STrZ.
SEBD: Disturbance in the SEB.

Unclarified is the cause of Jupiter's Red Spots. One hypothesis is that they dredge up chemicals from the deep changing in colour when exposed to sunlight.

For example, during a transit through the central meridian, the preceding (P) end of a prominent (1), dark (D), stripe (STRK) is measured. The code to be chosen according to the preceding list is: DP1 STRK.

There exist certain decidedly long-lived features on Jupiter, including the Great Red Spot, the Gray Veil, and also, according to recent observations, some white oval spots, as observed, for instance, in the STB. These form the category "WOS." In general, a series of observed features appears to be more stable than one would expect for purely atmospheric phenomena.

The first observation of the GRS dates from 1664 and is ascribed to the English physicist R. Hooke. The central column of the GRS extends 8 km above the cloud level and receives its energy from some long-lived internal source. The age of the GRS is estimated to be over 10^5 years. It is not fixed with respect to the surface (Figs. 128 and 129), but moves westward in the atmosphere at a rate of $0°.5$ per day. The reddish color of the GRS sometimes changes dramatically.

Another feature of the GRS is that its visibility alters: from time to time it becomes particularly obvious, as in the years 1880, 1927, 1936, 1960–1961; at other times even a practiced observer has to search for it, as in 1882, 1920, 1956–1957. The observer will find it worthwhile to investigate the following:

"On February 24, 2006, I made the biggest discovery of my life! I discovered that the white spot Oval BA of Jupiter has turned red!" The amateur Cristopher Go of Cebu, Philippines attracted notice to the morphing of the white oval known as BA into a miniature version of the Great Red Spot (GRS). See the webpages http://jupiter.cstoneind.com/, and http://www.christone.net/astro.

1. The speed of rotation of the Jovian atmosphere
2. Motions and changes in the GRS, the Gray Veil, and the WOS
3. Motions and changes in latitude and width of the bands and zones
4. Motions, occultations, and transits of the bright moons and their shadows

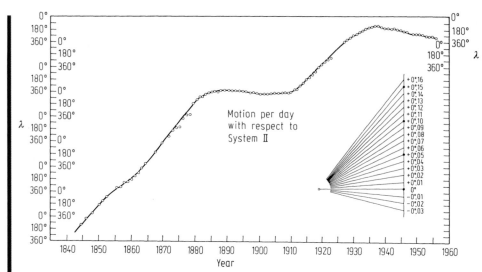

Fig. 128. Motion of the GRS on Jupiter during the period 1840–1950. After Löbering

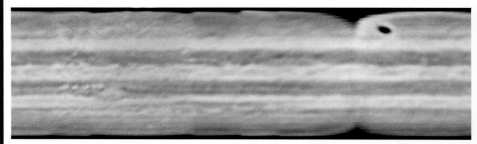

Fig. 129. Jupiter in opposition on April 3, 2005. Photographed during April 3 and 4, with the 10″ refractor of the Volkssternwarte München. Photos combined to a cylindric projection of Jupiter's atmosphere (see Fig. 105, p. 438). On the *left*, the Great Red Spot (GRS), and the shadow of Galilean satellite Io at the *top right*. From *Sterne und Weltraum*, July 2005, p. 62. Photograph by B. Gährken

Within the frame of visual observations, there exist the following possibilities for tackling the above tasks:

1. To represent the state of the visible surface by individual graphs (black-and-white or color) or overall maps, perhaps using color filters.
2. Observation of central meridian transits of bright and dark spots and other well-bounded objects. This may be done by visual estimates or, for especially well-defined features, by micrometer measurements.
3. Intensity estimates of bright and dark objects, in conjunction with color estimates using filters.
4. Determination of the width of bands and zones by estimation or by micrometer.

Features of low contrast become more apparent when observed through color filters. Use a light green filter (e.g., Wratten 58) or a Wratten 80A filter to boost the reddish brown markings. Yellow and orange glasses (e.g., Wratten 12 and 15) supports the visibility of subtle details (e.g., the bland polar regions). There are some specialized planetary filters on the market designed to boost the contrast of subtle markings beyond what is possible with standard filters (e.g., Baader Planetarium contrast Booster or the Variable Filter System by Sirius Optics).

A description of atmospheric features on Jupiter and Saturn and of their time variations necessarily includes accurate positions, since only by so doing can the motions and interactions of such features be identified. Haug and Kowalec [82] mention the various possibilities:

Positions may be derived from sketches, provided the observer was careful in graphing and timing. Stencils of Jupiter and Saturn are readily measured with commercially available overlays which contain coordinate lines at 10° intervals. The resulting latitudes and longitude differences against the central meridian are usually uncertain by at least ±5°.

More precise, and well-suited to studying motions of atmospheric features, is the timing of transits of an object (or a distinct point thereon) through the central meridian. This requires merely a good watch, checked against a time signal. For each transit, it is advisable to measure three readings (1/4 min each):

t1: Transit possibly occurred
t2: Transit probably occurred
t3: Transit certainly passed

The adopted average t_m is computed using $t_m = (t_1 + 2t_2 + t_3)/4$, giving double weight to the presumably closest estimate t_2. A cross-wire eyepiece oriented so that one wire marks the central meridian will be of advantage. A trained observer can obtain a substantial number of positions during an observing season of Jupiter or Saturn, and with a precision (±1° or 2°) adequate for determining motions. Extended objects such as the Jovian GRS can be measured during one night at several points: the preceding edge (P), the center (C), and the following edge (F) (Fig. 125).

Estimating transits over the central meridian is for most planetary observers the only way to achieve reasonably reliable longitude data. Such time estimates can be inserted in between regular observations (graphing) and thus permit one to test the graph for correctness in longitude.

The precision in longitude, and even more so in latitude, is improved by using a filar micrometer (with at least one fixed wire and one parallel, movable wire).

Measuring atmospheric features on Jupiter and Saturn in this way requires more time and experience, but is not restricted to the central meridian, and trained observers may attain a precision better than ±1°.

Whenever seeing permits, the planetographic latitudes of the north and south edges of zones and belts should be regularly measured, as well as the longitudes of well-defined features (GRS, white oval spots, dark bars). The following procedures are schematized in Fig. 130.

The zero reading at the superposition of the two wires is determined first, and subtracted from all subsequent readings. The equatorial diameter d is measured with the wires parallel to the central meridian (at right angles to the bands on Jupiter and Saturn). Objects are then measured in longitude (x) with the fixed wire at the left (E) edge of the disk, and the moving wire approaching the point to measure alternately from left

Working with filters you need more light. Therefore filters work well only with medium and large-sized instruments (6 in. and more). Filters may help to settle down the atmospheric seeing and to reduce blurring due to chromatic aberration and atmospheric color dispersion.

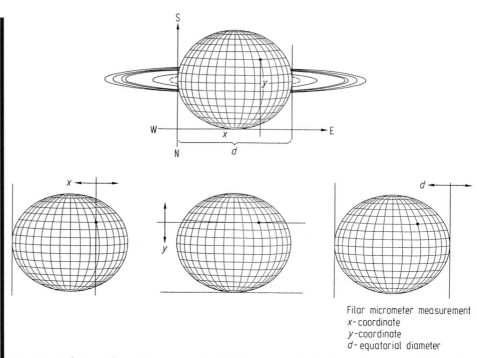

Filar micrometer measurement
x- coordinate
y-coordinate
d- equatorial diameter

Fig. 130. Definition of quantities measured with a filar micrometer (Jupiter and Saturn). After Kowalec

See the Jupiter sections
of the websites for ALPO
(http://www.lpl.arizona.edu/alpo)
or the BAA
(http://www.britastro.org/main)
for more on Jupiter observing
and how to make useful observations.

To analyze drawings, pho-
tographs and measurements
see the chapter "Jupiter" by
H.J. Mettig and others in *Planeten
beobachten* [78].

Hans-Jörg Mettig, Germany, runs
the Database for Object Positions
on Jupiter at http://home.t-
online.de/home/h.j.mettig.

and from right. Rotation the micrometer by 90°, so that the wire now parallel the equator, points are then measured in latitude (y) with the fixed wire at the lower (N) edge of
the disk, and again readings are repeated. The polar diameter need not be determined as
it is computed from the equatorial diameter. Record the time (to 1/10 min) at the beginning and end of each measurement. Needless to say, readings without good timings are
useless as the information on longitudes is lost.

Visual estimates of the jovigraphic latitudes of bands and zones have been performed
from systematic sets of estimates; motions and changes have been deduced. Again, a micrometer may be useful (see Fig. 131).

Intensity estimates may be made with the aid of a scale which runs from +5 (or 5 B)
equal to brilliant white, through 0 which is equal to neutral gray, to −5 (5 D) equal to
deep black.

These observations can be supplemented by recording colors. Observers recorded
the following tints: (1) red (magenta to bright orange), (2) brown, (3) blue (spots and
garlands), (4) yellow (zone areas), (5) white, (6) gray (apparently at the highest level).
Confirmation of color perceptions may be obtained by filter observations and by comparison with color and spectral photographs.

Photography The substantial apparent diameter of Jupiter favors photographic work
even with small- and medium-sized optics. Photographs showing the most prominent
bands and zones can be obtained with just a 4-in. telescope. Also, the image of the Great
Red Spot should not be too difficult, provided the spot is dark enough. Besides focal

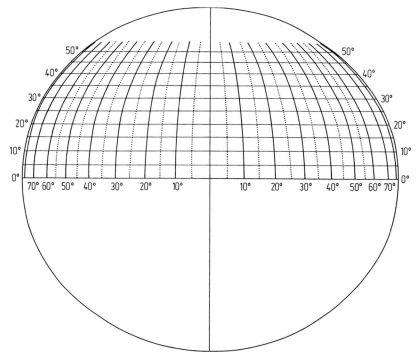

Fig. 131. Coordinate grid for measuring the positions and sizes of features in the Jovian atmosphere. Original from S. Cortesi, Locarno-Monti

photography, the eyepiece projection method is also well-suited for Jupiter (see Chap. 6, where hints are given).

No longer are amateur photographers of Jupiter and other planets limited to locations with excellent atmospheric seeing when using webcams and a specialized software. A new line of high-speed digital cameras is capable to record high-quality images of planets with modest telescopes. An example is the Lumenera SkyNyx 2-0 camera of Canada, capable of capturing video clips at up to 60 frames per second.

The tasks of visual observation as outlined on p. 458 can, in principle, be tackled photographically as well. Within a working group, it will usually be desirable to perform photographic and visual observations simultaneously. On the other hand, the solitary observer will find that taking photographs immediately after a visual drawing is of help in checking the graph and the positions determined. Similarly, filter photographs can supplement corresponding filter-visual observations.

Photographic studies are well-suited for measuring positions of distinct spots, for finding latitudes of bans and zones, and also for determining the positions of the Jovian moons. But position measurements from photographs are not without problems; in particular, the location of the true limb is made difficult owing to the strong limb darkening effect of the planetary atmosphere.

For information on how to process planetary images, especially Jupiter, Saturn and Mars read the report by Donald C. Parker, *Sky and Telescope* January 2007 issue, p. 129.

For information of SkyNyx 2-0 camera visit http://www.astromeccanica.it (Europe) or http://www.astrovid.com (USA).

Visit http://www.astromanie.de, the homepage of the German observer and photographer of Jupiter Ralf Gerstheimer, for more information about imaging the Great Red Spot and "Red Spot Jr.".

The photographic limb darkening effect causes substantial errors when determining latitudes. It is preferable to use a computed polar diameter, reconstructed from the measured equatorial diameter (the bright equatorial zone) and the known oblateness.

Photometry Measurements of individual bright or dark objects can be done photometrically, either by estimates (see above) or by area photometry (see Chap. 8). The apparent brightness of the moons I and II is somewhat variable. Observations also suggest that I (Io) has a slightly higher apparent brightness than usual immediately after re-emergence from the planet's shadow and for about 15 min thereafter. The polar caps of Io exhibit a reddish tint which has been confirmed by color photographs. By making comparative magnitude estimates of moons which do not differ significantly in brightness and occasionally comparing with nearby stars, the observer will gain much valuable experience.

Photoelectric photometry is well-suited for the above-mentioned tasks, as well as for following occultations and eclipses of the moons. Those phenomena which occur close to the bright disk of the planet, however, must be excluded because of the interference of the scattered planetary light. Not to be forgotten is integrated photometry of the planet and its satellites, with the goal of, for instance, finding evidence for relations between solar activity and brightness changes in planets and satellites. For details on photoelectric photometry, see Chap. 8.

Satellites of Jupiter From the results of the Voyager missions, one can assume that Jupiter has at least 39 satellites. Most of these moons are beyond the reach of amateur equipment. The more interesting moons from the amateur standpoint are I–IV, which were discovered by Galileo centuries ago. Together with the Earth's moon and Saturn's satellite Titan, they are the largest and most massive moons in the Solar System. Their apparent diameters and apparent magnitudes are given in Table 16.

The shadows of the moons on the disk provide good test objects for the quality of 3-in. apertures. To see the moons as disks requires a 5- or 6-in. aperture. When transiting the planetary disk, moons I and II (Io and Europa) appear bright on a gray background, III (Ganymede) gray on a brighter background and IV (Callisto) almost as dark as a shadow.

Observers equipped with telescopes of 30-cm aperture and over may try to find surface shadings or details on Ganymede. This satellite is somewhat larger than the planet Mercury; Io and Callisto are somewhat smaller, but still larger than the Earth's moon. The Galilean satellites are thus comparable in size to the smaller terrestrial planets. The Pioneer and Voyager missions have made possible the construction of detailed surface maps of the four large moons. Voyager discovered eight active volcanoes on Io and a crust of frozen sulfur and sulfur dioxide. The surfaces of the other satellites are most likely covered with ice formations. Ganymede is assumed to consist of half water and half ice.

Table 16. Apparent diameters and magnitudes of the four Galilean moons of Jupiter

Moon	Diameter	Magnitude
I (Io)	1″.05	5.m43
II (Europa)	0.87	5.57
III (Ganymede)	1.52	5.07
IV (Callisto)	1.43	6.12

The motions of the moons and their frequent transits, occultations, and eclipses provide an interesting demonstration of orbital motion. Such phenomena occur regularly for moons I to III, and for moon IV when the Jovian axis has a low tilt, which occurs when Jupiter is around right ascensions 8 h or 20 h. At these points, the lines of sight and of shadow fall in the Jovian equatorial plane, which coincides with the orbital plane of the satellites. It is at these times that mutual occultations and eclipses of the moons can occur (Fig. 132). During such an occurrence, an observer using a small telescope will notice brightness changes, while a large instrument will capture the actual event, for instance the eclipse of III by I. Also, the moons III and IV can then approach each other closely, resulting in a combined magnitude of +4.8, which at moons III's elongation can then be seen by keen-eyed observers without a telescope.

The albedos of the Jovian moons are markedly different: highest for I and II at over 0.6, lowest for IV at 0.2. During transits across the Jovian disk, moons I and II appear

Fig. 132. Triple solar eclipse on Jupiter. Shadows of the Galilean satellites Ganymede (*1*), Io (*2*), and Callisto (*3*). The satellites Ganymede (*blue spot*) and Io (*white spot*) are also seen on this image, which was taken by Hubble Space Telescope in the infrared (HST/NASA)

bright against a darker background, which moon III is glaring and IV could almost be mistaken for the shadow of a moon.

The four large moons provide the astronomer with an opportunity to "rediscover" Kepler's third law by observations, taken either photographically or by micrometer, of the relative positions of the moons with respect to Jupiter on several consecutive nights.

13.6.5.5 Saturn

Visibility Saturn is one of the most interesting and beautiful objects for the observer, owing in particular to its famous ring system. The observable atmosphere shows similarities to that of Jupiter. The advice for observing Jupiter in general also holds for Saturn. As a superior planet, Saturn is observable all night during opposition. Its visibility, beginning with the conjunction with the Sun, is as follows:

Day 0:   Conjunction with the Sun
Day 18:   Saturn appears in the morning sky
Day 125:   Saturn begins retrogression
Day 189:   Saturn in opposition with the Sun
Day 253:   Retrogression ends
Day 360:   Saturn disappears from evening sky
Day 378:   Next conjunction

A good almanac will supply the observer with the following data:

1. Right ascension and declination
2. Apparent equatorial diameter (ranging between 15″ and 21″)
3. Apparent polar diameter
4. Apparent magnitude (between +0.9 and −0.6)
5. Position angle of axis of rotation
6. Latitudes of Earth and Sun with respect to Saturn's equatorial plane
7. Distance from Earth (in AU)
8. Phase angle (or deficit of illumination, from $=0''.00$ fully illuminated to a maximum of $0''.04$)
9. Light-travel time from Saturn to Earth
10. Longitude of the central meridian in both rotation systems I and II
11. Time of transit
12. Orientation and opening of the rings
13. Phenomena of the brighter satellites

In addition to the geocentric planetary coordinates, some almanacs also contain heliocentric coordinates.

The apparent brightness depends upon not only the continuously varying distance from the Earth but also (and to a much larger extent) on the changing position of the rings. Saturn's equatorial plane, which contains the rings, is inclined against its orbit by $26°$ and against the ecliptic plane by $28°$. This inclination must certainly be taken into consideration, as the observer will view only the northern side of the ring system with respect to the line-of-sight during the observing season. The phase angle never exceeds $6°$ and can be neglected.

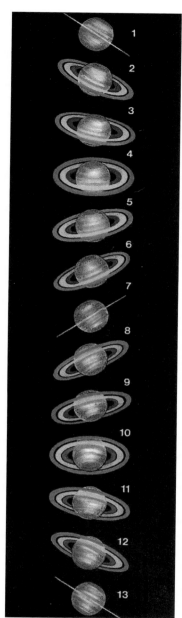

Fig. 133. The ring system of Saturn. Ringplane crossing: *1* rings edgewise to Earth 1995; *2 + 3* south face of rings on view 1996–2001; *4* rings south face widest open 2002; *5 + 6* south face of rings on view 2003–2008; *7* rings edgewise to Earth 2009; *8 + 9* north face of rings of view 2010–2015; *10* rings north face widest open 2016; *11 + 12* north face of rings of view 2017–2022; *13* rings edgewise to Earth 2023

Information on phenomena and events in the atmosphere, on the surface and within the ring system of Saturn were provided by US missions Pioneer 11, Voyager 1 and Voyager 2 between 1979 and 1981. In 1997, NASA's Cassini spacecraft was launched and entered Saturnian orbit in July 2004. The mission discovered a fascinating planetary world.

A small part only is visible with telescopes of amateur astronomers. But nevertheless the task remains thrilling!

Bright zones and dark bands similar to those in the atmosphere of Jupiter are observed on Saturn (see nomenclature for Jupiter). However, the features, in particular the bright spots and clouds, are less well-defined than they are on Jupiter. Occasionally more prominent features appear. A good example was the Great White Spot first noticed in September 1990 and which subsequently developed into a quite conspicuous object in the northern part of the EZ (see Fig. 134). D. Parker described it as "the brightest thing ever seen on the planet, a quarter of the diameter of the disk, and as bright as a Martian polar cap."

The first known Great White Spot was detected in December 1876 by American astronomer Asaph Hall in Washington, DC. The next one was found in June 1903 by E.E. Barnard with the 40-in. refractor at Yerkes Observatory, near Williams Bay, Wisconsin. The third and fourth were both found by eagle-eyed amateurs: in August 1933 by Will Hay in England, and in March 1960 by J.H. Botham in South Africa. All of these spots were seen in the northern hemisphere of Saturn, those in 1876 and 1933 at about the same latitude as the present one, while the two others were farther north at +40° (1903) and +58° (1960).

Most planetary astronomers agree that the Great White Spots are upwellings from the lower atmosphere, whereby large clouds move upwards and become visible when they penetrate the uppermost, hazy layers. They resemble the towering cumulonimbus clouds often seen in the earth's atmosphere. However, the lifting mechanism is not yet known; one possibility is that their upward motion is due to the release of heat by condensation of water vapor, perhaps in combination with strong updrafts from sublimating ammonia grains.

The spots become longer as the clouds are carried along by strong winds in the upper atmosphere (for global map of Saturn see Fig. 135). Eddies and whirl patterns undoubtedly develop because of the different wind velocities at different latitudes, but because of their smaller size they are very difficult to observe from Earth. This may imply that the spots, perhaps in particular those which have emerged more recently, are actually gigantic storm centers, such as the Giant Red Spot on Jupiter, which has now been visible for almost 400 years.

Micrometer measurements and transit times of suitable objects across the central meridian to determine the rotation are therefore difficult. Analysis of the observed data indicated that there are two rotation systems; since 1969, the Association of Lunar and Planetary Observers (ALPO) has published tables of the central meridians:

System I   (NEB, EZ, SEB) $10^h14^m13^s.08$ (= 844° per day)
System II (middle latitudes) $10^h 38^m 24^s.2$ (= 812° per day)

Fig. 134▸ The Great White Spot on Saturn 1990. This series of three exposures from the ESO La Silla Observatory shows the development of the spot over a period of two weeks. Between October 8 and 16, 1990, the Spot grew significantly in size; it became longer and probably also brighter. After October 16, it rapidly expanded to reach all the way around the equator of the planet. The images were made in blue light in mediocre seeing conditions with three different instruments, all equipped with CCD detectors. Photograph courtesy of the European Southern Observatory

Great White Spot on Saturn

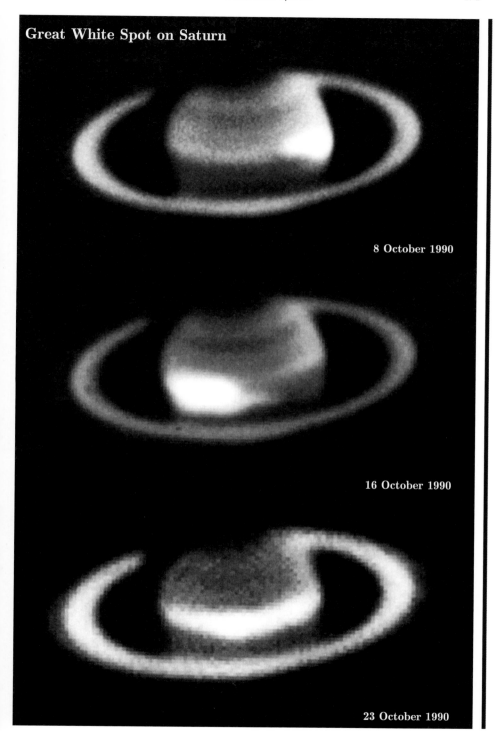

8 October 1990

16 October 1990

23 October 1990

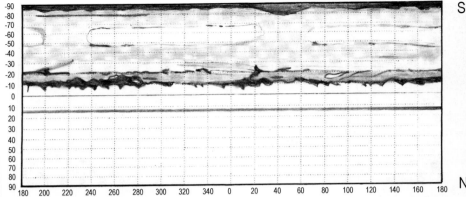

Fig. 135. Global map of the planet Saturn as obtained in April 1–8, 1974. Cylindrical projection distance true at the equator; flattening of 1:10 is allowed for. Instruments used were 6- and 12-in. refractors of the Wilhelm-Foerster Observatory, Berlin. Observers were Wolfgang Anklam and others

In 1980, the International Astronomical Union (IAU) slightly changed System I and added a System III, derived, as in the case of Jupiter, from radio observations:

System I   (IAU) Period of rotation $10^h\ 14^m\ 0^s.00$
System III (IAU) Period of rotation $10^h\ 29^m\ 59.^s42$

When observing Saturn's low-contrast markings it may be especially helpful to use color filters (see section on Jupiter p. 456). Because the more dimmed Saturn's image light-colored filters should be preferred.

Visual Observations For information relating to atmospheric phenomena on Saturn, the corresponding text on the planet Jupiter (Sect. 13.6.5.4) should be consulted, as in principle it also applies to Saturn. However, the number of the bright and dark objects perceivable with amateur instruments is significantly less. The intensities of the objects also are usually so low that micrometer measurements and central meridian transits are more difficult.

Figure 136 shows Saturn's rings and divisions before Voyager. The space missions discovered numerous new sections and divisions in the ring systems (see Fig. 137).

When the rings are fully opened, Cassini's Division, the dark dividing line between the outer ring A and the middle ring B, can be seen even in a 3-in. telescope. The visibility

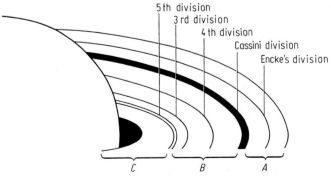

Fig. 136. Rings of Saturn (*A*, *B*, *C*) and the ring divisions known before the Voyager missions

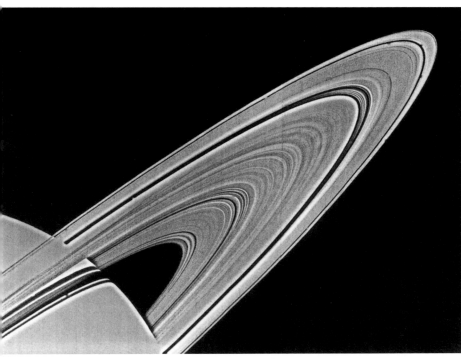

Fig. 137. Voyager I photograph taken on 1980 November 06 at a distance of 8 million km from Saturn, showing copious detail in the ring system

of other rings and divisions also depends on the degree of opening. When open widely enough, the rings can be seen to differ distinctly in brightness. Intensity estimates may be tried. In all graphs it should be remembered that the shadows cast by the planet onto the rings and vice versa must be correctly represented. Observed deformations of these shadows appear to be optical illusions. However, thickenings of the ring system as seen with the rings edge-on coincide with zones of elevated brightness. They also indicate that the brightness of the ring structure is far from uniform. It is also quite interesting to observe the "transparency" of the ring system when Saturn occults a star.

Photography Published reports have suggested that photography of Saturn be carried out using instruments with apertures of 20 cm and larger. Good results have indeed been obtained with instruments in this size range. Imaging Saturn with the introduction of webcams and software that automatically records and combines thousands of pictures made a revolution concerning photography by amateurs. For more information see Sect. 13.6.5.4, and Chap. 6.

Photometry Suitable features for photometric studies include, in addition to the bright and dark surface detail (see Sect. 13.6.5.4 on Jupiter), the brighter moons and the ring system. In particular, photometry can be performed in conjunction with photographs and with color filters of known transmissivity (see Chap. 6). Reasonable results, however, require apertures over 25 cm and a total focal length over 20 m.

Table 17. The brightest satellites of Saturn

Moon		Magnitude
II	Enceladus	11.7
III	Tethys	10.3
IV	Dione	10.5
V	Rhea	9.7
VI	Titan	8.3
VIII	Iapetus	10–12, variable

The Cassini spacecraft solved the variability of Iapetus: Due to the moon's slow rotation dust collects more on one side, the dark side, and heat it up. Any reflective ices vaporize and migrate to the other side, the bright one.

Dust coats Saturn's largest moons with a frosty layer. Titan is the lone exception caused by its thick atmosphere. Discoveries come from flyby encounters with icy moons.

The Satellites of Saturn As a result of the space missions, 60 Saturnian moons are now known (as of May 2007). Several of them of especial interest for the amateur observer are given in Table 17.

The *Astronomical Almanac* also includes data on the fainter moons I (Mimas) and VII (Hyperion). With the ring system seen edge-on, the brighter Saturnian moons show similar phenomena (eclipses, etc.) as those of Jupiter (see Sect. 13.6.5.4). But even the transit of the brightest and largest moon Titan and the observation of its shadow on the planet require an aperture of at least 20 cm.

Of interest are the apparent magnitudes of the moons, which may vary; visual photometry either by direct estimates or by photometer is within the reach of amateurs. It is well-known that Iapetus exhibits a light variation with an amplitude of about 2 mag. The period of variation corresponds to the orbital period of $79^d\,2^h$. The maximum brightness is observed near western elongation, the minimum near eastern elongation. Evidence for a relation between solar activity and brightness changes of Titan and Rhea requires long-term and very precise photometry, obtainable only by photoelectric means. Seasonal changes of 0.01 or 0.02 mag for Titan have been reported.

13.6.5.6 Uranus

Visibility Uranus was discovered in 1781, by F.W. Herschel. With a diameter of 50,800 km, Uranus is the third largest planet in the Solar System, revolving about the Sun in 84 years. Owing to its extreme distance, the planet's apparent diameter remains between $3''$ and $4''$, limiting its interest for telescopic observation.

On January 24, 1986, Voyager 2 passed Uranus at a distance of 93,000 km. The pictures obtained by the probe reveal an unstructured atmosphere, with a veil of haze surrounding the planet. Filter photographs showed clouds in the lower layers of the atmosphere moving with speeds of 100 m/s, about the same as the jetstreams on Earth 9 km above the surface, whereas the westerlies on Saturn attain speeds of up to 480 m/s near the equator. While the upper layers of Uranus' atmosphere consist of hydrogen and some helium, the lower parts are composed of methane and other hydrocarbons.

Strangely enough, the atmosphere over the dark pole of Uranus is found to be warmer than over the insulated pole. This may be explained by heat accumulated over the 42 years during which the now dark pole was exposed to the Sun. The rotation axis of the planet is almost exactly in the orbital plane (Fig. 138). Since the inclination of the planet against the ecliptic is low, Uranus' equator is almost perpendicular to the ecliptic. The passage of Voyager 2 permitted direct measurements of the rotation period. Data from radio

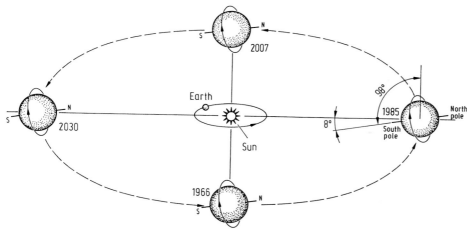

Fig. 138. Configurations of the planet Uranus relative to the Sun. The rotation axis is inclined 98° from the orbital axis

observations yield a period of $17^h239\pm0^h009$. Using measurements of the magnetic field, a rotation period of $17^h29\pm0^h20$ was derived. Uranus also displays a kind of polar aurora.

During the processing of observations from the 1977 occultation by Uranus of the star SAO 158687 (magnitude +8.8) in Libra, it was discovered that the planet is apparently surrounded by five narrow, thin rings; the total breadth of the ring system was found to be 7100 km. The ring system is in the equatorial plane of the planet, and its lower edge lies 17,700 km above the surface. Whether or not this system reaches the density of Saturn's rings still needs to be ascertained.

Four more rings had been detected even before the Voyager flyby. Again, it is assumed that all rings are in the equatorial plane. Table 18 shows the radii of the nine rings now known.

Between the ϵ and δ rings, Voyager 2 found a tenth ring, which has been given the preliminary designation 1986 U1R. This ring lies 50,040 km from the center of the planet. There are indications that other rings may exist outside the ϵ ring. By contrast with Sat-

Eighteen years since the Voyager flyby we observe a dynamic and complex atmosphere of Uranus with clouds varying in size, brightness and longevity. The major reason is the planet's extreme seasonal cycle.

Table 18. Radii of the rings around Uranus

Ring (Code)	Radius (km) (±30 km)
6	41,980
5	42,360
4	42,663
α	44,844
β	45,799
η	47,323
γ	47,746
δ	48,432
ϵ	51,697

urn, whose rings consist largely of golf ball-sized particles, the particles in the rings of Uranus have diameters of mostly about 1 m. This may be explained by assuming that the particles which make up the Uranus rings are composed of a sturdier material than that of their Saturnian counterparts, and are thus more resistant to the continual collisions.

Observations Some observers had reported that telescopes with apertures larger than 25 cm are large enough to show dark shadings (e.g., the "dark equatorial band"). No indications of such structure were found on photographs taken in 1970 by Stratoscope II,

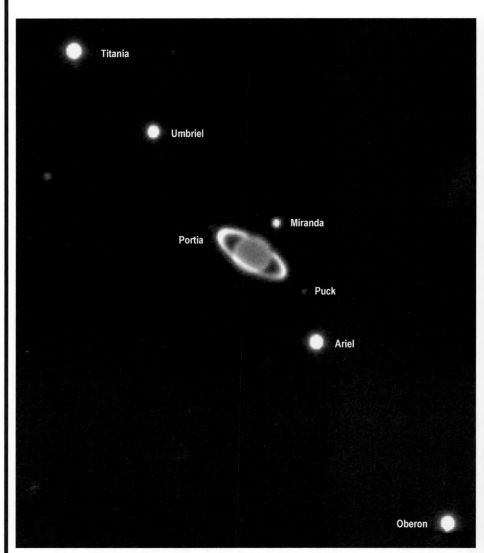

Fig. 139. Uranus with its ring system and the brighter satellites. Image by the Very Large Telescope of the European Southern Observatory (ESO)

a 91-cm telescope flying 2 km above the Earth's surface. These photographs, however, show a very small oblateness and a distinct limb darkening which is even stronger than that of Jupiter.

Uranus shows a pronounced variation of brightness with total amplitude 0.9 mag. This change has geometric as well as physical causes and may be monitored with amateur equipment, either by photometer or by the well-known Argelander "step method" (Chap. 8). While viewing Uranus, the observer may also note the greenish coloring of the planet.

Photography of the two outer moons Titania and Oberon (magnitude +14) is within reach of the amateur. These moons were discovered by W. Herschel in 1787. Ariel and Umbriel were discovered in 1851 by Lassell while observing with a 61-cm reflector on the island of Malta; their brightness is around magnitude +16. In 1948, G.P. Kuiper discovered Miranda at magnitude +17, and finally, the Voyager 2 pictures showed ten new moons. Uranus now has a total of 27 known moons (Fig. 139).

13.6.5.7 Neptune

Visibility Neptune was discovered in 1846 by J.G. Galle, who had followed the theoretical predictions of U. Leverrier. Despite its linear diameter of 49,000 km the apparent diameter of Neptune reaches only $2''.5$, and is, with a mean opposition magnitude of +7.75, below the naked-eye limit of visibility. Neptune revolves in 164 years about the Sun; its oppositions reoccur every 367 days. It has an extremely low orbital eccentricity of 0.009.

In 1989, the space probe Voyager 2 reached Neptune, approaching it on August 25 to within 29,240 km. The most important discovery in the atmosphere was the Great Dark Spot (GDS), a long-term stable cyclone with high white clouds ("cirrus") at the edges. It is comparable with the GRS on Jupiter. Voyager 2 found several other, less prominent bright and dark spots, as well as a band structure, the brightest zone being at latitude 20°S, and dark bands at 6° to 25°N and at 45° to 70°S. Compared with Uranus, Neptune shows a dynamical atmosphere (see changes in brightness in Fig. 141); the existence of a yet unexplained internal heat source is conjectured. Fast westerly winds reaching up to 300 m/s were recorded.

A clear periodicity of Neptune's radio radiation over several rotations around Voyager's closest approach, $16^h 13^m \pm 4^m$, is interpreted as the rotation period of the core of the planet. Its magnetic field was measured as 0.13 gauss, which is weaker than that of other planets (Jupiter 4.2, Earth 0.3, Uranus 0.2 gauss).

Voyager 2 confirmed the ring system previously suspected from observations of occultations of stars. The Earth-based measurements had suggested rather incomplete arcs, whereas Voyager reported closed rings like those around Jupiter, Saturn, and Uranus. The radii of the four rings known so far are given in Table 19.

As for the other Jovian planets, an extended disk of very fine dust surrounds Neptune in its equatorial plane, causing up to 300 impacts per second on the Voyager craft as it passed through it.

To the known moons Triton (discovered by Lassel in 1846) and Nereid (Kuiper 1949), Voyager 2 added six more. Triton's diameter was measured as 2760 km, and pictures of it were quite surprising as they revealed a wealth of surface detail. Resembling Mars and

Long-term monitoring of Uranus by ALPO members reveals a slight, slow dimming of 6 to 8% since 1991 (visit http://www.lpl.arizona.edu/alpo).

CCD images of Uranus visit http://astrosurf.com/pellier/planeturanus and the Astro-imaging webpage of the Texas astrophotographer Ed Grafton at http://www.ghg.net/egrafton.

See also VdS-Journal für Astronomie III/2007, page 86 (in German) www.vds-astro.de

A challenge to photograph (CCD camera) are the moons Umbriel, Ariel and Miranda (Fig. 140).

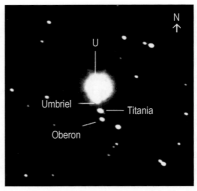

Fig. 140. View of some of Uranus's moons

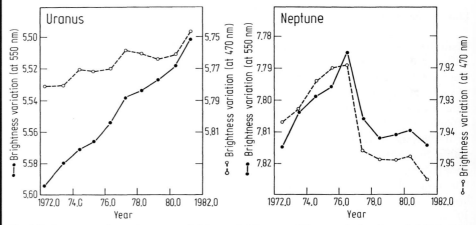

Fig. 141. Long-term change in apparent brightness in Uranus and Neptune during the period 1972–1982. Observations from the Lowell Observatory at 470 nm and 550 nm. Adapted from R.M. Genet (ed.), *Solar System Photometry Handbook*, Willmann-Bell, Richmond VA 1983

the moons Enceladus (Saturn) and Ariel (Uranus), no other body in the Solar System appears to show such a variety of surface formations. Triton's atmosphere was found to consist of a layer of nitrogen-methane haze.

Table 19. Radii of the rings around Neptune

Ring	Radius (km)
1989 N3R	41,900
1989 N2R	53,200
1989 N4R	53,200–59,000
1989 N1R	62,900

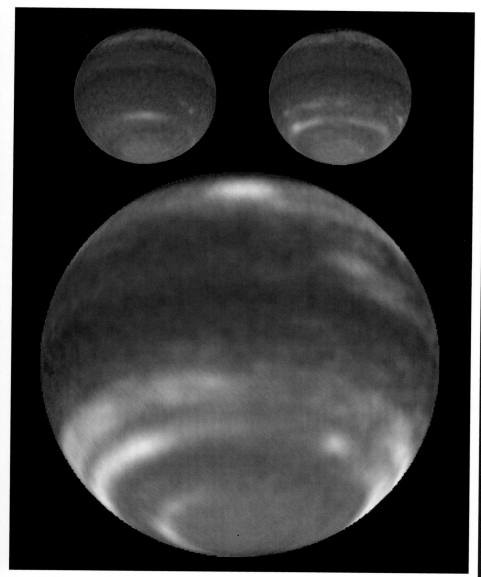

Fig. 142. Neptune with cloudy structures around the poles. Images from the Hubble Space Telescope 1996, 1998, and 2002 (HST/NASA)

Observations The amateur observer can see the planetary disk with magnifications over 300×, but atmospheric detail is ruled out. Similar to Uranus, the color of Neptune is observed to be bluish-green.

With telescopes of 10 in. and more equipped with a fast webcam a planet imager should be able to record cloudy features of Neptune.

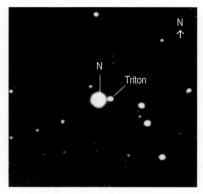

Fig. 143. Neptune's moon Triton

Photometric monitoring of the planet may have some scientific merit. The apparent brightness varies between +7.5 and +8.0; a rotational variation is suspected. The satellite Triton reaches +13.6 and is, at a diameter of 4000 km, one of the largest moons in the Solar System (Fig. 143). As with the brighter of Uranus's moons, attempts to photograph Triton may prove successful.

13.6.5.8 Pluto (Dwarf Planet)

Visibility The outermost Pluto was found on photographs in 1930 by Clyde Tombaugh. The opposition brightness is about +13.8, the apparent diameter about 0″.3. Even at high magnification, this dwarf planet does not look much different from a star.

Observations Locating Pluto has recently become a popular "sport" among some amateur astronomers. This may be achieved visually or photographically using a telescope with an aperture of at least 20 cm. The moon Charon was discovered in 1978. Opposition brightness is +15.5, and the distance from Pluto max. 1″.

13.6.6 Further Reading

Alexander, A.F.O'D.: The Planet Saturn. Faber and Faber, London (1962)

Alexander, A.F. O'D.: The Planet Uranus. A History of Observations, Theory, and Discovery. Faber and Faber, London (1965)

Bagenal, F., Dowling, T., Mckinnon, W. (eds): Jupiter. The Planet, Satellites and Magnetopshere. Cambridge University Press, Cambridge (2004)

Baum, R., Smith, R.W.: Neptune's Forgotten Ring. Sky Telescope **77**, 610 (1989)

Baumm, J.L.: Unusual activity on the night side of Venus 1986 September. JBAA **97**, 1 (1986)

Beatty, J.K.: Magellan at Venus: First Results. Sky Telescope **80**, 603 (1990)

Beebe, R.F.: Queen of the Giant Storms, Sky Telescope **80**, 359 (1990)

In August 2006 the General Assembly of the International Astronomical Union (IAU) in Prague resolved that Pluto is no longer a planet but a "dwarf planet" by a new definition and is recognized as a prototype of a new category of transneptunian objects. More informations see Sect. 13.7

Bell, J.: Postcards from Mars: The First Photographer on the Red Planet. Penguin, New York (2006)

Benton J.L.: The 1983–84 and 1985–86 Western (Morning) Apparition of Venus: Visual and Photographic Observation. JALPO **33**, 93 (1987)

Benton J.L.: The 1984–85 Eastern (Evening) Apparition of Venus: Visual and Photographic Observation. JALPO **34**, 1 (1989)

Binzel, R.P.: Photometry of Asteroids. Solar System Photometry Handbook. Willmann-Bell, Richmond, VA (1983)

Brahic, A., Hubbard, W.B.: The Baffling Ring Arcs of Neptune. Sky Telescope **77**, 606 (1989)

Cattermole, P., Moore, P.: Atlas of Venus. Cambridge University Press, Cambridge (1997)

Cruikshank, D.P.: The Ashen Light of Venus. Astronomical Society of the Pacific Conference Series, vol. 33, San Francisco (1992)

Dobbins, A., Parker, D.C., Capen, C.F.: Observing and Photographing the Solar System. Willmann-Bell, Richmond, VA (1988)

Gehrels, T.: Jupiter. The University of Arizona Press, Tucson (1976)

Gehrels, T.: Shapley, M. (eds.): Saturn. The University of Arizona Press, Tucson (1984)

Hanlon, M.: The Red Mars: Spirit, Opportunity, Mars Express and the Quest to Explore the Red Planet. Carroll and Graf, New York (2004)

Heuseler, H., Jaumann, R., Neukum, G.: Zwischen Sonne und Pluto. Die Zukunft der Planetenforschung: Aufbruch ins dritte Jahrtausend. BLV, München (1999)

Kargel, J.S.: Mars. A Warmer, Wetter Planet. Springer/Praxis, Berlin Heidelberg New York (2004)

Martinez, P. (ed.): The Observer's Guide to Astronomy, 2 volumes. Cambridge University Press, Cambridge (1994)

Maunder, M., Moore, P.: Transit: When planets cross the sun. Springer, Berlin Heidelberg New York (2000)

McBride, N., Gilmour, I. (eds.): An Introduction to the Solar System. Cambridge University Press, Cambridge (2004)

McFadden, L.-A., Weissmann, P.R., Johnson, T.V. (eds.): Encyclopedia of the Solar System. Academic, New York (2007)

Miner, E.D.: Uranus, the Planet, Rings and Satellites. Ellis Horwood, New York (1990)

Moore, P.: The Planet Neptune. Ellis Horwood, New York (1988)

H.M. Nautical Almanac Office (ed.): Planetary and Lunar Coordinates 2001–2020. Willmann-Bell, Richmond, VA (2001)

Oivarez, J.: Jupiter's Best Show in Twelve Years. Astronomy **15**, 64 (1987)

O'Meara, S.J.: Jupiter's North Equatorial Belt Erupts. Sky Telescope **79**, 94 (1990)

Parker, D.C., Dobbins, T.A.: The Art of Planetary Observing-II. Sky Telescope **74**, 603 (1987)

Peebles, C.: Asteroids: A History. Smithsonian Institution Press, Washington, DC (2000)

Peek, P.M.: The Planet Jupiter. Faber and Faber, London (1958)

Philips, J.L.: The Ashen Light of Venus. Sky Telescope **75**, 250 (1988)

Price, F.: The Planet Observer's Handbook, 2nd edn. Cambridge University Press, Cambridge (2000)

Rogers, J.H.: The Giant Planet Jupiter. Cambridge University Press, Cambridge (1995)

Roth, G.D.: The System of Minor Planets. Faber and Faber, London (1962)

Roth, G.D.: Compendium of Practical Astronomy, vol. 2, Earth and Solar System, chap. 19. Springer, Berlin Heidelberg New York (1994)

Schmadel, L.D.: Dictionary of Minor Planet Names, 4th edn. Springer, Berlin Heidelberg New York (1999)

Schober, H.J.: Asteroids. In: Hearnshow, J.B., Cottrell, P.L. (eds.) Instrumentation and Research Programmes for Small Telescopes. International Astronomical Union (IAU), Paris (1986)

Sky and Telescope: A Mars Observer's Guide. Sky Telescope **75**, 516 (1988)

Sobel, D.: Die Planeten. Berliner Taschenbuch, Berlin (2005)

Stern, A., Mitton, J.: Pluto and Charon. Ice Worlds on the Ragged Edge of the Solar System. Wiley, Weinheim (2005)

Standage, T.: The Neptune File. Walker and Co., New York (2000)

Strom, R.G.: Mercury, The Elusive Planet. Cambridge University Press, Cambridge (1987)

Weintraub, D.A.: Is Pluto a Planet? A Historical Journey through the Solar System. Princeton University Press, Princeton (2006)

Current observational tasks and possibilities for cooperation are subjects of reports in the *Journal of the British Astronomical Association*, the *Journal of the Association of Lunar and Planetary Observers* (USA), *Astronomy, Sky and Telescope* and *Sterne und Weltraum*. Advanced observers will find more information in *Icarus*, an international journal of solar system studies.

13.7 Asteroids and Kuiper Belt Objects

H. Boehnhardt

13.7.1 The Asteroid Belt and the Kuiper Belt

In between the planetary orbits there are two regions of enhanced occurrence of minor bodies, i.e., the Asteroid Belt and the Kuiper Belt. The former is located between the orbits of Mars and Jupiter at 2.1–3.3 AU, the latter defines the currently known outer edge of the giant planet region and extends from 35–50 AU. The common properties of both belts are: they contain small bodies compared to planets with maximum sizes of about 1900 (Asteroid Belt) and 2400 km (Kuiper Belt). Both regions of minor bodies are ecliptic oriented, although a wide scatter of inclinations are found. Most bodies therein move in close to circular or moderately eccentric orbits, while a few subpopulations have higher eccentricity and orbit arcs outside of the main regions of the respective belt.

The exploration of the Asteroid Belt started with the discovery of asteroid (1) Ceres by Guiseppe Piazzi on January 1, 1801. After an initially slow increase of the number of asteroids discovered in the nineteenth century, the searches for these small bodies have improved and are nowadays done very successfully by automatic telescopes leading to more than 450,000 objects with well-established orbits in the catalog of the International Astronomical Union (IAU). The Asteroid Belt is also called the "Main Belt." The objects in the belt are called "asteroids" since they appear like stars (though moving) in the telescopes. Since asteroids are small objects in planet-like orbits, they are sometimes also called "planetoids," which corresponds better to their physical nature than the widely accepted naming as asteroids.

The first Kuiper Belt object (KBO) was discovered in 1930, at that time without really knowing, since Pluto, the object in question, was considered the ninth planet at those times and decades after. Thus, the real detection of the Kuiper Belt happened only in 1992, when Dave Jewitt and Jane Luu in Hawaii found a very slowly moving object, preliminarily cataloged by the IAU as 1992 QB1. This object was far beyond the distance of Neptune. Its discovery did not happen by chance, but was the result of a dedicated search for such objects. Fifteen years later more than 1100 KBOs have been discovered. In 2006 the IAU "degraded" Pluto from planet status to that of a "dwarf planet," i.e., Pluto, and some more KBOs, fall now in the same category of bodies like 1 Ceres and several other asteroids in the Main Belt. Different names for the belt of KBOs are in use: The "Kuiper Belt", honoring Gerard Kuiper who speculated about the existence of bodies beyond Neptune in the 1940s and 1950s of the last century, and naming as the "Edgeworth–Kuiper Belt" acknowledges the somewhat earlier contributions of Kenneth Edgeworth to the speculative topic of Kuiper. The objects therein are frequently also addressed as transneptunian objects (TNOs), since the majority have orbits beyond the orbit of Neptune.

13.7.2 Naming Convention and Spacecraft Missions to Asteroids and Kuiper Belt Objects

13.7.2.1 Naming Convention
The designation and naming conventions for asteroids and KBOs are the same. When discovered and if observed during at least two nights, the new object receives a preliminary designation that contains the year of the discovery, the

Asteroids and Kuiper Belt objects do exist beyond the edges of the belts, but at much lower spatial density. Somewhat wider limits for the Asteroid Belt can be set at 1.8 and 4.0 AU, although the inner edge is not very sharp. For the Kuiper Belt the inner and outer extensions are less well-defined, but could be set to Neptune's orbit at 30 AU and the transition region to the hypothesized inner Oort Cloud, possibly at around 2000 AU from the Sun, respectively.

Roman (capital) letter for the half month in the year (starting with A for January 1–15) and another Roman (capital) letter in the sequence of discoveries during the respective half month (starting with A for the first discovery and omitting letters I and Z). In the case of numerous (more than 25 for the number of letters in the Roman alphabet) discoveries per month, optional numbers are added to the object designation indicating the number of times the second letter has been repeated in that half-month period, e.g., 1992 QB1, the first TNO discovered after Pluto, was the 27th object found in the period of August 16 to 31 1992. If the orbit of the asteroid or KBO is "secure," i.e., enough accurate astrometric positions have resulted in an accurate orbit determination, the object receives its object identification number (set in parenthesis). This designation "counts" in arabic numbers the minor bodies that are supposed to be found back easily (starting with (1) Ceres as the first one). Object names are assigned by the IAU Committee on Small Body Nomenclature to numbered asteroids and KBOs. The discoverer may propose a name for the object according to certain rules set by the IAU (see http://www.ss.astro.umd.edu/IAU/csbn/).

13.7.2.2 Space Missions to Asteroids and Kuiper Belt Objects Up to now, in total eight asteroids were visited by spacecraft, and two more missions are on their way to encounter four more asteroids over the coming years. One spacecraft heads out for a flyby to the Pluto-Charon system in the Kuiper Belt (Table 20). Most of the asteroid visits were flyby missions except two, NASA's Near-Shoemaker mission to asteroid (433) Eros and Japan's Hayabusa mission to asteroid (25143) Itokawa that went into orbits around their target bodies. Both missions performed also a landing at the respective asteroids. The Hayabusa spacecraft is even expected to return some material to Earth that was collected from the surface of the asteroid. The Dawn mission is on its way to investigate from gravitationally bound orbits two dwarf asteroids in the Main Belt, i.e., (4) Vesta in 2011 and (1) Ceres in 2015. NASA's New Horizons mission was launched in 2005 and aims, after a swingby at Jupiter in 2007, at the exploration of the Pluto-Charon system in 2015. After the close flyby at this multiple object, there is a chance that the mission may visit one or two other KBOs yet to be discovered beyond Pluto's distance from the Sun.

Like for comets, spacecraft exploration is a key element for progress in the scientific investigation of minor bodies in the Solar System. Earth-based observations, however, provide the global framework of our current and future knowledge on these bodies, since the results of the few spacecraft targets can be put into context of the whole variety of small objects in the planetary system.

13.7.3 Orbital, Physical Properties, Formation and Evolution of Asteroids and KBOs

13.7.3.1 Asteroid Orbits The asteroids have mostly only slightly elliptical and in general ecliptic-oriented orbits between 2 and 3.5 AU. In this distance range, an extended orbit domain exists that is long-lasting over the lifetime of the Solar System (4.6 billion years). Outside of this distance range, there are asteroid-like bodies found in close to circular stable orbits at about 4.5 (the Hilda group, named after the most prominent representative) and at about 5.2 AU (the Trojans). Moreover, a variety of escapees from the Asteroid Belt exist in widely random orbits reaching into the distance range of the terrestrial planets.

Note that recently speculations on potential "intruders" in the asteroid family were raised, since dormant or extinct comets may be almost indistinguishable from asteroids apart from some dynamical orbit properties – see chapter on comets.

Table 20. Past and future spacecraft missions to asteroids and Kuiper Belt objects. Mission status refers to June 2008. Information on the missions listed below can be found via the weblink http://solarsystem.nasa.gov/missions/index.cfm

Year	Comet	Mission type	Spacecraft
1991	(951) Gaspra	Flyby	GALILEO[a]
1993	(243) Ida	Flyby	GALILEO[a]
1997	(253) Mathilde	Flyby	NEAR SHOEMAKER[b]
1999	(9969) Braille	Flyby	DEEPSPACE 1[c]
2000–2001	(433) Eros	Orbiting, descent	NEAR-SHOEMAKER[b]
2002	(5535) Annefrank	Flyby	STARDUST[d]
2005–2007	(25143) Itokawa	Orbiting, landing, sample return	HAYABUSA[e]
2006	(132524) APL	Flyby	NEW HORIZONS[f]
2008	(2867) Steins	Flyby	ROSETTA[g]
2010	(21) Lutetia	Flyby	ROSETTA[g]
2011	(4) Vesta	Orbiting	DAWN[h]
2015	(1) Ceres	Orbiting	DAWN[h]
2015	(134340) Pluto	Flyby	NEW HORIZONS[f]

[a] After the asteroid flybys the Galileo spacecraft headed for a close exploration of the Jupiter system, the main target of the mission. Galileo discovered the first binary system (951 Gaspra and Dactyl) in the asteroid belt.
[b] Near-Shoemaker flew by asteroid (253) Mathilde and, thereafter, it went into orbit around asteroid (433) Eros. Although not foreseen in the original mission plan, the spacecraft landed on the surface of (433) Eros at the end of its science mission.
[c] After its flyby at the asteroid (9953) Braille, the Deepspace 1 spacecraft encountered comet 19P/Borrelly in 2001. Due to operational problems no images of the asteroid were transmitted to Earth.
[d] The Stardust spacecraft made a distant flyby of asteroid (5535) Annefrank and continued to an encounter with comet 82P/Wild2 in 2004. During its comet flyby and during cruise in interplanetary space it collected dust particles for return to Earth in 2007.
[e] Hayabusa was the second spacecraft that went into orbit around an asteroid. It managed to land on the surface of asteroid (25143) Itokawa with the aim to take samples of surface material and return them to Earth. By the time of writing, it is unclear whether the surface material sampling was successful. The spacecraft is expected to arrive back on Earth in June 2010.
[f] New Horizons passed by asteroid (132524) APL at a very large distance. After a flyby at planet Jupiter it continued to the main target, the Pluto-Charon system in the Kuiper Belt, where it is expected to arrive in July 2015. Beyond this flyby the spacecraft will hopefully be able to continue towards one or two further yet to be discovered KBOs. During most of the mission to Pluto, the location of potential new KBO targets in the sky is close to the Galactic center which complicates search programs for the new targets. However, as of 2012, i.e., still in time for proper targeting after the Pluto flyby, the chances to find suitable KBO targets will improve significantly.
[g] The main target of the Rosetta mission is comet 67P/Churyumov–Gerasimenko. While passing through the Asteroid Belt, the spacecraft will visit two asteroids, (2867) Steins and (21) Lutetia.
[h] The DAWN mission aims at visiting of two large bodies in the Asteroid Belt, (1) Ceres and (4) Vesta. Both asteroids will be investigated over a few months from close orbits around the bodies.

Kirkwood Gaps The number statistics of the semimajor axes of main belt asteroids displays a non-uniform distribution with remarkable gaps at well-defined distances from the Sun. In these gaps, the so-called Kirkwood gaps, only very few asteroids are found. The Kirkwood gaps comprise orbits with revolution periods of small integer number ratios with the revolution period of Jupiter, i.e., for instance 1:2 for the Hecuba gap, 1:3 for the Hestia gap, 2:5, 3:7 and many more. The gaps mark resonant orbits with Jupiter

Most of the asteroids ejected from the Asteroid Belt end up colliding with the Sun or with one of the planets. Although the risk for collision with Earth is low – about 1 per 10^6 years for an object a few 100 m in size – the destruction potential of such an encounter is extremely high: For instance a 1 km size asteroid may excavate a crater of about 10 km diameter on Earth. Impacts of about 100 km size bodies are responsible for the large basins on the Earth facing hemisphere of the Moon.

and the asteroids therein are exposed frequently to the same gravitational disturbance introduced by the massive planet. As is shown by dynamical calculations, the orbits of objects close to the gaps are thus very unstable and are exposed to large changes in their orbital elements even to the level that the orbits become very elliptical leading the objects well outside of the belt, either in the distance range of the terrestrial planets or into the region of the gas giants. Again through gravitational interactions with planets, these escapees can be permanently "extracted" from the Asteroid Belt and may become planet orbit crossers like for instance the Amor group (with semimajor axis $a > 1$ AU and perihelion $q > 1$ AU), the Apollo group ($a > 1$ AU, $q < 1$ AU), the Aten group ($a < 1$ AU, $q > 1$ AU). A bit surprisingly, but not unexpected from the dynamical modeling, there are two resonant orbits, the Hilda group at 3:2 resonance with Jupiter and the Trojans at 1:1 resonance with Jupiter, that are well-populated by asteroids. The Trojans, preceding or following Jupiter in its orbit around the Sun at an average angular distance of about 60°, orbit the Sun at the distance of Jupiter and represent the examples for the existence of stable orbit solutions in the Solar System of the three-body problem in celestial mechanics. Recently, a number of faint Mars and Neptune Trojans were discovered while for other planets objects in 1:1 resonance orbits were not found up to now [83].

The Trojans are located in the Lagrange points L_4 and L_5, the region of equilibrium between the gravitational forces of the Sun and Jupiter and the centrifugal forces from the orbital motion around the Sun. L_4 and L_5 each form a triangle of about equal legs together with the Sun and Jupiter. There are three more equilibrium points L_1 to L_3, all located on the connecting line between Sun and Jupiter. Only L_4 and L_5 allow stable long lasting orbits for objects in so-called libration drift orbits around the equilibrium locations.

Collision Families Within the belt there are collision families of asteroids (also called Hirayma families), that contain a larger number of objects with very similar orbital elements, in particular semimajor axis, eccentricity and inclination. Examples are the Eos, Flora family, Themis family and Koronis family families. These families are believed to be produced by mutual collisions and disruption of parent asteroids in the belt: The collision families represent the left-over pieces of the disastrous events that may have happened millions to billions years back in time. However, the dynamical families dissolve over the lifetime of the Solar System[84, 85].

Special Asteroid Groups As mentioned in the section on comets (Sect. 13.8), there are asteroid-like objects with some characteristics similar to comets. First, there is a relatively large population of objects with asteroid-like appearance, however, in orbits typical for comets. The similarity to cometary orbits is indicated by the Tisserand parameter (see Sect. 13.8). This population is possibly an armada of dormant or extinct comets that are unable to develop significant coma and tail activity as is typical for comets when getting close to the Sun. Another very interesting yet small population is the group of main belt comets. These objects (three were found by early 2008) move around the Sun in orbits typical for main belt asteroids; however, they eventually display comet-like activity, i.e., dust tails and comae.

Fig. 144a–g ▶ Spacecraft images of main belt asteroids. Note that the images are not to scale. **a** (951) Gaspra imaged by GALILEO. **b** (243) Ida imaged by GALILEO. The satellite Dactyl appears as small body close to the right edge of the image and is also seen in the insert panel on a larger scale. **c** (253) Mathilde imaged by NEAR-SHOEMAKER. **d** (132524) APL imaged by NEW HORIZONS. The asteroid is hardly resolved due to the large flyby distance of the spacecraft. **e** (433) Eros imaged by NEAR-SHOEMAKER. All previous images courtesy of NASA. **f** (25143) Itokawa imaged HAYABUSA. Courtesy of JAXA. **g** (5535) Annefrank imaged by STARDUST. Courtesy of NASA

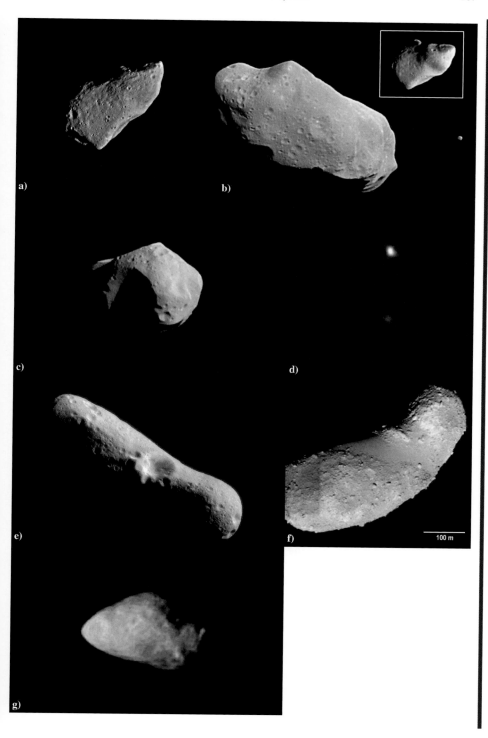

a)

b)

c)

d)

e)

f)

100 m

g)

Small asteroids of a few mm/cm to m in size may appear as bright meteors or extraordinary bolides in the sky when entering the Earth's atmosphere.

The first ever satellite of an asteroid was found in images of (951) Gaspra, taken by the GALILEO spacecraft during its flyby in 1991.

It is still considered a miracle why asteroid (4) Vesta seems to be metal-rich and the typical representative of a metallic core body, while the larger (1) Ceres seems to be a primitive and mostly unaltered asteroid with likely original body constitution. NASA's DAWN mission – see Table 20 – targets for a visit of these two bodies in order to verify and help interpreting the diversity of two of the largest asteroids in the belt.

The funny name "Cubewanos" is derived from the preliminary designation of the first KBO discovered (after Pluto), i.e., 1992 QB1, that also happens to be a classical disk object. It goes without saying that the Plutinos are named after Pluto, the most prominent object in this resonance.

13.7.3.2 Physical Properties of Asteroids

Asteroids display a wide range of sizes from a few meters, which might not be the lower limit, to about 950 km for the largest ones. The big bodies (above a few 100 km in size) like (1) Ceres or (4) Vesta are rather spherical in shape, while smaller objects tend to be of irregular shape (see Fig. 144). Due to the scattering cross-section and/or due to albedo variations on the surface, one can measure the rotation period of asteroids through visible and infrared light curves. Recently, more and more binary asteroids are detected through radar echo and/or light curve analysis techniques. The albedo of asteroids depends on the widely unknown surface composition and varies depending on the spectral type between about 5 and more than 40%. Objects considered to be "primitive" are usually darker, while "evolved" objects are among the brighter ones.

Visible and infrared spectra are used for classification of the surface taxonomy of asteroids, using some wide and mostly weak absorption features from silicate (pyroxene and olivine), metallic (Fe, Ni) and organic (CH bond) species. Asteroids of type A, R, S, and V have silicate absorptions and their surfaces could consist of basaltic material similar to the mantle of a differentiated body (like Earth). E-, M- and T-type asteroids display metal absorptions, occasionally also signatures of water hydration. They may come from the interior (core) of an evolved and/or differentiated body. Asteroids of type C, D, G and P have featureless spectra or display wide bands in the infrared that are usually attributed to carbon-rich material. They are considered little or not at all evolved and may contain the most original and unmodified material from the formation period of the Asteroid Belt. It is noteworthy to mention that the primitive objects are found more in the outer part of the Asteroid Belt, while the evolved, silicate and metal-rich objects are more abundant in the inner and central part of the belt. There is no doubt that many asteroids have experienced evolutionary changes after formation and over the lifetime in the Solar System. In parts, this may have happened during the early period, when the Sun passed through the T-Tauri phase and may have heated up the material at the inner edge of the Asteroid Belt. At least larger objects may also be able to develop enough heat in their interior and start melting, also supported by radioactive heating for instance by ^{26}Al. The heating and melting may have initiated composition changes and differentiation of the body structure. Despite some great similarities in orbit and spectral properties, a unique link between the asteroid taxonomy and the abundant meteorite samples could not yet be established, although it is out of question that such links do exist [86].

13.7.3.3 KBO Orbits

Most of the KBOs move around the Sun in ecliptic-oriented orbits beyond Neptune, inside about 50 AU. The dynamical "core" population is the group of the so-called Cubewanos or classical disk objects (CDOs) that have low eccentricity orbits with semimajor axes between 40 and 48 AU. The Plutinos are KBOs in the dynamically stable 3:2 resonance with Neptune. Apart from this abundant resonance population there are some weaker resonances in the Kuiper Belt that seem to be populated by KBOs as well. Beyond about 50 AU, objects in low eccentricity orbits are not found; instead, the objects in the scattered disk have highly eccentric orbits of widely different semimajor axes and partially of high inclination. Some of the scattered disk objects (SDOs) reach several 100 AU solar distance. However, the perihelion of their orbits falls close to Neptune's orbit which suggests that this population is scattered from the Kuiper Belt by the giant planet. Dynamical calculations confirm this interpretation and indicate rather short lifetimes (order of 10–100 million years) of the objects in the scattered disk. The detached disk of KBOs has similar orbit characteristics as the scattered disk. However, the perihelia are at solar distances beyond the outer edge of the main Kuiper Belt suggesting that their existence may originate from the belt, but a series of scattering processes involving first Neptune, and possibly also a passing star may have "extracted" them in the peculiar orbits completely detached from the Kuiper Belt. Objects in the detached disk are more difficult to find since faint, and they should be less abundant than the CDOs. Last, but not least, there are the Centaurs objects, moving around the Sun in eccentric orbits between the Kuiper Belt and the inner planetary system. The Centaurs represent the inward scattered KBOs and provide a continuous influx of cold and icy bodies for final ejection or capture by Jupiter. The latter process recruits new members of the Jupiter family of comets (Sect. 13.8) [87].

13.7.3.4 Physical Properties of KBOs

Because of their faintness—a consequence of their sizes, albedos and solar distances—physical observations of KBOs are difficult and require the best and largest telescope facilities in the world. Size and albedo measurements revealed that the largest objects are a few 100 km to a bit more than 1000 km in radius. Fainter KBOs should be smaller in size down to around 30–50 km, equivalent to the approximate detection limit of 8–10-m class telescopes. The albedo of the KBOs is usually low and amounts to a few to some ten percent. Only (134340) Pluto and (136199) Eris and some other very large objects in the Kuiper Belt display extraordinarily high albedo suggesting that their surface reflectivity is significantly different from that of the other KBOs. Fresh ice layers (water, methane, nitrogen; see the (134340) Pluto and Charon spectra in Fig. 145) from recent resurfacing are likely to be responsible for the bright albedos of the large KBOs. Double-peak light curves measured in some KBOs, allow the determination of the rotation periods, which range from a few hours to several days, and suggests slightly non-spherical shape of the bodies. Binary systems seem to be relatively abundant in the Kuiper Belt. Their existence is explained to be due to capturing or collision of objects.

Measurements of photometric filter colors in the visible indicate a wide range of surface reflectivity ranging from almost solar to very red (KBOs are among the reddest objects known in the Solar System). This is confirmed by reflectance spectroscopy of the objects where mostly featureless spectra of almost constant slope were detected. It is noteworthy to mention that two distinctly different color populations were found among the classical disk objects, i.e., neutral to moderately red population and very red ones. The former KBOs seem to be also larger in size and they are in slightly excited orbits at inclinations above about 4°. This dichotomy among the Cubewanos may suggest different body constitution and possibly also different origin for the two populations (despite both are now found at about the same distance from the Sun).

Infrared spectra have revealed the existence of ices on the surface of several KBOs, namely of H_2O and CH_4. CH_4 ice seems to dominate the reflectance spectra of the largest KBOs, while water ice is found in the smaller ones (Fig. 145). This could be seen as an indication that resurfacing may take place on the larger bodies, possibly driven by intrinsic activity that may produce temporary atmospheres when the objects gets closer to the Sun and that freeze out when it recedes again [88].

(136199) Eris, an object larger than (134340) Pluto, is in the scattered disk and was found close to its aphelion passage at 97 AU.

(134340) Pluto was the first object in the Kuiper Belt for which a satellite, Charon, was found in 1978. In 2005 two more small moons, Nix and Hydra, were detected such that Pluto-Charon represents the first multiple system in the Kuiper Belt.

A thin, most likely temporary atmosphere was detected in (134340) Pluto through occultation measurements of Charon and of stars. The atmosphere is expected to disappear in one or two decades from now. Pluto's surface is mostly covered by N_2, CO and CH_4 ices, while Charon has a water ice surface.

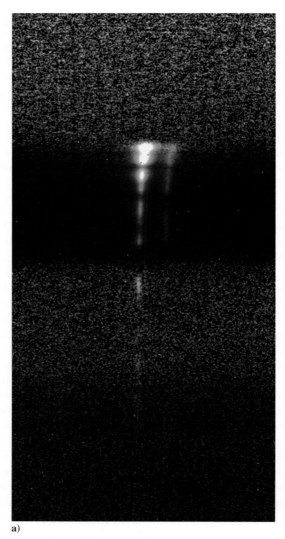

a)

Fig. 145. Surface spectra of Pluto and Charon. The close binary KBO is resolved by the adaptive optics instrument NACO at the ESO Very Large Telescope (VLT). **a** Near-IR spectra of Pluto (*left spectrum*) and Charon (*right spectrum*) covering the wavelength range from, *top* to *bottom*, 1 to 5 μm. Due to the faintness of the objects in L (Charon) and M bands (Pluto and Charon) the object spectra are hardly detectable in a single exposure. The spectra appear curved, in particular at the short wavelength ends, due to differential atmospheric dispersion. The gaps in the spectra along the wavelength range are mostly due to light absorption in the terrestrial atmosphere.

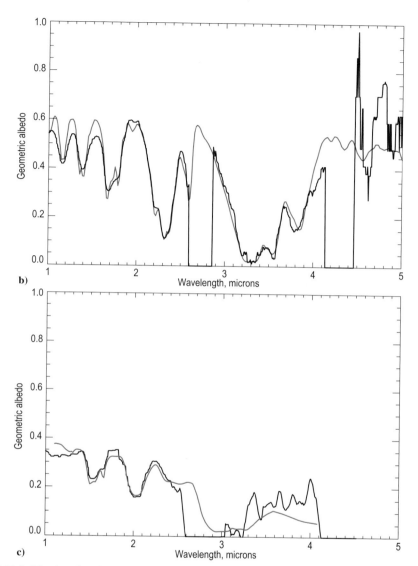

Fig. 145. b Wavelength-calibrated reflectivity spectrum of Pluto extracted from the two-dimensional spectrum of the *top* panel. The *dark line* shows the measured surface reflectance of the object, the *red line* provides a model spectrum trimmed to fit mostly the L band part. Most of the absorptions in the spectrum are due to methane ice on the surface of the dwarf planet except the one at 4.6 μm, which could be due to carbon monoxide ice and nitriles. Note that the sharp edges in the measured spectrum at 2.6–2.85 and 4.15–4.35 μm are caused by completely opaque absorption of light in the terrestrial atmosphere. **c** Wavelength-calibrated reflectivity spectrum of Charon extracted from the two-dimensional spectrum of the *top panel*. *Dark and red lines* as for *middle panel*. All absorptions in the spectrum are due to water ice on the surface of Pluto's moon. Note that the sharp edge in the measured spectrum at 2.6–3.0 μm is due to opaque absorption of light in the terrestrial atmosphere (2.6–2.85 μm) and the strong absorption of water ice of Charon's surface (2.85–3.0 μm) [89]

13.7.3.5 Formation and Evolution of the Asteroid Belt and of the Kuiper Belt The ecliptic-oriented geometries of both the Asteroid and the Kuiper Belts, are clear indications that both belts formed from the protoplanetary disk, though at different distances and possibly from bodies of different constitution. The Asteroid Belt was formed from silicate and metal-rich, but ice-poor material. Only in the more distant regions at the outer edge of the belt, asteroids may have been able to accrete volatile ices and organic material. The asteroids may thus resemble the best available planetesimals that formed the terrestrial planets. After the formation of the Main Belt at its current distance range, it got diluted and sculpted over time by Jupiter's gravity: Namely through the Kirkwood gaps, the majority of asteroids escaped, and they still do, from the belt over the lifetime of the Solar System such that the current belt may contain only a small fraction of the original mass. The early dilution process has possibly contributed to the heavy bombardment on Earth and on the terrestrial planets in the first few 100 million years of their existence. During this period of time the chances for importing organics and water to the planets should have been high. The existence of collision families and the asteroid size distribution suggests that mutual collisions play an important role for the bodies in the belt. In fact, it is very likely that the size distribution of the small to medium large bodies therein is dominated by collisions and only the largest objects in the belt may have kept their original sizes from formation time.

Not for sure, but very likely, the Kuiper Belt represents the outer part of the original formation disk of the planetary system around the Sun. It has accreted from a mass of some 10 to 100 Earth masses a sample of icy objects that themselves have been the planetesimals for the giant planet formation. Like for the Asteroid Belt, estimations of the current mass of about 0.2 Earth masses in the belt indicate a massive mass loss of the belt over the lifetime of the Solar System. In fact, this process is still on-going as is demonstrated by the detection of objects in the scattered and detached disks and of the Centaur population. Through the latter, also the supply of short-lived Jupiter family comets is guaranteed. Recent calculations of the orbital dynamics in the outer planetary system provide a new revolutionary scenario for the Kuiper Belt: It may have formed much closer to the Sun, possibly at about 2/3 of its current distance. During the clean-up process of the planetary formation disk by the giant planets, the outer planets Saturn, Uranus and Neptune may have migrated outward, and may have swept along the original population of planetesimals in the distance range, represented by the small and red Cubewano population. The bluish and moderately red Cubewanos got into the belt by a separate resonance transport process from regions closer to the giant planets.

There is no doubt that both the Asteroid and the Kuiper Belts are remnants from the formation disk of the planets, one at the inner and one at the outer edge of the giant planet region. Nowadays, both belts resemble very thin debris rings, much more diluted than the debris disks detected around other stars. The objects therein are of planetesimal size and may contain the most pristine and least altered material from the formation period of the planets.

13.7.4 Observing and Measuring Asteroids and KBOs

Observing and measuring asteroids and KBOs are challenges of two different kinds: Asteroids are much easier to find in the sky since their visible brightness can be quite bright.

(1) Ceres can be as bright as 7 mag and a large number of asteroids are well within the reach of amateur telescopes allowing for even quantitative measurements of the objects. KBOs are much more difficult to observe, since mostly due to their distance from the observer on Earth they are much fainter. Pluto-Charon, the brightest KBO, is about 14 mag, the next brightest objects are 17 mag and fainter and the majority of objects is well beyond 20 mag in brightness. Given their faintness we will not outline special observing and measurement techniques for KBOs. On one side, from the schematics point of view, they are very similar to those of asteroids. On the other side, amateur observations of KBOs may be restricted to only a few objects and in the first place restricted to astrometry and maybe light curve measurements. It should also be noted that a significant part of the detected KBOs are possibly "lost," since the few available astrometric measurements do not allow accurate orbit determination making recoveries of the objects some months after the last detection already very difficult and impossible if the objects are not observed for a year or more.

13.7.4.1 Search, Finding and Astrometry of Asteroids

The preparation of asteroid observations is not very much different from what is described for comets: One needs an accurate ephemeris and good finding charts of the area where the object is expected in the sky. It will be very useful to indicate the expected motion direction in the finding chart in order to ease the object detection at the telescope. Identifying the objects is done by comparison with the expected background stars; there should be at least one object more in the field than seen in the finding chart, and this object should be at the position and moving in the predicted direction expected for the asteroid. Some care should be taken if the asteroid is close to its "stagnation points" in the sky, i.e., during the period when it changes motion direction before and after opposition. During these time periods the motion rate can be slow or even close to zero such that identification by motion becomes more challenging. Compared to comets that are sometimes unreliable in keeping the brightness prediction and the expectations in their appearance, asteroids will appear as a (slowly moving) star at about the predicted brightness.

In photographic or CCD camera images exposed over a few minutes at sidereal tracking speed, the asteroid motion will produce a trailed image of the object which eases the a posteriori detection (although it reduces the limiting magnitude of detection). It is recommended to ascertain the finding of the object by exposing an image series in order to have more than a single detection and to perform the confirmation of detection through object blinking (Sect. 13.8), at least for CCD images. Blinking of images centered on the star background also allows an efficient search for new or unexpected moving objects in the field-of-view of the telescope.

When finding a new object, it is important to verify that the object is not in the catalog of known asteroids. Using an accurate astrometric position of the object or at least an approximate one, the check is easily possible through the IAU webpage: http://scully.cfa.harvard.edu/~cgi/CheckCMT. Entering the observing time and measured position of the moving object, the computer program lists all known objects (asteroids, KBOs, comets) within a given distance to the target position. If no known candidate object is found, the object can tentatively be considered to be "new." One can compute its approximate ephemerides using the above-mentioned web references in order to further evaluate the object identification. One could even try to observe the possible can-

Ephemerides can be obtained from NASA's Jet Propulsion Laboratory JPL at http://ssd.jpl.nasa.gov/horizons.cgi or through Harvard Center for Astrophysics CfA in Cambridge/USA at http://www.cfa.harvard.edu/iau/MPEph/MPEph.html. Finding charts can be produced via the All Sky Digital Sky Survey, for instance at http://archive.stsci.edu/cgi-bin/dss_form, at http://archive.eso.org/dss/dss, or through http://www.nofs.navy.mil/data/FchPix/cfra.html.

Up-to-date information on unusual and critical asteroids that need to be measured can be found at http://www.cfa.harvard.edu/iau/Ephemerides/CritList/index.html and http://www.cfa.harvard.edu/iau/lists/Others.html. Information on near-Earth objects and which one to observe with highest priority can be obtained from the IAU bureau webpage http://www.cfa.harvard.edu/iau/NEO/TheNEOPage.html or from the European Space Agency http://spaceguard.esa.int/SSystem/SSystem.html. Reports on observations of interesting objects can be uploaded directly to the respective information centers.

didate again using the new ephemerides. In the case of newly discovered asteroids, it is useful (1) to re-observe the object (in particular if only a single detection is available), and (2) to report it without delay to the IAU bureau through the respective discovery form (http://www.cfa.harvard.edu/iau/DiscoveryForm.html). In order to enable recovery observations of hitherto unknown objects for which a few astrometric positions are available, the IAU bureau also offers a webtool that allows calculating a quick and approximate short-term ephemerid table using http://www.cfa.harvard.edu/iau/MPEph/NewObjEphems.html.

The success rate for new asteroid detections by amateur observers has decreased considerably over the past decade, since a number of sky patrol projects (LINEAR, the Lincoln Near-Earth Asteroid Research of the Massachusetts Institute of Technology (MIT); LONEOS, the Lowell Observatory Near-Earth-Object Search of the Lowell Observatory; SPACEWATCH, the Spacewatch Project of the Lunar and Planetary Laboratory of the University of Arizona) have taken up duties, scanning the whole visible sky frequently with 1-m-size telescopes and wide-field CCD cameras. With the start of operation of the Pan-STARRS telescopes (the Panoramic Survey Telescope and Rapid Response System of the Institute for Astronomy of the University of Hawaii) the domain for asteroid search by professional observatories will even be strengthened in the near future.

Amateur observers can provide very valuable contributions to asteroid research by follow-up astrometry of newly discovered objects and of those with uncertain orbits as well as for near-Earth asteroids. The basic principles of astrometry of moving objects are addressed in Sect. 13.8 on comets. Measuring asteroid positions should be even easier than doing astrometry of comets, since the former do not show a fuzzy appearance as the latter frequently do. The only complication for performing accurate position measurements could be the trail length of the asteroid. Shorter trails will allow more accurate measurements, but may be more limiting in brightness. Thus, the respective image series should make use of different exposure times. Besides that, one should obtain at least two good astrometric measurements of the object. Orbit determination of asteroids is performed for instance by the IAU bureau, NASA's Jet Propulsion Laboratory or, for near-Earth objects, by the Spaceguard Foundation on behalf of the European Space Agency (ESA).

13.7.4.2 Optical Characterization of Asteroids

Optical characterization of asteroids is an interesting research field that benefits from contributions of active amateur observers. Light curve measurements allow the determination of rotation periods, shape information and scattering properties of the surface when available over a wider orbit arc. Spectroscopy and polarimetry, although interesting, are challenging with small telescopes (below a 20-cm aperture) and will have difficulties measuring objects not yet observed with larger telescopes by professional astronomers.

Visual Brightness of Asteroids Visual brightness estimates by telescope-aided eye results in quick, but only first-order guesses on the asteroid brightness. Useful quantitative results nowadays come from CCD camera images and traditional aperture photometry using existing computer software. Time series of images allow the coverage of major, if not the whole rotation period of the object. The duration of the exposure series depends on the expected rotation period, the integration time and on the brightness of the asteroid. Depending on the telescope size and object magnitude, the brightness can be measured to better than 0.05–0.1 mag. With some luck the rotation period follows from a single night observations, i.e., when the body is asymmetric and rotates in a few hours. Otherwise and if multiple night observations are available, it is required using a numerical period search routine for data analysis. The evaluation of the results can become a complicated task depending on the data volume and structure and might be better done by expert scientists.

Rotation Light Curve A double-peak light curve with more or less symmetric half-period profiles is indicative for an irregular shaped body, while a single-peak light curve argues more in the direction of dominant albedo variations on the surface. Nonetheless, the light curve profile should be taken only as first guess on the reason of the variability. However, since smaller asteroids tend to be irregular in shape, shape-dominated light curve variations are more likely. Detailed shape reconstruction of the asteroid and the determination of the rotation axis require complete light curve measurements from several arcs along the orbit—and of course a sophisticated light scattering model of the surface, a task again for scientists to do. However, the provision of useful data can come also from amateur measurements. Multi-revolution coverage may even allow following the temporal evolution of the rotation period and axis. From a practical point of view, observations of fast (periods of a few hours) rotating objects are simpler and provide most likely better quality results. Long-period variations require accurate data calibration over several nights.

> Changes of the rotation period and axis orientation are caused by internal effects in the body, close passages to planets and radiation pressure effects. Models of the latter, for instance, predict a dichotomy in the rotation period distribution of small asteroids, which should rotate preferably very quickly (orders of a few ten minutes to a few hours) or very slowly (orders of several days).

The light curve amplitude is directly related—depending on the form of the light curve—to the main axis ratio of the body or the albedo difference on the surface:

$$a/b = 10^{0.4(m_{min} - m_{max})}$$

with a and b being the large and small axis of the body (that is thought to be approximated by a two-axes ellipsoid here), m_{min} and m_{max} being the minimum and maximum brightness of the object determined from the light curve. In principle, the same formula applies for the estimation of the albedo ratio in an albedo-dominated light curve, in that case "a" and "b" representing the albedo values of the bright and dark surface parts, respectively.

In very lucky cases one may even measure the light curve of a binary system which otherwise is only possible when the objects gets really close to Earth and/or by adaptive optics at large aperture telescopes.

Absolute Magnitude, Phase Function, and Opposition Effect An interesting behavior can be seen in opposition light curves of asteroids, i.e., the object brightens above expectations when getting to small (<5–$10°$) phase angles α. This is called opposition brightening. Expectation is an increase of the observed magnitude with decreasing distance from the Sun (r) and from Earth (Δ) and with phase angle α:

$$m = H(1,1,0) + 5\log_{10} r\Delta + \phi(\alpha) .$$

The phase angle α is the angle between the Sun and the observer as seen from the asteroid, and it is a standard output of many ephemeris programs. $H(1,1,0)$ is the absolute magnitude of the asteroid, i.e., its brightness as seen at 1 AU from the Sun and Earth and for $\alpha = 0°$. As defined by the IAU, the phase function $\phi(\alpha)$ is a bit more complicated, though still simplifying mathematical expression as

$$\phi(\alpha) = -2.5 \log\left[(1 - G)^* \Phi_1 + G^* \Phi_2\right]$$

with G as function parameter. In the formula above Φ_1 and Φ_2 are given by

$$\Phi_1 = \exp\left[-3.33^* (\tan \alpha/2)^{0.63}\right]$$
$$\Phi_2 = \exp\left[-1.87^* (\tan \alpha/2)^{1.22}\right].$$

G can be determined from a fit of the measured magnitude of the asteroid over phase angle α [90].

$H(1, 1, 0)$ is related to the size and the albedo of the asteroid. Unfortunately, it is not possible to determine the two parameters, size and albedo, from the visible measurements alone. If the albedo is known (for instance from measurements in the thermal infrared), one can estimate the equivalent radius of the body. The latter is the radius of a sphere that has the same light scattering cross-section as the asteroid observed.

In order to detect the opposition effect, it is required to measure the mean (i.e., properly averaged over the light curve profile) object brightness over a wider phase angle range, typically from 20–30 to 0° (or at least to the smallest phase angle possible). The

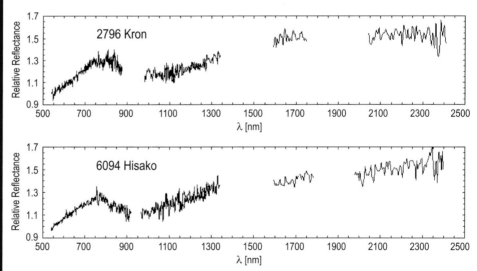

Fig. 146. Surface spectra of asteroids (2796) Kron and (6094) Hisako. The spectra show relative reflectance versus wavelength from about 600 nm to about 2400 nm, i.e., from the visible into the near-IR region. The gaps in the spectra are due to atmospheric extinction and instrumental properties. Both objects are S-type (silicate-type) asteroids. The spectra show a steep increase in the reflectance in the visible range and a flatter slope in the near-IR. The absorption between about 750 and 1300 nm indicates the presence of olivine and pyroxene-type minerals at the surface of the objects at different mixing ratios with olivine being more abundant in (2796) Kron compared to (6094) Hisako. Courtesy of A. Nathues (Max Planck Institute for Solar System Research)

brightening of the reflected light for small phase angles is due to two contributions, i.e., the shadowing effect and due to the so-called coherent backscattering. The former is simply described by the mutual shadows of the surface grains when illuminated by the Sun. The latter is an interference effect between light reflected by neighboring grains in the same direction. For non-zero phase angle positive interference of the light is small, while close to zero phase angle light amplification by interference may occur depending on the mean distance of the grains. By measuring the opposition effect it is thus possible to learn about the grain properties, i.e., the typical size and albedo.

13.7.4.3 Spectroscopy and Polarimetry of Asteroids Both spectroscopy and polarimetry are used by professional astronomers in order to constrain the surface composition and constitution of asteroids. In the visible wavelength range, spectroscopy allows the measurement of mostly the silicate composition of the surface at least for some objects. In particular, characteristic and wide absorption dips with minima at 900 and 1000 nm indicate the presence of pyroxene and olivine at the surface, respectively (Fig. 146). Polarimetry focuses on the light scattering properties and allows constraining the surface and grain albedo of the object.

For amateur telescopes spectroscopic and polarimetric measurements of asteroids are difficult, not so much in the technical performance, but due to the limited light flux of the objects. Moreover, most objects within reach for 30–50 cm telescopes are already measured and new surprising results are thus unlikely. If these types of observations are of interest, one can follow the approaches described in the section on comets (Sect. 13.8).

13.7.5 Further Reading

Barucci, M.A., Boehnhardt, H., Cruikshank, D.P., Morbidelli, A.: The Solar System Beyond Neptune. University of Arizona Press, Tucson (2008)

Belton, M.J.S., Morgan, T.H., Samarasinha, N.H., Yeomans, D.K.: Mitigation of Hazardius Comets and Asteroids. Cambridge University Press, Cambridge (2004)

Binzel, R.P., Gehrels, T., Matthews, M.S.: Asteroids II. University of Arizona Press, Tucson (1989)

Bottke, W.F., Cellino, A., Paolicchi, P., Binzel, R.P.: Asteroids III. University of Arizona Press, Tucson (2002)

Edberg, S.J., Levy, D.H.: Observing Comets, Asteroids, Meteors and the Zodiacal Light. Cambridge University Press, Cambridge (2008)

McFadden, L.A., Weismann, P., Johnson, T.: Encyclopedia of the Solar System, 2nd edn. Academic, Burlington (2007)

13.8 Comets

H. Boehnhardt

13.8.1 Naming Convention and Spacecraft Missions to Comets

Comets—named as "hairy stars" after the Greek word κομήτης, meaning "who carries long hair"—receive a lot of attention, both in the public when shining bright in the sky exhibiting an extended coma and tail, and in science since they are considered having preserved in the best possible way the original material from the formation period of our own Solar System. The comet phenomenon stands for a solar system object that moves around the Sun in an elliptical orbit and that consists of a solid nucleus, a gaseous and dusty coma, and three tails, i.e., a neutral gas tail, an ion tail and the dust tail. From these six different parts of the comet phenomenon, the nucleus and the neutral gas tail are usually not detectable even for well-equipped observers, while the gas and dust coma as well as the ion and dust tail are recognizable and even measurable with relatively simple devices by interested amateurs (see Fig. 147).

13.8.1.1 Designation and Naming Convention for Comets Since 1995 a new international numbering scheme and naming convention for comets were adopted by the International Astronomical Union (IAU). Periodic comets (periods usually up to a few hundred years), i.e., comets with a well-defined revolution period, are designated with a continuous integer number followed by letter "P/" and the name(s) adopted by the IAU for the discovers, for example, 1P/Halley or 67P/Churyumov–Gerasimenko or 196P/Tichy (for one of the first periodic comets discovered in 2008). Other comets, in particular those in highly elliptical, parabolic or hyperbolic orbits, receive a leading "C/" followed by the year of the discovery and an alphabetic letter designator for the half month when it was discovered (for instance "A" for first half of January, "B" for the second one and so on). Since usually more than one comet is discovered during the respective half month, the comets are numbered in sequence of their discoveries. At the end of the comet identification sequence, the name, adopted by the IAU for the discovers, is listed (for instance C/1995 O1 Hale–Bopp or C/2008 A1 McNaught). A maximum of three "discoverer" names are given by the IAU for a single comet. If the comet is permanently lost the letter designation is changed to "D/" (for instance D/1993 F2 Shoemaker–Levy 9 or 3D/Biela). Components of comets that have split are annotated by letters like for instance D/1993 F2-F for the sixth identified fragment of comet D/1993 F2 Shoemaker–Levy 9.

13.8.1.2 Space Missions to Comets Up to now, in total six comets were visited by spacecraft missions (Table 21). These missions returned a wealth of measurements which provided answers to many important and basic questions on comets. One of this mission (Stardust to comet 81P/Wild 2) has even returned cometary dust to Earth. However, the results have posed new questions for cometary research, now to be addressed by future spacecraft projects and by Earth-based observations. Currently, there are three spacecraft on their ways to three comets (Table 21) with the Rosetta mission being the most complex and comprehensive challenge in exploring these primitive objects from the early days of the Solar System. So far, all missions were fast flybys which provide only short-term and spatially restricted opportunities for the exploration of these cosmic vagabonds. Rosetta will change this situation, since in 2014 it will go into orbit around

A different designation scheme for comets was in place before 1995. The conversion from the old to the new designation scheme can be done via the webpage http://www.cfa.harvard.edu/iau/CometDes.html.

For information on the Rosetta mission see the weblink http://www.esa.int/SPECIALS/Rosetta/index.html.

Fig. 147. Comet C/1995 O1 Hale–Bopp in April 1997. The *bright cometary head* contains the coma of gas and dust emitted by the nucleus. The *dust tail* is slightly curved and appears white to yellowish due to the reflected sunlight. The *plasma tail* is straight and is bluish due to the light emission of CO^+ ions. Adapted from Wikipedia webpage on comet C/1995 O1 Hale–Bopp

comet 67P/Churyumov–Gerasimenko and will follow its path and evolution over more than 1 1/2 years from about 4.5 AU solar distance to perihelion at about 1.2 AU and beyond. In late 2014, when the comet has approached about 3 AU to the Sun, the Rosetta orbiter will release the 100-kg lander spacecraft Philae which is supposed to land on the surface and to perform in situ analysis and exploration of the physical conditions, composition, structure and processes of the cometary surface.

There is no doubt that spacecraft exploration of comets allowed and certainly will allow considerable progress in cometary and solar system science. Earth-based observations, however, provide the global framework of our current and future knowledge on comets, since the results of the few spacecraft targets can be put into context with the whole family of cometary objects.

Information on the missions listed below can be found via the weblink http://solarsystem.nasa.gov/missions/index.cfm.

Table 21. Past and future spacecraft missions to comets

Year	Comet	Mission type	Spacecraft
1985	21P/Giacobini–Zinner	Flyby	ICE
1986	1P/Halley	Flybys	GIOTTO[a], VEGA 1, VEGA 2, SAKIGAKE, SUISEI, ICE
1992	26P/Grigg–Skjellerup	Flyby	GIOTTO[a]
2001	19P/Borrelly	Flyby	DEEPSPACE 1
2004	81P/Wild2	Flyby, dust sample return	STARDUST[b]
2005	9P/Tempel 1	Flyby, impact	DEEPIMPACT[c]
2010	103P/Hartley 2	Flyby	DEEPIMPACT[c]
2011	9P/Tempel 1	Flyby	STARDUST[b]
2014–2015	67P/Churyumov–Gerasimenko	Orbiting, landing	ROSETTA[d]

[a] The GIOTTO spacecraft flew by two comets, i.e., 1P/Halley and 26P/Grigg–Skjellerup.
[b] The STARDUST spacecraft has visited its first comet 81P/Wild2 and is on its way to the second target (9P/Tempel 1) that will be encountered in 2011. A re-entry capsule has returned to Earth dust samples collected during the first comet flyby.
[c] The DEEPIMPACT mission performed a cratering experiment at 9P/Tempel 1 hitting its nucleus by an impactor. After the event the mother spacecraft that flew by the nucleus, continued its interplanetary travel towards another flyby at comet 103P/Hartley 2 in 2010. Comet 85P/Boethin, that was originally foreseen as second flyby target could not be recovered in time for proper mission planning.
[d] In 2004 ROSETTA was launched to the target comet 67P/Churyumov–Gerasimenko where the spacecraft orbiter will deliver in 2014 a lander (named PHILAE) for in-situ experimenting on the surface of the nucleus.
Note: NASA's CONTOUR mission, not listed in the table, was launched in 2002 and got lost during the injection maneuver for the interplanetary travel to meet comets 2P/Encke and 73P/Schwassmann–Wachmann 3. Also not listed in the table are NASA's GALILEO mission to Jupiter that managed to image directly the impacts of the fragments of comet D/1993 F2 Shoemaker–Levy 9 at this planet in 1994 and ESA's ULYSSES spacecraft that encountered the tail of comet C/1996 B2 Hyakutake in 1996.

13.8.2 Cometary Orbits and Comet Reservoirs

13.8.2.1 Cometary Orbits Comets usually have very extended, non-circular orbits around the Sun. Most of them are very to highly elliptical with eccentricity $e < 1$, clearly different from 0, i.e., the circular form; some are slightly hyperbolic ($e > 1$). The catalog of cometary orbits (Green 2005; [91]), issued by the Central Bureau of Astronomical Telegrams (CBAT) of the IAU, also lists a large number of comets with parabolic orbits ($e = 1$). The parabolic orbit approximation is used in cases where the orbit determination does not allow to distinguish between an orbit that is gravitationally bound to the Sun ($e < 1$) or not ($e > 1$). This situation can occur when the orbital arc of the object that is covered by position measurements, is short and/or when the astrometric measurements of the object have low accuracy as is the case for many of the comet apparitions in ancient history. From the more than 2000 comets registered until the end of 2007, none has an eccentricity much larger than 1 (maximum $e < 1.0005$), meaning that up to that year no "real" interstellar comet was found. Thus, all comets in that list are to be considered as long-lasting members of the Solar System. The objects in hyperbolic orbits appear to be affected by temporary non-gravitational forces, the so-called recoil effects due to out-

Interstellar comets originate from the formation disk of other stellar system and are ejected from the gravitational domain of the central star as is the case with some of the very distant comets of our own Solar System.

gassing of the nucleus when approaching the Sun. These forces modify the orbits from highly elliptical forms (though with $e < 1$) to slightly hyperbolic paths ($e > 1$) while the comet is passing close to the Sun.

13.8.2.2 Dynamical Classes of Comets

Based on the orbital elements and dynamical calculations, comets are classified in three dynamical classes, i.e., the short-period comets, the long-period comets and the Oort Cloud comets. The short-period comets have ecliptic-oriented, low-inclination prograde orbits with revolution periods of a few years. The majority of short-period objects belongs to the so-called Jupiter family of comets (JFC), since the evolution of their orbits is controlled by this planet. Hence, most JFCs have orbital periods of 5–6 years and make close approaches to Jupiter's orbit. By the end of 2007, 198 objects are cataloged as short-period comets.

The Oort Cloud comets (named after Jan Oort, a Dutch astronomer who identified this class in 1950 based on a statistical analysis of a sample of about 50 objects) have semimajor axes of around 10,000 to some 100,000 AU, almost uniformly distributed in inclination and perihelion orientation. The number of Oort Cloud comets larger than 2 km in diameter, is estimated to be more than 10^{12} objects, amounting in total to several 10 Earth masses. They are distributed in a spherical cloud of comets towards the border of the Solar System. The Oort Cloud comets are gravitationally bound to the Sun, but their orbit dynamics is influenced by gravitational interactions with neighboring stars, molecular clouds in the spiral arms of our galaxy and by the mass concentration towards the galactic center. The vast majority of Oort Cloud comets will never be observed from Earth, since their orbits do not reach into the planetary system. Usually, only large enough objects (i.e., a few kilometers in size) that get closer to the Sun than about 10 AU, have a chance to be detected from Earth.

The long-period comets are intermediate to the short-period ones and the Oort Cloud comets with orbits of semimajor axis between about 10 and about 10,000 AU with some preference for prograde, ecliptic-oriented orbits. However, they are not very much influenced either by the planets (namely Jupiter) or by the disturbing forces acting on Oort Cloud comets. Most of them are Oort Cloud objects that are "stranded" as long-period comets due to singular interactions with planets and/or a peculiar scattering from the Oort Cloud.

13.8.2.3 Comets in Peculiar Orbits and Other Comet-Like Objects

In recent years three peculiar classes of objects were identified: (1) the so-called sun-grazing and SOHO comets and two object types that might be closely related to comets; (2) a population of objects in short-period, elliptical orbits typical for comets, but with asteroid-like appearance (i.e., no coma and tail are detected); and (3) a few objects in the main asteroid belt, i.e., with typical asteroid-like orbits, but intermittent comet-like appearance (Main Belt comets).

The sun-grazing comets get very close to the Sun and are nowadays mostly discovered through searches in coronographic images of the SOHO solar observatory mission. Although these "sun-grazers" have long been known, SOHO has populated this family with a large (more than 1400 by early 2008) number of objects. The sun-grazer comets are in highly eccentric orbits forming three dynamical subfamilies. It is speculated that these families may actually have resulted from break-ups and/or dissolution of a larger parent body. Many sun-grazing comets do not survive the passage close to the Sun.

The population of asteroid-like comets is identified through the so-called Tisserand parameter T_J, a quasi-preserved constant of motion of a small body in the gravitational domains of the Sun and of Jupiter. The most prominent object in this population is asteroid (4015) Wilson–Harrington, first seen with a tail in 1949 and thus cataloged as comet 107P/Wilson–Harrington. These objects are considered "dormant" comets, i.e., objects, the continuous activity of which, has ceased or not yet started since the body may be covered by a protective surface crust.

The second small group of objects has orbits indistinguishable from those of main belt asteroids. Their comet-like appearance seems to be temporary. At least in one case, asteroid (7968) Elst–Pizarro also cataloged as comet 133P/Elst–Pizarro, the cometary appearance is "repetitive" since it occurred during close to perihelion passage when the object developed a dust tail in 1996, 2003 and 2007. During the rest of its path around the Sun, no activity is displayed.

From dynamical calculations it is expected that "dormant" (i.e,. inactive) cometary nuclei may also exist among the large group of near-Earth objects (NEOs). However, identification of such cometary nuclei is difficult and up to now no firm detection of a comet-like NEO is reported.

13.8.2.4 Comet Reservoirs and Formation Regions The Solar System contains two major reservoirs of comets, i.e., the Oort Cloud of comets and the Kuiper Belt region. The Oort Cloud is a continuous provider of comets that are scattered into the planetary system by the disturbing gravitational forces that also have shaped the cloud at the border of the Solar System. However, it is excluded that the Oort Cloud is also the region where the comets therein were formed (this would require an unrealistically large and dense initial formation disk of our Solar System). It is currently believed that Oort Cloud comets were formed in the region of the outer planets, i.e., in the distance range of the gas giants Jupiter, Saturn, Uranus and Neptune. During the clean-up period of the planetary disk at the end of the planet formation process, gravitational scattering by these planets has ejected a huge number of comets in the outskirts of the Solar System and even beyond into interstellar space. A large number of the scattered comets were captured in the Oort Cloud by the shaping forces mentioned above. Considering the number of stars in our galaxy and the likely similar planetary formation processes, an enormous number of interstellar comets like those from our own Solar System should exist. It may thus be surprising that among the more than 2000 comets cataloged there is no interstellar comet which undoubtedly came from another star. Such an interstellar comet can easily be identified, since the eccentricity of its orbit should be significantly different from meaning larger than 1. On the other side, these interstellar travelers may easily escape our attention since our "horizon" for comet detection is not very large, i.e., maybe 5 to 10 AU from the Sun, nowadays the typical discovery distance for comets [92].

The Kuiper Belt is considered the provider of most of the short-period comets. These objects are thought to have formed in the belt, being stored there since. Due to gravitational interaction with Neptune (close encounters or resonance motions) some of them are scattered out of their original orbits and can be injected into orbits that may bring them in the neighborhood of one of the other gas giants in the outer Solar System, where

T_J is expressed as $T_J = a_j/a + 2((1 - e2)a/a_j)^{1/2}\cos i$, with a and a_j as the semimajor axes of the object and of Jupiter, respectively, i as the inclination of the object with respect to Jupiter's orbit and e as its eccentricity.

yet another scattering (inward or outward) may occur. Finally, when arriving at Jupiter, they can be captured as a short-period comet or alternatively they are scattered outward again, leaving the sphere of influence of the planets. Dynamical calculations have shown that, statistically, the gravitational cascading of a Kuiper Belt object towards Jupiter orbit may take on the order of 10 million years. Moreover, the simulations indicate that only Jupiter is capable of recruiting a significant number of short-period comets, the so-called Jupiter family of comets JFC, which reside under its gravitational leadership. Kuiper Belt objects cascading towards the inner Solar System have eccentric orbits and, if discovered, they are classified as so-called Centaur objects (asteroid (2060) Chiron being the first and best-studied Centaur) [93].

13.8.3 The Nucleus, the Coma, and the Tails

13.8.3.1 The Nucleus: the Comet Machine Cometary nuclei are small irregularly shaped bodies with sizes between a few hundred meters to a few ten kilometers (Fig. 148). The surface albedo, i.e., the percentage of sunlight reflected at the surface, amounts to only a few percent. Hence, cometary nuclei are among the darkest bodies in the planetary system. Nucleus rotation periods of a few hours to days were measured, in some cases with indications—still to be verified—of an excited rotation motion, i.e., systematic, though complex drifts of the rotation axis of the body. The surface morphology, known from spacecraft imaging (Fig. 148), displays a wide range of landforms: flat and mottled terrains, mesas, crater-like structures, riffs, etc., still awaiting conclusive and comprehensive interpretations. Various hypotheses exist about the constitution of the nucleus interior ranging from a uniformly mixed body to a rubble pile structure. The composition of the nucleus material is concluded from observations of dust and gas during the phase of activity of comets when being close to the Sun: Water is most abundant (about 80%) among the ices of volatile species followed by CO and CO_2 and various minor, partly also organic, compounds. The non-volatile components consist of silicates like olivine, pyroxene, fosterite, and compounds of C, H, O, N that are likely to form organic species. The mass ratio between non-volatile and volatile material varies from 0.1 to 10 among the objects measured. Hence, cometary nuclei are frequently described as "dirty iceballs" or "icy dirtballs."

13.8.3.2 Cometary Activity The nucleus is the "machine for the activity" that creates the overall cometary phenomenon, the coma and the tails of dust and gas. Due to the eccentric orbits the interior of the nucleus remains deeply frozen (temperatures of $-200°C$ and below are expected from theoretical considerations) all around the orbit and also over its complete lifetime. Only the upper surface layers (a few meters) warm up due to the heating by the sunlight when the comet gets closer to the Sun. During perihelion passage in Earth distance, temperatures of $0°C$ and above may be common at the surface of cometary nuclei. Similarly, there is also a day-night cycle of the surface temperature.

The surface heating allows embedded volatile ices to evaporate and to leave the nucleus forming the gas coma. The evaporation process is called sublimation, since the molecules change from the solid directly to gaseous phase. It is expected that most of the released gas is produced in subsurface layers, since ices seem to be extremely rare at

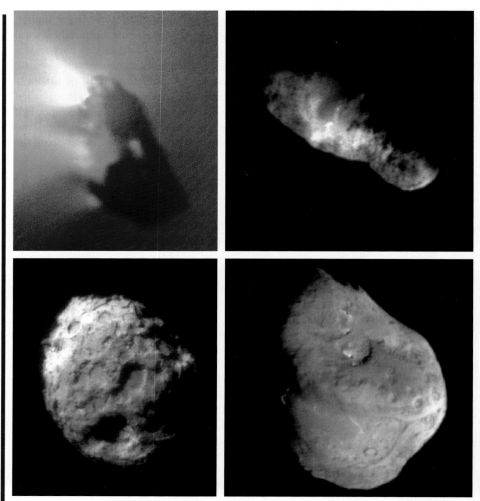

Fig. 148. Images of cometary nuclei obtained with flyby missions. (*Top left*) Comet 1P/Halley by ESA's GIOTTO mission. (*Top right*) Comet 18P/Borrelly by NASA's DEEPSPACE1 mission. (*Bottom left*) Comet 82P/Wild2 by NASA's STARDUST mission. (*Bottom right*) Comet 9P/Tempel 1 by NASA's DEEP-IMPACT mission. Courtesy of ESA for GIOTTO image, and courtesy of NASA for DEEPSPACE1, STARDUST and DEEPIMPACT images

the cometary surface. The gas stream from the surface lifts off dust grains and accelerates them away from the nucleus thus forming the dust coma. Although small grains of micron to submicron size are by far the most abundant dust emitted from the nuclear surface, most of the solid mass is actually lost in larger grains. Theoretical calculations indicate that in a highly active comet like 1P/Halley, grains of up to a few decimeter in size can be dragged away from the surface into the coma. During expansion the gas and dust coma overlap and form, together with the nucleus, the head of the comet. The acceleration zone of the dust due to the expanding surrounding gas ranges to about 10 nuclear radii. Further away from the nucleus the coupling of the gas and dust motion gets rapidly weaker. The overall activity depends on the nucleus composition (i.e., the amount of volatile gas and the dust-to-gas mixing ratio) and its variation across the surface, on the distance from the Sun and on the rotation of the nucleus (day-night cycle). Non-uniform and inhomogeneous distribution of volatile material is very likely, since regions of enhanced activity are seen in close-up images of cometary nuclei. Nucleus activity is usually strong when the comet is close to the Sun. Depending on the presence of volatile ices in the upper surface layers, it might cease during aphelion passage, also for short-periodic comets, or it continues out to solar distances beyond 20 AU if very volatile ices (for instance CO or CH_4 with low sublimation temperature) exist close to the surface.

13.8.3.3 The Cometary Coma
The cometary coma is fed by the nucleus activity, i.e., by the gas and dust released from the nucleus when heated by the Sun. During expansion the gas and dust is exposed to Sun illumination and to the particles from the solar wind. The coma mix of gas and dust represents a very thin temporary and unbound atmosphere around the nucleus that extends to several 10,000 to 100,000 km from the nucleus.

Dust grains absorb and scatter the sunlight, i.e., the grains heat up and shine in the infrared wavelength region beyond 3–5 µm; at the same time they reflect a smaller part of the sunlight giving the dust coma a close to solar color (i.e., yellowish) appearance. Light absorption and scattering imposes an acceleration force on the grains mostly radially away from the Sun. This so-called radiation pressure modifies the expansion tracks of the dust in the coma and pushes it away from the Sun into the direction of the dust tail.

Sunlight is also interacting with the gaseous species in the coma: On one side it splits up parent molecules into various daughter species, for instance H_2O into OH and H or into the single-charged H_2O^+ ion and a free electron. On the other side and at the same time, it excites molecules in a fluorescence process, i.e., the molecule absorbs light and is, for a short while, in an excited state after which it emits the light again returning into the original state. The light absorption and re-emission is very specific for the various atoms and molecules, i.e., it happens in very characteristic wavelength ranges. Hence, the re-emitted light from the coma gases shows the same characteristic spectrum and is thus very instrumental for the identification of the gaseous parent and daughter species in the cometary coma. The light emitted by the coma gas appears as emission lines and/or emission bands on top of the solar-like continuum that originates from the dust-reflected sunlight (Fig. 149). In the outer part of the coma neutral molecules get in contact with charged particles of the solar wind such that further reactions can occur (so-called charge exchange reactions).

13.8.3.4 The Cometary Tails
Comets can display tails, i.e., the dust tail, the ion or plasma tail, and the neutral gas tail (Fig. 150). The solar radiation pressure pushes the coma dust away into anti-solar direction, causing an extended linear or curved dust tail. The dust tail usually points away from the Sun. However, under special circumstances it can be projected as seen from Earth into the direction of the Sun. A sunward-pointing dust tail is called an "anti-tail." Like the dust coma, the dust in the tail region reflects solar light and thus it shines with close-to-solar colors. The charged gaseous species in the coma (ionized by sunlight or solar wind particles) interact with the solar wind in the

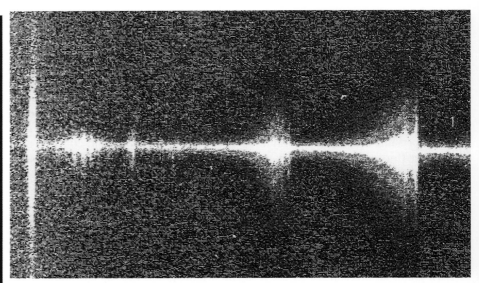

Fig. 149. Long-slit spectrum of comet 10P/Tempel 2 in the visible wavelength range between about 380–540 nm (*left to right*). The slit orientation (in *vertical direction*) is across the coma along the Sun-tail direction with the nucleus region close to the *center position*. The molecule emission appears as bright vertical bands, i.e., from *left to right* CN, C_3, CN, C_2 and C_2. The dust in the coma center produces a solar-type spectrum (*central horizontal band*). Courtesy of European Southern Observatory (ESO)

Fig. 150. The sodium, ion and dust tails of comet C/1995 O1 Hale–Bopp. The sodium tail appears narrow and straight in the left image (50° *left from top*), the ion tail is weak and has a streamer structure in the right image (45° *left from top*). Both images show the diffuse curved dust tail pointing towards the top of the frames. The sodium and the ion tails shine in emission lines of different wavelengths and are imaged through special narrowband filters. The dust tail reflects the full solar spectrum and is detectable through both filters. Courtesy of Instituto de Astrofisica de Canarias

outer part of the coma, namely with the magnetic field that is embedded therein. Since physical laws do not allow easily the motion of charged particles perpendicular to magnetic field direction, the cometary ions have difficulties to cross the solar wind magnetic field lines, while they can easily move in parallel direction. The solar wind flow at Earth distance is mostly away from the Sun. Thus, cometary ions will follow this flow, forming a long, relatively narrow and straight ion tail that points approximately opposite to the direction to the Sun. Since the solar wind contains turbulences and waves, the cometary ions follow the corresponding local field lines of the solar wind which can cause surprising geometric structures in the ion tail (streamers, kinks, waves, clouds). Like the coma gas, the ions in the ion tail shine in specific emission lines and/or bands, in the visible wavelength range mostly in the blue CO^+ emission and the red H_2O^+ emission bands. Both the dust and the ion tail can extend to large distances from the cometary nucleus (several million kilometers to even an astronomical unit AU and more) until they disappear in interplanetary space.

Last but not least, also the neutral atoms experience a repelling force from sunlight when shining in fluorescence light. In particular for sodium atoms, this force is very strong, blowing them rapidly out of the gaseous coma environment into anti-solar direction. Thus, a linear extended neutral gas tail can form. Since atoms that are very sensitive to such repelling forces (for instance sodium, Na) are not very abundant in comets, the neutral gas tail is very thin and faint. In fact, it was imaged for the first time during perihelion passage of the bright comet C/1995 O1 Hale–Bopp in 1997.

13.8.4 Observing and Measuring Comets

Nowadays, comets are observed over a wide wavelength range, i.e., from X-rays, over the ultraviolet, the visible and near-infrared, the thermal infrared to submillimeter and radio wavelengths. Depending on the wavelength range, different physical phenomena in comets can be studied. Still the visible wavelength range from about 370 to 1000 nm is by far the most important domain for research. It is this wavelength region where most amateurs and laymen observe comets and provide very valuable contributions to cometary science. These contributions complement the work of professional astronomers and scientists, for instance, in searching and discovering comets in the sky, measuring astrometric positions for orbit determination, estimating and measuring the brightness of the objects, as well as monitoring and documenting the optical appearance of comets.

13.8.4.1 Orbit Predictions and Finding Charts
The most essential information that is needed for successful comet observations are proper and accurate predictions of the comet position with time, i.e., so-called ephemerides. They list for given time stamps the position of the object in the sky, i.e., its right ascension and declination for the equinox of the chosen coordinate system (for instance astrometric or apparent coordinates) plus some complementary information like visual magnitudes, etc.

A useful complement to ephemerides are called finding charts, i.e., maps of the sky region where the comet is located showing also the location of background stars. Finding charts are particularly useful for fainter targets, since they allow to compare the stars in the circumference of the comet in the map to the actual constellations observed. Since comets are moving targets, a good finding chart provides, apart from the star background, orientation and scale that is adapted to the telescope field-of-view (maybe $1.5\times$ larger); it also provides the motion direction of the comet, or even better it plots the track of the object over time.

Comet ephemerides can be obtained through the HORIZONS tool of NASA's Jet Propulsion Laboratory JPL (http://ssd.jpl.nasa.gov/horizons.cgi) or through the IAU office at the Harvard Center for Astrophysics CfA in Cambridge/USA (http://www.cfa.harvard.edu/iau/MPEph/MPEph.html). Ephemerides for selected comets are found through several weblinks (for instance at http://www.cfa.harvard.edu/iau/Ephemerides/Comets/index.html).

Basic finding charts can be produced semiautomatically using the Digital Sky Survey DSS, for instance through the web addresses at the Hubble Space Telescope Science Institute STSCI (http://archive.stsci.edu/cgi-bin/dss_form), at the European Southern Observatory ESO (http://archive.eso.org/dss/dss); the US Naval Observatory catalogue of stars USNO (http://www.nofs.navy.mil/data/FchPix/cfra.html) also provides finding charts for the background stars.

For information on the various search projects see the respective webpages:
http://www.ll.mit.edu/mission/space/linear for LINEAR,
http://asteroid.lowell.edu/asteroid/loneos/loneos.html for LONEOS,
http://spacewatch.lpl.arizona.edu for SPACEWATCH,
http://pan-starrs.ifa.hawaii.edu/public for Pan-STARRS.

13.8.4.2 Searching and Discovering Comets

In the past, i.e., until the mid-1990s, searching for and discovering comets was the domain of individual observers, both professional and amateurs, who spent an enormous amount of time and patience in order to find a fuzzy moving spot in the sky, thereafter named after the discover(s). Nowadays, comet discoveries come routinely from professional search and patrol programs using 1-m-class telescopes to find and catalog small bodies in the Solar System, i.e., asteroids, near-Earth objects and comets. Looking through the naming list of recently discovered comets indicates the search programs in question: LINEAR, the Lincoln Near-Earth Asteroid Research of the Massachusetts Institute of Technology (MIT); LONEOS, the Lowell Observatory Near-Earth-Object Search of the Lowell Observatory; Spacewatch, the Spacewatch Project of the Lunar and Planetary Laboratory of the University of Arizona. In the very near future, Pan-STARRS, the Panoramic Survey Telescope and Rapid Response System of the Institute for Astronomy of the University of Hawaii will join here and will possibly provide a significant share of discoveries. The advantages of these search programs are telescope aperture size, wide-field imaging, all-sky coverage at short repetition cycles, very good observing sites and automatic computer-aided search and detection techniques for moving objects. The detection limit of the comet search programs is currently around 18–20 mag now and will reach 21 mag and fainter soon.

Amateur searches and discoveries are getting rarer, but are still possible as is documented for instance in the webpage "Catalogue of Comet Discoveries" (http://www.comethunter.de). Large sky coverage is essential for success. Visual searches for comets require good memory of the star constellations to faint magnitudes and well-trained recognition skills to spot a diffuse object that could represent a (new) comet. Although by using wide-field and fast f-ratio binocular or telescope optics, the human eye can develop a remarkable efficiency in pattern recognition for comet search, the detection depth is at the same time very much limited by the inability of the eye "to integrate longer." Here, telescopes equipped with wide-field cameras and CCD detectors, provide significant improvement. Before the era of automatic patrol telescope searches started in the 1990s, a trained amateur observer had to spend on the average about 1000 hours of night sky searching in order to discover—with great skills and even greater luck—a new comet that was not observed before.

The search using CCD imaging requires some planning, systematic observing and a careful, possibly computer-aided inspection and analysis of the images. The planning should aim for good sky coverage; the systematic observing should try to implement the respective exposure series, and the computer-aided inspection and analysis should provide the confidence in the detection. Traditionally, comet detections are done via two criteria: object motion and object appearance. The comet motion can be found by aligning on the computer two or more images of the same sky region such that the stars fall on the same screen pixels and subsequent blinking of the images in sequence of their exposure, i.e., displaying the individual images automatically one by one. Blinking frequencies could be around 1 s for a first inspection, a bit longer (5–10 s) for a deeper and more careful investigation. The images should be taken such to allow the object to have moved in between the exposures by at least more than the typical seeing, i.e., usually and with the exception of very distant objects integration times of a few to 15 min are sufficient. Image series of the same field during a single night are advantageous over series taken during different nights, simply because the former usually allow the detection of the object in more than a single exposure. However, also the latter is worthwhile, since it allows a complementary search through archived images. Search areas should avoid the so-called region of stagnation of the object motion, i.e., areas in the sky where the motion rate of objects is close to zero. The stagnation regions can be coarsely estimated for objects close to the ecliptic assuming circular orbits of different semimajor axes. Since short-periodic comets tend to have ecliptic-oriented orbits, they will mostly move at low ecliptic latitudes except when getting close to Earth. Long-periodic and Oort Cloud comets will appear all over the sky with about equal probability.

After detection it is important to determine the astrometric position (i.e., right ascension and declination for the mid-exposure time of the image; see Sect. 13.8.4.3) of the moving object for as many images as possible, but at least two exposures to allow an estimation of the motion rate of the object. The position measurements can also be used to perform a first check whether the object is known in the databases of the IAU Central Bureau of Astronomical Telegrams at Cambridge, USA. The IAU tools provide a list of Solar System objects (planets, asteroids, comets, Kuiper Belt objects) within a given angular distance to the astrometric position where the potentially new object is found in the sky. In order to cross-check motion rates of potentially new objects, it is advisable to run short-term ephemeris for the observing nights in order to get predictions of the astrometric positions including uncertainties, the motion rates and visual magnitude. Candidates for new discoveries should be reported immediately to the IAU Central Bureau for Astronomical Telegrams using the webpage http://www.cfa.harvard.edu/iau/DiscoveryForm.html. Note that one should provide apart from various (minimum two) astrometric positions and rough magnitude estimates of the object also a description of the optical appearance of the comet, like coma size, tail extension and direction, peculiarities like coma and tail structures. For those observers, who believe to have found a new comet or recovered a known object that has not been seen for a long time, and who are unable to perform accurate astrometric measurements of the object themselves, it is possible to report approximate positions to the IAU bureau. The bureau will then try to get independent sightings and astrometric measurements of the comet from other sources. Approximate positions of the object can be obtained by applying traditional visual position measurement techniques as described in [94] or in the minimum by plotting the comet positions in the star chart and determining the approximate right ascensions and declinations by interpolating between star positions. Although less accurate than required, if reported rapidly, they will help finding the object and to allow accurate astrometric positions to be obtained, thus securing data needed for the orbit determination of the comet. It is important to report accurate timings of two or more positions of the comet in order to allow a rough estimation of the motion direction for subsequent recovery observations.

A confirmed discovery of a new comet, in exceptional cases also its recovery, is reported in an IAU Circular by the IAU Central Bureau of Astronomical Telegrams giving proper references to the discoverers and the discovery circumstances. The IAU Circulars are the official channel for first and rapid announcements of such discoveries and are distributed by the IAU bureau to the astronomical community via e-mail and in printed version. Note that the IAU bureau reserves the right to check and verify discovery noti-

This check can be done basically online through the Internet by using the Web address http://scully.cfa.harvard.edu/~cgi/CheckCMT or http://scully.cfa.harvard.edu/~cgi/CheckMPC.

A proper weblink for a useful tool is http://www.cfa.harvard.edu/iau/MPEph/NewObjEphems.html.

The IAU Circulars can be found on the Internet at http://www.cfa-harvard.edu/iau/cbat.html. The circulars provide different kind of important information on opportunity observations and detections like comets, asteroids, planetary satellites, variable stars, novae and supernovae and more. Access requires a user account and is password protected which can be obtained from the IAU bureau by payment of the respective access fee.

fications before publication and that the cost for the publication might be charged to the person who issued the discovery note.

Useful contributions from amateurs could also come from searches for so-called lost comets, i.e., objects which appeared in the past, but were not seen at the expected time and at the predicted place in the sky for a long time (i.e., several apparitions) despite good visibility windows in the past. Examples are comets 85P/Boethin or long-periodic comets last seen decades (or centuries) ago like P/1994 J3 Shoemaker 4. Recoveries provide valuable improvements of the orbit knowledge of these comets. Another rewarding contribution of new discoveries, now even without performing one's own night-time observations, is the search for comets in the coronographic images of the SOHO solar mission. References can be found at the SOHO comets webpage (http://ares.nrl.navy.mil/sungrazers/) including instructions for searches and reporting. The comets to be found are of a special type, i.e., so-called sun-grazing comets (see Sect. 13.8.4.3). The family of SOHO comets is very abundant and is found when shining up while passing perihelion close to the Sun within a few solar radii. Most objects do not survive perihelion and orbit determination as well as the analysis of their physical properties is difficult, thus any new contribution may have the chance to make a difference. As of 2008 sun-grazer comets will also be found in images of NASA's solar observatory STEREO.

13.8.4.3 Astrometry and Orbit Determination of Comets

Astrometric measurements of comet positions in the sky are very useful, in particular if they extend over a long orbital arc and are performed with sufficient (i.e., subarcsecond, ideally 0.1–0.2) accuracy. Such measurements are used to compute and to refine the orbital elements of the objects including an estimation of the non-gravitational forces due to coma activity. While orbit calculations are mostly the domain of professional astronomers, astrometric measurements can be performed by educated amateurs with a reasonable investment of effort in learning how to perform the job properly, by devoting the time for the required observations and for the subsequent measurements and, last but not least, by investing the money to get the proper tools for doing a good job. Visual and/or photographic observations, that were competitive techniques of cometary astrometry in the last millennium, are described in [94]. Nowadays, telescopes with fast and wide-field optics equipped with CCD cameras allow better quality and more reliable astrometric measurements. Key specifications of the observing equipment are: telescope aperture of 15 cm and larger, in order to reach faint magnitudes (minimum 15 mag); CCD camera field-of-view of 10 arcmin and more, in order to cover enough background stars for the determination of the so-called astrometric parameters of the images; CCD pixel resolution of 1.5 arcsec and better, in order to reach acceptable position accuracy. The integration times of the comet images have to be a compromise between exposure depth (to reach faint targets) and not over-exposing the comet and the background stars (since this will make the determination of the object brightness centers during the astrometric reduction difficult or even impossible and in any case less accurate). Yet another compromise has to be made by the observer, i.e., whether to guide the telescope during the exposure at the stellar rate or at the motion rate of the comet. For brighter comets with a well-defined coma center, stellar tracking is preferable in order to reach optimum astrometric accuracy. A sky field with a fainter comet inside might be better exposed with differential tracking at the ob-

A guide to minor body astrometry can be found on the IAU bureau website at http://www.cfa-harvard.edu/iau/info/Astrometry.html. Dedicated astrometry software is also listed therein.

ject speed. An important issue for the quality of the measurement is the timing accuracy of the exposure, namely to be able to record the mid-exposure time to (at least) 1 s accuracy. Although a bit simplifying, for the actual observing it is important to reiterate the classical statement: One observation is no observation, two observations show the truth and the capabilities. Hence, it is essential to obtain two and more exposures for astrometric measurements of the same comet per night or at least per visibility. The IAU bureau who collects the astrometric measurements of comets, requires at least two astrometric measurements per object from a single object and/or observer (independent whether professional or amateur): a prerequisite for recognizing measurements as firm detections.

The actual measurements of the comet positions are done on the computer with the help of astrometry programs. Some commercial software packages for processing of astronomical CCD images allow basic astrometric reductions. The main tasks of the astrometric software packages is to find the moving object, usually through blinking or comparison of exposure series of the same field, to extract and identify the pixel positions of the objects (background stars and moving targets) and to relate them to catalog stars with accurate positions, to determine the astrometric parameters of the exposure like image scale and optical distortions, and last but not least to estimate the right ascension and declination of the comet and other moving objects in the field-of-view.

Reporting of the astrometric results is done via e-mail to the IAU bureau at mpc@cfa.harvard.edu. This office collects all position measurements of solar system bodies (i.e., not only of comets). A fixed format is required for reporting the astrometric measurements that is described in the Web reference mentioned above.

The orbit determination of comets is performed by experts who use the available astrometric positions of the respective objects for an estimation of its orbital elements for a certain epoch, i.e., for instance of the perihelion distance, the eccentricity, the ascending node, the argument of perihelion, the inclination and the perihelion transit time together with the non-gravitational parameters. The result fits the object measurements the best (for given model assumptions adopted by the modeler). A useful output for both the analyst and the observer are the residuals that describe the difference between measured and calculated position of the object using the best-fit orbit. Residuals of a good observer scatter around the calculated positions of an object within the measurement uncertainties in a statistical manner, i.e., without systematic trends (like for instance being always a bit off to the west or south). The astrometric measurements of comets that are reported to and accepted by the IAU bureau, are published in the Minor Planet Circulars (MPC) referencing the observing station, observers and measurers as well. The MPCs provide also recent determination of orbital elements of comets and other minor bodies.

The respective information can be found at http://www.cfa.harvard.edu/iau/MPCSummary/MPCsummary.html.

13.8.4.4 Optical Characterization of Comets

Optical characterization of comets provides a wide field for monitoring the specific, mostly unpredictable behavior of comets while passing around the Sun. Here, amateur astronomers can also contribute to scientific research since professionals, despite having much better technical equipment in principle available, usually lack frequent and continuous access to telescopes and instruments such that only short-term observing runs spread over longer time intervals can be accomplished. Amateur observers are able to complement the scientific observations of comets performed at larger observatories. The observers' monitoring can focus on the comet brightness, the coma activity, the tail appearances and on comet splitting events. Some simple numerical tools and procedures (photometry and structure enhancement techniques) will help enabling profound and high-quality detections and contribute more accurate measurements for the physical interpretation and scientific modeling of the comets (although the latter will be usually in the hands of professional researchers).

Visual Brightness and Overall Activity of Comets The most straightforward estimation that can be achieved by an observer is that of the cometary brightness. Usually when active, comets do not appear as stellar objects due to the presence of a surrounding diffuse coma. Hence, the traditional brightness estimation by eye through comparing the comet with neighboring stars, needs a small modification: defocusing of the telescope, to let the stars appear with similar average "surface brightness" as the comet such that one can compare them by guessing relative brightness differences. The brightness ratio method requires some training to reach reliable and good quality results. Well-experienced visual observers can reach an accuracy in the brightness estimation certainly better than 0.5 mag, although systematic differences among individuals, also depending on the optical equipment used, may exist. Furthermore, visual estimations of the observer can characterize the diameter of the coma, the condensation degree in the coma center and the extension of the dust tail, in favorable circumstances also of the ion tail (while the neutral gas tail remains mostly undetectable to human eyes). The coma diameters and tail extensions are estimated in degrees, usually by comparison to known distances between close-by background stars. An important interfering aspect for this estimation is the brightness of the background sky. It is thus not surprising that observers estimating in full moon sky or from inside big cities with lots of light pollution or with small telescopes, find smaller extensions than colleagues who observe under dark sky conditions using larger and fast f-ratio optics. Estimations of the condensation degree in the coma center use a 10 grade scale from planet-like (condensation degree = 9) to diffuse without clear brightness center (condensation degree = 0).

See for instance the comet sections of the British Astronomical Association and the Society for Popular Astronomy (http://www.ast.cam.uk/%7Ejds), of the Unione Astrofili Italiani (http://comete.uai.it), of the Vereinigung der Sternfreunde (http://kometen.fg-vds.de), of the Sociedad de Observadores de Meteoros y Cometas de España (http://www.somyce.org), of the Nederlandse Kometen Vereniging (http://www.kometen.nl) to mention at least a few. The above mentioned references are also good sources to get the latest brightness information on comets since they keep their archives very much up-to-date.

Visual brightness estimates of comets as well as other information obtained by visual observations from amateur observers, are collected in a data archive and are published in the International Comets Quarterly (ICQ) journal (http://www.cfa.harvard.edu/icq/icq.html). The format for reporting of the estimates is to be respected (for details see the ICQ webpage). Moreover, there are also amateur associations and groups observing, sharing, collecting and evaluating visual brightness estimations of comets.

The visual brightness of the comet reflects the activity status of the coma, here mostly that of the production of dust grains. Monitoring the total coma brightness over a wide orbit arc, best achieved in a combined effort by many observers, allows the reconstruction of the apparition light curve of the comet and the determination of some photometric parameters from it. Photographic brightness estimations are more difficult, since the sensitivity properties of the photographic emulsion used modifies the results which requires extensive and careful evaluation.

In the simplest and ideal case (note that there are also other formulas applied), the total coma brightness m_{coma} follows closely the mathematical relationship:

$$m_{coma} = M_{comet} + 5 \log_{10} \Delta + n \log_{10} r ,$$

where M_{comet} is the intrinsic brightness of the comet scaled to Sun (r) and Earth (Δ) distances of 1 AU. Δ and r of the comet can be found in ephemerides, m_{coma} is a set of coma brightness estimations. Linear regression least-square fitting—for instance available in office software tools on standard personal computers—allows the simultaneous estimation of M_{comet}, the intrinsic brightness of the comet as a measure of nuclear activity, and of n, the activity parameter of the comet that describes its increase or decay with solar distance. Both, M_{comet} and n, are characteristic for the individual comet. It is however noted that for an individual comet they may vary from apparition to apparition and even throughout a single apparition. It is also known that the intrinsic brightness M_{comet} of a comet can decay with time (usually over several apparitions), a phenomenon that is explained by the gradual exhaust of the driving gas for the activity and/or by crust formation on the nucleus that prevents gas emission from its surface.

Coma Photometry, the A$f\rho$ Parameter and the Dust Production of Comets With the availability of CCD cameras (even equipped with filters similar to the broadband filter set BVRI used in professional astronomy) and larger aperture telescopes, the more advanced observer is able to perform calibrated and quantitative measurements of the coma photometry and to achieve estimations of the cometary dust production. Further requirements are reasonably good sky conditions that allow flux calibration of the comet images taken through broadband filters.

Flux calibration implies that the filter raw images are bias, flat-field and sky-level corrected, and are flux calibrated through reference stars in the field-of-view. Integrating the coma flux in a circular aperture centered on the central brightness peak, allows to estimate a quantity called $Af\rho$, which is a measure of the dust-reflected sunlight and is proportional to the total effective light-scattering cross-section of all dust grains in the measurement aperture. In $Af\rho$, A stands for the average grain albedo, f for the filling factor, i.e., the percentage of the aperture cross-section that is indeed filled by dust grains, and ρ for the radius of the measurement aperture at the distance of the comet. From the measured coma flux F_{coma} one can derive the $Af\rho$ value according to

$$Af\rho = 4\Delta^2 r F_{coma}/(\rho F_{sun})$$

where Δ (in centimeters) and r (in AU) are the Earth and Sun distance, respectively, F_{sun} is the flux of the Sun at 1 AU in the filter band used for the comet exposure and the aperture radius ρ of the coma flux measurement is to be taken in centimeters at the distance of the comet. Strictly speaking, the $Af\rho$ measurement should use comet images taken through filters the transmission range of which does not contain (many and strong) emission bands of molecules of the coma gas. Broadband R and I filters have some contamination by weak gas emissions (mostly from NH_2), but are for many comets reasonably well-suited for measuring $Af\rho$ of the dust. Under the assumption that the cometary dust is produced continuously at a rather constant level and that it follows roughly a uniform, homogeneous and isotropic expansion scheme in the coma, it can be shown that $Af\rho$ is independent of ρ, meaning that its value is independent of the aperture size used for the measurement. In practice, the above-mentioned conditions are never strictly fulfilled, which results in some aperture-size dependency of the $Af\rho$ measurements of the coma. It is thus recommended to estimate $Af\rho$ for some reference aperture sizes comparing to 5000, 10,000 and 20,000 km at the distance of the comet. The

For a simple method of photometric reduction see for instance the paper by H. Mikuz and B. Dintinjana at http://www.astrovid.com/technical_documents/Comet_Photometry.pdf and some special instructions found at the ICQ website http://www.cfa.harvard.edu/icq/cometphot.html or through a full photometric reduction as for instance described by C. Sterken and J. Manfroid in *Astronomical Photometry*, published by Springer Press in 1992.

aperture radius ρ at the distance of the comet can easily be obtained via the geometric relationship

$$\rho = \Delta \times \tan(\alpha/2)$$

A database and descriptions of $Af\rho$ measurements of comets were recently published by the CARA project, a group of amateur astronomers within the Unione Astrofili Italiani (see the project webpage at http://www.cara-project.org/).

where α is the diameter of the aperture in the sky (in degrees) and Δ is the Earth distance of the comet at the time of the observation (in centimeters). Using wider apertures for the $Af\rho$ measurements will show effects from the radiation pressure on the dust, while for the smaller diameters the dust expansion is subjected less to modifications by the solar radiation. Measurements through very small apertures, however, may be affected by the atmospheric seeing.

The $Af\rho$ measurements (in centimeters) allow a very rough estimation of the dust production rate Q_{dust} of the comet (in kilogram/second) through the empirical approximation

$$Q_{dust} = Af\rho/C$$

with C being a constant (typical value of C is 9 s/kg cm). The dust production rate determined this way is at best an order of magnitude value for the amount of dust produced by the comet. In particular, it refers to the dust that is measured in the aperture by its reflected sunlight, i.e., it does not consider properly grains that are smaller than the typical wavelength of the reflected light and grains that are very much larger than about 100 μm. Both types of grains do not reflect very much sunlight and are thus difficult to be detected in the visible wavelength range. It is also noted that the C constant in the Q_{dust} formula above is badly determined and may be wrong by several factors.

The best and most accurate results are obtained using mathematical period estimation routines, which may not be easily available to amateur observers. Although a possibility for amateur research, the task is a difficult one and the risk for wrong estimations is high.

Rotation Light Curves and Nucleus Sizes of Comets Using an exposure series taken over a longer time interval can be used to explore whether the comet shows regular periodic variability. Here, contrary to the brightness profile with solar distance that can be obtained from visual estimates of the object magnitudes, a quantitative measurement of the coma light is required. In practice, small aperture measurements of the comet and a number of reference stars in the field-of-view with constant brightness are to be done and the relative brightness change of the comet has to be plotted and analyzed with respect to the set of reference stars. Great care must be taken to avoid introducing artificial brightness variations through variable atmospheric seeing affecting the aperture measurements. The brightness variability can be used to estimate the rotation period of the comet.

The estimation of the rotation motion of cometary nuclei is more easily possible through a measured light curve of the bare nucleus signal. When the comet is active and surrounded by a coma, it is virtually impossible for a ground-based telescope to detect the nucleus signal "on top" of the coma light. The "great disturber" for such a detection is the turbulence of the terrestrial atmosphere that produces seeing and the latter smears out the tiny nucleus signal among the much brighter seeing disk of the very coma center. Detection of the bare nucleus signal becomes possible when the coma activity has ceased, i.e., when the comet is further away from the Sun. Given the small sizes (order of a few kilometers or below) of most comets, the nucleus magnitude is very faint and usually beyond what is measurable through standard amateur telescopes. The proof of

absence of coma is another hurdle here, since it requires analysis of even deeper exposures. However, in very exceptional and fortunate cases (for instance when the comet
passes extremely close to Earth) it may be possible to measure the sunlight reflected by
the nucleus surface without coma or after coma removal. Using the measured nucleus
brightness $m_{nucleus}$ and assuming a canonical value for the nucleus albedo $A_{nucleus}$ (usually 4%), one can estimate the nucleus radius $R_{nucleus}$ (in meters) as

$$R_{nucleus} = 1.49610^{11} 10^{[0.2(M_{sun} - m_{nucleus} + 5\log(r\Delta) + b\phi)]} .$$

In the formula above M_{sun} stands for the filter brightness of the Sun at 1 AU, r and Δ
are the Sun and Earth distances of the comet measured in AU and $b\phi$ is a factor that
corrects for light losses due to the phase function of the nucleus, i.e., it accounts for an
incompletely illuminated nucleus of the comet as seen by the observer. The phase angle ϕ
(in degrees) can be obtained from the comet's ephemerides for the time of observations;
the empirical constant b is given in mag/deg and amounts to about 0.03–0.06 mag/deg for
some comets that could be measured. In the very, very lucky case that even the variability
of the nucleus signal can be measured, one can even assess—aside from the rotation
period of the nucleus—its nucleus shape and/or albedo variability over rotation phase.

Coma Appearance   Although at first sight the cometary coma may appear fuzzy and
rather uniform, a closer look may show interesting and time variable phenomena that
are indications of the on-going and changing activity of the cometary nucleus. Detection
of these phenomena is possible thanks to the high and linear sensitivity range of modern
CCD detectors, nowadays frequently used at medium-sized amateur telescopes.

The visual appearance of the coma is dominated by dust-reflected sunlight, hence
most of the coma phenomena addressed below are caused by cometary dust. Gas features are frequently related, but do not appear prominently except through special filters
that have transmission windows tailored to the wavelength intervals of the various emitting coma gases (CN, C_3, C_2, NH_2 in the visual wavelength range). At a first glance, the
decrease in brightness with increasing distance from the coma center is recognizable. On
the average the coma surface brightness falls off with $1/\rho$ and the total aperture brightness increases with ρ, the projected nucleus distance in the coma at the distance of the
comet (see p. 513). By integrating the coma light in concentric rings of increasing radius ρ and centered on the brightness peak in the coma, one can verify this relationship
and in many cases it applies well at least for comets at highest activity level close to the
Sun and despite all special features discussed below. Deviations indicate non-constant
and non-uniform material outflow in the coma.

The coma isophotes, i.e., lines of equal brightness in the coma, appear rather circular
close to the coma center and elongated towards the tail direction further away (usually
outside of about 10,000–20,000 km projected nucleus distance). Systematic deviations
from a smooth isophote pattern are due to inherent geometric structures, caused by localized material condensations in the coma. Various structures are known, i.e., jets, fans,
shells, clouds, arcs and arclets. Due to the low contrast of these features (a few percent)
compared to the bright coma background, they are not easily seen in normal coma images. However, some simple numerical processing of the images can enhance the coma
structures above background, thus enabling easy identification by the observer. Various enhancement techniques can be applied and, depending on the nature of the coma

Surprisingly, in the past coma
phenomena were also well-
recorded in the detailed drawings
made by experienced visual observers. It was mostly the usage
of photographic observations
with their rather narrow and
non-linear sensitivity range that
has questioned the reality and
correctness of the phenomena in
the drawings. CCD technology has
introduced a great revival here
and has opened a wide area for
contributions from well-equipped
and somehow skillful amateur
observers who can now follow
the time-variable, sometimes
unexpected features in the coma
of a comet over a long time
interval.

structures, they are more or less suitable with success, although there is no guarantee that a single method will enhance all coma structures equally well. The most popular techniques are unsharp masking, radial renormalization, and various numerical filtering methods. The common goal of these enhancement methods is to remove as much as possible the general coma background and at the same time to amplify any kind of deviations from it. It is advisable to apply different techniques to the same data set of coma images, since each method has its strengths and weaknesses depending on the geometry of the coma structures.

A description of the Larson–Sekanina method for structure enhancement can be found in *The Astronomical Journal*, vol. 89, 1984, pp. 571–578.

Unsharp masking uses a smoothed or rescaled version of the original image for subtraction or division of the latter. Smoothing has to be done with different box sizes, rescaling with different radial amplitudes. Radial renormalization uses the ring-aperture averaged radial coma profile for subtraction or division of the original exposure. A more complex enhancement technique is the Larson–Sekanina method that subtracts smoothed, shifted and rotated images from the original ones. Numerical filtering methods are available as part of astronomical data reduction packages. Adaptive Laplace filtering and wavelet transformations have been proofed to be very useful for coma structure enhancement. The afore-mentioned techniques have to be applied with great care, since they may produce different types of artifacts that could be misinterpreted as coma features. Yet another reason to apply more than one technique to the images! For instance, unsharp masking and radial renormalization, also the Larson–Sekanina method, are sensitive to centering errors between the two images that are subtracted or divided. Decentering between the two images can produce artificial structures close to the coma center. Numerical filtering may pick-up and enhance also trailed background objects in the coma or produce bull-eye structures around the coma center and/or around small localized brightness enhancements like for instance bad pixels or cosmic rays. Figure 151 shows examples, the isophote representation of the coma light as it appears in the comet image from the telescope and the inherent coma structures after some numerical processing of the data.

Figure 151 also provides examples of various structures found in cometary comae. The interpretation of the features requires some experience, and it is not straightforward to conclude on a possible interpretation just from some obvious similarity with structures seen elsewhere. Usually, one should apply numerical coma structure modeling in order to achieve a qualitative (and quantitative) understanding of the features and their interpretation, a process only a few experts in the world perform with success. Nevertheless, amateur images of comets can provide, with some numerical treatment, useful information on coma structures and may allow at least some qualitative understanding of the various features with respect to their nature.

Jets Different forms of coma jets are known, i.e., straight and curved (also called spiral) ones. Jets are usually connected to the central brightness peak in the coma, the location of the cometary nucleus. The "foot point" of the jet is believed to be related with a surface area of enhanced activity on the rotating nucleus. The rotation causes temporal changes in the position angle of gas and dust emission from the active region and maps into the geometric appearance of the jet (i.e., position angle, curvature). Hence, coma jets change appearance with the nucleus rotation which is recognizable easily in a series of exposures taken over time. Recurrent coma jets allow to constrain the rotation period

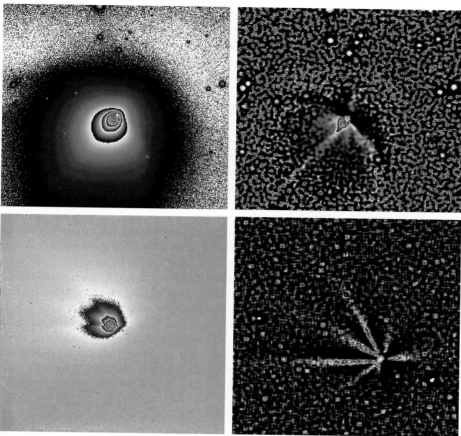

Fig. 151. Numerical enhancement of different coma structures. (*Top panels*) Component B of Comet 73P/Schwassmann–Wachmann 3. The *left panel* shows the isophote representation of the inner coma. The slightly asymmetric distribution of dust-reflected sunlight is an indication of inherent coma structures, which become obvious after numerical enhancement using adaptive Laplace filtering as shown in the *right panel*. The long central streamer of the dust tail extends in anti-solar direction while a short weak jet is pointing into the sunward coma. The structures sideways of the tail streamer are arclets from fragmentation events of the nucleus. (*Bottom panels*) Comet C/1995 O1 Hale–Bopp. The *left panel* shows very irregular isophotes in the inner coma. After some numerical enhancement this asymmetry appear as a porcupine pattern (*right panel*) of straight "jets," in fact the edges of coma fans produced by a number of active regions on the rotating nucleus. Courtesy of European Southern Observatory (ESO)

on the nucleus and the number of jets gives an idea of the number of active regions on the surface.

Fans Fans are sector structures of increased brightness in the coma. They are believed to be caused by an active region on the rotating nucleus that produces an emission cone of material in the coma. The cone when viewed side-on appears as a sector or fan of increased coma brightness. It frequently displays relatively sharp edges and a smoother and more uniform surface brightness within the sector area. The fan geometry gives a very rough idea on the orientation of the projected rotation axis (which should usually fall in the position angle range of the fan) and on the latitude of the active region on the nucleus (which may not be too far from 90° minus half the opening angle of the fan). The appearance of coma fans changes only slowly with time. Apart from some geometric information on the nucleus and its active regions, it does not allow constraining the rotation motion in detail. From numerical modeling of the emission cone of active regions one can conclude that a coma fan represents the side-on view, while curved jets (resembling spirals) provide the along-axis view of the same phenomenon.

Shells and Arcs   Shells and arcs are connected to jets and their day-night activity modulation over the rotation period of the nucleus. During the daytime period the active region on the nucleus is active, while at night it shuts down its activity due to the lag of solar heating of the driving gas that resides in the frozen ices of the nucleus. Hence, the jet displays a "switch-on-switch-off" pattern over a full nucleus rotation. With the new sunrise over the active region a new jet starts to evolve into the coma, while the jet material from the previous rotation has moved further away from the nucleus and may appear as an arc-like or shell-like structure depending on the range of position angle covered by the jet activity, in any case disconnected from the central brightness peak. A pattern of repetitive shells and arcs that can be traced to the same active region(s) on the nucleus allows the determination of the rotation period of the nucleus and of the projected expansion speed of the material emitted into the coma. Older shells and arcs disperse in width and extension, and sooner or later they disappear in the general coma background due to reduced surface brightness. Their geometry may also be influenced by the action of the solar radiation pressure. Shells and arcs are more frequently seen in the sunward coma hemisphere since they are related with nucleus activity driven by solar heating.

Clouds   Clouds can be seen occasionally in cometary comae. They have a smooth compact appearance, in some cases resembling small cometary comae embedded in the overall coma background. Depending on the motion speed of these clouds, that is directly measurable at least in projection on the sky, one can conclude the following: Fast moving clouds with velocities of the order of 100 m/s are caused by short-term localized activity of the nucleus, so-called activity puffs. Slowly moving clouds are more related with fragments in the coma that were once emitted from the nucleus and are now traveling close to its parent body through the coma, developing its own activity and an independent coma. Clouds related with fragments appear more frequently on the tailward side of the nucleus and their tracks tend to fall along the direction of the projected velocity vector of the comet in the sky (the latter can be obtained from comet ephemerides). Clouds from activity puffs tend to show up in the sunward coma hemisphere.

Arclets   Arclets are temporary phenomena in the cometary coma that are related with fragmentation events of the nucleus. When parent and daughter nuclei drift away from each other, they develop their own comae which both expand and, in the intersection surface between the two bodies, the expanding gases collide such that an enhancement of material is produced. With further separation of the subnuclei, the gas collision zone gets wider, but also weaker, and it gradually disappears with time within a few days after the fragmentation event. Arclets are recognized (in coma images of comets) as symmetric, slightly curved arcs, mostly in the tailward hemisphere of the coma. They indicate a recent (a few days old) fragmentation event, and one may expect the appearance of individual fragments a few further days or weeks after disappearance of the arclets.

13.8.4.5 Tail Observations

As already mentioned, from the three tail phenomena of comets only the dust and plasma (also called ion) tails are usually observable with amateur telescopes. The much weaker neutral gas tail is hardly detectable even with specialized equipment at larger telescopes. Observations of tail phenomena require wide-field and fast f-ratio optics, since the dust and plasma tails are fainter than the central coma of the comet and they cover a wider area in the celestial sky. In the cases of very bright comets the tail may even extend over a major part of the celestial hemisphere seen by the observer.

While the comet is further from the Sun both the ion and the dust tail orient very much along the Sun-comet direction, usually pointing away from the Sun. Then, plasma and dust tails overlap with each other and they are difficult to distinguish by appearance. When getting closer to the Sun and in particular for a more top-on view of the cometary orbit as seen from Earth, the ion and dust tail may appear separated in the sky (i.e., extending at different position angles with respect to the nucleus). Unfortunately, this favorable viewing geometry is mostly only the case for long-period comets with high inclination orbits or for short-period ones when they get very close to Earth.

Ion Tails (Fig. 150) are known to be visible only when the comet is closer (<1.5 AU) to the Sun. They shine with bluish light, caused by CO^+ molecule emissions of the tail plasma, and the ion tail position angle remains close to anti-solar direction. Bright comets can develop a rich variety of ion tail structures from so-called narrow tail rays that are connected to the comet head and fold in towards the tail axis with time, over "clouds," "knots," "waves" and "kinks" in the plasma tail itself. These temporary phenomena are all related with localized mass loading in the tail, caused by nuclear activity, and its interaction with the solar wind and its magnetic field. By measuring projected distances of these structures from the nucleus, it is possible to determine their velocity in the plasma tail and even their acceleration by the solar wind interaction along the tail axis. The repetitive appearance of clouds in the ion tail of 1P/Halley even allowed to conclude on the rotation motion of the nucleus. So-called disconnection events describe the separation of the plasma tail from the comet head and the subsequent formation of a new ion tail from the coma. It is believed that disconnection events are caused by passages of the comet through a reversal zone of magnetic polarity in the solar wind. The old tail is kept embedded in the magnetic field environment of the "old" solar wind plasma, while the "following solar wind plasma" carries opposite magnetic polarity that imposes opposite gyration of the cometary ions along the frozen-in magnetic field lines of the solar wind.

Dust Tails Dust tails appear yellowish since they are seen due to solar light scattering at cometary dust grains when moving away from the comet into interplanetary space. The diffuse appearance of the dust tail reflects the rather uniform distribution of the (mostly micron-sized) grains in the tail. When the comet is far from the Sun (>2 AU), the dust grains in the tail move mostly radially away from the Sun, driven by the solar radiation pressure. Closer to the Sun, the grain motion in the tail also reflects the emission velocity of the dust along the orbit that is dominated by the orbital motion of the nucleus. The grain motion "bends" away from radial direction into the orbital plane of the comet. Thus, when the comet is near perihelion, the dust tail can point by a significant amount out of anti-solar direction. Unfortunately, for short-period comets this situation that is in principle advantageous for dust tail analysis, is counteracted by the typical side-on projection geometry as seen from Earth, since these comets have low inclination with respect to the Ecliptic. In this case, the dust tail projects mostly along the Sun-comet line and tends to overlap with the ion tail of the comet.

The dynamical structure of the dust tail is characterized by so-called synchrones and syndynes. The former (i.e., the synchrones) represent the location of dust grains in the tail that were released from the nucleus at the same time, but are subject to different solar radiation pressure due to different size and/or material properties. Syndynes are lines of location of dust grains that are exposed to the same radiation pressure force, but are emitted at different times from the nucleus. Both, the synchrones and the syndynes patterns allow a complete dynamical description of the dust tail of a comet. Dust tails are found to contain grains of a few micron in size and emitted over a few weeks. In particular, when seen top-on, the dust tail can tell a long and detailed story on the dust properties (syndynes) and temporal activity (synchrones) of the comet along its orbit. More specifically, temporary short-term changes in dust emission activity along the orbit may appear as so-called streamers in the dust tail, i.e., they fall along synchrones that can be traced back to the cometary nucleus. Regular changes for instance from rotation motion and related variable activity due to different illumination of major active sources on the nucleus can produce a regular pattern of dust streamers in the tail. Dust features following syndynes in the tail are hardly seen which may indicate that the grain size spectrum may not change too much over the phase of activity represented by the dust tail. Nonetheless, the maximum extension of the dust tails allows to constrain the solar radiation sensitivity of grains in the tail that may even exceed two times the solar gravity: These grains, if surviving long enough, are literally blown out of the Solar System by the radiation pressure of the Sun.

Furthermore, three special phenomena may be seen in cometary dust tails once in a while: dust striae, an anti-tail and a neck-line phenomenon. Dust tail striae are diffuse, narrow and straight streaks superimposed on the diffuse tail background light. Striae are believed to be remnants from fragmentation processes of major boulders in the dust tail region of comets. These sometimes house-size fragments from break-up events that may have occurred long ago (months to years and even centuries) accompany the main nucleus along its orbit and only dissolve into a swarm of dust grains when getting closer to the Sun. Anti-tails can occur by projection when sections of the distant dust tail of a comet are seen sunward as projected in the sky of the observer on Earth. This type of an anti-tail is due to the special viewing geometry of an otherwise normal dust tail, when the Earth passes through the orbital plane of the comet. The neck-line phenomenon describes a narrow dust tail feature that can point away or towards the Sun (in the latter case it may be called an anti-tail as well) and is produced by larger dust that was emitted about 180° away from the actual orbit position of the comet and is passing through the orbital plane of the comet by the time of observation. Like the projection anti-tail, neck-line structures in comets are detectable when the Earth is passing through the orbital plane of the comet.

The heavy dust grains emitted from the nucleus have long lifetimes in the neighborhood of the comet. However, with time their orbits get distorted slightly due to gravitational interaction with the planets and due to radiation pressure effects, thus forming the so-called *dust trails*. Dust trails are the sources of many meteorite streams that are related with comets (for instance the Perseid stream is related to dust trails of comet 109P/Swift–Tuttle).

13.8.4.6 Comet Splitting Observations Comets are known to split. The physical reasons of the nucleus break-up are widely unknown except for one case, i.e., comet D/1993 F2 Shoemaker–Levy 9 that split due to tidal forces during a close passage to Jupiter in 1992. The statistical likelihood that an individual comet splits is a few percent in 100 years. The consequences of the nucleus splitting are manifested usually in three phenomena: brightness outbursts, arclets, and fragments. As for many other phenomena in comets, nucleus break-ups can occur at any time. In fact many events are first noticed by amateur observers while monitoring the objects in the sky [95].

Brightness outbursts, sudden increases towards temporary maxima followed by possibly slower decays in the visual light curve of comets, are known to be associated with comet splitting which produces apparently lots of gas and dust at a time. However, the sudden brightening of the coma light is not necessarily a safe indicator for a nucleus break-up, since there are cases of major outbursts known where none of the other break-up indicators are noticed and instead some temporary coma features occurred as described on p. 515. However, it was possible to associate brightness outbursts with the break-up events of cometary nuclei. In these cases, outbursts seemed to be the immediate, though indirect "indicators" of splitting events, since it could be shown that they were measured within a few hours after the splitting.

An arclet, a coma phenomenon described also on p. 515, may appear within a few days after the break-up event. Arclets disturb the regular smooth coma isophotes, but their full geometric structure may only be detectable after some coma structure enhancement of the respective comet images. They are thought to represent a temporary phenomenon that occurs due to the formation of a collision zone between the expanding material released by the two subnuclei departing from each other after the break-up. Arclets last for one to a few days only, showing up as symmetric narrow slightly curved arcs perpendicular to the separation track of the fragments and diffusing out and fading away quickly when the fragments achieve larger distances with time. Typical separation speeds of fragments are of the order of a few meters per second. Up to now, arclets are only reported from images taken with larger professional telescopes. However, they should be detectable in well-exposed amateur images as well and, given the frequency of occurrence of nucleus splitting events, amateurs should have a good chance to find them in their comet images.

In most cases fragments produced by a nucleus splitting event, can only be seen weeks to months after the actual break-up. They usually show up in the tailward side of the coma and roughly follow the direction of the projected velocity vector of the comet in the sky. A fragment appears as a little minicomet, i.e., it displays a small coma and sometimes even a tail, both "on top" of the stronger coma and tail of the parent comet. Nucleus fragmentation events are believed to produce swarms of pieces, possibly of different size, and in fact in some cases a real armada of fragments was found. However, it is very difficult to follow individual fragments of the swarm with time since they change brightness and fade away while new ones show up on short time scales. Some nucleus break-ups produce also one or more major fragments that survive for a very long time, developing more and more into comets of their own and independent from the parent nucleus like for instance components B and C of comet 73P/Schwassmann–Wachmann 3 (Figs. 151 and 152). By measuring the separation of fragments from the parent nucleus with time, either through relative or absolute astrometry of the components, it is possible to simulate the dynam-

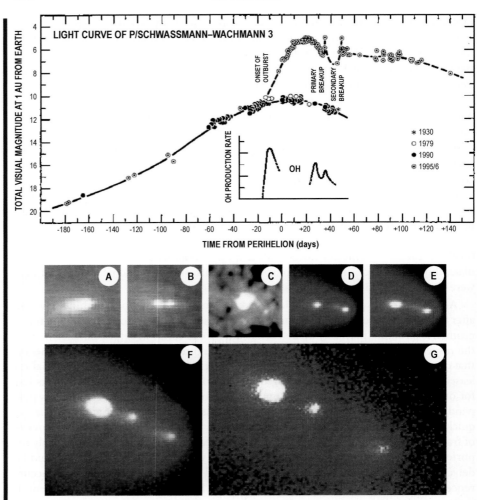

Fig. 152. The nucleus splitting of comet 73P/Schwassmann–Wachmann 3 in 1995. (*Top panel* = Plot) The light curve of the comet compiled from visual brightness estimations of amateurs, shows a large outburst starting a few days before perihelion passage followed by a moderate decay with two short outburst peaks around 30 and 50 days post-perihelion. The insert panel shows the detailed OH molecule production of the comet as measured through radio telescopes around perihelion. The OH production traces the increased emission of water molecules by the nucleus during the break-up event in 1995. (*Middle and Bottom panels* = Images A–G) A sequence of images of the comet showing the inner dust coma on October 30, 1995 (**A**), December 2, 1995 (**B**), December 12, 1995 (**C+D**), December 14, 1995 (**E**), January 7, 1996 (**F**) and January 31, 1996 (**G**). Two nuclei are detectable in the image of December 2, while with time yet a third nucleus appears in the coma. In panels (**D**) to (**F**) the components of 73P/Schwassmann–Wachmann 3 are **A**, **B**, and **C** from *right to left*. Courtesy of European Southern Observatory (ESO)

ics of the post-break-up phase of the pieces and to extrapolate the actual date and some physical parameters of the event. Indeed, amateur observers can contribute much to the monitoring of comet splitting and fragment separation.

13.8.4.7 Spectroscopy of Comets Spectroscopy is a major tool of professional planetary research for the exploration of comets, for studying their nature and the physics and chemistry behind them. It covers the widest possible wavelength range accessible to ground- and space-based observations, i.e., from X-ray to UV, visible, infrared, submillimeter to the radio domain. Visible spectroscopy of comets is the oldest and also most established technique available to amateur observers and interested laymen. All others require specialized and rather expensive instrumentation that is usually not affordable for individuals. Hence, we do not describe the respective observing techniques here, and instead we focus on a brief description of visible spectroscopy of comets.

Visible spectra of comets are frequently obtained when the comets are active and bright. Due to the dispersion of the light over a wider wavelength range covered by the detector of the spectrograph, the flux from the comet has to be high enough, and for this reason visible spectroscopy with telescopes of 1 m aperture or smaller may be restricted to comets of visible brightness of 6 mag and brighter. Spectra of cometary nuclei are sparse, since due to the faintness of the nuclei they require the light collecting power of large telescopes (4–10 m class or the Hubble Space Telescope).

Visible spectroscopy of comets can address the coma and the tails, the former being of highest interest followed by spectroscopy of the ion tail, while spectroscopy of the dust tail is relatively unimportant. The coma spectra of comets display emission bands of some cometary molecules, superimposed on a slightly reddened solar spectrum from dust-reflected sunlight (Fig. 149). The most prominent spectral features result from CN, C_2, C_3, and NH_2 gas in the coma, all products from other, partially unknown mother species, sublimating off the cometary ices and destroyed by the solar radiation while expanding into the coma. These gases start to appear in the coma when the comet gets closer than 2.5–3 AU to the Sun, and their emission bands get more prominent the closer the comet comes to the Sun. Unfortunately, the most abundant gas species in comets, H_2O, cannot be measured directly by visible spectroscopy.

The dust-reflected light usually dominates the flux in the visible spectrum of the coma center. Its slightly reddened color results from the reflection of sunlight at the dust surfaces and is indicative of the dust material and grain sizes. As already mentioned, mostly grains of micrometer to several tens of micrometers in size, contribute to the visible light of a comet, while dust of smaller and larger dimensions is usually not recognizable in the visible spectra of comets.

Spectra of the ion tail show emission lines from positively charged molecule ions, like CO^+ in the blue and H_2O^+ in the red part of the spectrum. Depending on the visibility of the ion tail (i.e., whether or not it overlaps by projection with the dust tail in the sky) the ion emission bands are superimposed on a weak dust continuum (or not). The dust tail shows a slightly reddened solar spectrum in the visible without noticeable spectral absorption or emission features.

The scientific goal of cometary spectroscopy is the determination of production rates and mixing ratios of the detectable species through flux measurements of their emission bands. This requires, apart from the usual calibration exposures as known for CCD image reduction (bias, flat-fields here from a continuum lamp), wavelength and flux calibration and the removal of a "reddened" solar spectrum. The emission band fluxes can be translated into a number for the amount of molecules in the coma region seen through the

It goes without saying that aside from spectroscopy, the respective imaging techniques are also applied.

slit of the spectrograph. From that the amount of gas released by the nucleus is derived via a physical model.

In the simplest approach the so-called 'Haser model' is applied, a semiempirical mathematical description of the cometary coma.

The spectrograph for the visible wavelength region consists of a slit unit, a collimator, a disperser (grating or prism), a camera and a detector device (usually a CCD). The spectrograph is mounted to the telescope and may need some alignment and focusing in order to be ready for cometary observations. It is advantageous if the slit can be oriented across the cometary coma at an observer-defined angle, and slit orientation along the Sun-comet position angle in the sky are preferred. However, in order to study specific localized phenomena in the comet, it may be required to put the slit on different areas in the coma and also with different slit orientations. It is obvious that good telescope tracking and guiding at the speed of the comet's motion in the sky is required, since the exposure times for spectroscopy are long (easily 30 min and more in order to achieve sufficient signal level in the comet spectrum).

13.8.4.8 Polarimetry of Comets

Polarimetry of comets is not very popular even among professional planetary scientists, since it is considered a difficult measurement and not very rewarding for the scientific outcome compared to other observing techniques. Light with randomly distributed vibrational planes and phases is called unpolarized, light with a predominant direction of vibration is called linearly polarized.

Besides linear polarization also circular polarization is claimed to be measured from cometary dust. However, these measurements are difficult and may be the domain of professional astronomers.

In comets linear polarization is caused by the light scattering of the dust in the coma and tail and due to fluorescence of the light-emitting gases in the coma. The polarization of the dust depends on the phase angle of the light scattering, i.e., the angle ϕ between the Sun, the comet and the observer. The flux measured through the polarizer is usually highest when measured in the light scattering plane and is lowest when measured perpendicular to it. The linear polarization of the dust increases with increasing phase angle ϕ up to 20–30% at ϕ of about 90°, and it decreases again for even larger phase angles. Polarization from the fluorescence of coma gas is widely independent from the phase angle ϕ. Following this experience and considering that the coma light is dominated by reflected sunlight, polarimetric measurements of comets are usually considered to constrain dust properties, although only in a very global manner. Over the years, some statistics on comet polarization was established that indicates the general behavior of dust polarization described above. And although some scientists have postulated two or even three different types of comets based upon their polarization properties, this classification scheme may be questionable, since it does not properly consider the usually unquantified contamination of the results due to the coma gas.

The polarization degree P and the predominant direction of polarization ψ (the position angle of maximum polarization) are the aim of the measurements, accomplished with the help of a polarizer device (for instance a polarization foil) mounted in a turnable frame in front of the telescope optics or close to the detector. Since the polarizer absorbs at least half of the light from the comet and a high signal level is required in order to achieve sufficient accuracy in the flux measurements in particular for comets at smaller or very large phase angles, comet polarimetry may only be possible for brighter objects (above 6 mag). Moreover, when using a polarizer device only, it is important to measure the comet in clear to photometric sky conditions and at similar airmass for the various exposures required (see below). Otherwise artifacts from weak absorptions due to faint atmospheric clouds and/or even the transparent atmosphere may influence the results.

Depending on the measurement goals, two or three exposures with the polarizer at two or three different position angles are required. For a simple determination of the polarization degree P it is sufficient to orient the linear polarizer with its polarization direction along and perpendicular to the precalculated scattering plane of the dust-reflected sunlight from the comet, i.e., the position angle of the Sun-comet radius vector projected into the sky (a value that is calculated by standard ephemeris programs). Here, the assumption is that the light scattering follows theoretical expectations. The polarization degree P is estimated from the fluxes $I_\parallel$ and $I_\perp$ measured parallel and perpendicular to the light scattering plane, respectively, i.e.,

$$P = (I_\parallel - I_\perp)/(I_\parallel + I_\perp) \,.$$

If one dares to determine the angle ψ of the linear polarization in the sky, in total at least three exposures with polarizer angle settings different by $45°$ each, will suffice. The measured fluxes I_1, I_2, I_3 allow the estimation of the polarization degree P and the polarization angle ψ as

$$P = \left[(I_1 - I_3)^2 + (2I_2 - I_1 - I_3)^2\right]^{1/2}/(I_1 + I_3)$$
$$\tan 2\psi = (2I_2 - I_1 - I_3)/(I_1 - I_3) \,.$$

It is important not to confuse the sequence of flux measurements I_1, I_2, I_3 for the different polarizer angles. The quadrant of ψ follows from the different signs in the second equation above. By observing standard stars with known polarization angles using the same technique, the observer's system can be related to the standard system. In order to remove the so-called instrumental polarization that may be inherent in the data due to the equipment used, one should measure an unpolarized standard star for comparison.

13.8.5 Further Reading

Festou, M., Keller, H.U., Weaver, H.: Comets II, Univ. Arizona Press, Tucson (2005)

Huebner, W.F (ed.): Physics and Chemistry of Comets, Springer Press, Heidelberg (1994)

Kronk, G.W., Marsden, B.G.: Cometography, Cambridge Univ. Press, Cambridge (1999)

Kronk, G.W.: A Catalog of Comets, 1800–1899, Cambridge Univ. Press, Cambridge (2003)

Kronk, G.W.: A Catalog of Comets, 1900–1932, Cambridge Univ. Press, Cambridge (2007)

Levy, D.H.: David Levy's guide to observing and discovering comets, Cambridge Univ. Press, Cambridge (2003)

McFadden, L.A., Weismann, P., Johnson, T.: Encyclopedia of the Solar System (2nd edition), Academic Press, Burlington (2007)

13.9 Meteors and Bolides

F. Schmeidler[†]

13.9.1 General Information on Meteors

The International Meteor Organisation (IMO) defines fireball as a meteor brighter than magnitude −3. A bolide (see Sect. 13.9.3.2) is defined as an exploding meteor, although it may be a fireball too.

The 1989 founded International Meteor Organisation (IMO) is the worldwide base of amateur work. See the website http://www.imo.net.

Various software packages are available related to meteors and meteor data processing.

The term *meteor* is used to describe the brilliant phenomenon which occurs when a material body from space, called a *meteoroid*, penetrates the Earth's atmosphere; the resulting fragment reaching the surface is called a *meteorite*. Observation of meteors is an area of astronomy which, at least in principle, continues today to provide opportunities for scientifically useful work by amateur astronomers.

One must also distinguish between the small "shooting stars," or meteors, and the large fireballs, also called *bolides*; the latter fall with substantial development of light and heat, and are often accompanied by thunderous noise. There is no sharply defined boundary between meteors and fireballs; as an approximate guideline, a meteor is considered a fireball when it reaches at least the apparent brightness of Venus (about −4$^{\rm m}$). The German word "Sternschnuppe" is reminiscent of the snuffs falling off candle wicks. Not until about 1800 was it clearly recognized that all meteors are the result of cosmic bodies entering the Earth's atmosphere and glowing in it. The explanation of the problems associated with these bodies has been handled primarily by professional astronomers, as amateur observations in most cases were not performed in a sufficiently systematic manner. With well-planned observing, however, amateur astronomers may supply useful contributions to scientific progress in this area.

There exists also a co-operation between professionals and amateurs in the Pro-Am Working Group of the Commission 22 of the International Astronomical Union (IAU). See website at http://www.iau.org.

The scientific interest in meteors pertains to data on their origin and physical constitution. The first question is tackled primarily by determining the spatial path of the meteor, the second by studying the luminous phenomenon of the fall; only for the large fireballs is there occasionally the opportunity to investigate a piece of the meteorite afterwards in the laboratory.

Only part of the desired results for meteor observations can be obtained from the records of a single observer. Many questions require observations by at least two observers at different locations. Suggestions for the planning and evaluation of observations are the subject of this chapter.

13.9.2 Methods of Meteor Observation

Meteors may be observed visually (with or without optical aid), photographically, or using radio-astronomical, spectroscopic, and video instruments. The photographical aspects are separately detailed in Chap. 6, the radio-astronomical in Chap. 9, and spectroscopic aspects in Chap. 7.

The magnification of binoculars should be in the range of 70 × 50 to 10 × 50 and the apparent field-of-view within 50° to 60°.

13.9.2.1 Visual Observations Observations by single observers as well as simultaneous observations from several different stations require the following important data:

1. Three points of the path are desired, including the beginning and end, if possible. It is best to state these points relative to fixed stars, for instance: "midway between α and β Ursa Majoris," or "1/4 of the line from α Cygni to α Lyrae," or a similar description.

In case the points cannot be placed relative to stars (i.e., for a bright fireball during the daytime), estimates relative to the Sun or Moon may be made. Comparison with prominent terrestrial markings or estimates of apparent altitude and azimuth can also be used, but are less dependable. If the beginning point was not seen, the point where the meteor was first caught should be stated, of course noting this fact. All data are recorded immediately in order to minimize the risk of memory errors.

2. Time and place of observation. Of course, the error of the clock or watch used needs to be precisely determined, a task best done by comparison with a broadcast time signal. The place of observation is most suitably described by geographic longitude and latitude.

3. The duration of the phenomena. It is here that the largest errors occur. Only by substantial practice can good estimates be reached! Poor data on the duration are scientifically just as useless and confusing as good results are in demand. Observers lacking a stopwatch may test their memory scale of time by performing an experiment in which a watch is repeatedly checked, for instance, for the time needed to slowly count: "1001, 1002, 1003, . . ." and so on. After practicing this for some time, one becomes accustomed to a certain counting rhythm, which then can be repeated for each individual case with closely matching speed.

4. The luminous phenomenon. Estimating the maximum brightness of a meteor requires considerable experience. It is suggested that the brightness be stated not in magnitudes but rather as compared with stars at about the same altitude (because of differential extinction). Typical descriptions could be: "as bright as β Aquarii," or "somewhat fainter than γ Aquilae but distinctly brighter than β Aquilae."

A luminous trail often remains visible for some time along the path: in this case data on intensity and duration of afterglow are very useful. Statements on the color of the phenomena are occasionally possible if the event is bright.

5. Acoustic data. Large meteors are sometimes accompanied by the sound of thunder. The time between the light and sound phenomena should be recorded as accurately as possible.

6. Weather data, in particular the state of cloudiness during the observation, are needed in every case.

Of course, observations will frequently aim at a particular program item, and the other features listed here will then be omitted or attended to with less effort. It all depends upon the specific target of observation.

In any event, the observation of meteors requires much patience. A certain sector in the sky is selected to be monitored for a long time, and all meteors appearing there are recorded. It makes little sense to try to survey the whole sky in the hope that a meteor falling anywhere would be detected owing to its brightness and its motion. Using optical aids has the advantage of reaching fainter meteors, but the disadvantage that it limits the field-of-view; which aspect weighs heavier depends upon the purpose of observing. Under all circumstances, of course, the optical instrument used, such as binoculars, should have as large a field-of-view as possible.

Practice has shown that experienced observers can fix the end point of a meteor track in the sky to within a few degrees. Often, instead of a statement giving the beginning and end of the apparent path of a meteor, the pertinent information is plotted onto a star map. In principle, there is no objection to this but in practice it seems to be less accurate.

13.9.2.2 Photographic Observations The target of photographic observations of meteors is essentially the same as for visual ones, though the circumstances will cause some modifications in detail. First, few amateur astronomers own instruments of high enough light-gathering power and a large field-of-view, as will be needed for the observation of meteors. Also, a good tracking mechanism must be provided. Commercial photographic cameras can be used with success only in exceptional cases, since the small linear scale on the negative will reveal only a few meteor trails, if any. Two principal objections connected with the rapidity of a meteoric event can be raised against the photographic method per se:

<aside>
For the detection of fireballs or bolides observers use an assembly of cameras and an all-sky mirror. See the IMO website at http://www.imo.net.
</aside>

1. The photograph shows only the brightest meteors.
2. It is based toward those meteors with the slowest apparent motions.

The first difficulty has been largely overcome by the so-called Super-Schmidt camera, constructed by Baker. There is principally no remedy against the second difficulty.

An advantage of photographic observations is the facility to measure the speed of meteors with suitable mechanical devices. If, by some means, perhaps using a rotating sector in front of the objective or the plate, the light is chopped at regular intervals (e.g., every 1/10 second), then the meteor will appear as a light trail which is broken at discrete intervals. Counting the interruptions and measuring their separations permits the determination of the duration and speed of the meteor. Of course, construction of such a device requires a reasonably well-equipped mechanical workshop.

<aside>
For hints to construct an automatic meteor camera see VdS *Journal für Astronomie* (in German) *Zeitschrift der Vereinigung der Sternfreunde*, no. 17, II/2005, p. 75 (http://www.vds-astro.de).
</aside>

When arranging photographic monitoring simultaneously from two different positions, both cameras are pointed to regions of the sky so that their lines of view intersect in the average height of meteors (about 80 to 100 km). One may consider 40 to 50 km the most favorable distance between two photographic observing posts.

In summary, it may be said that photographic observation of meteors is advised only when suitable optical and mechanical tools are available.

13.9.2.3 Radio Astronomical Observations The detection of meteors at radio wavelengths is possible because the air along the path of the meteor becomes more strongly ionized than elsewhere, and a radar beam can then be reflected by the trail of residual ionization. This fact was noted already before and during World War II, and thereafter utilized for the systematic exploration of meteors.

<aside>
Radio observations are the way to count meteors during Full Moon, when skies are clouded, and also during daytime.

Software for meteor detection and visualization of observations can be found at http://radio.meteor.free.fr

Programs for modeling and analyzing nearly any kind of antenna in its actual operating environment can be found at http://www.eznec.com
</aside>

Besides the recording of meteors, radar observations permit the determination of two quantities which are principally inaccessible to a single visual or photographic observation. These are the direction of the path of the meteor, or *radiant*, and its speed. The direction can be measured by the fact that a radar echo is received only when the beam meets the meteor orbit more or less perpendicularly. The meteor's speed is measured by small changes in the amplitude of the radar vibrations, which depend on the speed of the target.

13.9.2.4 Observation by Television

In recent years, meteors have also been observed using video cameras and television monitors. The process is in principle the same as in photography, except that the photographic camera is replaced by a television camera. According to results obtained thus far, the limiting magnitude reached is somewhat better than the corresponding photographic limit. On the other hand, the mentioned disadvantage of photographic data, i.e., that slow meteors are recorded preferentially, occurs to the same or even greater extent for television cameras.

13.9.3 Special Aspects of Observations

The following section addresses the specific questions which are expected to be answered by meteor observations. A summary of the present scientific knowledge is given first for meteors and later for bolides.

13.9.3.1 Meteors

There is a major distinction between *sporadic meteors* and those belonging to a *meteor shower*. The physical processes of ionization and incandescence, which are, of course, the same for both cases, are not yet fully explored. It is now assumed that air friction plays a relatively minor role and that the primary process is the collisional ionization of air molecules. Most meteors appear at 100 to 150 km above the ground and expire at about 50 to 100 km. The mass of most meteorites is under 1 gram, or comparable to a grain of sand.

One of the most directly accessible data is the *meteor frequency*, defined as the number of events noted per hour. It is much larger than would be expected; on a clear night an observer without optical aid can see an average of 8 to 10 meteors. In addition, there are distinct periodicities with time of day and time of year. Usually the best time to see meteors is when the apex of the Earth's motion (i.e., the point toward which the Earth is traveling in its orbit about the Sun), is high above the horizon. All meteors appearing on the sphere near the apex move through space in the direction opposite that of the Earth, and therefore their velocities relative to Earth will tend to be quite high. The apex always lies in the ecliptic 90° west of the Sun, and so most meteors appear in the early morning hours; moreover, the fact that the apex is highly elevated above the horizon during autumn results in appreciably more meteors seen at that time than in spring. These rules are modified depending on what actual speed the meteors have and whether or not their space motions exhibit a preferential direction. Answers to both questions are not yet scientifically clear, and thus amateur observers can still contribute usefully by simply counting meteors, so long as they are not averse to working until dawn!

The determination of the spatial path of meteors requires at least two observers, located at different sites. If both observers see the same meteors and have noted the requisite data, then the parallactic displacement gives the altitude above Earth's surface, and the speed gives the path in space. The necessary formulae are compiled in Sect. 13.9.4.

The question of meteoroid velocities in particular has in recent years given rise to noticeable differences of opinion. The orbit determination (see Sect. 13.9.4) gives the heliocentric velocity at the instant of penetration into Earth's atmosphere. If it is larger than 42 km/s, then the meteoroid had a hyperbolic orbit; otherwise it was an elliptical one. It has sometimes been claimed that a large number of meteoroids have hyperbolic orbits;

Actual observations by the European Group "Arbeitskreis Meteore" (AKM) available at http://www.imo.net/video/metrec.

MetRec is a software package for the automatic detection and analysis of video meteors, see http://www.imo.net.

For video meteor observations to investigate fainter meteors see Chap. 6.

these meteors thus should have come from outside the Solar System. More recent radio astronomical measures, however, have always indicated elliptical orbits. It is thus far unclear whether the discrepancy results from systematic errors of either visual or radio astronomical observing methods. Thus, the reality of "interstellar" meteoroids following hyperbolic orbits has been neither proven nor refuted. A decision awaits additional observations.

At specific times during the year, there appear unusually large numbers of meteors whose paths, when extrapolated backward, meet approximately in one point of the celestial sphere, the so-called *radiant*; the rule of perspective shows that these meteoroid follow parallel orbits in space and are members of a shower. The best-known meteor shower is that of the Perseids, which appears in early to mid-August. For some of these showers it has been proven that their paths coincide with the orbit of a specific comet. Thus, the particles which become meteors are in reality remnants of cometary matter. Table 22 provides data on the more prominent meteor showers.

Among the showers listed, the Draconids and the Leonids are *periodic showers*, which means that rich meteor displays occur only when the Earth crosses that particular part of the parent comet's orbit where matter is strongly concentrated. For the Draconids (Comet Giacobini–Zinner) this occurs every 6.5 or 13 years. Regarding the Leonids, particularly rich showers occurred in the last century every 32 or 33 years; the shower was subsequently absent for some time, but in November of 1966 it produced one of the most brilliant displays ever seen, also in 1998 till 2002.

The most impressive of all meteor showers are the Perseids, which may produce nearly 100 meteors per hour. The Geminids are also rich, but the frequently inclement weather in December impedes their observation. The tabulated dates of beginning and end are averages and may vary by a few days either way for individual appearances.

To best observe these phenomena, the following points should be carefully attended to:

1. It is useful to make new observations of sporadic meteors as well as of the known showers. The results (e.g., regarding richness) may change within a few years in an unexpected fashion.
2. The known showers should, if possible, be observed anew each year.

Of special interest are the Geminids for the source seems to be the minor planet (3200) Phaeton with a period of 1.5 years. Some astronomers look at Phaeton to be an inactive planet.

For a summary of prominent showers between 1998 and 2006 by J. Rendtel see *Kosmische Feuerwerke. Die großen Meteorströme der letzten Jahre* (in German), *Sterne und Weltraum* November 2007, p. 66–71.

The activity of the 2002 Leonids: Arlt, R., Krumov, V., Buchmann, A., Kac, J., Verbert, J.: Bulletin 18 of the International Leonid Shower, WGN, *The Journal of the IMO*, vol. 30, p. 205 (2002).

Table 22. Some prominent meteor showers

Stream	Dates	Radiant α	δ	Richness	Associated Comet
Quadrantids	Jan 01–Jan 04	230°	+50°	High	No comet known
Lyrids	Apr 19–Apr 23	273	+31	Low	Comet 1861 I
η Aquarids	Apr 28–May 16	340	0	Medium	Halley's Comet?
δ Aquarids	July 22–Aug 10	344	−15	Medium	No comet known
Perseids	July 27–Aug 18	40	+55	High	Comet 1862 III
Draconids	Oct 10 only	267	+56	High	Comet 1900 III
Orionids	Oct 15–Oct 25	94	+14	Low	Halley's Comet?
Taurids	Oct 26–Nov 22	54	+15	Low	Encke's Comet?
Leonids	Nov 15–Nov 17	151	+22	Low	Comet Temple 1866 I
Geminids	Dec 06–Dec 16	113	+32	High	No comet known
Ursids	Dec 21–Dec 23	217	+76	Medium	Comet Tuttle 1939 k?

3. The data from observations on numbers of meteors per hour increase in value, when also the brightnesses of these meteors can be stated.
4. Observations at zenith distances of over 70° and in moonlight are best avoided.
5. In the southern hemisphere of the Earth, the data collected so far are very scant. Observers who have the occasion to travel to southern latitudes may serve science considerably by observing meteors even without using any special instruments.

Radio-astronomical observations have also shown that daytime showers exist; they appear only in the daytime sky, and therefore are inaccessible to most other observing methods. Most of these showers occur during the months May and June. Reports by observers or groups thereof, having made systematic meteor observations and now state their results, are often found in various astronomy magazines.

13.9.3.2 Bolides A falling body of large dimensions will penetrate deep enough into the Earth's atmosphere to be heated on the outside by friction, and thus originates the brilliant event known as a *fireball* or *bolide*. In many cases, the meteorite, as it is suddenly heated, bursts into several pieces. After the speed has been sufficiently braked by friction, the visible meteor phenomenon terminates and the fragments fall to the ground. Among the meteorites found, the two groups of *stony* and *iron meteorites* are distinguished (Fig. 153). The latter are composed almost entirely of iron which, on the other hand, is not completely absent in the stony meteorites.

The methods and problems of observing fireballs are in principle the same as for the fainter meteors. In practice, however, the rarity and unpredictability of fireballs makes still greater demands on the observer's patience, and chance plays a much greater role.

Of particular interest are the acoustic phenomena which sometimes accompany the bolide. Depending on the height and the location of the observer relative to the path of the fireball, the sound may be quite different.

The meteorite fragments, which can have masses of several kilograms, may possess very darkened and crusty surfaces. Be warned, however, as not every stone with such characteristics is a meteorite! In each specific case, the decision as to whether or not a particular chunk found is actually a meteorite can be made only by a specialist. A characteristic feature of metallic meteors are the interlocking crystal patterns, the so-called *Widmannstätten patterns*, which become prominent after the stone has been cut, polished, and etched with acid. But there also exist iron meteorites without such patterns. The International Meteor Organization (IMO) publishes the IMO journal with actual

Regarding the difference between a bolide and a fireball see p. 526.

A formidable bolide and fireball was observed over Peru on September 15, 2007. See report with the photography of the crater in *Sterne und Weltraum*, December 2007, pp. 20–23 (http://www.suw-online.de).

Fig. 153. Iron meteorite from Cabin Creek, AR, USA. After Berwerth [96]

information for amateur observers (http://www.imo.net). National groups are also publishers of information sheets, e.g., in Germany the Section of Meteor Observes of the Vereinigung der Sternfreunde (VdS) with the news service "Meteoros" (http://www.vds-astro.de).

13.9.4 Orbit Determinations of Meteoroids

The spatial path of a meteoroid is always a conic section with a focus at the Sun. As the meteoroid approaches the Earth, this path is converted into a hyperbola with a focus at the Earth's center. The direction of motion naturally approaches the zenith, that is to say, the meteor falls more steeply than it would in the absence of Earth's gravity. The direction from which the meteor comes, i.e., its radiant, thus suffers a *zenith attraction*.

Only a small portion of the hyperbolic orbit, with a focus at the Earth's center, is actually seen. This portion can in practice be approximated by a straight segment and thus a circular segment on the celestial sphere. The task of orbit determination is to reconstruct from observations of these orbit segments the original orbit in space.

13.9.4.1 The Path Within the Atmosphere   As a first step in calculating the orbit of a meteoroid, the geometric elements of motion within the atmosphere need to be found. The necessary formulae are presented here without proof; they have been taken from Bauschinger [97], who provides rigorous mathematical derivations. The following mathematical quantities are introduced:

$$\rho = \text{Radius of Earth}$$
$$\lambda, \varphi = \text{Geographic longitude and latitude of the observer}$$
$$h, A = \text{Apparent height and azimuth of the endpoint of the orbit}$$
$$\alpha, \delta = \text{Right ascension and declination of the endpoint}$$
$$\alpha', \delta' = \text{Right ascension and declination of any other point of the orbit}$$
$$z = \text{Linear height of the endpoint above the ground}$$
$$A_0, D_0 = \text{Right ascension and declination of the radiant}$$

For the measurements from every observing site, the auxiliary quantities J, N, K and i are computed from the relations

$$\sin J \sin(\lambda - N) = \sin A \sin \varphi \,,$$
$$\sin J \cos(\lambda - N) = \cos A \,,$$
$$\cos J = \sin A \cos \varphi \,,$$
$$\tan i \sin(\alpha' - K) = \tan \delta' \,,$$
$$\tan i \cos(\alpha' - K) = \frac{\tan \delta - \tan \delta' \cos(\alpha - \alpha')}{\sin(\alpha - \alpha')} \,. \tag{20}$$

Then, the unknowns x, y, X, and Y are determined using

$$x \sin N \sin J - y \cos N \sin J + \cos J = 0 \,,$$
$$X \sin K \sin i - Y \cos K \sin i + \cos i = 0 \,. \tag{21}$$

Data from the two sites give the unknowns from two pairs each of Eqs. (21); if data are available from more than two observing sites, the most probable values of the unknowns can be computed by the method of least squares.

The following formulae are then evaluated:

$$\cot \chi \cos \eta = x, \cot D_0 \cos A_0 = X ,$$

$$\cot \chi \sin \eta = y, \cot D_0 \sin A_0 = Y ,$$

$$\sin^2 \frac{s}{2} = \sin^2 \frac{\varphi - X}{2} + \cos \varphi \cos \chi \sin^2 \frac{\lambda - \eta}{2} ,$$

$$z = 2\rho \sin \frac{s}{2} \frac{\sin(h + s/2)}{\cos(h + s)} , \tag{22}$$

and the coordinates A_0, D_0 of the radiant and the height z above ground for the termination point of the orbit are found. The linear length of the orbit segment for the two points observed is arrived at using three auxilliary quantities σ, τ, and d and the relations

$$\cos \sigma = \sin \delta \sin \delta' + \cos \delta \cos \delta' \cos(\alpha - \alpha') ,$$

$$\cos \tau = - \sin D_0 \sin \delta - \cos D_0 \cos \delta \cos(A_0 - \alpha) ,$$

$$(\rho + z)^2 = \rho^2 + d^2 + 2\rho d \sin h ,$$

$$l = \frac{d \sin \sigma}{\sin(\sigma + \tau)} . \tag{23}$$

Finally, the speed of the meteor is found as the quotient of the length l divided by the duration in which that segment was passed.

In many cases, the radiant can be adequately found by graphical analysis. The apparent paths as seen respectively from the two observing sites are traced out on a star map. The traces are extrapolated backwards until they intersect, the point of intersection being the radiant.

13.9.4.2 The Orbit in Space The directly observed quantities relating to the motion of the meteor in the atmosphere refer to the path described under the additional influence of Earth's gravity. From it, the spatial orbit which the meteor followed while still at considerable distance from Earth can be deduced.

There are essentially two influences to be considered, namely the elevation of the apparent radiant by zenith attraction, and the increase of the linear speed. There is also a shift of the radiant by diurnal aberration, but this effect is almost always negligible. Again, the relevant formulae are reported without proof. A detailed derivation is found in the book by Porter [98]. The following quantities (all in units of km/s) are introduced:

V = Speed of meteor in its original orbit

V_1 = Original speed of meteor relative to Earth

V_E = Speed of Earth in its orbit

V_2 = Observed speed of meteor in the atmosphere

The laws of celestial mechanics show that

$$V_1^2 = V_2^2 - 125 \ . \tag{24}$$

Thus, V_1 can be found. This serves to compute the amount of zenith attraction of the radiant:

$$\tan \frac{1}{2} \Delta \zeta = \frac{V_2 - V_1}{V_2 + V_1} \tan \frac{1}{2} \zeta \ . \tag{25}$$

This gives the amount by which the observed zenith distance of the radiant ζ must be increased. The radiant thus corrected is still only an apparent one as it corresponds to the direction of motion of the meteor relative to the Earth, and not of absolute motion of the meteor. The book by Porter [98] should be consulted for instructions on finding the true radiant. To determine the original speed V, the simple relation

$$V^2 = V_1^2 + V_E^2 - 2 V_1 V_E \cos \beta \cos(\lambda - \lambda_A) \tag{26}$$

can be used, where λ, β are the ecliptic coordinates of the apparent radiant corrected for zenith attraction and λ_A is the ecliptic longitude of the apex. It is linked with the ecliptic longitude $\lambda_\odot$ of the Sun by the relation

$$\lambda_A = \lambda_\odot - 90° + 57^m.6 \sin(\lambda_\odot - 102°.2) \ . \tag{27}$$

From the heliocentric velocity of the meteor thus found, the semimajor axis of the orbit can be derived using the well-known relation,

$$V^2 = k^2 \left(\frac{2}{r} - \frac{1}{a} \right) \ , \tag{28}$$

so that a can be found from V. The term r is the Earth–Sun distance, which, with adequate precision, can be taken as constant. If r and a are expressed in astronomical units (AU), and k^2 is set equal to 1, then V is found in units of the Earth's mean orbital speed, which is $29.8 \, \text{km/s}^{-1}$. In the parabolic limit ($a \rightarrow \infty$) this equation also gives the speed of $\sqrt{2} V_E = 42 \, \text{km/s}^{-1}$ as was mentioned in Sect. 13.9.3.1.

A numerical example of the determination of an orbit has been published by Schmitz [99]. Suggestions on the visualization of orbits of meteor showers and the comets which spawn them are given by Zeuner [100].

13.9.5 Further Reading

Beech, M.: Meteors and Meteorites, Origins and Observations. Crowood, Marlborough, Wiltshire (2006)

Bronshten, V.: Physics of Meteorite Phenomena. Reidel, Dordrecht (1983)

Hapgood, M., Rothwell, P., Royrwik, O: Monthly Not. R. Astron. Soc. **201**, 569 (1982)

Heide, F.: Kleine Meteoritenkunde. Springer, Berlin Heidelberg New York (1934)

Hoppe, G.: Die Sterne **58**, 352 (1982)

Jenniskens, P.: Meteors Showers and their Parent Comets. Cambridge University Press, Cambridge (2006)

Kichinka, K.: The Art of Collecting Meteorites. Bookmasters, Mansfield, OH (2005)

Knöfel, A.: Die Sterne **61**, 356 (1985)

Lovell, A.C.B.: Meteor Astronomy. Oxford University Press, Oxford (1954)

Schippke, W.: Sterne und Weltraum **20**, 287 (1981)

Sky and Telescope **52**, 391 (1976)

Sterne und Weltraum **14**, 247 (1975)

Whipple, F.: Sky and Telescope **8** (1949)

13.10 Noctilucent Clouds, Aurorae, and the Zodiacal Light

Ch. Leinert

The extended celestial phenomena which are the subject of this chapter in a sense are easy objects: all of them are observable with the naked eye, and their beauty makes a lasting impression. On the other hand, at mid-latitudes where most of the readers of this book may live they are observed either with difficulty or only rarely. In the following we will in turn describe and try to explain the phenomena. For all of them, beautiful images can easily be found on the Web with their name as search address.

13.10.1 Noctilucent Clouds

On summer evenings at moderately high geographic latitudes, bright silvery-white clouds with a cirrus-like appearance are occasionally seen above the northern horizon (Figs. 154 and 155) between the end of civil and astronomical twilights. Their occurrence at this late time gave them the name "noctilucent" (shining at night), which is somewhat misleading, as they are not visible during the night. These tenuous formations, which diminish

Fig. 154. Noctilucent clouds in the form of bands with some billow patterns, photographed by Viktor Veres in Budapest (latitude 47°) on June 15, 2007. Image courtesy of NASA

Fig. 155. Noctilucent clouds in the form of a veil, including a billow pattern. Photographed by Nathan Wilhelm in southern Sweden (latitude 59°) on July 14, 1986

See *Met. Mag.* vol. 20, 133 (1885).

the light of stars shining through them by much less than 1%, represent by far the highest clouds known to occur in the Earth's atmosphere.

The first recorded sighting of noctilucent clouds took place on June 8, 1885 by T.W. Backhouse in Kissingen, Germany. It was not mere coincidence that this discovery occurred shortly after the gigantic eruption of the Krakatau volcano in 1883, in which enormous quantities of dust were ejected into the stratosphere and distributed around the globe within a few weeks. This resulted in spectacularly bright and colorful twilight phenomena, which led to an increase in observations of the evening sky.

Systematic observations of noctilucent clouds began as early as the year of discovery and lead to a determination of the heights of the noctilucent clouds from simultaneous observations at different sites [101]. The surprisingly high value of 82 km has been confirmed by modern measurements to within less than 1 km. At these heights, the air pressure is only 10^{-5} that of the surface, and therefore no meteorological phenomena would have been expected.

Noctilucent clouds are transient phenomena, lasting usually just a few hours. Compared with their horizontal extent of 100 to 1000 km, their thickness between 0.5 and 2 km is extremely small. The different patterns and waves in noctilucent clouds have sizes of 1 to 10 km. According to the appearance of these patterns, noctilucent clouds are classed into four main types: *veils* (Type I), *bands* (Type II), *billows* (Type III), and *whirls* (Type IV).

See the old, but still informative, B. Fogle and B. Haurwitz, Space Science Rev. 6, 278 (1966).

A cloud usually shows several of the physical shapes simultaneously, e.g., waves resembling fish bones often appear across bands. A very readable overview on noctilucent clouds, unfortunately still the most recent one, can be found in [102].

13.10.1.1 Visibility To see noctilucent clouds one has to be in the right place at the right time, both with respect to the formation of the clouds and to the geometry of observation.

The latter is shown in Fig. 156. The noctilucent clouds are far too weak to be visible in the daytime. Not until twilight is far advanced, with the Sun at least 6° below the horizon, will they distinctly stand out against the now darkened sky. Thus, these clouds can in principle appear along the arc HNC, which at the beginning may still extend past the zenith of the observer. Below point N the sky is still too bright, which leaves the arc NLC. This range follows the sinking Sun toward the horizon until, with the Sun 16° below the horizon, sunlight has been cut off from the layer at which the noctilucent clouds occur. The most favorable conditions (i.e., sky already dark and visibility range still large) occur for positions of the Sun 9° to 14° below the horizon. The clouds thus seen are a few hundred kilometers distant.

The visibility of noctilucent clouds depends strongly on the geographic latitude. They never have been seen north of 77° (Greenland, Spitsbergen) and not south of 45° (Montreal, Lyon), with very few exceptions like an observation at Logan in Utah (latitude 42°) in June 2003. At lower latitudes, the clouds do not form; at higher latitudes in summer, when clouds form, the "midnight sun" keeps the sky too bright. The best conditions for

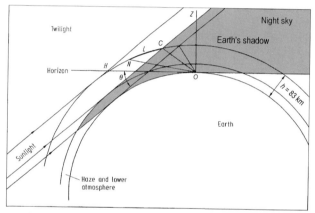

Fig. 156. Geometry of the observations of noctilucent clouds. The letters refer to the following: O observer, Z zenith, C beginning of obscuration of cloud layer by haze and the lower atmosphere, Θ depression of Sun below the horizon. NLC is the range in which noctilucent clouds can be seen

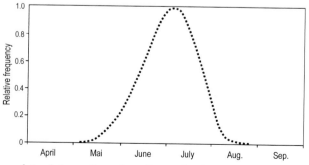

Fig. 157. Frequency of visual observations of noctilucent clouds during various seasons in the northern hemisphere. The maximum in early July is normalized to unity

observing noctilucent clouds are for latitudes between 50° and 65° and during the summer months (Fig. 157).

During the first half of July, the appearance of noctilucent clouds can be expected on more than half of the clear nights at northern latitudes of 55° to 60°. They do not occur during the winter. These relations hold for the corresponding latitudes and seasons in the southern hemisphere.

13.10.1.2 The Larger Picture The key to the formation of noctilucent clouds lies in their height of 83 km. This is only 5 km below the *mesopause*, the coldest part of the atmosphere, which takes its lowest temperatures during the summer months, when they often drop below 140 K. Water vapor, although comprising only a few ppm (parts per million) of the total atmosphere at these heights, then is oversaturated 100-fold and efficiently condenses as ice onto the condensation nuclei, which mainly consist of "smoke" ablated from infalling meteoroids. The presence of these tiny ice particles, about 10 nm in size, shows itself by radar echos. The particles grow while sinking with tens of centimeters per second, reaching a diameter of typically 80 nm and up to 150 nm before they enter the warmer regions below 83 km where they quickly sublimate. It is only these larger particles near this lower boundary which scatter sunlight sufficiently strong to be seen as noctilucent clouds.

Satellite experiments in the 1970s and 1980s (Orbiting Geophysical Observatory *OGO-6* and Solar Mesosphere Explorer *SME*) revealed that the polar caps of the Earth are often overlaid by a fine cloud layer at high altitude during their respective summers. The frequency of occurrence of this layer was found to increase dramatically with geographic latitude, from 20% at 55° to over 45% at 65°, and even 80% at 75°, so that in the polar regions the layer may be a permanent feature during this season. The height of the cloud layer of 83 km, its thickness of less than 3.5 km, the particle size of 0.14 μm, and the density of about 100 particles per cm^3 agreed so well with the values obtained for noctilucent clouds that the two are now considered as one and the same phenomenon, seen from above in the ultraviolet (0.2–0.3 μm) during daytime as polar mesospheric clouds or from below after sunset as noctilucent clouds. The noctilucent clouds are just the visible ragged low-latitude edge of the summer mesospheric polar caps. The larger part of the phenomenon remains hidden to the ground-based observer in the brightness of white nights and of the midnight sun.

In situ studies of the small noctilucent cloud particles by rocket experiments are notoriously difficult. Lidar observations (light detection and ranging) of noctilucent clouds are not subject to the limitations of visual observations and have confirmed the nature of noctilucent clouds as part of the polar mesospheric cloud complex.

13.10.1.3 An Indicator of Global Change? Noctilucent clouds are influenced by the solar activity cycle: during maxima the yearly number of noctilucent cloud events is reduced to half of its undisturbed value. This can in part be explained by heating of the mesopause region by the then increased solar Lyα flux, but neither the large amplitude of the effect nor its time lag of about one year with respect to the solar activity are yet understood. Figure 158 shows a similar effect in the cloud brightness.

Another trend was announced in the 1990s: an increase in the rate of occurrence of noctilucent clouds by a factor of ≈ 2 over the preceding two decades. It was hypothesized that this might reflect the increasing concentration of greenhouse gases in the

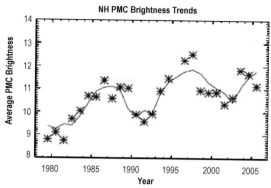

Fig. 158. Increase in the brightness of northern hemisphere polar mesospheric clouds over the years 1979 to 2006 as measured by the Solar Backscatter Ultraviolet (SBUV) satellite instruments for latitudes north of 74° (from M. DeLand et al. 2007). Note also the solar cycle effect on cloud brightness—maxima of solar activity occurred in 1980, 1990, and 2001

higher atmosphere. At these heights, methane gets dissociated and finally sets free water molecules, while CO_2 overall acts as a cooling agent at the region of the mesopause. In principle, these effects could favor the occurrence of noctilucent clouds. It was a tempting idea that such a conspicuous phenomenon as the noctilucent clouds could serve as an indicator of atmospheric changes. However, the claim of this positive trend may have been premature. An analysis of 43 years of noctilucent cloud observations (1964–2006) showed no significant increase in noctilucent cloud events with time. The sightings of the very weakest displays get more numerous, but this could be due to improved observing. Satellite observations confirm the absence of a significant change for the latitude range typical for noctilucent cloud observations (50–65°) but find a trend at high latitudes (see, e.g., Fig. 158).

At present, the stability of the phenomenon, e.g., the absence of a change in the height of noctilucent clouds over the last century, finds increased attention, including the questions why the mesopause region does not follow the cooling trend of the regions above and below it, or why its water vapor content presently shows a decrease instead of the predicted increase. The AIM satellite (Aeronomy of Ice in the Mesosphere) launched in April 2007, will monitor and analyze the occurrence of mesospheric cloud displays over both polar regions for the period of two years (Fig. 159).

13.10.1.4 Observing Noctilucent Clouds Visual observations by amateurs have the value that they have been performed in the same way over a much longer time span than those obtained with modern techniques. If carefully documented, they can help to ascertain the long-term trends discussed above. An excellent guide "Observing Noctilucent Clouds" has been prepared by Michael Gadsden and Pekka Parviainen on behalf of the the International Association of Geomagnetism and Aeronomy (IAGA). It gives, apart from a general introduction to the subject, definitions and photographs demonstrating the various types of noctilucent clouds. And it provides clear and detailed practical suggestions on how to perform noctilucent cloud observations and where to report them. Typical exposure times with focal ratio f/2.8 and ASA 400 are 1–10 s. But the main driver for observing noctilucent clouds probably always will be the beauty of the phenomenon.

See S. Kirkwood et al., Annales Geophysicae 25, 1 (2007) and M. Deland et al., JGR 112, D10315 (2007).

http://aim.hamptonu.edu

See http://www.iugg.org/IAGA/, IAGA Guide electronic form (September 2006).

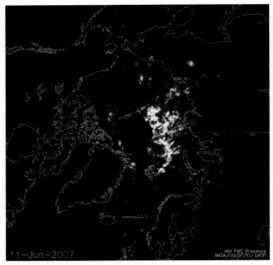

Fig. 159. Distribution of polar mesospheric clouds over the northern arctic regions on June 11, 2007 as seen from satellite AIM. For the *black circle* around the pole measurements were not possible. Image courtesy of NASA

13.10.2 Aurorae

According to reports by numerous observers, the sporadic event known as *aurora* is the most magnificent phenomenon to be witnessed in the night sky with the unaided eye. Aurorae occur most frequently at high latitudes, usually lasting from one to two hours. In the middle and southern parts of Europe and of North America, they are only rarely seen. The typical appearance of aurorae in middle latitudes is different from what is shown in Figs. 160 and 161, the display being more diffuse and the dominant color being red (Fig. 162). Due to the symmetry of the Earth's magnetic field, aurorae occur in the same way in the northern and southern polar regions, most often simultaneously.

13.10.2.1 Origin Attempts to explain aurorae are diverse and as old as the handed-down observations of the phenomenon, which date back to antiquity. Historical overviews can be found in [103, 104]. The first attempt of a scientific explanation was the theory proposed in 1708 by Suno Arnelius of Sweden, which held that the auroral light was sunlight reflected by ice particles high in the atmosphere. As it turned out, this is actually the explanation of noctilucent clouds.

For a tabulated history, see: http://pwg.gsfc.nasa.gov/Education, chronology.

With current understanding, the occurrence of aurorae is always linked to solar activity and resulting disturbances of the geomagnetic field, as shown, for instance, by magnetic deflections of up to a few degrees on the compass, and they are created by complex solar-terrestrial interactions (see Sect. 13.10.2.5). Among the currents of up to a million amperes which cause the change in the magnetic field, one flows through the ionosphere at heights of about 100 km and connects back on both sides along the geomagnetic field lines high into the Earth's magnetosphere. The existence of such field-parallel currents was postulated as early as 1908 by the Norwegian researcher Kristian Birkeland, but was generally accepted only after direct satellite measurements. The current in itself does not

Fig. 160. Spiral shaped aurora, photographed by Jacek Stegman in Kiruna, Sweden on February 17, 1985. In the *foreground* are city lights, in the *background* a band of clouds covers part of the aurora. Exposure time 10 s with aperture ratio f/2 and 200 ASA

Fig. 161. Aurora with distinct ray structure. The rays often move so quickly that they are difficult to display photographically. Photo courtesy of Klaus Rinnert

yet suffice also to cause an aurora. But in parallel a strong acceleration of electrons and ions happens at a height range of several thousand kilometers, and to energies of up to 10 keV and more. In the magnetic sheets where the current of this loop is directed outward, these fast electrons move downward and penetrate into the atmosphere down to about 100 km above the surface, where the atoms and molecules which inhabit the high atmosphere are excited to luminescence, which is then seen as the aurora.

The rays which often appear in auroral displays occur because electrons, as electrically charged particles, are forced to follow the magnetic field lines. This also accounts for the

Fig. 162. Arc of an aurora over the moonlit Earth in the region of the Tropic of Capricorn. The horizon appears doubled: the *upper, yellowish line* originates from luminescence, called *airglow*, in the high atmosphere at an altitude of 90 to 100 km. The auroral arc, about 1000 km long, extends distinctly higher, and assumes in its upper region *a reddish coloring.* Stars are visible close to the true Earth horizon. NASA photograph made on Spacelab 3 by Don L. Lind

distinct "drapery" structure of some auroral events. Changes in the acceleration region and in the tail of the magnetosphere are responsible for the often fast changes in auroral displays.

13.10.2.2 Apparent Shapes Aurorae can be classified into a relatively small number of typical shapes. Photographic examples of these are given in the *International Auroral Atlas* [105] produced by the International Union of Geodesy and Geophysics in 1963. Structureless displays are generally called *glow, patch* refers to a localized brightening, and *veil* to a background sometimes underlying other forms of aurorae. The more pronounced shapes are shown in Fig. 163. Homogeneous arcs and bands are the most quiescent features, the others showing larger variations in brightness and more rapid motions. The lower boundary of an auroral display is usually sharply defined. The most frequent colorings are listed in Table 23, the greenish-white coloring being the normal case. The connection between color and height of the aurora can be understood from the constituents of its spectrum.

13.10.2.3 Spectrum and Color The major constituents of the high atmosphere are atomic oxygen, molecular nitrogen and molecular oxygen. The first two essentially determine the spectrum of the aurora (Fig. 164). The lines of atomic oxygen dominating the spectrum are *forbidden*, which means that the excited atom requires an appreciable length of time before emitting light and thus will do so only when not colliding during that period with other molecules, as this would cost it the excitation energy. For the red lines at 630 nm, preferentially excited because of the lower excitation potential, this time interval is 110 s. Collisions will occur with this frequency at a height of 200 km, and therefore below this level this line emission is increasingly inhibited. For the green line at 557.7 nm, the time interval is 0.74 s. Here, the weakening occurs below 95 km. In this low

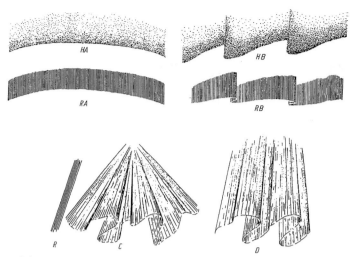

Fig. 163. Typical shapes of aurorae: *HA* homogeneous arc, *RA* rayed arc, *HB* homogeneous band, *RB* rayed band, *R* rays, *C* corona, *D* drapery, a particularly developed form of a rayed band

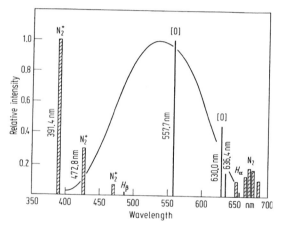

Fig. 164. Spectral features in aurorae in the visible range. The *square brackets* indicate forbidden lines, while molecular bands are *hatched*. The *solid line* shows the sensitivity of the eye

region, then the nitrogen bands around 670 nm will dominate the color which otherwise are outshone by the green line. This accounts for the lower red border of some aurorae. Similarly, the red color at high altitudes occurs because the red oxygen emission dominates there. A violet-purple color results from the *permitted* (i.e., not forbidden) lines of N_2^+ which—with the Sun not too far below the horizon and the aurora extending to above the shadow—are enhanced by direct excitation by sunlight.

Aurorae are observed over a large range of heights, but mostly, as measured by the well-defined lower boundary, occur in the range 100 to 150 km. The lower edge is determined by the energy of incident electrons: the higher their energy, the lower the boundary (see Table 23).

Table 23. Classification of most common aurorae according to color

Type	Description	Base height	Emission	Electron energy
b	Red lower border	90 km	N_2	30 keV
c	Green or whitish	100 km	[O] 557.7 nm	10 keV
d	Red	200 km	[O] 630.0 nm, 636.4 nm	1 keV

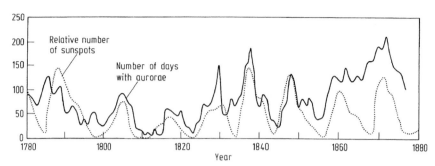

Fig. 165. Relation between incidence of aurorae and solar activity. After Sophus Tromholt (ca. 1880)

13.10.2.4 Observability At any moment, the probability of seeing an aurora is highest in a nearly circular ring with about 20° radius around the geomagnetic pole, the so-called *auroral oval*, which touches the northeast of Canada and also the north of Norway. Towards more southern locations, there is a steep decline in the frequency of aurorae, being reduced to one or two events per year at the latitudes of central Europe and central US.

Images and many links are found at: http://www.geo.mtu.edu/ weather/aurora.

The geomagnetic pole at 79.7° N and 71.8° W (2005) in northwestern Greenland is defined by the symmetry axis of the magnetic dipole field of the Earth. The arrangement of auroral phenomena with respect to this axis again shows their close linking with the magnetosphere. Their strong correlation with solar activity has been known since the nineteenth century (Fig. 165). As seen in the last several solar cycles, the highest incidence of aurorae occurs around or about one year before sunspot maximum, which may be expected to occur again in 2011/2012. Spectacular aurorae often occur after strong solar flares and in particular after the so-called *coronal mass ejections*.

With such violent events as triggers, the aurorae can extend far south. Such great aurorae occurred in March 1989 and October 2003, and the aurora of April 6/7 2000 gave a fine display in southern England and was visible even down to Portugal and the southern US (Fig. 166).

13.10.2.5 Auroral Oval and Connection to the Magnetosphere In perfect vacuum, the geomagnetic field would display the typical dipole structure as shown by a bar magnet in a classroom experiment. The solar wind, which arrives from the Sun at a speed of 400 to 700 km/s, is a plasma of electrons and protons with a density of typically 10 to 15 particles cm^{-3}. It deforms the Earth's magnetosphere, compressing it on the daytime side and dragging some of the field lines with it into a paraboloidal *tail* over 100 Earth radii. These become "open" field lines, which extend into interplanetary space without linking

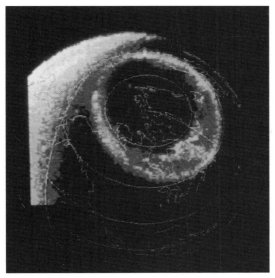

Fig. 166. Northern auroral oval, recorded on November 08, 1981, 03^h UT with the Dynamics Explorer I satellite. Geographic coast lines have been added for orientation to this false color picture. In the ultraviolet at 135 nm, where this image was taken, the brightness of the auroral oval is comparable with that of the sunlit atmosphere. The image was made 2 hours after the arrival of a shock wave in the solar wind and shows the broadening of the auroral oval towards lower latitudes on the night side. Photo courtesy of L.A. Frank

to the Earth again at the ends (Figs. 167 and 168). The magnetic field lines of the *boundary layer*, intermediate also between closed and open magnetic field regions, originate in the daytime half of the auroral oval. On the night side, a fraction of the electrons and protons of the passing solar wind plasma moves under the influence of electric and magnetic forces toward the plane of symmetry, such that a zone of increased density is formed beyond several Earth radii, the *plasma sheet*. This region separates open magnetic field lines of opposite direction in the northern and southern part of the tail and connects to the nighttime side of the auroral oval, with a moderate current flowing down along those field lines under quiet conditions.

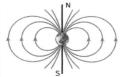

Fig. 167. The Earth's magnetic field

Strong shock waves in the solar wind, as they appear after eruptive events on the Sun, can lead to a strong compression of the Earth's magnetic field and much more dramatic events. The power of the dynamo constituted by the Earth's magnetosphere and the passing solar wind is greatly enhanced. Disturbances in the Earth's magnetic field show the onset of a *magnetic storm* with a duration of about 1 day. The plasma sheet gets strongly loaded with plasma, and large amounts of plasma are flung down towards the ionosphere. The currents through magnetosphere and ionosphere may increase 100-fold and reach 1 to 2 million amperes. Then, voltages build up along the the magnetic field lines which accelerate part of the electrons flowing down from the plasma sheet to energies of thousands of electronvolts: auroral displays occur. This sequence, called a *substorm* usually lasts from 1/2 to 2 hours and can repeat several times during a magnetic storm event. During strong magnetic storms, the plasma sheet is also pushed closer to the Earth, connecting to magnetic field lines of lower latitudes: the auroral oval on the night side widens

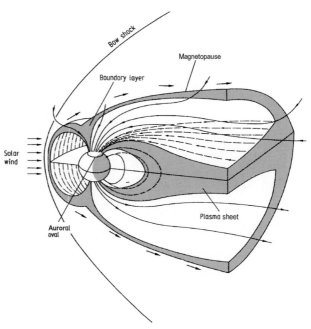

Fig. 168. Three-dimensional cut through the magnetosphere, showing the connection of the auroral oval (the circle in the polar region) to the magnetospheric structures. The magnetopause, the outer skin of the boundary layer, separates the plasma attached to the earth from that of the passing solar wind (*arrows*)

giving rise to auroral displays at the populated mid-latitudes of North America and Europe, and sometimes even to the south of these continents. The weather-like variability of the of the solar wind, which is the energy source, (one speaks of "space weather"), accounts for the the irregular and sporadic nature of the whole phenomenon. Auroral events are connected to strong magnetic fluctuations and precipitation of energetic particles. Because of the disturbances this may produce in satellite electronics, radio transmission, and by induced currents in power lines on the ground, short-term predictions for the probable occurrence of aurorae are made available on the basis of observed solar activity. These can also be used by the aurora observer. A more detailed and complete account of aurorae, well-readable, well-illustrated, and with many practical hints for the amateur observer can be found in the book by Neil Bone [106].

13.10.2.6 Observing Aurorae Aurorae are best seen around local midnight. As for noctilucent clouds, visual and photographic observations are still of some value since they can be compared to the observations in the times before space exploration, and thus allow to study long-term trends and changes. Reports can be sent to organizations like the Aurora Section of the British Astronomical Association, the Aurora and Solar Section of the Royal Astronomical Society of New Zealand, the Finnish association Ursa or the Arbeitskreis Meteore (in Germany). The first of these webpages, like the book of Neil Bone [106], describes in detail how such observations should be performed and documented, e.g., with 400 ASA sensitivity and focal ratio f/2.8 exposure times are 30–60 s for weak diffuse aurorae, 10–30 s for moderately bright aurorae, and 2–10 s for the bright ones, where longer exposures anyway would smear the finer structures of these moving displays. The webpage "Meteoros" has a graphical compilation of information pertaining to the prediction of aurorae. It goes without saying that taking individual photographs of these splendid displays will be very satisfying in itself.

http://www.britastro.org/aurora
http://www.rasnz.org.nz
http://www.ursa.fi/english.html
http://www.meteoros.de/polar/polare.htm

13.10.3 The Zodiacal Light

Under favorable viewing conditions, the faint, essentially colorless, pyramid-shaped *zodiacal light* may be seen above the western horizon in the evening about one hour after sunset (Fig. 169) or above the eastern horizon in the morning about one hour before sunrise. As its name implies, the zodiacal light extends along the ecliptic (Latin: zodiacus), which is the projection of the Earth's orbital plane onto the celestial sphere. The brightness of the zodiacal light increases sharply toward the Sun, and, at an angular distance of 30° outshines the brightest parts of the Milky Way. It is seen most distinctly when its cone makes a steep angle with the horizon. Such is always the case in tropical latitudes, but at mid-latitudes this condition holds only in late February or early March for the evening sky, and in October during the morning; also, the Milky Way during these seasons is far enough away so that it does not interfere. When he discovered the zodiacal light on March 18, 1683 Giovanni Domenico Cassini [107] also arrived at the explanation which is still valid: it is created by sunlight reflected by myriads of small particles orbiting the Sun as mini- or microplanets: the *interplanetary dust* in present terminology. The investigation of zodiacal light saw a renaissance when it became clear that the thermal emission of the interplanetary dust grains produces a veil of infrared emission which constitutes an unwanted foreground to views into the more distant universe, and

Fig. 169. The zodiacal light above the western horizon, photographed by the author on May 13, 1983 one hour after sunset from the peak of Mauna Kea (4200 m). The setting crescent Moon in the evening twilight is 19° above the Sun. Venus, located at the *top of the light cone*, is 43° from the Sun. Exposure time 90 s with focal ratio f/3.5 and focal length $f = 28$ mm on Kodak CF 1000

that any *exozodiacal light* around other stars, with its infrared emission, may mask the signal of Earth-like planets orbiting them.

Because of increasing light pollution, the zodiacal light is practically no longer visible in densely populated regions at middle latitudes. At low latitudes ($< 40°$) and at sites sufficiently distant from big cities and at sufficient altitude (at least 2000 m), the zodiacal light is easy to see if one knows where to look for it. Figure 169 was taken under such conditions; the eye, with its contrast enhancement, would perceive the cone as narrower and longer.

13.10.3.1 The Brightness Distribution Photometric measurements have shown the zodiacal light to extend over the entire sky, its brightness decreasing with increasing distance from the Sun and from the ecliptic. At $90°$ from the Sun in the ecliptic, the brightness has dropped to 1/10 of its value at $30°$. At the ecliptic pole, the zodiacal light is still fainter by another factor of three. However, at the point opposite the Sun, there appears a slight brightening which is barely detectable by the unaided eye. This is the *Gegenschein* (Fig. 170), also termed "counterglow," and was first detected by Brorsen in 1876. It is due to the enhanced backscattering of sunlight by the interplanetary dust particles. At smaller angles, the brightness of the zodiacal light continues to increase towards the Sun and joins (not visible from the ground in the twilight) into the corona, where it forms one of its constituents, the so-called *F-corona*. The zodiacal light is strongly polarized. This does not come from the Sun but originates in the reflecting particles; it therefore permits conclusions regarding the optical properties of the scattering dust grains. The color of the zodiacal light corresponds nearly to that of the Sun, with a slight reddening which the eye cannot perceive. Since very small particles would scatter the blue part of the light much more strongly, it may be inferred that the typical interplanetary dust par-

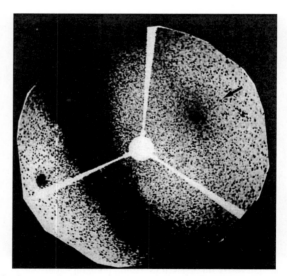

Fig. 170. Wide-angle photograph of the night sky with a spherical mirror taken from La Silla, Chile. The Gegenschein, which is visible in the tenuous light bridge of the zodiacal light *to the right of* the stronger Milky Way band, is marked with an *arrow*. Photo courtesy of W. Schlosser, Bochum

ticles are at least as large as the wavelength of the scattered light, i.e., they have diameters of at least 1 μm. Actually, their effective size is 10–30 μm and exceeds by far that of the submicron-sized interstellar dust particles.

A review of the topic at [108], and comprehensive book [109].

13.10.3.2 The Spatial Distribution of Interplanetary Dust The brightness distribution of the zodiacal light reveals how interplanetary dust is distributed in space. For instance, the concentration of the light cone toward the ecliptic leads quite directly to the flattening of the dust cloud shown in Fig. 171 and described approximately by an ellipse with an axis ratio of 1:5. To express this concentration in a different fashion, 1/2 AU above the Earth the space density of dust has dropped to 1/4 of its value in the ecliptic. The orbits of the interplanetary dust grains themselves are not as strongly concentrated as the resulting spatial distribution; their average inclination to the ecliptic is 32°. The strong increase in zodiacal light brightness towards the Sun mostly results because the spatial density of the interplanetary dust also increases toward the Sun. From measurements made by the space probes Helios 1 and 2 (in the inner Solar System) and Pioneer 10 and 11 (in the outer Solar System) in 1972–1980 it was learned that the density $n(r)$ of the interplanetary dust increases toward the Sun as $n(r) \sim r^{-1.3}$, approximately inversely with the distance. The space density, however, is extraordinarily low—1.0×10^{-19} kg m^{-3} near the Earth's orbit. The light scattering power of interplanetary dust corresponds to only about a dozen particles with a typical size of 10 μm per cubic kilometer. That the zodiacal light is visible at all is only due to the fact that along a line-of-sight the contributions of a great many particles add up. For instance, roughly 10^{24} particles contribute to the light cone seen in Fig. 169. Viewed from the distance of another star, the zodiacal light could hardly be detected next to the Sun which is about 10 million-fold brighter in the visual and almost 100,000-fold brighter than the zodiacal light at 10 μm. Nevertheless, the zodiacal light still outshines the radiation of the Earth by a factor of about 100 in the infrared. This deserves serious consideration when searching for the light of a distant Earth around another star.

The total mass of the interplanetary dust cloud can be estimated to 10^{16} to 10^{17} kg, barely more than the mass of a large comet. Its plane of symmetry is tilted by a few degrees with respect to the ecliptic, 3° in the inner Solar System, and somewhat less in the outer parts. This plane of symmetry coincides neither with the orbits of the largest planets Jupiter and Saturn, which have lesser inclinations, nor with the plane of the solar

Fig. 171. A model for the distribution of interplanetary dust perpendicular to the ecliptic, which is shown as *solid line*. The position of the Earth is indicated *right of the center* by a bar. The Sun would be in the *center of* the central dust-free zone, the latter being displayed five times too large for the sake of clarity

equator, the inclination of which is higher. Thus, it is not yet clear what kind of forces have determined the plane of symmetry of the interplanetary dust cloud.

13.10.3.3 Properties of Interplanetary Dust Particles
A dust particle hitting the surface of the Moon, which is not shielded by an atmosphere, leaves, after impact and evaporation, a microcrater whose diameter indicates the mass of the incident particle and hence its size. Studies of such microcraters have shown, that the main component of interplanetary dust responsible for the zodiacal light are particles in the range 1 to 100 µm, but most frequently 10 µm. (These would qualify also as "dust" in the terrestrial sense). Particles with diameters less than 1 µm are more numerous but optically insignificant. For these smallest particles, the solar radiation pressure exceeds gravity, so that they are, on a short timescale (about 1 year), "blown" out of the Solar System. They are called β-meteoroids (β is an often-used symbol for the force ratio of radiation pressure to gravity).

The first solid knowledge on the structure of the particles was obtained by D.E. Brownlee, who in the 1970s succeeded in collecting interplanetary dust grains in the stratosphere from balloons and high-flying planes. Those with sizes of less than 50 µm survive the braking upon infall into the Earth's atmosphere without damage. Their interplanetary origin was recognized from the helium enrichment in their outer layers. Many of them show a rather loose structure composed of pieces of about 0.1 µm in size (Fig. 172). Most of them are very dark owing to a carbon content of a few percent.

13.10.3.4 Origin of the Zodiacal Light
Particles revolving near the Earth's orbit with speeds of about 30 km s^{-1} receive the sunlight not radially but slightly shifted toward the front. This *aberration of light* gives rise to a force component of the radiation pressure which is opposite to the direction of particle motion, and braking it. This so-called *Poynting–Robertson effect* drives the particles ever closer to the Sun until they evaporate by the intense heat, or are destroyed by collisions, so that they leave the system

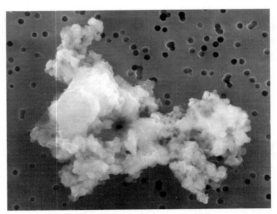

Fig. 172. Interplanetary dust particle 10 µm in size collected in the stratosphere with a U2 aircraft. It is composed of numerous small (≈ 0.1 µm) grains. It contains glass, carbon, and silicate minerals. It is typical in the sense that both more compact and more fluffy interplanetary dust particles have been found. NASA electron microscope image

as β-meteoroids. The lifetime of a particle until evaporation is computed from its radius s (μm), its density ρ (g cm^{-3}), and the semimajor axis a (AU) of the original orbit as $t_{PR} = 700\,s\rho a^2$. For a particle in the Asteroid Belt with size 1 to 10 μm, the lifetime is 10^4 to 10^5 years; this can be shortened by collisions. The resulting loss to the interplanetary dust cloud is 1 to 10 tons of solid material per second. This means that a continued presence of the zodiacal light can only have been maintained by a continuous resupply of dust particles, as it seems unlikely that the zodiacal light phenomenon exists only for the eyes of the present-day observer.

See S. Wyatt and F. Whipple, Astrophys. J. 111, 134 (1950).

Comets were originally held to be the source of interplanetary dust. Their dust tails are direct evidence that they can release large amounts of dust. Fred Whipple, who proposed the now accepted model of comets as "dirty snowballs," suggested that a large comet, like comet Encke, could have set free some ten thousand years ago most of the dust we see today in interplanetary space. However, it remains doubtful, if comets can supply dust at a sufficient average rate to maintain equilibrium.

In recent years, asteroids have increasingly be considered as an important source which, by mutual collisions, could release dust and pieces of rock, which again by further collisions would finally be ground down to dust. This found support by the infrared satellite *IRAS*, which observed bands of increased zodiacal light brightness close to the ecliptic and 10° off on both sides. These loci match the three asteroid "families" Themis, Coronis, and Eos, whose members have orbits similar to each other and consequently will suffer collisions more frequently and could supply the dust seen in the dust bands. Similarly, an increased infrared brightness of zodiacal light in the "wake" of the Earth has been interpreted by a circumsolar dust ring along the Earth's orbit resulting from gravitationally trapped asteroidal particles (Fig. 173). Quantitative modeling led to the conclusion that at least one third if not the majority of interplanetary dust particles would

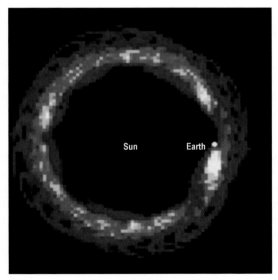

Fig. 173. Computer simulation of a ring of asteroidal dust particles along the Earth's orbit (from Dermott et al. 1994). Evidence for a similar dust ring along the orbit of Venus was recently found in the data of the Helios spaceprobes

See S. Dermott et al. Nature 369, 769 (1994).

be of asteroidal origin. For the time being, it appears that both, comets and asteroids, are important sources for the origin of interplanetary dust.

Acknowledgement. The author thanks Uwe Berger, Franz-Josef Lübken, Kristian Schlegel and Ulf von Zahn for expert advise and reading of the manuscript, and Monika Dueck for patient and efficient help with the acquisition of literature.

13.11 The Terrestrial Atmosphere and Its Effects on Astronomical Observations

F. Schmeidler[†]

13.11.1 General Remarks on the Atmosphere

The atmosphere of the Earth is a source of numerous unfavorable effects on the performance of astronomical observations for professional as well as amateur observers. These influences can be largely eliminated by, for instance, observing from a mountain top or from an airplane, rocket, or spacecraft. But such opportunities are normally not available to the average observer, and thus the impact of the atmosphere upon astronomical observations requires serious consideration. In this regard, there should be a careful distinguishing between factors which are strongly weather dependent and those which remain about the same regardless of weather conditions.

The atmospheric influence on radio observations will not be discussed here; it has been dealt with in Chap. 9. While the effects in the radio range bear some similarities to those in the optical range, they are nevertheless in many respects very dissimilar.

Information about the global atmospheric research program (THORPEX) can be found online at: http://www.wmo.int/thorpex/mission.html. From which the following is quoted: "Emphasis on ensemble prediction, and global-to-regional influences on weather forecast skill. Interactive forecast systems 'tuned' for end users using targeted observations called on in 'sensitive areas', adaptive data assimilation, grid computing."

13.11.2 Weather-Dependent Phenomena

This section will concern itself primarily with the constraints imposed by various weather patterns on the possibilities of optical astronomical observations. Obviously, no optical work can be performed with an overcast sky (although radio observations are still possible), but there exist other factors which may preclude astronomical observations even under cloudless or slightly cloudy skies.

Information about weather, clouds and climate can be found online at: http://www.wetter.de http://www-imk.physik. uni-karlsruhe.de/~muehr/ wetter.html

13.11.2.1 Assessment of Weather Patterns Factors evident from weather maps , such as the geographic distribution of highs and lows provide some guidance as to the observing conditions on a given night, but of course not an unambiguous correlation. Weather patterns may have substantially different local effects at places less than 50 km apart, owing to the influence of, for instance, water surfaces (e.g., lakes and rivers) and upslope or downslope mountain winds. This also affects (locally) the total annual average number of clear nights. On a larger scale it is the seasonal distribution of clear nights, which is usually fairly constant with time. Large parts of western Europe as well as California, for instance, experience, as a rule, the best observing conditions in late summer and autumn, the poorest in late winter, while much of the central and eastern US has the worst season in the late spring and early summer.

The German Weather Service uses the following classification of large-scale weather patterns, as graphed in Fig. 174, characteristic for much of Europe. (Note that T in this case stands for the German for *low, tief*.)

W   Westerlies
BM   High-pressure ridge over central Europe
HM   High centered over central Europe
SW   Southwest flow (High over southeastern Europe, Low over North Atlantic)
NW   Northwest flow (High over western Europe)
HN   High over Arctic Ocean
HB   High over Great Britain
N   North flow (High over North Sea, Low over eastern Europe)
TrM Low pressure trough over central Europe
TM   Closed Low over central Europe
TB   Closed Low over Great Britain
TrW Trough over western Europe
S   South flow (Low over western Europe, High over eastern Europe)
SE   Southeast flow (High over eastern Europe, Low over Mediterranean)
HF   Closed High over Fennoscandia
HNF Closed High over Fennoscandia and Arctic Ocean
NE   Northeast flow (High bridge from Azores to Fennoscandia)
Ww   Angular westerly flow

BM, HM, and HB are always anti-cyclonic over central Europe (index a), while TrM, TM, TB, and TrW are always cyclonic (index z). The other patterns may have either isobaric curvature.

Statistical data show that none of these patterns guarantees or prohibits cloudless skies, but there are correlations. Most favorable are HM and BM, as might be expected, whereas certain cyclonic flows like NWz and SEz are found least promising. A count of observing nights in Munich, Germany, has led to the following groups with regard to astronomical promise:

- Favorable: HM, BM, Wa, HNa, SWz, TB, HB, NEa
- Average: SWa, Na, SEa, HFa, HNFa, TrW, HFz, Ww
- Unfavorable: All other patterns

An instance of local effects is here the high incidence of clear weather in association with certain westerly flows (Wa, SWz, TB) as a consequence of downhill foehn winds north of the Alps.[1]

---

[1] Analogous maps for the eastern US (specifically the mid-Atlantic seaboard) are added in Fig. 175.

(1) *Warm Atlantic (Bermuda) High off-shore, or extending inland*: frequently good seeing but poor transparency owing to accumulating moisture.

(2) *Rear of cold High, normally traveling eastward*: usually indicates fair weather, but only for short duration.

(3) *Traveling cold front*: on its rear side, the transparency (outside the cloud range from the Great Lakes) is often excellent, the seeing invariably poor for about 36 hours.

(4) *Stationary front along shore* (particularly when held in place by a Bermuda High): expectation of some interval of inclement weather owing to probability of secondary lows forming.

(continued on page 557)

Information about weather charts can be found at:
http://www.maps.ethz.ch/cat_int7t.html#klima

European Organization for Meteorological Satellites (EUMETSAT) :
http://www.eumetsat.de

National Oceanic and Atmospheric Administration, USA (NOAA) :
http://www.noaa.gov

Information about the global weather situation can be found on the Internet at:
http://www.wmo.ch
http://www.weatherbase.com
http://www.worldclimate.com

For information about clouds and their formations see:
http://www.wolkenatlas.de

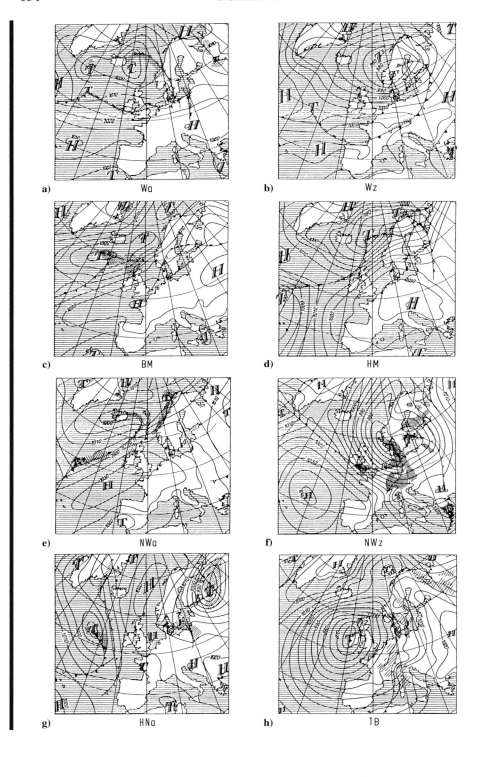

a) Wa

b) Wz

c) BM

d) HM

e) NWa

f) NWz

g) HNa

h) TB

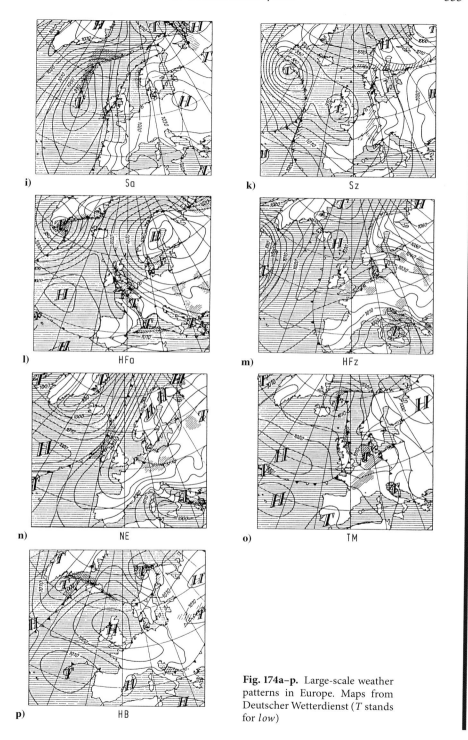

Fig. 174a–p. Large-scale weather patterns in Europe. Maps from Deutscher Wetterdienst (*T* stands for *low*)

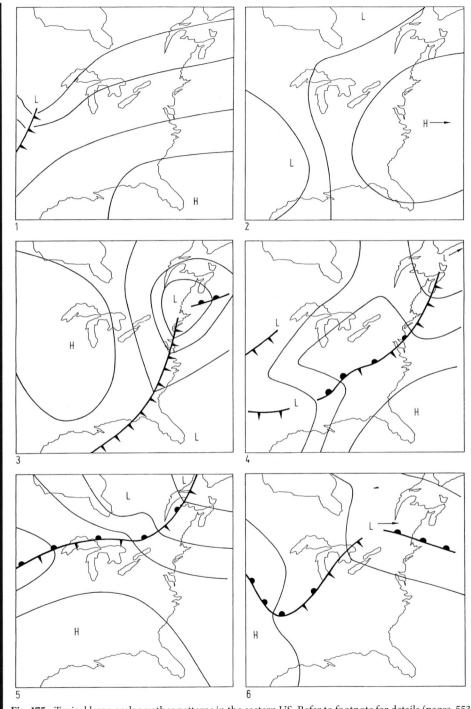

Fig. 175. Typical large-scale weather patterns in the eastern US. Refer to footnote for details (pages 553 and 557)

13.11.2.2 Atmospheric Turbulence and Scintillation
A certain amount of turbulence is present even in cloudless skies, and causes some vibration of an image in the telescope. This unsteadiness can be so strong that it is often perceived by the naked eye as the "twinkling" phenomenon in bright stars. It is caused by small, local density and temperature differences in the atmosphere effecting changes in the refraction index. When substantial air motions are added, the variation can be quite rapid, thus causing the images to flicker wildly. Essentially the same effect can be observed on a hot summer afternoon above the ground, and in particular above hot asphalt.

Turbulence brings about a change in both the apparent brightness and position of a star. One thus distinguishes between (1) intensity scintillation, or simply scintillation, and (2) directional scintillation, or seeing, although the term seeing is often used to include both phenomena. Directional scintillation may manifest itself as either a displacement of the entire image around a mean position, or as a diffuse apparition of a nonmoving image. Some authors summarily state that the displacement dominates for small apertures, while the fuzziness or diffuseness prevails at large apertures. Not all observers concur with this assessment, however, and the experiences are too diverse for all to be placed over a common denominator.

Just how much the atmospheric unsteadiness impedes astronomical observations depends on the specific purpose of the observing program. Most affected are studies which depend crucially upon resolving power, such as measurements of close double stars and visual observations of planetary surface details; many nights are completely useless for work of these kinds. Unsteadiness can be an especially bothersome impediment for astronomical photography in which long focal lengths and high magnifications are needed to produce detailed pictures of the Sun, Moon, or planets. Photographic work at short to moderate focal lengths (under 2 m), on the other hand, is less sensitive to poor seeing, as are photometric observations. For observations without a telescope (e.g., of meteors and the zodiacal light), atmospheric scintillation has practically no influence.

The degree of turbulence is influenced by various factors. It is obviously larger near the horizon than near the zenith, but even this almost trivial rule is not always complied with; there are nights in which the unsteadiness increases very little or not at all from the zenith to the horizon, and there are nights when the increase is quite large. Unambiguous criteria for these cases are thus far unknown.

On a statistical average, the turbulence depends quite strongly on the time of day or night. The high degree of unsteadiness around noon is as readily understood as the observed minima of scintillation before sunrise and after sunset. What is not expected is the secondary maximum of unsteadiness that numerous observers concordantly find around midnight, but sometimes earlier or later. It is caused by a corresponding secondary max-

Some observers who have worked with telescopes of different sizes have noted that images suffer greater distortion with larger apertures, because more simultaneous turbulence elements in the beam contribute to the diffraction. This is compensated by the higher image brightness, as faster eye perception and shorter recording time cut the time variation of turbulence. In a nutshell, the advantage of larger apertures is most pronounced in good seeing and in poor transparency. The most detrimental scintillation frequencies are in the range about 1 to 5 Hz, but this depends on the surface wind speed: the stronger the wind, the higher the contributing frequencies.

(5) *Westerly weather* (particularly when supported by a High bridge over the southern states): generally good, often stable (unless a strong pressure gradient at the polar front to the north triggers cyclonic motion).

(6) *Warm-air sector of northern traveling Low* (usually small pressure gradients in the region): frequent, but least predictable regarding cloudiness.

These samples should enable observers to recognize the most typical patterns and how they correlate with cloud cover under other weather regimes. See also E. Palmen and C.W. Newton, *Astronomical Circulation Systems*, Academic Press, New York 1969, for a more comprehensive description of weather patterns, and H.H. Lamb, in *The English Climate*, English Univ. Press, London 1964

imum of atmospheric turbulence at that time, following the effects of atmospheric mass exchange. Consequently, the visual observer regularly finds that the image quality in the telescope gradually improves in the hours after sunset, but deteriorates again one or two hours before midnight.

Individual cases may differ substantially from these averages. As always in meteorological problems, there are no sure-fire rules, but rather only crude guidelines as to when to reckon with favorable or unfavorable conditions. As a rule, wind (even a weak one) is associated with a noticeable degree of unsteadiness, but this self-evident correlation also has numerous exceptions. High humidity almost always improves the image quality to the extent that observations through ground fog often make very steady images—just so long as the star is visible at all!

The seasonal variation of unsteadiness is very different, depending on locale. For instance, observers in northern Germany report the best seeing during the months of spring, while in southern Germany, experience shows the late summer and fall (August through October) to be the best months, without any readily apparent cause for this difference. The aforementioned large weather patterns show only a weak correlation with scintillation. In brief, anti-cyclonic situations (when the wind flow is curved clockwise in the northern hemisphere) are more conducive to favorable weather conditions than are cyclonic ones. An evident rule of thumb is that the stabler the large-scale state of the atmosphere, the better the astronomical image quality.

Finally, there are various local, microclimatic influences which can modify the situation entirely, even over short distances, such as nocturnal heat from an urban climate, forests protecting a nearby town from wind, or sloping terrain causing turbulent winds.

Overall, the factors affecting image quality on a given night are so complex that even with extensive experience in a particular locale, it is not possible to make reasonably good predictions for more than 50% of the cases.

13.11.2.3 Halos, Rainbows, and Other Optical Phenomena This section deals with features which are technically not of an astronomical nature, but which are naturally so often noticed by observers that a short description of their basic features is warranted. A *halo* is (usually) seen around a bright source (e.g., the Sun or Moon) as a ring-shaped apparition of light, which originates when light is refracted or reflected by ice crystals in the atmosphere. Only when these crystals are present at a certain level in the atmosphere can the halo arise, which means that the phenomenon is connected with specific meteorological patterns. The kind, distribution, and shapes of ice crystals give rise to a variety of halo types. Also, a preferential orientation must be present, because a random distribution of crystal axes would distribute the light isotropically and thus not produce any definite luminous phenomenon.

A lot of information can be found in text books and popular science books (see "Further Reading" on p. 566)

A faint ring of radius 22° around the Sun or Moon is most frequently observed as the halo, and its inner edge is distinctly reddish. Sometimes, the so-called large halo with a radius of 46° appears instead. The so-called horizontal circle lies parallel to the horizon through the position of the Sun. Its intersection with the halo proper sometimes generates particularly bright spots of light called *sun dogs*.

Cirrus clouds are the most evident indicators of the requisite ice crystals in the atmosphere, and thus halos are most often observed in the presence of this type of cloud.

The phenomenon of the *rainbow* is much more widely known than halos. It is based on similar optical principles, but in this case the light is refracted by water droplets instead of ice crystals. Since water in liquid particles has only one symmetric shape (i.e., spherical), the rainbow does not show the confusing variety of appearances, but rather has only two shapes: the *primary rainbow* surrounds the point opposite the Sun with an angular radius of 42°, while the less frequently seen (and much fainter) *secondary rainbow* possesses a 51° radius. Both rings are best seen with the Sun at the horizon.

The distinctly visible color phenomenon in rainbows gave rise to the proverbial expression of "all the colors of the rainbow." The main bow has red on the outer and violet on the inner side, and the secondary bow reverses this order. Weak secondary bows are occasionally seen joined to one or both of these bows.

The appearance of rainbows requires, of course, the illumination of water droplets in the atmosphere by the Sun or Moon. This may happen when, for instance, during or after a rainstorm the Sun emerges from a small opening among the clouds. Even under cloudless skies, rainbows can be seen near sprinklers or water fountains. Rainbows can be caused by moonlight too, but these usually appear colorless since their low brightness does not permit the human eye to discern the colors even though they are present.

Under suitable conditions with respect to kind, distribution, and density of clouds, a few other phenomena such as the *corona* and the *glory* can appear because of the Sun or Moon. The latter are colored rings of small diameter around the point opposite the Sun. There is some fascinating information on all the phenomena described here in papers by B. Albers [110] and C. Leinert [111, 112]. See also [113, 114, 115].

13.11.3 Permanent Atmospheric Phenomena

In contrast to the weather-dependent optical phenomena discussed thus far, there are several atmospheric influences which are always present in any weather situation whether or not the sky is clear. Their effects should, if necessary, be considered in astronomical observations.

13.11.3.1 Refraction Like any physical medium, atmospheric air has a refractive index, so that light rays entering the atmosphere from empty space suffer a small but measurable change in direction. Of course, this fact is significant only if the direction of the incident light (i.e., the position of the celestial body at the sphere) is of importance. The amount of refraction, which always causes a celestial light source to appear elevated above its true position relative to the horizon, depends on both the height or zenith distance of the source and on the color of the light. When two neighboring celestial objects are, owing to a small difference in zenith distance, subject to different amounts of refraction, the result is termed *differential refraction*. The fact that light from a source at a sufficiently large zenith distance is, owing to the wavelength dependence of the refractive coefficient, spread out into a spectrum is termed *atmospheric dispersion*.

The general refraction naturally vanishes in the zenith and increases toward the horizon. The rule of increase has been the subject of detailed theories. At zenith distances down to around 75°, the amount of refraction is practically independent of the atmospheric stratification, that is, of the prevailing density and temperature decrease with altitude (this is known as the *law of Oriani and Laplace*). An approximation adequate

Color images as well as general information on atmospheric optics can be found on the Web:
http://www.atoptics.co.uk
http://www.meteoros.de
http://mintake.sdsu.edu/GF/
http://www.polarimage.fi
http://www.allthesky.com/
atmosphere/atmosphere.html

for most observations can be derived from the simplifying assumption that the Earth's atmosphere extends with constant density to a certain upper level, and thus a light ray from space suffers a one-time refraction at that level. Consequently, the direction of the ray after refraction is observed from the ground as the apparent zenith distance z; it was $z + R$ before refraction, where R is the amount of refraction. According to the law of refraction with μ as the refractive coefficient of atmospheric air,

$$\sin(z + R) = \mu \sin z ,\qquad(29)$$

where R is a small angle so that $\cos R = 1$ and $\sin R = R$ can be substituted. Expanding the left side gives

$$\sin(z + R) = \sin z \cos R + \cos z \sin R ,\qquad(30)$$
$$R = (\mu - 1) \tan z .\qquad(31)$$

This is the well-known rule that refraction is proportional to the tangent of the zenith distance. The constant of proportionality is very close to $1'$, which gives the rule of thumb: The amount of refraction in arcminutes equals $\tan z$.

Only at large zenith distances, or when high precision is needed, does this rule break down. More accurate theoretical work shows refraction to be expressible as a series of odd powers of $\tan z$. Detailed tables or precise calculation of refraction include terms down to the thirteenth power of $\tan z$.

The amount of refraction also depends on air pressure and temperature since the refractive index is a function of these quantities. If R_0 is the "mean" refraction, valid for $0°C$ and $760\,\text{mm Hg} = 1013\,\text{mb}$, then

$$R = \left(\frac{b}{1013}\right)\left(\frac{1}{1 + T/273}\right) R_0 ,\qquad(32)$$

where b is the atmospheric pressure in millibars (mb) and T the temperature in $°C$.

The mentioned expansion of the refraction in powers of $\tan z$ fails entirely near the horizon where $\tan z \to \infty$. Calculating the horizontal refraction is thus a specific problem in any refraction theory, into which the vertical stratification of the atmosphere also enters. Observations reveal that horizontal refraction amounts to about $34'$. A star apparently just setting is actually already one-half degree below the horizon. This fact results in an increase in the length of the day. The beginning and end of day are defined by the rising and setting of the upper limb of the solar disk. As the apparent radius of the Sun is $16'$, then at the moment of sunrise the center of the Sun's disk is in reality $34' + 16' = 50'$ below the horizon. In mid-latitudes, this increase in the length of the day amounts to about 8–12 min, but in the polar regions it is substantially greater.

Another consequence of refraction (in this case, differential refraction) is the apparent elliptical shape of Sun and Moon when near the horizon. The lower limb of the disk, being closer to the horizon, suffers a stronger "lifting" by refraction, which reduces the apparent angular diameter of the Sun. Of course, this phenomenon occurs at any altitude of the Sun, but only near the horizon is the effect sufficiently pronounced to become conspicuous to the eye.

Extraordinary atmospheric conditions may cause strong distortions of the solar/lunar disks. Anomalous refraction can also lead to *mirages*, which are sometimes observed in desert regions.

13.11.3.2 Extinction The attenuation or extinction of light from space by the Earth's atmosphere has two physically very different causes: (1) light is partially absorbed by the air molecules (i.e., converted into heat), and (2) partially deflected from its direction (i.e., scattered). The total light loss is thus a combination of contributions from absorption and scattering. The amount of extinction can be expressed in terms of the *extinction coefficient*, or sometimes by its complement, the *transmission coefficient*.

The extinction coefficient states the percentage of vertically incident radiation (star at the zenith) which is absorbed. The transmission coefficient gives the percentage of radiation reaching the ground. Both depend on the degree of *opacity* of the atmosphere, and also very much on the wavelength of the light. Strictly speaking, the definition given is valid only for monochromatic light, but only a very small error is incurred in the case of polychromatic light by assuming a simple *mean* extinction or transmission coefficient over the relevant range. The transmission coefficient of the atmosphere under average conditions and cloudless skies is about 0.8; in other words, a star near the zenith loses about 20% of its brightness. This is valid for the visual spectral range; in the photographic (blue) range, the extinction is usually higher by a factor 1.5 or 2, and is also much more dependent on the specific atmospheric status. The extinction also depends somewhat on the color of the star.

Of course, the amount of extinction increases toward the horizon. To a very good approximation, the attenuation of starlight is proportional to the secant of the zenith distance. One does not normally express a measured stellar magnitude corrected entirely for atmospheric influences, but merely states the magnitude which the star would have at the zenith. Thus, if p is the transmission coefficient, then the amount of "reduction to zenith" which must be applied is

$$\Delta m = -2.5 \log p (\sec z - 1) . \tag{33}$$

The factor $-2.5 \log p$ is about $0^m.25$ to $0^m.30$ in the visual range and $0^m.40$ to $0^m.60$ in the photographic range. This reduction is subject to factors which are rarely controllable, and therefore it is advisable to limit photometric measurements to small zenith distances (if possible, smaller than $30°$). The measurements should be arranged such that stars at about equal zenith distances are compared so that only a small extinction difference needs to be determined. As expected, the atmospheric extinction at high elevations is greatly reduced. The numbers given here are for low altitudes of around a few hundred meters above sea level; for mountain-top observatories, the amount of extinction may be as much as ten times lower, but the exact amount must be determined for each location.

Measurements by Arsenijevic [116] showed that substantial seasonal variations in the extinction occur on the island Hvar in the Adriatic Sea. Whether or not this is due to a one-time, local effect or a more widespread phenomenon has not yet been determined. In any event, there is good reason to regularly check the extinction coefficient when making photometric measurements.

The fact that extinction is *selective* (i.e., wavelength dependent) gives rise to a number of well-known phenomena. Short-wave light is attenuated much more than long-wave light. For this reason, the Sun, Moon, and stars appear much redder near the horizon than at the zenith. Another consequence of this is the blue color of the daytime sky, caused by scattering of sunlight by atmospheric dust. The red portion of direct sunlight is scattered very little, the blue part quite strongly. Thus, the light which dust particles at large angular distances from the Sun receive and then scatter toward the observer is predominantly of blue color. If there were no atmosphere or one without scattering, the sky would appear deep-black in the daytime, and only at the exact locations of Sun and Moon would direct light be seen at all.

The mathematical rules governing how extinction depends on wavelength differ for the various physical processes involved. The simplest is *Rayleigh's law* of scattering by air molecules, which gives the amount of the light lost by scatter as inversely proportional to the fourth power of the wavelength ($\sim \lambda^{-4}$). Since the atmosphere contains not only air molecules but also dust particles of various sizes and properties, the actual situation is more complex and also time variable. By and large, the extinction caused by dust can be taken as proportional to λ^{-1} to $\lambda^{-1.5}$. The net result is a combination of both the air and dust contributions in a manner which varies from case to case.

13.11.3.3 Twilight The phenomenon of twilight is also a consequence of light scattering in the atmosphere, and can also be treated in conjunction with the extinction. Even after evening twilight ends, the sky is not perfectly dark, but retains a small, nocturnal brightness of its own. Without atmospheric scatter of light, the day would pass immediately into dark night; indeed, this transition takes place very rapidly in the desert, where the air normally contains very little dust.

The phenomenon of twilight occurs after sunset and before sunrise, when no direct sunlight reaches the observer, but does reach the upper layers of the atmosphere where it is scattered in different directions by air molecules; after one or more scatters, some fraction of this radiation reaches the eye of the observer at the ground.

There are actually several definitions of twilight. *Civil twilight* is defined as the time interval during which it is possible to read comfortably under cloudless skies. It ends or begins with the Sun 6° below the horizon. *Astronomical twilight* begins and ends with the altitude of the Sun at $-18°$, characterizing the point at which no trace of scattered sunlight is seen. The term *nautical twilight*, which is limited by a solar altitude of $-12°$ and characterized by the visibility limit of the horizon at sea, is also occasionally heard mentioned. Naturally, the actual duration of twilight is much shorter under cloudy skies. On the occasion of exceptionally clear weather, the Earth's shadow may be glimpsed during evening twilight in the eastern sky as a rather sharp boundary between dark and slightly brightened sky, and similarly during the dawn twilight in the western sky.

At mid-latitudes, civil twilight lasts about 30 to 40 min. It is longest in summer and winter, and shortest in spring and fall. Astronomical twilight varies little in length during fall, winter, and spring. Its duration is about one and one-half hours at latitudes of 30° and two hours at 50°, but it increases at higher latitudes during summer. Latitude 49° marks the beginning of the zone of the so-called white nights, where, during a certain period around the summer solstice, the astronomical twilight lasts all night as the Sun is never lower than $-18°$. The farther north the observer, the longer this time interval, for example, it lasts:

- At latitude 49° from June 11 to July 3
- At latitude 52° from May 21 to July 23
- At latitude 55° from May 9 to August 5
- At latitude 58° from April 29 to August 15

The transition to latitudes of the white nights is so sharp that even at 48° (e.g., Munich, Vienna, Vancouver), almost nothing is seen of twilight at midnight. In some years or on certain nights, the situation may be different depending on the amount of dust in the atmosphere.

A number of interesting color phenomena, which under favorable circumstances can be seen during twilight, will not be dealt with here because of their exclusively meteorological significance. Moreover, satisfactory explanations do not yet exist for most of them. Twilight is, of course, of great importance to the astronomer, as it indicates when the stars will become visible (and for how long) so that he can begin (and later terminate) his observations. The exact starting and stopping times depend on the purpose of the observation, which dictates at what sky brightness this is possible. Generally speaking, photometric measurements require perfectly dark skies, while positional measurements may be made just as soon as the sought star can be seen.

This also raises the question of visibility of planets and stars during the daytime. The unaided eye can find the planet Venus at greatest brilliancy ($-4^m.5$) when the position is exactly known. Fainter objects are invisible during the daytime without telescopic aid, and the often-heard myth that stars can be seen during daylight hours from the bottom of a deep well is wholly false. When using a telescope, on the other hand, the background brightness of the sky will interfere less with the visibility of stars the larger the objective aperture and the smaller the focal ratio. Thus moderately bright stars can be seen during the daytime in the telescope. Under excellent conditions and full daytime brightness, a 2-in. telescope reveals stars of magnitude 1, while a 4-in. brings out nearly magnitude 3 stars. Conditions are considerably better on mountain stations. However, the observation of planets with substantial disks (Mars, Jupiter, Saturn) is pointless in the daytime because the available light, though substantial, is distributed over a relatively large area. In the case of Venus or Mercury, observability with the telescope in the daytime depends very much on its phase, and also, of course, on the magnification chosen. It is worth mentioning that even the Moon is difficult to observe in the daytime with a telescope.

13.11.3.4 The Brightness of the Night Sky
The faint residual brightness of the sky after the end of astronomical twilight consists of several contributions, and—even excluding the part caused by artificial illumination—its intensity is subject to time- and weather-dependent variations. The mean background brightness of the night sky corresponds roughly to that of a star of magnitude 22^m per square arcsecond. This places a practical limit on the ground-based observation of objects with faint surface brightnesses, although it is significant only for very large telescopes.

The night sky brightness can be traced to the following causes:

(a) Residue of twilight
(b) Recombination light of air molecules
(c) Zodiacal light
(d) Faint stars and nebulae
(e) Scattered light of the sources (b), (c), and (d)
(f) From time to time, aurorae or noctilucent clouds

The causes mentioned under (c) and (d) are of an extraterrestrial nature, while the others lie in the atmosphere itself. The entire night background has a slightly reddish tinge, but this color is of course too weak for the eye to perceive. For medium- and small-sized telescopes, the sky background does not interfere with observations.

13.11.3.5 The Polarization of Sky Light The diffuse scattered light of the sky is polarized to a greater or lesser degree. In other words, the direction of electromagnetic vibration at right angles to the propagation of the light is not random but rather oriented in a somewhat preferential direction. Although this fact is practically insignificant for most astronomical observations, it should be included in a complete discussion of the optical phenomena connected to the Earth's atmosphere.

The cause of the polarization of the sky background light is the scatter by air molecules. The laws of optics show that, during the processes of reflection, refraction, and scattering, a certain portion of the light waves becomes polarized. The percentage of polarized compared with total light, the so-called *degree of polarization*, depends on circumstances such as the angle of incidence or the particle size. While the human eye cannot directly distinguish polarized from unpolarized light, some animals, such as bees, have been provided with this faculty by nature. The human eye, however, can see the polarization effects by viewing the incident light through a polarization filter. When rotated, the brightness of unpolarized light remains the same in any position of the filter, while for polarized light a maximum brightness occurs in that position where the vibration direction in the filter coincides with that of the incident light. The degree of polarization of diffuse sky light depends on the position of the region in the sky relative to the Sun. In a cloudless sky, there are a few (usually three) positions free from polarization; all three points lie in the vertical through the Sun, and are:

1. The *Arago Point*: 20° above the point opposite the Sun
2. The *Babinet Point*: 10° above the Sun
3. The *Brewster Point*: 15° below the Sun

The Arago Point is above the horizon only when the Sun is rather low. The numbers of heights for the three points given are averages and may vary with opacity and other atmospheric parameters. The highest degree of polarization (usually some 60 to 80%) is reached at the point of the vertical through the Sun and 90° from it. Here, too, the degree of polarization and, to a lesser extent, the position of maximum polarization depend on atmospheric opacity. In general, the variation of polarization in the sky can be described by combining two rules:

1. The degree of polarization increases with the distance from the Sun up to 90° and then decreases again.
2. The degree of polarization diminishes toward the horizon.

13.11.3.6 The Apparent Shape of the Celestial Sphere The fact that the celestial vault appears to the human eye not as a perfect hemisphere but rather flattened has nothing to do with the atmosphere, but it is worth briefly mentioning here. The observer has the impression that the horizon is more distant than the zenith point. Of course, this is not a genuine phenomenon, because nowhere does the sky as such have a well-defined "distance" from the observer; in principle, the view everywhere reaches into infinity. The phenomenon is of a physiological nature, and is connected with the structure of the human eye. This is easily recognized: an observer lying prostrate on the ground sees the zenith point of the sky much farther away than when standing upright.

An important consequence of this property of the human eye is that angles of altitude are usually substantially overestimated. Although the amount of error varies from case to case, the figures presented in Table 23 may be taken as averages.

Thus the height of an object not too far above the horizon may be overestimated by up to 20°. The effect is smaller at night than in the daytime.

Because of this subconscious overestimation of altitudes near the horizon, the Sun and Moon when rising and setting appear larger than when at the zenith. The same holds true for constellations. It is also a consequence of the properties mentioned of the human eye that stars seem to appear near the zenith even at true zenith distances of 10° to 20°. The observer should take this effect into account when estimating the altitudes of stars.

Table 23. Comparison of the true with the estimated altitude

True Altitude	Estimated Altitude
0°	0°
15°	30°
30°	50°
45°	65°
60°	75°
75°	84°
90°	90°

13.11.4 Site Selection for Astronomical Observations

The problem of site selection for major observatories has been the subject of discussions among astronomers for many decades. These studies were occasioned by the construction of a substantial number of large telescopes, whose high costs were justifiable only by optimum utilization in exceptionally dark, tranquil skies. As such, this problem is rarely encountered by observers who use small telescopes. Nevertheless, if an astronomer is considering several different sites at which to locate a telescope or an observatory, it might be prudent to weight the final decision according to those criteria which were gained by site testing for large telescopes. Only atmospheric criteria will be mentioned here, but other practical considerations, for instance, transportation accessibility, might also play an important role.

Of primary importance is unquestionably the frequency and kind of cloudiness at the site considered to accommodate astronomical instruments. This information may be obtained by the nearest meteorological station, but it should be kept in mind that neighboring locations often have substantial differences in cloudiness, particularly in mountainous regions. It is thus advisable to perform the crucial observations at the site itself. Also, the low cloudiness rate by itself does not make a site particularly suited for observations; the air steadiness under average conditions should also be carefully examined. Finally, the amount of dust in the atmosphere plays a role by dimming the starlight.

Although individual situations will depend on various factors difficult to separate, efforts toward comprehensive evaluation of numerous experiences have been made. At best, there exist only some crude guidelines. Of course it is desirable in any event to avoid the nearness of urban areas and industry. On the other hand, to build observatories on high mountain tops, as is often recommended, does not necessarily prove advantageous for every purpose: while the atmospheric transparency on mountain tops is superior to that in low-lying regions (very much so in the infrared owing to the lesser H_2O content), high winds and large local and temporal temperature differences usually also increase the air turbulence.

In view of the above considerations, the best compromise of avoiding the atmospheric dimming as well as its inherent unsteadiness seems to be a smooth plane located at a high altitude above sea level. Indeed, the astronomical site tests made in Chile and in south-western Africa seem to verify this statement. Another possibility which has proven very favorable is a site surrounded on several sides by water (e.g., an island). The thermal inertia of large bodies of water forestalls the buildup of strong temperature gradients in the air which would generate much turbulence.

Additional information on the influence of local atmospheric conditions on astronomical observations is provided in an article by M.F. Walker [117], which also gives some details on physical processes in the atmosphere that may impede the observations. The article also presents some useful hints on how the use of a few simple accessories can permit at least a qualitative testing of the situation.

13.11.5 Further Reading

Minnaert, M.G.J.: Light and Color in the Outdoors (original 1937, new English issue). Springer, Berlin Heidelberg New York (1993)

Tape, W.: Atmospheric Halos. American Geophysical Union, Washington, DC (1994)

Schlegel, K.: Vom Regenbogen zum Polarlicht. Spektrum, Heidelberg (1995)

Vollmer, M.: Lichtspiele in der Luft: Atmosphärische Optik für Einsteiger. Spektrum/Elsevier, Heidelberg (2006)

Lynch, D.K., Livingston, W.: Color and Light in Nature, 2nd edn. Cambridge University Press, Cambridge (2001)

Roth, G.D.: Die BLV Wetterkunde. blv, München (2009)

References

Section 13.1

1. Beck, R., Hilbrecht, H., Reinsch, K., Völker, P. (eds.): Solar Astronomy Handbook. Willmann-Bell, Richmond, VA (1995)
2. Reinsch, K., Beck, R., Hilbrecht, H., Völker, P. (eds.): Die Sonne beobachten. Sterne und Weltraum, Heidelberg (1999)
3. Taylor, P.O.: Observing the Sun. Cambridge University Press, Cambridge (1991)
4. Kiepenheuer, K.O.: Solar Site Testing. In: Rösch, J.: Site Testing, IAU Symposium No. 19, p. 193. International Astronomical Union, Paris (1962)
5. Müller, R.: Sichtbedingungen bei Sonnenbeobachtungen. Sterne und Weltraum **1**, 170 (1962)
6. Bray, R.J., Loughhead, R.E.: Sunspots. Chapman and Hall, London (1964) and New York (1979)
7. Wilson, P.R.: The Structure of a Sunspot. Solar Phys. **3**, 243 (1968)
8. Solanki, S.: Sunspots: An Overview. Astron. Astrophys. Rev. **11**, 153 (2003)
9. McIntosh, P.S.: The Classification of Sunspot Groups. Solar Phys. **125**, 251 (1990)
10. Giovanelli, R.G.: Secrets of the Sun. Cambridge University Press, Cambridge (1984)
11. Waldmeier, M.: Sunspot Numbers and Sunspot Areas. Astronomische Mitteilungen der Eidgenössischen Sternwarte Zürich, **358** (1978)

12. Künzel, H.: Hinweise für die heute übliche Zählweise von Sonnenflecken zur Bestimmung der Relativzahl. Astronomie und Raumfahrt **14**, 121 (1976)

13. Beck, R.: Eine neue Definition der Sonnenfleckenrelativzahl. Sonne 2, **1**, 56 (1977)

14. Pettis, H.S.: Eine systematische Studie von Sonnenflecken. Saturn (Astron. Arbeitsgem. Paderborn) **11** (1978)

15. Malde, K.I.: Klassifikationswerte: eine neue Messung der Sonnenaktivität? Sonne 36, **9**, 159 (1985)

16. Keller, H.U.: A-Sonnenfleckenbeobachtungen von bloßem Auge. Orion **38**, 180 (1980)

17. Keller, H.U.: Der Sonnenfleckenzyklus No. 21 von bloßem Auge registriert. Orion **44**, 154 (1986)

18. Waldmeier, M.: Neue Eigenschaften der Sonnenfleckenkurve. Astr. Mitt. Eidgen. Sternw. Zürich **133**, 105 (1935)

19. Bendel, U., Staps, D.: Kurz- und mittelfristige Sonnenfleckenprognose mit der P17-Mittelung. Sterne und Weltraum **19**, 180 (1980)

20. Gleissberg, W.: Die Häufigkeit der Sonnenflecken. Akademie-Verlag, Berlin (1952)

21. Solanki, S.K., Usoskin, I.G., Kromer, B., Schüssler, M., Beer, J.: Unusual Activity of the Sun During Recent Decades Compared to the Previous 11,000 Years. Nature 431, 28 (2004)

22. Karkoschka, E.: Neue Relativzahl-Mittelung. Sonne 9, **3**, 33 (1979)

23. Sheeley, N.R.: Polar faculae: 1906–1990. Astrophys. J., **374**, 386 (1991)

24. Tandberg-Hanssen, E.: The Nature of Solar Prominences. Kluwer Academic Publishers, Dordrecht (1995)

25. Völker, P.: Die Protuberanzenbeobachtung des Amateurs. VdS-Nachrichten, **19**, 14 (1970) in Sterne und Weltraum, **9** (1970)

26. Waldmeier, M.: Ergebnisse und Probleme der Sonnenforschung. Akademische Verlagsgesellschaft, Leipzig (1941)

27. Roth, G.D.: Compendium of Practical Astronomy. Springer, Berlin Heidelberg New York (1994)

28. Stetter, H.: Protuberanzenaktivität: Ergebnisse 20-jähriger visueller Beobachtung. Sonne 118, **30**, 54 (2007)

29. Völker, P.: Die H-α-Relativzahl: Ein Beobachtungsprogramm für das PST. Interstellarum 14 (57), 36 (April/May 2008)

30. Kiepenheuer, K.-O.: Die Sonne. Springer, Berlin Heidelberg New York (1957)

31. Bruzek, A., Durrant, C.J.: Illustrated Glossary for Solar and Solar-Terrestrial Physics. Reidel, Dordrecht (1977)

32. Stix, M.: The Sun: An Introduction, chap. 7. Springer, Berlin Heidelberg New York (2002)

33. Howard, R.: Solar Rotation. Annual Review of Astronomy and Astrophysics, vol. 22, pp. 131–155. Annual Reviews, Palo Alto, CA (1984)

34. Balthasar, H., Stark, D., Wöhl, H.: The Solar Rotation Elements i and Ω derived from Recurrent Single Sunspots. Astron. Astrophys. **174**, 359 (1987)

35. Zerm, R.: Bestimmung der differentiellen Rotation der Sonne. Astronomie und Raumfahrt **22**, 127 (1984)

36. Joppich, H.: Differentielle Rotation 2005, Sonne 117, **30**, 20 (2006)

37. Balthasar, H., Vàsquez, M., Wöhl, H.: Differential Rotation of Sunspot Groups in the Period 1874–1976. Astron. Astrophys. **155**, 87 (1986)

38. Möller, M.: Synoptische Karte der Rotation 2030. Sonne 114, **29**, 55 (2005)

39. Möller, M.: Butterfly diagram. Sonne 122, **32**, 66 (2008)

40. Hammerschmidt, S.: Schmetterlingsdiagramm. Sonne 11, **11**, 11 (1987)

41. Yallop, B.-D., Hohenkerk, C.Y.: Solar Butterfly Diagram 1874–1976. Solar Phys. **68**, 304 (1980) and Sonne 19, **5**, 96 (1981)

42. Jahn, C.H.: Eigenbewegungen einer Sonnenfleckengruppe. Sterne und Weltraum **25**, 340 (1986)

43. Pfister, H.: Spezielle Eigenbewegungen in Sonnenfleckengruppen. Astron. Mitt. Zürich **342** (1975)

44. Pfister, H.: Klassifikationsschema für Eigenbewegungen. Sterne und Weltraum **28**, 598 (1989)

45. Mehltretter, J.P.: On the Proper Motion of Small Pores in Sunspot Groups. Solar Phys. **63**, 61 (1979)

46. Brauckhoff, D., Delfs, M., Stetter, H.: Polfackeln – Neuland für den Amateursonnenbeobachter. Sonne **10**, 114 (1986)

47. Vogt, O.: Positionsbestimmung von Sonnenflecken. Sterne und Weltraum **16**, 58 (1977)

48. Treutner, H.: Sonnenpositionsfotografie. Sonne 4, **1**, 141 (1977)

49. Junker, E.: Drei Wege zur richtigen Sonnenfleckenposition. Sonne 47, **12**, 87 (1988)

Section 13.2

50. Guest, J.E., Greeley, R.: Geology of the Moon. Wykeham, London (1977)

51. Kopal, Z.: The Moon. D. Riedel, Dordrecht (1969)

52. Kopal, Z.: Physics and Astronomy of the Moon. Academic, New York (1962)

53. Kuiper, G.P., et al.: Photographic Lunar Atlas. University of Chicago Press, Chicago (1960)

54. Schwinge, W.: Fotografischer Mondatlas. Johann Ambrosius Barth, Leipzig (1983)

55. Westfall, J.E.: Lunar Incognita: Completing the Map of the Moon. Sky Telescope **67**, 284 (1984)

56. Classen, J.: Die Sterne **50**, 157 (1974)

57. British Astronomical Association (ed.): Guide to Observing the Moon. Enslow, Hillside, NJ (1986)

58. Price, F.W.: The Moon Observer's Handbook, p. 309. Cambridge University Press, Cambridge (1988)

59. Westfall, J.E.: Atlas of the Lunar Terminator. Cambridge University Press, Cambridge (2000)

60. MacRobert, A.M.: Three Lunar Challenges. Sky Telescope **77**, 520 (1989)

61. North, G.: Observing the Moon: The Modern Astronomer's Guide. Cambridge University Press, Cambridge (2000)

62. Kopal, Z., Carder, R.W.: Mapping of the Moon. D. Reidel, Dordrecht (1974)

63. Fauth, P.: Die Sterne **33**, 158 (1957)

64. Roth, G.D. (ed.): Planeten beobachten. Spektrum Akademischer, Berlin (2002)

65. Hopmann, J.: Die Genauigkeit der Angaben von relativen Höhen auf dem Monde. Neue Werte für 163 Punkte. Mitteilungen der Sternwarte Wien **12** (19), (1965)

66. Zimmermann, O.: Astronomisches Praktikum, 5th edn. Sterne Weltraum, München (1995)

67. Hedervari, P.: Lunar Photometry. In: Genet, R.M. (ed.) Solar System Photometry Handbook, pp. 4–8. Willmann-Bell, Richmond, VA (1983)

68. Viscardy, G.: Atlas-Guide photographique de la Lune. Ouvrage de référence à haute résolution. Masson, Paris (1985)

69. Duerbeck, H.W., Hoffmann, M.: Compendium of Practical Astronomy, vol. 1, chap. 8. Springer, Berlin Heidelberg New York (1994)

70. Böhme, D.: Polarimetrie für Amateurastronomen. Sterne Weltraum **25**, 544 (1986)

Section 13.3

71. Espenak, F.: Fifty Year Canon of Solar Eclipses: 1986–2035, NASA Reference Publication 1178 Revised (1987)

72. Harrington, P.S.: Sonnen- und Mondfinsternisse beobachten. Spektrum Akademischer, Heidelberg (2002)

Section 13.4

73. Riedel, E.: Sternbedeckungen durch den Mond. Sterne und Weltraum Basics 1. Spektrum der Wissenschaft, Heidelberg (2002)

74. Kretlow, M.: Sternbedeckungen durch Kleinplaneten im Jahr 2003. VdS J **11** (2003)

75. Leinert, C.: Mit Mondbedeckungen auf der Jagd nach jungen Doppelsternen im Taurus. Sterne und Weltraum **6** (1992)

Section 13.5

76. Sposetti, S., Tilger, B.: Die ASTRA-Satellitenfamilie. Sterne und Weltraum **6** (1998)

Section 13.6

77. Löbering, W.: Jupiterbeobachtungen von 1926 bis 1964. Nova Acta Leopoldina. Johann Abrosius Barth, Leipzig (1969)

78. Roth, G.D. (ed.): Planeten beobachten. A Practical Guide for Amateurs in German. Spektrum Akademischer, Berlin (2002)

79. Zmek, W.P.: Rules of Thumb for Planetary Scopes I. Sky and Telescope July, 91 (1993)

80. Zmek, W.P.: Rules of Thumb for Planetary Scopes II. Sky and Telescope September, 83 (1993)

81. Dragesco, J.: High Resolution Astrophotography. Cambridge University Press, Cambridge (1995)

82. Haug, H., Kowalec, C.: Sterne Weltraum **24**(11), 600 (1985)

Section 13.7

83. Greenberg, R., Scholl, H.: Resonance in the asteroid belt. In: Gehrels, T. (ed.) Asteroids, pp. 310. Univ Arizona Press Tucson (1979)

84. Kozai, Y.: The dynomical evolution of the Hirayama families. In: Gehrels, T. (ed.) Asteroids, pp. 334. Univ Arizona Press Tucson (1979)

85. Bendjoya, Ph., Zappala, V.: Asteroid family identification. In: Bottke, W.F., Cellino, A., Paolicchi, P., Binzel, R.P. (eds.) Asteroids III, pp. 613. Univ Arizona Press, Tucson (2002)

86. Cellino, A., Bus, S.J., Doressundiram, A., Lazarro, D.: Spectroscopic properties of asteroid families. In: Bottke, W.F., Cellino, A., Paolicchi, P., Binzel, R.P. (eds.) Asteroids III, pp. 633. Univ Arizona Press, Tucson (2002)

87. Morbidelli, A., Brown, M.E., Levison, H.F.: The Kuiper Belt and its primordial sculpting. In: Davies, J.K., Barrera, L.H. (eds.) The first decadal review of the Edgeworth–Kuiper Belt, pp. 1. Kluwer Acad. Pub., Dordrecht (2004)

88. Brown, M.E.: The largest Kuiper Belt objects. In: Baruchi, M.A., Boehnhardt, H., Cruikshank, D.P., Morbidelli, A (eds.) The solar system beyond Neptune, pp. 335. Univ. Arizona Press, Tucson (2008)

89. Protopapa, S., Böhnhardt, H., Herbst, T.M. Cruikshank, D.P., Grundy, W.M., Merlin, F., Olkin, C.B.: Surface characterization of Pluto and Charon by L and M band spectra, pp. 365. A&A 490 (2008)

90. Meeus, J.: Astronomical algorithms, pp. 230. William Bell inc. Pub., Richmond (1998)

Section 13.8

91. Green, D.W.E. (ed): Catalogue of cometary orbits 2008. Central Bureau for Astronomical Telegrams, Cambridge (2008)

92. Dones, L., Weisman, P.R., Levison, H.F., Duncan, M.J.: Oort Cloud formation and dynamics. In: Feston, M.C., Keller, H.U., Weaver, H.A. (eds.) Comets III, pp. 193. Univ. Arizona Press, Tucson (2004)

93. Morbidelli, A., Brown, M.E.: The Kuiper Belt and the primordial evolution of the solar system. Feston, M.C., Keller, H.U., Weaver, H.A. (eds.) Comets III, pp. 171. Univ. Arizona Press, Tucson (2004)

94. Häfner, R.: Comets. Compendium of Practical Astronomy, vol. 2, p. 261, Springer-Verlag (1994)

95. Boehnhardt, H.: Comet splitting observations and model scenarios. Earth, Moon, and Planets, **89**, 91 (2002)

Section 13.9

96. Berwerth, F.: Ann. Naturhist. Hofmuseum Wien **27** (1913)

97. Bauschinger, J.: Die Bahnbestimmung der Himmelskörper, 2nd edn., p. 587. Engelmann, Leipzig (1928)

98. Porter, J.G.: Comets and Meteor Streams, p. 81. Chapman and Hall, London (1952)

99. Schmitz, B.: Sterne Weltraum **18**, 9 (1979)

100. Zeuner, H.: Sterne Weltraum **23**, 256 (1984)

Section 13.10

101. Jesse, O.: Die Höhe der leuchtenden Nachtwolken. Astronon. Nachrichten **140**, 161–168 (1896)

102. Gadsden, M., Schröder, W.: Noctilucent Clouds. Springer, Berlin Heidelberg New York (1989)

103. Eather, R.A.: Majestic Lights: the Aurora in Science, History, and the Arts. American Geophysical Union, Washington, DC (1980)

104. Brekke, A., Egeland, A.: The Northern Light from Mythology to Space Research. Springer, Heidelberg New York Berlin (1983)

105. International Union of Geodesy and Geophysics: International Auroral Atlas. Edinburgh University Press, Edinburgh (1963)

106. Bone, N.: Aurora: Observing and Recording Nature's Spectacular Light Show. Patrick Moore's Practical Astronomy Series. Springer, Berlin Heidelberg New York (2007)

107. Cassini, G.D.: Découverte de la lumière céleste qui paroist dans le zodiaque., Memoires de l'Academie Royale des Sciences Tome VIII (1666–1699), p. 119. Compagnie des Libraires, Paris (1730)

108. Leinert, C., Grün, E.: Interplanetary Dust. In: Schwenn, R., Marsch, E. (eds) Physics of the Inner Heliosphere, p. 207. Springer, Berlin Heidelberg New York (1990)

109. Grün, E. et al.: Interplanetary Dust. Springer, Berlin Heidelberg New York (2001)

Section 13.11

110. Albers, B.: Sterne Weltraum **12**, 137 (1973)

111. Leinert, C.: Sterne Weltraum **25**, 18 (1985)

112. Leinert, C.: Sterne Weltraum **25**, 136 (1985)

113. Greenler, R.: Rainbows, Halos, and Glories. Cambridge University Press, Cambridge (1980)

114. Meinel, A., Meinel, M.: Sunspots, Twilights, and Evening Skies. Cambridge University Press, Cambridge (1983)

115. Trickler, R.A.: Introduction to Atmospheric Optics. Elsevier, New York (1971)

116. Arsenijevic, J.: Publications de l'Observatoire astronomique de Belgrade, **33**, 36 (1985)

117. Walker, M.F.: Sky Telescope **71**, 139 (1986)

14. Stars and Stellar Systems

14.1 The Stars

K.P. Schröder

14.1.1 Stars by Position and Magnitude: Atlases and Catalogs

In astronomy, stars are identified and cataloged by their apparent position in the sky, given in the equatorial coordinates *right ascension* α and *declination* δ (see Chap. 2). In addition, every star is primarily characterized by its *visual magnitude* m_V, which represents the apparent visual brightness of the star—or more accurately—which is a logarithmic measure of the stellar flux F_{vis} perceived by the observer's eye, relative to the flux of a comparison star. "Flux" here means the energy of the light entering the eye, per receiving area and second.

Most first-time users of a star catalog are puzzled by the fact that the numerically smallest magnitude values actually refer to the brightest and boldest-plotted stars. This historic convention stems from the Greek idea of a celestial hierarchy with six distinguishable levels (magnitudes), where brightness reflected importance. Hence, the very brightest stars are of first magnitude (mag), while the sixth mag is populated by countless, apparently unimportant stars close to the perception-limit of an unaided eye. Since the human eye has a logarithmic response, a factor of 2.5 in brightness (F_{vis}) is equally significant to any two stars (a) and (b), which differ by a full magnitude, regardless of their brightness. The definition formula

$$m_V\ (a) - m_V\ (b) = -2.5 \times \log_{10} \frac{F_{vis}\ (a)}{F_{vis}\ (b)} \ ,$$

reflects this logarithmic response and the ancient historic background of the magnitude scale: If star (a) is 2.5× (apparently) brighter than (b), this makes a difference of 0.4 on the logarithmic scale, and m_V (b) becomes numerically larger (which means fainter) than m_V (a) by 1.0.

What looks complicated at first does offer some convenience to the observer: For stars of similar brightness, for example, a 0.05 difference in (apparent) magnitude m_B is equivalent to (nearly exactly) a 5% difference in the fluxes. On the other hand, a still humble difference of 5.0 mag already represents a factor of 100 in apparent brightness.

The precession of the orbit of Earth and (to a much smaller extent) stellar proper motions both impose slow changes on all stellar coordinates. Hence, every star catalog and atlas must refer to a specific epoch, which currently is 2000.0. Popular examples are the *Sky Catalogue 2000.0* [1] and Wil Tirion's matching *Cambridge Star Atlas*, and *Sky*

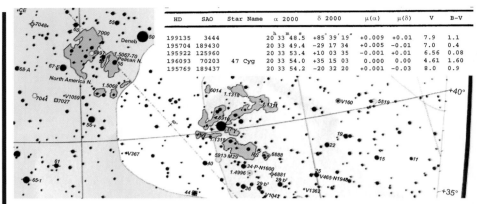

HD	SAO	Star Name	α 2000	δ 2000	μ(α)	μ(δ)	V	B–V
			^{h} ^{m} ^{s}	$^\circ$ ' "				
199135	3444		20 33 48.5	+85 39 19	+0.009	+0.01	7.9	1.1
195704	189430		20 33 49.4	−29 17 34	+0.005	−0.01	7.0	0.4
195922	125960		20 33 53.4	+10 03 35	−0.001	+0.01	6.56	0.08
196093	70203	47 Cyg	20 33 54.0	+35 15 03	0.000	0.00	4.61	1.60
195769	189437		20 33 54.2	−20 32 20	+0.001	−0.03	8.0	0.9

Fig. 1. Small cuts from *Sky Atlas 2000.0* by W. Tirion [2], and the *Sky Catalogue 2000.0*

Atlas 2000.0 [2] (see http://www.SkyandTelescope.com and Fig. 1). These books are essential tools for any preparation, and they provide accurate orientation during the actual observing session, when used with an eyepiece with a red light.

The modern computer-user may prefer a planetarium program to a printed star atlas. An excellent and even free example is *Cartes du Ciel*. Advantages of the PC-based approach are the ability to zoom into any specific region of the night sky, thus to adjust the display to the field-of-view of your telescope or viewfinder, to get essential object data by mouse-clicking, and to have planets and brighter minor planets shown at their actual positions. The obvious disadvantage is your dependence on a laptop while observing under cold, moist and dark outdoor conditions. Several issues must be resolved, if a planetarium program is to replace a printed star atlas in your observing nights: (1) extra power supply, perhaps by a good car battery and voltage converter, (2) a deep-red filter-foil for the laptop screen to keep the observer's night vision uncompromised and as a courtesy to any fellow observers, and (3) precautions to keep your laptop dry and sufficiently warm, e.g., by means of a box, in which keyboard and screen are kept under transparent plastic.

Several modern digital star catalogs, which provide precision coordinates for any observing date, are available for free download. They are precision tools for doing astrometry with digital (or digitized conventional) astrophotos (see Chap. 6). Noteworthy are:

1. The Hipparcos Catalogue. It only reaches to mag 8–9 but offers an incredible milliarcsecond (0.001" or mas) precision.
2. The UCAC2 of the US Naval Observatory, accomplished by ground-based CCD imaging. With a still impressive 20 mas positional accuracy, it reaches much deeper and includes 48 million stars down to mag 10–14.
3. The Hubble GSC (Guide Star Catalogue). The largest compilation of stellar positional data and still better than, at least, 1" (arcsecond).

The image-processing software AstroArt, e.g., allows to load and use the GSC directly from its astrometric routines. It then computes positions in α and δ of any unknown object in your photo with respect to the GSC stars recorded around it. This is an extremely powerful tool for observers of comets and minor planets, who wish to determine orbital parameters. Hence, any student or modern amateur astronomer can produce very useful and precise astrometric data by just a few mouse-clicks!

Stellar magnitudes m_v are a measure of the apparent stellar brightness in visual light, on an inverted, logarithmic scale.

The excellent planetarium program *Cartes du Ciel* is freely available in French and English language from http://www.astrosurf.com/astropc/cartes.

Customized datasets of digital star catalogs are freely available online from:
http://archive.ast.cam.ac.uk/hipp/hipparcos.html or
http://vizier.u-strasbg.fr (Hipparcos)
http://vizier.u-strasbg.fr (UCAC2)
http://archive.eso.org/gsc/gsc (GSC)

14.1.2 Stellar Colors and Spectral Types

Color photographs of rich star fields and of star clusters reveal that hot stars are blue, while cooler stars appear red. At the same time, spectra taken of such different stars show remarkably different absorption-line patterns. Hence, observational astronomy has developed well-defined standards to quantify stellar colors and spectral types, in order to empirically derive the "effective temperature" of a star from the quality of its radiation.

In order to measure a stellar color, its apparent magnitude must be measured in at least two different spectral windows (see Chap. 8 for technical details). Since the response of the human eye is strongly in favor of yellow-green light, "visual" magnitudes m_V are obtained by the deliberate use of a yellow-green (V) filter. Blue and near-UV filters, likewise, are required to measure m_B (B) and m_U (U). Today's photometric standard sets of such UBV filters (e.g., Optec, Inc., USA, see http://www.optecinc.com) are available at reasonable prices. Their transmission curves (see Fig. 2) are consistent with the UBV filters used by H.L. Johnson (Lick Observatory) in the early 1960s. For this, changed spectral detector response has been considered, too, i.e., modern solid-state detectors are more red-sensitive than the phototubes of Johnson's days. The *Johnson color system* of our days now also includes R (red) and I (infrared) filters. In addition, old sources often refer to a *photographic magnitude* m_{pg}, which crudely represents the blue-green spectral response of historic (orthochromatic) B&W photographic emulsions.

The logarithmic nature of the magnitude system conveniently allows to represent the ratio of, say, blue over visual flux of any given star, simply by a difference in magnitude. Hence, stellar colors are measured in terms of U–B, B–V, V–R, R–I. If a star emits more flux towards bluer light, then the bluer magnitude in any of the above differences is numerically smaller and, hence, the resulting color index is negative. Cool stars, the Sun included, have positive color indices. The zero-point of all indices is defined by ordinary (main sequence) stars of spectral type A0, which have an effective temperature of nearly 10,000 K (see Table 1).

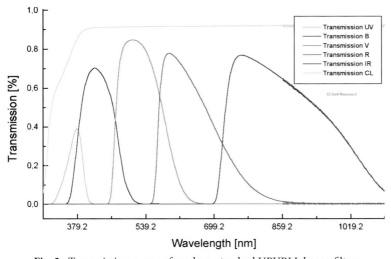

Fig. 2. Transmission curves of modern, standard UBVRI Johnson filters

B–V is the most common color index, listed by all modern star catalogs. But very hot stars are better distinguished by their U–B index, cool stars better by V–R or R–I. The color indices can easily be measured by any amateur or student equipped with a CCD, filter set and catalog of reference stars. However, the measured color is not always the original stellar color; often, interstellar reddening has a non-negligible effect (see Sect. 14.1.6).

The total stellar flux, received over the *entire* spectrum, suggests the (ideal) concept of a *bolometric magnitude* m_{bol}. The difference between m_{bol} and m_V is called the *bolometric correction* (BC). In practice, limited spectral detector response and imperfect transparency of atmosphere and optics render any direct measurements of m_{bol} impossible. But computer models of the spectral energy distribution of a particular star allow the professional astronomer to give BC as a function of temperature or spectral type. Obviously, BC differs from 0.0 the farther away the maximum of the spectral energy distribution is from yellow-green light. Notable examples are O stars (maximum in UV) and M stars (maximum in R), see Table 1.

As mentioned above, stellar effective temperature is reflected not only by color, but also by specific, temperature-sensitive characteristics of the absorption-line spectrum, resulting in different spectral types. These are defined by the *Harvard classification scheme*. Organized by temperature, the Harvard sequence of spectral classes reads, from hot ($\approx 50,000$ K) to "cool" (≈ 3000 K): O-B-A-F-G-K-M. Due to a historic convention, classes O to A are also called "early-type," K and M "late-type." Each class is further divided into subclasses 0 to 9. The Sun, with an effective temperature of 5780 K, is of spectral type G2, an A0 star has nearly 10,000 K. Due to the extremes of the sequence, however, it is difficult to assign exact effective temperatures. Also, effective temperature depends a bit on the luminosity class (see below). Most stars, however, are ordinary "main sequence" stars, to which Table 1 can be applied.

When inspecting different stellar spectra (see Fig. 3), we find that cool stars have overwhelmingly rich absorption line spectra, cast mostly by neutral (with type K stars) or singly ionized metal atoms (types G and F). The coolest stars (M) even show molecular lines. All these absorbers have a very complex atomic structure and offer millions of different frequencies to absorb photons. By contrast, very few and very different absorption lines appear with hot stars. Above about 9000 K (spectral type A), most metals become doubly ionized and do not contribute much line absorption to the optical spectrum. The most prominent absorption lines, the Balmer series (i.e., H_α, H_β, H_γ, etc.), are due to excited neutral hydrogen, found at their strongest with spectral type A0. In the hottest stars (types O and early B) hydrogen, too, is ionized. Hence, the now nearly featureless spectra contain only a few lines of singly ionized and/or highly excited neutral helium.

Table 1. Approximate effective temperatures, $(B-V)_0$ colors (unreddened) and BCs for main sequence stars of different spectral type

Type	O5	B0	B5	A0	A5	F0	F5	G0	G5	K0	K5	M0
T_{eff}/K	45,000	30,000	16,000	9800	8500	7300	6500	6000	5700	5200	4400	3850
$(B-V)_0$	−0.33	−0.30	−0.17	−0.00	+0.15	+0.30	+0.44	+0.58	+0.68	+0.81	+1.15	+1.40
BC	−3.9:	−3.0:	−1.44	−0.15	−0.02	−0.01	−0.03	−0.10	−0.14	−0.24	−0.66	−1.2:

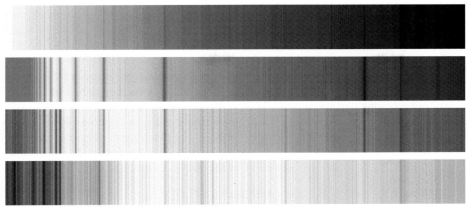

Fig. 3. Stellar spectra from deep blue light (*left*) to green (*right*), showing spectral types O5 (*top*), A1, F5, and K0 (*bottom*)

By these characteristics, spectral classes are well-defined, assessed easily, and stellar temperatures can be derived. A more detailed description of stellar spectra and spectral classes can be found in the book by Kaler, *Stars and their Spectra* [3], and we also refer the reader to Chap. 7.

14.1.3 Fundamental Physical Quantities of Stars

We now take a look at the fundamental stellar quantities, their physical meaning and relations. This allows us to characterize a star in absolute terms. For a much more detailed approach, we refer to the book by Böhm-Vitense, *Introduction to Stellar Astrophysics* [4].

- R_*: For main-sequence stars like the Sun, the definition of a *stellar radius R* is quite trivial; the photosphere is a fairly well-defined, thin layer. Not so for supergiants like Betelgeuse (α Ori): their photospheric extent reaches a significant fraction of the stellar radius! But apart from such notable exceptions, the simple-minded concept of a stellar radius is very adequate. If the distance is known, R_* can be obtained by means of angular diameter measurements. But the only absolute (i.e., distance-independent) measurements of stellar radii are provided by eclipsing binaries (see Sect. 14.3) with spectroscopically known orbital velocity v. For a central eclipse, the duration of each partial phase divided by v simply gives the diameter of the secondary. Similarly, the duration of totality plus one partial phase, together, yields the diameter of the primary.
- M_*: The *stellar mass M* is a well-defined and important quantity: It determines how stellar evolution proceeds. Nevertheless, M_* can be measured only from the orbital velocities of spectroscopic binary stars. To do this, their orbital inclination i must be known. Again, this mostly leaves us with eclipsing binaries, where i evidently is $\approx 90°$.
- T_{eff}: A star has no surface and, hence, no surface temperature. And there is no uniform photospheric gas temperature, either. Hence, the "*effective temperature*" of a star is a somewhat artificial construction. Physically, it is just a measure of the stellar surface flux F_* (the total radiation energy emitted per stellar surface area and second). It is *defined* by the equation $T_{\text{eff}} = (F_*/\sigma)^{1/4}$, with $\sigma = 5.67 \times 10^{-8}$ W m^{-2} K^{-4} as the

Both, stellar color and spectral type indicate the effective temperature of a star.

A popular phrase to remember the odd sequence of spectral classes, in the order from hot to cool, goes: "Oh, Be A Fine Girl, Kiss Me!"

You can easily familiarize yourself with the full wealth of stellar spectra by using the excellent CLEA student training software. It comes with a detailed user-manual and is available for free from http://www3.gettysburg.edu/marschal/clea/speclab.html.

Stefan–Boltzmann constant. This definition equation is modeled on the simple physics of a perfect "blackbody radiator," where the radiation is in full thermodynamic equilibrium (TE) with the surface temperature. But just a hot, blackbody would not do: this stringent condition is fulfilled only *inside* a hot oven or cavity with very thick walls of even temperature, and within which every surface area receives as much radiation as it emits. Stellar surfaces, by contrast, are far from TE, because they radiate to all sides without receiving any radiation back! Nevertheless, the effective temperature of a star does give us about the right idea of the true gas temperatures in its photospheric layers.

– L_*: The *luminosity* of a star is its total energy output per second, integrated over the whole electromagnetic spectrum and the whole stellar surface. Hence, L_* is related to R_*, F_* and T_{eff} by simple geometry and by the above definition equation for T_{eff}:

$$L_* = 4\pi R_*^2 F_* = 4\pi R_*^2 \sigma T_{eff}^4 .$$

From this equation we can immediately get some very basic insight: A luminous star must either be very large, or very hot, or both. And cool stars must even be extra large to be luminous, hence, known as "cool giants" or "supergiants."

14.1.4 Distance, Luminosity and Absolute Magnitude

Getting reliable distances is a fundamental problem in observational astrophysics. After all, the derivation of such fundamental stellar quantities as luminosity and radius depends on an exact knowledge of the distance d. The only direct measurement of d is by *trigonometric parallax*. It is obtained from the apparent motion of a star in the sky against a distant stellar background, caused by the orbital motion of Earth. The value π is the respective angular radius in arcseconds of the projected inclined circle described by a sufficiently near star in the course of a year. By definition of the convenient distance unit "parsec" (pc), we obtain the stellar distance from

$$d[pc] = 1/\pi[''] .$$

In the popular unit of light years (ly), 1.0 pc is equal to 3.2615 ly. Before the Hipparcos satellite (i.e., before the mid-1990s), precise trigonometric parallaxes reached only to about 20 pc. Even today, we still depend on secondary (i.e., not direct) distance measurements beyond $d \approx 200$ pc.

This brings us to the dilemma with all objects of unknown distance: In principle, a star could appear bright to us, because it is indeed very luminous, *or*, equally well, because it just happens to be very near (but modest). Would life not be much easier (for the astronomer), if all stars had the same distance? With this idea in mind, the concept of an *absolute magnitude M* was born. M is the would-be magnitude of the star if it could be "moved" to the convenient distance of 10 pc. Now, with all stars thought to be at the same distance of 10 pc, their bolometric absolute magnitudes M_{bol} are simply a (logarithmic) measure of their true luminosities L_*, and for any two stars (a) and (b) we can write (compare with Sect. 14.1.1):

$$M_{bol}(a) - M_{bol}(b) = -2.5 \times \log_{10} \frac{L_*(a)}{L_*(b)} .$$

In fact, the Sun would be just a fairly insignificant star of $+4.74^m\,(=M_{bol,\odot})$, if seen from a distance of 10 pc. But we can use the Sun as a reference star and relate the absolute magnitude of any star with a luminosity given in solar units of $L_\odot = 3.85 \times 10^{26}$ W:

$$M_{bol} = 4.74 - 2.5 \times \log_{10} \frac{L_*}{L_\odot}\ .$$

In real life, of course, all stars have their specific and very different distance d. The apparent brightness, or rather the flux F_{vis} received on Earth, is diluted with d^{-2}. Hence, absolute and apparent magnitudes are, for any (same) star and color, related by the so-called *distance modulus*:

$$m - M = 5.0 \times \log_{10} \frac{d/pc}{10pc}\ .$$

Hence, if the distance d is known, M_{bol} (and, subsequently, L_*) can be determined from m_V by, first, the bolometric correction (to get m_{bol}, see Sect. 14.1.2) and then subtraction of the above distance modulus.

Nearly a century ago, it was noted that the sharpness of spectral lines is an indicator of stellar luminosity. Consequently, the Harvard scheme defines *luminosity classes* (LC), in addition to spectral types. Obviously, there is a large range of stellar luminosities: Ordinary giants (LC III, narrow lines) have $M_V \approx 0$, $L_* \approx 10^2 L_\odot$, while supergiants (LC I, very sharp lines) have $M_V \approx -5$, $L_* \approx 10^4 L_\odot$. All main sequence (MS) stars, the Sun included, have LC V, with distinctly broader lines compared to giant stars. Here, M_{bol} depends strongly on the spectral type, since hotter MS stars are much brighter.

Hence, from a full assessment of luminosity class and spectral type of a distant star, it is possible to estimate the absolute magnitude and luminosity. In that case, the distance modulus $(m - M)$ can be used the other way round: to find the (spectroscopic) distance of the star.

14.1.5 The Hertzsprung–Russell Diagram and Stellar Evolution

Around the year 1913, Einar Hertzsprung and Norris Russell independently thought out a very useful way to present stars by their most important and spectroscopically observable physical quantities, L_* and T_{eff}. Their diagram has since become the roadmap for stellar astrophysics and is known as the HR diagram, or HRD. The original form of the HRD is still used by observers: It plots $\log L_*$ in absolute magnitude (or luminosity class), brightest stars above, over spectral type or color index (most commonly B–V), from hot (left) to cool (right). Figure 4 shows the observed HRD of the solar neighborhood, with all stars with M_V brighter than 4.0 and $d < 50$ pc. The theoretical HRD has, correspondingly, $\log L_*$ over $\log T_{eff}$ (i.e., with an inverted x-coordinate).

From the relation between L_*, R, and T_{eff} presented in Sect. 14.1.3, it is clear that the cool supergiants must be found in the top-right corner of the HRD, hot supergiants in the top-left corner, "brown" (cool) dwarfs in the bottom-right corner, and "white" (hot) dwarfs in the bottom-left corner of the HRD, while the main sequence (MS) runs from upper-left (hot, luminous, massive stars) down to bottom-right (cool, low-luminosity, low-mass stars).

Absolute bolometric magnitudes M_{bol} are a logarithmic measure of stellar luminosity L_*.

By definition, absolute magnitude M and apparent magnitude m coincide if the star is at a distance of 10 pc. Otherwise, the "distance modulus" $m - M$ must be applied.

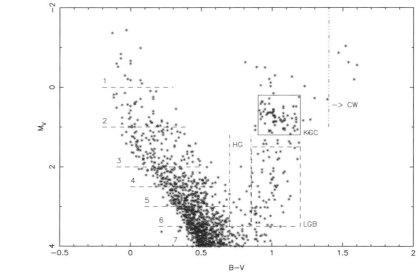

Fig. 4. Observed HRD of all stars within 50 pc distance and $M_V < 4.0$, according to Hipparcos parallax data. Indicated are the MS (1–7), K giant clump (KGC), lower RGB (LGB) and stars with cool winds (CW)

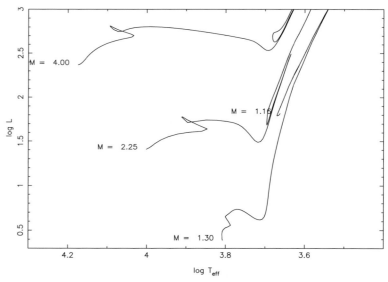

Fig. 5. Theoretical HRD, covering about the same stellar quantities as above, with evolution tracks for the initial stellar masses of 1.30 (AGB starting with 1.16), 2.25 and 4.00 $M_\odot$

Because the basic idea about any star comes so intuitively from just its location in the HRD, this diagram has become the natural choice to visualize stellar evolution: In it, all changes of the basic physical quantities appear in terms of a track, as the star evolves and ages (see Fig. 5). The stellar "life" begins on the left edge of the MS, where stars find their equilibrium right after the ignition of central hydrogen burning. The position on the MS is set by the stellar mass, and the MS actually is a monotonous mass sequence from large (top) to small (bottom). Most of the stellar lifetime (between 75% and 90%) is spent on the MS, which makes it a very populated region in the HRD. But change is inevitable, when all central hydrogen has been converted into helium. In the Sun, the hydrogen sufficiently near the center (i.e., about 10% of the entire solar hydrogen) will last around 10 billion years; half of it has been spent already.

In general, stellar lifetimes become much shorter with increasing mass, because massive stars are so much more luminous. On the MS, $L_* \propto M_*^{3.5} \propto M_*^{4.0}$, so nuclear burning lifetimes $\tau_N \propto M_*/L_*$ must decrease approximately with $\tau_N \propto M_*^{-2.5} \propto M_*^{-3.0}$. To most MS stars, we can also apply the proportionality $T_{eff} \propto M_*^{0.6}$ and $R_* \propto M_*^{0.7}$. These empirical (and approximate!) relations are observed with eclipsing (well-detached) binary stars, for which we have precise knowledge of their masses and other quantities (see also Sects. 14.1.3 and 14.3).

Post-MS stellar evolution is complex and depends a lot on the initial stellar mass, and a little on the chemical composition. Low-mass stars like the Sun (i.e., around $1\,M_\odot$) slowly develop a hydrogen-burning shell around a now inactive helium-core. In this phase, a low-mass star first very gradually expands to become a cool giant. It then brightens and slowly moves up in the HRD on the *red* (or first) *giant branch* (RGB). For the Sun, this whole phase will last about 2 billion years, of which the last half billion years will be spent climbing up the RGB. Due to the shrinking of the inactive He-core in this stage, pressure in the outer H-burning shell steadily increases, and so does, subsequently, its energy output (i.e., the luminosity) on the RGB. The very low surface gravity of stars high on the RGB, which results from the considerable expansion, makes such a giant gradually lose a noticeable fraction of its mass, by means of a "cool wind."

The rise on the RGB ends abruptly, when, finally, the growing and still contracting He-core has reached a mass of about $0.5\,M_\odot$ and is now dense enough to start He-burning. Although this is an explosive event, the "core He-flash," it is not visible to observers outside the star. The surrounding stellar structure absorbs all the extra energy supplied in the short time of the He-flash. Rather, the star adjusts to a less cool and less bright equilibrium state (observed as a K giant of LC III). By contrast, very old, low-metallicity stars—like the ones found in globular clusters—are hotter and more compact in this phase. They form, at about the same luminosity of $M_{bol} \approx 0.6$, the *horizontal branch*, as seen in color-magnitude diagrams (see Sects. 14.1.6 and 14.4) of globular clusters.

After the central He has all been burnt, the stellar core becomes inactive again and shrinks once more. As on the RGB, this causes the now two burning shells (one inner of He, one outer of H) around the core to produce more luminosity again, and the star ascends on the *asymptotic* (or second) *giant branch* (AGB). The Sun will have done this already 130 million years after its He-flash.

The evolution of more massive stars (in the range of 2 to $8\,M_\odot$) is similar, but with several characteristic exceptions: (1) The H-shell-burning phase on the RGB is much shorter. (2) The central helium-burning phase starts smoothly, without flash. (3) It is more luminous and (4) becomes the longest-lasting post-MS phase. (5) The more massive He-burning giant is also hotter and so performs a "blue loop" in the HRD. (6) The subsequent AGB giants become more luminous and lose more mass by cool winds, as their initial mass increases.

The really massive ($>8\,M_\odot$) stars are extremely fast (on astronomical timescales) in eating through their nuclear fuel: in much less than 100 million years. When they run out of energy, these stars are blue supergiants. Since their burnt-out cores are too massive to find a stable state without energy production, these stars explode as supernovae (mostly of type II), triggered by the implosion of their core. Unfortunately, we cannot give a more detailed picture here of the complex and very interesting physics of stellar evolution; the interested reader may be referred to a dedicated textbook by Prialnik, *An Introduction to the Theory of Stellar Structure and Evolution* [5].

14.1.6 Star Clusters

The main sequence (MS) is not a cooling-sequence, as thought a century ago, but it is a sequence according to mass: Stars with large masses settle into a very luminous H-burning state, while low-mass stars are found low on the MS.

For stellar astrophysics, star clusters provide two huge advantages: (1) all its member stars have about the same distance from us, and (2) all cluster stars have about the same age! The first advantage means, that the same distance modulus can be applied to all cluster stars. Consequently, we may start with a simple color-magnitude diagram (CMD, see Fig. 6), which already looks like an observed HRD, except that the magnitude scale still requires, as a whole, to be shifted by the distance modulus. In fact, after a correction for interstellar absorption and reddening (see below), the cluster-MS can be matched with the MS of an ordinary HRD (as of Fig. 4), and the distance modulus is simply read off the y-scale, as the actual difference between CMD and HRD.

The second advantage, i.e., all cluster stars are of the same age, has given astrophysics a basic, empirical understanding of stellar evolution, long before the first computer models became available in the 1960s: CMDs of young clusters show a fully populated MS up to the hottest O stars. Towards older ages, MS stars do not disappear evenly, but the MS rather seems to "burn down" in luminosity like a candle—the lower the mass and luminosity, the longer the MS stars remain. This is visible proof of the MS-relation between lifetime and mass, as derived in Sect. 14.1.5. Hence, the cluster age can be derived from the location of the tip of the cluster MS, the "turn-off point" (as here the cluster stars turn off the MS to become giants).

The only tricky problem is, how first to correct for the right amount of interstellar absorption A_V and the related reddening or *extinction* E(B–V). This makes a star cluster shift in both coordinates of its CMD, in V and in B–V, often quite significantly—in particular in distant clusters, or clusters in the dust-rich galactic plane. In the latter case, A_V may easily reach several magnitudes per kiloparsec, instead of an average $\approx 1^m$/kpc, elsewhere.

The extinction (or reddening) is proportional to the interstellar absorption: usually, $A_V = 3.0$ to $3.2\,$E(B–V). The factor may vary a little with the specific properties of the interstellar dust in the line of sight. Hence, if a few cluster stars are spectroscopically well-classified, then it is possible to compare their observed color (B–V) with what it should be, according to the spectral type ((B–V)$_0$ in Table 1). The resulting difference gives E(B–V), from which A_V is determined. But this approach requires a lot of additional observational effort, or to rely on published spectral classification.

There is, actually, a simple and purely photometric method to determine the interstellar extinction and the corresponding absorption towards a cluster: In a CMD, the MS appears as a pretty straight line and it gives us no clue to the correction required. But this is different in a *two-color diagram* (2CD), which plots U–B over B–V (see Fig. 7).

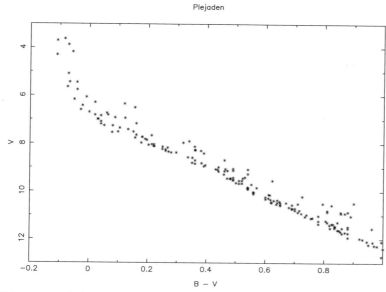

Fig. 6. Color-magnitude diagram of the Pleiades, which are almost unaffected by interstellar extinction. The MS is straight, except for its bright, blue end

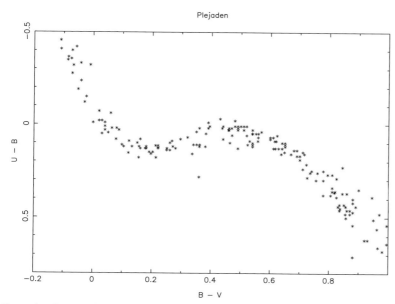

Fig. 7. Two-color diagram (U–B versus B–V) of the Pleiades. Note the characteristic double-wiggle of the MS (see text)

Here, the MS describes a double-wiggle, of which the turning points have the unreddened colors $(B–V)_0$ of 0.20 and 0.50. Hence, by comparison of a standard MS with that of the actual cluster in a 2CD, the extinction is simply the resulting difference of the B–V scales. The only requirement is an additional photometric measurement of each cluster star in the U filter.

CMDs like Fig. 6 can be produced by any student or amateur astronomer in just half a night's work, followed by one afternoon of data reduction. This work requires only a moderate telescope size (8″ to 16″ aperture) and a CCD with a sufficiently linear response, e.g., CCDs with no ABG (no antiblooming gate), and makes a very instructive project (see Chap. 8 for an introduction into photometry in general). Furthermore, stellar evolution models from the scientific literature may be checked out to find those which match the cluster stars at the turn-off point, and to derive the cluster age. This is a nice student project to gain experience in both observational practice *and* theoretical background.

For more information on star clusters, i.e., morphology and kinematics, see Sect. 14.4.

A rich source of UBV photometric data of a large number of galactic clusters is freely accessible at http://www.univie.ac.at/webda/ (the former "Geneva Cluster Database"). This site also holds, e.g., interstellar extinction data for many clusters.

14.2 Variable Stars

H.W. Duerbeck

14.2.1 Introduction

The subject of variable stars, or simply variables, originally referred to just a handful of stars noted for substantial changes in brightness: the occurrence of a "new star" in the constellation Cassiopeia in 1572 was the trigger of this field of research.

Present-day astrophysical studies have demonstrated that brightness changes are only part of the variability picture. The definition of variability in a star has been generally extended to include time variability of any parameter, especially the luminosity, temperature, and radius. With the development of radio and satellite astronomy, the changes refer more generally to a time variation of radiation over the entire electromagnetic spectrum from radio to X-rays.

Complex structures as stars can never be "absolutely constant." At sufficiently high resolution, every star displays some variability. For example, active regions in the surface layers of the Sun form, evolve and decay in an 11-year cycle. During times of high activity, both the ultraviolet radiation and the particle emission of the Sun show substantial and mostly short-period changes, while the solar magnetic field and the shape of the corona change more slowly. The solar constant, i.e., the flux arriving at $1\,m^2$ on the Earth's surface, changes at the 0.1% level. Another example: stars surrounded by planets can show eclipses which cause brightness drops of 2% or less.

Within the framework of this article, variable stars are defined as stars whose light changes are of intrinsic nature and are dependent on the particular structure of the stellar interior, which are mostly related to particular stages of stellar evolution. Variable stars form several distinct classes which are discussed in this chapter. Besides these classes, the "eclipsing binary stars," which are also counted among the variables, will be discussed in more detail in Sect. 14.3 on binary stars.

We will focus here on an up-to-date introduction to the various classes and types of variable stars.

14.2.1.1 Initial Discoveries The history of variable star research began in 1572 with the appearance of Tycho's Supernova. The indisputable change in the "translunar" realm shook the Aristotelian notion of the universe and triggered vehement philosophical discussions. In 1595, Fabricius discovered the "miraculous star in Cetus" which is known as Mira Ceti. Originally thought to be a new star, its periodic nature was soon uncovered. Further discoveries followed, and by the early eighteenth century, four "new stars" and four variables were known to exist. By 1800, seven more variables had been found.

Although this was a short list, with only 15 objects, it provided a good cross-section of all major classes of variables. After 1800, the number of identified variable stars grew, but still at a slow rate. In 1900, 700 variable stars were known. The most recent printed catalogue lists discoveries up to 1982, and comprises 28,484 variables. Additional variable stars have been discovered, e.g., by the Hipparcos satellite (5665 variables), and by ongoing sky surveys (e.g., All Sky Automated Survey (ASAS), 30,000 variables), and new space projects (MOST, COROT).

14.2.1.2 Nomenclature, Catalogs and Classification Variable stars are designated according to a somewhat cumbersome convention: brighter stars, such as β Lyrae, γ Cassiopeiae, δ Cephei, or 44 Bootis, which already have a name or number, receive no further designation if found to be variable. Other variables are coded by Latin capital letters in the sequence of discovery in each constellation. First, the letters R, S, T, ..., Z are used in succession, then RR, RS, RT, ..., YZ, ZZ, and finally from AA, AB, AC, ..., BB, BC, ... to QY, QZ, adding the constellation name in the genitive form. Examples include T Coronae Borealis, RR Lyrae, or HZ Herculis. This coding scheme is sufficient for 334 variables in each constellation. For additional variable stars, the sequence number preceded by the letter "V" for variable, is used, and thus, beginning with V335, the number of available designations is infinite.

This system is not followed for variables in globular clusters and extragalactic star systems, which usually carry only an ad hoc designation.

Newly discovered variables first receive a provisional coding. After the variability has been examined, the final coding is given by the Astronomical General Catalogue of Variable Stars Research Group at Sternberg Astronomical Institute, Moscow, Russia, on behalf of the International Astronomical Union. This institution is in charge of maintaining the *General Catalogue of Variable Stars* (GCVS).

The first edition of the GCVS appeared in 1948. The fourth edition, in five volumes, was edited by P.N. Kholopov and coworkers [6]; the first three volumes list the variables according to constellations, from Andromeda to Vulpecula, the fourth (N.N. Samus, editor [7]) gives reference tables, the fifth one extragalactic variables [8]. In view of the rapidly growing number of variable stars, an eventual fifth edition will presumably be available only on the World-Wide Web.

The growing number of discoveries gradually revealed the wide variety of variable star types with different mechanisms of variability. The development of astrophysics permitted a better understanding of the structure and evolution of stars and hence also of the causes of variability. In recent decades physically meaningful classification schemes

GCVS: The fourth edition of the General Catalogue of Variable Stars (1985–1988), in five volumes, lists variable stars in the Milky Way discovered until 1982. The final volume lists extragalactic variable stars. It contains a catalog of 10,979 variable stars in 35 stellar systems (including the Magellanic Clouds, the large galaxies in Andromeda and Triangulum, etc.), as well as a catalog of extragalactic supernovae with 984 confirmed or suspected supernovae. An updated electronic version is available at: http://www.sai.msu.su/groups/cluster/gcvs/gcvs/intro.htm. A catalog of variable stars in globular clusters of the Milky Way (1997) is kept by C. Clement, and can be downloaded from: http://www.astro.utoronto.ca/~cclement/read.html. New lists of names are published from time to time in the Information Bulletin on Variable Stars of Commission 27 (Budapest): http://www.konkoly.hu/IBVS/IBVS.html.

have been worked out, which nevertheless undergo revisions from time to time. Ordering principles usually depend on the physical nature of the variability (e.g., pulsations of a star or eruptions in the outer layers of the atmosphere), and on the evolutionary status (e.g., young stars, or evolved stars with compact components, etc.).

The most recent classification of variable stars has been elaborated on by N. Samus and coworkers in 2006, and will be used, perhaps again after some improvement, in forthcoming editions of the GCVS. It contains six basic types of variability: pulsating, eruptive, rotating, cataclysmic, eclipsing, and X-ray. The following list closely follows this classification, as well as Table 1 in Duerbeck and Seitter (p. 129 in [9]) although some types are grouped together.

Classification of variable stars:
http://www.sai.msu.su/groups/
cluster/gcvs/gcvs/iii/vartype.txt

A. *Pulsating variables* (Sect. 14.2.2)

1. δ Cephei, W Virginis stars, and BL Bootis stars
2. RR Lyrae stars
3. α Cygni stars
4. PV Telescopii stars
5. β Cephei variables
6. Slowly pulsating B stars
7. δ Scuti stars
8. γ Doradus stars
9. RV Tauri stars
10. Mira stars and OH/IR variables
11. Semiregular variables
12. Irregular or red variables
13. Rapidly pulsating hot subdwarfs
14. ZZ Ceti stars (pulsating white dwarfs)

B. *Eruptive variables* (Sect. 14.2.3)

1. S Doradus stars
2. Wolf–Rayet stars
3. Variable Be stars (γ Cassiopeiae stars)
4. R Coronae Borealis stars
5. UV Ceti stars (flare stars)
6. L dwarfs
7. Young (nebular) variables of all types
8. Protoplanetary nebulae

C. *Rotating variables* (Sect. 14.2.4)

1. α_2 Canum Venaticorum stars, or magnetic variables, also with rapid oscillations
2. Ellipsoidal variables
3. Reflection variables
4. RS Canum Venaticorum stars
4. FK Comae stars
5. BY Draconis stars (spotted stars)
6. Pulsars

D. *Cataclysmic variables* (Sect. 14.2.5)
1. Supernovae
2. Classical novae and helium-flash novae
3. Recurrent novae
4. Symbiotic novae and Z And stars
5. Dwarf novae (UG, UGSS, UGZ, UGSU, UGWZ, UGER)
6. Nova-like objects at permanent outburst or with rare fadings

E. *Eclipsing variables* (Sects. 14.2.6 and 14.3)
1. Eclipsing Algols (E)
2. Binaries in shallow or deep contact (EC)
3. Stars showing eclipses by planetary companions (EP)

F. *X-ray variables* (Sect. 14.2.7)
1. X-ray variables
2. X-ray novae with low-mass companions
3. X-ray novae with high-mass companions
4. Low-mass X-ray binaries
5. High-mass X-ray binaries
6. Microquasars

These variables will be discussed below. Some objects fit into two or even three different classes. Ellipsoidal stars, for instance, may be treated as "failed" eclipsing variables, but they are included here among rotating variables. Novae and dwarf novae, as well as X-ray variables, are in all likelihood close binaries, and many of them exhibit eclipsing light variations. The above classification sorts the variables through a somewhat subjective judgement as to what the primary characteristic feature of an observed variable is.

14.2.2 Pulsating Variables: Generalities

A major group of physically variable stars are the objects which show periodical, or at least cyclic, brightness changes between a minimum and a maximum level, usually without a well-defined "rest" magnitude. Sometimes, these brightness oscillations have a regularity which is comparable in precision to that of a quartz clock (S Cephei), but sometimes of much lower precision, but still fairly regular (Mira Ceti), and sometimes barely identifiable as cyclical (RV Tauri). Such stars are called pulsating variables. Periodic brightness changes are caused by a periodic expansion and contraction of the star, usually in the radial direction. Besides the brightness variation due to the change in stellar surface area as well as temperature, the radial velocity of the star (averaged over the visible surface layers in motion) varies with the same period, giving an unambiguous proof of the pulsation process. It is also possible that the radial motions of stellar layers at a given time are, e.g., not all directed outward, others may be directed inward, and those in between are at rest (an analogous example is an oscillating string, vibrating at the first overtone). Such stars are said to pulsate in the first overtone. Stars may pulsate in several overtones, which can also interfere with each other.

Besides radial motions, other motions of stellar layers can occur. For instance, surface waves moving around the star, or torsional-type deformations of the stellar body may occur, generally resulting in light variations of low amplitudes and many periodicities. These are called non-radial pulsations or oscillations.

Radially Pulsating Stars The first theory of pulsations was given by A. Ritter in 1879, who established a relation between the pulsation period and the mean density ($P\rho^{1/2} \approx$ const). Only in the 1920s, was the pulsation explanation accepted.

The important parameters of these stars change as they pulsate: the observed brightness variations with time result from periodic changes in radius and effective temperature.

The effective temperature is the temperature of a blackbody which radiates the same total amount of energy from the same surface area. The total amount of radiation emitted per second (i.e., the luminosity) by a star is $L = 4\pi R^2 \sigma T_{\text{eff}}^4$ where $\sigma = 5.67 \times 10^{-8}$ J m^{-2} K^{-4} s^{-1} is the Stefan–Boltzmann constant.

From the observed spectral type as well as from the color, changes in the surface temperature can be estimated while the rate of change in radius dR/dt is derived from the radial velocity curve. Upon integration, the latter yields the difference $R_{\max} - R_{\min}$. The change in luminosity (total radiation per second) can be calculated from these data. It is usually smaller than the brightness change in the visual or photographic ranges. This difference is particularly striking in the long-period Mira variables. During their cycle, the radiated total energy changes by at most a factor 2, while the change in the visual brightness may reach 7 or 8 magnitudes, or a factor of about 1000. In these low-temperature stars the maximum intensity of radiation lies in the infrared, and at lower temperatures additional strong absorptions from molecules influence the spectral energy distribution. Dust formation may also play a role.

In all radially pulsating variables, the lowest surface temperature corresponds to a brightness minimum. Figure 8 shows the change of various stellar parameters during one pulsation cycle of δ Cephei. The brightness change follows this temperature variation rather closely, whereas the maximum radius (see the ΔR curve) does not occur at the time of maximum light, but rather halfway between maximum and minimum on the descending light branch. This is to be expected since the luminosity depends on the fourth power of the temperature, but only on the square of the radius.

Stellar pulsations may be seen as an analogy to a simple pendulum which, in the idealized case of no irreversible loss of mechanical energy by friction, will oscillate indefinitely. As in the case of an everyday pendulum a certain amount of mechanical energy of the star is always dissipated into the surrounding environment. This damps the motion and continuously diminishes its amplitude until the oscillation ceases at the equilibrium state. Most stars are stable against small perturbations, insofar as all vibrations are damped around the equilibrium point. In pulsating variables, such vibrations are evidently not damped, because the vibrating layers receive some energy at appropriate phases.

The vast majority of pulsating stars act as a heat engine. The net work done by a stellar layer during one cycle of pulsation is the difference between the heat flowing into the gas, and that leaving it. For "driving," the heat must enter the layer during the high-temperature part, and leave during the low-temperature part. Just as a spark plug in an automobile engine fires at the end of the compression stroke, the driving layers of a pulsating star must absorb heat around the time of maximum compression.

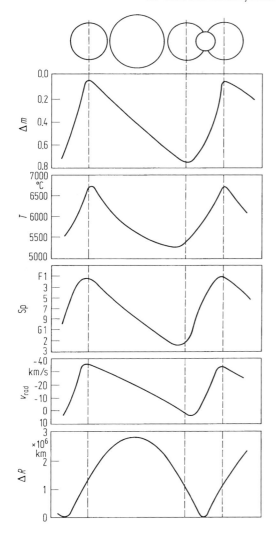

Fig. 8. Periodic variation of brightness, surface temperature, spectral type, radial velocity, and radius during one pulsation cycle of δ Cephei

The absorption of radiation by matter is described by the opacity κ (kappa), which depends on the density and temperature of the stellar material. In most regions, the opacity decreases with compression, thus the above mechanism does not work.

The special condition of exciting and maintaining stellar pulsations were identified by S.A. Zhevakin in the 1950s. A "valve" mechanism can operate at the partial ionization zones, layers of the star are partly ionized. In such a zone, work on the gas is partly used to produce further ionization, rather than raising the temperature. The compression causes an increase in density, but the temperature is hardly increased. This causes an increase in opacity. Similarly, during expansion, the temperature does not decrease so much, since the ions recombine with electrons and release energy: the density change dominates again, and opacity decreases during expansion.

Summing up, the stellar layer can absorb heat during compression, is pushed outward and releases the heat during expansion, and falls back to begin another cycle. This is called the κ-mechanism.

Main ionization zones in stars are: (1) a broad zone where the ion of neutral hydrogen (H I → H II) and the first ion of helium (He I → He II) occur at $1 - 1.5 \times 10^4$ K. (2) The second zone involves the second ionization of He (He I → He II) and occurs at about 4×10^4 K (= He II partial ionization zone).

If a star is too hot (7500 K), the ionization zone will be located near the surface, and there is not enough mass to drive the oscillation efficiently. This is the blue (hot) edge of the instability strip in the HR diagram. If a star is too cool, the onset of convection in the outer layers may damp the oscillation. The red (cool) edge of the instability strip seems to be the result of this. Figure 9 shows those regions where the various types of variables are found. Pulsating variables occur predominantly in the region surrounded by a dashed line (instability strip) with the following four most prominent types:

1. RR Lyrae stars with periods under 1 day
2. W Virginis stars with periods between 1.5 and 25 days (both these and RR Lyrae types are old, metal-poor Population II stars with masses around $1 M_\odot$)
3. δ Cephei stars with periods between 2 and 40 days (massive stars of Population I with 10 to 20 $M_\odot$)
4. Mira stars or long-period variables with periods from 150 to 500 days, stars of older Population I and Population II

The diagram illustrates that a strip of instability or pulsations exists. After the main-sequence phase, stellar evolution takes place above the main sequence in the giant and supergiant regions. Depending on the mass, stars may reach the instability strip and leave it again after 10^6 to 10^7 years. Massive stars may cross the strip several times.

Most pulsating variables are giants or supergiants which are suffering substantial mass loss, primarily in the form of a strong stellar wind with a low terminal velocity (≈ 10 km s^{-1}). Pulsations enhance the mass loss.

Non-Radially Pulsating Stars Non-radial stellar pulsations occur when parts of the stellar surface move outward, while other parts contract. Physical variables like pressure and temperature follow the expansion and compression of different regions inside or on the surface of a star.

The wave equation for non-radial pulsations contains a radial number (k, equivalent to that for radial pulsations, giving the number of nodes within the stellar interior). In addition, the angular variation is represented in terms of spherical harmonics, and characterized by two integers, l and m. The term l is a non-negative integer (number of nodal circles where the stellar radius does not change), and m is equal to any of the $2l + 1$ integers between $-l$ and $+l$. The term $|m|$ gives the number of circles passing through the poles of the star (the remainder of circles are parallel to the stellar equator). Simple radial oscillations of the entire surface layer of a star are expressed by $l = 0$, while $l = 1$ represents dipole oscillations (when the hemispheres vibrate separately), $l = 2$ the quadrupole oscillations (with four separately vibrating spherical segments), and so on. Visualizations can be seen on the Internet (see margin), and are also depicted in Fig. 10.

Visualization of non-radial oscillations:
http://www.asteroseismology.org/ ("animations")

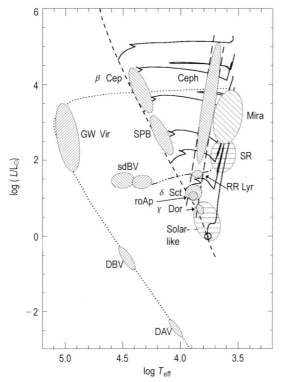

Fig. 9. Schematic Hertzsprung–Russell diagram (logarithm of luminosity (in solar units) versus effective surface temperature) illustrating the location of several classes of pulsating stars. The *dashed line* shows the zero-age main sequence, the *continuous curves* are selected evolution tracks (from *bottom* to *top*) at masses 1, 2, 3, 4, 7, 12, 20 $M_\odot$, the *dot-dashed line* is the horizontal branch (evolved, helium-core burning stars of Population II), and the *dotted curve* is the white dwarf cooling curve. Abbreviations include: GW Vir; DOV, DAV, DBV = ZZ Ceti stars; sdBV rapidly pulsating hot subdwarfs or extreme horizontal branch stars (with kind permission of J. Christensen-Dalsgaard)

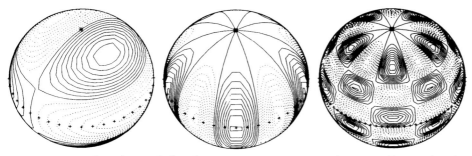

Fig. 10. Contour plots of non-radially pulsating stars. Positive contours (e.g., expanding areas) are indicated by *continuous lines*, negative contours by *dashed lines*. The stellar axis has been inclined by 45° towards the viewer. The equator is shown by "++++". The following cases are illustrated (from *left* to *right*): $l = 3$, $m = 1$; $l = 5$, $m = 5$; and $l = 10$, $m = 5$ (with kind permission of J. Christensen-Dalsgaard)

Non-radial pulsation may be seen as sound waves traveling horizontally as well as radially through the star. Since pressure provides the restoring force for sound waves, pulsation of this type is known as *p-modes* (pressure modes). Like in the case of radial pulsation modes, most of their motion occurs near the stellar surface. The other class of non-radial pulsation are the *g-modes*. These oscillations are related to the buoyancy of the stellar gas. Small blobs of material undergo pressure and density variations, e.g., when a gas bubble is displaced from its equilibrium position upward by a small distance. In a non-convective stellar region, it will be pushed back and will oscillate like a weight on a string. The g-modes involve significant motions of gas in the stellar interior.

Non-radial oscillations evidently affect the overall brightness of the star, although their amplitudes are much smaller than those of radial pulsations. The same is true for radial velocity variations, although the profiles of spectral lines are influenced quite noticeably by non-radial oscillations.

DCep, CW, BLBoo

14.2.2.1 δ Cephei Stars, W Virginis Stars, and BL Bootis Stars

Cepheids are radially pulsating, high-luminosity (classes Ib-II) variables with periods in the range of 1 to 135 days and amplitudes from several hundredths to 2 mag in V. The spectral type at maximum light is F; at minimum, the types are G–K. The longer the period of light variation, the later is the spectral type. The maximum of the surface-layer expansion velocity almost coinciding with maximum light.

δ Cephei stars are the "classical Cepheids," comparatively young objects of Population I that have left the main sequence and evolved into the instability strip of the Hertzsprung–Russell (HR) diagram, they obey the well-known Cepheid period-luminosity relation and belong to the young disk population. A subgroup are the DCEPS, δ Cephei variables having light amplitudes $<0^m.5$ in V and almost symmetrical light curves; as a rule, their periods do not exceed 7 days. They are probably first-overtone pulsators and/or are in the first transition across the instability strip. Another subgroup are the the Beat Cepheids, CEP(B), displaying the presence of two or more simultaneously operating pulsation modes (usually the fundamental tone with the period P_0 and the first overtone with P_1). The periods P_0 are in the range from 2 to 7 days, with the ratio $P_1/P_0 \approx 0.71$.

W Virginis stars are pulsating variables of the galactic spherical component (Population II) with periods of approximately 0.8 to 35 days and amplitudes from $0^m.3$ to $1^m.2$ in V. For an equal period value, the W Virginis variables are fainter than the δ Cephei stars by $0^m.7 - 2^m$. The light curves of W Virginis variables for some period intervals differ from those of δ Cephei variables either by amplitudes or by the presence of humps on their descending branches, sometimes turning into broad flat maxima. W Vir variables are present in globular clusters and at high galactic latitudes.

Cepheids and other pulsating stars have achieved a particular significance in astronomy, since they follow certain period-luminosity relations. Knowing the length of the pulsation period leads directly to the average luminosity or absolute magnitude of a star. The period-luminosity relation (PLR) had been discovered before radial pulsations became generally accepted as the cause of the characteristic light variations: Henrietta Leavitt at the Harvard Observatory had noted in 1912 that the apparent brightness (at both maxima and minima) of Cepheids in the Small Magellanic Cloud was systematically higher for stars with longer periods. The relation was found to have the form $m = a + b \log P$, and it soon became evident that a similar relation holds for absolute magnitudes, since all stars of the SMC have approximately the same distance from the solar system. The slope of the line in the $(m, \log P)$ or $(M, \log P)$ diagram is simply determined and yields results which are about the same for the SMC, the LMC, M 31, M 33, and the Milky Way. To calibrate the PLR, the distance of at least one Cepheid must be determined by an independent method. Various attempts in the twentieth century lead to a continuous refinement of such relations.

Knowing the PLR, the direct observation of the (average) apparent magnitude of a star and its period then leads to its distance, after taking into consideration the effect of interstellar absorption $A(r)$:

$$m - M = 5 + 5 \log r - 5 + A(r) \, .$$

The difference $\mu = m - M$ is also called the distance modulus. Since δ Cephei stars are supergiants of high absolute luminosity, they can be observed in galaxies of the nearer regions of the universe. In 1923 Edwin Hubble, using the 2.5-m telescope at Mount Wilson, discovered Cepheids in the galaxies M 31, M 33 and others, and determined their distances by means of the period-luminosity relation. Thus he proved conclusively that these and some other "nebulae" are in reality extragalactic star systems.

Period-luminosity relations exist for RR Lyrae, δ Cephei, and W Virginis stars, and to some extent also for Mira stars.

Recent research yielded the following relations for δ Cephei stars in the visual and infrared:

$$< M_V > = -2.24 \log P - 1.28 \, , \tag{1}$$
$$< M_I > = -2.98 \log P - 1.74 \, , \tag{2}$$
$$< M_K > = -3.36 \log P - 2.28 \, , \tag{3}$$

where P is expressed in days (Fig. 11). The $<>$ symbol indicates that the brightness is averaged over the light curve. (Usually the light curve is expressed in intensities rather than in magnitudes, and the intensity averaged over the period.)

The W Virginis stars follow a somewhat different relation to both slope and zero point, although the scatter around the median line is somewhat larger. W Virginis stars are fainter than "classical" Cepheids by $1^m.5$ for periods of 2.5^d, and by $2^m.5$ around 20^d:

$$< M_V > = -1.64 \log P + 0.05 \, , \tag{4}$$
$$< M_I > = -2.03 \log P - 0.36 \, , \tag{5}$$
$$< M_K > = -2.41 \log P - 1.00. \tag{6}$$

For RR Lyrae-type stars, discussed in Sect. 14.2.2.2, an average absolute magnitude that is nearly period-independent, is found:

$$\bar{M}_V = +0.54 \, , \tag{7}$$
$$\bar{M}_K = -0.63 \, . \tag{8}$$

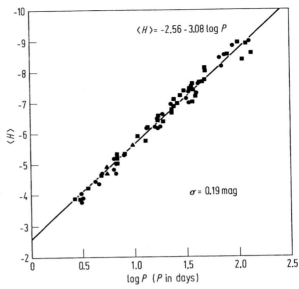

Fig. 11. Period-luminosity relation for Cepheids in the infrared (H-band at 1.6 μm). This graph includes data from Cepheids in the Milky Way and in the Magellanic Clouds

BL Bootis stars are "anomalous cepheids," stars with periods characteristic of comparatively long-period RR Lyrae variables, but with considerably higher absolute magnitudes. The prototype is BL Boo, the variable star 19 in the globular cluster NGC 5466.

RR

14.2.2.2 RR Lyrae Stars Variables of the RR Lyrae type are radially pulsating giant A5-F5 stars having amplitudes from $0^{m}.2$ to $1^{m}.5$ in V. Cases of variable light curve shapes ("Blazhko effect") as well as variable periods are known. Traditionally, RR Lyrae stars are sometimes called short-period Cepheids or cluster-type variables. The majority of these stars belong to the older population of the Galaxy; they are present, sometimes in large numbers, in some globular clusters, where they are known as pulsating horizontal-branch stars. Like Cepheids, maximum expansion velocities of surface layers for these stars practically coincide with maximum light.

Previously, the following subgroups were in use:

RR(B)   RR Lyrae variables showing two simultaneously operating pulsation modes, the fundamental tone with the period P_0 and the first overtone, P_1. The ratio P_1/P_0 is approximately 0.745.

Catalog and identification of RR Lyrae stars brighter than $12^{m}.5$: http://www.astro.uni-bonn.de/~gmaintz/aa3230-05MZ.pdf

RRab   RR Lyrae variables with asymmetric light curves (steep ascending branches), periods from 0.3 to 1.2 days, and amplitudes from $0^{m}.5$ to 2^{m} in V.

RRc   RR Lyrae variables with nearly symmetric, sometimes sinusoidal, light curves, periods from 0.2 to 0.5 days, and amplitudes not greater than $0^{m}.8$.

ACyg

14.2.2.3 α Cygni Stars These variables are supergiant stars of spectral types Bep Ia to Aep Ia; supergiant stars of types O and F may also belong to this group. The superposition of many non-radial pulsations with similar periods cause light curve changes of the order of $0^m.1$, which appear at first glance irregular. Cycles from several days to several weeks are observed.

14.2.2.4 PV Telescopii Stars

These variables are supergiants of spectral type Bp (p for peculiar) with weak hydrogen lines and enhanced lines of helium and carbon. They pulsate with periods of 0.1 to 1 day, or vary with an amplitude $\approx 0^m.1$ during a time interval of 1 year.   **PVTel**

These stars, also called extreme He stars or hydrogen-deficient carbon stars, show small amplitude light and radial velocity variations, due to pulsation. They appear to be related to the R CrB stars (see Sect. 14.2.3.4). Because of their higher surface temperatures, dust formation, which manifests itself by brightness drops in R CrB stars, seems to be suppressed.

14.2.2.5 β Cephei Variables

β Cephei stars, also called β *Canis Majoris stars*, form a small, sharply distinguished group of pulsating early-type giants. Their periods range from 0.1 to 0.3 days, their visual amplitudes are very small. Their pulsation proceeds usually in one radial and one or more non-radial modes, it is multiperiodic and their amplitudes exhibit a "beat effect." The small amplitudes (below $0^m.1$) render photometric studies difficult; the radial velocity variations usually range between 20 and 100 km s^{-1} and are more readily measured. The equatorial rotational velocities are not very large, usually well below 100 km s^{-1}.   **BCep**

β Cephei stars are located in the HR diagram far away from the "classical" instability strip. Their spectral types are between B0 and B2, and luminosity classes are IV (subgiants) and III (giants). β Cephei stars lie about one magnitude above the main sequence, where the core hydrogen burning presumably ends and the evolution toward the giant region begins for stars with masses in the range 10 to 15 $M_\odot$.

The pulsation instability is driven by the opacity of iron atoms in layers with temperatures between $1–2 \times 10^5$ K. The metal abundance obviously plays a role, since in the metal-poor Magellanic Clouds a deficiency of such stars is observed.

14.2.2.6 Slowly Pulsating B Stars

B stars of luminosity classes V to III often show periodic light variations of small amplitude. The slowly pulsating B stars are main-sequence stars with spectral types B3 to B9. They oscillate in gravity modes with periods of the order of days, excited by the κ-mechanism in partial ionization zones of iron-group elements at temperatures near 200,000 K.   **SPB**

14.2.2.7 δ Scuti Stars

δ Scuti stars have short periods (below 0.3 days) and small amplitudes (a few $0^m.01$); they show multiperiodic behavior. With spectral types between A5 and F2, they are located in the vicinity of the RR Lyrae stars in the HR diagram. Their luminosity class ranges from V to III; about 30% of stars in this region appear to be δ Scuti stars. They often occur in open clusters and in binary systems. An interesting exception is the star SX Phe ($P = 0^d.055 = 79^m$), which distinctly shows low-metallicity and subdwarf features. Perhaps this star represents an unexplored class of Population II δ Scuti stars.   **DSct**

δ Scuti stars have normal masses and chemical compositions. They are located in the downward extension of the δ Cephei instability strip. Their pulsational mechanism seems similar to that of the Cepheids. About 20% of all δ Scuti stars show rather large

Online catalog of δ Scuti stars:
http://www.iaa.es/~eloy/
dsc00.html

GDor

light curve amplitudes of up to $0^m.7$. Their light curves may then look similar to those of some RR Lyrae variables. Multiple periodicities occur quite frequently in δ Scuti stars; in addition to the basic (radial) oscillation of period P_0, in many cases the first overtone of period P_1, and sometimes even the second overtone P_2 can be clearly identified. The ratios P_1/P_0 are near 0.77. Hotter stars pulsate in the first overtone, cooler stars in the fundamental mode.

14.2.2.8 γ Doradus Stars γ Doradus stars are stars of spectral type F0 to F2 (at or beyond the red edge of the δ Scuti instability strip), on or near the main sequence (luminosity classes V or IV). They show multiple periods from several tenths of a day to about 2–3 days. Amplitudes usually do not exceed $0^m.1$. They are multiperiodic, and appear to pulsate in high order (m) low spherical degree (l) non-radial g-modes.

Together with the δ Scuti and the SX Phoenicis stars, they are found near the main sequence in or near the Cepheid instability strip in the HR diagram. The radial velocities are nearly in phase with the light curves, contrary to the findings in δ Scuti stars.

Online catalog of γ Doradus stars:
http://www.chjaa.org/aiying/
gdorlist.html

RV, RVb

14.2.2.9 RV Tauri Stars The RV Tauri stars (Fig. 12) are pulsating yellow supergiants, whose light curves are characterized by alternating deep and shallow minima, with amplitudes of $3-4^m$ in V. They form a rather small group with less than 100 known members. For some time these stars were thought to be typical representatives of the semiregular variables, even though they show enough regularity to be considered a subgroup of "regular" pulsating variables. In fact, both by their periods of from 30 to 150 days and by their spectral types ranging from F-G at maximum and K-M at minimum, they occupy a place somewhere between Cepheids and Miras.

The characteristics of the light curves for RV Tauri stars are twofold. Besides the alternate succession of a deep and a less deep minimum, there is in some systems an apparently sporadic long-term light modulation, appearing about once or twice in 10 years: the amplitudes of the deep minima decrease and simultaneously the overall brightness diminishes distinctly. Objects showing this characteristic belong to the subgroup RVb, if the amplitude of the secondary wave is at least equal to that of the main variation.

RV Tauri stars do not form a uniform group; some of them belong to Population II, as occasionally they are found in globular clusters. Others, however, show strong metallic lines and thus belong to Population I. RV Tauri stars are in their final transition from the asymptotic giant branch (AGB) to the white dwarf stage.

M

14.2.2.10 Mira Stars and OH/IR Variables Mira-type stars, sometimes called long-period variables (LPV), have spectral types M, C, and S with emission lines, and light variations from $2^m.5$ to 10^m in V; infrared variations are substantially smaller. The well-pronounced periodicity lies between 80 and 1000 days. The light curve of the Mira variable R Dra is reproduced in Fig. 3 of Chap. 8.

Mira stars follow period-luminosity relations; in the period range $200 < P < 500$ days, they are (for V, at maximum, for K and bolometric magnitudes, averaged over the light curve):

$$M_V = 0.004 \log P - 2.6 , \tag{9}$$

$$< M_K > = -3.47 \log P + 1.0 , \tag{10}$$

$$< M_{bol} > = -2.34 \log P + 1.3 . \tag{11}$$

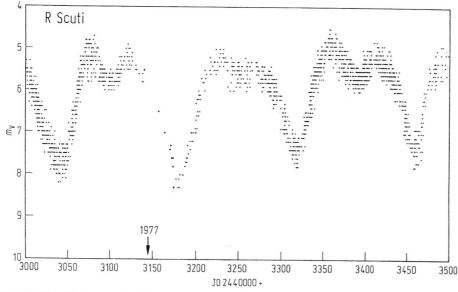

Fig. 12. Visual light curve for R Scuti covering the period 1976–1977, serves as a typical example for the small class of RV Tauri stars. The observations were provided by the AAVSO

Mira variables have very distinctive spectra, with types M 6 to M 9 at minimum and M 3 to M 7 at maximum. In general, shorter periods are correlated with earlier spectral types, but this relation is not very tight. About 10% of Mira-type variables belong to the classes S or C. Type C is distinguished by a high carbon abundance; S-type stars show other chemical peculiarities (e.g., zirconium instead of titanium). These are side branches off the temperature sequence at the types K and M.

These stars always show emission lines, and this property is so essentially that all Me-type giant stars can be assumed to be pulsating variables. The presence of emission lines, which occur in particular at certain phases in the light curve, may be explained by outward-moving shock waves in the stellar atmosphere.

The large amplitude of the light curves and the emission lines, as well as their relatively high luminosity, contribute to the large number of known variables. At present, about 6000 Mira stars are known, constituting about one-quarter of all stars contained in the new GCVS. Miras represent the most frequent type of intrinsic variability.

The Mira-type giants have absolute visual magnitudes around -2^m, and corresponding bolometric magnitudes are around -4.5. Assuming an effective surface temperature of 3000 K, then $M_{bol} = -4.5$ leads to an enormously large stellar radius of about 1.5 AU. If this structure has a mass of 1 $M_\odot$, the corresponding mean density suggests a period around 500 days. Therefore, the stellar parameters are consistent to within an order of magnitude. In practice, one must also consider that a relatively large fraction of stellar mass, perhaps as much as one-half (at least 0.5 $M_\odot$) is concentrated in a compact, dense core which, according to current interpretation, is composed of carbon and oxygen, and has originated in an earlier stage of post-main sequence evolution via helium and carbon burning. The Miras are in the second giant stage, characterized by hydrogen and helium-burning in alternate shells. This corresponds in the HR diagram to a position on the steeply sloped AGB. The enormous envelope surrounding the tiny core, which is barely larger than a white dwarf, loses mass at a substantial rate, between 10^{-8} and 10^{-4} $M_\odot$, and could be completely lost within a few 10^6 years.

The galactic distribution and the statistics of motions of Mira variables indicate a fairly broad spread in population membership and hence in the ages of these objects. Mira variables with periods below 200 days may be counted among the Population II stars, while long-period ones belong to Population I.

A class of variable stars closely related to the Miras group are the *OH/IR* stars, which can be regarded as *infrared Mira stars*. These stars were first noted as "molecular" radio sources through their 1612 MHz (18.6 cm) microwave radiation of the OH radical. The intensity of the radiation point to the existence of a special "pumping" mechanism in the circumstellar envelope. It is called the *maser effect*, and consists of a complex interaction of stimulation and emission in the electron shells of atoms and molecules.

Infrared measurements, particularly those with the IRAS satellite, also show strong radiation sources, but as yet no corresponding objects have been found in the visible range. Since many Mira stars also show maser effects, it seemed reasonable to assume that late-type stars of this new class, owing to strong mass loss, develop a massive circumstellar dust envelope at some distance from the star. This shell absorbs the short-wavelength portion of the radiation almost entirely, is heated to 500 to 700 K and radiates intensely in the infrared range.

Studies show that the stars "behind the dust envelope" are substantially more luminous than ordinary Mira stars; they lie at the top of the AGB. The absolute bolometric luminosities lie in the range -4 to -8. These stars also pulsate with periods ranging up to 5 years. They are longer than for Mira variables, and the masses may also be higher: they are expected to range from 2 to 6 $M_\odot$, or even as high as 8 $M_\odot$.

The infrared Mira-type stars are evidently in a stage of very rapid evolution, and at this point there seems only one direction indicated for them: when, after suffering a nearly catastrophic mass loss the "last" envelope has been ejected, only the hot stellar core remains as the central star of a planetary nebula.

SR, SRd

14.2.2.11 Semiregular Variables Related to the class of regularly pulsating stars is the large class containing semiregular and irregular variables. Semiregular variables pulsate, although not in a strictly cyclic fashion. Even in irregular variables, the light variation originates at least partly in pulsation-like changes of the volume and temperature, which occur apparently erratically.

The semiregular variables are late-type giants or supergiants, with spectral types M, C, S, for which a kind of average period or cycle can be defined. The periods lie between 20 and 2000 days, and can be broken by temporarily occurring light curve irregularities; amplitudes are usually small, $1-2^m$ in V. The most recent GCVS lists about 3000 SR variables. Together with the irregular variables, which are also late-type giants and supergiants, they form a broad class of objects also often called red variables.

In previous editions of the GCVS, an object is placed into one of the subclasses of semiregular variables based on the following definitions:

SRa   "Fairly regular" light curve with amplitude usually less than 2^m; periods lie between 35–1200 days. They are similar to Miras, but with smaller amplitudes.

SRb Less regularity, but periodicity still discernible; periods lie between 20–2300 days.

SRc Like SRa,b but the variables are presumably quite massive supergiants. Periods lie between 30 and several 1000 days; amplitudes around 1^m.

SRd Collective class for semiregular variables with spectral types earlier than M, S, and C. Periods lie between 30–1100 days; amplitudes between $0^m.1$ and 4^m.

SRs Pulsating red giants with short periods (several days to months), probably high overtone pulsators. The prototype is AU Aur.

In the most recent classification, only the classes SR and SRd are kept, i.e., semiregulars of spectral types M (including C, S) and variables with spectral types earlier than M.

The most famous representative of the SR stars is the bright red supergiant α Orionis (= Betelgeuse). This star was previously considered a typical irregular variable; its more correct classification is SRc. The slow brightness variations of about $0^m.2$ to $0^m.3$ seem to indicate a cyclic pattern with about 1200 days as the length of the cycle. The brightness variations, however, are more probably connected with pulsation-type changes of the radius; the few interferometric measurements of its diameter are not very consistent, varying from $0''.04$ to $0''.06$.

14.2.2.12 Irregular or Red Variables This inhomogeneous category comprises primarily late-type giants or supergiants which vary slowly and with only small amplitudes (normally only a few $0^m.1$) and without a detectable periodicity. The spectral type M prevails, as in the case of semiregular variables. Since the third edition of the GCVS, they are labelled by "L" with the following subdivisions: **L, Lb**

Lb Irregularly variable giants, mostly of spectral type M (also C, S), but occasionally of types G and K.

Lc Virtually all late-type supergiants with irregular fluctuations in brightness and radius, perhaps also in color. The class Lc is closely related to the semiregular subclass SRc, and possibly its continuation to less regular variations. It will, apparently, be eliminated in a new variable star classification.

14.2.2.13 Rapidly Pulsating Hot Subdwarfs These stars are very rapidly pulsating hot stars (RPHS). Typical periods are hundreds of seconds, amplitudes are within several hundredths of a magnitude. In the HR-diagram, they are extreme horizontal branch stars (EHB): low-mass stars with a He-burning core, surrounded by a H-rich shell. They pulsate in non-radial p-modes (EC14026 stars, $T = 28,000$–$36,000\,\mathrm{K}$, $P \sim$ a few hundred seconds) or in non-radial g-modes (PG1716 stars, $P = 1$–2 hours). The κ-mechanism is caused by partly ionized iron. **RPHS**

14.2.2.14 ZZ Ceti Stars: Pulsating White Dwarfs These stars are white dwarfs of spectral type DA, DB and DO (D = degenerate is the prefix indicating a white dwarf). They lie on the downward extension of the Cepheid instability strip, with a similar driving mechanism. Unlike some δ Scuti stars, they pulsate in non-radial g-modes. Multiple periodicities of about 30 to 1500 s occur almost without exception. The light variations are, with rare exceptions, of very small amplitude, $0^m.001$ to $0^m.2$ in V, for instance $0^m.012$ for ZZ Ceti. The multiple periodicities cause the light curve to change from one cycle to the next. In almost all cases, two or more nodes interfere and produce beats. **ZZ**

ZZ Ceti stars are confined to narrow ranges in color and temperature. DA stars with $T \sim 10{,}000\,\mathrm{K}$ (ZZ Ceti); DB stars with $T \simeq 30{,}000\,\mathrm{K}$ (V777 Herculis); DO stars with $T \sim 100{,}000\,\mathrm{K}$ (GW Virginis).

Some ZZ Ceti stars occur in nova, nova-like or dwarf nova systems (e.g., DQ Her), see Sect. 14.2.4. Some central stars of planetary nebulae (PNNV) also show such light variations.

14.2.3 Eruptive Variables

Eruptive variables are stars varying in brightness because of violent processes and flares occurring in their chromospheres and coronae. The light changes are usually accompanied by shell events or mass outflow in the form of stellar winds of variable intensity and/or by interaction with the surrounding interstellar medium. This class includes the following types of stars:

SDor

14.2.3.1 S Doradus Stars S Doradus is an extremely bright member of the Large Magellanic Cloud, an A5 supergiant with absolute magnitude $M \approx -10$. At about 10-year intervals, it becomes fainter by about 1^{m} for a few months. In 1964, the star dimmed by $2^{\mathrm{m}}.4$, then became 1^{m} brighter, and has since oscillated around the new magnitude. Occasionally, the often-observed (and perhaps related) stars P Cygni and η Carinae are counted among the S Doradus-type stars, although these stars at a time behaved like very slow novae: P Cygni in 1600 reached a maximum brightness of $+3$, η Carinae in 1843 a spectacular -1. The corresponding absolute magnitudes were in fact near -11 for P Cygni, and around -12.5 or even higher for η Carinae, whose distance is better known. Presently, P Cyg and η Car have absolute magnitudes of about -8 and -4.5, respectively, which makes them still fairly bright stars. They are presumably related to the *Hubble–Sandage* objects, which are extremely bright, often irregularly variable, single stars in neighboring galaxies. We see here possibly the early evolution of very massive stars (with over $100\,M_\odot$, and perhaps as much as $150\,M_\odot$), which cannot arrive at a "normal" equilibrium state, and therefore suffer very strong mass loss.

WR

14.2.3.2 Wolf–Rayet Stars Wolf–Rayet stars are evolved, massive, hydrogen-poor early-type stars with broad emission features of He I and He II as well as C II-C IV, O II-O IV, and N III-N V. Their photometric variability is small and extremely complex. Irregular light changes with amplitudes up to $0^{\mathrm{m}}.1$ in V are probably caused by non-stable mass outflow from their atmospheres, i.e., random small density enhancements embedded in their massive winds. Some also show eruptions (theory suggests that they are unstable to pulsation), some others are eclipsing.

A satellite study of WR123, a WN8 star, shows no variability on short timescales, but variations with timescales between hours and 10 days.

Be

14.2.3.3 Variable Be Stars Be stars—stars of spectral type B with emission lines—are closely related with the so-called shell stars. In the HR diagram, they are found between late O an early A types, near the main sequence at luminosity classes V to III. They are characterized by equatorial rings, disks, and spherical envelopes of gas which generate emission and absorption lines.

The character of these lines (intensities, ionization and excitation, Doppler shifts, and peculiar line profiles) depend on physical parameters of the circumstellar material (density, temperature, flow velocities) and on the geometrical aspect (e.g., pole-on or equatorial view of the star). The emissions and shell absorptions are superposed upon a normal photospheric absorption spectrum distinguished by rather broad line profiles. The Doppler broadening is caused by the rapid rotation of the star. The shell spectrum often varies with time, the intensity and profiles of shell lines may change, sometimes within a few days. Examples are γ Cas and ζ Tau.

As concerns the photometric behavior, the majority of Be stars are photometrically variable. In previous editions of the GCVS, such stars were designated as GCas (γ Cassiopeiae-type) variables. However, the light variations of γ Cassiopeiae, which undergoes strong shell phases and is photometrically very active, is not typical for this group. Quite a number of Be stars show small-scale variations not necessarily related to shell events. In some cases, the variations are quasi-periodic, λ Eri is an example. If a Be variable cannot be readily described as a GCas star, its variability should be typified as Be.

14.2.3.4 R Coronae Borealis Stars

R Coronae Borealis stars (Fig. 13) are low-mass yellow supergiants. The small but remarkable group has about 30 known members and shows an uncommon type of light variation: normally the brightness is nearly constant, showing small variations of at most $0^m.1$ or $0^m.2$. At irregular intervals (a few years on the average), this is interrupted by deep minima which may reach 7 or 8^m. The decline in such minima is quite steep and occurs within a time span of about 4 to 6 weeks. The ascent is slower, and may be accompanied by irregular fluctuations or secondary minima. The prototype of these variables (R CrB) was discovered in 1795. Observations spanning 180 years, many of which carried out by amateurs, show that these minima occur totally irregularly.

R CrB stars are supergiants of spectral types late F or early G. They are hydrogen-poor (with very weak Balmer lines) and carbon-rich, indicating that they are in a far advanced evolutionary state and have suffered extremely high mass losses. They have shed their original outer atmospheric layers, and now consist of a helium-core in which the helium-burning into carbon and oxygen occurs, and a thin hydrogen-poor extended atmosphere. The products of helium-burning have reached the photosphere. This scenario explains why these stars, in spite of their high luminosities, have masses of only about 1 $M_\odot$; their original mass could easily have been 5 to 6 $M_\odot$. Some experts have suggested that these stars are evolving into helium stars or into central stars of planetary nebulae.

The pronounced minima of R CrB stars are caused by the ejection of circumstellar matter, probably in the form of jets into various directions. The carbon-rich atmosphere could also explain the large depth of the minima, since the opacity of the ejected shell is particularly high when the carbon at larger distances from the star no longer occurs as gas but in the form of "graphite flakes" or as a soot-like, amorphous condensate. The intense infrared emission seems to support this picture: dust and soot particles are heated and re-emit the absorbed light at long wavelengths.

R CrB stars are located in the HR diagram in the instability strip or close to it, and some of them also show pulsations (e.g., R CrB, UW Cen, S Aps, and very clearly RY

RCB

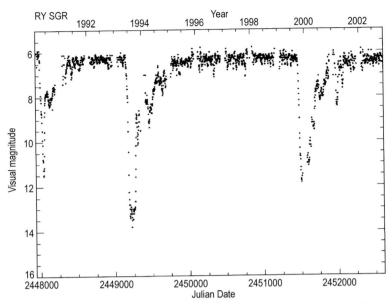

Fig. 13. Visual light curve for RY Sagittarii from 1991 to 2003. Three declines with different depths, durations and intervals in between are visible. A 39-day pulsation is marginally visible at maximum light. The data were provided by the AAVSO

Sgr). The periods are always about 35 to 40 days. The variations in brightness and color of RY Sgr illustrate the semiregular nature of the pulsational variation.

14.2.3.5 UV Ceti Stars UV Ceti stars or flare stars (Fig. 14) are dwarf stars of spectral types K to M, which show flares, emission lines in the visual and UV spectra, and X-ray emission from a hot corona (temperatures up to 10^7 K). Flares are rapid light outbursts. During these eruptions, which occur at irregular intervals of typically a few days, the visual brightness of the star increases within a few seconds to minutes by from 1 to 6^m, and then declines within a few minutes to 2 hours back to normal minimum level. On the average, a brightness gradient of about $0^m.05$ to $0^m.1$ per second is reached during ascent to maximum; in some cases it can be substantially larger. Fast photometry has revealed very rapid "spike" flares during a normal flare. Figure 14 shows an example of a spiked flare.

The amplitude of a flare increases with decreasing wavelength, and therefore is larger in the blue and ultraviolet than in red light. During eruptions, the continuum radiation in the blue and ultraviolet regions becomes stronger. Also, numerous emission lines (particularly of H and He) appear, which are much weaker or absent at minimum.

Owing to the low luminosity of very late-type dwarf stars, which are located slightly above the lower main sequence in the HR diagram, UV Ceti stars can be detected only in the local neighborhood of the Sun. Presumably they are evolutionarily young stars which, despite an age of about 10^9 years, have not reached the main-sequence state since the evolution at a mass of only a few tenths of the Sun's mass is correspondingly slow. Similar flare activities are observed in the very young T Tauri stars, which are found

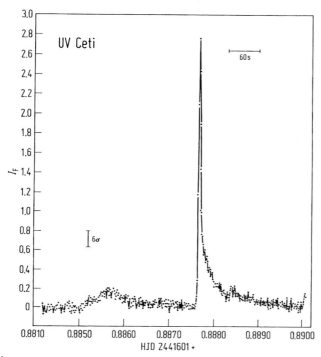

Fig. 14. Photoelectric measurement in white light with no filter of a spiked flare of UV Ceti, recorded with an integration time of 1 s

at somewhat earlier spectral classes (F to K) in the HR diagram, also above the main sequence. Some flare stars are also BY Draconis variables (see Sect. 14.2.4.5).

The physical process responsible for flare activity appears to be related to various mechanisms of solar activity and also with eruptive phenomena within the chromosphere and corona of the Sun (solar flares, surges, prominences). The flares can be linked with complex magnetic field structures undergoing rapid changes with time and space. In this connection, the emission of relativistic electrons and photon scattering by the inverse Compton effect could cause short-term enhanced electromagnetic radiation.

14.2.3.6 L Dwarfs Class L dwarfs get their designation because they are cooler than M stars and L is the remaining letter alphabetically closest to M. Some of these objects have mass large enough to support hydrogen fusion, but some are of substellar mass and do not, so collectively these objects should be referred to as L *dwarfs*, not L *stars*. They are very dark red in color and brightest in the infrared. Their atmosphere is cool enough ($T = 1300$–2000 K) to allow metal hydrides and alkali metals to be prominent in their spectra. Examples are VW Hyi (cool component), and 2MASSW J0746425+2000321 (binary).

Indications of variability have been found in red and infrared color bands, although not in a consistent way. Rotation and clouds ("weather") may be the cause for the ob-

The *Compton effect*, which provided an early proof of the particle nature of light, consists of a collision between a quantum of light and an electron of low energy or at rest, transferring part of the photon energy to the electron. The inverse Compton effect involves a collision of a photon in the radio or visible range with a *high-energy* (relativistic) electron, in which case the photon receives additional energy at the expense of the electron.

dL

served variations. Periods of several hours and variations of up to $0^m.01$ seem to be typical. As often, the question has to be asked where the borderline between variable and non-variable stars lies, and that between stars and "brown dwarfs" has to be drawn.

14.2.3.7 Young Irregular Variables

Variations in photometric and spectroscopic properties are found in evolutionarily young stars in the pre-main sequence phase. While previous editions of the GCVS had introduced a quite sophisticated system of classification, the recent revision has only four classes: irregular variables (I) whose association with star-forming regions is not established without doubt, "classical" T Tauri stars (CTT) with typical emission-line spectra (see below), weak-line T Tauri stars (WTT) that lack strong emission lines in their optical spectrum, and lack both strong stellar winds and circumstellar accretion disks, and Herbig Ae-Be stars (HAeBe), which are more massive T Tauri stars with spectra type F or earlier.

In the years between 1920 and 1930, such objects, which display irregular brightness variations, were found in the Orion Nebula. By about 1950 many other related objects, which are located slightly above the main sequence and in the subgiant region on the HR diagram, have been identified. They show spectral types from O to M. Such stars are found primarily in groupings, termed associations, within dark or emission nebulae, often in the vicinity of massive, young O- and B-type stars (OB associations). A subgroup is called T Tauri-type stars, after their prototype. These objects are young, solar-type protostars characterized by distinct spectroscopic features. Groupings of protostars in star-forming regions are generally termed T associations, after the important class of T Tauri stars.

The first attempts to find a cause for the photometric and spectroscopic irregularities centered upon the process of accretion of interstellar matter by normal stars. Comparison with theoretical evolutionary models soon showed that these objects had to be extraordinarily young stars, prior to reaching the main sequence. The irregularity of the photometric variations and spectroscopic features is due to the complexity of the process by which newly formed stars accrete interstellar matter.

These irregularly variable stars were previously called Orion variables or nebular variables, since they are connected with interstellar, diffuse gaseous nebulae. There are similar objects which are apparently not connected with nebulosity; they were designated as RW Aurigae stars. Among them are early spectral types (O, B, A), as well as middle- to late-types (F to M). In most cases, the absorption line spectrum is superposed by emission lines (for example, Hα), which is interpreted as evidence for circumstellar shells and for inflow and outflow of plasma (P Cygni line profiles). The irregular brightness variations cannot be unambiguously characterized. Brightness variations of the order of one or more magnitudes on timescales of several tens of days are observed against temporarily constant brightnesses, with dips lasting from one to a few days, short-term eruptions for hours or minutes, as well as quasi-periodic changes with cycle lengths in days.

The T Tauri stars are named after the prototype located in the star-forming Taurus-Auriga dark cloud complex. The spectral types of the objects with masses between about 0.3 and 3 $M_\odot$ range from F to M. In the HR diagram, they are found about 2 or 3^m above the main sequence, at absolute magnitudes between +7 and +3. T Tauri stars are classed on the basis of spectroscopic features, the criterion being the presence of the following spectral lines:

- Emission lines of the Balmer series and of Ca II (the H- and K-lines)
- Fluorescence emission lines of Fe I ($\lambda\lambda$ 406.3, 413.2 nm)
- Forbidden emission lines of neutral oxygen ([O I] ($\lambda\lambda$ 630.0, 636.3 nm) and of singly ionized sulfur ([S II] $\lambda\lambda$ 406.8, 407.6, 671.6, 673.0)

A large subgroup (~40%) of T Tauri stars are the *YY Orionis stars* (e.g., S CrA), with clear spectral indication for the accretion of matter from the surrounding circumstellar gas and dust cloud (the so-called inverse-P Cygni line profiles, emissions with red-shifted absorption components).

FU Orionis stars (Fuors) are characterized by gradual increases in brightness by about 6^m in several months, followed by either almost complete constancy at maximum that is sustained for long periods of time or slow decline by $1-2^m$. Spectral types at maximum are in the range Ae–Gpe (with Hα emission). After an outburst, a gradual development of an emission spectrum is observed and the spectral type becomes a later type. These variables probably mark one of the evolutionary stages of T Tauri-type variables, as evidenced by an outburst of one member, V1057 Cyg, but its decline ($2^m.5$ mag in 11 years) occurred immediately after maximum brightness was reached. All presently known FU Ori variables are associated with reflecting cometary nebulae.

Analogs of T Tauri stars in the higher mass range (2–8 $M_\odot$) and B spectral type pre-main sequence stars, are called *Herbig Ae/Be stars*. These stars are pre-main sequence stars—young (< 10 Myr) stars of spectral types A and B. They are still embedded in gas-dust envelopes and may be surrounded by circumstellar disks. Their spectra show hydrogen and calcium emission lines. They are named after the American astronomer George Herbig, who first distinguished them from other stars in 1960.

Sometimes Herbig Ae-Be stars show significant brightness variability. It is believed to be due to clumps (protoplanets and planetesimals) in the circumstellar disk. In the lowest brightness stage the radiation from the star becomes bluer and linearly polarized (when the clump obscures direct star light, scattered disk light relatively increases—it is the same effect that causes the blue color of our sky).

14.2.3.8 Protoplanetary Nebulae Protoplanetary nebulae (PPN) are objects in transition from the AGB to the planetary nebula (PN) phase (other post-AGB objects are called RV Tau stars, with spectral types G-K and periods 50–150 days, and the UU Her stars, with spectral type F). The spectral energy distribution of these objects shows a characteristic double peak, the first maximum in the visible and near-infrared (reddened stellar photosphere) and the other in the mid-infrared (re-emission from circumstellar dust).

Almost all PPN vary in light and radial velocity (in phase). Typical amplitudes of $0^m.15$ to $0^m.40$ are observed. The variability is due to pulsation; it has no simple periodic form, but varies in amplitude and period, or maybe multiple periods exist. A relation between period and spectral type is indicated: Late types (G) show longer periods (80–150^d), middle (A-F) middle periods (25–100^d), and early (O-B) short timescale variability (< 10^d).

14.2.4 Rotating Variables

Rotating variables are stars with non-uniform surface brightness and/or ellipsoidal shape, whose variability is caused by axial rotation with respect to the observer. The non-uniformity of surface brightness distributions is caused by the presence of spots or by some thermal or chemical inhomogeneity of the atmosphere caused by a magnetic field whose axis is not coincident with the rotation axis. These stars are subdivided into the following types:

ACV, roAp, ACV+roAp

14.2.4.1 α_2 Canum Venaticorum Stars (Magnetic Variables), also with Rapid Oscillations

In the spectral range between late B to early F, numerous stars are found with sometimes quite pronounced spectral peculiarities:

1. Metallic lines have a strength unusual for A-type stars (*metallic A* or *Am-type* stars).
2. Lines of certain elements usually not showing high intensity dominate the spectrum: He, Mn, Cr, Eu, Sr, Ir, Y, and also Si, in particular some lines in the vicinity of 420 nm (*peculiar A* or *Ap-type stars*).

About 15 to 20% of all A-type stars belong to the subclass Am, and 2 to 5% to subclass Ap. Many Ap stars, especially the later-type, cooler member of the sequence (the Si-Cr-Eu-Sr stars), are variables in a threefold fashion:

– Strong, variable magnetic fields occur, with strengths of several 10 to 10^4 gauss.
– Spectra are variable, both with respect to occurrence and strengths of lines as well as to radial velocities.
– There are small brightness variations of several $0^m.01$ up to $0^m.2$, with the polarization also varying.

These changes usually have the same period, but in a few cases the photometric and spectroscopic variations follow exactly one-half of the magnetic period. Apart from rare exceptions, the periods are in the range of 5 to 9 days. The light curves are approximately sinusoidal.

Brightness changes in the Ap stars, and especially the prototype α_2 CVn, are a manifestation of stellar rotation. The decisive role is played by the strong magnetic field, which causes a concentration of certain elements at the stellar surface. The magnetic axis does not usually coincide with the rotation axis, this configuration is termed "oblique rotator." The Earth-based observer finds that the magnetic field and the distribution of the "peculiar" elements vary. The strength of line absorption in the photosphere (blanketing) thus varies, too, and this is a cause of the small brightness variations. Variations of several $0^m.1$ are possibly also caused by additional starspot activity.

Rapidly oscillating Ap stars (roAp stars) are non-radially pulsating stars with periods between 6 and 12 min, and amplitudes less than $0^m.01$ in V. These variations are superimposed on the light variations due to rotation.

This theory of the oblique rotator requires a redistribution of radiation emitted by the stellar surface, so that the energy output is reduced in certain spectral regions and enhanced in others. The brightness variation in the far-ultraviolet is larger and runs in a phase counter to that in the optical range. The near-UV, on the other hand, is a sort of "neutral" range, with scarcely any changes observed.

It has sometimes been suggested that Am stars, which may be related to Ap stars, could show a similar light variation. The very small amplitude, however, is an impediment to obtaining a convincing proof of such variability.

SX Arietis-type variables are main-sequence stars of spectral types B0p-B9p with variable-intensity He I and Si III lines and magnetic fields. They are sometimes called helium variables. Periods of light and magnetic field changes (about 1 day) coincide with rotational periods, while amplitudes are approximately 0.1 mag in V. These stars are high-temperature analogs of the ACV variables, and are now counted among the ACV stars.

14.2.4.2 Ellipsoidal Stars Close binary systems which are characterized by the presence **Ell** of a strong ellipticity effect (ellipsoidal form on the stars). The light curves are sinusoidal, with $P_{var} = 1/2P_{orb}$; maximum brightness coincides with the quadratures. Eclipses may be absent. Typical brightness fluctuations can reach up to $0^m.35$ in V.

The brightest ellipsoidal variable is Spica (α Virginis), another famous one is b Persei.

14.2.4.3 Reflection Variables Close binary systems which are characterized by the **R** presence of strong reflection (re-radiation of the light of the hot star illuminating the surface of the cooler companion). The light curves are sinusoidal, with $P_{var} = P_{orb}$, maximum brightness coincides with the passage of the hot star (often a hot white dwarf) in front of the companion. Eclipses may be absent. Examples are BE UMa and MT Ser.

14.2.4.4 RS Canum Venaticorum Stars RS Canum Venaticorum variables are close **RS** binary stars of spectral type G or K. They are rapidly rotating because of tidally enforced synchronism in relatively short-period binaries. ("Short" means days, weeks or months in the case of luminosity classes V, IV and III.) The stars have active chromospheres which result in variations in their observed luminosity. The period of variations is, in general, close to the rotational period of the binary system. Some systems exhibit light variations because they are eclipsing binaries. A typical brightness fluctuation in such a star is $0^m.2$.

14.2.4.5 FK Comae Stars FK Comae variables are rapidly rotating giants with non- **FKCom** uniform surface brightnesses, which have spectral types G-K with broad H and K Ca II and sometimes Hα emission. They may also be spectroscopic binary systems. Their periods of light variation are up to several days and equal to their rotational periods, amplitudes are several $0^m.1$. It is possible that these objects are endpoints of the evolution of contact binary systems (W UMa systems), where the system has evolved into a giant, and orbital angular momentum has been transferred to the stellar envelope.

14.2.4.6 BY Draconis Stars ("Spotted" Stars) BY Draconis stars are cool stars of spec- **BY** tral types K and M, with noticeable variability due to cool spots and rotation. Their spectra show emission lines due to chromospheric activities. The stars can be single or binary stars. Rapid rotation is a consequence of a star's young age, or a tidally enforced synchronism in an orbit with a short period. Quasi-periodic light changes with periods from a fraction of a day to 120^d and amplitudes from several hundredths to $0^m.5$ in V are observed.

Many BY Draconis stars also show flares. The differentiation between spotted stars and flare stars (UV Ceti stars; see Sect. 14.2.3.5) remains somewhat problematic. The variability of both types is due to activity in the near-surface layers of late-type main-sequence stars. In BY Dra-type stars the observed phenomena originate primarily in the photosphere, in flare stars the activity is located mainly in the chromosphere and possibly in the corona.

In the case of starspots, the dominant feature is a temperature-dependent diminution of radiant intensity. The spot areas are, as on the Sun, about 500 to 1000 K cooler than the surrounding stellar photosphere. At photospheric temperatures of only 3500 to 4000 K, this means a local reduction of surface brightness in starspot regions by 55 to 70%. However, these spots must be far larger than the solar ones, covering at least 30 to 40% of the visible surface in order to cause measurable photometric effects of some tenths of a magnitude (sunspots seldom cover more than 0.3% of the surface, even at times of maximum activity). As in the case of the Sun, large spot regions have lifetimes of several stellar rotations, and thus generate nearly periodic light curves.

After early attempts to explain stellar variability by starspot formation, the idea was again taken up by G. Kron who suggested that the apparently periodic disturbances in the light curve of YY Gem (Castor C) could be interpreted in this way. Recognition of other objects as spot stars soon followed: BY Dra, CC Eri and XY UMi.

The light curves sometimes show strong changes and even the occasional disappearance of variability, which could be explained by the termination of an activity center, or by motions of spots. Should a spot move across the pole, it would change its "longitude" by 180°, and thus cause a characteristic jump in phase. Such jumps have actually been observed in BY Dra as well as in CC Eri. One important qualitative result seems to have stood the test of time: strong starspot activity occurs most frequently in the polar regions of spotted stars, in contrast to the case of the Sun, where the spots prefer lower and middle latitudes.

Spotted stars also show flare-like eruptions, but it is worth noting that no relation has thus far been found between starspots and flare activity. This is not compatible with the notion of extended, active regions which exhibit a wide range of activities on the Sun, and which stretch from layers immediately below the photosphere up into the inner corona.

The overwhelming majority (nearly 80%) of known BY Dra stars as well as flare stars occur in moderately close binary systems. It is possible that the duplicity causes the activity, or at least contributes to it (perhaps tidal effects). In all RS CVn-type binary stars, starspot activity has been convincingly proven from the light curves and from distorted line profiles in the spectrum. Some authors assign the RS CVn stars as a subclass of the BY Dra types, but the dissimilar evolutionary states of these stars make the different arrangement in the present text preferable.

An amateur observer with photoelectric equipment can contribute much to useful research on BY Dra stars. The essential requirements are high accuracy and an extended series of observations, over at least a few months, or better still over one or two years.

Psr

Catalogs of pulsars are available on the Internet: http://cdsarc.u-strasbg.fr/viz-bin/Cat?VII/189 (catalog of 706 pulsars) http://www.atnf.csiro.au/research/pulsar/psrcat/ (catalog of 1789 pulsars)

14.2.4.7 Pulsars Radio pulsars show variability almost exclusively in the decimeter and meter ranges of the electromagnetic spectrum. Only in a few cases, pulsars can also be observed in the visible range. The most famous one is CM Tauri, the optical counterpart of the Crab pulsar. The amplitude of the light pulses can reach 1^m in V.

The discovery of radio pulsars in 1967 was carried out with a radio telescope with high time resolution. The first source with rapid and apparently very regularly pulsed radiation, originally called CP 1919 (Cambridge pulsar at $\alpha = 19^h 19^m$) had a pulse period of $1^s.34$. Figure 15 shows the schematic radio pulse profiles of this object; the form is typical of many radio pulsars. Its period increases by $P = +0.1165$ nanoseconds per day. The pulsar is now designated PSR 1919+21, which includes the declination information.

Currently almost 1800 radio pulsars are known, and an approximate distance from the Sun has been determined for most of these. Most of the pulsars were found in large search programs at the radio observatories of Jodrell Bank (England), Green Bank (West Virginia, USA), Arecibo (Puerto Rico, USA), and Molonglo (Australia).

The characteristic feature in the curve is a main pulse which lasts about 1/10 of the period. Periods range from 1.5 ms to about 4 s. Often, a lower interpulse occurs. The clockwork-like mechanism underlying this stable, periodic pattern is the rapid rotation of a compact object. Such extremely short periods are possible only for neutron stars, which have radii of 10 to 15 km and masses lying somewhere between about 1 and 2 $M_\odot$. The corresponding densities from 10^{14} to 10^{15} g cm^{-3} lie in the range of that for atomic nuclei. These extraordinary objects also have very strong magnetic fields of up to 10^{12} gauss and extended magnetospheres of high-energy charged particles accelerated by enormous electric fields of up to 10^{12} V. The interaction of these relativistic particles with the magnetic field generates the observed radio emission, or synchrotron radiation, which is strongly focused by the magnetosphere into a beam. The observed pulses occur when this beam of radiation, which rotates with the star, sweeps past the Earth (the lighthouse effect).

The energy for this radiation is supplied by the kinetic energy of the continuously decelerating rotation of the neutron star; the periods of these objects increase without exception. After a few million years, the neutron star will rotate so slowly that it will no longer emit measurable radio radiation. Perhaps as many as 10^9 of such "silent" neutron stars populate the Galaxy, since the birthrate can be estimated as about 1 pulsar per 30 to 40 years, or even 10 to 20 years. It should be emphasized, however, that neither is there currently a satisfactory explanation for the conversion of kinetic rotational energy into

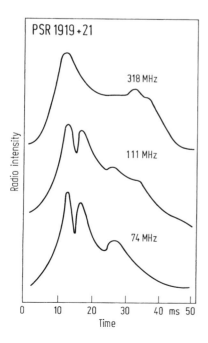

Fig. 15. Integrated radio pulse profiles of PSR 1919+21 at three different radio frequencies. The pulse period is $1^s.34$. Many radio pulsars show pulses with similar profiles

energy in the magnetosphere or into radio radiation, nor is the beaming mechanism of the radiation entirely understood.

14.2.5 Cataclysmic Variables and Other Stars with Outbursts

Cataclysmic variable stars (CVs) have experienced much attention in recent decades. "Explosive" CVs show outbursts caused by thermonuclear processes in their surface layers (novae) or deep in their interiors (supernovae). "Nova-like" CVs are those variables that show outbursts caused by a rapid energy release in the surrounding space (UG-type stars, see below) and also objects not displaying outbursts, only other variations (flickering, eclipses), but resembling explosive variables at minimum light by their spectral or photometric characteristics.

The majority of explosive and nova-like variables are close binary systems. Often the hot, compact component of the system is surrounded by an accretion disk formed by matter lost by the other, cooler, and more extended component (Fig. 16).

Cataclysmic variables derive their name from the Greek "kataklysmos," which means catastrophe or inundation, and refers to the physical process causing the eruption of these objects. The sudden outburst of brightness and related peculiar spectroscopic and photometric properties can be explained by effects of the evolution-driven mass transfer between the two components of a close, interacting binary system. The analysis of such systems was advanced particularly in the last 50 years, through the development of high time-resolution photometry, and space-based UV and X-ray astronomy. Owing to the large number of cataclysmic variables and to the opportunity to study important processes of interaction, these systems are significant sources of information for scrutinizing theories of evolution of close binaries.

The primary symptom of almost all cataclysmic systems, are brightness outbursts observed irregularly or only once, and which liberate large amounts of energy within very short timespans. The immediate and most conspicuous effect of such eruptions is an increase of visual brightness by 2 to 19 magnitudes within hours to months, and a subsequent, slower decline of the brightness to the original level (before eruption) in timespans of days to decades, depending on the class of objects.

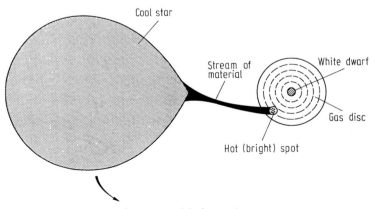

Fig. 16. Schematic model of a cataclysmic system

The diversified patterns of cataclysmic variables are classified according to physical characteristics. The primary parameter here is the energy freed during an eruption. To distinguish the various subgroups, it must be ascertained whether the observed eruption was a one-time event or whether the eruptions repeat regularly in certain cycles, or irregularly. Another feature of classification is the timescale in which the outburst proceeds. In particular, the time interval between reaching visual maximum brightness and that instant at which the brightness has declined by 2 or 3 magnitudes (the t_2 and t_3 times, respectively) is often a characteristic scale for novae. Besides these general distinguishing features, more precise classification includes individual photometric and spectroscopic properties, which can now be determined with the aid of highly sensitive detectors in photometry with high time resolution, and by spectroscopic measurements in X-ray, UV, visual, and IR ranges. The cataclysmic variables can be subdivided into supernovae (SN Ia, SN Ib, SN II; this class is somewhat removed from the following ones), novae (subgroups fast (Na), slow (Nb), very slow (Nc), recurrent (Nr) novae), dwarf novae (DN) (subgroups of UGSS, UGSU, UGWZ, UGER and UGZ systems, each named after its prototype, see Sect. 14.2.5.6), and nova-like variables (NL), which also comprise several subgroups (UX, VY, AM, DQ, and AM CVn systems, see Sect. 14.2.5.5).

14.2.5.1 Supernovae The two "new stars" in 1572 and 1604 were typical representatives of *supernovae* in the Galaxy. In reality, a supernova is quite the opposite of a new star; it is the catastrophic explosion of a highly evolved star which then becomes totally restructured or even destroyed in the process. At the time of maximum light, supernovae may reach absolute magnitudes of -18 to -20, which translates into about 10^4 times brighter than ordinary novae. Hence, the luminosity becomes comparable with that of an entire galaxy.

During a supernova explosion, atomic nuclei of "heavy" elements (i.e., heavier than hydrogen and helium) are produced in large amounts. In particular, all heavy elements with atomic numbers larger than 209 (bismuth) are formed exclusively in supernovae (lighter ones also through the slow neutron capture process in AGB stars). Some proportion of these heavy elements are dispersed by the supernova explosion into the interstellar medium so that later generations of stars become enriched by these elements. In this way, supernovae play a considerable role in the chemical evolution of galaxies and of the entire universe.

By mid-2008, almost 5000 supernovae had been discovered as telescopic objects in extragalactic systems; this number grows at a rate of about 500 per year, mostly because of the intensive search in cosmological supernova projects.

It has been ascertained that within historical times the following supernovae occurred in the Milky Way galaxy: SN 1006 in Lupus, SN 1054 (remnant is the Crab Nebula), SN 1181 (observed in China and Japan), SN 1572 (Tycho's Supernova), SN 1604 (Kepler's Supernova), and a supernova in the second half of the seventeenth century which was not optically observed but which left the radio source Cas A as a remnant. In neighboring galaxies, supernovae appeared in 1885 in M 31 and in 1987 in the Large Magellanic Cloud.

A supernova explosion results in a rapidly expanding cloud of gas with a mass in the range from a few tenths to several $M_\odot$. The initial velocity of expansion amounts to 10,000 to 20,000 km s^{-1}. This expanding gas collides with the interstellar medium (and possibly with pre-outburst circumstellar matter), and, by compressing and heating it,

A catalog and atlas of cataclysmic variables (up to Feb. 1, 2006) is available on the Web:
http://archive.stsci.edu/prepds/cvcat/index.html.
A catalog of cataclysmic binaries, low-mass X-ray binaries and related objects (RKcat Edition 7.10, July 1, 2008) is available on the Web:
http://www.mpa-garching.mpg.de/RKcat/.

SN

Catalogs of supernovae:
http://www.cfa.harvard.edu/iau/lists/Supernovae.html
http://www.sai.msu.su/sn/sncat/

generates a strong source of radiation with a characteristic spatial structure. The bulk of this radiation lies in the radio domain, and, as it is generated by high-energy (relativistic) electrons gyrating in the galactic magnetic field, its spectral distribution differs markedly from that of hot stellar or interstellar gas. This so-called synchrotron radiation is the trademark of a supernova remnant (SNR), and the negligible absorption of radio radiation in space enables astronomers to detect the remnants virtually everywhere in the Galaxy. They are very sharply concentrated toward the galactic plane and show a peak in the direction of the galactic center.

The best-known SNR is the Crab Nebula, which is associated with the "guest star" observed by the Chinese in 1054 AD. In 1968, a short-period pulsar (P = 0.033 s)—a rapidly rotating neutron star—was discovered in the Crab Nebula. A neutron star can be yet another result of the supernova outburst arising from the collapse of the core of a massive progenitor star. About 80% of the SNRs belong to the "shell-type" class. The Crab, with its entire volume filled by emission of mainly synchrotron radiation, is a rather rare representative of the so-called "filled-center remnants." About 15% of the SNRs are of the "combination-type," with shell features and partly filled central parts; such nebulae may evolve ultimately into shell-like objects. In a Crab-like nebula, a pulsar is needed to support the strong radiation, while in the combination-type nebula the presence of a pulsar is probable but not necessary. Cas A and also the remnants of Tycho's and Kepler's supernovae are, for instance, very young, 300–400 year-old shell-type remnants, apparently containing no energizing pulsars, while the large, circular nebula in the constellation Cygnus, called the Cygnus Loop, is a typical old remnant, perhaps as old as 10^5 years.

The absence of observed pulsars in the vast majority of SNRs is not surprising, and is in fact explicable by several valid arguments. In cases where the original star was of low mass, it may have been completely destroyed in the explosion, while in cases of very massive progenitor stars the formation of a black hole, instead of a neutron star, is to be expected.

The ultimate energy source of some supernova outbursts is the gravitational collapse of a star, and frequently, the burned-out core of a massive star. The liberation of gravitational energy can account for the prodigious energy output observed during the supernova outburst, to an order of magnitude: 10^{43}–10^{44} J in the optical region, and 10^{44}–10^{45} J in the kinetic energy of the expanding gas. Moreover, a third form of energy output is indicated since the total gravitational energy of the collapse is near 10^{46} J. The theory of stellar collapse suggests that the rest of the liberated energy should take the form of neutrino emission. A very large number (about 10^{58} or more) of neutrinos and antineutrinos is generated, which represents a total energy of 1 or 2×10^{46} J.

The earliest optically observed phase of a supernova is the originally steep increase in luminosity, which corresponds to an expanding quasi-photosphere, an optically thick layer radiating like an expanding atmosphere of a gigantic star—much larger than a supergiant—and steadily cooling. These particular events are not unlike those of a nova eruption, except that the moving masses are four or five orders of magnitude larger and their expansion speeds 10 times higher. Since about 1940, two main classes of supernovae have been distinguished, differing in their spectra and also in their light curves. The two classes are referred to as type I and type II supernova. The type II objects are dominated by strong Balmer lines of hydrogen, both in absorption and in emission, showing marked P Cygni profiles. At an early phase the spectra bear somewhat of a similarity to nova spectra, except that the very broad lines indicate the order of magnitude higher velocity of the expanding gas. With some difficulty, other lines, all very broad, can be identified. Type I supernovae show no trace of hydrogen and the identification of the highly broadened, overlapping emission lines, can be assigned to several metals.

Type I supernovae can be divided into two subclasses according to their observed spectra. Thus, there are effectively three main classes:

Ia   Lacking both hydrogen and helium lines
Ib   Lacking hydrogen, but showing helium lines
II   Showing hydrogen (and also helium) lines

These spectral criteria characterizing the expanding shell, in combination with the observed distribution of the supernovae among different types of galaxies and their distribution within these galaxies, have enabled astronomers to gain remarkable insight into the nature of the progenitors, the stars which have undergone the supernova explosion.

Figure 17 shows the mean light curve for a type I supernova. In a somewhat later phase of development, 40 days after maximum brightness in type I, 80 to 100 days in type II, the light curves appear to enter a linear descent. Since the magnitude scale is logarithmic, a strictly linear descent means an exponential decline of the light, which is characteristic of radioactive decay. Calculations of nuclear reactions occurring in supernovae indicate that element synthesis reaches to ^{56}Ni and produces a large amount ($\approx 0.5 M_\odot$) of this unstable radioactive isotope. The repeated decay by electron capture and γ-photon emission leads to iron:

$$^{56}\text{Ni} \rightarrow {}^{56}\text{Co} \rightarrow {}^{56}\text{Fe} \quad \text{(stable)}.$$

The half-lives are 6 and 77 days, respectively. The latter is nicely compatible with the observed decline of the light curve.

The calibration of absolute brightness of the supernovae is dependent on the distance scale of the galaxies, that is, on the adopted value of the Hubble constant. On the other hand, supernovae of type Ia have turned out to be useful "standard candles" which serve to study the expansion of the universe and its possible deceleration or acceleration.

Only a few years ago it was generally accepted that the brightest supernovae are of type I, and that type II objects are 1.5 to 2^m fainter. With the recognition of the two classes comprising type I, the data may be subject to some revision. The following data are based on an early discussion by Kowal, as modified by van den Bergh and coworkers. Assuming a Hubble constant of $H = 75 \, \text{km s}^{-1} \, \text{Mpc}^{-1}$ it is found that

$$M_V(\text{max}) \approx -19.2 \quad \text{Type Ia} \tag{12}$$
$$\approx -17.7 \quad \text{Type Ib} \tag{13}$$
$$\approx -17.1 \quad \text{Type II}, \tag{14}$$

with an estimated scatter of $\pm 1^m$ in each case.

Type II supernovae occur almost exclusively in spiral galaxies, close to or in the spiral arms. Moreover, the large majority of these supernovae seems to be directly associated with giant H II regions, where star formation is going on. It is virtually certain that the progenitors are massive stars evolving very rapidly and which, after having completed

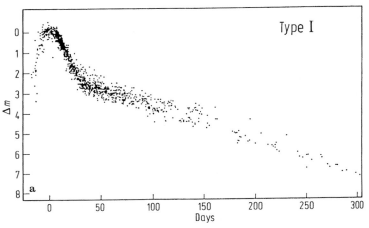

Fig. 17. Composite light curve of 38 type I supernovae. After Barbon, Ciatti, and Rosino

nuclear evolution, reached the stage of core collapse with most of their hydrogen-rich envelopes still in place. The minimum mass of this evolutionary path is about 8 $M_\odot$; the most massive stars (25–30 $M_\odot$ or more) may produce black holes in the supernova explosion.

Type Ib supernovae appear amidst younger stellar populations, but are less directly associated with recent or ongoing star formation. The lack of hydrogen, as indicated by the spectrum, leads to the working hypothesis that the progenitors may be Wolf–Rayet stars which, having lost their hydrogen-rich envelopes and reduced the originally high stellar mass to a value of perhaps 4 to 8 $M_\odot$, are still rich in helium. The mass loss can be the consequence of a "catastrophic" stellar wind or also mass exchange in close binary evolution. Type Ia supernovae may occur in older stellar populations lacking massive stellar objects. Normally, one would expect these relatively low-mass stars (after some mass loss) to terminate their evolution as white dwarfs, without explosion. One promising hypothesis is that the progenitors are, indeed, white dwarfs, specifically, accreting CO-white dwarfs (neither hydrogen nor helium present) in a *close binary system*. Under rather specific circumstances, the mass of a white dwarf can approach—without the intervention of a nova explosion—1.4 $M_\odot$, that is, the Chandrasekhar limit of stability. The accreted "burnable" material provided by the non-degenerate companion can be converted to C and O in short flare-ups of nuclear burning termed nuclear flashes, during which no mass loss occurs. Upon approaching a critical mass close to the limit, the star explodes in a complicated process that most likely does not produce a compact central remnant.

N, NHf

14.2.5.2 Classical Novae and Helium-Flash Novae Novae unexpectedly brighten up to 15^m in only a few days. One often finds a faint star, the pre-nova, on survey images taken before the nova explosion. After several decades, the nova has sunk back to its pre-outburst brightness and is now a post-nova. Detailed studies of post-novae show that their properties are similar to those of nova-like stars (see Sect. 14.2.5.5). Nova-like objects are thus potential novae.

Statistical considerations lead to the conclusion that a nova will erupt many thousands of times, although with intervals of several hundred to several hundred thousand years.

Novae which erupt every few decades are classified as recurrent novae (see Sect. 14.2.5.3). They are, however, somewhat different in structure from the classical novae.

The light curves of novae exhibit a bewildering diversity (Fig. 18). The classes Na, Nb, and Nc are characterized by typical t_3 times (the time for brightness decline from maximum) by 3^m, by some 10, 100, or 1000 days, respectively. Measured light curves of the fast nova V1500 Cyg (= Nova Cygni 1975), the slow nova DQ Her (= Nova Herculis 1934), and of the very slow nova HR Del (= Nova Delphini 1967) are shown in Fig. 18. Conspicuous differences occur primarily in the transition state following the first, rapid decline from maximum. The curve there may follow a more or less smooth path, or it may show periodic fluctuations of various amplitudes, or a deep brightness minimum may occur. The maximum itself is sharply pronounced in fast novae, while in slow novae it may be several hundreds of days broad and have a complex shape.

Amplitudes of most novae are around 11 to 12^m; extremely high values were observed in CP Pup, with over 16^m, and in V1500 Cyg, with over 18^m. These statistics include a selection effect which favors small amplitudes, as the magnitudes of most pre-novae are not known and particularly so for the large-amplitude cases. The mean absolute brightness at maximum M_V is between -8 and -6^m for slow novae and between -11 and -8^m for fast novae.

Novae can be recognized by their typical spectral features during an outburst. At maximum, the spectrum resembles that of a B-, A-, or F-type star with broad absorption lines, Doppler blue-shifted by the amount of several hundred $km\,s^{-1}$, and originating in an expanding pseudo-photosphere. With increasing expansion of the shell, the spectrum changes to that of a supergiant with a middle to late spectral type. On the long-wavelength side of the absorptions, broad emissions of hydrogen (Balmer lines) and singly ionized metals (e.g., Ca II, Fe II) appear at around zero velocity. While the absorption is formed only in the column of the expanding shell projected onto the source, which moves toward the observer (hence the blueshift), the emissions are produced by photons escaping from all parts of the optically thin envelope, and this leads to a more or less symmetric Doppler broadening of the unshifted emission lines.

Shortly after the appearance of the sharp absorptions (the principal spectrum), there appear much more blue-shifted absorptions (the diffuse enhanced spectrum), followed a short time later by another system of usually even broader absorption and emission lines (the Orion spectrum of strong He I, C II, N II, O II lines). Near the beginning of the transition state, the nova shell, due to its continuous expansion, has rarefied sufficiently so that the plasma density approaches values typical for planetary nebulae. Absorption lines and continuum thus disappear almost entirely, and the spectrum is dominated by strong emission lines characteristic of nebular spectra. At the beginning of this nebular stage, the forbidden lines are strongest in the visual and UV range (e.g., [O III] lines at 495.9, 500.7, and 436.3 nm, and lines of multiply ionized C, N, Ne, Si, or Fe). Ionized coronal lines (up to [Fe XIV]) are also observed. The nebular stage may last up to several years, and then passes gradually into the post-nova stage, where the system arrives back at the original pre-nova brightness.

While the photographic evolution of nova spectra has been studied especially by McLaughlin and Payne-Gaposchkin, modern digital spectra of many novae have been described and classified by R.E. Williams and coworkers. These papers should be consulted when it is intended to derive information from outburst spectra of novae.

Spectra of novae by R.E. Williams: http://articles.adsabs.harvard.edu/cgi-bin/get_file?pdfs/JAD../0009/2003JAD.....9....3W.pdf

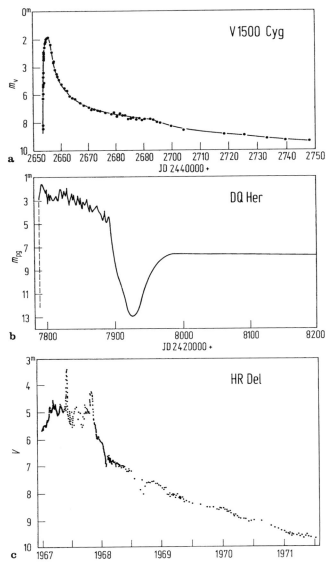

Fig. 18. Eruption light curves of novae. (**a**) Fast nova (Na) V 1500 Cygni. (**b**) Slow nova (Nb) DQ Herculis (1934). (**c**) Very slow nova (Nb) HR Delphini (1967)

The complex appearance and the spectral variations during the course of a nova eruption can, in principle, be modeled as an explosion in the central binary system which ejects a small fraction (10^{-4} to 10^{-5} $M_\odot$) of the total system mass in the form of the nova shell. A substantial fraction of the total explosive energy (10^{37} to 10^{38} J) goes into the kinetic energy of the shell, accelerating it to speeds of several thousand km s^{-1}.

Recently, the GCVS has introduced the concept of helium-flash novae. Unlike classical novae, the outburst occurs as a final helium-burning of an object which is on its way from the red giant to the white dwarf stage, the central star of a planetary nebula. The expansion occurs gently, the "nova" is very slow, the configuration is also named a "reborn giant." The expanding shell is extremely hydrogen-deficient and carbon-rich, and forms after several years a complete dust shell, so that the decline of the object is caused by dust (soot) obscuration.

Only V605 Aql (Nova Aquilae 1919) and V3443 Sagittarii (Sakurai's object) are well-established helium-flash novae.

14.2.5.3 Recurrent Novae

Recurrent novae are distinguished from classical novae, as their name suggests, by their repeated outbursts, with typical cycle lengths of a few decades, and by a smaller amplitude, averaging about 3 to 4 magnitudes less. It is generally accepted that classical novae also will show repeated eruptions, but at long time intervals of 10^4 to 10^5 years, so that only one outburst is historically documented in each case.

About one half of recurrent novae have evolved, late-type giants as secondary companions, contrary to classical novae. Presumably, the masses of the white dwarfs are quite near the Chandrasekhar limit of 1.4 $M_\odot$.

Only nine objects are known, the best known and investigated being T CrB (outbursts: 1866, 1946), T Pyx (1890, 1902, 1920, 1944, 1966), U Sco (1863, 1906, 1936, 1979), and RS Oph (1898, 1933, 1958, 1967, 1985). Others are CI Aql (1917, 2000), V394 CrA (1949, 1987), V3890 Sgr (1962, 1990), V745 Sco (1937, 1989), and IM Nor (1920, 2002).

Nr

14.2.5.4 Symbiotic Novae and Z Andromedae Stars

Symbiotic variables of the Z Andromedae type are close binaries consisting of a hot star, a giant star of late type, and an extended envelope excited by the hot star's radiation, showing emission lines of high ionization (e.g., He II, C IV, N V, O IV, Ne V). The combined brightness displays irregular variations with amplitudes up to 4^m in V. The observed periodic or quasi-periodic light variations typically have amplitudes of 1 to 3^m, and photometric and spectroscopic periods of several hundreds of days.

NSym, Sym

Substantial or longer lasting increases in luminosity often occur, and for this reason these objects were formerly also called "nova-like" variables, in contrast to the current definition. A well-known example of such an eruption is the active state in RR Tel, which lasted several years. This system increased its luminosity in the year 1944 by about 7^m. Other objects, such as AG Peg, also show brightness increases which are stretched over several decades, and these qualitatively resemble outbursts of extremely slow novae. Outside of active states, the light curves with their sinusoidal variations resemble those of Mira variables. The spectra of the late-type components are usually classified as M giants, and are compatible with that type of pulsational variability. In some objects, a strong infrared excess and radio emission indicate the presence of dense, extended circumstellar dust shells.

The working model for symbiotic binaries consists of a cool, late-type giant with a radius of about 100 $R_\odot$ and a hot subdwarf which is smaller and substantially hotter than the Sun. The entire system is surrounded by a gaseous shell formed by non-conservative mass transfer or by mass loss from the expanded giant component due to pulsations. The variability of the system is due to by the pulsations of the giant, variations in the mass accretion of the hot subdwarf, changes of density and structure in the circumstellar envelope, and possibly explosive H-burning on the hot component, like in the case of classical and recurrent novae.

14.2.5.5 Nova-Like Objects A heterogeneous group of objects partly of different physical structure and evolutionary status is comprised under the term nova-like systems. They are phenomenologically related to cataclysmic systems, but show no outbursts. As for dwarf novae at minimum and for ex-novae, random brightness variations from seconds to minutes, and with amplitudes of some hundredths to tenths of a magnitude are found. This phenomenon, called flickering, is caused by the mass transfer in the close binary. Some nova-like systems and dwarf novae also exhibit coherent, periodic, or quasi-periodic oscillations with amplitudes of some thousandths to hundredths of a magnitude, and with periods in the range from seconds to minutes. These periodic variations are interpreted as oscillations of the degenerate component of the binary.

The spectra of nova-like systems feature a continuum similar to that of hot, OB-type stars, which strongly increases toward the ultraviolet. As in dwarf novae, variable emission components are superimposed upon the broad Balmer absorption lines. Also observed, among others, are lines of He I, He II (especially at 468.6 nm) C III/N III (464 nm). Moreover, in longer-period systems ($P > 6$ hours), an absorption-line spectrum of late type (K, M) can be identified.

In recent years, detailed photometric, spectroscopic, and polarimetric studies have led to a systematic division of the various nova-like systems into the following subclasses:

1. *UX Ursae Majoris systems* These often have near-constant brightness over several decades. In addition to the usual flickering and statistical fluctuations of $0^m.1$ to 1^m within minutes and hours, low-amplitude coherent oscillations are in some cases observed. The spectra much resemble those of dwarf novae at minimum.

2. *VY Sculptoris systems* Also dubbed "antidwarf novae." These objects are found predominantly in so-called active phases which last for a few years when they have, apart from fluctuations by some tenths of a magnitude and the usual flickering, nearly constant mean brightness. These active states are sporadically interrupted by considerably shorter "inactive" phases, when the brightness diminishes by 3 to 5^m. A well-known object is TT Ari. The spectra of VY Scl stars in active phases can be compared with those of dwarf novae in eruption, or of Z Cam systems at stationary brightness, whereas the spectra in inactive phases are similar to those of dwarf novae at minimum, especially through the appearance of certain emission lines.

3. *AM Herculis systems* Also called "polars": their distinctive feature is a strong and time-variable circular polarization ($> 10\%$) of visual and IR radiation. The continuous intensity distribution in the optical range can be explained by synchrotron radiation emitted by electrons in magnetic fields on the order of 10^3 to 10^4 tesla. In the higher energy range, soft X-ray bremsstrahlung ($kT \approx 10$ to 100 eV, corresponding to temperatures of about 10^5 to 10^6 K), and hard X-rays ($kT > 10$ keV; $T \approx 10^8$ K) are found. Similar to VY Scl but contrary to dwarf novae in eruption, the prominent active phases show emission lines, particularly of hydrogen and helium in the visual range, and of multiply ionized elements in the UV.

4. *DQ Herculis systems* Also called "intermediate polars," they have photometric and spectroscopic properties similar to polars, i.e., long-lasting, active states are occa-

sionally interrupted by phases of lesser brightness ($\Delta m \approx 1$ to 3), and the spectra show strong emission lines. Brightness pulsations with amplitudes of some tenths of magnitude and long-term stable periods ranging from about 1 min to 1 hour are typical for these systems. The name "intermediate polars" originates from the comparatively weaker polarization of the optical radiation, which in the case of DQ Her is under 1% as compared with 20% in polars (AM Her systems).

5. *AM Canum Venaticorum systems* The two objects AM CVn and GP Com are eclipsing binaries with the shortest known orbital periods of 18 minutes and 46 minutes. respectively. These binaries, each presumably composed of two degenerate stars, show a pure helium emission-line spectrum. The relation with cataclysmic variables is photometrically indicated by flickering in the visual range, and by periodic, coherent oscillations with periods of a few minutes. Superposed on the rapid fluctuations are phase-dependent, ellipsoidal brightness changes owing to the orientational light variation of the binary components deformed by tidal action.

14.2.5.6 Dwarf Novae Once generally called U Geminorum stars, dwarf novae received their name from the similarity to the nova phenomenon, but the recurring outbursts have an amplitude several magnitudes less than in classical novae, and their spectroscopic appearance at outburst does not indicate a mass loss from the system (Fig. 19).   **UG**

The structure of these systems is described by the basic model of an interacting close binary, with a compact component (usually a white dwarf), a cool component (near the main sequence), and an accretion disk around the compact star (Fig. 20). The cause of the outburst is found in the accretion disk, which at mass transfer rates occurring in dwarf novae is unstable and switches from an optically thin state with low accretion to an optically thick state of high accretion. Dwarf novae can be subdivided by characteristic photometric properties and by eruption patterns into three groups:

1. *U Geminorum or SS Cygni systems* These show outbursts with a brightness increase by 2 to 8^{m} in one or two days, and with a subsequent decline to minimum in a few days or weeks. Such outbursts occur quasi-periodically and in similar shape with an average cycle length of about ten to several hundred days.   **UGSS**

2. *SU Ursae Majoris systems* These show, after several "normal" outbursts, a supermaximum which is about 1 to 2^{m} brighter and lasts several days longer than a normal maximum. An interesting phenomenon observed during supermaxima are periodic "superhumps," sharp intensity peaks superposed on the broad supermaximum, and recurring with a slightly variable photometric period which generally differs by only a few percent from the orbital spectroscopic period of the binary. A notable feature of SU UMa systems is their very short orbital periods, usually less than 2 hours. Supermaxima and superhump periods can be explained by a special elliptic shape of the accretion disk, which is formed by tidal action in systems with low mass ratios.   **UGSU**

3. *WZ Sagittae systems* This subset of the UGSU group is characterized by intervals between superoutbursts which are unusually long (decades), while normal outbursts are few (if at all present) and far between. Observations of the 1978 outburst of WZ Sge revealed superhumps, which are the defining characteristics of SU UMa type dwarf novae; thus WZ Sge is now considered the prototype for a subset of the SU UMa class. Other WZ Sge stars include AL Com and EG Cnc, which have superoutburst intervals of approximately 20 years.   **UGWZ**

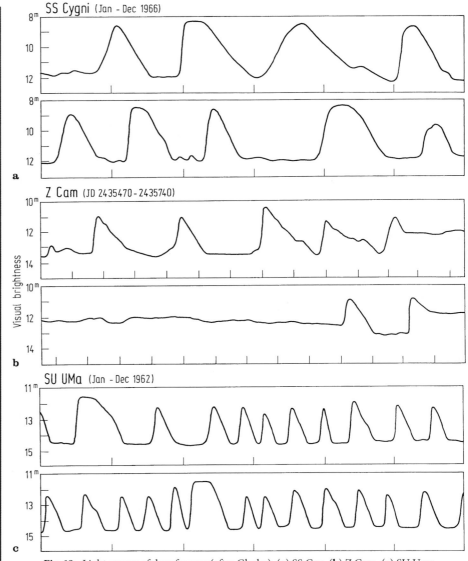

Fig. 19. Light curves of dwarf novae (after Glasby). (**a**) SS Cyg, (**b**) Z Cam, (**c**) SU Uma

The approximately 30-year supercycle length of WZ Sge indicates that this object is the most extreme of the SU UMa type stars. The factor determining the different timescales appears to be the mass-transfer rate. WZ Sge stars have a very low mass transfer rate, perhaps only 10^{12} kg/s. Given the slow rate of mass transfer, it takes decades to accumulate enough material for a superoutburst.

UGER

4. *ER UMa systems* A variety of UGSU in which the interval between superoutbursts is unusually short. ER UMa stars typically spend a third to a half of their time in superoutburst, with a supercycle (the interval between superoutbursts) of only 20

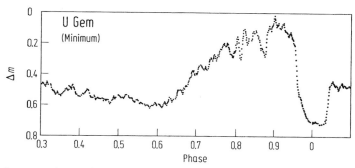

Fig. 20. High time resolution photometric light curve for the dwarf nova U Geminorum in minimum state. The deep depression in brightness at phase 0.0 is caused by the occultation of the hot spot by the secondary component. The much-reduced flickering during eclipse is also distinctly seen. After Warner and Nather

to 50 days. When not in superoutburst these stars show frequent normal outbursts, about one every 4 days.

5. *Z Camelopardalis systems* These show quasi-periodic outbursts often interrupted by longer lasting stationary brightness levels. The brightness may, shortly after time of maximum, remain constant at about 1/3 of the amplitude below maximum intensity for some days to many months, before the system then, in most cases, drops to minimum brightness.

UGZ

14.2.6 Eclipsing Variables

E, EA, EC, EP

Eclipsing-type variable stars are close binaries with the orbital plane oriented in space such that an observer on Earth sees mutual occultations of the two components. This gives rise to an apparent periodic variation in the integrated light of the system. During one revolution, the light curve typically shows a deeper *primary minimum* and a shallower *secondary minimum*. The primary minimum is caused by the eclipse of the hotter (and usually more luminous) component of the photometric binary by the cooler secondary star; one-half period later, the reverse configuration creates the secondary minimum. The ratio of depths of both minima is determined by the ratio of the surface brightnesses of the two components, which is determined essentially by the effective surface temperatures.

If the two stars are fairly distant from each other, and tidal and reflection effects are not very strong, the systems are Algol systems (EA). If they are close to each other, and thus in shallow or even deep contact, they are counted among the contact binaries (EC). In previous editions of the GCVS, the latter group was split into β Lyrae-type (EB) and W UMa-type (EW) binaries. All details are discussed in Sect. 14.3 on binary stars.

A newly introduced group (EP) contains stars showing eclipses by their planets.

A catalog of stars showing eclipses by exoplanets: http://exoplanet.eu/catalog-transit.php

14.2.7 X-Ray Variables

Since the first cosmic X-ray sources were found in 1963 by rocket experiments, optical counterparts were identified and studied. The question of the mechanism of energy generation was answered soon afterwards. It was supposed that stellar X-ray sources are interacting close binary systems in which transfer and accretion of matter onto a compact component causes emission of high-energy radiation, due to the large potential gradient.

While in cataclysmic systems the matter is collected by a white dwarf, which, as a degenerate object, has a mass of about 1 $M_\odot$ and a diameter of about 10^4 km (density $\approx 10^7$ g cm^{-3}), the much higher luminosity (by several orders of magnitude) in X-ray sources indicates that the accreting component is substantially more compact than a white dwarf. It was supposed from the very beginning that a neutron star (diameter ≈ 10 km, density $\approx 10^{15}$ g cm^{-3}) or a black hole is present here.

In the most recent version of the GCVS there have been attempts to classify the X-ray binaries into the following groups:

X

14.2.7.1 X-Ray Variables: Unclassified
Close binary systems that are sources of strong, variable X-ray emission and which do not belong to or are not yet attributed to any of the other types of variable stars listed in this section. One of the components of the system is a hot compact object (white dwarf, neutron star, or possibly a black hole). X-ray emission originates from the infall of matter onto the compact object or onto an accretion disk surrounding the compact object. In turn, the X-ray emission is incident upon the atmosphere of the cooler companion of the compact object and is reradiated in the form of optical high-temperature radiation (reflection effect), thus making that area of the cooler companion's surface an earlier spectral type. These effects lead to quite a peculiar complex character of optical variability in such systems.

XN

14.2.7.2 Nova-Like (Transient) X-Ray Variables with Low-Mass Companion
X-ray, nova-like (transient) systems containing, along with a hot compact object, a dwarf or subgiant of G-M spectral type. These systems occasionally rapidly increase in brightness by 4 to 9^m in V, in the visible simultaneously with the X-ray range, with no envelope ejected. The duration of the outburst may be up to several months (V616 Mon).

XMB

14.2.7.3 Nova-Like (Transient) X-Ray Variables with High-Mass Companion
X-ray, nova-like (transient) systems with an early-type supergiant or giant primary component and a hot compact object as a companion. Following the main component's outburst, the material ejected by it falls onto the compact object and causes, with a significant delay, the appearance of X-rays. The amplitudes are about 1 to 2^m in V (V725 Tau).

Xlm

14.2.7.4 Low-Mass X-Ray Binaries
A low-mass X-ray binary (LMXB) is a binary star where one of the components is either a black hole or neutron star. The other component, the donor, usually fills its Roche lobe and transfers mass to the compact star. The donor is less massive than the compact object, and can be on the main sequence, a degenerate dwarf (white dwarf), or an evolved star (red giant).

Catalog of low-mass X-ray binaries: http://webviz.u-strasbg.fr/viz-bin/VizieR?-source=J/A%2bA/469/807

A typical low-mass X-ray binary emits almost all of its radiation in X-rays, and typically less than one percent in visible light, so they are among the brightest objects in the X-ray sky, but relatively faint in visible light. The apparent magnitude is typically around 15 to 20. The brightest part of the system is the accretion disk around the compact object. The orbital periods of LMXBs range from ten minutes to hundreds of days.

Soft X-ray transients (often abbreviated to SXT) are also known as X-ray novae. Typical SXTs are usually very faint, or even unobservable, in X-rays and their apparent magnitude at optical wavelengths is about 20. This is called the "quiescent" state.

In the outburst state, the brightness of the system increases by a factor of 100 to 10,000 in both X-rays and optical wavelengths. During outburst, a bright SXT is the brightest object in the X-ray sky, and the apparent magnitude is about 12. The SXTs have outbursts with intervals of decades or longer, as only a few systems have shown two or more outbursts. The system fades back to quiescence in a few months. During the outburst, the X-ray spectrum is "soft" or dominated by low-energy X-rays, hence the name soft X-ray transients.

SXTs are quite rare, about 100 systems are known. A typical SXT contains a K-type subgiant or dwarf that is transferring mass to a black hole through an accretion disk. In some cases the compact object is a neutron star, but black holes are more common. During quiescence, mass is accumulated in the disk, and during outburst most of the disk falls into the black hole. The outburst is triggered as the density in the accretion disk exceeds a critical value. High density increases viscosity, which results in heating of the disk. Increasing temperature ionizes the gas, increasing the viscosity, and the instability increases and propagates throughout the disk. As the instability reaches the inner accretion disk, the X-ray luminosity rises and outburst begins. The outer disk is further heated by intense radiation from the inner accretion disk. A similar runaway heating mechanism operates in dwarf novae.

14.2.7.5 High-Mass X-Ray Binaries

HMXB

A high-mass X-ray binary (HMXB) is a binary star system that is strong in X-rays, and in which the normal stellar component is a massive star: usually an O or B star, a Be star, or a blue supergiant. The compact X-ray emitting component is generally a neutron star, black hole, or possibly a white dwarf. A fraction of the stellar wind of the massive normal star is captured by the compact object, and produces X-rays as it falls onto the compact object.

Catalog of galactic high-mass X-ray binaries: http://webviz.u-strasbg.fr/viz-bin/VizieR?-source=J/A%2BA/455/1165

In a high-mass X-ray binary, the massive star dominates the emission of optical light, while the compact object is the dominant source of X-rays. The massive stars are very luminous and therefore easily detected. One of the most famous high-mass X-ray binaries is Cygnus X-1, which was the first identified black hole. Be X-ray binaries are a subclass of high-mass X-ray binaries that consist of a Be star and a neutron star. The neutron star is usually in a wide highly elliptical orbit around the Be star. The Be stellar wind forms a disk confined to a plane often different from the orbital plane of the neutron star. When the neutron star passes through the Be disk, it accretes plenty of gas in a short time. As the gas falls onto the neutron star, a bright flare in hard X-rays is seen.

14.2.7.6 Microquasars

Xmq

A microquasar (or radio emitting X-ray binary) is a smaller cousin of a quasar. Microquasars are named after quasars, as they have some common characteristics: strong and variable radio emission, often resolvable as a pair of radio jets, and an accretion disk surrounding a compact object which is either a black hole or a neutron star. In quasars, the black hole is supermassive (millions of solar masses); in microquasars, the mass of the compact object is only a few solar masses. In microquasars, the accreted mass comes from a normal star, and the accretion disk is very luminous in the optical and X-ray regions. Microquasars are sometimes called radio-jet X-ray binaries to distinguish them from other X-ray binaries. A part of the radio emission comes from relativistic jets, often showing apparent superluminal motion.

Microquasars are very important for the study of relativistic jets. The jets are formed close to the compact object, and timescales near the compact object are proportional to the mass of the compact object. Therefore, ordinary quasars take centuries to go through variations a microquasar experiences in one day.

Noteworthy microquasars include SS 433, in which atomic emission lines are visible from both jets; GRS 1915+105, with an especially high jet velocity; the very bright Cygnus X-1; and the microquasar LS I +61 303, which has been discovered to emit very high-energy gamma rays.

14.2.8 A Guide for the Amateur to Objects and Literature

While the present section tries to give an overview over the complete field of variable stars, the amateur will perhaps find him or herself somewhat lost, if he or she plans to become active in the field. A general "hands on" introduction is provided on the web-pages of the American Association of Variable Star Observers. There are other amateur associations in other countries that try to combine the activities of amateurs. There are so many bright or faint, known and unknown variables in the sky that each amateur can find objects for own studies (with the eye, binoculars, telescopes, and any type of photometer) that will provide more insight on the ongoings in the sky.

However, it should be clear that there are also observational and instrumental limits. Noisy light curves or a poor coverage of a variable may mean a lot of observing time wasted on an unsuited object.

If visual observations with binoculars are carried out, light curves of pulsating stars or minimum times of bright eclipsing binaries are suitable objects for study. It should be noted, however, that eclipse timings have a much higher accuracy if photoelectric or CCD photometry is used. Observations of multiperiodic variables should only be carried out within world-wide campaigns, since due to the unavoidable gaps in single-observer light curves it is impossible to disentangle the various pulsational periods.

Period studies of pulsation or eclipsing binary stars, long-term monitoring of irregular (or simply poorly known) variables, the search for erupting novae in the Milky Way or supernovae in galaxies—there are plenty of fields where a serious amateur astronomer can carry out useful work.

The books by Good [10], North and James [11] as well as Percy [12] give useful introductions to the field, also concerning the planning and realization of observations. Others give a more detailed overview into the physics of variable stars.

Amateur associations for variable star observations:
http://www.aavso.org/ (American Association of Variable Star Observers)
http://www.britastro.org/vss/ (British Astronomical Association Variable Star Section)
http://cdsweb.u-strasbg.fr/afoev/ (Association Française des Observateurs des Étoiles Variables)
http://www.bav-astro.de/ (Bundesdeutsche Arbeitsgemeinschaft für Veränderliche Sterne e.V.)
More links to organizations and astronomical societies:
http://www.aavso.org/links.shtml

14.2.9 Further Reading

Bode, M.F., Evans, A. (ed.): Classical Novae. Cambridge University Press, Cambridge (2008)

Hellier, C.: Cataclysmic Variable Stars: How and Why They Vary. Praxis, London (2001)

Hoffmeister, C., Wenzel, W., Richter, G.: Variable Stars (transl. by S. Dunlop). Springer, Berlin Heidelberg New York (1985)

Unno, W., Osaki, Y., Ando, H., Shibahashi, H.: Non-radial Oscillations of Stars. University of Tokyo Press, Tokyo (1979)

Smith, H.A.: RR Lyrae Stars. Cambridge University Press, Cambridge (1995)

Sterken, C., Jaschek, C. (eds.): Light Curves of Variable Stars: a Pictorial Atlas. Cambridge University Press, Cambridge (1996)

Warner, B.: Cataclysmic Variable Stars. Cambridge University Press, Cambridge (1995)

Based on *Variable Stars* by T.J. Herczeg and H. Drechsel, Compendium of Practical Astronomy, Vol. 3, Chap. 25, Springer, Berlin, Heidelberg, New York (1994)

14.3 Binary Stars

H. W. Duerbeck

14.3.1 Introduction

Binary and multiple systems of stars first appeared as a topic of astronomical research more than two hundred years ago. In 1767, the Reverend John Michell suggested that double stars are not merely chance alignments, but revolve around each other under their mutual gravitational attraction. While the Mannheim astronomer Christian Mayer described "satellites of fixed stars" (i.e., faint visual companions of stars) in 1777, John Goodricke of York, England in 1782, found periodic brightness variations in Algol (β Persei), which he explained as eclipses caused by a planet orbiting the star.

While some of these early observers supposed that they had found "planets," it was soon clarified that we deal in both cases with binary stars, a term coined by William Herschel in 1802: Early micrometric observations of visual binaries established orbital displacements, supporting the conjecture that most of the double stars seen are not perspective random pairings (so-called optical double stars) but rather are physically connected systems (binary stars).

Also, the concept of mutual stellar eclipses was put forward as an explanation for the periodic brightness changes of certain apparently variable stars. These eclipsing binary stars (or photometric binaries) will also be discussed in this chapter.

14.3.2 Binary Stars: an Overview

Even the casual observer will certainly have encountered many fine double stars when sweeping the sky, and will have used them in various degrees of difficulty for optical and seeing tests (Fig. 21). As large telescopes, technologies, and theory progressed, binary stars became a rewarding target for high-precision astronomy; the literature shows that

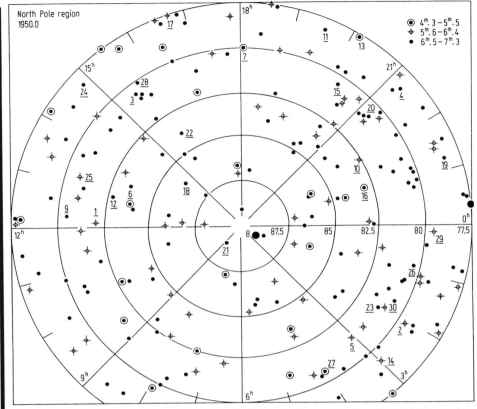

Fig. 21. Double stars in the north polar region which are suitable as test objects (see Table 2 number 1–30)

the research emphasis has shifted toward close (interferometric and photometric) systems and subsystems, and toward binaries containing low-mass, low-luminosity dwarf stars.

Visual binaries in particular are characterized by long periods of revolution. Their study depends on position measurements collected over many years, but the quantities to be measured are still minute enough that attention must be paid to the effects of random and systematic errors. Many of the archived data have only statistical significance until perhaps much later when a particular star receives special attention for one reason or another; investigators then appreciate finding measurements which date far back in time. Visual double stars have always been the domain of only a few researchers, indicating that experience and perseverance do play a role.

The two components of a binary star move in elliptical orbits around their common center of mass. The center of mass is not a visible point; what is actually observed is the relative orbit of one star with respect to the other, which is the sum of the component orbits and similarly elliptical. For the relative orbit, Kepler's third law states that

$$a^3 = (r\alpha)^3 = (M_1 + M_2)P^2 \,, \tag{15}$$

Table 2. List of 30 visual double stars which are circumpolar. The coordinates, apparent visual magnitudes of the components, and their separations in arcseconds are given

No.	α_{2000}	δ_{2000}	m	ρ	No.	α_{2000}	δ_{2000}	m	ρ
1	12^h09^m	$+82°0$	6.5–8.5	67″	16	22^h48^m	$+82°9$	5.0–8.7	3″5
2	0213	79.7	6.5–7.1	54	17	1643	77.5	6.1–9.4	2.9
3	1500	80.6	6.9–7.1	31	18	1433	86.2	7.0–10	2.4
4	2115	78.5	7.2–10	26	19	2244	78.4	7.5–9.3	2.3
5	0312	81.3	6.0–10	24	20	2114	80.2	7.8–8.5	2.0
6	1249	83.7	5.3–5.8	21.6	21	0929	88.8	7.1–10	1.7
7	1800	80.0	5.8–6.2	19.2	22	1549	83.8	7.6–8.1	1.4
8	0221	89.0	2.1–9.1	18.4	23	0210	81.3	6.9–9.1	1.2
9	1212	80.3	7.3–7.8	14.4	24	1453	78.3	6.5–10	1.2
10	2210	82.6	7.0–7.5	13.7	25	1312	80.6	6.3–10	1.0
11	1929	78.3	7.6–8.3	11.3	26	0119	80.6	8.0–8.0	0.9
12	1255	82.5	6.9–11	10.5	27	0409	80.7	5.7–6.4	0.9
13	2008	77.7	4.4–8.4	7.4	28	1539	80.1	7.4–8.2	0.7
14	0306	79.3	5.8–9.0	4.6	29	0010	79.5	6.8–7.1	0.5
15	2015	80.5	6.8–11	4.0	30	0150	80.7	7.8–8.1	0.3

where a is the orbital semimajor axis in astronomical units (AU), α the corresponding quantity in seconds of arc, r the distance of the pair from the Sun in parsec, M_1, M_2 the masses in solar units, and P (mutual) period of revolution in years. All stellar masses are determined by this formula. It also indicates that the great majority of binaries separated in small telescopes must have large true distances between their components, corresponding to periods of millenia and longer. Only in those binaries which are relatively close to the Sun, the orbital displacement will become conspicuous over a few decades or, in a few cases with periods less than 100 years, within a few months or a year.

As already mentioned, apart from these physically associated objects, there also exist *optical double stars* whose components are not related but merely appear close together owing to perspective along the line of sight; such pairs are without scientific interest. A physical binary system is identified by its curved orbital motion (in contrast to the rectilinear relative motion of optical components), or by a common proper motion in the sky in cases where the relative motion is too slow to distinguish its curved or linear shape. Bright, close pairs are almost certainly physical; among stars brighter than magnitude +8 with separations under 5″, optical pairs are rare exceptions, as observations and also statistical considerations show. Optical objects dominate at separations over 20″, but there are still many known pairs and groups of common proper motions which have much larger separations.

Binary and multiple systems are quite abundant in the solar neighborhood. It is estimated that roughly 85% of stars (excluding red dwarfs) are components of binary systems, thus leaving the truly single stars as a 15% minority, and even they are likely to have originated within groups and escaped thereafter. The triple and higher-multiple systems—also by no means rare—show components normally at very unequal mutual distances, such as a close pair with a third star ten or more times farther out, or two closer pairs circling each other in a very wide orbit. These patterns of distances are vital for the long-term stability of the systems. Near-equal component separations, like the Trapezium in the Orion Nebula, are found almost exclusively among young stars.

This type of arrangement is evidently unstable and dissolves quickly—even faster than do open clusters. The higher incidence of single stars among M-type dwarfs (perhaps 40%) is likely to have two causes: first, these red dwarfs were born with far too little angular momentum to form close binaries, and second, as very low-mass objects, they were preferentially ejected from the unstable original groupings.

Orbit dimensions and periods of binaries occur over a vast possible range. The closest pairs revolve with their surfaces practically in contact or within joint envelopes around both stars, while the widest pairs are separated so far that their gravitational bond barely exceeds that of the general gravity field of the Galaxy. Binaries substantially wider than α and Proxima Centauri (separated by 13,000 AU) will not remain bound over their lifetime. The widely varying orbital scales usually require separate observational techniques: a subdivision into the *visual* (long-period) *binary* (including interferometric binaries), the *photometric* (eclipsing short-period) *binaries*, and the *spectroscopic binaries*.

Mass determinations by Eq. (15) require that the distance a is known in kilometer or in astronomical units (AU), freed from the orbital inclination i and split into the relative orbits. This leaves two possibilities:

1. *Visual binaries*, in which the orbit (defined by a and P) is supplemented by astrometric studies of parallax and mass ratio (by making measurements relative to several unrelated background stars in the field). The parallax usually contributes the largest relative error and may be replaced by spectroscopic measurements, if available, of the radial-velocity amplitude of at least one component.
2. *Eclipsing binaries*, in which radial-velocity amplitudes of both components are required to find the scales of relative and absolute orbits; the distance r from the Sun does not enter, so its uncertainty does no harm.

The *mass-luminosity relation* (MLR) for main-sequence stars is:

$$\log L_{\mathrm{bol}} = 3.8 \log M \qquad\qquad \text{for } 0.5 < M < 2.5$$
$$\log L_{\mathrm{bol}} = 2.6 \log M - 0.3 \qquad \text{for } 0.1 < M < 0.5 \ , \qquad (16)$$

where M and L are the mass and luminosity in solar units. The slope of 2.6 for low-mass stars is also approximately followed by stars with masses over 2.5 $M_\odot$, but here it is less well defined owing to the high incidence of evolving, over-luminous stars. The evolution can be calculated through stellar models as a set of luminosity *isochrones* (loci of equal age). The change of energy generation on the main sequence is slow enough that in many cases the zero-age line (arrival at the main-sequence equilibrium) still holds. Data on distances and masses, however, show a substantial number of these stars to be incipient subgiants, even if not yet spectrally classified as such.

Bodies below a mass limit around 0.08 $M_\odot$ cannot tap the resource of nuclear energy, and—lacking hydrostatic equilibrium—they never become proper stars. These objects are termed *brown dwarfs*. Since substellar objects are characterized by mass, which here does not relate to luminosity, they cannot be identified by their brightness. Brown dwarfs are found by infrared observations and by stellar coronagraphs, used to block out the light a nearby more massive binary companion.

The frequency of stellar masses peaks at about one-fourth of the solar mass (absolute visual magnitude ≈ 12); stars with masses around 0.1 $M_\odot$ are less common and substellar masses are probably quite rare. Although not much is known quantitatively about the origin of the planetary system, its genesis (the accretion of very small bodies under the influence of a central gravitational force) is utterly different from that of binary and multiple stars, and inferences from one process to the other are not valid.

14.3.3 Features of Visual Binaries

The position of the secondary (fainter) star relative to the primary at the celestial sphere is given by two numbers (Fig. 22): the separation ρ, given in arcseconds, and the position angle p counted in degrees from north through east, south, west from $0°$ to $360°$. These polar coordinates can be most easily be obtained by means of a filar micrometer. On the other hand, photographic plates or CCD images give rectangular coordinates x (positive toward north) and y (east), both in arcseconds. This requires that the imaging device is properly calibrated in direction and plate scale. The conversion formulae are

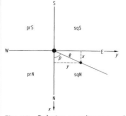

$$\Delta\delta = x = \rho\cos p \qquad \text{(Difference in Dec)}$$
$$15\cos\delta\Delta\alpha = y = \rho\sin p \qquad \text{(Difference in RA)} . \qquad (17)$$

Or, conversely,

$$\tan p = y/x \quad \text{and} \quad \rho = \sqrt{x^2 + y^2} . \qquad (18)$$

Fig. 22. Relative coordinates and quadrants for a binary star in the field

The axes divide the field into four "quadrants" referred to in Fig. 22.

From the total magnitude m_{tot} and the magnitude difference Δm of a pair, the component magnitudes m_A and m_B can be found and vice versa. Table 3 relates Δm with the amount $m_A - m_{\text{tot}}$, by which the primary star is fainter than the combined light.

Visual binaries are named with a discoverer's code and a number. Most objects accessible to small telescopes in the northern hemisphere are already numbered in the oldest systematic catalogs (before 1850), those by W. Struve (Σ) and O. Struve (OΣ); numerous other codes added since need not be detailed here. Often used is the ADS number after Aitken's compilation catalog of double stars (north of $\delta - 30°$ and known before 1927), and the coding now preferred is by the coordinates for the year 2000. Thus, Castor (α Geminorum) can be found under any of the designations Σ 1110, ADS 6175, and 07346 N3153 (coordinates 7^h34^m6, $+31° 53'$ for the equinox 2000.0). Within a system the components are distinguished by a capital letter, generally according to brightness: A = primary star, B = companion, C, D, and so on for more distant components, if present.

Example:
Data for Castor are $m_{\text{tot}} = +1.54$ and $\Delta m = 0.96$.
Table 3 gives $m_A - m_{\text{tot}} = 0.37$, so the primary is $+1.91$ and the companion $+2.87$. Equally bright components thus are 0.75 magnitudes fainter each than the combined light, whereas at $\Delta m > 3$ the companion contributes almost nothing to the total light.

Table 3. Relation between magnitude difference Δm of a pair and the amount $m_A - m_{\text{tot}}$, by which the primary star is fainter than the combined light

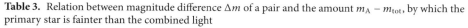

Δm	$m_A - m_{\text{tot}}$	Δm	$m_A - m_{\text{tot}}$	Δm	$m_A - m_{\text{tot}}$
$0^m.0$	$0^m.75$	$1^m.0$	$0^m.36$	$2^m.0$	$0^m.16$
0.2	0.66	1.2	0.31	3.0	0.07
0.4	0.57	1.4	0.26	4.0	0.03
0.6	0.49	1.6	0.22	5.0	0.01
0.8	0.42	1.8	0.19	6.0	0.00

The last compilation to appear in print was the *Index Catalogue* (IDS) in 1961. The *Washington Double Star Catalogue* (WDS) is available at: http://ad.usno.navy.mil/wds/ (see also G.L. Wycoff et al. [13]).

The examination of stars in the Durchmusterung catalogs (to about tenth magnitude) for duplicity has been largely completed; most of the recent discoveries are very close (interferometric) pairs or stars fainter than ninth magnitude.

The color and magnitude differences seen in visual binaries basically correspond to the Hertzsprung–Russell diagram. Main-sequence pairs may have components of equal magnitudes and colors, or unequally bright stars showing the corresponding color difference. The combination of a red or yellow giant with a white main-sequence star also occurs (with the well-known contrast illusion that the white star looks blue) and occasionally also with a degenerate companion. For larger magnitude differences the irradiation normally prevents a reliable determination of the color and spectrum of the companion, and even its magnitude estimates may be quite uncertain. In an evolutionary sense, visual (and many spectroscopic) pairs are wide enough to be non-interacting, that is, the stars affect each other only through orbital gravity, and not by mass transfer. The familiar color-luminosity picture does not hold for many components in eclipsing binaries, where deformation of stellar shapes, mass transfer, mass loss, and unusual evolutionary patterns are observed.

14.3.4 Micrometer and Visual Observations

Analog and digital bifilar micrometers are manufactured by Van Slyke Instruments, 12815 Porcupine Lane, Colorado Springs, CO 80908 (http://www.observatory.org/bfm.htm).

The analog bifilar RETEL micrometer is manufactured by D. Dolle, RETEL Ltd, 37 Banbury Road, Nuffield Industrial Estate, Poole, Dorset BH17 7UG (http://www.webbdeepsky.com/notes/dsretel.html).

Screw micrometers are an eighteenth century invention and are still in use. Although their research function has ceased for most observational applications, their use for visual double stars remains. The best-known type of micrometer is the *filar micrometer*, which consists of a frame located in the focal plane of a telescope, at right angles to the optical axis and rotatable around the axis, and a slide moved by a long precision screw along the frame.

Two commercial micrometers have movable wires driven by engineering micrometers [14]. They also allow an analog or digital dial readout of filar settings (Fig. 23).

In the filar micrometer, the slide and frame each carry a thin wire at right angles to the screw (Fig. 24); spares and cross wires may be added as desired. The wires *a* and *b* should be closely adjacent in depth, but should not touch when passing; the shortfocus eyepiece is focused on the middle distance so that both wires look equally thick. In order to protect the wires, the instrument is mounted in a closed case in which the eyepiece is

Fig. 23. Example of an analog (*sideward view, left*) and a digital (*top view, right*) bifilar micrometer. Image courtesy Van Slyke Instruments

inserted, the latter preferably movable on a small, independent slide. In earlier versions of the micrometer, the screw is connected with a divided drumhead to read parts (1/100) of a revolution, and a scale to count full revolutions, so that the displacement of the movable wire can be precisely found. Modern micrometers have an analog or digital dial gauge for determining the position of the movable wire.

The position circle to read the orientation of the micrometer is divided into degrees (counterclockwise) and is firmly attached to the telescope tube. A circle of about 20 cm diameter with sharp divisions suffices to read tenths of degrees. The rotating parts of the device (case, frame, and index on the circle) should be rigidly connected, and the connection screw/slide should be free of backlash. Also needed is a faint, adjustable source of illumination. Some observers prefer directly illuminated, bright wires, while others choose a faint (preferably red) field illumination in which the wires appear black.

The degree of difficulty with which a particular double star can be observed with a given telescope depends (apart from atmospheric conditions and eye sensitivity) on three stellar properties: the separation ρ, the magnitude difference Δm, and the combined magnitude. It should be emphasized that the often quoted rules about limiting magnitudes, seeing disks, magnifications, etc., do not apply to the present task.

The well-known Dawes formula states that the resolving power R for separating doubles (of adequate brightness and with equal components) is given by

$$R = 115/D \,, \tag{19}$$

where R is the resolution in arcseconds, and D the telescope aperture in meters. In closer pairs the stellar disks touch and merge into an oblong shape. The Dawes limit agrees closely with the half-width of diffraction images predicted by theory, but for larger apertures (over 0.5 to 1.0 m, depending on conditions) the gain in resolution is less than calculated; as more turbulence cells simultaneously in the beam affect the image. The magnification used is normally the highest that atmospheric conditions will permit, except for faint components. For modest apertures (less than the 0.5 or 1.0 m quoted), experienced observers prefer to use "over-magnification." (Since the eyepiece frame is close to the plane of the wires, beware of wire damage when changing eyepieces!) Stepping down the

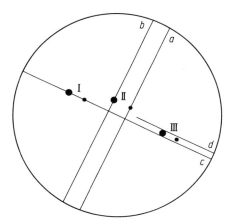

Fig. 24. Measuring with the filar micrometer (see text)

aperture is useful when the gain of better image definition and reduced glare outweigh the loss in resolution, that is, for bright stars observed with large-aperture instruments.

Difficulties and systematic errors occur in close pairs when the images are adjacent, and when the space between the wires is less than the wire thickness and therefore hard to estimate. Less-experienced observers should avoid separations below twice the Dawes limit. In the other direction, toward wide pairs, the precision decreases when high magnification limits the convenient, simultaneous viewing of the stars—this occurs at about 10^4 arcsec divided by the magnification. Systematic errors are also much larger for extended objects such as planetary disks and features.

The resolution is fully exploited only for near-equal components. Even $\Delta m = 1$ mag is much more difficult and also substantially increases the risk of systematic errors. The impediment by Δm can be included in the separation by an empirical "scale of difficulty"

$$C = 0.22\Delta m - \log \rho . \tag{20}$$

When a 0.4 m instrument, for instance, reaches $\rho = 0''\!.3$ at the Dawes limit, then $C = 0.5$, and the corresponding limit at $\Delta m = 3$ will be around $\rho = 1''\!.4$.

The seeing scale is very different from that in other observations, since the visual observer concentrates on the bright nucleus of the turbulent disk and not on the total spread of light. The author has often estimated the seeing as $0''\!.2$, while the photometrist at the telescope next door recorded $1''$ or $1''\!.5$.

Observations are also impeded when the stars are too faint for convenient direct (foveal) viewing. Under good conditions, an aperture of 0.6 m may just reach components of thirteenth magnitude, but the "convenient" limit (beyond which the uncertainty of measuring increases) is at magnitude +11.

For an aperture of 25 cm these numbers become 2 magnitudes lower. The usual guidelines on how faint a star a given telescope will show are irrelevant here, as there is an enormous difference between a barely visible point of light and one whose position can be measured.

The measuring procedure is illustrated in Fig. 24: a fixed wire (c) is rotated until positioned so as to join the centers of the stellar disks (position I), and the pointer at the position circle read; the setting is repeated a few times. Separation is measured with a pair of wires perpendicular to the line joining the stars; the fixed wire (a) is placed over star 1, and the movable wire (b) then brought to bisect star 2 (position II), whereafter the wire positions are exchanged. The difference of the two drum readings corresponds to twice the separation. This also is repeated. Settings of angle may also be done with a close pair of wires (c, d in position III) bracketing the double star in the middle, or the crosswire (c) can be dispensed with by using (a) with a subsequent rotation by 90°. The trick of the trade is to make the settings quickly (to avoid eye fatigue), yet precise, and to catch the fleeting moments of steady images. The line joining the eyes is parallel (or perpendicular) to that joining the stars; observations made with a slanted orientation of the eyes risks large errors in the position angles.

A pair is measured on two or more nights of the same year, so that comparison of the results shows the amount of random scatter. The observing record includes seeing, magnification (when using more than one eyepiece), and the quadrant of the secondary star according to Fig. 22 so that 180° confusion is avoided. A separation estimate may be added to safeguard against gross reading errors. Some experienced observers also record magnitude estimates; catalog data on brightness are often unreliable for faint objects.

The measurements are simply reduced by the following procedure:

1. Calculate average of angle readings minus zero point of circle = position angle.
2. Find average of double-separation readings multiplied by half the screw value = separation in arcseconds.
3. Observing time is converted into fractions of the year. (A half-year is added to the numbers in the *Astronomical Almanac*, Section B, in order to transfer from the middle to the beginning of the year.)

The zero point of the circle is the reading when the wire used for position angles points exactly north. It is found from the east-west direction, by placing the measuring wire in repeated attempts so that an equatorial star trails exactly (to ±0°.1) along the wire across the field. These readings are performed at the beginning and end of every observing session; experience will show how far the constancy of the zero point can be relied upon. Systematic position angle errors will be caused by defective grading or centering of the circle; they can be checked by reading differences between auxiliary pointers at various angular positions. The usefulness of a field-reversing prism in front of the eyepiece (to find subjective errors by comparing readings with and without prism) is doubtful.

In case of the direct reading of the screw setting, the screw value (the angular equivalent of the screw pitch in the focal plane) must be determined accurately, but it is then known once and for all. The interval of sidereal time required to travel a certain number of screw revolutions is noted and then converted into arcseconds by the factor $15 \cos \delta$. Star transits around $\delta = \pm 75°$ are recommended; closer to the pole the motion is so slow that the scintillation is much more annoying. The screw value may also be found by measuring distances (total, or only in declination) between very widely separated stars with well-known positions (e.g., the Pleiades), allowing for differential refraction. For comparison, the screw value is computed:

$$\text{Screw value} = 206265'' \times \text{screw pitch (mm)}/\text{effective focal length (mm)} . \qquad (21)$$

14.3.5 Interferometric, Speckle, Photographic and CCD Observations

The quest for higher resolving power has brought interferometry into the technical arsenal used to study binaries. Monochromatic light of wavelength λ falls through two slits and is subsequently recombined to form a pattern of interference fringes which, in the case of double stars, disappears (or reaches minimum visibility) in a specific position: the slits are lined up with the position angle and are $\lambda/2\rho$ apart, which gives the angular separation ρ. The enormous light loss experienced by interferometry has—at least until the recent arrival of high-speed detectors—severely limited its astronomical applications, and the shortage of the requisite large-telescope time has also contributed to the low productivity of interferometric investigations.

The small turbulence elements of the air, which are 10 to 20 cm in size and have timescales around 1 to 0.1 s, dissolve the image into rapidly changing spots, or speckles. The trained eye subconsciously perceives an instantaneous image in the spots but this ability is hampered, as noted above, for large apertures, and thus for close pairs, by the increased number of simultaneous air cells in the image. The so-called speckle methods purport to undercut the detrimental frequencies of scintillation by still shorter

Most data came from Finsen [15], who skillfully constructed and operated an eyepiece interferometer, reaching reliable measures down to 1/2 of the Dawes separation limit of his modest refractor. A simple apparatus has been described by Kerschbaum [16].

integration times of 1/30 to 1/100 s. The focal image is magnified by microscope optics, recorded by high-speed CCDs, which have replaced the previously employed image intensifiers and cameras, and the data fed into a computer, which makes the image of the binary re-emerge from the morass of hundreds of speckles. The closest pairs at separations down to about 0$''$03 have been measured in this way; in particular, many objects known only as spectroscopic pairs became resolved and were subsequently processed to obtain masses. Several (somewhat different) devices of this type have been successfully applied during the last decade, and are expected to soon become the dominant measuring technique; their operation requires technical support, and their use is feasible only for large observatories. In addition, the correct link of the composite speckle image with the binary star position requires a series of careful calibration measurements, including a safeguard against spurious "companion" speckles.

Compared with visual measurements, photographic observations of double stars permit a higher precision, as one plate accommodates multiple exposures of an object which can then be measured carefully and repeatedly. However, few objects exist which are photographically well-separated and yet worthy of regular observation, and so the technique is little practiced. Exceptions are some nearby stars covered by parallax programs, and the common proper-motion pairs and streams for studies of moving groups. Owing to the low quantum efficiency of photographic emulsions, more scintillation accumulates during the requisite exposure time; the seeing disk—much larger than in visual work—limits the resolution much more than does the plate graininess.

The angular separation in radians is given by the linear separation distance in the focal image divided by the effective focal length. Converted into arcseconds: 1 mm on the plate = 206$''$265 divided by the focal length in meters. Well-defined images, correctly exposed and made in good seeing, may be measurable down to a separation of 0.07 mm, but troublesome adjacency effects arising from both development and measurement of the images occur below about 0.15 mm. Thus, even for long-focus telescopes, good photographic resolution is above ten times the visual Dawes limit.

The use of a CCD camera appears to offer more advantages than photographic studies (see also Chap. 8). Multiple exposures can be carried out subsequently, without the possibility of overlaps with other stars; the spatial stability of a CCD is better than that of photographic emulsion, and photographic adjacency effects do not occur.

Owing to the very specialized and long-term nature of astrometric observations, they are not treated in a special chapter in this book; in the present context, a few explanations will do. Proper motions of stars and binaries (and sometimes mass ratios of the latter) are found from plate measurements relative to unrelated comparison stars in the field whose positions and proper motions are already known from transit-circle catalogs; this requires long-focus coverage over considerable time spans, usually decades. Images of larger reference star fields rely on precise orthogonality of the plate and the optical axis.

Measurements of wide star pairs have to be corrected for differential refraction; this also applies when such measurements are used to determine the image scale or the visual micrometer screw value. The atmosphere lifts the lower star more strongly; per arcmin, the altitude difference is thus diminished by about 0$''$04 at 40° altitude, and by 0$''$07 at 30°. Reduction of reference star fields by standard formulae of photographic astrometry (Sect. 3.3) removes the linear part of differential refraction; the non-linear remainder is negligible except at unusually large zenith distances or hour angles. The atmospheric

With a suitable image processing software, a standard stellar intensity profile can be constructed from the undisturbed stellar images on the CCD frame (see Sect. 8.4.3). By fitting this profile to a blended binary star profile, precise relative positions and the magnitude difference of the binary star components can be derived.

A 20-cm plate covering an angle of 1°, but slanted by 0.5 mm on one side, will displace the optical center relative to the geometric center by about 1 μm, and hence the central star will be shifted relative to comparison stars near the edges at the level of measuring precision.

spectrum and part of chromatic aberration in refractors are eliminated by suitable color filters, a double star of substantial Δm can be exposed through gratings so that side images of the overexposed primary star appear.

14.3.6 Orbital Elements and Ephemerides

The position of the companion at any other time is then determined by the first (law of elliptical motion) and second (law of equal areas) of Kepler's laws. Depending upon the orientation of the orbit with respect to the line of sight, the observed, projected ellipse differs from the true orbit by projection, which is defined by three angles.

Seven quantities, called *orbital elements*, determine the orbit of a visual binary star. They are defined in analogy to the elements of planetary orbits:

a = Semimajor axis of the true orbital ellipse in arcseconds.
e = Numerical eccentricity, which fix the size and shape of the orbit.

The relative motion is expressed by

P = Period of revolution in years.
T = Time of passage through periastron (the point of smallest true distance between the components).
i = Inclination, the angle between the orbital plane and the tangential plane at the sphere, counted from 0° to 180°. The projection factor $\cos i$ is taken as positive ($i < 90°$) when the motion is toward increasing position angles (direct, or counterclockwise); $i = 90°$ corresponds to edge-on viewing.
Ω = Node, the position angle of intersection between the true and tangential planes. Usually $\Omega < 180°$ is stated; the other node is 180° opposite. Precession changes all position angles, and also the node, by the amount $+0°.557 \sin \alpha \sec \delta$ per century. Measurements from different times are reduced accordingly in order to refer to a common equator.

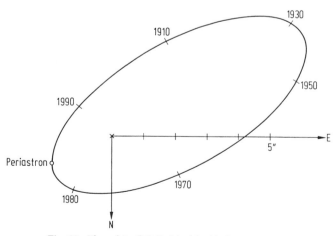

Fig. 25. The orbit of 70 Ophiuchi with $P = 88.3$ years

ω = Angular distance from node to periastron in the true orbit, counted in the direction of motion.

The *Thiele-Innes* elements serve to describe the orbit in rectangular coordinates. They correspond to the group (a, ω, Ω, i), and are composed in the following way:

$$A = a(\cos \omega \cos \Omega - \sin \omega \sin \Omega \cos i),$$
$$B = a(\cos \omega \sin \Omega + \sin \omega \cos \Omega \cos i),$$
$$F = a(-\sin \omega \cos \Omega - \cos \omega \sin \Omega \cos i),$$
$$C = a(-\sin \omega \sin \Omega + \cos \omega \cos \Omega \cos i) . \tag{22}$$

An ephemeris can be computed from the elements which gives the positions of the binary at any time t (Fig. 25). To optionally determine the polar coordinates p and ρ or the rectangular coordinates x and y, the following sequences of easily programmable formulae are employed:

$$360°/P(t - T) = M = E - e \sin E; \tag{23}$$
$$\tan v/2 = \sqrt{(1 + e)/(1 - e)} \tan E/2, X = \cos E - e,$$
$$r = a(1 - e^2)/(1 + e \cos v), \quad Y = \sqrt{1 - e^2} \sin E; \tag{24}$$
$$\tan(p - \Omega) = \tan(v + \omega) \cos i, \quad x = AX + FY,$$
$$\rho = r \cos(v + \omega)/\cos(p - \Omega), \quad y = BX + GY . \tag{25}$$

The conversion proceeds via an auxiliary angle E to coordinates in the true orbit (v and r, or X and Y, respectively) by Eqs. (24). Projection into the corresponding coordinates of the apparent orbit is done by Eqs. (25), which contain the direction cosines. Equation (23) is called the Kepler equation, expressing the law of areas, and is solved iteratively (see p. 35 in [17]). With degrees as units, the term $e \sin E$ is multiplied by 57.296 in order to obtain the same unit.

New observations sometimes suggest that a new orbit or a substantial improvement over a previous result may be obtainable, and that is usually done by the observer.

The latest catalog is the *Sixth Catalogue of Orbits of Visual Binary Stars* [18], which lists 2024 orbits of 1888 systems.

The available orbits are mostly those of close pairs; separations in the range $0''.1$ to $1''$ dominate, periods are mostly between 10 and 400 years, and distances from the Sun rarely exceed 200 pc. Many orbits are known with high precision, while others are still quite tentative—especially the long-period ones which may need another century or so to affirm by complete coverage. Differential corrections to initial elements are readily computed from a good and complete body of observations.

14.3.7 Features of Eclipsing Binaries

The orientation of the orbit of a binary system, as seen from Earth, may be such that its components pass in front of one another during their mutual revolution. This gives rise to an apparent periodic variation in the integrated light of the system. During one revolution, the light curve typically shows a deeper *primary minimum* and a shallower *secondary minimum*. The primary minimum is caused by the eclipse of the hotter (and usually more luminous) component of the photometric binary by the fainter secondary star; one-half period later, the reverse configuration creates the secondary minimum. The ratio of depths of both minima is determined by the ratio of the surface brightnesses of the two components, which is determined essentially by the effective surface temperatures. the periodic brightness variation due to the mutual eclipses reveals its binary nature. As is the case for genuine variable stars, the discovery and study of these *eclipsing binaries* lies in the realm of photometry. These binaries are listed in the *General Catalogue of Variable Stars* (see Sect. 14.2).

Eclipses are unlikely when the distance between the stars is large compared with their diameters. Thus, photometric binaries tend to be mostly short-period systems, and the components preferentially the larger stars (subgiants and giants) of advanced evolutionary status. Periods are typically a few days; they range from an extreme 27 years for the supergiant ϵ Aurigae down to nearly 1^h for the closest pairs of faint dwarfs. Since they appear as single stars in the telescope, orbital motion can be studied spectroscopically by the Doppler shift of stellar lines. Depending on the brightness ratio of the two components, double-lined or single-lined *spectroscopic binaries* are observed.

In parentheses, we note that for inclination angles between 0° and about 66° or more, no eclipses occur. In such cases, spectroscopic duplicity can still be found. In some cases, the deformation of stellar surfaces because of tidal interaction also leads to a weak photometric variability: such objects belong to the class of *ellipsoidal variables*. Since the inclination angle is difficult to establish for such systems, they are of minor interest to the astrophysicist.

Owing to the importance of obtaining the parameters of evolving stars and for their comparison with theory, several hundreds of eclipsing binaries have been carefully studied through two-color and multicolor light curves, spectroscopic orbits, and computer modeling of the photometric and spectroscopic phenomena.

14.3.8 Structure of Close Binaries

The revolution of the two stellar components around the common center of gravity proceeds generally on elliptical orbits with the radius vector

$$r = \frac{a(1 - e^2)}{1 + e \cos \theta} ,$$
(26)

where a is the semimajor axis of the relative orbit, e the orbital eccentricity, and θ the phase angle (true anomaly).

For non-circular orbits, the distance between the two stars varies between the maximum $a(1 + e)$, called *apastron*, and the minimum $a(1 - e)$, which is the *periastron*. The special case of a circular orbit $(e = 0)$ is approximately fulfilled in many close binaries. The period P of revolution is related to the scale of the orbit (semimajor axis a) and the masses $M_{1,2}$ of the components via Kepler's third law:

$$P = 2\pi \left[\frac{a^3}{G(M_1 + M_2)} \right]^{1/2} ,$$
(27)

where G is the universal gravitational constant.

A *close* binary is any system in which the distance between the components is of the same order as their radii. Owing to the mutual influences on structure and evolution of the components, such a system is also called an *interacting binary*. The most important interacting process is the mass exchange between components, which is responsible for peculiar properties of several classes of variable stars (e.g., W UMa systems, cataclysmic binaries, X-ray binaries). Generally, eclipsing binaries serve as a source of information for stellar sizes and masses.

If periodic Doppler shifts of spectral lines of both components can be observed, the ratio of radial velocity amplitudes K_1/K_2 found from the observed radial velocity curves $V_{1,2}(t)$ gives the mass ratio:

$$M_2/M_1 = K_1/K_2 . \tag{28}$$

When the orbital inclination i is known from the analysis of the light curve, with a_1 and a_2, $a = a_1 + a_2$ is derived and the two masses can be found.

Often, only the lines of the more luminous component are seen in the spectrum. Then a relation between the two masses and the angle of inclination i, the so-called *mass function*, is available:

$$F(M_1, M_2, i) = \frac{(M_2 \sin i)^3}{(M_1 + M_2)^2}$$
$$= \frac{4\pi^2}{G} \frac{(a_1 \sin i)^3}{P^2} . \tag{29}$$

For relatively well-separated eclipsing binaries, i is usually close to $90°$ and in any event lies above a certain lower limit, so that a good estimate of M_2 can be obtained from a value of M_1 assumed from the spectral type.

Kepler's third law and the motion of the components on closed orbital ellipses holds strictly only for centrally condensed spherical stars (sufficiently well-approximated with point masses) with gravitational potentials proportional to r^{-1}. Tidal interaction in close pairs leads to a deformation of the stars and a modification of the potentials, which, in an eccentric orbit, causes a rotation of the apsides (the direction of the semimajor axis). Through the viscous damping of stellar oscillations, tidal forces cause a decay of the orbital eccentricity, and a circular orbit is reached in the stable state of minimum total energy at constant total angular momentum, on a timescale that is short compared with the evolutionary age of the system. Thus, most close systems show circular orbits and synchronous rotation, i.e., the periods of rotation and of revolution of the two stars have the same value.

The closer the pair, the more pronounced is the departure of the shape of the two stars from spherical symmetry. With decreasing distance between the mass centers, the influence of tidal interaction increases proportional to a^{-3}, and also the angular speed of the bound rotation increases, in accordance with Kepler's third law, in proportion to $a^{-3/2}$, which increases the centrifugal action in proportion to a^{-2} and hence the rotational flattening of the stars. The three-dimensional shape of the surface of each stellar component corresponds to a surface of constant pressure and constant density, and hence a surface of constant potential. It is therefore called an equipotential surface. The spatial shapes of equipotential surfaces in close binaries are described by the Roche model. Three important assumptions permit a simple description of the shapes of the equipotential surfaces:

– The mass distributions in both stars can be approximated by point masses.
– The orbit is circular.
– The rotation of the stars is synchronous with the orbital motion.

In a corotating Cartesian coordinate system with its origin at M_1 ($M_1 \geq M_2$) where the x-axis is along the line joining M_1 and M_2 and the z-axis is perpendicular to the orbital plane, the potential ψ, composed of the gravitational potentials of M_1 and M_2 and of the centrifugal action at a point $P(x, y, z)$, is given by

$$\psi = -G\left(\frac{M1}{r_1} + \frac{M_2}{r_2}\right) - \frac{\omega^2}{2}\left[\left(x - \frac{M_2 a}{M_1 + M_2}\right)^2 + y^2\right],\tag{30}$$

where $r_1 = (x^2 + y^2 + z^2)^{1/2}$ and $r_2 = ((a-x)^2 + y^2 + z^2)^{1/2}$ are the distances of P from M_1 and M_2, respectively. The angular velocity $\omega = 2\pi/P$ is equal to the Keplerian value $[G(M_1 + M_2)]^{1/2}a^{-3/2}$.

The *Roche equipotential surfaces* $\Omega = $ const are connected with the potential ψ via

$$\Omega = \frac{a\psi}{GM_1} + \frac{q^2}{2(1+q)}.\tag{31}$$

Normalizing the potential by putting $a = 1$, Ω depends only on the mass ratio $q = M_2/M_1$, and not upon the individual masses.

An intersection of equipotential surfaces ($\Omega = $ const) with the orbital plane ($z = 0$) is shown in Fig. 26. Near the mass center of each star, the equipotential surfaces are nearly spherical since the gravitating influence of the other star on its shape is negligible. With increasing distance from the the star, the potential of the other component causes a distortion toward it; the line joining the stellar centers forms the symmetry axis of the approximately ellipsoidally deformed body. The effect becomes stronger the more the surface of the star approaches the so-called inner critical Roche surface, which is the largest closed equipotential surface around each component, and thus defines the maximum possible volume in a detached configuration.

In addition to the tidal action, the synchronous rotation of the stars also leads to a distortion owing to the centrifugal force; the stars are flattened on the rotation axis perpendicular to the orbital plane. The equipotential surfaces graphed in Fig. 26 include the centrifugal potential. The outermost surface graphed is the largest closed envelope of the system, and thus is called the *outer critical Roche surface*. With the aid of the equipo-

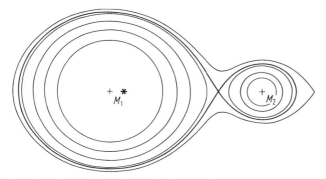

Fig. 26. Intersection of equipotential surfaces with the orbital plane of a binary system with mass ratio $q = M_2/M_1 = 0.1$. The mass centers of the two components are marked by plus (+) signs, the center of mass of the system with an asterisk (*). Some equipotential surfaces representing detached configurations, the figure-eight inner critical surface, and the outer critical surface are shown

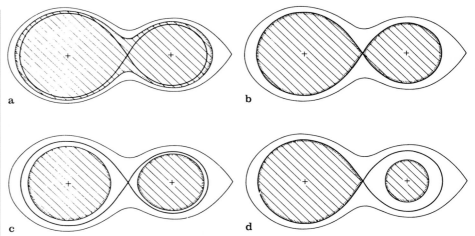

Fig. 27. Different configurations of close binary systems. The equipotential structure graphed for the mass ratio $q = M_2/M_1 = 0.5$. **(a)** Over-contact system. **(b)** Contact system. **(c)** Detached system. **(d)** Semidetached system

tential surfaces, a classification of various configurations of closed binary stars can be carried out (see Fig. 27):

1. Both stars are smaller than their maximum volumes bounded by the inner critical surface; this is a *detached system* (Fig. 27c).
2. One of the stars fills its Roche limit and the system is *semidetached* (Fig. 27d).
3. Both stars just fill their maximum volumes, thus making a *contact system* (Fig. 27b).
4. *Over-contact systems* (Fig. 27a) have both components further expanded so that matter overflows the inner Roche limit and fills part of the space between the inner and outer Roche surface where the matter is bound to the system but cannot be associated with either of the two components.

Over-contact systems can be characterized by a parameter known as the *degree of contact f*:

$$f = \frac{\Omega_i - \Omega}{\Omega_i - \Omega_0} \quad 0 \le f \le 1 , \tag{32}$$

where Ω, Ω_i and Ω_0 are the values of the potentials for the actual stellar surface and for the inner and outer critical surface, respectively. When the stars expand beyond the outer Roche surface, material is lost from the system.

14.3.9 Classification of Light Curves

In the past, light curves of eclipsing binaries were classified phenomenologically, into one of three classes:

1. Algol-type (EA)
2. β Lyrae-type (EB)
3. W UMa-type (EW)

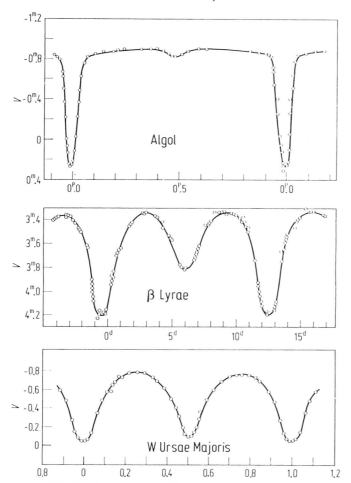

Fig. 28. Light curve of Algol (*top*), β Lyrae (*middle*), and W Ursae Majoris (*bottom*), the prototypes of eclipsing binaries of types EA, EB and EW, respectively. After Tsesevich [19]

Figure 28 shows the light curves of the prototypes, which are, however, not all typical representatives of their classes. Algol (β Persei) is actually a semidetached system with a red subgiant as the contact component, whereas many members of the photometric class EA are widely separated binaries composed of main-sequence components. Algol-type light curves show a pronounced primary minimum and—depending on the system geometry, inclination, and luminosity ratio of the components—a more or less deep secondary minimum. Eclipses may be total, in which case a constant level of brightness is observed for some time before and after the center of a minimum. Between eclipses, the luminosity changes only slowly, as tidal deformations are quite negligible for these widely separated stars, but the reflection effect (heating) is often noticeable.

β Lyrae also is not at all typical of the class EB. This extraordinary system is semi-detached, containing a massive, but spectroscopically unidentifiable, secondary component. Many EB systems, however, are close, yet detached, binaries with two main-sequence components nearly filling their Roche lobes and showing strong interactions. Tidal deformation and reflection cause a pronounced, continuous variation of brightness with orbital phase, which is distinctly noticeable outside of eclipses owing to the changing direction of view onto the stars.

W UMa, however, is a typical EW-type system. Both components fill their Roche lobes and form an over-contact configuration. Strong interactions cause the large brightness variation between the minima, which here, in contrast to β Lyrae systems, are about equally deep owing to the nearly equal surface temperatures of the components.

As we have seen, Algol is not a "typical" EA system, but rather a semidetached one; β Lyrae is a very complicated system and certainly not a good prototype for the EB class. The classification in the 4th edition of the *General Catalogue of Variable Stars* (GCVS) has specified the description of the binary, first by giving the light curve type (EA/EB/EW), and then on information based on the status of the filling of the Roche lobe, on the spectrum, or the luminosity class:

- RS: RS CVn systems with Ca II emission in the spectrum and a distortion wave in the light curve
- WD: White dwarf in the system
- WR: Wolf–Rayet star in the system
- AR: AR Lac systems composed of two subgiant components not filling their Roche lobes
- KE/KW: Contact binary with early or late spectral type
- D, DM, DS, DW: Detached system with main sequence component(s), with subgiant component
- DW: Like KW, but not in contact
- SD: Semidetached system

Since detailed analyses are necessary for a proper classification along these lines, another more straightforward classification has been suggested for the forthcoming GCVS: E, EA, EC (the latter one comprises the "contact" systems, i.e., those with equally deep minima, and continuously changing light during eclipses). Another group, EP, comprises stars eclipsed by planetary companions.

14.3.10 Analysis and Synthesis of Light Curves

An analysis of the light curves of eclipsing binaries which attempts to determine system parameters requires a realistic physical model and a suitable mathematical method of solution. Graphical solutions with nomograms (the Russell–Merrill method), and the Fourier analysis proposed by Kopal, have come out of use (see [20, 21] for an introduction to these techniques). Nowadays, numerical methods are generally preferred, which were pioneered by Wilson and Devinney [23, 22]. With some guessed starting values, a synthetic light curve is computed and compared with the observed one. Kallrath and Linnell [24] have combined the Wilson–Devinney model with a mathematical algorithm (SIMPLEX) which yields a rapid and unambiguous convergence.

The Roche model is the physical basis for calculations. In addition to tidal and rotational deformation, the mutual irradiation of the components (which gives rise to the reflection effect), limb darkening, and gravitational darkening are allowed for. The radiation flux in the direction of the observer is computed as a function of system parameters and orbital phase, and the theoretical light curve thus obtained is compared with observations. Corresponding iterative variations of parameters are obtained by fitting the computed light curve to the observed one by a least-squares solution.

14.3.11 The Zoo of Eclipsing Binary Systems

More than one-half of all stars are members of binary and multiple systems. Visual (i.e., optically separated) binaries with known absolute orbits and particularly spectroscopic binaries which simultaneously show eclipses are the primary source of knowledge on stellar masses and radii. Moreover, the two components of a binary system are of equal age, and thus offer the opportunity to test the theories of stellar evolution from, for example, the observed luminosity ratio. The evolution of the individual binary components is analogous to that of single stars only when the system is sufficiently separated that the structure and evolution of one component is not affected by the other. The structure and evolution are, however, mutually influenced for most close binaries.

The various classes have traditionally been based on the shape of the light curve, while the physical structure of the system was a priori unknown, and was deduced only by more detailed analyses of photoelectric measurements. The light curves of close detached and of semidetached systems do not differ appreciably in shape but show a smooth transition from one to the other, so that distinction between detached and semidetached systems is not always possible on the basis of the light curve alone. In the following, we list a few important types.

14.3.11.1 Ellipsoidal Variables
Systems with tidally deformed components nearly filling their Roche lobes may exhibit a sinusoidal, orientation-dependent light variation, even if the components do not undergo mutual eclipses. The orientational light variation is caused by the continuously changing aspect along the line of sight. Owing to tidal deformation, the projected stellar surfaces vary in size with the orbital phase. The radiating areas are largest at orbital phases $\phi = 0.25$ and 0.75 (quadrature), and smallest near conjunctions $\phi = 0.0$ and 0.5. Hence, the brightness of the system, which is proportional to the radiating area, varies in a sinusoidal shape with a period one-half that of the revolution. The ellipsoidal light variation is enhanced by the mutual irradiation of the components (reflection effect), which plays a significant role in very close systems.

14.3.11.2 RS CVn Systems
The stars of RS CVn-type are regarded here as a subclass of eclipsing variables, but one with peculiar properties, presumably conditioned by evolution. RS CVn stars, which are characterized by F- or G-type primary components and periods of from 5 to 12 days, are detached systems which should show little interaction. Both the strongly perturbed light curve as well as the Ca II emission in the spectrum indicate very strong photospheric and chromospheric activity in these stars. A noticeable hump in the light curve migrates with time toward decreasing phases, and is interpreted as being due to starspot activity. The continuous phase shift is caused by one

Binary Maker 3 Light Curve Synthesis Software (for Windows, Mac, Linux and Sun Solaris platforms) is available from Contact Software, 725 Stanbridge St., Norristown, PA 19401-5505, USA (see http://www.binarymaker.com)

The program *Nightfall* (for LINUX) for synthetic light curves can be obtained at: http://www.hs.uni-hamburg.de/DE/Ins/Per/Wichmann/Nightfall.html.

The program *PHOEBE* (PHysics Of Eclipsing BinariEs, for Linux, Mac and Windows) for synthetic light curves can be obtained at: http://phoebe.fiz.uni-lj.si/.

Another site (with a Microsoft Windows program) is: http://www.midnightkite.com/binstar.html.

binary component of inhomogeneous surface brightness rotating nearly, but not precisely, synchronously with the orbital period. The RS CVn also emit radio radiation and X-ray emission, indicating enhanced coronal activity.

14.3.11.3 Semidetached Systems Algol is a particularly good example of an interacting, semidetached system. This class was first recognized in connection with the so-called Algol paradox. From about 1940 onward, it was found that the low-mass components of Algol systems had much larger luminosities and radii than normal main-sequence stars of the same mass, and hence appeared to be in an advanced evolutionary state. However, the higher-mass companions of these subgiants are still on the main sequence, in spite of the rule that more massive stars evolve faster than less massive ones. The observed findings seemed to contradict the fact that binary star components have a common origin and are of equal age. It was recognized that the originally more massive components in such systems had in the course of evolution lost a large fraction of their mass. In Algol systems, this matter was transferred to the companion star or partly lost from the system. Therefore, the mass ratio was inverted and the now lower-mass components are more evolved than the more massive ones, while still remaining on the main sequence.

Semidetached systems often show signs that mass exchange is currently still taking place. For instance, high-resolution spectroscopy in the optical and UV ranges show the existence of plasma rings, disks, or circumstellar envelopes rotating around the hotter and less-evolved star which accretes matter. Another indication of mass transfer can be derived from the change in the orbital period, which may occur in a secular, continuous, or discontinuous fashion; for example, U Cephei showed a period change of about 20 seconds in 100 years.

14.3.11.4 Contact Systems Over 850 eclipsing contact systems are known, and they seem to be the most abundant type of close binary. EW-type light curves have primary and secondary minima of about equal depth, indicating that both components are similar in surface brightness and temperature, but they usually are a pair of stars with different masses and radii. An extreme example is the W UMa-type system AW UMa, with a mass ratio 1:12 and a ratio of radii 1:3.5. Spectroscopic studies of these short-period objects (typical periods of 5 to 15 hours) normally show two very similar stars of the lower main sequence; the spectral type is always compatible with the more massive component's size or mass.

The continuous light variation outside eclipses indicates a strong tidal deformation of the two stars. They have a common surface at an equipotential surface located between the inner and outer critical Roche surfaces. The two stellar nuclei embedded in this common envelope have different temperatures in the inner layers, but the energy flow entering the common envelope is distributed by currents and turbulent mixing, causing a nearly homogeneous temperature of the photosphere. The outer shells of W UMa stars are convective; also the components rotate considerably faster than do the corresponding single stars with the same spectral type. The dynamo effect thereby gives rise to chromospheric and coronal activity in the outer atmospheric layers, as plasma flows along closed magnetic loops originating within the stellar interior. This leads to the emission of lines and continuum in the UV and X-ray regions.

W UMa systems are subdivided into A-types, where the primary minimum corresponds to the occultation of the more massive, larger component by the smaller companion (transit minimum), and W-types, where the primary minimum is an occultation minimum when the smaller star is totally eclipsed by the larger one. Since the smaller star has a relatively higher surface brightness and higher temperature than the more massive one, the total brightness of the system during occultation is less than at transit.

14.3.12 Outlook

Studies of both visual binary stars as well as eclipsing binary stars are rewarding fields of activity for amateur astronomers. There are many visual binaries which have been neglected over many years, and recent measurements of distances and position angles are lacking. Recent automatic sky surveys for variable stars and related objects have a lot of eclipsing binary stars still requiring more detailed studies (periods, period changes, multicolor light curves). Amateurs equipped with suitable telescopes and auxiliary equipment will encounter a lot of interesting work that needs to be done.

Based on *Binary Stars* by W.D. Heintz, Compendium of Practical Astronomy, Vol. 3, Chap. 26, Springer, Berlin, Heidelberg, New York (1994)

14.4 The Milky Way and Galactic Objects

K.-P. Schröder

14.4.1 Galactic Structure and Visual Appearance of the Milky Way

The view of the divided band of the summer Milky Way, with its bright star clouds between Scutum and Sagittarius, is one of the finest naked-eye sights in a dark, clear night (Fig. 29). It resembles the telescopic view of an edge-on galaxy like NGC 891 or NGC 4565, and that is for a good reason: The galactic structure of our Milky Way is very typical for a spiral galaxy (see Fig. 30).

The milky perception stems from the myriads of unresolved stars in the galactic disk beyond the solar neighborhood. The surface brightness of the Milky Way to a naked-eye visual observer is equivalent to the light of 150 to 500 stars of tenth magnitude spread out evenly over one square degree.

Most stars, especially the young, massive ones, open clusters and interstellar matter (see Sect. 14.4.2) are concentrated towards the galactic plane and form the "thin" galactic disk with its spiral arms. It has a radius of about 17 kpc and a thickness of only about 1 kpc, and it surrounds a 5 kpc thick, central galactic "bulge." The galactic disk appears to us as the band of the Milky Way, which is inclined against the celestial equator by 62.9°. A good deal of it is divided along its middle by dark, dusty clouds of condensed interstellar medium (ISM) nearest the galactic plane.

All galactic objects of the thin disk orbit the galactic center with a very small inclination against the galactic plane. The center itself is physically marked by a black hole of four million solar masses, seen as a radio source (Sagittarius A) at $\alpha(2000) = 17^{h}45^{m}40.0^{s}$, $\delta(2000) = -29°00'28''$, but hidden in visual light behind thick interstellar (IS) dust clouds. The Sun, at a distance of about 8 kpc from the galactic center, has

Fig. 29. The summer Milky Way makes a spectacular view for an observer in the northern hemisphere, rising from the southern horizon towards the zenith. Photo by Ralf Raab, Germany (wide-angle lens 28/5,6, 45 min exposure)

Fig. 30. With a superwide field-of-view, the Milky Way appears as an edge-on galaxy. Composite-photo by Eckhard Slawik, Germany.

an orbital velocity of 220 km s^{-1} and circles the galactic center in, roughly, 240 million years.

Stars older than about 10 billion years keep larger average distances (≈ 1 kpc or more) from the galactic plane by having more inclined orbits. They form the "thick" galactic disk, while globular clusters (GC) and the very oldest stars of our galaxy (summarized under the term Population II, by contrast to Population I for the thin disk stars) populate the galactic "halo." It has an approximate diameter of 50 kpc and is surrounded by a sparsely populated galactic "corona." The halo density increases towards the galactic bulge, and its objects are the only surviving remains of the very first phase of galaxy formation about 14 billion years ago. They reveal the very metal-poor chemical composition of the early galactic ISM (see also Sect. 14.4.5).

The stellar component of the Galaxy is mostly made of field stars, but a significant fraction of the younger stars is found in open clusters (see Sect. 14.4.4). These disintegrate in the course of some hundred million years from gravitational interaction with their galactic environment as they orbit in the galactic disk. Hence, we may conclude that most, if not all stars, where born in open clusters.

Fig. 31. The Antares region shows IS dust in both ways: as absorbing *dark clouds* and as reflection nebulae around the brightest stars (i.e., in *orange* around *red supergiant Antares*). Photo by Klaus-Peter Schröder, Germany, 2.8/100 mm, 30 min exposure on Kodak Gold 100, push-processed, in landscape format

The glow of gas heated to around 10–20,000 K temperature is found in the vicinity of the youngest galactic clusters and individual hot stars, which emit powerful, ionizing UV radiation. Some of the most notable examples are still-active star-forming regions (see below and Sect. 14.5.3). The photoionized and recombining hydrogen atoms produce strong Balmer line emission. The prominent H_α ($\lambda = 656.3$ nm) emission leads to the red appearance of all emission nebulae in color photographs. But the visual observer rather perceives the blue-green light emitted in the H_β line ($\lambda = 486.1$ nm), because human night-vision lacks red-sensitivity.

If, by contrast, the UV light of a bright star is not sufficient to ionize the IS gas, or evaporate the IS dust surrounding it, then we can observe a "reflection nebula." Here, in the absence of the glow of hot hydrogen gas, the star-light scattered at (or "reflected" by) the dust is visible. A popular example is the faint nebulosity around the Pleiades, but a much easier object is M 78 (see also Sect. 14.5.3). The interstellar dust grains have a typical size of 1 micron (1/1000 mm), which makes them scatter blue light better than red light (see Sect. 14.4.2), hence the typical blue hue of most reflection nebulae. The rare exception of an orange reflection nebula occurs, however, if the illuminating star itself is very red, like Antares (see Fig. 31). In other rare cases (e.g., the flaming star nebula IC 405 in Auriga, Fig. 32) we can see a mixture of reflection and emission nebulosity.

A special class of emission nebulae are planetary nebulae (PN). Either nearly spherical or of a bipolar shape, they are found to surround a "dying" central star. Popular examples are M 57 and M 27. These relatively slowly expanding circumstellar (CS) shells are not to be confused with the chaotic and fast-expanding remnants of supernovae (SN) like the famous M 1.

A specific catalog, which would list all these different galactic objects together, does not exist. However, most visually observable galactic objects, i.e., nebulae and star clusters, have an NGC number; they are listed in the well-known *New General Catalogue*, published in 1888 by Dreyer. In addition, there are a number of extended and faint galac-

Fig. 32. IC 405 and IC 410: reflection nebula mixed with HII emission. Photo by Gerald Rehmann, Austria, Schmidt-camera 9″ f/2, hypersensitized TP6415, RGB 45/90/120 min exposures.

tic nebulae which were discovered on deep photographs. These mostly have an IC number, which indicates they are listed in one of the two supplements to the NGC, the *Index Catalogues*. Younger findings, discovered on the Schmidt plates of the Palomar Observatory Sky Survey, have a POSS identification. A noteworthy photographic collection of galactic nebulae is the more recent *Atlas Galaktischer Nebel* by Neckel and Vehrenberg [25]. Case studies of a number of galactic objects (among others), illustrated with excellent photographs, are offered by Stoyan et al. *Atlas of the Messier Objects* [26].

The great splendor and large variety of the galactic objects provides a never-ending motivation for visual observers and astrophotographers, alike. As different as they are, there is one common aspect: the life-cycle of the galactic mass! This cycle starts with (1) the cold, dark molecular clouds of the ISM, which eventually undergo gravitational collapse and fragmentation. The next stage (2) is a star-formation region of cool dust and hot gas mixed around a still partially enshrouded, newly born star cluster. On average, about 10 $M_\odot$ are converted into new stars every year in the whole galaxy. As then open cluster (3) age, they disintegrate and release their members into the galactic disk to become ordinary field stars (4). Most stars, after having exhausted their central hydrogen, become red giants (see Sect. 14.1.5) and develop CS shells of cool gas, in which dust particles can form. Because of the very low surface gravity, a considerable fraction of the stellar mass is lost to the ISM in form of a "cool stellar wind." Finally, the extremely hot stellar core becomes exposed, and its intense UV radiation turns the surrounding CS shell into a PN (5). Only the most massive stars, those with over 8 solar masses, recycle their material by a more spectacular process—a SN explosion.

14.4.2 The Interstellar Medium

Already a naked-eye view of the Milky Way reveals the most prominent structural characteristics of our host galaxy. Most of the cool, dusty interstellar (IS) matter appears to be concentrated in dark, opaque clouds. The smaller ones, around 10 pc in size and ≈ 10 to $100\,M_\odot$ in mass, contrast like blots of black ink against their starry background. They are, typically, $100\times$ more dense than the average cool IS matter (see below). Over 200 years ago, the famous Sir William Herschel mistook these prominent dark clouds for voids in the Milky Way. On the other end of the scale, over 100 pc large and with masses of up to several thousand $M_\odot$, we find huge molecular clouds like OMC-1 in Orion. The latter covers a region of over $10°$ in our winter sky. The well-sheltered interiors of these large IS clouds cool down to only 10 K, and the only observational insight comes from a multitude of low-energy lines at IR, mm and radio wavelengths of more than 60 different, well-detected molecules. A prominent example is the 2.6 mm line of the very abundant CO molecule. Hydrogen, too, mostly assumes its molecular form of H_2 in these cool molecular clouds.

However, dark IS clouds do not occupy more than a few percent of the galactic disk volume. Rather, there is a lot of widely distributed cool IS matter, which is not so obvious at first glance. More exposed to the stellar radiation field than are the interiors of the molecular clouds, the distributed cool ISM typical temperature lies higher, around 100 K. Its average density amounts to only about $1\,\mathrm{atom/cm^3}$, with only one dust particle of ≈ 1 micron in a cube as large as 50 m on one side!

Nevertheless, over the vast interstellar distances, dust-related IS reddening (extinction $E(B–V)$) and absorption (A_v) of visual light both reach considerable values. Extinction results from selective absorption and is well-described by the Mie theory of light, scattered at particles comparable in size to the wavelength. It shows that the probability of an interaction increases towards shorter wavelengths (i.e., for blue light) and, as a result, a larger fraction of red light can pass the IS dust grains in its way. That changes the observed colors B–V of stars: $B–V = (B–V)_o + E(B–V)$. For open clusters, where $E(B–V)$ should normally be about the same for all stars, it can be determined from two-color diagrams (see Sect. 14.1.6).

Along radial lines of sight through the galactic disk, the IS absorption (A_v) amounts to many magnitudes in visual light, blocking most or all light of any extragalactic object. This leads to the apparent, nearly complete absence of galaxies in the Milky Way fields. On average, the A_v is about $1^m/\mathrm{kpc}$ distance, with an extinction of $E(B–V) = \approx A_v/3.2$. However, there is a large variation in A_v in the Milky Way, even on small angular scales.

Despite these measurable effects of the IS dust component, it in fact amounts to only about 1% of the IS mass. It follows that 99% of the IS matter consists of gas, mostly hydrogen! But IS gas cannot be traced so easily. There is an IS line absorption in stellar spectra, most notably the sharp IS components in the NaI (neutral sodium) D-doublet at 589 nm. But this observation requires very high spectral resolution (i.e., several 10^4 or more), normally not available to small observatories.

An excellent way to study the cool hydrogen gas in the Milky Way, which these days may be in reach of advanced amateur and college observatories, is offered by the 21-cm line radio emission. This rare transition in the fine-structure of cool, neutral hydrogen was first predicted from theory in 1944 by van der Hulst and then detected in 1951. Despite a very low probability of this transition, a sufficiently large signal is obtained by the sheer amount of galactic hydrogen, but without the line being self-absorbing. This allows radio astronomers to look through the whole galactic disk, and to identify different spiral arms by line contributions with different radial velocity. These characteristics make the

From our point-of-view, about 8 kpc away from the galactic center, wide-angle photos of the Milky Way show an edge-on spiral galaxy.

Young stars and the interstellar matter (ISM) are sharply concentrated towards the galactic equator and map out the spiral arms of the galactic disk.

The very different types of galactic objects document the mass life-cycle in every galaxy: stars are formed from ISM and later, stellar mass-loss is fed back into the ISM.

21-cm line an excellent marker of the spiral structure of our galaxy. Its detection requires a paraboloidal antenna of 3 m or more (i.e., a large, old and retired satellite dish), provided a sensitive receiver with a good signal-to-noise ratio is available (see Chap. 9 for an introduction into the techniques of radio astronomy and first, simple applications).

But about three-fourths of the IS volume is *not* filled with cool IS matter. Instead, this space is occupied by hot (50,000 to half a million K) to superhot (several million K) and very thin IS gas. In fact, the Solar System happens to be located in such a "hot bubble." The two very different temperature regimes of the IS medium must assume a pressure balance to coexist. And indeed, with $P \propto \rho \cdot T$, the cool but dense IS regions have about the same pressure P as the very hot (large T) but very thin (small ρ) IS gas in between.

The hot IS gas is heated by shock waves from SN explosions and replenished by the hot winds of blue supergiants and WR stars (see Sect. 14.4.3). Under these conditions, all IS dust is evaporated. That leaves the hot IS regions absolutely transparent and causes its large directional variability. It even creates a few "windows": openings in the veil of IS absorption with a relatively free view of the space beyond the galactic disk—like the direction towards the galaxy NGC 6939, at the border of Cygnus with Cepheus. Other good examples are the famous "Baade's windows," which give glimpses of the galactic bulge.

The global energy density of the galactic IS medium is composed almost in equipartition of (1) the stellar radiation field, (2) the kinetic energy by the motion of individual clouds, (3) the cosmic rays and (4) the galactic magnetic field. It is quite remarkable (but still lacks explanation), that these so very different sources of energy contribute nearly evenly—on the order of 1 eV per cm^3 (1.6×10^{-4} J/km^3) each—to the global IS energy density.

14.4.3 Galactic Nebulae

Certainly, the galactic nebulae show the widest variety in shape, size and appearance of all galactic objects. As mentioned in the previous section, the interstellar medium can assume very different temperatures, and a nebula can either show scattered star light (reflection nebula), or be glowing by the emission of hydrogen recombination lines (emission nebula), caused by photoionization from intense stellar UV radiation. Furthermore, we find very different individual types of galactic emission nebulae, which each represent a very different state of the galactic mass cycle. And the various forms of IS clouds prior to forming a star demonstrate nicely the process of concentration and fragmentation: the smaller the cloud, the denser it is (for details, see below).

In this section, we focus on those object categories, which qualify for visual observation and deep-sky imaging, approximately in order of a decreasing average physical size. Hence, with only a small telescope, any student or amateur observer or astrophotographer can for herself or himself collect vivid evidence of the dynamic processes in the Galaxy and the mass cycle mentioned in Sect. 14.4.1. Table 4 summarizes the physical quantities of several representative objects mentioned in this section.

For the most extended galactic nebulae, small telescopes (or binoculars) with short focal lengths and a wide field-of-view are the obvious choice. A very dark rural site and a night with good transparency are the most important requisites for a successful visual observation. To cope best with the very low surface brightness of extended galactic

The cool, dusty component of the ISM is noticed by its absorption and reddening extinction, and in the form of reflection nebulae.
Large regions of very hot and thin IS gas separate the cold, dark and molecular clouds and provide "windows" into the veil of IS absorption.

Table 4. Selection of representative galactic nebulae (*HII* HII region, *HB* hot bubble, *SNR* SN remnant, *SFR* compact star-formation region, *RN* reflection nebula, *PN* planetary nebula), their EP2000 equatorial coordinates, apparent size of the object (in brackets: size of whole complex), and fundamental physical quantities (distance d and diameter D)

Object	Type	RA (h m)	δ (o ′)	Size	d/pc	D/pc
NGC 281	HII	00 53.0	+56 37	30′	3000	25
NGC 1435	RN	03 46.3	+23 57	30′	130	1
M 1	SNR	05 34.5	+22 01	6′ × 4′	1900	3
M 42	cSFR	05 35.4	−05 27	90′ × 60′	400	10
IC 434	HII	05 40.9	−02 28	90′	400	10
M 78	RN	05 46.7	+00 03	8′ × 6′	400	1
NGC 2437-39	HII	06 31.9	+04 56	70′	1500	30
NGC 2261	RN	06 39.2	+08 44	2′	800	0.5
M 8	cSFR	18 03.8	−24 23	45′ × 30′	1300	17
M 17	cSFR	18 20.8	−16.11	30′ × 20′	1800	15
M 57	PN	18 53.6	+33 02	1′	700	0.25
M 27	PN	19 59.6	+22 43	8′ × 4′	350	0.8
NGC 6888	HB	20 12.1	+38 21	18′ × 13′	1400	7
NGC 6992	SNR	20 56.4	+31 43	60′ (2.7°)	770	36
NGC 7000	HII	20 58.8	+44 20	120′ (5°)	600?	50
NGC 7293	PN	22 29.6	−20 48	16′ × 12′	200	0.9
NGC 7635	HB	23 20.8	+61 13	15′ × 8′	2200	9

nebulae, some observers seek remote mountain sites. Altitudes around 2000 to 2500 m (6000 to 8000 ft) are the best choice for avoiding the haze of the lower atmospheric layers without experiencing reduced eye sight, as caused by the lack oxygen at higher altitudes. Modern photography of emission nebulae, however, employing a cooled CCD and narrow-band filters , can produce astounding results even under suburban skies (read more in Chap. 6).

The home-bound visual observer will appreciate the high surface brightness of compact emission nebulae (some PNs in particular). With the use of a narrow-band filter for visual observation, known as UHC or OIII-filter, some of these objects withstand urban light pollution very well. The band-pass of an UHC filter is centered on the green [OIII] emission line (λ = 500.7 nm, characteristic for the denser gas in PNs, see below), and most transmission curves include the H$_\beta$ line (λ = 486.1 nm). However, the compact size of such surface-bright PNs demands higher visual magnification, and their photography requires a longer focal length and very accurate guiding. Hence, access to a larger, well-mounted telescope is a big advantage here.

14.4.3.1 HII Regions HII regions are extended hydrogen regions of low surface brightness, which are glowing predominantly in the light of Balmer emission lines. Best examples are the North America Nebula (Fig. 33), its neighbor the physically connected Pelican Nebula, the California Nebula, and the rather difficult to observe glow, which underlines the dark cloud that forms the famous Horsehead Nebula (IC 434 in Orion, Fig. 34). In these cases, extended IS clouds of moderate density are in reach of the photoionizing UV radiation of a powerful young star, mostly of a very early spectral type (O9–O5).

Fig. 33. The popular North America Nebula (NGC 7000) is a prototype galactic HII emission nebula. Photo by Norbert Mrozek, Germany, 4″ f/4.5 Newtonian, DSLR, exposure 3 × 30 min at 800 ASA

Most of these nebulae represent the the wider neighborhood of a star-formation region, like NGC 7000, or a part of it, like IC 434. Other objects are just less compact star-formation regions, like NGC 281 in Cassiopeia (Fig. 35) and the Rosetta Nebula (NGC 2437-39) around the star cluster NGC 2244. The latter extends over about 30 pc and contains gas and dust of several thousand solar masses. There is no physical difference but a gradual transition to the more compact star-formation regions discussed below.

The energetic UV photons beyond $h\nu = 13.60\,\text{eV}$ (below $\lambda = 91.2\,\text{nm}$) of the young star(s) keep most of the hydrogen in the HII region ionized and heat its gas. At the same time, due to free electrons being attracted to the hydrogen ions by Coulomb forces, there is a permanent recombination rate. The recombined hydrogen atom is then, usually, still in an excited state (i.e., energy level with $n \neq 1$, see Fig. 36). But within a short decay time of $\approx 10^{-8}$ s, one or several photons are emitted and the hydrogen atom reaches the ground level, until it will again be ionized. Very often, one of the Balmer-line transitions down to the intermediate energy level of $n = 2$ becomes part of the recombination cascade. That contributes to the line emission of the HII region in, e.g., the red H_α line ($\lambda = 656.3\,\text{nm}$) and blue-green H_β line ($\lambda = 486.1\,\text{nm}$).

Fig. 34. Popular but faint Horsehead Nebula IC 434. Photo by Bernd Koch, Germany, C11 at f/6.1, exposure 120 min on hypersensitized Fujicolor SHG 400

Photons generated by transitions directly down to the ground level ($n = 1$), forming the Lyman series (see Fig. 36), are so energetic that they are all part of the UV spectrum. On the other hand, transitions between higher energy levels (of bound electron states) of the hydrogen atom have much smaller energies (e.g., the Paschen series down to level $n = 3$). These and the respective frequencies are easily calculated in terms of the numbers n_l, n_u of the lower and upper level involved in the transition:

$$h\nu = \Delta E = (1/n_l^2 - 1/n_u^2) \cdot 13.60\,\text{eV} \ .$$

These transitions are found in the IR and (between bound electron states with large n) in the radio spectrum; the Balmer series presents the only hydrogen lines in visible light.

The size of an HII region is either limited by the reach of the ionizing radiation, or by the extent of the IS cloud, or both. Typical values range from about 10 to about 50 pc, typical densities from 10 to 100 atoms/cm^3. The line emission of the gas in the HII region is its main cooling mechanism, which balances the heating by photoionization. In addition, the free electrons in the HII region produce a noticeable radio-continuum emission.

Fig. 35. NGC 281 in Cassiopeia. Photo by Andreas Röhrig, Germany, R200SS, MX916 cooled CCD, LRGB, total exposure-time 135 min

14.4.3.2 Compact Star Formation Regions In contrast to extended HII regions, this category represents the compact star-formation region around an emerging, very young star cluster. These bright emission nebulae result from a compact IS cloud of several hundred to a few thousand solar masses, which recently suffered a gravitational collapse—often triggered by the shockwave from a nearby supernova.

But even without external triggering, any IS cloud becomes unstable to gravitational collapse, if it exceeds a critical mass M_J (the Jeans mass) which depends on the gas temperature T and density ρ. Alternatively, a stable cloud of given density and mass can become unstable, if it cools below the related critical temperature. The relevant relation was derived by Jeans almost a century ago, comparing the (stabilizing) gas pressure with the internal gravitational forces of the IS cloud. In convenient units, the Jeans criterion reads:

$$M_J > \left(\frac{T}{50\,\mathrm{K}}\right)^{3/2} \cdot \left(\frac{\rho}{10^{-23}\mathrm{g\,cm^{-3}}}\right)^{-1/2} \cdot 10^3 M_\odot \ .$$

The subsequent process of fragmentation and star formation runs fastest with the most massive stars, which form within about a million years. Hence, in the present day, these

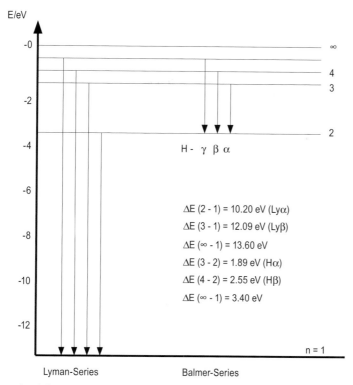

Fig. 36. Energy level diagram of hydrogen and its most common transitions of the Balmer and Lyman series

The diagram shows:

E/eV axis with markings -0, -2, -4, -6, -8, -10, -12

Energy levels: ∞, 4, 3, 2, n = 1

H - γ β α

$\Delta E \,(2 - 1) = 10.20 \text{ eV (Ly}\alpha)$

$\Delta E \,(3 - 1) = 12.09 \text{ eV (Ly}\beta)$

$\Delta E \,(\infty - 1) = 13.60 \text{ eV}$

$\Delta E \,(3 - 2) = 1.89 \text{ eV (H}\alpha)$

$\Delta E \,(4 - 2) = 2.55 \text{ eV (H}\beta)$

$\Delta E \,(\infty - 1) = 3.40 \text{ eV}$

Lyman-Series   Balmer-Series

Fig. 37. M 8, the Lagoon Nebula. Photo by Klaus-Peter Schröder, Germany, 10″ Newtonian at f/5, Hutech-modified Canon EOS 300D, exposure 6 × 1 min at 1600 ASA

now provide an intense UV radiation field, which photoionizes the surrounding nebulosity. Hydrogen gas, often glowing quite intensely in the Balmer lines (see Fig. 36), is mixed with dark structure—unconsumed remains of the formerly cool, dust-rich dark molecular cloud.

Most prominent examples are the Lagoon Nebula M 8 (Fig. 37), the Omega Nebula M 17, and the Orion Nebula M 42, with diameters between 10 and 20 pc. These compact star-formation regions have gas masses between several hundred to several thousand solar masses, and gas densities here reach several hundred to several thousand atoms/cm^3. A more evolved stage is represented by M 16, where a lot of the gas has already been expelled by the radiation pressure of the young star cluster NGC 6611.

The radio intensity $I_{cSFR,r}$ received from such a compact, star-formation region (cSFR) is $I_{cSFR,r} \propto V_{cSFR} \cdot n_e^2$ (with n_e being the density of the free electrons). Hence, for radio astronomers, $I_{cSFR,r}$ of a cSFR is a measure of the electron density inside—as long as the distance and, hence, the true volume V_{cSFR} of the nebula is known.

The fragmentation process within a compact star-forming region creates very small (<0.1 pc), compact, dense and dust-rich protostellar clouds and "proplyds," protostellar dust disks. These objects each contain an embryonic star and are so dense (10^4 to 10^5 atoms/cm^3) and opaque (at optical and UV wavelengths) that their inside conditions are in stark contrast to the surrounding energetic UV radiation and hot gas. Nevertheless, smaller examples may be facing evaporation.

In general, the dust inside protostellar clouds cools very efficiently by sending out far-IR radiation and so maintains a low temperature around 100 K. Hence, these objects are detected as far-IR bright dots with dedicated satellites, especially in M 8 and M 42. In addition, some foreground examples appear as dark blobs against their bright emission nebula background in deep exposures in optical wavelengths, and are in reach of modest observatory equipment.

In the formation of an individual massive protostar or protobinary, abundant angular momentum together with hot stellar radiation and mass-loss result in a bipolar outflow during the final stages of star formation. The outflow has an orientation perpendicular to the opaque dust-disk around the protostar and creates a small (<1 pc) bipolar nebula, a Herbig–Haro (HH) object. The outflowing gas is excited and ionized by the very young, massive and hot central star. In some cases, the backward outflow is hidden by the opaque dust-disk and we see only the facing outflow in form of a cometary nebula. Today, a larger number of HH objects is known, often found in the outskirts of large, extended galactic star-formation regions.

14.4.3.3 Bubbles and SN Remnants There are a few very pretty, visible examples of hot, round bubbles in the process of formation, which are filled with the thin, hot IS gas mentioned (see Sect. 14.4.2). Such a bubble has been blown up over hundreds of thousands of years by the hot wind of a Wolf–Rayet star. This most extreme form of a very hot, massive and luminous blue supergiant with 30 to 60 $M_\odot$ develops a highly ionized wind, several 10^5 K, with velocities of about 2000 km s^{-1}. A quite considerable mass-loss rate can remove a substantial fraction of the giant's mass within just a few hundred thousand years.

Well-known objects are NGC 6888, about 7 pc in size, surrounding WR 136 (Fig. 38), and the Bubble Nebula NGC 7635 (Fig. 39, about 9 pc diameter). The visible shell is made of the IS material swept up and compressed by the hot winds. It has an intermediate temperature of around 20,000 K, at which it radiates in the same recombination lines of hydrogen (and other elements) as ordinary HII regions do (see above).

Fig. 38. The gas shell of NGC 6888 was formed by the hot wind of a superluminous Wolf–Rayet star. Photo by Wolfgang Promper, Austria, 4.0/300 mm telelens, SXV-H9 cooled CCD, H_αRGB, exposures 180/15/15/15 min

Fig. 39. The Bubble Nebula NGC 7635. Photo by Wolfgang Promper, Austria, C8 at f/6.5, SXV-H9 cooled CCD, H_αRGB, exposures 180/20/20/20 min

Remnants of violent SN explosions, on the other hand, show hot stellar debris of several solar masses, driven out violently and chaotically, with truly large expansion velocities of 10,000 to 20,000 km s^{-1}. Very hot (several millions of degrees K) gas, detectable with X-ray satellites, is mixed with cooler and denser filaments, which are visible in optical line emission. In addition, relativistic (very fast, $v \approx c$) electrons spinning in the strong magnetic field of the SNR (SN remnant) emit synchrotron radiation. Its characteristic spectral distribution $I_\nu \propto \nu^{-0.6\pm0.3}$ can be observed in radio frequencies, in some strong cases (i.e., a compact SNR) even in the optical spectrum.

A famous example of a young and compact ($\approx 3\,\mathrm{pc}$) SNR is M 1, related to the SN reported by Chinese astronomers in the year 1054. This object contains a 16^m pulsar (the fast-spinning neutron star remnant of the collapsed supergiant) which is in reach of visual detection by $16''$ to $20''$ telescopes. Modern CCD photography and digital image processing has put it in reach of amateur or college observatories to measure the expansion rate of this SNR. The difference image of two high-resolution photos of M 1, taken with the same instrumentation and a time-difference of about a decade, reveals slight displacements of the expanding filaments.

An aged and very expanded example of a SN remnant is, by contrast, the Veil Nebula in Cygnus (NGC 6992 and other fragments). In this case, a considerable fraction of the visible nebula already consists of swept-up, cool IS matter. The Veil Nebula has now reached a physical size of 38 pc and is about 15,000 to 20,000 years old. A lot of mechanical momentum and heating is injected into the ISM by SN remnants and the hot winds of Wolf–Rayet stars. In large star-formation regions, the combined effect of many young, hot supergiants and a large SN rate can form the source of a hot galactic superwind, which emerges from the galactic disk like a fountain.

Fig. 40. M 78 (NGC 2068, here shown with NGC 2071/2067/2064) is one of the brightest reflection nebulae in our night sky. Photo by Andreas Röhrig, Germany, R200SS, MX916 cooled CCD, LRGB, total exposure-time 135 min

Fig. 41. Hubble's Variable Nebula (NGC 2261) is illuminated by the star embedded in its tip. Photo by Wolf Bickel, Germany, 60 cm f/5 Newtonian, THX 7395 CCD, RGB, exposures 1.5/9/10 min

14.4.3.4 Reflection Nebulae The same Mie-scattering, which preferably removes blue photons from the line of sight through thin IS dust (see Sect. 14.4.2), can be seen in action in cool IS clouds, which are not too dense and envelope stars that are not hot enough to ionize the gas around them, or evaporate (yet) the dust of the IS cloud. Typical sizes are 1 pc. Under these specific physical conditions, we can see the scattered star light in emission against a dark background, normally with a blue tint.

Prominent examples are the extended nebulosity around the Pleiades, in particular NGC 1435 around Merope—a thin IS cloud in an accidental encounter with the cluster— and the much denser cloud of M 78 (Fig. 40), scattered remains of the IS cloud which only recently (less than a few million years ago) created the blue, luminous, young stars in the center of M 78.

Of course, there are also examples of close coexistence of an emission nebula with a reflecting nebula in one and the same IS cloud complex, like M 20. And the above-mentioned reflection nebula around the red supergiant Antares defies the rule by not having a blue tint; rather, it is of a yellow-orange color. In this case, the light of Antares is too red to be turned into a blue hue by Mie-scattering.

A very peculiar example of a reflection nebula is Hubble's Variable (NGC 2261). Here, circumstellar (CS) material enshrouds the star R Mon (see Fig. 41), and its appearance shows some variation on timescales of years and decades. This is, in fact, only a projection onto the outer, sun-facing CS envelope: Patterns of stellar light and shadows are cast by inner, opaque dust clouds, which orbit R Mon on timescales of 10 years.

14.4.3.5 Planetary Nebulae Planetary Nebulae (PNs) are a very special kind of galactic nebula, which come with a fascinating variety of shapes. Many PNs are quite compact (with diameters typically under 1 pc) and have a high to moderate surface brightness, which makes them easier to observe under light-polluted conditions. But only a few PNs like M 57, the popular Ring Nebula (Fig. 42), actually exhibit a spherical shape and have

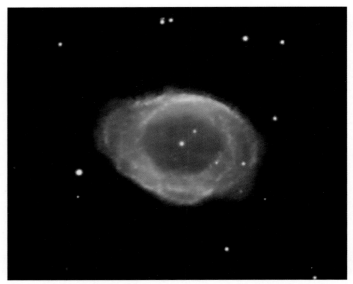

Fig. 42. M 57, the Ring Nebula. Photo by Stefan Heutz, Germany, C11, SXV-H9 cooled CCD, LRGB, exposures 28/6/6/6 min

Fig. 43. M 27, the Dumbbell Nebula. Photo by Peter Bresseler, Germany, C14 at f/7, ST9E, RGB 1:1:1.8, total exposure 1100 sec

a remote resemblance of a planetary disk. In fact, there are many more examples with an obvious bipolar shape, like the popular Dumbbell Nebula M 27 (Fig. 43). Altogether, about 3000 galactic PNs are known, about 200 of which are well-studied.

The compact form and high surface brightness of PNs correctly indicates gaseous shells of a relatively high density, in the range of 10^2 to 10^4 atoms/cm^3. With temperatures around 10^4 K, these gas densities generate some characteristic emission lines in addition to the Balmer lines, which are so common with all galactic nebulae (i.e., HII regions, see above). Many of the specific PN emission lines are "forbidden" in a quantum mechanical sense, the most intense example is [OIII] λ 500.7 nm of doubly ionized oxygen. Special narrow-band filters (found under the names UHC or OIII-filter) centered on this line are very efficient in enhancing the contrast between a PN shell and a light-polluted background: an ideal aid to urban observers and photographers. The other intense, forbidden PN emission line, [NII] λ 654.8 nm of ionized nitrogen, is very close in wavelength to the red H_α line. Both are outside (i.e., too red) the spectral sensitivity of human night-vision. All other PN emission lines are either outside the visually perceivable spectrum, or of a much lower intensity, or both, to be of any importance for the visual observation or imaging of PN shells.

PN emission lines are excited by collisions with free electrons in the ionized gas of the PN shell. Some line-strength ratios are very good density indicators and allow direct (i.e., distance-independent!) measurements of the PN gas density. It only requires emission-line spectra of a modest spectral resolution, but a good linear CCD response is needed, as obtained with moderate exposures on non-ABG CCDs (ABG is the antiblooming gate). For a detailed description of emission-line diagnostics in astrophysics, we may refer to the book of Osterbrock, *Astrophysics of Gaseous Nebulae and Active Galactic Nuclei* [27].

The visible PN shell is surrounded by a cool CS (circumstellar) shell, which is observed by the far-IR emission of its dust. The PN itself is excited and ionized by an ultrahot, ultracompact central sdO (subdwarf O) star, the now exposed core of a former tip-AGB red supergiant (for more details on stellar evolution, see Sect. 14.1.5). The cool wind of that supergiant has gradually been building up the outer, cool, CS shell. During the final $\approx 30{,}000$ years of its tip-AGB phase, the mass-loss of the supergiant develops into a cool "superwind." It is driven by radiation pressure on the now abundantly forming dust grains, in combination with an extremely low surface gravity—of about 10^{-5} of that of the Sun. The "superwind" mass-loss rates climb to 10^{-4} $M_\odot$ per year and gradually remove the outer layers of the supergiant. In this way, PNs are testimony to the recycling and reinjection of stellar material into the IS medium.

Eventually, the ultracompact, hot giant core becomes exposed. This is the moment, in which the visible PN is born: The hot ($\approx 20{,}000$ K) UV radiation of the now naked core and some hot, fast (but thin) wind ploughs into the cool CS gas, heats and ionizes it, and makes it glow. At the same time, the outer edge of a PN still expands with the speed of the former cool "superwind," typically 10–30 km s^{-1}, but the inner edge is often found to expand faster, pushed by the hot wind and ionizing radiation from inside.

Density profiles derived from the far-IR emission of the dust in the outer CS shells of PNs reveal the final mass-loss history of the former tip-AGB supergiant, including some fluctuations of the mass-loss rate. From such observations, we also know that most objects in their final ≈ 5000 years have had relatively fewer mass-loss via their poles of rotation than in their equatorial regions. Consequently, the hot, fast wind is faster in eating through the polar regions of most PN shells, creating a bipolar PN shape, or a hallow cylinder of glowing PN gas. A number of quite extreme cases of bipolar symmetry, however, may be related to an interference with the orbital angular momentum of a secondary companion of the central star.

The Internet is a rich source of excellent astrophotos of all kinds of galactic nebulae. Mostly produced by skilled amateurs, many of these are of professional quality. Here are just a few examples of some "must-see" sites:
http://www.spiegelteam.de/
http://home.tiscali.de/heutz_st/
http://panther-observatory.com/
http://btlguce.digitalastro.net/
http://homepage.univie.ac.at/
peter.wienerroither/pwafo0d.htm

Some professional sites offer public archives of instructive astrophotos, as well:
http://hubblesite.org/gallery/
http://www.eso.org/esopia/
http://www.aao.gov.au/images.html
http://www.noao.edu/image_gallery/
http://www.noao.edu/image_gallery/
WIYN/images.html

Suitable progenitor stars of PNs have initial masses from a little over $1\,M_{\odot}$ to $\approx 8\,M_{\odot}$. The more massive stars end by a supernova explosion, while less massive stars (probably including the Sun itself) do not become luminous enough on their tip-AGB to produce a final "superwind," without which a PN cannot be formed.

With continuing expansion of the PN shell, gas density and PN emissions decrease, since the now diluted intensity of the ionizing UV radiation field from the central star decreases as wel. A good example of an evolved PN is the Helix Nebula, NGC 7293. The expansion process gives PNs a maximum lifetime (i.e., as a visible PN) of about 20,000 years; most objects are about (or less than) 10,000 years old.

14.4.4 Open Clusters

Open clusters of stars are formed in compact star-formation regions in the galactic disk and its spiral arms (for examples see Figs. 44 and 45). After several million years, all remains of the IS host-cloud have been dispersed by the strong UV radiation and hot winds of the most massive, hottest cluster members. Cluster masses (in $M_{\odot}$) and number of cluster stars range from several hundred to several thousand. Usually, the more massive clusters can withstand disintegration for longer. This process is driven by tidal interaction with other objects in the galactic disk (occasionally) and with the galactic center (all the time). It results in a gradual loss of cluster stars across the *tidal limiting radius* of

$$r_{\mathrm{t}} \approx D_{\min} \cdot \left(M_{\mathrm{cl}}/2M_{\mathrm{gal}}\right)^{1/3},$$

which sets an upper limit to the physical cluster radius, depending on the cluster mass M_{cl} (over the mass of the galaxy, M_{gal}) and the minimum distance $D_{\min}$ of the cluster to the galactic center on its (eccentric) orbit. Obviously, once a cluster has lost the majority of its stars, its tidal limiting radius shrinks. That accelerates the cluster losses and eventually dissolves it.

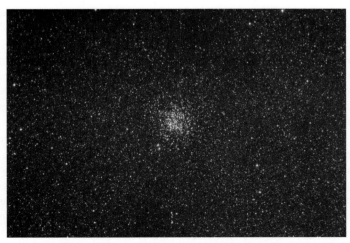

Fig. 44. Rich open cluster M11 (d = 1.9 kpc, size = 7 pc, Trümpler class: II2r). Photo by Andreas Rörig, Germany, R200SS, ST10XME, RGB-exposures 11/10/13 min

Fig. 45. Rich open cluster M 38 ($d = 1$ kpc, size = 5 pc, Trümpler class: II2r) and 3× smaller NGC 1907 250 pc behind it. Photo by Andreas Rörig, Germany, R200SS, Canon 300D, exposure 7 × 5 min

The process of tidal disintegration limits the dynamical age of any open cluster, but its dispersed stars can live much longer, depending only on their mass (see Sect. 14.1.5). The oldest open clusters have reached an age of 2 to 3.5 billion years (e.g., M 67). This is still a far cry from the age of the average globular cluster (see Sect. 14.4.5). On the other hand, ordinary open cluster ages range between only 50 to 500 million years.

Open clusters not only vary in their mass, their age-related composition by bright and faint stars, but also in their apparent compactness, which in part depends on the distance to the cluster. In the hope to derive distances from the visual appearance of open clusters, Robert J. Trümpler set up a classification scheme in the 1930s, which is still in use today, in terms of three parameters for:

1. Degree of concentration: I (strong), II (some), II (none, but cluster still appears detached from background), IV (none, and cluster not well-detached from background)
2. Range of brightness: 1 (most stars have the same brightness), 2 (a medium range of brightness), 3 (some bright stars mixed with many faint stars)
3. Richness: p (poor, < 50 stars), m (medium, 50 to 100 stars), r (rich, > 100 stars)

In addition, Trümpler used standard comments like "e" for "elongated" or "n" for "contains nebulosity." However, the application of his scheme seems to depend too much on personal judgement and telescope size. And with the inherent diversity among open clusters (typical diameters range from 3 to 20 pc), the only use of the Trümpler classification is a characterization of the visual appearance.

It rather requires photometric studies in terms of CMDs (color-magnitude diagrams) and 2CDs (two-color diagrams), assisted by modern stellar evolution models, to reveal the distance, extinction and age of an open cluster (for details, see Sect. 14.1.6). Because of their limited dynamical age, open clusters are usually found close to the galactic plane. Their younger examples are very good markers of the galactic spiral arms. Hence, distance and extinction studies of open clusters are very important for our quantitative, three-dimensional picture of the Galaxy.

Many open clusters are fairly easy targets, even for modest telescopes and sites with some light pollution. Many individual stars are visible already in a modest telescope, and some clusters are well-resolved even with binoculars. Good color perception, or simple color photographs already distinguish the red giants of a cluster from bright, blue or white main-sequence (MS) stars. Furthermore, it may be instructive to compare the personal impression of the cluster (visually, or on photographs) with the Trümpler class stated in catalogs.

14.4.5 Globular Clusters

Only globular clusters (GC) are massive enough to withstand tidal disintegration over the whole lifetime of the Galaxy. By contrast to the relatively short-lived open clusters discussed above, most GCs have lost but a fraction of their member stars since their formation in the earliest phase of the Galaxy. GCs are as old as the galactic halo, and they predate the formation of the galactic disk and most of the enrichment of the IS medium with heavy elements. Hence, there is a distinct gap between globular clusters and the oldest open clusters in mass, age, and in the abundance of heavy elements ("metallicity"). In addition, GCs show a completely different spatial distribution: They form a spherical cloud around (and significantly concentrated towards) the galactic center.

Typically, globular cluster masses range from 100,000 to over one million solar masses, with at least as many stars (see Figs. 46–48). Typical sizes of GCs are 15 to 100 pc, with varying degrees of concentration. The GC ages seem to have a real spread (i.e., apart from the spread caused by the uncertainties of the age determination) from 11.5 to 13.5 billion years. GC metallicities range from a sizeable fraction of the solar abundances (young GCs, formed close to the galactic center, where heavier elements were cooked up in stars first and fastest) to less than 1/200 of the solar metallicity (old GCs, formed far from the galactic center).

The very different age and metallicity is reflected on the properties of GC stars and on the appearance of typical CMDs of GCs, by contrast to those of open clusters. Hence, there are no GC stars left with more than 1 solar mass, due to the large cluster age, and all the brighter MS stars are completely absent. Furthermore, the low-mass GC stars take a very long time to move up on the RGB which, hence, is a prominent feature of the GC CMD, hardly distinguishable in population from the MS. The low metallicities in GCs create quite blue He-burning stars, at the same luminosity ($M_v \approx 0.7$) but at significantly

Fig. 46. Globular cluster M 22. Photo by Herrmann von Eiff, Germany, 6″ Apo at f = 825 mm, Canon 300D

higher effective temperatures than the K giant clump known from old open clusters. In the GC CMD, these mostly blue stars form the "horizontal branch" (HB), and their apparent magnitude is a good, first distance indicator.

At a distance of 7 kpc, at which we see several prominent GCs, HB stars are of the fifteenth magnitude—well within reach of amateur photography. On enhanced color photos, they contrast well with the red RGB giants of similar and higher (up to 3 mag brighter) luminosity. The brightest RGB stars of a GC can already be resolved visually (depending on the distance of the GC) with 4″ to 8″ telescopes, using magnifications of up to 25 times per inch of aperture. In general, good seeing and well-collimated optics are very important factors in resolving GCs.

The dynamical evolution of GCs shows differences with open clusters, too. While most GCs are so massive that they are, despite their old age, comfortably far away from tidal disintegration, they face the probability of a dynamical instability of their core, the "core collapse." About 15% of the galactic GCs appear to have undergone a core collapse: driven by an accelerating rate of close stellar encounters, kinetic energy is transferred from the core into the outer regions of the GC, leaving behind a tightly star-packed core region, about 0.5 pc in diameter. The average distance between two stars in such a collapsed GC core can be reduced to as little as 0.01 pc; hence, it may not exceed much the

Fig. 47. Globular cluster M 13. Photo by Stefan Heutz, Germany, C11 at f/7, SXV-H9 cooled CCD, RGB, exposures 5/3/5 min

Fig. 48. Globular cluster ω Cen. Photo by Klaus-Peter Schröder, Germany, 10″ Newtonian at f/5, Canon 300D, exposure 5 × 40 sec

size of the Solar System! Prominent examples are M 15, M 30 and M 70, but their actual collapsed cores are very small, just a few arcseconds, and only large telescopes can resolve them.

Nevertheless, GCs resemble the open clusters in the shape of the radial star-density profile: In principle, all clusters can be well-described by classical King profiles. The much higher star densities of GCs, however, make them appear much more like spheres, which gave them their name.

Like open clusters, GCs come with a varying degree of concentration. In the late 1920s, Shapley and Sawyer introduced a simple classification scheme of 12 classes, from I for very concentrated towards the core, to XII for extremely loose and scattered. Examples are found for nearly every class.

14.5 Extragalactic Objects

J.V. Feitzinger

14.5.1 Introduction

One objective of this chapter on galaxies is to present a compact survey of the world of galaxies. The other aim is to arouse interest and to give hints for the many possibilities to do successful extragalactic amateur observations.

The building blocks of the observable universe on the scale of several megaparsecs (Mpc) are galaxies, stellar aggregations similar to the Milky Way. The hierarchical structure of the universe continues on an even grander scale as galaxies are ordered together into clusters of galaxies, the latter often connected together in the shape of chains or garlands. Between these chains of clusters of galaxies, there exist enormous voids.

Statistically, galaxies show a very large scatter with respect to their external morphologies, and this is also true of their physical parameters such as masses, diameters, rotational velocities, and luminosities. The energy output sometimes varies enormously from one galaxy to the next, and this fact allows astronomers to distinguish normal galaxies from the energetic active galaxies and quasars.

Galaxies are entities which inhabit the space exterior to the Milky Way system. The distances to them are therefore much larger than to objects within the Milky Way Galaxy. The remoteness of these objects generally means that they are necessarily faint and of small angular extension, a fact that the neophyte observer should keep in mind in order to prevent major disappointments. In the northern hemisphere, the Andromeda Nebula is the most suitable galaxy for the beginning observer, while in the southern hemisphere the Large Magellanic Cloud (LMC) is the best choice.

14.5.2 Catalogs and Pictures

Burnham's *Celestial Handbook* by R. Burnham (1977) [32] is, for good reason, subtitled *An Observational Guide for the Universe Beyond the Solar System*. This work provides a list of galaxies which are accessible to the amateur astronomer. Using data on apparent brightness and apparent diameters, one can decide which galaxies can easily be reached with a given telescope. As a good first exercise, one should try to photograph some of the 39 bright extragalactic objects in the Messier list. In any event, Burnham's Celestial Handbook is one the most detailed lists of nearby northern and southern hemisphere objects. Individual names have been given to only a handful of galaxies, such as the Andromeda Galaxy and the Large Magellanic Cloud. Otherwise, galaxies are known only by a catalog number and name. For example, the Andromeda Galaxy is M 31 (object number 31 in the Messier list), also NGC 224 (object number 224 in the New General Catalogue), where the Triangulum Galaxy is listed as M33, and NGC 598.

The general literature on galaxies and the different aspects of structure, dynamics and evolution can be found in many textbooks and in overwhelming richness on the Internet (search for *galaxies lesson*). Here we offer only some basic hints for text books: Carroll and Ostlie [28], Feitzinger [29], Jones and Lambourne [30] and the compendium *Sterne und Weltraum*, vol. 8 of the Lehrbuch der Experimentalphysik [31].

An updated version of the NGC and IC (Index Catalog) catalogs, last update April 2007, is given by W. Steinicke (http://www.klima-luft.de/steinicke/ngcic/rev2000/Explan.htm#3). A Deep Sky Database is stored at http://virtualcolony.com/sac/.

The most complete printed catalog of galaxies for the northern skies is the *Uppsala General Catalogue of Galaxies* by Nilson (1973) [33], which employs information obtained using large Schmidt telescopes with apertures of 1 m and larger. For the southern skies, there is the *ESO/Uppsala Survey of the ESO (B) Atlas* by Lauberts [34]. The *Third Reference Catalogue of Bright Galaxies* by de Vaucouleurs et al. [35] compiles the data, including types, integrated magnitudes, and diameters, of the 23,024 brightest galaxies.

Picture galleries of galaxies are offered in the following references: *The Hubble Atlas of Galaxies* by Sandage (1961) [36] and the *Revised Shapley—Ames Catalogue* by Sandage and Tammann (1981) [37]. The latter includes, along with the pictorial material, an extensive catalog section, while the Hubble Atlas describes the individual galaxies (189 pictures) in some detail. Galaxies with peculiarities, such as distortions or tails caused by tidal forces, are pictured in H. Arp's Catalogue of Peculiar Galaxies (1966) [38]; Arp and Madore (1987) [39] have also compiled a corresponding catalog for the southern sky. A compilation of printed galaxy atlases and catalogs is given by Madore (1999) [40].

The present chapter could have easily been embellished with spectacular photographs of galaxies gained with giant observatory telescopes. However, it was felt that it would be much more appropriate to provide pictorial material which had been obtained using equipment and techniques which are generally accessible to amateurs. The illustrative portion of the present chapter therefore uses photographs of galaxies obtained only by amateur astronomers using amateur instruments. The reader can easily find many and diverse examples of fine galaxy photographs by leafing through a few issues of the monthly magazines *Astronomy, Astronomy Now, Sky and Telescope, Sterne und Weltraum, Astronomie Heute* and *Astronomie und Raumfahrt*, and others. The beginner should first study some of these volumes and then try to duplicate the practical procedures presented therein. This will certainly spare the observer much disappointment and unnecessary du-

Approximately since the year 2000 the electronic services via the Internet present nearly every galaxy catalog and the most comprehensive galaxy compilations. More than 1.5 million objects are compiled in the Star Atlas Pro-Astronomical Catalog data (http://www.skylab.com.au/astronomical_catalogues.html). The hitherto greatest data base is the NASA Extragalactic Database (NED), freely available for use by the astronomical community (NED, http://nedwww.ipac.caltech.edu). NED is an ongoing project that organizes a broad range of published extragalactic data into a computer-based central archive.

Fig. 49. M 51/NGC 5194 Sbc I–II. 16″f/8 hypergraph ST10XME with Astronomic filter set Type II, red filter RG 610 IR-blocked; LRGB (L=6 × 900 s; R = 3 × 500 s; G = 3 × 650 s; B = 3 × 1500 s; RGB bin2). Photo courtesy of Bernd Flach-Wilken

plication. The illustrations are here compiled in Figs. 49 through 56. Completeness with respect to galaxy types, instruments, photographic materials or electronic devices (CCD cameras) used has not been attempted. These pictures will hopefully provide incentive for the amateur's own work and to illustrate what is currently possible. They are put at

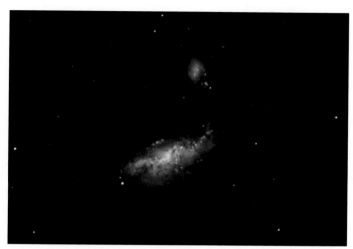

Fig. 50. NGC 4485/NGC 4490 Sb tidal disturbed and Scd III peculiar 15″f/8 Newtonian for luminance and a 16″f/8 hypergraph for color; SBIG ST10XME, AO7 and Astronomic filter set Type II; LRGB (L = 4 × 900 s; R = 3 × 600 s; G = 3 × 600 s; B = 3 × 600 s). Raw FWHM was used for L-channel ~ 2.4″. Photo courtesy of Volker Wendel and Bernd Flach-Wilken

Fig. 51. M 82/NGC 3034 S/Irr activity 15″f/8 Newtonian & 16″f/8 hypergraph; SBIG ST10XME with AO7 and SBIG STL6303 with AO-L; AstroDon filters for colors RGB (L = 5 hours in total (seeing about 2.8–3.0″), Hα = 12 hours in total (6 nm AstroDon Hα filter); R = 6 × 600 s; G = 6 × 600 s; B = 6 × 600 s). Photo courtesy of Volker Wendel and Bernd Flach-Wilken

Fig. 52. NGC 891 Sb on edge 15″f/8 Newtonian ST10XME, AO-7 and AstroDon filter set; LGB (L = 16 × 600 s; R = 3 × 600 s; G = 2 × 600 s; B = 2 × 600 s; RGB bin2). FWHM 1.8–2.1″. Photo courtesy of Volker Wendel

Fig. 53. M 81/NGC 3031 Sb I–II 15″f/5,4 Newtonian with Paracorr corrector; SBIG ST10XME, CFW8. Processed with wavelets and color saturation; LRGB (L = 6 × 600 s; R = 5 × 600 s; G = 2 × 600 s; B = 3 × 600 s; color channels bin 2). Near the lower frame some resolved parts of a Holmberg galaxy are visible. Photo courtesy of Volker Wendel

our disposal by B. Flach-Wilken and V. Wendel, a world-wide leading group in galaxy photography (http://www.spiegelteam.de/).

14.5.3 Classification of Galaxies

The richness of shapes of galaxies has always presented a challenge to observers. How can the various types be compiled into a scheme? The purpose of any classification system is to identify typical features in order to then progress to a deeper understanding of

Fig. 54. NGC 5850/NGC 5846 SBb I–II and SO/E2 Celestron 14 at f/5,9; ST10XME with standard CS filter set; LRGB (L = 6 × 600 s bin 1; R = 2 × 600 s; G = 2 × 600 s; B = 2 × 600 s; RGB bin2). Seeing below average caused the bloated stars, shot in Namibia. Photo courtesy of Volker Wendel and Bernd Flach-Wilken

Fig. 55. M 61/NGC 4303 Sc I–II 16″f/8 hypergraph; SBIG STL6303; LRGB (L = 5 × 1200 s; R = 3 × 450 s; G = 3 × 450 s; B = 3 × 900 s; RGB bin2). Shot in Namibia under very dark conditions. Photo courtesy of Bernd Flach-Wilken and Volker Wendel

the systemic properties. One scheme of classification which is crude, but best suited for introductory purposes, was suggested by E. Hubble, and redefined by A. Sandage and Tammann (1981). It distinguishes three classes of galaxies: *ellipticals, normal spirals*, and *barred spirals*. The scheme is illustrated in Fig. 57.

For elliptical systems, the number n attached to the symbol "E" expresses the ellipticity:

$$n = 10(b - a)/a ,$$

where a and b are the semiaxes of the ellipse.

Fig. 56. NGC 4449 Sm IV Irr 15″f/4,6 Newtonian @ f/8 with APO-Barlow; ST10XME with AstroDon filter set; LRGB (L = 9 × 600 s; R = 3 × 600 s; G = 3 × 600 s; B = 4 × 600 s; RGB bin2). Seeing conditions variety from 2.0–2.2″ during luminance imaging. Photo courtesy of Volker Wendel

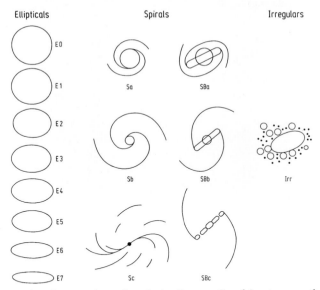

Fig. 57. The Hubble classification scheme for galaxies. Between E and S systems are the transition type S0 galaxies. The sequence a, b, c, d, m from tightly wound to more open, loose, and disrupted structures

For spiral, or "S" galaxies, the letters *a, b, c, d, m* express a sequence of increasing openness of the spiral arms and/or of decreasing brightness of the central body relative to the disk. There is a gradual decrease in the continuity of the spiral pattern from *a* to *m*. When spiral arms are attached to a bar-like structure, the galaxy is classified as an "SB."

Table 5. Apparent relative frequencies of galaxies in galaxy clusters

Type	%
E + SO	60.0
Sa + Sba	10.0
Sb + SBb	12.0
Sc + SBc	8.0
Irr	3.0
Other	7.0

Completing the classification are the irregular systems, which show no rotational symmetry. In between elliptical and spiral galaxies are the SO systems; galaxies so classified contain a disk component but lack spiral structure and significant amounts of interstellar matter ($\leqslant 2\%$).

The luminosity class of a spiral galaxy can be estimated from the morphology of its spiral structure. Systems with well-developed, large-scale spiral arms are assigned luminosity class **I**, and systems with disrupted and ill-defined arms are classed **V**. The latter are also called *dwarf galaxies*. These are, on the average, smaller in brightness, diameter, and mass than normal galaxies by one or two orders of magnitude. Dwarf galaxies also exist among elliptical systems. The frequency percentages of the various galaxy types in galaxy clusters are shown in Table 5.

Interacting and Peculiar Galaxies The class of irregular galaxies is not to be confused with peculiar galaxies. Peculiar features could include warping of the systems, ring structures, matter bridges, or matter ejection in the form of gas clouds. Such features may also be identifiable in certain wavelength ranges, for instance, the radio luminosity is unusually high at the centers of some galaxies. These peculiar galaxies, also called active galaxies, including objects such as quasars, Seyfert galaxies, BL-Lac objects, and radio galaxies, is somewhat flexible. The physical causes for the departures from the average (see Sect. 14.5.6). However, the limits of the classification presented here are properties of the corresponding galactic systems involving internal instabilities and star-forming processes on large scales.

Tidal actions can also cause such peculiarities in galaxies. Distortions, warps, bridges, tails, or the triggering of bursts of vigorous star formation are mutually generated in pairs or multiples of galaxies. Galaxies undergoing tidal interaction show an enormous variety of shapes, and the reader should consult atlases of galaxies [38, 36] to get an idea of the many diverse features of disturbances.

Close groups of galaxies provide a natural laboratory for experiments on tidal interactions. Observations of these phenomena and attempts to explain them provide insight into the workings of these multicomponent star systems. Figure 49 shows the interacting system M 51 (NGC 5194/5), while Fig. 50 shows NGC 4485/90. Enormous amounts of interstellar or intergalactic matter have disturbed the galaxy M 82/NGC 3034 (Fig. 51); it is currently being debated whether or not matter has recently been ejected by violent star formation and supernova explosions.

14.5.4 The Structure of Galaxies

A galaxy is composed of stars, interstellar matter (gas and dust), high-energy cosmic particles (electrons and protons) tied to interstellar magnetic fields and dark matter. These components organize themselves via equilibriums between attractive and centrifugal forces, and according to the energy densities of the cosmic magnetic fields, galactic rotation, the turbulent motion of matter, and the entire radiation field. The three galactic constituents, interstellar matter, cosmic particles, and magnetic fields are distributed primarily in the plane of symmetry, the so-called disk, of spiral galaxies. A disk component is distinctly absent in elliptical systems, although they may well contain substantial amounts of interstellar matter. The dark matter occupies the halo of the galaxies.

Most galaxies contain a distinct nucleus. It can be the locus of active phenomena such as mass ejection, strong radio radiation, or X-ray emission. The nucleus is embedded in a central bulge, onto which the disk can be superimposed. Elliptical galaxies, lacking a disk component, resemble in shape the central spheroidal structures of spirals. Spiral galaxies may be illustrated by imagining a discus with a small spheroid the size of an egg clamped onto its center; a sand grain representing the nucleus lies at the center of the spheroid. The discus-shaped part can itself be structured by spiral arms. In barred spiral galaxies, the rotation spheroid of the central body is partially replaced by a cigar-shaped spheroid.

Elliptical galaxies are larger than the central spheroids of spiral galaxies; on the scale of the discus-egg-sand-grain model, an elliptical galaxy would correspond approximately in size and shape to an (American) football. Spiral galaxies are evidently also surrounded by a spherical halo. It extends up to 100 kpc from the center of the galaxy. The halo contains mainly Population II stars concentrated in globular clusters and some tenuous gas. The halo is sometimes referred to as "stellar halo" in order to distinguish it from the "dark matter halo." Its existence is inferred from the orbital velocities of stars and gas around the centers of galaxies and from the rotation curves. In this model, its size would be represented by a basketball. The basic components of a galactic system are not composed of stars of equal ages. Stellar populations of various ages, chemical compositions, and velocity distributions are the subunits of multicomponent galaxies. The stellar populations can also be found intermingled with each other. Basically two populations can be distinguished: Population I consists of young stars with ages substantially less than 10^9 years, a mean velocity dispersion of about $20 \, km \, s^{-1}$, and a heavy-element content of less than 4%; also belonging to Population I is the interstellar medium of gas and dust. Population II consists of stars with ages on the order of 10^{10} years, a very low abundance of heavy elements, and a mean velocity dispersion of over $40 \, km \, s^{-1}$. Figure 58 illustrates the population components of a galaxy in schematic form. Population II dominates in the central bulge and, with decreasing strength, in the disk. It is steadily distributed over the entire galaxy, and also contains globular clusters. Population I stars have a discontinuous distribution, but are essentially concentrated in the disk. Stars, dust, and gas may appear preferentially along spiral arms, and the disk may appear to have a dearth of Population I objects around its center. This population has a steeper outward decline than does the older Population II objects.

The fact that different star groups have various ages means that they also possess different average colors. Population I objects are on the average hotter, and hence bluer,

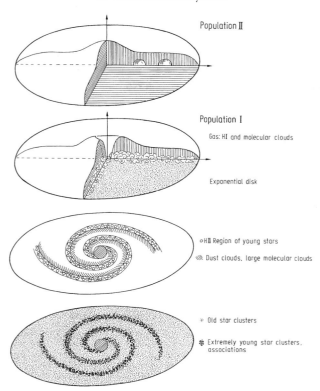

Population II

Population I

Gas: HI and molecular clouds

Exponential disk

o HII Region of young stars

Dust clouds, large molecular clouds

* Old star clusters

Extremely young star clusters, associations

Fig. 58. Cutaway view of a galactic system showing its components. The graphs are to be visualized as symmetric to the central plane. (*Top*) Two parts are hollowed out as spiral arms. (*Bottom*) The uniform distribution of older clusters illustrates the steady transition to Population II objects. Globular clusters, as the oldest Population II constituents are distributed with spherical symmetry with respect to the entire system, in the same way as the dark matter halo

than Population II objects, which are redder. A qualitative separation of the two primary populations can be achieved by observations in different wavelength regions. This can be done by red or blue filters. Radio astronomical methods can register the gaseous interstellar matter.

14.5.4.1 General Properties of Galaxies In order to deduce the properties of galaxies, their distances must first be known. True diameters, masses, and luminosities always depend on distance, as do the corresponding stellar parameters. The apparent angular size of a galaxy on the celestial sphere may be appreciable when the system is located near the Milky Way and very small when it resides far off in the depths of the Universe, irrespective of the true size.

After the distance has been ascertained, the measured quantities for a galaxy must be corrected for absorption and inclination effects. Also, it is necessary to decide up to which isophote of surface brightness the extension of a galactic system should be counted. As a boundary, the Holmberg radius, which is defined as the major axis of the isophote of apparent photographic magnitude +26.5 per square arcsecond, is often used.

With the angular diameter defined and the distance to the system known, the linear diameter is then determined.

The casual observer will certainly find it instructive to make diameter comparisons on a galaxy. Photographs taken on various days under different atmospheric conditions and also using different wavelength regions (using filters) and varying exposure times, will readily show that apparent differences in diameter can be found in a particular galaxy.

The range of linear diameters of galaxies extends from 0.1 to 50 kpc, with spiral systems falling mostly between 10 and 30 kpc, and irregulars between 5 and 20 kpc. Dwarf elliptical galaxies typically have the smallest diameters. The Milky Way Galaxy is a large spiral system with a diameter of about 36 kpc.

Galaxies do not have a sharp boundary. As with the definition of the diameter, the overall apparent brightness of a galaxy is defined in terms of the Holmberg radius. The sum of the individual brightness contributions within this radius forms the total apparent magnitude. Once the apparent magnitude has been determined, the absolute magnitude is obtained with the aid of the distance and two corrections. Absorption within the Milky Way itself must be considered, and the so-called K-correction applied. The latter is meant to guarantee that the same spectral range is always used in determining magnitudes; irrespective of the cosmological redshift a system may have (see Sects. 14.5.5 and 14.5.7). A galaxy's redshift may bring short-wavelength (UV) portions of the spectrum into the range defined for photographic observations, thus altering the isophote.

The absolute magnitude M and the system radius R in kiloparsec are closely related via the formula,

$$\log R = -3.94 - 0.246M .$$

The absolute calibration of luminosity classes (I to V) has the following mean values: V ~ −15, IV ~ −18, III ~ −19, II ~ −20, I ~ −21 in blue absolute magnitudes.

Typical galaxies range between $M = -15$ to -21 in absolute magnitude. The range is larger for elliptical than for spiral systems. Again, the dwarf systems occupy the lower end of the brightness distribution, whereas some large deviations in the higher classes can be explained by peculiar cases. The overluminous giant elliptical galaxies ($M = -3.3$) observed at the centers of clusters of galaxies are explained as mergers of several galaxies.

Brightness and Color Surface brightness distributions can be used to deduce the radial brightness distribution of a galaxy. The patterns of surface brightness follow a more or less universal law of luminosity which is characteristic for both ellipticals and spirals. The occurrence of a uniform law suggests that the star systems are structured according to similar "blueprints." The physical interpretation is that they should be in the same or at least in similar dynamical states. For elliptical systems, the empirical rule is

$$\log I(\rho) = -3.33(\rho^{1/4} - 1) , \tag{33}$$

where $\rho = R/R_e$. R is the radial coordinate and R_e the effective radius of the isophote enclosing 50% of the total light. Then $I(\rho)$ gives the surface brightness measured along the major axis as a function of distance from the center.

In spiral systems the above equation is augmented by a second term

$$I_2(R) = I_0 e^{-\alpha R} , \tag{34}$$

which describes the exponentially declining brightness distribution in the disk. I_0 measures the brightness at the center, and α is the inverse scale of the brightness gradient. The quantity $1/\alpha$ can lie between 1.5 and 5 kpc, depending on galaxy type.

The central bulge of a spiral galaxy is characterized by Eq. (34), which applies to nearly all elliptical systems. All disk systems show an exponential brightness decline in the radial direction, which means that the star population in the disk is always in a similar dynamical state. Within the Hubble classification of galaxies, the ratio of disk to central brightness is used as a classifying criterion. The distance at which the transition from the elliptical to the exponential brightness occurs is a quantitative measure of the light contributions in radial brightness distributions; the more advanced the type of the galaxy (c, d, m), the more inward are the transitions. Brightness distributions in the bars of barred (SB) galaxies follow the $R^{1/4}$ law. Elliptical galaxies also may occasionally show disk structures. These, however, are either small, compact components deeply embedded in the spheroidal body, or else extended disk-shaped shells caused by tidal effects from neighboring galaxies.

From the apparent ellipticities of spheroidal components, the true frequency distribution of flattening can be computed. If a and c are the true major and minor axes, respectively, then the ellipticity is

$$e = 1 - c/a = 1 - q \, .$$

A representative average is $e = 0.36$ for elliptical and $e = 0.75$ for spiral galaxies. The value $q = c/a$ decreases along the Hubble sequence from the E- to the S-type systems, and so does the ratio of brightness of central body to disk. Oblateness is not entirely conditioned by rotation. Since oblateness as a basic dynamical property cannot change on timescales of $< 10^9$ years, such differences between E and S galaxies must have been imposed at the time when galaxies were formed. A galaxy cannot evolve from one type into another, external triggering is needed. Galaxy-type transformation is only possible by the merger of many smaller or a few similar galaxies. The history of a galaxy in terms of the merger events, which have led to the formation of that galaxy, can be represented by a merger tree.

Surface photometry of a galaxy can be done practically in any color range (e.g., U, B and V). Depending on the wavelength chosen, different brightnesses are obtained which characterize the dominance of the different populations. The nuclear regions of spiral systems are reddish in color, as are elliptical systems throughout. From Sa to Sm, the outer parts exhibit a progressively bluer color as a result of stronger star-forming activity in the more irregular-type galaxies. A two-color diagram (Fig. 59) shows the galaxies arranged according to mean colors in a broad band above the typical stellar main sequence.

Stars and Gas The fact that diverse stellar populations can coexist in space indicated early on that the distribution of matter in galaxies is not like a well-mixed dough. The various constituents of different ages, chemical compositions, and motion patterns are found to preferentially inhabit distinctly different regions of galaxies. Of course, galaxies consist of the same parts as the Milky Way. The entire great varieties of objects—normal stars, variables, novae, supernovae, planetary nebulae, clusters and the gaseous constituents—are found in galaxies, although the extreme values of the parameters of these constituents may have systematic differences. One striking example is the set of *young* (blue) globular clusters in the Large Magellanic Cloud.

The integral properties of galaxies are summed from the features of the constituents. The correlation of color indices with the morphological classification (Fig. 59) reflects the different mixture ratios of the populations. Galaxies become increasingly blue from type Sa to Sm, which expresses a steady increase in young stars and their maximum radiation output in the blue spectral range. Also increasing from Sa to Sm is the absolute luminosity of the brightest stars in the spiral arms and the size and number of HII regions. In addition, the percentage of gas and dust of the total mass increases, reaching about 15% to 20% in Sd and Sm systems, compared to 5% to 10% in Sa galaxies. Elliptical galaxies are not totally devoid of interstellar matter, but the mass contribution is less than about 3%.

Interstellar matter is strongly concentrated toward the central plane when the galaxy has a dominating central body. In cases of widely extended arms and the absence of a distinct central part, the plane of symmetry of the galaxy is enveloped in interstellar matter. Neutral atomic hydrogen in S-type systems has a typical extension of more than 2 Holmberg radii. Molecular gas components (CO, H_2) are concentrated predominantly in the optically visible galactic disk. Ring-shaped distributions can also be seen in molecular hydrogen as well as in atomic hydrogen.

More than 90% of the continuous radio radiation is generated within the disk of the spiral galaxy. It is radiated from free electrons in the interstellar medium, with the electrons traveling at high velocities along interstellar magnetic field lines. The magnetic fields are tied to interstellar matter, and it has been established from studies of a few nearby galaxies that they follow the spiral arms. The radio luminosity of the disk and nucleus or central bulge is proportional to optical luminosity.

The 21-cm radio line of neutral atomic hydrogen provides information on the distribution, mass fraction, and motion pattern (rotation curve) of gas, and hence on the

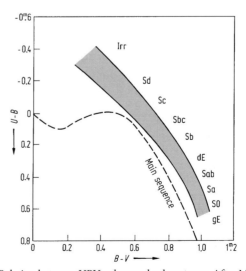

Fig. 59. Relation between UBV colors and galaxy types. After Mitton [41]

internal-motion features of the entire galaxy. Every part of the electromagnetic spectrum yields information on certain parts of the galaxy. For example, infrared radiation may stem from the dust component of interstellar gas, or the X-ray emission from gas at a temperature of several 10^6 K in the central region of the galaxy. Composing all of this information, an overall picture of the galaxy may be stepwise constructed.

An optical spectrum recorded at moderate dispersion from an entire galaxy yields a composite (average) of the contributions of various populations, and, depending on which population dominates, is characteristic for the type of galaxy. Galaxies may be assigned integrated spectral classes resulting from the blending of various spectral types. The sequence of spectral types runs parallel with an increase in younger population components from E to Sm.

Masses and the Mass-Luminosity Ratio As is the case for stars, the mass of a galaxy is a fundamental parameter. The mass distribution then describes the structure and determines the kinematics and dynamics of the system. Mass determinations always require a theoretical model of the system according to which the calculations are laid out. The simplest model is the assumption of a point mass undergoing purely circular motion in the outer part of a galaxy around the total mass concentrated at the center. The actions of gravitational and centrifugal forces then balance and lead to a mass calculation.

Typically more than 70% of the mass of a galaxy resides in *dark matter*. Dark matter can only be detected through its gravitational attraction. Dark matter appears neither to emit nor absorb electromagnetic radiation. Some fraction is made up of baryons, but most is believed to be composed of something else. Dark matter is distributed in an approximately spherical volume surrounding the luminous parts of a galaxy. The luminous parts of galaxies occupy the highest density part of the dark matter halo and are held in place by the gravity of the dark matter.

There are seven methods which can be used to determine the mass of a galaxy: (1) from the rotation curve, (2) from the width of the 21-cm emission of neutral hydrogen, (3) from the velocity dispersion of stars, or (4) of galaxies in a cluster of galaxies, (5) from the motion of a globular cluster or a dwarf elliptical system accompanying a large primary galaxy, (6) from the motion of a pair of binary galaxies and (7) by gravitational lensing. The methods (5) and (6) in principle are the same as used for mass determination in binary stars. The methods of velocity dispersions employ the Virial Theorem, which states that twice the amount of the average kinetic energy equals the gravitational energy when the system is considered invariant in time. The average kinetic energy is then determined from the average velocity dispersion.

Most recently, the mass determination of galaxies has used with high accuracy the correlation between absolute magnitude and total width of the 21-cm emission of neutral interstellar hydrogen. The line profile carries the information on the inner status of motion of the system, which, however, is governed by the mass distribution and, hence, by the total mass of the system. This correlation is also usable for distance determinations.

The rotation curve of a galaxy relates the rotational velocity of the system, as determined by the mass distribution, with the radius of the galaxy. The total mass of the galaxy can be calculated from the rotation curve, assuming circular orbits about the center, and

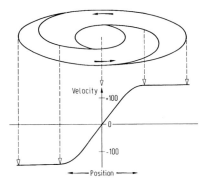

Fig. 60. The schematic rotation curve of a spiral galaxy shows a steep ascent in the inner part and a flat line of constant velocity in the outer part

Table 6. Basic parameters for different types of galaxies

	Sa/SBa	Sb/SBb	Sc/SBc	Sd/Sm	Irr
M in blue magnitudes	−17 to −23	−17 to −23	−16 to −22	−15 to −20	−13 to −18
Mass in solar masses	10^9 to 10^{12}	10^9 to 10^{12}	10^9 to 10^{12}	10^8 to 10^{10}	10^8 to 10^{10}
Diameter in kiloparsec	5 to 100	5 to 100	5 to 100	0.5 to 50	0.5 to 50
V_{max} in kilometer per second	160 to 370	145 to 330	100 to 305	40 to 80	40 to 70
Mass/luminosity (without dark matter)	6.5	4.5	2.5	1	1

by superposing various mass distributions. Figure 60 illustrates the schematic run of a rotation curve. The flat rotation curves of galaxies can only be understood by a spherical dark matter halo.

The *mass-luminosity ratio* combines two physical parameters. The luminosity is photometrically observed and the mass derived, as was stated previously, from kinematical data via a model. The mass-luminosity ratio describes how much luminosity per unit mass is developed in the system. Table 6 lists some typical masses and luminosities according to galaxy type and other basic galaxy parameters. Underlying this connection is evidently a link between both the proportion of interstellar mass and the total luminosity with the type. The more interstellar matter present, the greater the number of stars which can form. Large numbers of young, hot stars in turn increase the blue luminosity. This explanation makes plausible some of the differences between types of galaxies.

14.5.4.2 Formation of Structures in Galaxies The significance of the rotation curve for a galaxy has already been pointed out on p. 679. It is readily understood that the large-scale motion in a galaxy, namely the rotation, has significant influence on the formation of structures, be they bar- or spiral-shaped. Rotation curves of galaxies are largely flat; after a steep increase in the inner part of a galaxy, the rotation speed does not drop but instead levels off to some constant value. This simple observed result has far-reaching consequences. If a galaxy does not exhibit a decline in rotational speed in its outer parts, as would be expected for a Keplerian orbit, an additional mass component must exist. Such a component has thus far not been detected in the optical, infrared, or radio astronomical ranges. This additional mass component is called the *dark halo*, and it affects nearly all properties of a galaxy except its visual appearance.

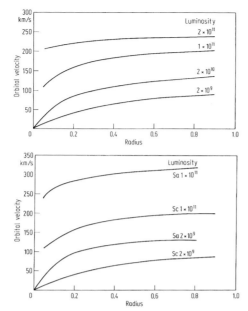

Fig. 61. Schematic rotation curves of Sc galaxies in different luminosities; radii normalized so that the edge of optical disk = 1 (*upper picture*). Comparison of Sa and Sc galaxies. At equal luminosities, the Sa systems show higher rotational velocities. Adapted from Rubin [42]

In a real galaxy, the rotational velocity remains high, the local mass density decreases but slowly, the integral mass continuously increases, and so does the mass-luminosity ratio. This result is caused by the influence of the halo component, which increases radially outward, combined with the brightness decline in the disk. What kind of matter composes the dark halo? Only one thing is currently certain: this matter must have an M/L ratio of 100 or higher.

The maximum rotational velocity of a galaxy, represented by the mean of the flat part of the rotation curve, is correlated with the morphological type of the galaxy. The slower the rotational velocity, the more open the spiral structure. It dissipates in the slowly rotating Sm spirals ($V_{max} = 70$ km s^{-1}) showing only some spiral arm filaments.

Figure 61 illustrates that luminosities for a given spiral type (here Sc) range over a factor of 100, and that the luminosities increase with the rotational speed of the galaxy. A comparison between Sa and Sc galaxies shown that, at equal luminosity, the rotational speed of a Sa system is always higher, so they must contain more mass per unit luminosity and volume than Sc galaxies.

In contrast to spiral galaxies, the structure of elliptical systems (including globular clusters and the central bulges of galaxies) is determined by the velocity dispersion of the individual stars. Elliptical systems rotate nearly not, if at all, than the degree of ellipticity might suggest. V_{max} is less than 100 km s^{-1}. Velocity dispersions of the stars composing the systems differ in the directions of the three major axes. Indeed, theory and observations reveal that elliptical galaxies are being triaxial galaxies with three axes of different lengths. The relation between shape (oblateness), central velocity dispersion σ, and luminosity L is a basic connection of dynamical and photometric parameters in elliptical systems:

$$L \sim \sigma^n \quad 4 < n < 6 .$$

The mean values of velocity dispersions are between 200 and 300 km s^{-1}.

Lacking pronounced rotation, elliptical systems yield only traces of rotation curves, and these are certainly not suitable for mass determinations. In such cases, the velocity dispersion in the center or its change among major axes may be used to calculate the mass. Velocity profiles measured along a major axis can thus serve as a surrogate for the rotation curve and are the most important observed data for describing the dynamics of such systems, and hence for interpreting the structure of elliptical galaxies. The typical decrease in the velocity dispersion from the inner to the outer regions of an elliptical galaxy is by a factor of two.

Spiral Structure Because of differential rotation, any accumulation of matter within the galactic disk experiences a shear. The more distant a star is from the center of a galaxy, the longer is the period of revolution about the center; the inner parts overtake the outlying regions. Shear causes spiral structures to be trailed as a natural consequence of galactic rotation. This fact, however, actually presents a major dilemma: such rotation would quickly wind up any spiral structure so that it would effectively disappear. What mechanism prevents this winding up? Two theories purport to provide an explanation: the density-wave theory, and the theory of self-propagating, stochastic star formation. The latter describes very well the more flocculent spiral structure; the former explains the grand design of spiral arms.

The density-wave theory begins with the assumption that in the disk a spiral-shaped symmetric perturbation of density forms and then rotates rigidly with fixed velocity. This can be illustrated by analogy with a moving construction area on a highway. The flow of cars (stars) condenses in a particular region, and this region slowly moves along with the progress of the construction work. The cause of these jam-ups in the traffic flow (or star flow) is the narrowed road. In nature, an external force (perhaps tidal action of a neighboring galaxy) or instability within the disk itself may create such a mass concentration, which then acts as a region of enhanced gravitational attraction on the surrounding gaseous interstellar medium. The interstellar gas which is pulled into the region will be decelerated and compressed, thus initiating intense star formation in the cool molecular clouds. The outcome, in the form of young stars and HII regions, then manifests itself as a visible spiral arm. Thus it is understood why spiral arms consist so predominantly of young, recently formed stars and cool interstellar clouds (optically visible as dust lanes).

The difficulty of this theory lies in providing an explanation for the origin of the stellar density wave and for its lifetime. As is true of any wave, the density wave propagates at some typical velocity, moving outward until it exits the galactic disk. The spiral structure disappears on a timescale of a few 10^8 years, which corresponds to just one galactic rotation period and needs a permanent stimulation. The density-wave theory requires a perturbation in the disk density as well as an interaction of the interstellar medium with this disturbance in order to explain the spiral structure. By contrast, the theory of self-propagating star formation is much simpler; here, it is the interaction of interstellar medium with the rotation of the galaxy which shapes the structure and produces more chaotic, short spiral arm filaments.

This theory begins with the notion that the formation of structure (the origin of a spiral arm) is linked to the random event of star formation. This idea is observationally founded upon spatially propagating star birth.

The formation of stars is a process of extremely low efficiency. Only a few percent of the mass of a cool interstellar cloud will condense into stars, the rest becoming heated and dissipated by the newly formed hot stars. The leftover gas, which constitutes at least 85% of a cloud, is then shifted into adjacent regions, causing the gas already existing there to condense, triggering more star formation, or it can cool, merge with other clouds, and thus form the birthplaces for later generations of stars. It has previously been mentioned that the angular speed of a galaxy decreases with increasing distance from the center, and that this causes shears. The rotating galaxy necessarily generates, via shears in stellar nurseries, a transient spiral structure. The spiral pattern changes with time as it is hallmarked by continuously forming stars. The pattern itself, however, is self-propagating, as the new stars continue to be born.

Assuming that each star-forming cell is a region with an extent of 200 pc, and imagining that a galaxy is subdivided into such cells, and then each star-forming cell will, over the course of time, come into contact with various other neighboring cells by way of the differential rotation of the galaxy. The spark of star birth then can spill over into the neighboring cells. Intense star formation in one cell increases the probability of such activity in neighboring cells. The probability for generating stars depends, of course, on the contents of the cell, i.e., on its previous history. If the cell has not experienced star formation for a long time, it will contain many cool, dense molecular clouds, and will be quite ripe for star birth activity. If, on the other hand, a cell has recently produced fresh stars, it will require some time to "recover" before the birthing process can begin anew.

The processes described above outline a self-propagating chain-reaction of star formation which changes the contents of a star-forming cell with time. Also, the encroachment into neighboring cells triggers a pattern of structure. This encroachment, physically called a percolation process, enhances the probability that additional star formation will occur.

The spiral arm may appear to possess a filamentary and disjointed structure, and may have a short coherence length. Or, it can form continuously with large coherence lengths and present a large-scale pattern. Coherence lengths are the regions of contiguous connections, and are determined by the dynamics and thermodynamics of the interstellar medium. Not only stellar morphology but also the distribution of the various gaseous components (neutral hydrogen, molecular clouds, and the hot coronal phase of the interstellar medium) can be generated in models and compared with observations. Finally, Fig. 62 displays the time evolution of a spiral pattern [43]. Though the substructures change, the large-scale shape is preserved over a long time.

The overlap of density wave structures and the stochastic behavior may be a solution for interpreting spiral arm features. The density waves generate the large scale features; the stochastically propagating star-formation processes generate the more filamentary and flocculent spiral arm appearance. In the case of Sc–Sm/SBc–SBm galaxies the more stochastically working processes govern the rudimentary spiral structure.

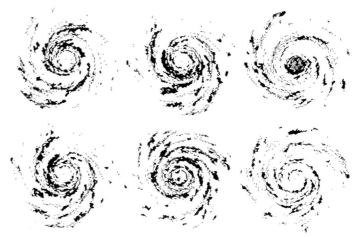

Fig. 62. Evolutionary sequence of a galaxy model at time steps 152, 154, 155, 161, 175 and 231 ($\times 10^7$ year). Star-forming regions are graphed by *dots* whose sizes decrease with age. The *smallest dots* represent a maximum age of 10^8 year

Cosmic Cycles and Energy Equipartitions The basic question raised by a classification of galaxies is what it signifies physically. The effects of two different processes, namely the formation as well as the evolution of galaxies, will here be linked. The basic correlations associated with the morphological classification will help clarify this connection.

The continuously decreasing ratio of central bulge diameter to disk diameter from elliptical to spiral systems is tied to the origin of galaxies. The elliptical central bulges of various extents had very early on undergone star formation, together with the final shaping of the galaxy.

The second primary correlation is the increasing influence of interstellar matter and of young stars upon the sequence of classification. For spirals, the decrease of maximum rotational speed along the sequence provides a further correlation. Here, also, certain initial conditions seem to have been impressed upon the galaxies since the time of their formation.

The difference between E and S systems is the result of certain dynamically different initial states by merging processes which shaped their basic morphologies. Within the framework of these premises this always means first a chemical evolution. The interstellar medium is continually enriched with heavy elements (atomic numbers > 2) as new generations of stars are born and die. As is the case for spiral structure, the shaping constituents continuously change and are rejuvenated. The prerequisite for this cosmic cycle—interstellar matter, star birth, evolution, and death, return of matter into the interstellar medium—is the equilibrium of energy sources and energy reservoirs which govern the evolution of spiral galaxies. The structure of E galaxies is predominantly governed via relaxation by stellar dynamical energy equipartition.

Energy sources include the energies of differential rotation (E_{rot}), of cosmic radiation (E_{cos}), and of stellar radiation (E_{rad}). The interstellar medium harbors three kinds of energy: (1) the thermal energy of gas particles (E_{thr}), described by the kinetic temperature T, (2) the turbulence energy (E_{tur}) of regions moving relative to one another, and ultimately (3) the magnetic energy (E_{mag}) coupled with the hot gas. The sources for the radiation energy are gravitation, which, to begin with, initiates the collapse of a gas cloud to form stars, and nuclear fusion in stellar interiors. The radiation emitted into intergalactic space constitutes the energy loss of a galaxy and the blow out of interstellar material.

Typical energy densities E of the various mechanisms are:

$$E_{rad} = 7 \times 10^{-13} \text{ erg cm}^{-3} = 7 \times 10^{-20} \text{ J cm}^{-3}$$
$$E_{cos} = 10 \times 10^{-13} \text{ erg cm}^{-3} = 10 \times 10^{-20} \text{ J cm}^{-3}$$
$$E_{rot} = 7 \times 10^{-13} \text{ erg cm}^{-3} = 7 \times 10^{-20} \text{ J cm}^{-3} .$$

The energy densities in the interstellar medium amount to:

$$E_{thr} = 4.5 \times 10^{-13} \text{ erg cm}^{-3} = 4.5 \times 10^{-20} \text{ J cm}^{-3}$$
$$E_{tur} = 3 \times 10^{-13} \text{ erg cm}^{-3} = 3 \times 10^{-20} \text{ J cm}^{-3}$$
$$E_{mag} = 3.5 \times 10^{-13} \text{ erg cm}^{-3} = 3.5 \times 10^{-20} \text{ J cm}^{-3} .$$

There exists a striking equality between these energies:

$$E_{thr} \sim E_{tur} \sim E_{mag} \sim E_1$$
$$E_{rad} \sim E_{cos} \sim E_{rot} \sim E_2$$
$$E_1 \sim E_2 .$$

These equalities seem to contain the secret of structure formation and of cyclic processes in galaxies. The interstellar medium is described as a system of energy reservoirs, couplings, feeding and releasing mechanisms (Fig. 63). Stellar and cosmic radiations contribute, via absorption, to the increase in the thermal radiation. The differential galactic rotation feeds energy into the turbulence via the friction between different gas clouds. Turbulence and thermal energy are coupled by way of spatial differences of densities, velocities, and temperatures. Equality is reached through the conversion of turbulence energy, which appears as an increase in the mean thermal energy. On the other hand, local heating (e.g., through radiation or the formation of new stars) causes inhomogeneities in the temperature distribution and hence gradients in pressure and velocity, which in turn increase the turbulent energy.

Magnetic energy can be converted, through turbulence, via an extension of field lines. Conversely, turbulence is generated when field lines are curved so much at the local level that the magnetic pressure exceeds the gas pressure. Of course, energy also escapes from the system via radiation, and it is assumed that, owing to the expansion of the Universe, the photon density in intergalactic space does not increase substantially, because the photons are simultaneously reduced in energy through that expansion. Also gas is blown out into the intergalactic space.

The equality of the energy reservoirs in galaxies (except distorted and active galaxies, of course) and thus the operation of long-term evolutionary and cyclic processes can be achieved in the following way: the time constants of the coupling mechanisms between energy reservoirs must be very much smaller than those in the energy feeding and releasing processes. Turbulent friction and imbalances in temperatures, pressures,

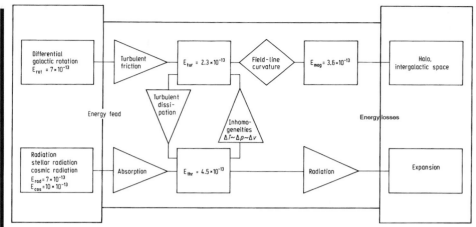

Fig. 63. Scheme of energy balance in a galaxy. The mean energies are given in erg cm^{-3}

and velocities will be smoothed out almost instantaneously, so that the energy reservoirs almost always maintain equal energy contents. If the energy contents of the reservoirs are all equal, there must also be equality in the contents of the energy sources. Per unit time, no more energy can flow into a system than flows out. Equipartition between energy sources is coupled via the origin and evolution of stars. Rotational energy and the rate of star formation seem to be proportional to one another, and star formation provides the energy source of stellar radiation and, ultimately, via the deaths of massive stars, for cosmic radiation. Stars exploding as supernovae feed, through interstellar magnetic fields, the energy source of cosmic radiation. The two named equalities are the most difficult to understand and also the least investigated.

14.5.5 Determination of Distances

A basic prerequisite for an understanding of the structure of the Universe is the determination of distances to galaxies. A typical description of the state of the Universe could be derived solely from the knowledge of cosmic distance scales. In the determination of distances, primary and secondary methods are distinguished. The primary procedures come from studies of the Milky Way Galaxy, and have been calibrated on objects contained within it. Secondary methods employ the extragalactic systems themselves, and have been calibrated on nearby galaxies. The principle of determining distances is to observe a class of objects, either from the Milky Way or from other galaxies, whose brightness or geometrical size is known. This further assumes that the physics of these objects is the same everywhere in the Universe. A system of scale must always be constructed first, for instance by measuring the distances and diameters of globular clusters in the Milky Way. Such a scale can then be linked to the values measured for globular clusters in other galaxies.

The connection between angular extension α and the distance D of an object follows the simple relation

$$D = R/\tan \alpha \, ,$$

where R is the calibrated diameter of the object. The connection with magnitudes is affected by the relation between the apparent magnitude m and the absolute magnitude M of the object,

$$\log D = 0.2(m - M) + 1 ,$$

where absorption effects must be allowed for. At very large distances, a correction depending on the particular model of the Universe employed also enters.

Primary distance indicators can be any of the following:

- The brightest stars (supergiants)
- Cepheids, using their period-luminosity relation, or other kinds of variable stars
- The brightest stars in globular clusters
- Integrated brightness or diameters of globular clusters
- Novae and supernovae
- Planetary nebulae
- Diameters of HII regions

Counted as secondary distance indicators are:

- Overall brightness of galaxy types (large range)
- Brightest member of a duster of galaxies
- Diameters of galaxies
- Luminosity classes of galaxies
- Surface brightness of E galaxies
- Supernovae

A correlation between absolute magnitude and maximum rotational velocity of a galaxy has proven to be a valuable method of distance determination. The width Δv_o of the 21-cm line of the total neutral hydrogen in a galaxy, as measured by radio observations, is proportional to the galaxy's absolute magnitude in blue light according to the relation

$$M_B = -8.2 \log \Delta v_o .$$

This line width again is proportional to $2 V_{\max}$ and thus depends on the galaxy type. Large radio telescopes permit the measurement of Δv_o fairly easily and quickly.

Combined with the knowledge of the apparent magnitude of the galaxy, a distance determination then becomes possible.

The various methods of distance determination have different reliabilities and limiting ranges. A combination of various methods will result in determining dependable distances, which then can be used to fix the value for the Hubble parameter. Figure 64 illustrates the distance ladder.

All of the procedures for finding distances contribute to the determination of the parameter H, named after E. Hubble, which connects the general recessional motion of the galaxies (expansion of the Universe) with their distances. The larger the distance of a galaxy, the higher is the velocity of recession. The measured redshift of spectral lines is interpreted as a Doppler effect due to a receding motion. This velocity v is proportional to the distance D via

$$v = HD .$$

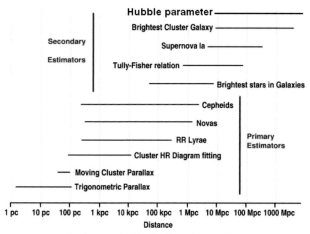

Fig. 64. The distance ladder leads to the Hubble parameter

The parameter H was determined by many methods and is $72\,\mathrm{km\,s^{-1}\,Mpc^{-1}}$.

As the largest cosmic distances are measured via redshift, that is, from the relative line shift $z = \Delta\lambda/\lambda$ from one or several spectral lines in the spectrum of an extragalactic object, it is this latter quantity which is often quoted in the literature instead of the distance; z is well-defined, while the deduced distance depends on the chosen value of H. The relation between the recessional velocity and the distance follows from the Doppler equation in its simplest (non-relativistic) approximation:

$$v = cz,$$
$$v = c\Delta\lambda/\lambda = HD$$
$$z = HD/c\,,$$

with c the speed of light in a vacuum and $z = \Delta\lambda/\lambda$ the redshift. This linear relation holds to $z = 0.25$, approximately 1000 Mpc. At greater distances the relation becomes non-linear.

14.5.6 Active Galaxies and Quasars

The observing methods of amateur astronomers reached in the last years the faint class of active galaxies.

The class of *active galaxies* comprises objects of various types, all of them showing an energy output which cannot be explained by normal evolution and radiation of stars. This energy output is limited to a very small region at the centers of these objects. Among active galaxies counted are Seyfert galaxies, radio galaxies with active nuclei, BL-Lac objects (named after the first-known object of its kind, which was originally classed as a variable star), LINER galaxies (low-ionization nuclear emission line region), and quasars (QSOs, quasi-stellar objects).

A galaxy falls into one of these various group if at least one, but preferably two, of the following criteria apply:

1. A very compact nuclear region, much brighter than the cores of galaxies with the same morphological type.
2. Emission lines in the nuclear region, indicating an origin different from the radiation mechanisms operating in normal stellar atmospheres.
3. Continuum radiation and/or emission-line radiation which is variable with time.
4. Continuum radiation from the nuclear region is non-thermal.

LINER systems show properties (2) and often (1) or (4), while BL-Lac objects exhibit features (1), (3), and (4). Many Seyfert galaxies, radio galaxies, and quasars comply with all four criteria.

Two groups of objects, namely Seyfert galaxies and quasars , will be considered here in more detail. Seyfert systems account for 1% of all bright spirals. The very bright nuclei of these galaxies exhibit all four of the characteristics mentioned. Luminosities of their nuclear regions range from 10^{35} to 10^{38} J s^{-1}. Two different classes are distinguished: Seyfert-1 systems show very broad, forbidden emission lines indicating speeds of around 5000 km s^{-1} in their nuclei. Seyfert-2 systems exhibit emission lines (both forbidden and permitted) with line widths ranging from 300 to 1000 km s^{-1}. The gas generating these emission lines lies at various distances from the central energy source and consequently displays various states of excitation and velocity. Seyfert galaxies are relatively nearby and therefore quite accessible to most observers. In Seyferts a first glance can be cast on what is the central powerhouse of galaxy activity.

By contrast with Seyferts, quasars are among the most distant objects in the observable universe. With their energy output in the range 10^{45} to 10^{48} erg s^{-1}, quasars represent the most luminous active nuclei, corresponding to absolute magnitudes in the range $-31 < M_V < -24$. Quasars appear as star-like points on photographic plates, but under very good seeing and on very deep photographic exposures, a galaxy disk can be photometrically detected around some quasars. A quasar is the nucleus of an elliptical galaxy or of a spiral galaxy of extremely low apparent magnitude. The smallest redshifts in quasars are about $z = 0.1$, while the largest measured values of z are currently 6.5 and up. The brightest and nearest quasar is 3C 273 at the 1950 coordinates $(\alpha, \delta) = (12^{\mathrm{h}}26^{\mathrm{m}}33^{\mathrm{s}}, +02°20')$. It has an apparent magnitude $m_v = 12.8$ and a redshift of $z = 0.158$.

The spectral features, such as emission lines, and the continuous energy distribution at optical, infrared, radio, X-ray, γ-ray wavelengths need to be explained by a uniform model for active galaxies. This model could then also provide information on the masses and the evolution of massive galactic nuclei. The nucleus contains an extreme mass-rich black hole with 10^8 to 10^{10} M$_{☼}$ (solar mass). Matter flowing into the black hole liberates the energy required to power an active galaxy. In quasars, about 10 M$_{☼}$ per year must be consumed, while for Seyfert galaxies, only about 0.1 M$_{☼}$ per year is required to account for the radiated energy.

The different types of active galaxies are explainable by the different orientation angles under which we can see the galactic nuclear regions, i.e., the black hole center. If the active galaxy is orientated perpendicular to our observing direction we identify the activity as a BL-Lac object, at 45° as a Seyfert 1 or quasar activity and at 90° as Seyfert 2 or radio galaxy activity. The different observed activity levels can presumably be distinguished in terms of various distributions of matter around a black hole (accretion disk, hot and cool gas clouds, steep or flat density gradients) and the rate of mass influx.

Such active nuclei within their mother galaxies are also sources of radio astronomically and partially optically identified jets and of the symmetrically structured, far-extended radio emission regions. Most of the extended radio sources are symmetric with an optically visible galaxy or a quasar. The jets reach into the extended radio emission. The extension of these radio structures ranges from 100 kpc into the megaparsec range. Optical jets are 3 kpc to 400 kpc long. The symmetry and the straight-line arrangement of the jet, measured from the innermost core to the outermost radio lobes, holds to within a few degrees. The jets are often found to be continuously curved, and the extended radio tails are displaced relative to one another. This suggests that the central source of activity rotates, and this motion is reflected in the ejecta. Taken together with estimates of the lifetime (about 10^5 years) of the radio-emitting regions, this fact suggests that active galactic nuclei eject matter and constantly resupply energy. Only in this fashion can these enormous cosmic structures be created and maintained.

14.5.7 The Universe

The entire cosmos is hierarchically structured. Stars are ordered into stellar associations, clusters, and galaxies. Galaxies are ordered into groups and clusters, and even clusters of galaxies are apparently linked up with a still higher order. Field galaxies, i.e., single or isolated stellar systems, exist only in low numbers.

An agglomerate of galaxies must comply with three conditions in order to be ranked as a cluster:

1. There should be no connection between apparent brightness and redshift (distance). If such a relation does exist, then the group in reality consists of randomly aligned objects in space.
2. The number density per unit area in the group should, from a statistical standpoint, be clearly above random fluctuations of surrounding galaxy numbers.
3. The crossing time of the cluster diameter for the individual galaxies of the possible group must be less than 1/3 of the age of the universe, so that a stable equipartition of kinetic energy within the group has had time to form. This energy distribution guarantees a good mixture in the group. The central region is often occupied by a large dominant galaxy.

The Milky Way is a member of the *Local Group*, which consists of about 41 members, including three large galaxies (the Milky Way, the Andromeda Nebula, and M 33) and 37 dwarf galaxies. The Local Group is apparently a small appendage of a local supercluster, whose center lies in the Virgo cluster of galaxies. If the Local Group is taken as a representative sample, then the overwhelming majority of galaxies in the Universe must be dwarf objects, which at larger distances will be beyond the reach of observation.

In clusters of galaxies, as in the case of groups, the compliance with the criteria given must also be carefully adhered to in order to identify true clusters. The best known and nearest clusters of galaxies are the Coma and Virgo clusters. The Virgo cluster has about 3000 certain members, and lies at a distance of 17 Mpc; the Coma cluster has 11,000 members (estimated) and a distance of 85–90 Mpc. Dwarf galaxies are not included in the membership numbers.

Clusters of galaxies show diverse structures, with the central concentration being the primary distinguishing criterion. Those clusters with many members show a strong concentration of the brighter members toward the center. The richer and more compact a cluster of galaxies, the higher the fraction of elliptical systems it contains. On the other hand, the more loosely packed a cluster is assembled, the more spiral galaxies it contains. These distributions are connected with internal interactions of galaxies within the cluster. During collisions between galaxies, interstellar matter is swept out of the systems, a frequent occurrence in compact clusters.

The intergalactic gas in these clusters was first identified by its X-ray emission. Since it has been found to have nearly the same chemical composition as stellar and interstellar matter, it must come from the galaxies themselves. Its temperature of 10^7 K to 10^8 K corresponds to the kinetic energy of cluster galaxies from whose energy supply in motion and radiation the energy is fed into the intergalactic medium. Average relative velocities of galaxies in clusters are in the range of 1000 to 2000 km s^{-1}.

Dark matter appears in individual galaxies as well as in the clusters to which they belong. The mass as estimated from the state of motion of a cluster is considerably higher than that derived from the optical luminosity. The fact that the observed brightness underestimates the mass is explained by noting that a substantial part of the mass is contained in the invisible dark matter haloes of the galaxies. The mass distribution is approximately: galaxies 10%, cluster gas 10–25%, dark matter 70–90%; the mass-to-light ratio lies in the range 100 to 400.

Apparently, there are no spherical superclusters composed of well-defined clusters of galaxies. The latter appear rather to be arranged in chains. These chains form a kind of lattice with larger voids between its nodes. Superclusters thus are fluctuations on the largest size scale in the distribution of cosmic matter. The size of a typical supercluster is 35–55 Mpc, the diameters of voids are 55–65 Mpc. The spatial uniformity of the Universe is called the cosmological principle.

Observationally it has been shown that the galaxies in our local neighborhood ($z = 0.3 \sim 1500$ Mpc) have taken their appearance relative recently. In the distant universe galaxy morphology deviates systematically from that of nearby galaxies. By $z = 1 \sim 3000$ Mpc our galaxy classification becomes meaningless. Only more or less irregular systems with violent star formation and in the state of merging are observed. Indeed, galaxies evolve on cosmic scales. The birth and evolution of objects in the Universe is tied to its origin. The precursors of galaxies were created during a specific phase of expansion by the contraction of denser regions of dark matter and of pre-galactic gas clouds bounded in these regions. The degree of turbulence in these protogalaxies and subsequent and frequent merging processes has determined the type of galaxy which formed. The distribution of angular momentum resulting from the turbulence is the basic parameter for the later dynamical evolution of the galaxies, in particular with respect to interstellar matter and star formation.

Cosmologies are scientific theories regarding the large-scale structure of the universe, and purport to explain the structure of space, the origin, expansion, and age of the Universe, and the origin and evolution of the objects within it. Knowledge about the origin of the Universe is derived from the distribution of matter as accessible from the recessional motion and morphology of galaxies, from elemental abundances, and from the cosmic background radiation. The latter, the microwave background in the wavelength range 0.1 to 6 cm, arrives uniformly from all directions in the sky. It is interpreted as the remnant of blackbody radiation emitted during the initial phase of the Universe. This radiation originates 380,000 years after zero-time, the Big Bang . This investigation makes cosmology a precision science. This remnant of the radiation from the cosmic zerotime flooded into the expanding universe, and it now has the properties which would be expected after about 13.7×10^9 years, the age of our cosmos. Next to the recessional motion of galaxies, it is the most convincing evidence for an evolving universe.

The enormous fascination of present theories of cosmologies lies in the fact that an understanding of the origin of the Universe is also an aid in determining the physics of atoms, protons, neutrons, and electrons. The physical descriptions of elementary particles, which form the classical constituents of atoms, trace their roots back to the first seconds of the birth of the Universe.

14.5.8 Amateur Techniques and Projects

New electronic cameras, the CCDs, modern high-speed photographic films and cheaper greater telescopes brought extragalactic astronomy into the focus of amateur astronomy. The computer techniques of picture processing in connection with different filter registrations have opened entirely new avenues for galaxy studies. Automatically operating telescopes with electronic cameras and computer data storage are usable for long-term supervision and sky scanning.

The spectral ranges from the UV to gamma rays and those from the near-infrared to radio wavelengths are largely inaccessible to the amateur, but active working groups whose personal computers have sufficient storage capacity and printing equipment should not be discouraged, as they can obtain from professional scientists data on the inaccessible wavelength ranges. Thus, radio charts of nearby galaxies scaled to the optical photographs will allow the observer to work on many interesting astrophysical problems. Astronomical databases may also be of help in data acquisition, but here it will be necessary to provide evidence of serious intent.

Making surveys of supernovae, novae, and variables in galaxies in the Local Group is a virtually unending project. Here, the first virtue of the observer is perseverance and continuity. The same galaxies, photographic materials or electronic cameras, and reduction procedures should always be used in order to achieve success, which, however, may be realized only after years of patient work. Working groups of amateur associations can provide much and helpful instruction in this regard. This area of work will certainly expand in the years to come as optoelectronics become less expensive and more widely accessible.

A study of galaxies may approach various astrophysical concepts; such an investigation may address the following possibilities:

1. Testing various photographic materials or electronic cameras for speed and color, sensitivity and, in combination with the telescope, for resolution, color and sensitivity.
2. Observations of galaxies centered on different wavelengths through suitable camera/filter/film combinations (e.g., red and blue), which will qualitatively show the different populations in galaxies, the outskirts of galaxies and tidal deformations.

3. Dark cloud distribution in spiral arms or at galaxies seen edge on.

4. Variability of the center regions or star-forming regions and the search for variables, especially novae and supernovae.

5. Construct one's own catalog of galaxies, including, if possible, all morphological types.

One very fascinating observing project with great scientific value is the search for supernovae. CCD cameras on computer-controlled telescopes with well-adopted software for controlling and data reduction allows the steady observations of galaxies. Stars as faint as nineteenth magnitude become accessible. Such projects are done in many countries. They are also well-suited for astronomy working groups in schools, see for example, D. Muelheims, Astronomie AG, Overbach Schule, Jülich/Barmen (http://www.meade.de/index.php?id=332/).

An example is the detection of SN 2005cs in M51 by the amateur astronomer W. Kloehr (http://www.dsi-astronomie.de/bericht_EN.htm/).

In nearly every country special amateur working groups are busy on extragalactic astronomy devoted to deep-sky objects. The Internet helps to quickly find contacts. Also a multitude of excellent observer's handbooks for detailed questions can be consulted.

The primary requirements for extragalactic astrophotography are:

1. The night sky background should be as dark as possible, the atmospheric transparency quite good, and the air turbulence negligibly small. High contrast and rich detail are reached in adequate quality only at excellent sites and under favorable conditions.

2. For long-focus astrophotography, stable and precise tracking is required. Long exposures (one hour or more) demand accurate guiding. Very fine galaxy photographs will be obtained only with a mechanically stable instrument. This is also true for CCD astronomy and shorter observing times.

3. Precise focusing is always a problem in long-focus astronomy; the knife-edge method is recommended for satisfying results. During the night the focus should be reexamined after each exposure, as temperature and mechanical changes of the optics or of telescope flexure may affect the sharpness of the image.

What instrument is best suited for extragalactic astrophotography? Neither very large instruments nor expensive equipment is required to achieve success in deep-sky astronomy; telescopes with focal lengths in the range 750 mm to 1500 mm can produce images of high quality. All photos in this chapter were taken by amateurs using instruments which are small in comparison with professional telescopes. It is hoped that they will provide inspiration for the reader to make his or her own attempts in this realm.

Actually, any observer, irrespective of what size or kind of telescope he or she has, can attempt it. The best results, however, are obtained by using a stably mounted, long-focus instrument, and by taking long exposures with careful guiding. Respectable pictures can be secured with focal lengths of only $f > 1$ m. As the light-gathering power increases with the square of the aperture A, the brightness of extended objects is thus given by squaring the aperture ratio (A/f).

More information can be found under the following addresses:
http://www.supernovae.be/supernova.htm
http://astronomy.swin.edu.au/sao/guest/evans
http://www.amateur-astronomy.gr
http://myweb.tiscali.co.uk/tomboles/

Webb Society lists the most important deep-sky projects:
http://www.webbsociety.freeserve.co.uk

Students for the Exploration of Space (SEDS) are a universal server for astronomy:
http://www.seds.org

The National Deep Sky Observers Society (NDSOS) has headquarters in the USA:
http://www.cismall.com/deepsky/

Valuable books are by:
Steinicke and Jakiel [44]; Luginbuhl and Skiff [45]; Gordon [46]; Seip (2006); Schröder [47]

References

Section 14.1

1. Hirshfeld, A., Sinnot, R.: Sky Catalogue 2000.0. Cambridge University Press, Cambridge (1982)

2. Tirion, W.: Star Atlas 2000.0. Cambridge University Press, Cambridge (1981)

3. Kaler, J.B.: Stars and their Spectra. Cambridge University Press, Cambridge (1989)

4. Böhm-Vitense, E.: Introduction to Stellar Astrophysics, vol. 1 & 2. Cambridge University Press, Cambridge (1989)

5. Prialnik, D.: An Introduction to the Theory of Stellar Structure and Evolution. Cambridge University Press, Cambridge (2000)

Section 14.2

6. Kholopov, P.N., Samus, N.N., Frolov, M.S. et al.: General Catalogue of Variable Stars (GCVS), vol. I–V, 4th edn., vols. I–III. Nauka, Moscow (1985–1988)

7. Kholopov, P.N., Samus, N.N., Durlevich, O.V. et al.: General Catalogue of Variable Stars, 4th edn., vol. IV. Nauka, Moscow (1990)

8. Artyukhina, N.M., Durlevich, O.V., Frolov, M.S. et al.: General Catalogue of Variable Stars, 4th edn., vol. V. Extragalactic Variable Stars, Kosmosinform, Moscow (1995)

9. Duerbeck, H.W., Seitter, W.C.: Variable Stars. In Voigt, H.H. (ed.) Landolt-Börnstein New Series. Group VI, vol. 3B, p. 127. Springer, Berlin Heidelberg New York (1996)

10. Good, G.A.: Observing Variable Stars. Springer, Berlin Heidelberg New York (2003)

11. North, G., James, N.: Observing Variable Stars, Novae & Supernovae. Cambridge University Press, Cambridge (2003)

12. Percy, J.R.: Understanding Variable Stars. Cambridge University Press, Cambridge (2007)

Section 14.3

13. Wycoff, G.L., Hartkopf, W.I., Douglass, G.G., Worley, C.E.: Astron. J. **122**, 3466 (2001)

14. Argyle, B. (ed.): Observing and Measuring Visual Double Stars. Patrick Moore's Practical Astronomy Series. Springer, Berlin Heidelberg New York (2004)

15. Finsen, W.S.: Astron. J. **69**, 319 (1964)

16. Kerschbaum, F.: Ein visuelles Sterninterferometer im Eigenbau. Sterne Weltraum **4**, 383 (1999)

17. Heintz, W.D.: Double Stars. Reidel, Dordrecht (1978)

18. USNO 2006: The US Naval Observatory Double Star Compact Disc 2006.5[1]. Available from: http://ad.usno.navy.mil/wds/cd_request.html

19. Tsesevich, V.P. (ed.): Eclipsing Variable Stars. Wiley, New York (1973)

20. Binnendijk, L.: Properties of Double Stars: a Survey of Parallaxes and Orbits. University of Pennsylvania Press, Philadelphia (1960)

21. Kopal, Z.: Language of the Stars: a Discourse on the Theory of the Light Changes of Eclipsing Variables. Reidel, Dordrecht (1979)

22. Wilson, R.E.: Astrophys. J. **234**, 1054 (1979)

23. Wilson, R.E., Devinney, E.J.: Astrophys. J. **166**, 605 (1971)

24. Kallrath, J., Milone, E.F.: Eclipsing Binary Stars: Modeling and Analysis. Springer, Berlin Heidelberg New York (1999)

---

[1] Contains the *Washington Double Star Catalog* (WDS), the *Second Photometric Magnitude Difference Catalog*, the *Fourth Catalogue of interferometric Measurements of Binary Stars*, and the *Sixth Catalogue of Orbits of Visual Binary Stars*

Section 14.4

25. Neckel, T., Vehrenberg, H.: Atlas Galaktischer Nebel I–III. Treugesel, Düsseldorf (1985, 1987, 1989)

26. Stoyan, R. et al.: Atlas of the Messier Objects: The Deep Sky Highlights. Cambridge University Press, Cambridge (2008)

27. Osterbrock, D.E.: Astrophysics of Gaseous Nebulae and Active Galactic Nuclei. University Science Books, Sausalito, CA (1989)

Section 14.5

28. Carroll, B.W., Ostlie, B.A.: Modern Astrophysics. Addison-Wesley, Reading (2002)

29. Feitzinger, J.V.: Galaxien und Kosmologie. Kosmos, Stuttgart (2007)

30. Jones, M.H., Lambourne, R.J.A.: Galaxies and Cosmology, Cambridge University Press, Cambridge (2004)

31. Bergmann, Schäfer: Lehrbuch der Experimentalphysik, Vol. 8 Sterne und Weltraum. de Gruyter, Berlin (2002)

32. Burnham, R.: Celestial Handbook I–III. Celestial Handbook, Flagstaff (1977)

33. Nilson, P.: Uppsala General Catalogue of Galaxies. Uppsala Astron. Obs. Ann. **6** (1973)

34. Lauberts, A.: The ESO/Uppsala Survey of the ESO (B) Atlas. European Southern Observatory, Munich (1982)

35. de Vaucouleurs, G., de Vaucouleurs, A., Corwin, H., Buta, R.J., Patruel, G., Fouque, P.: Third Reference Catalogue of Bright Galaxies. Springer, Berlin Heidelberg New York (1991)

36. Sandage, A.: The Hubble Atlas of Galaxies. Carnegie Institution of Washington, Washington (1961)

37. Sandage, A., Tammann, G.A.: A Revised Shapley–Ames Catalogue of Bright Galaxies. Carnegie Institution of Washington, Washington (1981)

38. Arp, H.: Atlas of Peculiar Galaxies. California Institute of Technology, Pasadena (1966)

39. Arp, H., Madore, B.F.: A Catalogue of Southern Peculiar Galaxies and Associations, vols. 1, 11. Cambridge University Press, Cambridge (1987)

40. Madore, B.F.: In: Landoldt-Börnstein, Astronomy and Astrophysics, vol. 3c, p. 179. Springer, Berlin Heidelberg New York (1999)

41. Mitton, S.: Exploring the Galaxies. Scribner Sons, New York (1976)

42. Rubin, V.C.: Scientific American, **6**, June (1983)

43. Seiden, P.E., Schulman, L.S., Feitzinger, J.V.: Astrophys. J **253**, 91 (1982)

44. Steinicke, W., Jakiel, R.: Galaxies and How to Observe Them. Springer Berlin Heidelberg New York (2006)

45. Luginbuhl, C.B., Skiff, B.A.: Observing Handbook and Catalogue of Deep Sky Objects. Cambridge University Press, Cambridge (1998)

46. Gordon, B.: Astrophotography. Willman-Bell, Richmond (1985)

47. Schröder, K.P.: Praxishandbuch Astrofotografie. Kosmos, Stuttgart (2003)

Index

I

J